MATHEMATICAL AND PHYSICAL PAPERS

IN TWO VOLUMES

VOLUME I

MATHEMATICAL AND PHYSICAL PAPERS

BY

SIR JOSEPH LARMOR, Sc.D., F.R.S.

HON. F.R.S. EDIN.: HON. MEM. R. IRISH ACAD.: HON. MEM. ASIATIC SOC. OF BENGAL, MANCHESTER LIT. AND PHIL. SOC.: HON. FOR. MEM. U.S. NATIONAL ACAD. OF SCIENCES, AMERICAN ACAD. OF SCIENCE AND ARTS, BOSTON, AMERICAN PHILOSOPHICAL SOCIETY, PHILADELPHIA, WASHINGTON ACAD., R. ACCADEMIA DEI LINCEI, ROME, ISTITUTO DI BOLOGNA: CORRESPONDANT OF THE INSTITUTE OF FRANCE: LUCASIAN PROFESSOR IN THE UNIVERSITY OF CAMBRIDGE, AND FELLOW OF ST JOHN'S COLLEGE

VOLUME I

CAMBRIDGE
AT THE UNIVERSITY PRESS
1929

CAMBRIDGE
UNIVERSITY PRESS

University Printing House, Cambridge CB2 8BS, United Kingdom

Cambridge University Press is part of the University of Cambridge.

It furthers the University's mission by disseminating knowledge in the pursuit of education, learning and research at the highest international levels of excellence.

www.cambridge.org
Information on this title: www.cambridge.org/9781107536463

First published 1929
First paperback edition 2015

A catalogue record for this publication is available from the British Library

ISBN 978-1-107-53646-3 Paperback

PREFACE

MANY years ago an invitation came to the writer to adapt a collection of papers on electrodynamic theory into the form of a book for the Cambridge University Press. Some preparations were then made: but the material was felt to be incomplete, and its further development was delayed by administrative and other public duties that fell to his lot, and continued in various forms over twenty years. The project has now been resumed with a wider scope: about half of the present collection of papers is of electrical character, the other half being mainly General Dynamics and Thermodynamics including the dynamical history of the Earth, Formal Optics, and Geometry.

It may seem to be late to take up this thread: but among types of reasons that might be alleged, there is the circumstance that the history of progress of electrical theories towards a coherent system is still unwritten and possibly is hardly yet ripe for formulation. The systematic materials for it that are readily accessible are not plentiful: while every investigator bears the stamp of the domicile in which he has been brought up, of its contemporary literature and point of view with which he has been specially familiar. It might have been possible by condensations to reduce the electrical half of this volume by perhaps one-quarter of its extent. But, in deference to the advice that was open to him, the writer has abstained from that course. It was urged that out of regard for future historical interests, the order of succession in years should not be blurred. And, moreover, the governing motive for the production of memoirs in formal physical theory, the progressive clarification of the writer's own outlook, may be worth preserving.

In the present collection the names that recur most frequently are, after Faraday, those of Ampère, Helmholtz, Kelvin, Maxwell, also for the earlier time Stokes and Kirchhoff, for the later Rayleigh. They built the foundations for the present electric age: their theoretical message, the most fundamental since Newton, has now gone over largely, in simplified form, into the electrotechnic domain, where seizing in recent times on the practical possibilities opened up by the detection and identification of the free electron by Thomson, it is still the main guide in marvellous modern developments. The most interesting, even tantalizing, recent kaleidoscopic theoretical glimpses of an ordered under-world, emerging from an extensive field of correlations in atomic spectroscopy, in connection with the cardinal phenomenon of the disintegration of atoms as discovered and very closely systematized by Rutherford at the beginning of the century and later, form a special domain, still mainly in the exploring and assembling stage, which appears only incidentally, as regards its remoter ideal sources, in these papers.

The writer has been allowed to counteract any inconvenience arising from discussions recurring in modified forms, by placing index headings along the margins of the pages. It can be held that the practical amenities of an unsystematic collection of papers can thereby in various respects be much enhanced.

Many footnotes and nine appendices have been added. To distinguish them, the footnotes to the original papers are now indicated by numbers, the new ones by asterisks and other such signs. Other incidental additions are enclosed in square brackets. Minor improvements and corrections, especially in algebraic typography, have been freely introduced: but it is hoped that substantial changes have not been made without suitable indications. The collection will be completed in a second volume.

The electromagnetic system of units, employed in Maxwell's and the other earlier classical writings, has been adhered to throughout. It is the foundation of universal electrotechnic practice, as settled long ago internationally: and any introduction now, alongside it, of more recent unitary schemes (see Appendix VI), which though more simple are no whit more rational, would, as troublesome experience shows, only produce confusion. For similar reasons the notation is throughout that of Maxwell's *Treatise*.

As was the custom of the time, the earlier papers on Electrodynamics proceed largely by help of auxiliary conceptual models of dynamical type, representing the aethereal systems that are involved. Afterwards there came a period of wide repudiation of such aids to theoretical construction, on the ground that the legitimate province of theory cannot extend beyond bare description of the course of Nature by means of equations: in contrast with Kelvin's creed that he could never be assured of the internal coherence and consistency of any complex scheme of equations until he could construct some sort of model of a process which they would represent. Abstract mathematicians, with their close attention to existence theorems, ought to be in sympathy. Without Maxwell's provisional models there would have been no electric theory of light. Without a model of an electron (cf. p. 514, *infra*) there would be no dynamically coherent notion, however provisional, of how one electric source can affect another across a distance. Indeed it might be imagined that a deeper knowledge of the electron must be awaited before

Maxwell's pressure of light is fully understood: now traced to a convected momentum of radiation, entirely foreign to the circuital equations of the field as hitherto considered by itself, as Kelvin did not fail to note, yet present and interchangeable on Amperean principles where the rays meet the flux of electrons. A closer scrutiny shows (cf. *infra*, p. 655) that the Maxwellian scheme, as translated into electrons alone, is wholly valid though incomplete: his aethereal medium in which the electrons subsist refuses to pass so entirely into the background. The ultimate foundation, covering relativity and all else, is the specification of the Action throughout this medium: from its analysis in relation to sources comes the electron scheme. But there are other things outstanding, including the momentum of radiation, which can be reduced to that simpler formulation only as regards interaction with aggregations of electrons, not with electrons individually. The absence or latency of the complementary atomic form, the positive electron, in the actual world must however be a deeper and an urgent problem.

More recently, the practice of reasoning by aid of conceptual models of mixed geometric-mechanical type, now more provisional, has been recovering its credit; for no progress at all in theories relating to processes in the interior of the atom is feasible without their assistance. This new development, mainly after the lead of N. Bohr, is in a different plane, recalling in some degree Newton's effort towards fitting together the fragments of an orbital Lunar Theory from the empirical features revealed by the observations, and also the formal constructions of organic chemistry.

Acknowledgment is due, and is now tendered, to the Royal Society, the Royal Society of Edinburgh, the Royal Astronomical Society, the Cambridge Philosophical Society, the London Mathematical Society, the Manchester Literary and Philosophical Society, and the proprietors of the *Philosophical Magazine* and of *Nature*, for consent to the re-publication of material from their collections.

The writer has much pleasure in acknowledging the expert co-operation of the staff of the Cambridge University Press, which has resulted in the production of this handsome volume.

J. L.

CAMBRIDGE,
September 1927.

CONTENTS

I

PLANIMETRY ON A MOVING PLANE.

[*Nature*, Oct. 27, 1881.]

PROF. MINCHIN'S theorem in *Nature* [which is also M. Darboux's, namely, "If a plane, A, move about in any manner over a fixed plane, B, and return to its original position after any number of revolutions, all those right lines in the plane A which have enveloped glisettes of the same area, are tangents to a conic, and by varying the area of the glisette we obtain a series of confocal conics"] may be proved easily by considering the motion as due to the rolling of one closed curve on another back into its first position, their lengths being of course commensurable. If you measure y for the rolling curve from the straight line which forms the envelope, and x along that line, then the differential of the area between the envelope and the fixed curve is easily seen to be $ydx + \frac{1}{2}y^2 d\omega$, where $d\omega$ is the angle turned through by the rolling curve, and is equal to ds multiplied by the sum of the curvatures at the point of contact, which we shall call σ. The summation of the former part is a multiple of the area of the rolling curve, and therefore the same for all lines; that of the latter is half the moment of inertia of matter distributed over its perimeter with density σ, about the line in question. The result is therefore the well-known property of equi-momental ellipses. Similar reasoning, with the use of the property of the centre of inertia of a system, leads to the further result that when the perimeter of the envelope is of constant length, the line touches a circle, and different values of the constant correspond to concentric circles. In the same way by a property of the centre of inertia we may also prove immediately the known theorem that when the area traced out by a point is constant, the point lies on a circle, and different values of the constant correspond to concentric circles; and we may extend it to areas traced on a sphere.

Lines enveloping equal areas:

equal perimeters.

Points tracing equal areas.

2

ON CRITICAL OR "APPARENTLY NEUTRAL" EQUILIBRIUM.

[*Proc. Camb. Phil. Soc.* Vol. IV. Pt VI. (May 28, 1883).]

1. When a solid body is resting on a fixed surface, its equilibrium is stable when its centre of gravity is vertically below the centre of curvature at the point by which it rests, and unstable when vertically above it: when the two points coincide the equilibrium is often said to be apparently neutral, and its real character is discriminated by an analysis of the differentials of higher orders. It may be worth while to trace the origin of this peculiarity, and its practical effect on the nature of the equilibrium in cases which approximate to this critical condition.

Range of instability:

2. Let us take the case of a heavy body symmetrical about two principal planes through its axis AB (one of them the plane of the figure), and resting on a horizontal plane at A. The evolute of the section has a cusp at O, the centre of curvature corresponding to A. Let us suppose it to point downwards, so that the radius of curvature is a minimum at A, and let us suppose the centre of gravity G to be a very short distance above O. The position of the body is unstable, but a stable position exists in immediate proximity on each side, in which the tangents from G to the evolute are vertical. We see therefore, that when left free the body will oscillate at first round its upright position, and will finally settle down in one of these two slightly inclined positions. When G moves down to O, these two flanking stable positions come nearer to the upright position, and finally come up to it, so that the equilibrium is really stable. But there is this peculiarity, that its oscillations round the vertical are no longer approximately Simple Harmonic, but follow another law which we can easily investigate, and that they are executed with extreme slowness: and we can trace the change to this new law from the rocking motion which is compounded of oscillations round the two flanking stable positions alternately.

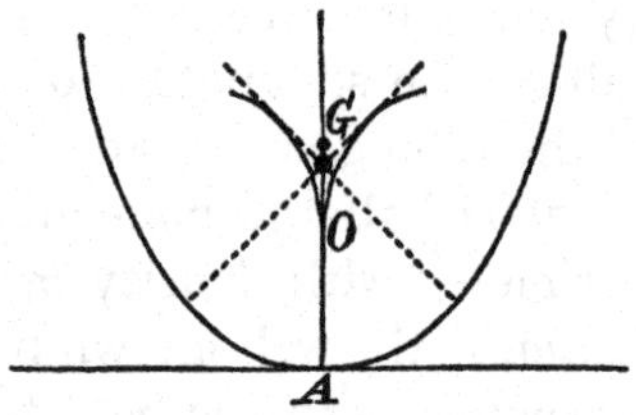

how it closes up.

Range of stability:

3. If the cusp pointed upwards, and G were a very short distance below O, we would have a vertical stable position flanked by two

very near positions of instability: and so, when G moves up to O, the vertical position becomes unstable. It is important, then, to bear in mind, that cases which satisfy the condition of stability, but are near to the critical case, are practically unstable for oscillations of any considerable amount, when the radius of curvature is a maximum at A.

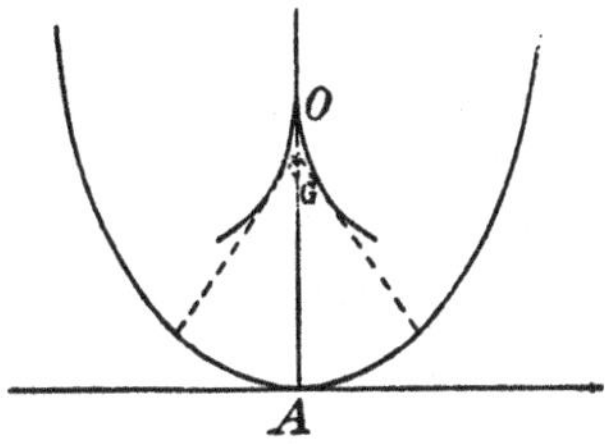

how it closes up.

4. These considerations clearly apply to all cases of critical or "apparently neutral" equilibrium, so that the determination of its real character carries with it the determination of the *practical* character of all other cases which approximate to that condition.

Range of stability of a ship:

5. In the case of a floating body this discrimination is easy. If we consider, as usual, oscillations in which the displacement is constant, the centre of gravity of the displacement traces out a surface called the Surface of Buoyancy, and we know that the tangent plane to this surface corresponding to any position of the oscillating body is always horizontal, and that therefore the resultant fluid pressure acts along the normal at its point of contact. The circumstances of the oscillation are therefore the same as those of a solid body with the same centre of gravity, but bounded by the surface of buoyancy, and rolling on a frictionless horizontal plane. The evolute of the section of the surface of buoyancy is the locus of the metacentre, and has been called the Curve of Stability: and a case of equilibrium which approximates to the critical condition will come under § 2 or § 3 according as this curve has its cusp pointing downwards or upwards. Now the radius of curvature of the surface of buoyancy is known by the ordinary theory to be equal to the moment of inertia of the corresponding plane of floatation about an axis through its centre of gravity perpendicular to the plane of displacement, divided by the volume of the displacement: and therefore the case comes under § 2 or § 3 according as this moment of inertia increases or diminishes as the degree of heeling increases, a criterion usually easy of application.

the criterion:

6. In a very numerous class of cases we can completely determine by geometry the curves mentioned above. Since any surface of the second degree may be derived by successive orthogonal projections, real or imaginary, from a sphere, it follows that the locus of the centre of gravity of a constant volume cut off from it by a plane is a similar and concentric surface. Hence for all this class of surfaces, which includes quadrics, cones, cylinders, pairs of planes, the surface of

buoyancy is similar to the bounding surface of the floating body. Now the radius of curvature of a parabola or hyperbola is least at the vertex, and that of an ellipse is least at the ends of the major axis, and greatest at the ends of the minor axis: hence for parabo- results for a class of cases. loidal and hyperboloidal surfaces, and cones, and wedges, the critical position is really stable, and for surfaces of which the section in the plane of displacement is elliptic it is stable or unstable according as the major or minor axis is vertical.

Further, if the shape near the water line come under any of these heads, the above conclusions clearly all apply, irrespective of the shapes of the other parts of the body.

Case of parabolic section.

7. We shall now investigate, as a typical case, the nature of the oscillations of a body whose section at the vertex approximates most closely to that of the parabola $y^2 = ax$, which rolls on a rough plane, and in which G is a very short distance c above O.

The equation of the evolute near the cusp is of the form

$$y^2 = \frac{16}{27a} x^3,$$

and when it is displaced through a small angle θ, the distance of the vertical tangent to the evolute from G is easily seen to be equal to

$$\frac{a}{4}\theta^3 - c\sin\theta,$$

to the third order in θ.

Thus, if κ represent the radius of gyration of the solid round an axis through its vertex, the equation of motion is

$$\frac{\kappa^2}{g}\frac{d^2\theta}{dt^2} = c\left(\theta - \frac{\theta^3}{6}\right) - \frac{a}{4}\theta^3,$$

to the third order in θ.

And the flanking positions of stable equilibrium are given by

$$\theta = \sqrt{12c/(3a+2c)}.$$

This equation is of the form

$$\frac{\kappa^2}{g}\frac{d^2\theta}{dt^2} = c\theta - m\theta^3,$$

where c is very small compared with m, which may be put equal to $\frac{1}{4}a$; and in fact we are investigating the character of oscillations under a restoring force proportional to the cube of the distance disturbed by a small force proportional to the distance.

Oscillations near the critical stage.

8. We have

$$\frac{\kappa^2}{g}\left(\frac{d\theta}{dt}\right)^2 = \tfrac{1}{2}m\beta^4 - c\beta^2 + c\theta^2 - \tfrac{1}{2}m\theta^4,$$

where β is the amplitude of the swing; and therefore

$$t = \int \frac{\kappa d\theta}{g^{\frac{1}{2}} \sqrt{A + c\theta^2 - \frac{1}{2} m\theta^4}}$$

$$= \frac{\kappa}{\sqrt{Ag}} \int \frac{d\theta}{\sqrt{(1 - p\theta^2)(1 + q\theta^2)}}$$

$$= \frac{\kappa}{\sqrt{Apg}} \sqrt{1 - k^2} \int \frac{-d\phi}{\sqrt{1 - k^2 \sin^2 \phi}},$$

where $\cos\phi = \sqrt{p}\,\theta, \quad k^2 = \dfrac{q}{p+q},$

$$A = \tfrac{1}{2} m\beta^4 - c\beta^2,$$

$$q - p = \frac{c}{A}, \text{ which is small,}$$

$$qp = \frac{m}{2A};$$

so that $q + p = \sqrt{\dfrac{2m}{A}} \left(1 + \dfrac{c^2}{4mA}\right)$, approximately.

Also $\dfrac{c}{m\beta^2}$ is a small quantity which we shall call e.

Hence $t = \dfrac{\kappa}{g^{\frac{1}{2}} \sqrt{A(p+q)}} \displaystyle\int \frac{-d\phi}{\sqrt{1 - k^2 \sin^2\phi}}$

$$= \kappa (mg\beta^2)^{-\frac{1}{2}} \left(1 + \frac{c}{2m\beta^2}\right) \int \frac{-d\phi}{\sqrt{1 - k^2 \sin^2\phi}}$$

$$= \kappa (mg\beta^2)^{-\frac{1}{2}} (1 + \tfrac{1}{2} e) \left\{ F\left(\frac{\pi}{2}, k\right) - F(\phi, k) \right\};$$

where $\cos\phi = \sqrt{p}\,\theta = \dfrac{\theta}{\beta}(1 - \tfrac{1}{2} e), \quad k^2 = \tfrac{1}{2}(1 + e).$

Now write $\cos\psi = \dfrac{\theta}{\beta}$, and use the results for differentiating F that are given in Cayley's *Elliptic Functions*, § 73. We find, on putting $\phi = 0$, that a quarter period of the oscillation of amplitude β is given by

$$T = \kappa (mg\beta^2)^{-\frac{1}{2}} (1 + \tfrac{1}{2} e) \left[F\left(\frac{\pi}{2}, \frac{1}{\sqrt{2}}\right) + e \left\{ E\left(\frac{\pi}{2}, \frac{1}{\sqrt{2}}\right) - \tfrac{1}{2} F\left(\frac{\pi}{2}, \frac{1}{\sqrt{2}}\right) \right\} \right]$$

$$= \frac{\kappa}{g^{\frac{1}{2}} m^{\frac{1}{2}} \beta} [F_1 + eE_1],$$

in which $F_1 E_1$ stand for the complete elliptic integrals of the first and second orders to modulus sin 45°, and $e = c/m\beta^2$.

Also

$$t = T - \frac{\kappa}{g^{\frac{1}{2}} m^{\frac{1}{2}} \beta} (1 + \tfrac{1}{2}e) \left[F + e(E - \tfrac{1}{2}F) - \tfrac{1}{2} e \sin\psi \cos\psi + \frac{e \cos\psi}{2 \sin\psi \sqrt{1 - \frac{1}{2}\sin^2\psi}} \right]$$

$$= T - \frac{\kappa}{g^{\frac{1}{2}} m^{\frac{1}{2}} \beta} \left[F + eE - \frac{e}{2\beta^2} \theta \sqrt{\beta^2 - \theta^2} + \frac{e\beta\theta}{\sqrt{2(\beta^4 - \theta^4)}} \right],$$

in which F, E are the functions of $\text{arc} \cos \frac{\theta}{\beta}$ to modulus $\sin 45°$, whose values can be taken at once from Legendre's tables.

9. The disturbance produced by the small term proportional to the distance is represented by the terms multiplied by e. And, in particular, if $e = 0$, so that the equilibrium is critical, we have

The critical oscillations.

$$T = \frac{\kappa . F\left(\frac{\pi}{2}, \sin 45°\right)}{g^{\frac{1}{2}} m^{\frac{1}{2}} \beta},$$

$$t = \frac{\kappa}{g^{\frac{1}{2}} m^{\frac{1}{2}} \beta} \left\{ F\left(\frac{\pi}{2}, \sin 45°\right) - F\left(\text{arc} \cos \frac{\theta}{\beta}, \sin 45°\right) \right\},$$

where $m = a/4$, $\beta =$ amplitude of excursion.

In this case, therefore, the period of an oscillation varies inversely as its amplitude, and is equal to

$$\frac{2\kappa}{g^{\frac{1}{2}} a^{\frac{1}{2}} \beta} \times 1{\cdot}85407.$$

If the plane on which the motion takes place be frictionless, κ will be the radius of gyration about an axis through the centre of gravity: and the same will be the case in the problem of hydrostatic oscillations.

10. In the case of one sphere resting on another it is very easy to form the equation of energy, and thence determine the value of m.

Body rolling on another.

In the case of one symmetrical body resting on another, whether on the summit or not, we have, if θ be the angle by which G overhangs the vertical through the point of support,

$$\frac{\kappa^2}{g} \frac{d^2\phi}{dt^2} = \left(\frac{d^3\theta}{ds^3}\right)_0 \frac{s^3}{6}$$

in the critical case; therefore

$$\frac{d^2 s}{dt^2} = \frac{g}{6\kappa^2} \left(\frac{1}{\rho} + \frac{1}{\rho'}\right)^{-1} \left(\frac{d^3\theta}{ds^3}\right)_0 s^3,$$

in which the value of $\left(\frac{d^3\theta}{ds^3}\right)_0$ is given by Mr Routh in the *Quarterly Journal of Mathematics*, XI, p. 106; and thus the value of m is determined. The nature of the equilibrium in the critical case appears to have been first discussed by Dr A. H. Curtis, *Q. J.* IX, p. 42.

11. The exception to this critical case again, is that in which five positions of equilibrium come together. Then the equation of motion is of the form **Higher critical circumstances.**

$$\frac{d^2\theta}{dt^2} = -\mu\theta^5.$$

Hence
$$t = \sqrt{\frac{3}{\mu}}\int\frac{d\theta}{\sqrt{\beta^6 - \theta^6}}$$
$$= \sqrt{\frac{3}{\mu}}\frac{1}{\beta^2}\int\frac{d\phi}{\sqrt{1 - \phi^6}}.$$

And on writing
$$\phi^2 = \frac{1}{1 + y^2},$$
$$y = 3^{\frac{1}{4}}\tan\frac{\psi}{2},$$

this reduces to (*vide* Bertrand's *Integral Calculus*)

$$t = \sqrt{\frac{3}{\mu}}\frac{1}{\beta^2}\frac{1}{3^{\frac{1}{4}}}\int\frac{d\phi}{\sqrt{1 - \sin^2 15^\circ \sin^2\psi}},$$

and a quarter period is given by

$$T = 2\frac{3^{\frac{1}{4}}}{\mu^{\frac{1}{2}}\beta^2} \times 1{\cdot}59814,$$

which varies inversely as the square of the amplitude.

3

ELECTROMAGNETIC INDUCTION IN CONDUCTING SHEETS AND SOLID BODIES.

[*Philosophical Magazine*, Jan. 1884.]

Electro-dynamic flow in unlimited sheets: Maxwell's image-system: Jochmann's approximation for spherical case:

I. The problem of electromagnetic induction in continuous conductors has often engaged attention. Maxwell in 1871 (*Proc. Roy. Soc.*, and "Electricity and Magnetism") gave the complete solution for the case of an infinite sheet, by means of a moving trail of images of the inducing disturbance. He refers to Jochmann (Crelle's *Journal*, 1866), who had investigated the steady currents that would be set up in an infinite plane or a spherical conductor, rotating uniformly in a field of force, on the supposition that the effect of this external field on the conductor is very great compared with that of the system of currents excited. This, we shall find, presupposes a velocity of all the parts of the conductor in the immediate neighbourhood of the seat of the disturbance which must not exceed a certain moderate limit, or else a correspondingly great resistance.

C. Niven's analytic solutions.

In the *Philosophical Transactions* for 1880 Professor Charles Niven has developed a complete mathematical solution of Maxwell's equations for infinite plane sheets and spherical sheets of any thickness; but he has not fully discussed any of the simpler cases. The generality of the solution necessarily complicates his expressions; and in what follows we shall, to secure compactness and completeness, work out *ab initio* the results that we require. The methods are chosen with a view to deducing the results directly from consideration of the physical quantities involved, with as little appeal to mathematical transformations as possible.

The solutions given below are all for cases that can be presented in a simple form. A principal object in obtaining them is to examine the circumstances of what may be called electromagnetic screens.

Electro-magnetic screening:

It has long been known that a plate of soft iron placed in front of a magnet will partially screen the space on the other side of the plate from the influence of the magnet, and that the screening increases with the thickness of the plate. Sir William Thomson has applied the effect in recent times by enclosing his marine galvanometer in a heavy soft-iron case to protect it from the magnetism of the iron of the ship. And lately Stefan has published a paper in

Wiedemann's *Annalen*, in which he investigates experimentally the diminution of the strength of a magnetic field which is produced in the interior of soft-iron cylinders, and compares the results with the indications of Poisson's theory of induced magnetization.

Now if we have a sheet of conducting matter in the neighbourhood of a magnetic system, the effect of a disturbance of that system will be to induce currents in the sheet of such kind as will tend to prevent any change in the conformation of the tubes of force cutting through the sheet. This follows from Lenz's law, which itself has been shown by Helmholtz and Thomson to be a direct consequence of the conservation of energy. But if the arrangement of the tubes *in* the conductor is unaltered, the field on the other side of the conductor into which they pass (supposed isolated from the outside spaces by the conductor) will be unaltered. Hence if the disturbance is of an alternating character, with a period small enough to make it go through a cycle of changes before the currents decay sensibly, we shall have the conductor acting as a screen.

from alternating field by conducting sheet:

Further, we shall also find, on the same principle, that a rapidly rotating conducting-sheet screens the space inside it from all magnetic action which is not symmetrical round the axis of rotation.

from steady field by spinning sheet:

The earth considered as a rotating body comes under this case; for the upper strata of the atmosphere are conductors of electricity, whether the conduction follows the law of Ohm or not, and therefore these principles show that this layer must more or less protect the interior from any external magnetic action, not symmetrical round the Earth's axis, that might exist outside it. In the case of a spherical sheet conducting according to Ohm's law, we shall find that the screening action depends upon the angular velocity multiplied by radius divided by the resistance of the sheet, so that defect of conductivity is fully made up for by a large radius.

***e.g.* atmospheric sheet around the Earth, is effective.**

We shall also find expressions for the magnetic moment of a copper sphere or spherical shell rotating in a magnetic field, and for the expenditure of power required to keep up the motion, and also for the rate of damping of the oscillations of such a body.

Spinning globes and discs in magnetic field.

Finally, we shall show how solutions may be obtained for the case of a rotating circular disc of conductivity not very large by neglecting the mutual action of the induced currents.

II. In the case of a conductor rotating steadily in a magnetic field, which is that with which Jochmann deals, a steady distribution of currents in space will ensue when the conductor is symmetrical about the axis of rotation; and the electromotive force along any line will be given, according to Faraday's rule, by the number of

No currents induced by a symmetrical field:

tubes of force of this steady field that are cut through by the line per second. Now when the magnetic field is symmetrical round the axis of rotation, the number of tubes enclosed in any closed moving circuit in the conductor will not alter at all, so that there will be no current round any circuit, and therefore no induced current whatever: the electromotive force along each open line will accumulate a statical electric charge at one end of it, so that the conductor will become electrified until the induced electromotive force is exactly neutralized by the statical difference of potentials. This conclusion holds whatever be the shape of the body.

the annulling slight electrification.

Taking cylindrical coordinates, if α, β, γ be the components of the magnetic force at the point $r\theta z$ in the directions of $dr, rd\theta, dz$ respectively, the electromotive forces between the ends of these elements of length will be $-\omega r\gamma dr$, 0, $\omega r\alpha dz$, where ω is the angular velocity, by Faraday's rule; and thus the electrostatic potential which neutralizes these will be

$$\psi = \int(\omega r\gamma dr - \omega r\alpha dz)$$

$$= \omega \int r\,(\gamma dr - \alpha dz).$$

From this the electrification of the conductor may be deduced at once. For example, taking a uniform field $\gamma = \gamma_0$ parallel to the axis of rotation, we have

$$\psi = C + \tfrac{1}{2}\omega\gamma_0 r^2,$$

and therefore there is a uniform volume density of electricity

$$-\frac{1}{4\pi}\nabla^2\psi = -\frac{\omega\gamma_0}{2\pi}.$$

We must add the proper surface distribution: for instance, if the conductor be a sphere, the outside potential which corresponds to the given value at the surface is

$$\psi_1 = C\frac{a}{\rho} - \tfrac{1}{2}\omega\gamma_0 a^5\,\frac{3\cos^2\phi - 1}{3\rho^3} + \tfrac{1}{3}\omega\gamma_0\,\frac{a^3}{\rho},$$

where ρ is the radial line, and ϕ its inclination to the axis. Thus the surface density

$$= \frac{1}{4\pi}\left(\frac{d\psi}{d\rho} - \frac{d\psi_1}{d\rho}\right)$$

$$= \frac{\omega\gamma_0 a}{8\pi}\left(-\tfrac{8}{3} + 5\sin^2\phi\right).$$

The arbitrary constant C allows us to superpose any free distribution. If the charge of the body was originally zero, we may give it such a value that the charge shall remain zero; but if the axis of rotation is uninsulated, the condition is that C is zero. The above agrees with Jochmann's results.

For the case of a flat disc rotating about an axis perpendicular to its plane in any uniform field of force, we may divide the force into two components—one parallel to the plane of the disc, which produces no induction, on account of the thinness of the sheet, and the other perpendicular to it, whose effect has just been estimated. Thus Arago's rotations will not occur in a uniform field.

Arago's rotations require a non-uniform field.

By connecting one terminal of a wire to the axis and making the other terminal rub along the circumference, in Faraday's manner, we utilize the difference of potential to produce a current in the external circuit.

Faraday's dynamo or motor.

Of course these static charges are of an exceedingly minute character. We have found ψ in electromagnetic units; and to reduce to electrostatic units we must divide by the reducing factor v, which is approximately equal to the velocity of light: the electric volume density is therefore $-\frac{\omega\gamma_0}{2\pi v}$ electrostatic units, and the surface density must also be divided by v to reduce it to electrostatic units.

Induced charges, insensible.

III. We now proceed to the case of a spherical conducting shell at rest in a magnetic field.

Spherical shell in changing magnetic field:

Let Φ be the current function in the sheet: the value of Φ at any point will be the strength of the equivalent magnetic shell at that point. It is well known (Maxwell, art. 670) that the magnetic potential in space not occupied by the sheet is $\frac{d}{dr}(Pr)$, where $-aP$ is the potential of a distribution of density Φ over the shell. The corresponding vector potential may be represented, on the analogy of the case of plane strata (Maxwell, art. 657), by components 0, $\frac{d\,(Pr)}{r\sin\theta d\phi}$, $-\frac{d\,(Pr)}{rd\theta}$ in the direction of dr, $rd\theta$, $r\sin\theta d\phi$. This is true, in fact, because Laplace's equation for P can be written in the form

the current function expresses its magnetic field:

and also the vector potential thereof:

$$\left\{\left(\frac{d}{dr}\right)^2+\left(\frac{d}{rd\theta}\right)^2+\left(\frac{d}{r\sin\theta d\phi}\right)^2\right\}(Pr)=0.$$

The given state of the external field we can express similarly in terms of an auxiliary potential function P_0.

The equations giving P, Q, R the components of the electromotive force at a point are (Maxwell, art. 598)

and the electric field:

$$P=-\frac{dF}{dt}-\frac{d\psi}{dr},$$

$$Q=-\frac{dG}{dt}-\frac{d\psi}{rd\theta},$$

$$R=-\frac{dH}{dt}-\frac{d\psi}{r\sin\theta d\phi};$$

and therefore, by Ohm's law, expressed for the components along $d\theta$ and along $d\phi$,

leading to equations of electric flow.

$$R\frac{d\Phi}{a\sin\theta d\phi} = -\frac{d}{dt}\frac{d(P+P_0)}{\sin\theta\, d\phi} - \frac{d\psi}{a d\theta},$$

$$-R\frac{d\Phi}{a d\theta} = \frac{d}{dt}\frac{d(P+P_0)}{d\theta} - \frac{d\psi}{a\sin\theta d\phi},$$

where a is the radius of the shell, and R is here its specific surface resistance.

These equations are satisfied by

$$R\Phi = -a\frac{d}{dt}(P+P_0), \qquad (1)$$

without the intervention of any electrostatic potential ψ.

To obtain a solution, let us suppose the external field to be given by

Solution for an alternating field

$$\frac{d}{dr}(P_0 r),$$

where $$P_0 = A_0 e^{\iota\kappa t}\left(\frac{r}{a}\right)^n Y_n,$$

Y_n being a spherical harmonic of the nth order, and $\iota = (-1)^{\frac{1}{2}}$.

Let the corresponding values of P be

$$P_1 = Ae^{\iota\kappa t}\left(\frac{r}{a}\right)^n Y_n \text{ inside;}$$

and therefore $$P_2 = Ae^{\iota\kappa t}\left(\frac{a}{r}\right)^{n+1} Y_n \text{ outside.}$$

We have $$\frac{dP_1}{dr} - \frac{dP_2}{dr} = -\frac{4\pi}{a}\Phi \text{ at the sheet;}$$

therefore $$\Phi = -A\frac{2n+1}{4\pi}e^{\iota\kappa t}Y_n;$$

and by (1), $$-A\frac{2n+1}{4\pi}R = -a(A+A_0)\iota\kappa.$$

Therefore, writing $$p = \frac{(2n+1)R}{4\pi a\kappa},$$

$$A = \frac{\iota}{p-\iota}A_0 = \frac{-1+\iota p}{1+p^2}A_0.$$

Thus, taking real parts, we find, for a field of force oscillating in intensity according to the harmonic law, given by

obtained:

$$P_0 = A_0\cos\kappa t\left(\frac{r}{a}\right)^n Y_n,$$

the intensity of the internal field due to the current sheet will be given by

$$P = A_0 \frac{-\cos \kappa t - p \sin \kappa t}{1 + p^2} \left(\frac{r}{a}\right)^n Y_n$$

$$= -A_0 \cos \alpha \cos (\kappa t - \alpha) \left(\frac{r}{a}\right)^n Y_n, \qquad (2)$$

where $\tan \alpha = p$.

For copper at 0° C., $R = 1640 \div$ thickness of the shell in centimetres.

This solution represents a new and opposite field, with strength reduced by the factor $\cos \alpha$, and phase of variation retarded by the fraction $\alpha/2\pi$ of a complete period. described.

For each harmonic term in the potential, the corresponding induced currents flow along the level lines of that term: for example, in a uniform field of force the currents circulate round the axis of the field in circles, and produce the same external effect as a simple magnet of moment $\frac{1}{2}Fa^3 \cos \alpha \cos (\kappa t - \alpha)$ at the centre, with its axis pointing in the stable direction along the lines of force, where $F \cos \kappa t$ is the strength of the inducing field, and $\tan \alpha = 3R/4\pi a\kappa$. Example of uniform field:

When the oscillations are produced by sound, as in the telephone, we find, on taking the number of vibrations as 350 per second, and $n = 2$ to roughly represent the case, that the values of p and κ are equal for a thickness of $\frac{1}{10}$ centimetre and a radius of 3 centimetres. In this case half the original field is suppressed inside the shell, and a new field of the same intensity as this half is added whose phase of oscillation is increased by a quarter period. By increasing the radius or the thickness, we soon make κ the most important term; but if the area of the original magnetic disturbance remain the same, $n = 2$ will not represent the circumstances when the radius is very large. At any rate, if the number of vibrations per second or the thickness of the sheet is increased, say fourfold, we have now $\kappa = 4p$, $\frac{16}{17}$ of the original field will be suppressed, and a new field of $\frac{4}{17}$ of the original with phase augmented by a quarter period will be introduced. numerical for telephone. Screening by sheet: leakage:

In the *Phil. Mag.* for 1882 Lord Rayleigh describes an experiment in which a coil conveying a current with a microphone-clock in circuit was placed close to another coil connected with a telephone, and the insertion of a thick plate of copper between the coils very considerably diminished the inductive effect in the telephone. test by Rayleigh.

IV. For the case of a solid sphere* or a shell of considerable Induction in solid globe:

* Problems of electric radiation from spheres, or collapse of the static fields of released charges, are included in the same type of analysis by substituting $\frac{K}{4\pi c^2}\frac{d}{dt} + \sigma^{-1}$, where $\frac{d}{dt}$ is ιp for period $\frac{2\pi}{p}$, in place of the conductance σ^{-1}.

thickness, the very remarkable conclusion holds that no external magnetic disturbance whatever can induce currents which do not circulate in concentric shells. In fact, it appears that no difference would be produced if the sphere consisted of concentric layers separated by infinitely thin non-conducting partitions. For, assuming such a distribution, and replacing it for the instant by its corresponding lamellar magnet, of strength Φ at the point considered, we have for outside space a corresponding magnetic potential $\Omega = \frac{d}{dr}(Pr)$, where P is the potential due to a distribution of matter whose density is $-\Omega/r$ throughout the shell. The components of the vector potential outside are, as before, o, $\frac{dP}{\sin\theta d\phi}$, $-\frac{dP}{d\theta}$. Inside the shell Ω clearly denotes the potential *of the magnetic induction* of this equivalent lamellar magnet, from which the vector potential inside is derived by the same formula. We have $\nabla^2 P$ equal to $-4\pi\Phi/r$ inside the shell, and equal to zero outside, by Laplace's equation.

the currents are spherically stratified, their stream function: and magnetic field, its vector potential. Magnetic induction. Equation of potential.

The equations which give the currents become

Equations of electric flow,

$$\sigma\frac{d\Phi}{r\sin\theta d\phi} = -\frac{d}{dt}\frac{d(P+P_0)}{\sin\theta d\phi} - \frac{d\psi}{rd\theta},$$

$$-\sigma\frac{d\Phi}{rd\theta} = \frac{d}{dt}\frac{d(P+P_0)}{d\theta} - \frac{d\psi}{r\sin\theta d\phi},$$

$$0 = \frac{d\psi}{dr},$$

where P_0 represents the influencing field, and σ is the specific resistance of the medium.

verified to be in strata.

These are satisfied, as before, by $\psi = 0$,

Key equation.

$$\frac{\sigma\Phi}{r} = -\frac{d}{dt}(P+P_0).$$

But

$$\nabla^2 P = -\frac{4\pi\Phi}{r};$$

Final single characteristic equation: with interfacial conditions.

therefore

$$\nabla^2 P = \frac{4\pi}{\sigma}\frac{d}{dt}(P+P_0), \tag{3}$$

which determines P to satisfy all the conditions of the problem*. The boundary conditions are simply that the current shall not become infinite at the surface, and therefore that the distribution Φ shall not have a finite surface-density; this requires that dP/dr as well as P shall be continuous at the boundary.

* An external field P_0 suddenly established produces a current system soaking in strata inwards from an initial current sheet for which $P = -P_0$.

To obtain a solution, let us assume a varying external magnetic field given by

Solution for inducing field of one harmonic term:

$$P_0 = A_0 e^{\iota\kappa t} r^n Y_n;\ \text{viz. } \Omega = (n + 1)\, P_0;$$

then
$$P = Q e^{\iota\kappa t}\, Y_n,$$

where Q is a function of r; and substituting in equation (3), we find

$$\frac{d^2Q}{dr^2} + \frac{2}{r}\frac{dQ}{dr} - \frac{n(n+1)}{r^2}\,Q = \frac{4\pi}{\sigma}\,\iota\kappa\,(Q + A_0 r^n);$$

therefore
$$\frac{d^2Q}{dr^2} + \frac{2}{r}\frac{dQ}{dr} + \left(\lambda^2 - \frac{n(n+1)}{r^2}\right) Q = -\lambda^2 A_0 r^n,$$

where
$$\lambda^2 = -\iota\,\frac{4\pi\kappa}{\sigma}.$$

The general solution of this equation is

$$Q = -A_0 r^n + L\,(\lambda r)^n \left(\frac{1}{\lambda r}\frac{d}{d\lambda r}\right)^n \frac{\sin \lambda r}{\lambda r} + M\,(\lambda r)^n \left(\frac{1}{\lambda r}\frac{d}{d\lambda r}\right)^n \frac{\cos \lambda r}{\lambda r}, \qquad (4)$$

forms for outside and for inside.

separable into two parts of which one vanishes at the centre, and the other term vanishes at infinity*.

In the case of a shell, the values of L, M are to be determined by means of two sets of boundary conditions.

In the case of a solid sphere, it is better for our purpose to proceed as follows.

Case of solid sphere:

Write
$$Q_1 = Q r^{\frac{1}{2}},$$

and the equation, wanting the right-hand side, becomes

$$\frac{d^2Q_1}{dr^2} + \frac{1}{r}\frac{dQ_1}{dr} + \left(\lambda^2 - \frac{(n+\frac{1}{2})^2}{r^2}\right) Q_1 = 0,$$

of which the solution for this case, when Q_1 is to be finite at the origin, is

$$Q_1 = A_1 J_{n+\frac{1}{2}}(\lambda r).$$

finite Bessel series:

Hence
$$Q = \frac{A_1}{r^{\frac{1}{2}}} J_{n+\frac{1}{2}}(\lambda r),$$

or, say,

$$Q = L\,(\lambda r)^n \left\{1 - \frac{(\lambda r)^2}{2\,.\,2n+3} + \frac{(\lambda r)^4}{2.4.2n+3.2n+5} - \dots\right\}.$$

* This form is more suitable when the medium is a dielectric: a disturbance then subsides in waves of type expressed in real form by

$$P = (\lambda r)^n \left(\frac{1}{\lambda r}\frac{d}{d\lambda r}\right)^n \left\{L\,\frac{\sin(\lambda r - nt)}{r} + M\,\frac{\cos(\lambda r - nt)}{r}\right\} Y_n,$$

where $\lambda^2 = \kappa_n{}^2/c^2$, now real, and Y_n is a spherical surface harmonic of degree n expressing its angular character. The principal type $n = 1$ works out simply: for example the free periods of a dielectric sphere can be found.

Thus the solution is

$$P = \{-A_0 r^n + A_1 r^{-\frac{1}{2}} J_{n+\frac{1}{2}}(\lambda r)\} Y_n e^{\iota\kappa t}$$

inside the sphere, and

$$P = \{-A_0 a^n + A_1 a^{-\frac{1}{2}} J_{n+\frac{1}{2}}(\lambda a)\} \left(\frac{a}{r}\right)^{n+1} Y_n e^{\iota\kappa t},$$

outside the sphere, where

$$-nA_0 a^{n-1} + A_1 \frac{d}{da}\{a^{-\frac{1}{2}} J_{n+\frac{1}{2}}(\lambda a)\}$$

$$= (n+1) A_0 a^{n-1} - (n+1) A_1 a^{-\frac{3}{2}} J_{n+\frac{1}{2}}(\lambda a);$$

or $$A_1 = \frac{(2n+1) a^{n-1} A_0}{\{(n+1) a^{-1} + d/da\} \{a^{-\frac{1}{2}} J_{n+\frac{1}{2}}(\lambda a)\}}. \quad (5)$$

solution completed. And by expanding the potential of the field in a series of harmonics of different orders, we obtain the general solution as the sum of the corresponding special solutions for each term.

Approximation: Now λr is usually very small, in so far as λ^2 involves frequency divided by specific resistance, so that we may write

$$Q = L(\lambda r)^n \left\{1 - \frac{(\lambda r)^2}{2(2n+3)}\right\},$$

when we have

$$P = \left[-A_0 r^n + B_1 (\lambda r)^n \left\{1 - \frac{(\lambda r)^2}{2(2n+3)}\right\}\right] Y_n e^{\iota\kappa t},$$

where $$B_1 = \frac{(2n+1) a^{n-1} A_0}{\left\{(n+1) a^{-1} + \dfrac{d}{da}\right\}\left\{(\lambda a)^n - \dfrac{(\lambda a)^{n+2}}{2(2n+3)}\right\}},$$

or $$B_1 \lambda^n = \frac{(2n+1) A_0}{2n+1 - \frac{1}{2}(\lambda a)^2}.$$

Thus, finally, for a sphere in a varying field of force given by P_0, for which $\iota\lambda^2 = \frac{4\pi\kappa}{\sigma}$ is a small quantity, the value of P inside is given by

$$-\left[-1 + \frac{1 + \dfrac{(\lambda r)^2}{2(2n+3)}}{1 + \dfrac{(\lambda a)^2}{2(2n+1)}}\right] A_0 r^n Y_n e^{\iota\kappa t},$$

or $$-\frac{\lambda^2}{2}\left(\frac{r^2}{2n+3} - \frac{a^2}{2n+1}\right) A_0 r^n Y_n e^{\iota\kappa t};$$

or, taking real parts only, if the field is given by

$$P_0 = A_0 r^n Y_n \cos \kappa t,$$

we have $$P = -\frac{2\pi\kappa}{\sigma}\left(\frac{r^2}{2n+3} - \frac{a^2}{2n+1}\right) A_0 r^n Y_n \sin \kappa t \quad (6)$$

inside the sphere; and

$$P = + \frac{4\pi\kappa A_0}{(2n+1)(2n+3)\sigma} \frac{a^{2n+3}}{r^{n+1}} Y_n \sin \kappa t \qquad (7)$$

outside the sphere.

Remembering that the magnetic potential is $\frac{d}{dr}(Pr)$, we see that the effect of the sphere is to introduce a new varying field with phase diminished by a quarter period, and which in other respects bears the same relation to the old as the electrostatic field produced by an uninsulated sphere of the same radius bears to the inducing field, when the strength of the former is diminished in the ratio **result described,**

$$\frac{n}{(n+1)(2n+1)(2n+3)} \frac{4\pi\kappa a^2}{\sigma}.$$

By combining the proper harmonic solutions, any possible case can of course be represented.

The form of the solution obtained in equation (4) suggests another way of looking at the problem. The first term gives a field equal and opposite to the inducing field at each instant; the following terms represent its decay. Since $\nabla^2 P_0 = 0$, that opposite field can only be due to a surface distribution of currents. Thus, if at each instant we suppose a system of currents to start in the superficial layer of the body which neutralizes for internal space the effect of the outside changes, the actual state of the body is that produced by these currents soaking into it and decaying by their own mutual action. The equation of decay (3) with P_0 omitted is the same as the equation for the diffusion of heat from the surface into the body, though the boundary conditions are different; the corresponding thermal diffusivity (conductivity divided by thermal capacity) will be $\sigma/4\pi$. The value of this for copper is about 130; and as the actual heat diffusivity for copper is about 1·2, we see that to have penetration into the solid *of the same order* in both cases the oscillations must be about 100 times quicker for the electrical case. **as a screening current sheet which soaks into the sphere,** **not unlike heat conduction:** **comparison.**

V. We now proceed to consider the practically important case of a conductor in motion in a magnetic field. If one considers any closed circuit in the conductor, it is clear that the electromotive force round it depends only on the change produced by its motion in the number of tubes of force that it encloses, and is therefore quite independent of whether the relative motion of the conductor and the field be ascribed to the conductor or to the magnetic field, or to both conjointly. Therefore the currents induced in the body are derived from the same equations, whether the axes are fixed, or are moving *in any manner, uniform or not*. But in the case of an unclosed circuit **Spinning conductor in steady field:** **effects are relative,** **so independent of frame of reference,**

there is a difference introduced in the value of the electrostatic

except for an electric potential:

potential. In fact such an open line which is at rest relatively to the moving axes is displaced across the field, owing to the motion of the axes: if one supposes the ends of the line in its former position (1)

its rationale:

and in a near position (2) at a very short distance from it, to be connected so as to form a closed circuit, the number of tubes of force on the positive side of the line will be diminished by the number which pass through this closed circuit supposed circulating in the direction of the isolated arrows. The diminution is therefore equal to the flux of the vector potential along (2) minus its flux along (1), together with its fluxes along the two lines of motion of the ends of the open line. Thus, when the line has moved from (1) to (2), we must suppose the potential at each end diminished by the flux of the vector potential along the line of motion of that end. Therefore in the equations for electromotive force we must include terms for the change of the rate of variation of this flux as we pass from point to point of the conductor; that is, instead of the true electrostatic potential Φ of the outside charges, we shall get from our equations $\Phi + \Phi'$, where Φ' is the scalar product of vector potential and velocity of the point supposed connected with the moving system of axes, and is therefore

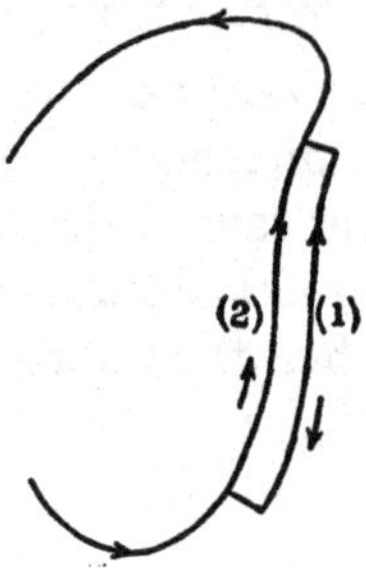

the induced electric potential.

$$-\left(F\frac{\delta x}{\delta t} + G\frac{\delta y}{\delta t} + H\frac{\delta z}{\delta t}\right).$$

We have thus deduced from first principles the result obtained by Maxwell analytically by transformation from his equations of the electromagnetic field.

The method that we have adopted also brings before us very clearly what it is on which the term Φ' really depends*. It will have different values according as we take one or another body in the system to be absolutely at rest; and as there is no criterion of absolute rest at all, so far as matter is concerned, we must conclude that the true value of Φ' is that derived from axes fixed with reference to some system or medium which is the seat of the electromagnetic action.

The aether as absolute frame.

We conclude, then, that when a *constant* electromagnetic system is moving through this medium in any manner, the effect produced by the relative motion is an electrostatic charge of the system of such character that its potential is Φ'. This static charge, however, itself exerts a magnetic effect by virtue of its motion; but it is easy

* It should be compensated by a charge in the body as *infra*, of insensible amount which ought to be independent of the frame of reference.

to see that this depends on v^2, and is therefore very minute. We know nothing of the relation of this medium to ponderable matter: the motion of the one probably affects that of the other, so that we could not get the actual value of such an electrification, but it seems that this electrification represents the total effect produced. It is to be noticed that our argument is independent of axes of coordinates altogether; and that for any motion of a constant electromagnetic system, the value of Φ' at any point is the scalar product of the vector potential and the relative velocity of the conductor with respect to the medium at the point.

We can now simplify the equations which give the electric currents when the conductor is in motion, as we can reduce it to rest, and solve the corresponding relative problem, where motion across the lines of force is replaced by a variation of the field itself.

(1) Let us first take the case of a spherical shell rotating with angular velocity ω in a field given, as before, by the function P_0. We found for the case of the shell at rest the equation

Spinning shell:

$$R\Phi = -a\frac{d}{dt}(P + P_0);$$

in this case it becomes $$R\Phi = -a\omega\frac{d}{d\phi}(P + P_0). \tag{8}$$

To obtain a solution, let us take a harmonic term of order n and type s, as follows:

in a tesseral harmonic field:

$$P_0 = A_0\left(\frac{r}{a}\right)^n Y_n^s e^{\iota s\phi},$$

$$P_1 = A\left(\frac{r}{a}\right)^n Y_n^s e^{\iota s\phi}, \quad P_2 = A\left(\frac{a}{r}\right)^{n+1} Y_n^s e^{\iota s\phi};$$

and therefore $$\Phi = -A\frac{2n+1}{4\pi}Y_n^s e^{\iota s\phi}.$$

On substituting, we have

$$-AR\frac{2n+1}{4\pi} = -\iota a\omega s(A + A_0);$$

therefore $$A = \frac{-A_0}{1+\iota p}, \quad = \frac{-1+\iota p}{1+p^2}A_0,$$

where $$p = \frac{(2n+1)R}{4\pi a\omega s};$$

Taking real parts only, we find that if

$$P_0 \propto A_0 \cos s\phi,$$

then $$P_1 \propto A_0\frac{-\cos s\phi - p\sin s\phi}{1+p^2}$$

lag of the induced field:

$$\propto -A_0\cos\alpha.\cos(s\phi - \alpha), \tag{9}$$

where $\tan\alpha = p$; and for the total inside field, we have

$$P_0 + P_1 \propto A_0 \sin\alpha \cos(s\phi + \tfrac{1}{2}\pi - \alpha);$$

which shows a lagging of $(\frac{1}{2}\pi - \alpha)/s$, while the intensity is reduced in the ratio of $\sin\alpha$ to unity, where $\alpha = \tan^{-1} p$ and so depends on the spin and the tesseral order.

In a uniform field, whose magnetic force is F_0, the potential due to the rotating shell will be

case of uniform field:

$$\frac{Fa^3}{2r^2} \frac{\cos\phi + p\sin\phi}{1 + p^2}, \quad p = \frac{3R}{4\pi a\omega};$$

result described: and the shell will therefore have the same outside effect as a simple magnet of moment $Fa^3/2\sqrt{1 + p^2}$, whose axis is inclined to the direction of the force at an angle $\tan^{-1} p$.

The opposing couple experienced by the rotating shell will therefore be the same as for this magnet, *i.e.* it will be

torque on the spinning shell:

$$G = F^2a^3p/2(1 + p^2);$$

driving power required, and the rate of expenditure of power required to keep up the rotation will be

$$G\omega.$$

compared with a spinning ring: (For a Delezenne's ring, of the same thickness and diameter, and 1 cm. in breadth, rotating in the same field, we have

$$\text{Current} = \frac{\pi a^2 \sin\phi}{2\pi a R} \omega F;$$

rate of expenditure of power in driving it

$$= \frac{\pi a^3 \omega^2 \sin^2\phi}{2R} F^2,$$

the mean value of which $= \dfrac{\pi a^3 \omega^2}{4R} F^2;$

and the couple $= \dfrac{\pi a^3 \omega \sin^2\phi}{2R} F^2,$

the mean value of which $= \dfrac{\pi a^3 \omega}{4R} F^2.$

ratio for ring increases with speed. The ratio of this effect to that produced in the shell is $\dfrac{3}{8}\dfrac{1 + p^2}{p^2}$, which is small for slow speeds, but increases without limit as the speed increases.)

We notice that if p is small,

$$P_1 = -A_0 \cos s\phi - A_0 p \sin s\phi;$$

Screening incomplete: so that the effect is (1) to neutralize the outside field throughout the interior of the shell, (2) to add on a new weak field equal to the former turned through an angle $\pi/2s$ in the negative direction and multiplied by p.

For moderate values of s, since R is large, p can be small only by a and ω or both becoming large. In particular, for a copper shell $\frac{1}{3}$ cm. thick, $R = 3 \times 1640$ about: thus, if in a uniform field of force ($s = 1$, $n = 1$) p is to be so small as $\frac{1}{5}$, we must have

$$a\omega = 5870,$$

which it is impossible to realize. In a possible case, say $\omega = 2\pi \times 20$ (which corresponds to 1200 turns per minute, an ordinary speed for the armature of a dynamo), $a = 10$ cm., thickness of shell $= \frac{1}{3}$ cm., we find $p = 1$ nearly; and the result is that inside the shell half of the field is rotated through a right angle, the other half remaining as before. If the thickness were 1 cm., we would have $p = 3$, the field inside would be diminished to $\frac{1}{10}$, and a new field of $\frac{3}{10}$ of the former turned through a right angle would be added on. its defect not small: concrete examples.

(2) For the case of a thick shell or a solid sphere, the equation for P_0 which, when the sphere is at rest, has been found to be Spinning thick shell:

$$\nabla^2 P = \frac{4\pi}{\sigma}\frac{d}{dt}(P + P_0),$$

now becomes

$$\nabla^2 P = \frac{4\pi}{\sigma}\,\omega\,\frac{d}{d\phi}(P + P_0). \qquad (10)$$

The solution can be conducted as above. Taking an external field,

$$P_0 = A_0 r^n Y_n^s e^{\iota s\phi},$$

then

$$P = Q Y_n^s e^{\iota s\phi},$$

where Q is a function of r; and substituting in equation (10), we find

$$\frac{d^2Q}{dr^2} + \frac{2}{r}\frac{dQ}{dr} - \frac{n(n+1)}{r^2}Q = \iota\frac{4\pi}{\sigma}\omega s(Q + A_0 r^n);$$

therefore

$$\frac{d^2Q}{dr^2} + \frac{2}{r}\frac{dQ}{dr} + \left\{\lambda^2 - \frac{n(n+1)}{r^2}\right\}Q = -\lambda^2 A_0 r^n,$$

where

$$\lambda^2 = -\iota\frac{4\pi\omega s}{\sigma}.$$

And we adapt our previous solution for Q by writing ωs for κ.

Thus, in the case of a solid sphere in a field of force given by spinning solid sphere.

$$P_0 = A_0 r^n Y_n^s \cos s\phi,$$

we have, by (6), when we neglect squares of $\frac{4\pi\omega}{\sigma}$,

$$P = -\frac{2\pi\omega s}{\sigma}\left(\frac{r^2}{2n+3} - \frac{a^2}{2n+1}\right)A_0 r^n Y_n^s \sin s\phi \qquad (11)$$

approximately inside the sphere, and

$$P = \frac{4\pi\,\omega s\,A_0}{(2n+1)(2n+3)\,\sigma}\frac{a^{2n+3}}{r^{n+1}}\,Y_n^s \sin s\phi \tag{12}$$

outside the sphere.

This solution may be expressed as before.

We have found that a magnetic potential

$$\Omega_0 = A_0 r^n\, Y_n^s \cos s\phi$$

Induced currents for tesseral field: generates a steady system of currents in the rotating conductor whose external potential is

$$\Omega_1 = A\,\frac{a^{2n+3}}{r^{n+1}}\,Y_n^s \sin s\phi,$$

where $$A = A_0 \frac{4\pi\omega s}{\sigma}\frac{n}{(n+1)(2n+1)(2n+3)}.$$

This potential is the same as would be produced by a distribution of magnetic matter over the surface of the sphere whose surface density D is given by

$$D = -A\,\frac{2n+1}{4\pi}\,a^{n+1}\,Y_n^s \sin s\phi.$$

resulting torque: The couple exerted by the system Ω_0 to retard the rotation is therefore

$$G = \iint D\,\frac{d\Omega_0}{d\phi}\,dS$$

$$= AA_0\,\frac{(2n+1)\,s}{4\pi}\,a^{2n+3}\,\pi\int_{-1}^{1}(Y_n^s)^2\,d\mu$$

$$= A_0^2\,a^{2n+3}\,\frac{\pi\omega s^2}{\sigma}\,\frac{n}{(n+1)(2n+3)}\,\frac{2}{2n+1}\,\frac{(n+s)!}{(n-s)!}$$

(Ferrers's *Spherical Harmonics*, p. 86), on taking

$$Y_n^s = \sin^s\theta\left(\frac{d}{d\mu}\right)^s P_n,$$

where $\mu = \cos\theta$, and P_n is the zonal harmonic of order n.

And by combining such conjugate harmonics in the proper manner, we obtain the general solution in the form

$$\Omega_0 = \Sigma\left(\frac{r}{a}\right)^n Y_n^s\,(A_n^s \cos s\phi + B_n^s \sin s\phi),$$

$$G = \frac{2\pi a^3}{\sigma}\,\omega\Sigma\,\frac{ns^2}{(n+1)(2n+1)(2n+3)}\,\frac{(n+s)!}{(n-s)!}\,(A_n^{s2} + B_n^{s2}) \tag{13}$$

and the rate of expenditure of power required to keep up the motion is

$$G\omega.$$

In a uniform field $n = 1$, $s = 1$; therefore if the magnetic force is F in the direction $\theta = \frac{1}{2}\pi$, $\phi = 0$, we have $F = -A_0$; result for uniform field:

$$\Omega_1 = -F\frac{2}{15}\frac{\pi\omega}{\sigma}\frac{a^5}{r^2}\sin\theta\sin\psi,$$

$$G = \frac{2}{15}\frac{\pi\omega}{\sigma}a^5F^2.$$

We see that, still neglecting squares of the small quantity $4\pi\omega/\sigma$, the rotating sphere behaves as a magnet of moment equivalent magnet.

$$\frac{2}{15}\frac{\pi\omega}{\sigma}a^5F$$

broadside on to the direction of the force, with its poles placed so as to oppose the motion.

VI. We shall next consider the damping of the torsional oscillations of a spherical conductor in a magnetic field. If $2\pi/\kappa$ denote the period of the oscillations, we have to write $e^{\iota\kappa t}\omega$ instead of ω in the equations. Damping of oscillations,

First, taking the case of a thin shell, we have of thin conducting shell:

$$R\Phi = -a\omega e^{\iota\kappa t}\frac{d}{d\phi}(P + P_0);$$

hence $$-AR\frac{2n+1}{4\pi} = -\iota a\omega s e^{\iota\kappa t}(A + A_0),$$

or $$A = \frac{-\iota a\omega s e^{\iota\kappa t}}{\iota a\omega s e^{\iota\kappa t} - (2n+1)/4\pi . R}A_0;$$

and taking the real part only, we have for the induced field, corresponding to $A_0 \cos s\phi$ and $\omega \cos \kappa t$, the real part of

$$\frac{-\cos s\phi - p(\cos\kappa t - \iota\sin\kappa t)\sin s\phi}{1 + p^2(\cos 2\kappa t - \iota\sin 2\kappa t)}A_0,$$

where $$p = \frac{(2n+1)R}{4\pi a\omega s}.$$

This

$$= \frac{-(\cos s\phi + p\cos\kappa t\sin s\phi)(1 + p^2\cos 2\kappa t) - p^3\sin\kappa t\sin 2\kappa t\sin s\phi}{1 + p^4 + 2p^2\cos 2\kappa t}A_0$$

$$= -\frac{(1 + p^2\cos 2\kappa t)\cos s\phi + p(1 + p^2)\cos\kappa t\sin s\phi}{1 + p^4 + 2p^2\cos 2\kappa t}A_0. \quad (14)$$

It is the part of this expression containing $\sin s\phi$ that furnishes the retarding couple. Thus, in a field whose potential is

$$\Omega = A_0\left(\frac{r}{a}\right)^n Y_n^s \cos s\phi,$$

we find $$\text{Couple} = A_0^2\frac{p(1 + p^2)\cos\kappa t}{1 + p^4 + 2p^2\cos 2\kappa t}s\frac{(n+s)!}{(n-s)!};$$ torque, for uniform field: and in the case where the force is uniform and equal to F_0 we find

$$\text{Couple} = a^2F_0^2 \frac{p(1+p^2)\cos\kappa t}{1+p^4+2p^2\cos 2\kappa t},$$

where now $p = 3R/4\pi a\omega$, and the angular velocity is $\dot{\theta}, = \omega\cos\kappa t$; and therefore we may write the couple

equation for damping.

$$= a^2F_0^2 \frac{\omega p(1+p^2)\dot{\theta}}{\omega^2(1-p^2)^2+4p^2\dot{\theta}^2}.$$

For a thin shell, p is large, and the damping is regular.

Solid sphere; Secondly, for a solid sphere,

$$\nabla^2 P = \frac{4\pi}{\sigma}\omega e^{\iota\kappa t}\frac{d}{d\phi}(P+P_0).$$

And, as before, we have approximately, if

$$P_0 = A_0 r^n Y_n^s \cos s\phi,$$

$$P = -\frac{2\pi\omega e^{\iota\kappa t}}{\sigma}\left(\frac{r^2}{2n+3} - \frac{a^2}{2n+1}\right) A_0 r^n Y_n^s \sin s\phi, \qquad (15)$$

the real part of which corresponds to an angular velocity $\omega\cos\kappa t$. In fact, to our degree of approximation the whole solution is the same as for rotation, with $\omega\cos\kappa t$ written for ω.

in uniform field: For a sphere whose moment of inertia about a diameter is I, in a uniform field whose magnetic force is F, we have the equation of motion

$$I\frac{d^2\phi}{dt^2} + m\phi + \frac{2}{15}\frac{\pi}{\sigma}a^5F^2\frac{d\phi}{dt} = 0, \quad m = \kappa^2 I;$$

therefore $$\phi = \omega e^{-\frac{\pi a^5F^2}{15\sigma I}t}\cos\kappa t,$$

when the damping is small. So that the vibrations decay owing to

rate of subsidence: electromagnetic action at logarithmic rate $\dfrac{\pi a^5F^2}{15\sigma I}$, which is equal to $\dfrac{F^2}{6\sigma d}$, where d is the density of the sphere.

If we write $\dfrac{F^2}{6\sigma d} = \lambda$, we have the general solution

$$\phi = \omega e^{-\lambda t}\cos t\sqrt{\kappa^2-\lambda^2};$$

becomes dead-beat: thus the motion will just cease to be vibrational when $\lambda = \kappa$, *i.e.* when

$$F^2 = 6\sigma d\kappa.$$

illustration. For copper, taking $a = 3$ cm., $\sigma = 1640$, $d = 8{\cdot}8$, and $\kappa = \frac{1}{10}$, which is thus our first order of small quantities, we find

$$F = 93, \text{ nearly},$$

or about 200 times the earth's magnetic force.

VII. It is interesting to look back over the solutions, and observe what part of the result is due to the mutual action of the induced currents; we shall then be able to form an estimate of the cases in which their mutual action may be neglected. Now the effect of neglecting this mutual action is merely to drop P in the formation of the potential functions. In the accurate solutions, for thin shells the part of the induced field which is directly opposed to the original field will disappear. This part is, for a solution of the nth order, the fraction $4\pi a\kappa/(2n+1)R$ of the other component; and the approximate solution would hold to the first order of this small quantity. In the case of thick shells, we would neglect λ^2 on the right-hand side of equation (3), and the solution then will be exactly the approximation to which we were conducted on neglecting squares of the above small quantity, as may be easily shown. Simplification by neglecting mutual induction.

For a bounded plane sheet we have, if the magnetic potential is $\frac{dP}{dz}$, and r, θ, z are columnar coordinates, Formulation for bounded plane sheet:

$$R\frac{d\Phi}{rd\theta} = -\frac{d}{dt}\frac{d(P+P_0)}{rd\theta} - \frac{d\psi}{dr},$$

$$-R\frac{d\Phi}{dr} = \frac{d}{dt}\frac{d(P+P_0)}{dr} - \frac{d\psi}{rd\theta},$$

of which the *general* solution is

$$R\Phi = -\frac{d(P+P_0)}{dt} + \chi, \tag{1}$$

where

$$\frac{d\chi}{rd\theta} = -\frac{d\psi}{dr}, \quad \frac{d\chi}{dr} = \frac{d\psi}{rd\theta};$$

and therefore

$$\frac{d}{dr}\left(r\frac{d\chi}{dr}\right) + \frac{1}{r}\frac{d^2\chi}{d\theta^2} = 0, \tag{2}$$

which shows that χ is a potential function in two dimensions. Also

$$2\pi\Phi = -\frac{dP}{dz}. \tag{3}$$

We have thus to solve (1) and (3), with χ so determined as to make Φ constant round the boundary of the sheet.

If we neglect self-induction, we have solvable when mutual induction negligible.

$$R\Phi = -\frac{dP_0}{dt} + \chi,$$

where χ is a two-dimension potential function, finite at the origin, and equal to a constant plus dP/dt at the boundary, the determination of which is a well-known problem in electrostatics.

In the *Phil. Mag.* for 1880 Professor Guthrie and Mr C. V. Boys give details of experiments on the couples experienced by Arago's C. V. Boys' experiments on torque.

discs in a rotating magnetic field; and in a second paper is given an account of the determination of the relative resistance of polarizable liquids from the couples experienced by the containing vessel in a uniform rapidly rotating magnetic field. It would be very interesting to determine [*infra*, Appendix, p. 29] the absolute value of the couple for a copper sphere in a uniform rotating field of known strength, and to determine the log. decrement of its torsional oscillations due to induction in a stationary field, and to deduce thence, by approximation from the general formulae, two values for the specific resistance of its material at the given temperature in absolute units, which should be consistent with themselves and with the known value. For a uniform thin spherical copper shell the calculation would be very simple, and at the same time the temperature conditions uniform. No absolute quantitative experimental verification appears to have yet been made for induction-currents in continuous media.

H. Hertz. Since writing the above, I have found that H. Hertz has given solutions of problems of rotation to a considerable extent identical with those discussed here (Inaugural Dissertation, Berlin, 1880, p. 93). I have not yet been able to procure his paper*; but the results, so far as indicated in a notice in *Phil. Mag.* Dec. 1880, agree with the above.

VIII. In the case of the Earth rotating round its axis we have, approximately,

$$a\omega = \frac{10^9}{2\pi}\frac{2\pi}{24 \times 60 \times 60};$$

therefore, in a uniform field ($n = 1$, $s = 1$),

$$p = \frac{3R}{4\pi a\omega} = \frac{2R}{10^5} \text{ approximately,}$$

Screening of Earth by rotating upper atmosphere. where R is equal to specific resistance divided by thickness, so that p could be small, and therefore the screening nearly complete with but a moderate amount of conducting power in the rarefied strata of the atmosphere.

Although the sheet of air is very thick, yet it is still so thin in comparison with the radius of the earth that the solution applies to the problem by the principle of similarity, assuming Ohm's law to hold for the current. Weighty evidence has been brought together from different experimenters by Edlund (*Phil. Mag.* 1880) in favour of the theory that much the greater part of the resistance experienced by the electric discharge in passing through vacuum tubes occurs at

* This inaugural dissertation "On Induction in Rotating Spheres" occupies pp. 35–126 in the English translation of Heinrich Hertz's *Miscellaneous Papers* edited posthumously by P. Lenard (Macmillan, 1896). See also p. 29 *infra*.

the surfaces of the electrodes, at which accordingly an opposing electromotive force may be supposed to act (*vide* C. F. Varley, *Proc. Roy. Soc.* 1871; E. Wiedemann, *Phil. Mag.* 1880). That being so, rarefied gases may well be fairly good conductors of electricity, even when they refuse to allow a spark to pass between electrodes, a view which is also adopted by Balfour Stewart in his theory of the magnetic variations mentioned below.

It is well known that disturbances of the terrestrial magnetic force are largely connected with changes of the sunspots. There is also a daily variation in the magnetic elements which follows the Sun, but lags behind him by about 40°. Lloyd and Chambers have shown that this is of such a character that it cannot be accounted for by a direct inductive magnetic action of the sun upon the Earth, and in particular that the lagging could not be so explained. Balfour Stewart (*Encyc. Brit.*, Art. "Meteorology") gives reasons for assigning the magnetic variations to the effect of currents actually produced in the upper conducting strata of the atmosphere by the circulation caused by the Sun's heating and tidal action. He also refers the large disturbances, amounting sometimes to $\frac{1}{100}$ of the total force, to the same cause; and as any cause capable of such large effects in a short time may also be assumed to be capable of producing the total magnetism of the Earth (cf. W. G. Adams, *B. A. Report*, 1880), he is inclined to look upon the Earth's magnetism as the accumulated effect of such action.

Sunspots affect Earth's magnetic field, indirectly.

Its variations referred to currents induced by circulation of upper atmosphere,

effect might accumulate like a dynamo.

Now, even leaving out of account its high temperature, we could not suppose the matter of the Sun to be so strongly magnetized as to produce direct effects at the distance of the Earth; but if the very considerable changes in the Earth's magnetism as well as in the long run its total amount, are to be referred to the effect of convection currents produced by the Sun's radiation, we must be prepared to admit that, owing to the enormous convection effects which take place in the Sun, combined with the great velocity at its surface due to its rotation in 26 days, convective currents of electricity on an enormous scale are possible in it whose magnetic effect might be felt at the Earth. It is easy to verify that, if the Sun were equivalent to a magnet with 13,000 times the Earth's intensity of magnetization (*i.e.* if it were magnetized rather more strongly than we can magnetize steel), it could produce a variation of $\frac{1}{100}$ of the Earth's total force*.

Sun as a magnet could not directly affect Earth:

nor Solar electric currents

* This inadequacy of direct solar magnetization to produce the magnetic disturbance at the Earth was already emphasized by Lord Kelvin, *Phil. Trans.* 1863, note attached to W. C. Chambers' memoir: *Math. and Phys. Papers*, Vol. v. p. 164; also later. The solar effect is now usually ascribed to streams of electrons from the Sun passing the Earth.

Atmospheric screen admits the axial components, diminishes and delays the rest.

Our object is to point out the effect of the conducting atmospheric layer on all such hypotheses. If its conductivity is sufficiently high, it will screen off all such external influences except the components parallel to the Earth's axis of rotation; and in any case it will partly screen them off, and at the same time produce a lagging behind the Sun of an amount determined above. On the other hand, if, adopting Sir C. W. Siemens's hypothesis, we were to suppose that the outer part of this conducting stratum did not participate fully in the rotation of the Earth, currents would be induced in it which would screen external space from the part of the Earth's magnetism which is not symmetrical round its axis, and the reaction of these currents would tend to keep the atmosphere rotating.

The sudden and very brilliant outburst of light that was observed on the Sun by Carrington in 1859 was shown to be accompanied well within an interval of 15 seconds by a very sudden and well-marked disturbance of the magnetometers, which, however, went on increasing afterwards (*vide* Balfour Stewart, *loc. cit.*). Now the light took about eight minutes to reach the observer; and it seems therefore that we have good warrant for concluding that this magnetic disturbance was not propagated from the Sun directly and practically instantaneously, as was shown long ago by Laplace to be the case with gravitation; but that either (1) it was an immediate indirect effect produced in the atmosphere by the sudden increase of radiation, or (2) it was an effect propagated from the Sun *with the same velocity as light.*

Galway,
Nov. 13, 1883.

Appendix.

ELECTROMAGNETIC INDUCTION IN SPHERES.

[*Philosophical Magazine*, April 1884.]

IN the *Philosophical Magazine* for January 1884, Prof. Larmor has published a paper containing highly interesting researches on the electromagnetic induction in continuous conductors, especially in spherical shells and in spheres. I hope I may be allowed to publish a few lines in order to show that the results of Prof. Larmor concerning the induction in spheres have been published some years ago by myself.

Previous investigations for spheres:

In a paper, "Ueber die Schwingungen eines Magnets unter dem Einfluss einer Kupferkugel,"[1] I have dealt with the electricity induced in a copper sphere by an oscillating magnet and with the influence of the induced currents on the motion of the magnet. In a second paper, "Einige Versuche über Induction in körperlichen Leitern,"[2] I have treated the same problem for a system of magnets. This includes the case of a rotating sphere and magnets at rest, the electromagnetic induction being the same whether the sphere is at rest and the magnet rotates round an axis, or the sphere is rotating round this axis and the magnet is at rest. Moreover I have developed the solution of the equations for the case of a sphere rotating in a uniform magnetic field, and for all the cases I have determined the couple of the induced currents on the moving magnets or on the rotating sphere. As Prof. Larmor has given his formulae expressed in Bessel's functions and in spherical harmonics, they can easily be compared with mine given in terms of the same functions. Indeed the results agree very well.

with experimental tests.

At page 21 of his paper Prof. Larmor says: "It would be very interesting to determine the absolute value of the couple for a copper sphere in a uniform rotating field of known strength, and to determine the log. decrement of its torsional oscillations due to induction in a stationary field, and to deduce thence, by approximation from the general formulae, two values for the specific resistance....No absolute quantitative experimental verification appears to have yet been made for induction-currents in continuous media." I have tested my formulae by experiment, and derived from these experiments values for the specific resistance of copper. For a first copper sphere I have

[1] Inaugural Dissertation, Göttingen, 1875. *Nachrichten der kgl. Ges. der Wissenschaften zu Göttingen*, 1875, p. 308.

[2] Wiedemann's *Annalen*, Bd. XI. p. 812 (1880). *Nachrichten der kgl. Ges. der Wissenschaften zu Göttingen*, 1880, p. 491.

found: Diameter $d = 92{\cdot}94$ mm.; specific weight $s = 8{\cdot}88$; specific resistance $\lambda = 1/444278{\cdot}10^{12}$. For a second sphere: $d = 59{\cdot}85$ mm.; $s = 8{\cdot}9$; $\lambda = 1/474074{\cdot}10^{12}$. These numbers for λ agree sufficiently well with those found for copper wires, as follows: Jacobi, $\lambda = 1/374116{\cdot}10^{12}$; Kirchhoff, $\lambda = 1/451043{\cdot}10^{12}$; W. Weber, $\lambda = 1/463382{\cdot}10^{12}$. For galvanoplastic copper $\lambda = 1/513144{\cdot}10^{12}$.

F. HIMSTEDT, Dr, Professor.

Freiburg i. B., *March* 1884.

I regret that I was unacquainted with Dr Himstedt's papers when preparing my communication referred to in the above. The first paper I have not at present access to; and I did not find it mentioned by any subsequent writer on the subject. On referring to the second paper, I find, in addition to the very interesting experimental verification, a discussion of the general solution for solid spheres. This latter, however, is of a much more analytical character than the investigation contained in my paper, in which one of my aims was to reduce the mathematical analysis to as narrow limits as possible, replacing it by a discussion of the physical phenomena. A comparison of the results shows agreement when a slip in my work is corrected.... [See footnote p. 26 *supra*.]

J. LARMOR.

Queen's College, Galway,
March 22, 1884.

4

ON LEAST ACTION* AS THE FUNDAMENTAL FORMULATION IN DYNAMICS AND PHYSICS†.

[*Proc. London Mathematical Society*, xv. (1884) pp. 158–184.]

(i) On the Immediate Application of the Principle of Least Action to Dynamics of a Particle, Catenaries, and other related Problems.

1. It is a matter of constant observation that different departments of Mathematical Physics are closely connected together, so that the solution of a question in one branch of the subject admits of being transferred into another branch, and serving as the solution of a corresponding problem there.

To facilitate the general discussion of these relations between different problems, it is desirable to express each in its simplest mathematical terms, so that it shall involve as little reference as possible to the detailed theory of the special subject. Now nearly every such question can be expressed as a problem of maximum or minimum; and this mode of statement probably conveys a clearer and more compact representation of the question, and an easier grasp of its mathematical relations as a whole, than any other. It at once places the question beyond the accident of the particular coordinate system employed. We should therefore expect that it would be a direct and easy method of arriving at dynamical analogies.

Minimal relations the easiest to grasp and explore.

The development of this method is the object of the following paper. Many of the examples given are more or less known, but there does not seem to have been any systematic attempt to group them together as illustrations of a single general principle.

2. The complete mathematical statement of every question in Dynamics is involved in either of the forms of the Principle of Least Action, when once the expressions for the potential and kinetic energies of the system are known. The principle admits of easy and elementary proof, and therefore forms a suitable foundation from

Thomson and Tait.

* In subsequent expositions the term *Minimal Action* is employed as involving the general idea, with the limitation that for long paths it may only be stationary instead of least.

† This general subject has been treated later in a systematic physical memoir by Helmholtz, with rediscovery of Routh's and Kelvin's developments in terms of latent momenta, as *infra*, (ii) § 3, p. 46, Crelle's *Journal*, 100 (1886), and in relevant historical notes going back to Leibniz, now reprinted in *Abhandlungen*, Vol. III. There is also a historical exposition in a lecture by M. Planck, now translated into English in *A Survey of Physics*, pp. 109–129.

which to construct dynamical theory. See, for example, Thomson and Tait's *Nat. Phil.* Vol. I. Art. 327.

Thermo-dynamics. The prominent place that this principle directly holds in general dynamical theory is illustrated by the attempts that have been made to place the Second Law of Thermodynamics on a dynamical basis. The investigations of Clausius, Boltzmann, and Szily have shown that the Principle of Least Action in Dynamics is in close relationship with that law. And this agrees with what we might have expected from the above; for the law must have some dynamical interpretations, and the only [scalar] dynamical relations hitherto discovered that are absolutely general and independent of coordinate systems of representation are the Principle of Energy, of which the First Law of Thermodynamics is a case, and the Principle of Least Action.

Hamilton. Hamilton pointed out the importance and began the development of the general method of reducing the solution of every dynamical question to the consideration of a single function (*Phil. Trans.* 1834–1835); and it has since been widely extended and carried on by others (cf. Cayley's *Report, Brit. Assoc.* 1857). What we now wish to point out is the ease and simplicity with which the mere statement of the principle, without any consequent analytical work, reveals the existence and nature of many analogies between different problems, whose relationship has been hitherto treated in a more or less isolated manner.

Analogies.

When we wish to compare the motion of one system with the equilibrium of another, we have the variational energy condition of equilibrium at hand, which may, if we please, be itself considered to be a case of the general principle.

3. In what follows we shall develop the relations between different problems relating to the motion of a particle, and the analogies between the motion of a particle and the forms of systems of rays and of catenaries that flow immediately from the minimum principle. The application of the method of inversion and of other methods of transformation flows naturally from it, and we are guided immediately to a large number of generalizations of solvable cases of motion that have formed the subjects of separate papers by Liouville and others.

Transformations in dynamics of a particle.

It may be observed that the method for brachistochrones, given, from Woodhouse [after J. Bernoulli], in the Appendix to Tait and Steele's *Dynamics*, affords also a very simple and elementary geometrical proof of the Principle of Least Action for a particle. For, drawing the system of equipotential surfaces of the field of force, the velocity of the particle between any two consecutive surfaces is invariable, as the energy of the motion is given. To determine that path which makes $\int v\,ds$ least, we take three consecutive points on the

Proof for a particle.

path, and find the position that the intermediate one must have on the equipotential through it, in order to make the variation of the integral zero. We shall be led to the ordinary law of centrifugal force in the orbit; and it follows that the actual orbit is the one possessing the stationary property.

In another paper the same principle will be applied to the discussion of problems connected with general solid and fluid systems, cyclic and acyclic, and of certain close analogies with questions relating to elastic wires and electric currents[1].

Application of the Method of Inversion.

4. When a particle moves in a given field of force, the form of its orbit is determined by the condition that $\int v\,ds$ satisfies the maximum-minimum condition, in which v is given by the equation of energy

Eulerian Action form for orbits; time eliminated.

$$\tfrac{1}{2}v^2 + V = E, \text{ a constant},$$

where V is the potential of the field of force.

Now, if we invert the orbit with respect to any origin, with radius of inversion k, $\int v\,ds$ is equivalent to $k^2 \int \frac{v}{r'^2}\,ds'$, where dotted letters refer to the inverse orbit; since, by Geometry,

Inversion of orbits.

$$ds : ds' = r : r' = k^2 : r'^2.$$

This new integral therefore satisfies the maximum-minimum condition, when taken along the inverse curve; and therefore this curve is the orbit of a particle in a field of force which makes

$$v' \propto v r'^{-2},$$

and therefore $$\tfrac{1}{2}v'^2 \propto (E - V)\, r'^{-4},$$

in which V is now to be expressed in terms of the dotted letters.

The value of V', the potential of the new field of force, is therefore given by

$$AV' + (E - V)\, r'^{-4} = \text{constant},$$

or, what is the same,

$$r'^2 (E' - V') \propto r^2 (E - V).$$

The velocity of projection in the new orbit is determined by the above value of v'.

We notice that, if the original law of force only contain powers of the coordinates higher than the inverse fifth, this velocity-condition will in general be that the particle be projected with the velocity from infinity. (See ex. 2, § 5.)

The ratio of the velocities at corresponding points of the related orbits is clearly inversely proportional to that of the radii vectores.

[1] The original paper has been divided into two, and some other changes have been made, at the suggestion of the referee.

5. The following are examples:

(1) With no forces acting, a particle constrained to remain on a fixed surface describes a geodesic with constant velocity; the curve corresponding to this geodesic on the inverse surface satisfies the relation $\delta \int r^{-2} ds = 0$, and is therefore a free path for a particle constrained to remain on the surface, under the action of a force varying as the inverse fifth power of the distance from the centre of inversion, and projected with the velocity from infinity.

Orbits on a sphere: On a sphere, the geodesics are great circles. Therefore particles projected on the surface of a sphere under the action of a force varying inversely as the fifth power of the distance from any fixed centre, and with the velocity due to approach from infinity, will describe circles on the sphere, which are its curves of intersection with the system of spheres passing through the centre of force and its inverse point. In particular, if the original sphere becomes a plane, we get Newton's result for free motion in a circle with a centre of force on its circumference.

on a ring-surface: On a cylinder of any form of cross-section, the geodesics make equal angles with the generators. Therefore on the ring of vanishing aperture, which is the inverse of such a cylinder, the paths of particles projected under the same conditions as before (the centre of force being now at the aperture) will be curves, each of which cuts all the meridians of the surface at the same angle. This result holds for any ring with circular meridians all passing through the same point; in particular, for a common anchor ring of vanishing aperture.

Again, lines of curvature invert into lines of curvature, and umbilici into umbilici; for the directions of the lines of curvature at a point are determined by the nature of the contact with spheres touching at the point, and spheres invert into spheres. Therefore we arrive at the result that, on any inverse of an ellipsoid, and in particular on Fresnel's Surface of Elasticity (in which case the centre of the surface is the centre of force), all particles projected from one umbilicus under the law of the inverse fifth power and with the velocity from infinity, will, when reflected by a line of curvature, pass through the other umbilicus, will then be again reflected through the first, and so on continually. If the orbit pass through the two umbilici without reflection, it will have the form of a curve winding round and round the surface, and passing through them once in each revolution; if it do not pass through either, it will be such a curve touching the two branches of a line of curvature, one surrounding each umbilicus, once in each revolution. All this, of course, results from the well-known properties of geodesics on an ellipsoid.

on inverse of an ellipsoid.

Further, the orbits of particles moving on a surface under this law of force with any velocity of projection whatever, become on inversion the orbits described under similar conditions on the inverse surface; while, with the particular velocity discussed above, they become geodesics.

(2) For free motion in a plane the application is easy. Thus, if for the original orbit, the force (repulsive) $= \mu r^n$, we have Compound law of force:

$$\tfrac{1}{2}v'^2 = \left\{E + \frac{\mu}{n+1}\left(\frac{k^2}{r'}\right)^{n+1}\right\} r'^{-4},$$

the centre of inversion being the centre of force; and therefore in the inverted orbit the force $= \mu_1 r'^{-5} + \mu_2 r'^{-n-6}$, where the values of μ_1, μ_2 are obvious. If $n > -5$, the velocity will be that from infinity; while, if the velocity in the original orbit was that from the origin when $n > -1$, or to infinity when $n < -1$, we have the further simplification $\mu_1 = 0$.

We may also invert with respect to a point not in the plane of the orbit, and so obtain the fields of force in which curves drawn on a sphere are free orbits with proper velocity of projection. There is no need to supply examples. on sphere.

6. *Transformation by Conjugate Functions.*

If a plane area is transformed by writing ξ, η for x, y, each element of length near a point is altered in the ratio λ to 1, where Conformal plane transformations.

$$\lambda^2 = \frac{d\,(\xi\eta)}{d\,(xy)} = \left(\frac{d\xi}{dx}\right)^2 + \left(\frac{d\xi}{dy}\right)^2 = \left(\frac{d\eta}{dx}\right)^2 + \left(\frac{d\eta}{dy}\right)^2,$$

for in the original case ξ, η were the coordinates, but they are now replaced by x, y.

Hence the relation $$\delta \int v\,ds = 0$$

becomes $$\delta \int \lambda v\,d\sigma = 0,$$

where $d\sigma$ is the corresponding element of arc of the transformed curve*. The transformed curve is therefore the orbit, with proper velocity, in a field of force given by

$$V' \propto (E - V)\,\lambda^2.$$

* This is confused. The original plane of the orbits was (ξ, η), which is transformed into (x, y): the original orbits were expressed by an Action formula $\delta\int v d\sigma = 0$ with v as a function of ξ, η, which transforms into an Action formula $\delta\int \lambda v ds = 0$, with λ and the same v now expressed in terms of x, y. The argument may be verified on the example from Newton, which follows: also on the next footnote, in which conversely the Action $\int v ds$ in the transformed plane is expressed in terms of an Action in the original plane. The recent relativity theories find their essential basis in such Action-correspondences, there regarded as merely analytical transformations of the same physical system. Cf. "Newtonian Time essential to Astronomy" in supplement to *Nature*, April 1927.

For instance, if the orbit be the straight line

$$x + ay = b,$$

we have $V = \text{constant},$

and therefore the curve $\xi + a\eta = b$

[on the plane x, y] can be described freely in a field $V' \propto \lambda^2$.

Ex. from Newton.

On making $\xi = n \log r$, $\eta = n\theta$, we find $\lambda^2 = n^2 r^{-2}$, and therefore the spiral $r = Ae^{-a\theta}$ is the orbit under the law of the inverse cube, with velocity from infinity, as Newton proved.

7. We have here also a ready method of inventing fields of forces, in which the motion can be integrated. For example, motion in a field whose potential is given by

$$V = f(x) + F(y)$$

Compound force:

can obviously be reduced to quadratures. The transformed curves are free orbits in a field whose potential is

$$V' = \lambda^2 \{f(\xi) + F(\eta) - E\},$$

with proper velocity, E being the total energy of the original orbit. But further, being given this law of potential, without *any* velocity-condition, and transforming back again, we arrive at the law

$$V = f(x) + F(y) - E + C\lambda^{-2},$$

where C depends on the velocity; so that, if λ^{-2} is of the form $\phi(x) + \psi(y)$, we can integrate without any restrictions as to initial velocity.

Liouville.

Liouville arrived at these results from a discussion of the equations of motion.

Case of two centres of force:

(1) He applies them to the case of motion under two centres of force, by using elliptic coordinates with these centres as foci; thus, if μ, ν represent the semi-axes of the confocals through a point, and $2b$ the distance between the foci, we have

$$r = \mu + \nu, \quad r' = \mu - \nu,$$

$$ds^2 = (\mu^2 - \nu^2)\{(\mu^2 - b^2)^{-1} d\mu^2 + (b^2 - \nu^2)^{-1} d\nu^2\},$$

so that λ^{-2} is here equal to $\mu^2 - \nu^2$, and we can reduce to quadratures motion under any law

$$V = (\mu^2 - \nu^2)^{-1}\{f(\mu) + F(\nu)\} = (rr')^{-1}\{f(r + r') + F(r - r')\}^1.$$

[1] (Thus $\int v\,ds = \sqrt{2}\int\sqrt{E - V}.ds = \sqrt{2}\int\sqrt{(E - V)(\mu^2 - \nu^2)}\,d\sigma$

$$= \sqrt{2}\int\sqrt{E(\mu^2 - \nu^2) - (f\mu + F\nu)}\,d\sigma,$$

where $d\sigma^2 = d\mu'^2 + d\nu'^2$, with μ' and ν' functions of μ and ν as later, and this comes under the first form of § 7.)

A case of this is easily seen to be

$$V = gr^{-1} + g'r'^{-1} + kR^2,$$

where R is the distance from the centre. (*Vide* Liouville, Vol. XI. 1846; or Prof. Cayley's "Report on Special Problems of Dynamics," *Brit. Assoc.* 1862.)

Further, if $\psi(x, y) = c$ be the equation to the orbit under the law $V = f(x, y)$ when the energy of the motion is E, we see that $\psi(\mu', \nu') = c$ will be the equation to the orbit under the law

$$V = (\mu^2 - \nu^2)^{-1} \{f(\mu', \nu') - E\},$$

where $\mu' = \int(\mu^2 - b^2)^{-\frac{1}{2}} d\mu, \quad \nu' = \int(b^2 - \nu^2)^{-\frac{1}{2}} d\nu,$

the velocity being such that

$$\tfrac{1}{2}v^2 + V = 0.$$

For instance, in the field, simple example:

$$V = A'(\mu^2 - \nu^2)^{-1} = A'(rr')^{-1},$$

particles projected with the velocity from infinity will describe oblique trajectories of the system of confocals.

(2) On taking $\xi = \log r$, $\eta = \theta$, we find $\lambda^2 = r^{-2}$, and we have the Jacobi. solution immediately for the law

$$V = r^{-2}\phi\left(\frac{x}{y}\right) + \psi(r),$$

with *any* velocity of projection, since $\lambda^{-2}\ (= x^2 + y^2)$ is of the proper form.

This law has been discussed by Jacobi and Bertrand from the equations of motion.

(3) Adopting the dipolar system of coordinates

$$\xi = \log r - \log r', \quad \eta = \theta - \theta',$$

we find $\qquad \lambda = b(rr')^{-1},$

and we have at once the equations in quadratures of the orbits in a More complex bifocal fields: field of force given by

$$V = \frac{1}{r^2 r'^2}\left\{f\left(\frac{r}{r'}\right) + F\left(\frac{r^2 + r'^2 - b^2}{2rr'}\right)\right\}$$

$$= \frac{1}{r^2 r'^2}\phi\left(\frac{r}{r'}\right) + \psi(rr'), \text{ say,}$$

with the velocity of projection which makes $\tfrac{1}{2}v^2 + V = 0$.

We may notice that, if $V \propto (rr')^{-2}$, particles projected with velocity simple case. from infinity will describe paths cutting all circles through the foci at equal angles.

These results might have been arrived at by the method of inversion.

Orbits on ellipsoid: (4) For motion on an ellipsoid, we have

$$ds^2 = \frac{(\mu^2 - \rho^2)(\mu^2 - \nu^2)}{(\mu^2 - h^2)(\mu^2 - k^2)} d\mu^2 + \frac{(\nu^2 - \mu^2)(\nu^2 - \rho^2)}{(\nu^2 - h^2)(\nu^2 - k^2)} d\nu^2,$$

where ρ, μ, ν are the principal semi-axes of the confocals through the point, and h^2, k^2 the constant differences of the squares of the semi-axes of each surface. Hence

$$\int v\,ds$$

becomes

$$\int v\,(\mu^2 - \nu^2)^{\frac{1}{2}}\,(d\theta^2 + d\phi^2)^{\frac{1}{2}},$$

where

$$\theta = \int \left(\frac{\mu^2 - \rho^2}{\mu^2 - h^2 . \mu^2 - k^2}\right)^{\frac{1}{2}} d\mu,$$

$$\phi = \int \left(\frac{\rho^2 - \nu^2}{\nu^2 - h^2 . \nu^2 - k^2}\right)^{\frac{1}{2}} d\nu.$$

compound field. Therefore, if the potential is of the form

$$(\mu^2 - \nu^2)^{-1}\{f(\mu) + F(\nu)\},$$

the equation of the trajectory can always be reduced to quadratures, by § 7[1]. By means of Hamilton's Characteristic Equation, Liouville has generalized this result so as to include free motion in space (*Journal*, Vol. XII. 1847). Liouville-Hamilton extension.

Sphere: polar field: (5) On a sphere with ordinary polar coordinates,

$$ds^2 = a^2 d\theta^2 + a^2 \sin^2\theta\, d\phi^2.$$

Hence

$$\int v\,ds$$

becomes

$$\int av \sin\theta \left\{\left(\frac{d\theta}{\sin\theta}\right)^2 + d\phi^2\right\}^{\frac{1}{2}},$$

and similar results may be stated.

stability included. This method may clearly be applied very widely; and it possesses the advantage of establishing a complete correspondence between the two systems of orbits, so that, for instance, the stability of either orbit is determined at once when that of the other is known.

From free orbit to brachistochrone: 8. *A curve of quickest descent* from one point to another in a given field of force, with given velocity of projection from the first point, is such that

$$\delta \int \frac{ds}{v} = 0,$$

and is therefore the same as a free orbit in a field which would make the velocity equal to v^{-1}; this latter implies a definite law of potential and a definite velocity of projection. The converse proposition is of course equally true. Examples of their application have been given at

[1] (I find that this method has been applied to the discussion of some cases of motion on an ellipsoid by Mr W. R. W. Roberts in *Math. Soc. Proc.* 1883, Vol. XIV. pp. 230–5.)

length by Mr Townsend in the *Quarterly Journal of Mathematics*, Vol. XV; this short and direct statement has already been given by Sir W. Thomson (*vide* Tait and Steele's *Dynamics*). We notice that the constrained orbit is not brachistochronous beyond the point at which the free orbit becomes unstable; so that simple observation as to whether a consecutive free orbit does or does not intersect the given one between the two fixed points, in Mr Townsend's examples, determines at once whether the corresponding brachistochrone is real, or only apparent, as corresponding simply to a maximum-minimum. true or apparent.

9. *For rays of light or sound*, Fermat's Principle of Least Time, From orbits to system of rays:

$$\delta\int v^{-1}ds = 0, \quad \text{or} \quad \int \mu ds = 0,$$

follows at once from the notion of waves. The analogy of this to the principle of least action for a particle has long been known, and indeed this case of the principle of action appears to have been suggested to Euler by the interpretation of Fermat's principle according to the corpuscular theory of light*. (Serret, *Bulletin de Math.* II. p. 97.) We can therefore apply to the case of rays all the transformations that have been indicated for the motion of a particle, and all the examples already given can be translated into solutions for the paths of rays in heterogeneous media by simply writing μ for v. The restriction as to velocity of projection that often affected the generality of the transformations is here rather an advantage, as it simplifies the law of the index of refraction. The following are simple examples: historical.

(1) By inverting the straight rays in a uniform medium, we find that, in a medium for which $\mu \propto r^{-2}$, the rays are circles passing through the origin. Ray systems: examples:

(2) If $\mu \propto \sqrt{\frac{2}{r} - \frac{1}{a}}$, the rays are ellipses with the origin as focus, the major axis of each is $2a$, and the rays can never pass outside an exterior caustic which is an ellipse with the radiant point and the origin as foci. (*Vide* Tait and Steele's *Dynamics*, Ch. IV.) caustics:

Further, if in these examples r denote the distance from an axis, the particle can have a constant velocity along the axis in addition to the orbital motion round it; and therefore for a medium distributed in cylindric layers round an axis we have for the rays helical curves, cylindric optical fields:

(1) On circular cylinders,

(2) On elliptic cylinders,

whose forms are completely determined. And we see also that the rays in (2) are bounded externally by a caustic elliptic cylinder, and

* Cf. the recent *Wellenmechanik* (de Broglie and Schrödinger) of the theory of quanta.

focal lines. that in (1) they all pass through a line parallel to the axes; and, as the conjugate foci for small pencils lie on the caustics, they are completely determinable, and each pencil has an infinite number of conjugate foci symmetrically placed.

By inverting this class of cases, we arrive at distributions of refractive index which lead to ring-shaped caustics, with corresponding properties.

From orbits to catenaries.

10. *For a chain of uniform density* hanging in a field of force whose potential is V, the figure of equilibrium is determined by the condition

$$\delta \int (V + \lambda)\, ds = 0,$$

where λ is a constant that has to be adjusted to the given length of the chain, according to the known method of the Calculus of Variations. There is here one condition more than in the case of the motion of a particle. We conclude, however, that the forms assumed by chains of different lengths attached to two fixed points are the trajectories of particles projected from one of the points so as to pass through the other in a field where

$$v = V + \lambda,$$

and the potential is therefore $-\frac{1}{2}(V + \lambda)^2$, λ being variable.

Newton's stability criterion for orbits.

11. When the chain assumes the form of a circle round a centre of force as centre, the determination of its stability is thus at once reduced to Newton's well-known determination of the stability of a circular orbit, by calculation of the apsidal angle. If the apsidal angle be imaginary, the circular form will be stable; if it be real, the circular form will only be stable for arcs shorter than the distance between two consecutive apses. For the condition of stability is that the potential energy be a minimum, and we know, by Jacobi's rule in the Calculus of Variations[1], that it can only be a minimum when taken to a limit short of the point in which the curve is met by any consecutive curve of equilibrium of the same length. As was to be expected, the value of the apsidal angle contains λ, and therefore depends on the length of the chain; however, the relation between λ and the radius of the circle is easily determined. In fact, following the analogy with dynamics of a particle, we have, if

$$u = r^{-1},$$

$$\frac{d^2u}{d\theta^2} + u = \frac{P}{h^2u^2} = \frac{V + \lambda}{h^2}\frac{dV}{du},$$

since P, the attractive force in the problem of the particle, is the rate of increase of the potential outwards, and is therefore the rate of

[1] See Routh's *Rigid Dynamics*, 3rd ed. p. 561.

decrease of $\frac{1}{2}v^2$ outwards from the origin. When the orbit is nearly circular, we have

$$u = a + x,$$

$$V \equiv \phi u = \phi a + x\phi' a,$$

where x is small; and h is a *constant* for which we may write the approximate value

$$h = vp = (\phi a + \lambda)\, a^{-1}.$$

Therefore $$\frac{d^2x}{d\theta^2} + x + a = \frac{ha + x\phi' a}{h^2}(\phi' a + x\phi'' a)$$

$$= \frac{a}{h}\phi' a + x\left\{\left(\frac{\phi' a}{h}\right)^2 + \frac{a\phi'' a}{h}\right\}.$$

Therefore, if we make $\phi' a = h$,

we have for x the oscillatory value

$$x = A \sin(\mu\theta + \alpha),$$

where $$\mu^2 = \frac{a\phi'' a}{\phi' a},$$

and the apsidal angle is π/μ.

For a force varying as the inverse nth power of the distance acting on the string, we have

Instability of chain-form in a field of force:

$$\phi u = Bu^{n-1},$$

$$\frac{u\phi'' u}{\phi'(u)} = n - 2,$$

so that, if $n < 2$, the apses will be imaginary; and we conclude that the circular form will be unstable for the law of the inverse square, or any higher inverse power[1]. Under a lower power, a string originally held in a circular or nearly circular form round the centre of force as centre will maintain that form; under any other power, it will depart from that form unless held fast at points whose distance is less than the distance between two consecutive apses. This appears to be at variance with Mr Townsend's statement (*Quarterly Journal of Math.* XIII. p. 238).

relieved by a set of point-supports.

12. When the force emanates from a fixed axis, the trajectory of a particle will be a curve, on a cylinder whose cross-section is the plane central orbit, such that the velocity parallel to the axis of the cylinder will be constant. When the cross-section is closed, the trajectory will be a helical curve. So also *one* of the catenaries for the

Helical catenary.

[1] (Mr R. R. Webb has given for the period of that vibration of the string in which it is divided into m complete waves, the expression $\frac{2\pi}{m}\left\{ag\frac{m^2+n-2}{m^2+1}\right\}^{-\frac{1}{2}}$ the law of force being $g\,(ar)^{-n}$ (*Cambridge Math. Tripos*, Jan. 1884); which confirms the above statement.)

corresponding law of force will be this helical curve; but any length more than one turn of it would usually be unstable.

The difficulty in establishing a strict analogy between the orbit of a particle and the form of a catenary arises from the fact that between two points an infinite number of catenaries of different lengths can lie, while only a definite number of orbits with given total energy can pass through them. This explains the appearance of the additional constant λ.

Loaded chain. The problem of determining what forces must be applied to a chain, or how it must be loaded, in order that it may retain a certain given form, is, of course, much less general than the inverse problem of the determination of the system of catenaries that are possible in a given field of force; and accordingly the analogy is easily enough stated for the more special problem, but in that form it applies only to the special curve considered.

The following, however, is a specimen of the kind of results that flow from the more general form of the analogy:

Adaptation of ray-caustics. Suppose rays diverging from a centre in a medium of varying index μ to touch a caustic; the paths of the rays will be the curves assumed by uniform strings, fastened to the fixed centre and wrapped round the caustic, in a field of force whose potential is equal to μ.

The provision that the string has to touch the caustic, here disposes of the extra degree of generality in the catenaries, and limits us to that *one* which passes through the two terminals, and has the same tangent as the ray at one of them.

Catenary of flexible electric current in a magnetic field: 13. A very interesting *series of natural catenaries* is included in the problem of finding the form assumed by a flexible string carrying an electric current in a given magnetic field. Here the energy condition leads to the solution that for equilibrium the variation of the number of tubes of magnetic force enclosed by the string should be zero, while for stability the number of tubes enclosed must be a maximum.

geodesic: It follows that all such catenaries are geodesic curves on the surfaces formed by the lines of force which intersect them; for, if not, then, by making them geodesics on wider surfaces, they can be made to include more lines of force.

Let us take the case in which the force is everywhere in a fixed direction. We know that it must then also be uniform. The catenary will be a geodesic on a cylindrical surface, and would therefore become straight when the surface is unrolled. Hence its projection on a cross-section of the surface will also be of constant length, and the cylinder

in uniform field is circular helical: will therefore include the greatest number of tubes of force (*i.e.* the greatest area) when it is circular. Thus the catenaries will all be circular helices; but it is clear that more than a single turn of the helix will be unstable, unless the string is confined by being enclosed in a non-conducting cylinder, or wrapped round one. Even if this cylinder do not lie along the lines of force, or if it is not circular, the portion of string in contact with it will still lie along a geodesic.

stability: with guiding cylinder.

Le Roux has observed the helical form in a platinum wire, rendered flexible by incandescence, and placed in the sensibly uniform magnetic field between two flat poles of an electro-magnet. Professor Stokes has deduced the helical form* analytically in connection with the explanation of the spiral discharges in a rarefied gas, along a magnetic field, observed by Spottiswoode. (*Vide* Chrystal, *Encyc. Brit.*, Art. "Electricity"; Spottiswoode, *Proc. Roy. Soc.* 1875.) We notice that, to explain the number of convolutions observed in the spirals, we must take also cognizance of the outer and less heated portions of the gas as performing the part of the constraining cylinder mentioned above.

Spiral electric discharge in gas:

the outer constraint.

The method which we have just employed shows also that, in a field of force radiating from a centre, the catenary will be a geodesic on a right circular cone; while, to obtain more than a single turn of the spiral, the string must be enclosed in or wrapped round a circular cone. This form of the catenary has been already deduced analytically by Darboux.

Radial field:

geodesic on cone.

A flexible electric current attracted by a very powerful straight current will, if lying in one plane, assume a form such that

Electric catenary in field of a straight current.

$$\cos\psi = 1 - A\log\frac{y}{a},$$

where ψ is the angle its direction makes with that of the straight current, at a point whose distance from it is y.

(ii) On the Direct Application of the Principle of Least Action to the Dynamics of Solid and Fluid Systems, and Analogous Elastic Problems.

1. In the preceding paper it has been observed that the most direct and compendious method of stating the mathematical conditions of a physical problem is to express it as a maximum or minimum relation, and that in dynamics this can always be accomplished by means of the Principle of Least Action. It is natural, then, to employ this principle in the transformation of dynamical problems, and in the investigation of analogies between different departments of the science; and it was so employed in connection with Dynamics of a Particle and related subjects.

The Action minimal equation:

We proceed to apply the same method to dynamics of general systems. There are two well-known forms of the minimum principle

* The result has been extended to spirals on a cone in a radial field, by Poincaré, as in fact already stated in next paragraph.

which we shall use, viz. that if T be the kinetic energy, and V the potential energy of the system, then, between any two fixed configurations,

its two principal types:

(1) $\delta \int (T - V)\, dt = 0$, when the time of motion is constant,

(2) $\delta \int T dt = 0$, when the total energy of the motion E $(\equiv T + V)$ is constant.

one orbital, without time but with energy conserved:

The second is the more usual form of the principle, and admits of direct elementary proof (cf. Thomson and Tait's *Nat. Phil.* § 327). The first, which is more immediately connected with the Lagrangian equations of motion, can be even more easily established. For, using Cartesian coordinates, we are to prove that

$$\delta \int \{\tfrac{1}{2}\Sigma m\, (\dot{x}^2 + \dot{y}^2 + \dot{z}^2) - V\}\, dt = 0.$$

in the other, energy comes into the variation along the path.

But with fixed initial and final configurations, and with the time limits of the integral fixed, this variation is, by the ordinary rule, when we vary the coordinates and not the time, equal to

$$\int \{-\, \Sigma m\, (\ddot{x}\delta x + \ddot{y}\delta y + \ddot{z}\delta z) - \delta V\}\, dt;$$

and the quantity inside the integral, being the virtual work of the applied forces opposed by the effective forces, is equal to zero, by d'Alembert's principle and the Principle of Virtual Velocities.

Ultimate foundations.

When the time of the motion is not supposed constant, we must vary both the coordinates and t: the process is not much more difficult (*vide* Routh's *Rigid Dynamics*, 3rd ed., footnote, p. 307), and shows that the variation of the time of passage between the fixed configurations introduces simply a term $-E\,\delta t$. In fact, the complete variation is

Complete variation effected for connected system of particles:

$$[(T - V)\, \delta t + \Sigma m\, \{\dot{x}\, (\delta x - \dot{x}\,\delta t) + \ldots + \ldots\}]$$
$$+ \int \left[-\, \Sigma m\, \{\ddot{x}\, (\delta x - \dot{x}\,\delta t) + \ldots + \ldots\} - \left\{ \frac{dV}{dx}\, (\delta x - \dot{x}\,\delta t) + \ldots + \ldots \right\} \right] dt,$$

since $\delta x - \dot{x}\,\delta t$, and not δx, is now the virtual displacement of the particle. In this expression the first term refers to the limits, at which δx, δy, δz are zero. The integral term is zero as before. Hence, if δt represent the variation of the time of passage, we have

$$\delta \int (T - V)\, dt = -\, E\,\delta t.$$

including both forms.

Again, if we now suppose that E is constant during the motion, we have

$$\delta \int (T + V)\, dt = \delta\, (Et) = E\,\delta t + t\,\delta E;$$

therefore, adding, $$2\delta \int T dt = t\,\delta E,$$

so that, if the energy of the motion is not varied,

$$\delta \int T\,dt = 0,$$

which is the second form of the principle.

Having once shown the truth of the first form, of course it only remains to transform T and V into any generalized coordinates, and the method of variation will then afford a proof of Lagrange's equations of motion.

Its far-reaching character.

2. In treating of the motion of a general system, it is important to find the forms assumed by these principles in the cases in which some of the coordinates do not explicitly appear in the expressions for the kinetic and potential energies, and in which the corresponding momenta are therefore constant. From this modified form the equations of motion with ignored coordinates, as developed independently by Routh and Sir W. Thomson, are immediate deductions.

Systems with latent momenta.

Kirchhoff's analogy between the form assumed by an elastic wire originally straight, when subjected to stresses at its ends, and the motion of a solid about a fixed point under gravity, is explained from this fundamental point of view; and an extension to the case in which the wire was originally of circular or spiral form immediately suggests itself. The brachistochronous character of the motion of a sphere in a fluid of the same density with any fixed boundaries or obstacles whatever, is pointed out.

Examples: dynamical:

The principle is then applied to the motion of solids in a fluid with cyclosis through apertures either in some of the solids or in fixed boundaries. A theorem of Kirchhoff's relating to circulation of fluid round thin rigid cores is proved immediately, and equations are deduced for the motion of the solids which, I have since found, agree with those given as the result of a much longer analytical process by Sir W. Thomson in 1872. They are also consistent with the result of a long direct hydrodynamical investigation by Dr C. Neumann, published in 1883.

solids in a fluid medium: with circulation.

3. In the very important case in which the expressions for the potential and kinetic energies do not contain some of the coordinates the corresponding momenta are constant, and therefore, by elimination, the velocities can also be made to disappear from the expressions, being replaced by these momenta. Lagrange's and Hamilton's equations of motion have been independently adapted to this case by Routh and Sir W. Thomson*, and applied to very important problems.

Wide physical extensions, including latent permanent motions:

* Also later by Helmholtz, "On Least Action," Crelle, 100 (1886): also earlier papers "Statics of Monocyclic Systems," arising out of thermodynamics, beginning 1884. All in *Collected Papers*, Vol. III.

(The text of §§ 3–4 has been improved in this reprint.)

When the variables are reduced in number by means of these equations of condition, the original functions of Hamilton no longer retain their minimum properties, and we shall investigate the functions to which they now belong. Following Routh's example, we shall at first proceed as if the momenta introduced were not necessarily constant.

The coordinates are thus divided into two groups; one group of velocities is retained, while the other is eliminated, being replaced by the corresponding group of momenta.

Let T, V (the kinetic and potential energies of the motion) be expressed in terms of a set of coordinates θ, ... and another set ψ, ..., and suppose we wish to obtain a form of the principle that shall involve the corresponding momenta Ψ, ... instead of the velocities $\dot\psi$, ... If $L = T - V$, the first form of the principle is that

$$\delta S \equiv \delta \int L\,dt = 0,$$

with time of motion given.

Now $\delta \int L\,dt = \int \left(\frac{dL}{d\theta}\delta\theta + \frac{dL}{d\dot\theta}\delta\dot\theta + \ldots + \frac{dL}{d\psi}\delta\psi + \frac{dL}{d\dot\psi}\delta\dot\psi + \ldots \right) dt,$

a variation of the time being unnecessary. In this we are to write Ψ for $\frac{dL}{d\dot\psi}$, ..., and to substitute for $\dot\psi$, $\delta\dot\psi$, ... in terms of Ψ, ... from these relations. We find from it

$$\int \left(\frac{dL}{d\theta}\delta\theta + \frac{dL}{d\dot\theta}\delta\dot\theta + \ldots \right) dt$$

$$= \delta \int (L - \Psi\dot\psi - \ldots)\,dt + \int \left(\dot\psi\,\delta\Psi - \frac{dL}{d\psi}\,d\psi + \ldots \right) dt.$$

But the first side of this identity is the variation of S produced by varying θ, $\dot\theta$, ... without causing any variation in ψ, $\dot\psi$, ..., and is therefore zero. Hence, with our new variables θ, $\dot\theta$, ..., ψ, Ψ, ... the second side is zero.

We shall write $L' = L - \Psi\dot\psi - \ldots$.

the generalized Lagrangian Action integrand obtained.

Since $L = T - V = \frac{1}{2}(\Theta\dot\theta + \ldots + \Psi\dot\psi + \ldots) - V$,

we see that L' differs from L by containing the difference between the part of the kinetic energy due to the momenta Θ, ... and the part due to the momenta Ψ, ... instead of the whole kinetic energy, and by being expressed in terms of different variables. We shall find that this observation will often very much simplify the determination of L', especially in cases of cyclical fluid motion[1].

Work expended on latent spins:

[1] (The total quantity of work required to impulsively generate the motion is independent of the order of operations; but the quantity consumed by each impulse depends on the order of succession in which the impulses are ap-

In case the ψ coordinates do not enter into T or V, we have Ψ, ... constants; and the variational equation of motion becomes simply

now not a pure quadratic in the velocities.

$$\delta S' \equiv \delta \int L'dt = 0,$$

with time of motion given. Therefore, in such steady motion, $\int L'dt$ is a minimum, for the given values of the steady momenta, and the given value of the time of motion, provided, as usual, that the interval be not so long as to make it only stationary.

System with latent spins:

The equations of motion are therefore, by the Calculus of Variations, (d/dt being total)

Lagrangian equations formally unaltered.

$$\frac{d}{dt}\frac{dL'}{d\dot{\theta}} - \frac{dL'}{d\theta} = 0,$$

$$\dots\dots\dots\dots\dots\dots\dots$$

and are exactly analogous to Lagrange's equations, except that L' now contains terms with $\dot{\theta}$, ... in the first degree.

If the motion is not steady, so that Ψ, ... are not constants, we have the additional equations

System specified by mixed velocities and momenta:

$$\frac{dL'}{d\Psi} = -\dot{\psi}, \dots,$$

$$\frac{dL'}{d\psi} = \frac{dL}{d\psi}, \dots.$$

These two sets of equations, the former of the Lagrangian, and the latter of the Hamiltonian type, were given by Routh under the name of modified Lagrangian equations.

Routh.

4. The other form of the variational equation, $\delta A = 0$, with the total energy E of the motion given, where $A = \int T dt$, may be reduced

Generalized Eulerian form of Action:

plied. The expressions in the text refer to the case in which all the impulses are simultaneously and equably applied to the system, so that the state of motion rises from zero to its final value, remaining always similar to itself. In the example of § 10, the motion is therefore to be supposed generated by the simultaneous and equable application of Θ, ... to the boundaries of the fluid, and Ψ, ... to the barriers.

If, however, the gyrostats possess initially their steady constant momenta Ψ, ..., the impulses that must be applied to the solids to which they are attached in order to start their motion are equal to $\frac{dL'}{d\dot{\theta}}$, ... respectively; as is found by taking the time integral of the modified equations of motion of this section, extended over the duration of the impulse. Thus we see that, in the case of § 10, the same impulses are required to start the solids, whether the fluid be circulating or not. But the amount of work required is different, being $\frac{1}{2}\rho\int(\phi_0+\phi_1)\frac{d\phi_0}{dn}dS$ in the former, and $\frac{1}{2}\rho\int\phi_0\frac{d\phi_0}{dn}dS$ in the latter case.—*Sept.* 4.)

their impulses.

to the new variables as follows. The time is now variable, so that the variation of θ with time unvaried, as the foundation on virtual work requires, now becomes $\delta\theta - \dot{\theta}\,\delta t$, and that of its velocity $\dot{\theta}$ becomes $\delta\dot{\theta} - \ddot{\theta}\,\delta t$: and therefore, as usual,

$$\delta\int L\,dt = [L\,\delta t] + \int\left(\frac{dL}{d\theta}(\delta\theta - \dot{\theta}\,\delta t) + \frac{dL}{d\dot{\theta}}(\delta\dot{\theta} - \ddot{\theta}\,\delta t) + \dots \right.$$
$$\left. + \frac{dL}{d\dot{\psi}}(\delta\dot{\psi} - \ddot{\psi}\,\delta t) + \dots\right) dt,$$

as we are now, for simplicity, taking $\psi, \dots$ absent and so $\Psi, \dots$ equal to $dL/d\dot{\psi}, \dots$, all constant. Therefore, rearranging, as $L' = L - \Psi\dot{\psi} - \dots$,

$$\delta\int L'dt = [L'\delta t] + \left[\Sigma\frac{dT}{d\dot{\theta}}(\delta\theta - \dot{\theta}\,\delta t)\right],$$

by the same process of variation (cf. p. 65 *infra*), as the final terms under the integral vanish as before; thus

$$\delta\int L'dt = [L'\delta t] + \left[\Sigma\frac{dT}{d\dot{\theta}}\,\delta\theta\right] - [(2T - \Psi\dot{\psi} - \dots)\,\delta t]$$
$$= -\,[E\,\delta t] + \left[\Sigma\frac{dT}{d\dot{\theta}}\,\delta\theta\right].$$

But the energy, now conserved, is $E = T + V$, therefore

$$\delta\int(T + V)\,dt = E\,\delta t + t\,\delta E,$$

and, adding this to the above result, we find

$$\delta\int(2T - \Psi\dot{\psi} - \dots)\,dt = t\,\delta E + \left[\Sigma\frac{dT}{d\dot{\theta}}\,\delta\theta\right].$$

with conserved energy. We conclude that $\qquad \delta A' \equiv \delta\int T'dt = 0,$

with conserved energy, which is unvaried, wherein

$$T' = T - \tfrac{1}{2}(\Psi\dot{\psi} + \dots),$$

i.e. where T' is equal to the part of the kinetic energy which is due to the momenta $\Theta, \dots$, but not to $\Psi, \dots$.

5. The values of the quantities under the integral signs in these modified formulae* can be easily calculated. Thus, following Routh (*Essay on Stability*, Ch. IV, Art. 21), let us write

$$T \equiv \tfrac{1}{2}T_{\theta\theta}\dot{\theta}^2 + T_{\theta\phi}\dot{\theta}\dot{\phi} + \dots + \tfrac{1}{2}T_{\psi\psi}\dot{\psi}^2 + T_{\psi\chi}\dot{\psi}\dot{\chi} + \dots,$$

and we have $T_{\psi\psi}\dot{\psi} + T_{\psi\chi}\dot{\chi} + \dots = \Psi - T_{\psi\theta}\dot{\theta} - T_{\psi\phi}\dot{\phi} - \dots,$

$$T_{\psi\chi}\dot{\psi} + T_{\chi\chi}\dot{\chi} + \dots = \mathrm{X} - T_{\chi\theta}\dot{\theta} - T_{\chi\phi}\dot{\phi} - \dots,$$
$$\dots = \dots$$

* The two principal forms of the Action are usually named Hamiltonian and Eulerian, the former because Hamilton based on the Lagrangian form a complete reconstruction of Analytical Dynamics. Hamilton designated the Lagrangian form as the Principal Function. *Phil. Trans.* 1833–4.

where we shall, for brevity, write the second sides

$$\Psi - X, \ \mathrm{X} - Y, \ldots.$$

Now

The various Action forms analytically worked out.

$$T = \tfrac{1}{2}T_{\theta\theta}\dot{\theta}^2 + T_{\theta\phi}\dot{\theta}\dot{\phi} + \ldots + \tfrac{1}{2}\dot{\psi}(\Psi + X) + \tfrac{1}{2}\dot{\chi}(\mathrm{X} + Y) + \ldots,$$
$$L' = \tfrac{1}{2}T_{\theta\theta}\dot{\theta}^2 + T_{\theta\phi}\dot{\theta}\dot{\phi} + \ldots - V - \tfrac{1}{2}\dot{\psi}(\Psi - X) - \tfrac{1}{2}\dot{\chi}(\mathrm{X} - Y) - \ldots,$$
$$T' = \tfrac{1}{2}T_{\theta\theta}\dot{\theta}^2 + T_{\theta\phi}\dot{\theta}\dot{\phi} + \ldots + \tfrac{1}{2}\dot{\psi}X + \tfrac{1}{2}\dot{\chi}Y + \ldots;$$

and, eliminating $\dot{\psi}$, $\dot{\chi}$, ... from each of these and the above set by a determinant, we find

$$T = \tfrac{1}{2}T_{\theta\theta}\theta^2 + \ldots - \frac{1}{2\Delta}\begin{vmatrix} 0, & \Psi + X, & \mathrm{X} + Y \ldots \\ \Psi - X, & T_{\psi\psi}, & T_{\psi\chi} \ \ldots \\ \mathrm{X} - Y, & T_{\psi\chi}, & T_{\chi\chi} \ \ldots \\ \ldots & \ldots & \ldots \end{vmatrix}$$

in which the determinant does not contain any terms in Ψ, X, ... of the first degree, as it is not altered by changing the signs of these quantities:

$L' = \tfrac{1}{2}T_{\theta\theta}\dot{\theta}^2 + \ldots - V +$ a symmetric determinant of the same kind, which does contain terms of the form $\dot{\theta}\Psi$, $\dot{\theta}\mathrm{X}$, ...:

$$T' = \tfrac{1}{2}T_{\theta\theta}\dot{\theta}^2 + \ldots - \frac{1}{2\Delta}\begin{vmatrix} 0, & X, & Y, & \ldots \\ \Psi - X, & T_{\psi\psi}, & T_{\psi\chi}, & \ldots \\ \mathrm{X} - Y, & T_{\psi\chi}, & T_{\chi\chi}, & \ldots \\ \ldots & \ldots & \ldots & \ldots \end{vmatrix},$$

which is the same as L' with the omission of V, of the quadratic terms in Ψ, X ..., and of half of each term of the first degree in $\dot{\theta}$, $\dot{\phi}$ ....

An illustration of these is Sir W. Thomson's expression that will be proved below independently for the kinetic energy in the case of solids in motion in a cyclically moving fluid, and the associated functions L' and T' for that problem.

6. As a first example of the application of the method, we will take Kirchhoff's analogy between the form of a bent elastic wire and the motion of a solid body round a fixed point.

Example: analogy of strained elastic wire to motion of solid pendulum:

If we consider a naturally straight elastic wire of any form of section, provided it be small and uniform, we can express its potential energy of deformation by the following known plan (Thomson and Tait's *Nat. Phil.* Arts. 593–97): Draw through each point of the central axis of the wire three rectangular axes in fixed directions in space, one of them in the direction of its length, and suppose them to be rigidly connected with the substance of the wire; then, when it is deformed, the directions of these axes will be altered, and the deformation at any point will be completely specified by the displacements of their directions per unit length measured along the wire; and, if the curvature be not so great as to prevent our assuming the stresses to be proportional to the strains produced by them, the potential energy of deformation per unit length will, by Green's theory of elastic solids, be a homogeneous quadratic function of these rates of variation of direction of the axes. That is, if we suppose a point to travel along the wire with unit velocity, and suppose a certain solid to rotate round a fixed point, so that three axes fixed in the solid shall be always parallel to the axes through the point moving along the wire, the potential energy of deformation per unit length will be equal to the kinetic energy of rotation of this solid.

Let one end of the wire be clamped, and the other end be acted on by a given system of forces, reducible to a force and a couple in the ordinary way. The total potential energy of the wire will assume the form of a line integral along the central section, if we replace the balancing stresses at its ends by equal balancing stresses at the ends of each element of it; for the virtual moment of the balancing couples due to a rigid displacement of the element will be zero, while that of the forces will be $P\delta s \sin\theta\, d\theta$, wherein P is the intensity of the force, and θ is the angle which ds makes with the constant direction of P. Hence the energy condition of equilibrium will be

$$\delta V \equiv \delta \int (T + P\cos\theta)\, ds = 0,$$

between the given terminal points, and with the *given constant length of the wire*, where T is the quadratic function before referred to.

But, for the rotating body of the analogy, the equations of motion may be expressed as the conditions that

$$\delta L \equiv \delta \int (T - V')\, dt = 0,$$

between the given initial and final configurations, with *constant time of motion*.

The two conditions will agree exactly, if

$$V' = -P\cos\theta,$$

and this is clearly secured by causing the constant force P to act on the solid in its constant direction at a point unit distance along that axis which is parallel to the direction of the wire. Kirchhoff's proposition follows, then, as an immediate consequence, that the motion of the solid round a fixed point under gravity bears to the deformation of the wire the relation already explained.

7. The analogy does not in this case throw much light on the question of stability, but we may discuss it by considerations similar to those employed in the case of a chain. For the stability of the wire it is necessary and sufficient that V should be an absolute minimum with the given conditions at the extremities, and the given length. But Jacobi's theorem before quoted shows that V is a mini-

limit of free length of wire for stability.

mum only up to the first point at which the curve is cut by a neighbouring curve satisfying the conditions. If the curve be of sinuous form and the wire only tied at the ends, it is therefore clear that it will not be stable if longer than half a wave-length, unless it be tied down at points within that distance; for the curve is intersected by an equal curve at points whose distance apart is half a wave-length. For a steel band, clamped at the ends, the limits are wider, but do not apparently amount to three-quarters of a wave, even in the comparatively favourable case in which it is confined to one plane, and therefore assumes one of the forms corresponding to the simple pendulum (figured in Thomson and Tait, and in Rankine's *Applied Mechanics*); for that case, however, the instability is practically removed by the passive friction of the plane surface on which the curve is formed, unless the displacement be so great that the disturbing force exceeds the frictional limits.

Experimental forms of stressed wire:

can be stabilized by slight passive friction.

8. If the elastic wire, instead of being straight, had been originally a helix, the three components of whose curvature are each constant, we should have had for the potential energy of bending, per unit length, a quadratic function of the component curvatures, each diminished by the corresponding constant, and the energy condition of equilibrium might be written down as before. We notice that it now contains terms of the first degree with constant coefficients in the expression for T; but these are exactly such terms as would be introduced into the modified T (§ 4) of the dynamical problem by fly-wheels mounted on the rotating body. So that we are prepared for this extension of Kirchhoff's proposition, viz.—That the motion of a gyroscopic pendulum (*i.e.* a solid rotating about a fixed point with a revolving fly-wheel mounted on an axis fixed in it) is exactly analogous to the form assumed by a wire originally helical, when deformed by any distribution of forces applied to its extremities. Indeed, an independent proof is now obvious; for the constant angular momentum of the fly-wheel is equivalent to three constant component angular momenta round the axes through the point of suspension fixed in the solid, in addition to the angular momenta that are due to the action of the forces, while in the wire there are the three constant component curvatures round the three axes, in addition to those caused by the action of the forces; and, as for the principal axes of the solid angular momenta round them are proportional to the angular velocities round them, and these latter are equal to the curvatures in the wire, we see that the analogy still holds.

Analogy extended to forms of wire initially curved:

now a-gyrostatic pendulum,

The solid of the analogy may also be replaced by a pendulum having annular cavities filled with fluid which is circulating round them, or by a pendulum in a fluid with apertures in it, through which

or other cyclic system.

the fluid is circulating; as the motion of each of these follows the same laws as that of the gyroscopic pendulum.

Ex. of gyrostat: various mountings:

A wire of uniform circular section corresponds to a solid symmetrical about an axis; and the fly-wheel is to be attached with its centre of gravity in the axis, and with its axis in such a direction that its component angular momenta round the axis and in a perpendicular direction are equal to the couples corresponding to the tortuosity and curvature of the helix, respectively. A helix with a few convolutions far apart corresponds to a slowly rotating fly-wheel, while a helix with many close convolutions corresponds to a rapidly rotating wheel, mounted nearly parallel to the axis, and the forms which it assumes represent the motion of the solid with this "gyrostatic domination." Under these latter circumstances, the helix bends and twists very much as if it were a straight wire; so that the solid moves with its axis in rapid gyration nearly at right angles to a mean position whose own motion does not very greatly differ in character from that of a solid moving freely round the fixed point under the same forces.

its motion indicates forms of strained wires.

Bent circular spring.

As a still simpler illustration of this extension, we notice that a spring of the form of a circular arc will, when deformed by end forces, in its own plane, assume one of the known forms of the elastic curve for a straight spring; and from such a wire we might obtain the curves due to a distribution of force *and couple* on each extremity of a straight spring, without having to apply the bending force at the end of a rigid arm attached to the spring (*cf.* Thomson and Tait).

Again, such a spring when bent out of one plane will correspond to the motion of a rotating body with a gyrostat attached to it, having its axis parallel to one of the principal axes through the point of suspension.

Action for solids moving in fluid, expressible with fluid ignored.

9. The solution of questions relating to the motion of solid bodies in fluids without cyclosis has been reduced to Lagrange's equations in Thomson and Tait (2nd ed. Art. 320). In this case the ignored coordinates are those corresponding to the individual particles of fluid, and the momenta that correspond to them are the momenta remaining in the fluid when the solids have been all reduced to rest. In the problem discussed by Thomson and Tait, these momenta are all zero; but, when the fluid circulates through apertures in the solids, the momenta are definite, and are linear functions of the cyclic constants of all the independent circuits.

Cyclic momenta.

Recalling the argument already employed, which is simplified now by the constancy of these momenta, we have

$$\delta S \equiv \delta \int L\,dt = \int\left(\frac{dL}{d\theta}\,d\theta + \frac{dL}{d\dot\theta}\,\delta\dot\theta + \ldots + \Psi\delta\dot\psi + \ldots\right)dt,$$

with given time of motion, and therefore

$$\delta S' \equiv \delta \int (L - \Psi \dot{\psi} - \ldots)\, dt = \int \left(\frac{dL}{d\theta}\, \delta\theta + \frac{dL}{d\dot{\theta}}\, \delta\dot{\theta} + \ldots \right) dt.$$

That is, the total variation of S' when expressed in terms of $\theta, \ldots$, $\dot{\theta}, \ldots$, and the constants $\Psi, \ldots$, is equal to that part of the variation of S which is due to variations of $\theta, \ldots, \dot{\theta}, \ldots$ alone, while $\dot{\psi}, \ldots$ remain constant; it is therefore zero. And, also as before, we see that

Form of Action for rings in fluid with permanent circulation.

$$S' \equiv \int L' dt,$$

where L' is obtained by subtracting from the total energy of the system twice that part which is due to the constant momenta.

Let us consider, however, in the first place, the case in which there is no cylcosis, and therefore $\Psi, \ldots$ are each zero. We have then for δS the same value as if there were no ignored coordinates; and the ordinary proof for rigid systems (§ 1) now suffices to deduce the other minimum theorem, that for given total energy of motion

Eulerian form for null momenta of the latent freedoms.

$$A \equiv \int T dt,$$

satisfies the minimum condition, between any two configurations of the motion.

For example, suppose we have a solid sphere moving in fluid enclosed in a boundary of any shape, and with any fixed obstacles immersed, but without circulation. The only coordinates that we are concerned with are those of the centre of the sphere; and T and E can be expressed in terms of them. But, if further, the sphere be of the same density as the fluid, or if no external forces act on the sphere, we have T constant, and the variational equation becomes $\delta \int dt = 0$ with given energy. We conclude that the sphere will move from one point to another in a shorter time than if it were guided by frictionless constraint so as to proceed by any other path, it being understood, as usual, that for paths beyond a certain length the minimum relation may cease to hold, and the time may be only stationary. The same proposition applies to cylinders moving in two dimensions. It follows, just as in the case of motion of a particle, that if equal spheres are projected from a point with equal energies in variable directions, their paths are everywhere intersected at right angles by a system of surfaces of equal action, which are here also surfaces of equal time.

Paths of free spheres in perfect fluid are brachistochronous, like rays.

10. Now let us discuss the case of perforated bodies with circulation through their apertures. The kinetic energy T of the motion has been shown by Sir W. Thomson to consist of a quadratic function $\mathfrak{T}$ of the component velocities of the solids, the method of calculating which is not affected by the existence of cyclosis, together with a

Energy for rings with cyclic flow.

quadratic function Ω of the cyclic constants $\kappa_1, \kappa_2 \ldots$, which latter clearly represents the kinetic energy of the motion that would remain after the solids are brought to rest. (*Vide* Lamb's *Fluid Motion*, Art. 120.) The momenta due to the ignored coordinates are those which contain $\kappa_1, \kappa_2 \ldots$; and the part of the kinetic energy due to these momenta (*i.e.* the work that would be done by the corresponding impulses in the instantaneous production from rest of the actual fluid motion, the amount of which is obtained at once by integrating over each barrier closing an aperture half the product of impulsive pressure $\kappa\rho$ and velocity $d\phi/dn$) is

Conserved momenta of the cyclosis: their separated energy.

$$\tfrac{1}{2}\rho\Sigma \int \kappa \frac{d\phi}{dn}\, d\sigma,$$

taken over the surfaces of all the barriers. (We draw dn outwards from a boundary, and in the direction of the circulation from a barrier.) But this expression is equal to the above-mentioned quadratic function Ω of $\kappa_1, \kappa_2 \ldots$, together with terms involving products of κ's, and velocity components of the solid; and these latter are exactly equal to

$$-\tfrac{1}{2}\rho\Sigma \int \phi_1 \frac{d\phi}{dn}\, dS,$$

taken over the surfaces of the immersed solids, where ϕ_1 is the part of the velocity potential due to cyclosis, and therefore $d\phi_1/dn$ is zero. These statements indeed all follow at once from Green's theorem which can be applied to each of the acyclic compartments into which the whole space is divided by the barriers; thus, if ϕ_0 is the part due to the motion of the solids, so that $\phi = \phi_0 + \phi_1$,

$$\text{kinetic energy of fluid} = \tfrac{1}{2}\rho \int \left\{ \left(\frac{d\phi}{dx}\right)^2 + \left(\frac{d\phi}{dy}\right)^2 + \left(\frac{d\phi}{dz}\right)^2 \right\} dv$$

$$= \tfrac{1}{2}\rho \left\{ \Sigma \int \phi \frac{d\phi}{dn}\, dS + \Sigma \int \kappa \frac{d\phi}{dn}\, d\sigma \right\};$$

but, for the whole surface of each of the compartments, we have

$$\int \left(\phi_1 \frac{d\phi_0}{dn} - \phi_0 \frac{d\phi_1}{dn} \right) dS = \int (\phi_1 \nabla^2 \phi_0 - \phi_0 \nabla^2 \phi_1)\, dv = 0,$$

and therefore, adding for all the compartments,

$$\Sigma \int \phi_1 \frac{d\phi_0}{dn}\, dS = 0;$$

or, in our previous notation,

$$\Sigma \int \kappa \frac{d\phi_0}{dn}\, d\sigma + \Sigma \int \phi_1 \frac{d\phi}{dn}\, dS = 0.$$

We have, therefore,

kinetic energy of fluid

$$= \tfrac{1}{2}\rho\Sigma\int\phi_0\frac{d\phi_0}{dn}\,dS + \text{the quadratic function } \Omega \text{ of } \kappa_1\kappa_2\ldots,$$

in which the coefficients of the quadratics are all functions of the coordinates.

The modified principal function is therefore

Form of Lagrangian Action for the immersed solids:

$$S' = \int L'dt,$$

$$\text{where } L' = T_0 + \tfrac{1}{2}\rho\left\{\Sigma\int\phi_0\frac{d\phi_0}{dn}dS + 2\Sigma\int\phi_1\frac{d\phi_0}{dn}dS\right\} - \Omega - V$$

$$= T_0 + \tfrac{1}{2}\rho\Sigma\int(\phi_0 + 2\phi_1)\frac{d\phi}{dn}dS - \Omega - V,$$

and T_0 is the kinetic energy of the solids by themselves. So that L' is now a non-homogeneous quadratic function in $\dot{\theta}, \ldots, \kappa_1, \ldots$. Its differential coefficient with respect to κ_r is the flux through the corresponding aperture with sign changed; its differential coefficients with respect to the velocities $\dot{\theta}, \ldots$ are the impulses required to start the motion of the solids in the circulating fluid.

a pure quadratic when no cyclosis.

It is obvious that these expressions are cases of the general expressions for L and L' which have been calculated above (§ 5) for dynamical systems.

The equations of motion are of the form

$$\frac{d}{dt}\frac{dL'}{d\dot{\theta}} - \frac{dL'}{d\theta} = 0,$$

....

and the determination of the motion is thus reduced to analysis.

11. If we suppose the immersed solids to be mere rings or cores of extreme thinness, round which the fluid circulates, while they offer no obstacles to its other motion, the terms of the first degree in L' will disappear, and we have

$$L' = T_0 - \Omega - V.$$

In other words, L' is the same as if there were no circulation, but the potential of the field in which the cores are placed were increased by Ω. Now Helmholtz has shown in his Memoir on "Vortex Motion" that Ω is equal to the potential energy* of electric currents, whose strengths are equal to the respective cyclic constants multiplied by ρ, and which circulate round the cores. We arrive, therefore, at the

Forces between skeleton rings with cyclosis:

analogy (partial) with electric currents:

* This quasi-potential energy is $-\Omega$, so that the forces are opposite instead of the same, a discrepancy in the analogy which Kirchhoff and Kelvin had emphasized.

result that such cores in a fluid appear to exert forces on one another which are the same [with signs changed] as would be exerted if they were carrying these electric currents, *i.e.* that forces equal and opposite to these would have to be applied in order to prevent their motion. The proof of this result for two cores is the subject of a paper by Kirchhoff (*Borchardt*, Bd. 71, 1869; *Gesamm. Abhandlungen*, pp. 404–16), who reduces the potential of the forces to Neumann's well-known form of the mutual energy of two electric currents.

12. The general equations of motion that we have obtained are in accordance with equations deduced from the fundamental equations of hydrodynamics by Dr Carl Neumann (*Hydrodynamische Untersuchungen*, 1883, pp. 1–90). He gives them as a correction to those in Thomson and Tait; but he appears to have misunderstood the reasoning and the limitations there given, his account of it being merely taken from a short notice in W. M. Hicks' Report on Hydrodynamics (*Brit. Assoc. Report*, 1882).

Kelvin's Hamiltonian method.

Since developing the form of solution just given, I have found that the same problem has been discussed by Sir W. Thomson himself, whose final equations agree with those here given (*Proc. Roy. Soc. Edin.* 1871–72; *Phil. Mag.* May 1873). His method there consists in forming the generalized Hamiltonian equations of motion in terms of the momenta and coordinates of the system (which are clearly exactly suited to the problem), and transforming them analytically into the Lagrangian form; though at bottom the same, the argument is perhaps hardly so compact as that which we have chosen.

Sir W. Thomson proceeds to consider the case in which there are moving solids as well as cores in the fluid. When there is a moving sphere, he points out that the additional terms in T are $\frac{1}{2}\mu q^2 + w$, where q is the velocity that the fluid would have if the sphere were absent, μ is its mass increased by half that of the fluid displaced, and w is the kinetic energy that would be destroyed if the fluid which would then occupy its position were solidified by the action of internal forces. When the sphere is very small, the latter part may be neglected. He applies this result to determine the conditions on which it depends whether a small spherule of the same density as the fluid will be sucked into the axis of a straight vortex, or will escape from it, and shows that its path is a Cotes' Spiral.

Motion of small bubble: round a straight vortex.

General electro-magnetic analogue:

Still neglecting w, it is clear that the forces which the cyclically moving fluid appears to exert on the sphere are the [opposite] as the corresponding electric currents would exert on a magnetic molecule with its axis pointing in the direction of the velocity q, and whose moment is $3q\rho/8\pi$ multiplied by the volume of the sphere; for, as regards the velocity produced by it at a point outside it, the sphere may be replaced by the small core which is the analogue in Kirchhoff's theorem of this magnetic molecule.

13. This method may also be applied to prove Kirchhoff's theorem itself. For each core may be replaced by a system of indefinitely small cores, forming a network, just as Ampère similarly subdivided a finite electric circuit. The mutual force exerted by two of these elementary cores is the same as if each was replaced by the corresponding sphere, as above; for the force which one exerts on the other is the effect of the motion which it propagates through the fluid, and at any distance great compared with the radius this motion is the same for the sphere as for the core. The resultant force exerted on either sphere can easily be found by integrating the pressures on different parts of it; and it follows that a force equal and opposite to this must be applied to keep the sphere from moving. This will be found to agree with Kirchhoff's rule: which can now be extended to the general case by summing up for all the elementary cores. extended to interaction of thin vortexes.

14. The employment of the modified form of the Principle of Least Action, instead of the Principal Function, leads us to a minimum theorem, which is rather simpler than that given above, but is not so easily transformed into equations of motion. It is as follows: Eulerian modified form for Action:

Of all the motions between fixed initial and final configurations that have the same total energy and the same cyclic constants, the actual free motions are those that possess the property that $\delta A' \equiv \delta \int T' dt$ is zero, where T′ is equal to the kinetic energy of the immersed solids together with

$$\tfrac{1}{2}\rho \int \phi \frac{d\phi}{dn}\, dS,$$

taken over their surfaces; or, in other words, T' is equal to the amount of work that must be properly applied at any instant to the solids (in the manner specified above, § 3, note) in order to start the actual motion. This integral $\int T' dt$ is therefore smaller for the actual motion than for any other neighbouring motion that might be produced in the same cyclically moving fluid by the introduction of frictionless constraints, which do not alter the energy. its minimal theorem.

15. *Sept.* 4. The application of Lagrange's equations to the motion of solids in a fluid without circulation, by Thomson and Tait in 1867, was justified first by Kirchhoff (*Borchardt*, 1870); his method was extended somewhat by Boltzmann (*Borchardt*, 1871), and appears from him in the German edition of Thomson and Tait. Another method was used independently by J. Purser (*Phil. Mag.* 1876). Historical.

Kirchhoff's hydrodynamic procedure sketched: The method of Kirchhoff admits of easy extension to the general theory of § 10. Neumann considered the question in this way. Thus, by § 1, we have, if we vary the limits, but not the time,

$$\delta \int_{t_0}^{t_1} L\,dt = \left[\Sigma m \left(\frac{dx}{dt}\,\delta x + \frac{dy}{dt}\,\delta y + \frac{dz}{dt}\,\delta z\right)\right]$$
$$= \left[\rho \int \left(\frac{d\phi}{dx}\,\delta x + \frac{d\phi}{dy}\,\delta y + \frac{d\phi}{dz}\,\delta z\right) d\tau\right]$$

taken at the limits t_0 and t_1.

Here δx, δy, δz are any displacements; let us now taken them to be consistent with a velocity potential, and therefore equal to $\frac{d\delta\phi}{dx}$, $\frac{d\delta\phi}{dy}$, $\frac{d\delta\phi}{dz}$, and we have, by Green's theorem,

$$\delta \int L\,dt = \left[\rho \int \phi \frac{d\delta\phi}{dn}\,dS\right],$$

the integral being extended over all the boundaries and barriers. If now we suppose the solids not to be virtually displaced, $d\delta\phi/dn$ is zero at each boundary. Hence

acyclic case: (1) if there is no circulation in the fluid, $\delta \int L dt = 0$ for given initial and terminal position of the solids, and given time of motion; and Lagrange's equations of motion therefore apply:

cyclic ring-solids. (2) if there is circulation,

$$\delta \int L\,dt = \left[\rho \Sigma \int \kappa \frac{d\delta\phi}{dn}\,d\sigma\right]$$
$$= \left[-\rho \Sigma \int \phi_1 \frac{d\delta\phi}{dn}\,dS\right],$$

by § 10; hence, if ϕ_1, the velocity potential of the circulation, is supposed unvaried, and therefore $\kappa_1, \kappa_2 \ldots$ constants, we have

$$\delta \int L' dt = 0,$$

where
$$L' = L + \rho \Sigma \int \phi_1 \frac{d\phi}{dn}\,dS,$$

for given initial and terminal positions of the solids, given circulations, and given time of motion.)

Appendix.

THE PRINCIPLE OF LEAST ACTION.

[Addition to Maxwell's *Matter and Motion*, ed. 2, 1920.]

THE great *desideratum* for any science is its reduction to the smallest number of dominating principles. This has been effected for dynamical science mainly by Sir William Rowan Hamilton, of Dublin (1834–5), building on the analytical foundations provided by Lagrange in the formulation of Least Action in terms of the methods of his Calculus of Variations (1758), and later (1788) but less fundamentally for physical purposes on the principle of virtual work in the *Mécanique Analytique*.

Hamilton.

Lagrange.

The principle of the Conservation of Energy, inasmuch as it can provide only one equation, cannot determine by itself alone the orbit of a single body, much less the course of a more complex system (thus §§ 107–112 above need some qualification). But if the body starts on its path from a given position in the field of force and with assigned velocity, the principle of energy then determines the velocity this body must have when it arrives at any other position, either in the course of free motion or under guidance by constraints such as are frictionless and so consume no energy. If W, a function of position, represents the potential energy of a body in the field, per unit mass, the velocity v of the body is in fact determined by the equation

The rôle of energy only partial:

$$\tfrac{1}{2}mv^2 + mW = \tfrac{1}{2}mv_0^2 + mW_0 = mE,$$

but determinative for a single body.

where the subscripts in v_0 and W_0 refer to the initial position; and mE is the total energy of the body in relation to the field of force, which is conserved throughout its path. Thus

$$v = (2E - 2W)^{\frac{1}{2}};$$

so that the velocity v depends, through W, on position alone.

Now we can propound the following problem. By what path must the body, of mass m, be guided under frictionless constraint from an initial position A to a final position B in space, with given conserved total energy mE, so that the Action in the path, defined as the limit of the sum $\Sigma mv\delta s$, that is as $\int mv\,ds$, where δs is an element of length of this path, shall, over each stage, be least possible? The method of treating the simpler problems of this kind is known to have been familiar to Newton: in the case of the present question, first vaguely proposed by Maupertuis[1] when President of the Berlin

leading to a minimal accumulated Action.

[1] The notion of an Action possibly with minimal quality, not merely passive inertia, as concerned in the transmission of Potentia or energy, is ascribed to Leibniz by Helmholtz in 1887.

[A system moving free and reversibly, without receipt or abstraction of energy from outside, always pursues the path requiring least accession of Action, that accession being expressed by the Vis Viva of the system continuing through time.]

Historical. Academy under Frederic the Great, the solution was gradually evolved and enlarged by the famous Swiss mathematical family of Bernoulli and their compatriot Euler: and finally, extended to more complex cases, it gave rise, after Euler's treatise of date 1744, in the hands of the youthful Lagrange (*Turin Memoirs*, 1758) to the Calculus of Variations, the most fruitful expansion of the processes of the infinitesimal calculus, for purposes of physical science, since the time of Newton and Leibniz.

Its geometric foundation for a single body, following Newton's methods

Let us draw in the given field of force a series of closely consecutive surfaces of constant velocity, and therefore of constant potential energy mW: and let us consider an orbit, $ABCD$... intersecting these surfaces at the points $B, C, D, \ldots$. We shall regard, in the Newtonian manner[1], the velocity as constant, say v_1, in the infinitesimal path from B to C, and constant, say v_2, from C to D: these elements of the path are thus to be regarded as straight, the field of force being supposed to operate by a succession of very slight impulses at B, C, D, ... such as in the limit, as the elements of the path diminish indefinitely, will converge to the continuous operation of a finite force.

If $\Sigma v \delta s$ is to be a minimum over this section $ABCD$... of the path, then by the usual criterion any slight alteration, by frictionless constraint, which would compel the body to take locally an adjacent course $BC'D$, ought not to alter the value of the Action so far as regards the first order of small quantities. Now, on our representation of the force as a rapid succession of small impulses, the change so produced in the value of this function of Action is equal to

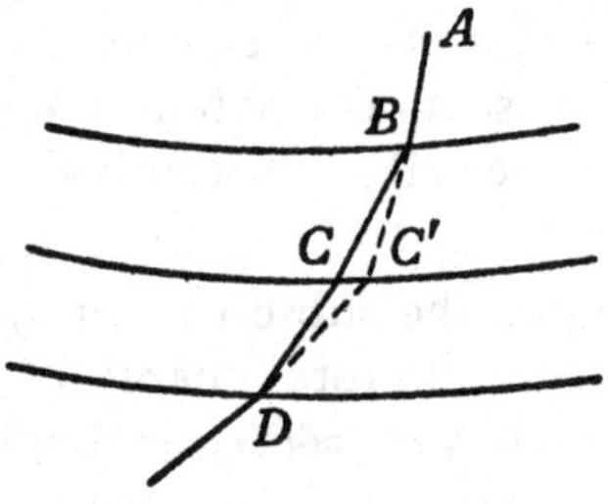

$$v_1 (BC' - BC) + v_2 (C'D - CD);$$

hence this must vanish, up to the first order. But $BC' - BC$ is equal to $- CC' \cos BCC'$, and $C'D - CD$ is equal to $- CC' \cos DCC'$. Thus the condition for a stationary value is that the component of v_1 along CC' is equal to the component of v_2 along the same direction, where CC' is any element of length on the surface of constant v, that is of constant W, drawn through C. This involves that the impulse which must be imparted to the body at C in order to change its velocity from v_1 to v_2 must be wholly transverse to this surface: or, on passing to the limit, that the force acting on the body must everywhere be in the direction of the gradient of the potential W. That is, whatever the form of this potential function may be, the succession of impulses

[1] Cf. *Principia*, Book I. Sec. II. Prop. I, on equable description of areas in a central orbit.

must be in the direction of its force; it is already prescribed by the form of v that they are of the amounts necessary to make changes in the velocity that are in accord with conservation of energy. These are just the criteria for a free orbit. Hence for any short arc of any free orbit the Action $m\Sigma v\,\delta s$ is smaller than it could be if the orbit were slightly altered locally owing to any frictionless constraint. The free orbit is thus describable as the path of advance that would be determined by minimum expenditure of Action in each stage, as the body proceeds: though this does not imply that the total expenditure of Action from one end to the other of a longer path is necessarily or always the least possible. This formula of stationary (or say *minimal*) Action, expressed by the variational equation

Restriction in statement.

$$\delta\int mv\,ds = 0, \quad \text{where } \tfrac{1}{2}mv^2 + mW = mE,$$

is by itself competent to select the actual free orbit from among all possible constrained paths*.

And generally, for any dynamical system having kinetic energy expressed by a function T of a sufficient number of geometrical coordinates, and potential energy expressed by W, it can be shown that the course of motion from one given configuration to another is completely determined by the single variational equation

Generalized statement: Eulerian,

$$\delta\int T\,dt = 0 \text{ subject to } T + W = E,$$

E being the total energy, which is prescribed as conserved, so that the variations contemplated in the motion must be due only to frictionless constraints.

Another form of the principle is that

$$\delta\int (T - W)\,dt = 0$$

provided the total time of motion from the given initial to the given final configuration is kept constant. This form is more convenient for analytical purposes because the mode of variation is not restricted to frictionless constraint; as conservation of the energy is not imposed, extraneous forces, which can be included in a modification of W,

Lagrange-Hamilton.

* A sheaf of trajectories starting from the same point, or from a surface orthogonal to them all, are orthogonal to a set of surfaces, those of equal Action.

Thus for a plane sheaf, if δn is an element of the normal to a path, $v\,\delta s$ being the same for adjacent paths between two surfaces of Action,

$$\frac{d}{dn}(v\,\delta s) = 0 \quad \text{or} \quad \frac{dv}{dn} = \frac{v}{R},$$

where R is the radius of curvature of the path; corresponding to the formula for a ray in optics $\frac{1}{\mu}\frac{d\mu}{dn} = -R$. The kinetic interpretation is $\frac{v^2}{R} = \frac{d}{dn}(\tfrac{1}{2}v^2) = -\frac{dW}{dn}$, which is the normal component of the force.

may be in operation imparting energy to the system. Constancy of the time of transit, which here takes the place of conservation of the energy, is analytically, though not physically, a simpler form of restriction. From this form the complete set of general equations of motion developed by Lagrange (see p. 64) is immediately derived by effecting the process of variation.

If T is a homogeneous quadratic function of the generalized components of velocity, $T^{\frac{1}{2}}dt$ is a quadratic function of infinitesimal elements of the coordinates: therefore the first form when expressed

Jacobian. (after Jacobi) as

$$\delta\int (E - W)^{\frac{1}{2}}\, (T^{\frac{1}{2}}dt) = 0$$

does not any longer involve the time. It thus determines the geometrical relations of the path of the system without reference to time; for a simple orbit it reduces to the earliest form investigated above.

In the modern discussions of the fundamental principles of dynamics, especially as regards their tentative adaptation to new regions of physical phenomena whose dynamical connections are concealed, this principle of variation of the Action, which condenses the whole subject into a single formula independent of any particular system of coordinates, naturally occupies the most prominent place.

Source of its power

As a supplement to Chapter IX these statements of the Principle of Action will now be established for a general dynamical system. This can be done most simply and powerfully by introducing the analytical method of Variations, invented by Lagrange as above mentioned.

Analytical development

The principle, as already deduced for the simplest case, relates to the forms of paths or orbits: if it is also to involve the manner in which the orbits are described the time must come in. The criterion of a free path was that $\delta\int v\,ds = 0$ with energy E_0 constant throughout the motion: it is the same as $\delta\int v^2 dt = 0$ under the same condition; or, writing T for the kinetic energy $\frac{1}{2}mv^2$, it is $\delta\int 2T dt = 0$ under the same restriction to constancy of the total energy.

by variation to adjacent paths:

Let us conduct the variation directly from this latter form, but now keeping the time unvaried,

$$\delta\int T\,dt = \delta\int_{t_1}^{t_2} \tfrac{1}{2}m\left\{\left(\frac{dx}{dt}\right)^2 + \left(\frac{dy}{dt}\right)^2 + \left(\frac{dz}{dt}\right)^2\right\} dt$$

$$= \int m\left(\frac{dx}{dt}\frac{d\delta x}{dt} + \frac{dy}{dt}\frac{d\delta y}{dt} + \frac{dz}{dt}\frac{d\delta z}{dt}\right) dt$$

in which d is the differential of x as the body moves along its orbit with changing time, but δx is the variation of the value of x as we pass from a point on the orbit to a corresponding point on the adjacent possible path that is compared with it. The introduction of different symbols d and δ to discriminate these two types of change is the essential feature of the Calculus of Variations: we have already used the fundamental relation $\delta dx = d\delta x$. Integrating now by parts, in order to get rid of variations of velocities which are not independent variations and so not arbitrary, we obtain

$$\delta\int T\,dt = \left| m\frac{dx}{dt}\delta x + m\frac{dy}{dt}\delta y + m\frac{dz}{dt}\delta z \right|_1^2 - \int\left(m\frac{d^2x}{dt^2}\delta x + m\frac{d^2y}{dt^2}\delta y + m\frac{d^2z}{dt^2}\delta z \right) dt;$$

in this the first term represents the difference of the values at the upper and lower limits of the integral, indicated by subscripts 2 and 1, which correspond to the final and initial positions of the particle. The second term is equal to

$$-\int (X\delta x + Y\delta y + Z\delta z)\,dt,$$

where (X, Y, Z) is the effective force acting on the particle m, as determined by the acceleration which the particle acquires.

We can extend this equation at once to any system of particles in motion under both extraneous and mutual forces. If there are no forces exerted from outside the system, but only an internal potential energy expressed by a function W, then the work of the internal forces of the system tends to exhaust this energy, so that

extended to a general intercon-nected system,

$$\Sigma\,(X\delta x + Y\delta y + X\delta z) = -\,\delta W,$$

and this holds good whether the algebraic equations expressing the constraints contain t or not, provided t is unvaried.

even with constitution changing if in prescribed manner,

Thus if T now represents the total kinetic energy, and all the forces are internal, we can write, for variation from a free path to any adjacent path by frictionless constraint, and with times unvaried,

$$\delta\int (T - W)\,dt = \left| \Sigma m\frac{dx}{dt}\delta x + \Sigma m\frac{dy}{dt}\delta y + \Sigma m\frac{dz}{dt}\delta z \right|_1^2.$$

Strictly, this result has been obtained for a system of separate particles influencing each other by mutual forces. It is natural to expand it to any material system consisting of elements of mass subject to mutual forces, thus including the dynamics of elastic systems. The ultimate analysis of the element of mass is into molecules or atoms in a state of internal motion: that final extension would include the dynamical theory of heat.

but not extending down to the irregular thermal residue.

We can now express all the coordinates x, y, z of the particles or

The independent degrees of geometrical freedom: expressed: which may change with time.

elements of mass in terms of any sufficient number of independent quantities $\theta, \phi, \psi, \ldots$ which determine the position and configuration of the system as restricted by its structure. Their number is that of the degrees of freedom of the system. The equations which express x, y, z in terms of them may involve t explicitly, for the equation of virtual work involves the displacements possible *at given time*; thus the new form of $T - W$ can contain t*. Then we can assert that when t is not varied, and the time limits t_1 and t_2 are therefore constant,

$$\delta \int_{t_1}^{t_2} (T - W)\, dt = 0$$

when the frictionless variation is taken between fixed initial and final positions of the dynamical system.

Kinetic potential of a system:

This quantity $T - W$ is the Lagrangian function L defining by itself alone the dynamical character of the system: the function $-L$ or $W - T$ is thus the potential energy W as modified for kinetic applications, and has been appropriately named by Helmholtz the kinetic potential of the system. Thus the particular case of a system at rest is included: for

$$\delta \int W dt \quad \text{or} \quad \int \delta W dt \quad \text{is equal to} \quad \delta W \int dt$$

static potential.

as W remains constant during the time: hence the equation of Action asserts in this case that

$$\delta W = 0,$$

which comprehends the laws of Statics in the form that the equilibrium is determined by making the potential energy stationary. For stability it must be minimum.

Again, as L is expressed as a function of the generalized coordinates $\theta, \phi, \ldots$ and their velocities,

$$\delta \int L\, dt = \int \left(\frac{\partial L}{\partial \theta} \delta\theta + \frac{\partial L}{\partial \dot{\theta}} \delta\dot{\theta} + \ldots \right) dt,$$

where $\dot{\theta}$ represents $\frac{d\theta}{dt}$ and $\delta\dot{\theta}$ is equal to $\frac{d}{dt}\delta\theta$: thus integrating by parts as before

$$\delta \int_{t_1}^{t_2} L\, dt = \left| \frac{\partial L}{\partial \dot{\theta}} \delta\theta + \frac{\partial L}{\partial \dot{\phi}} \delta\phi + \ldots \right|_1^2$$

$$- \int \left\{ \left(\frac{d}{dt}\frac{\partial L}{\partial \dot{\theta}} - \frac{\partial L}{\partial \theta} \right) \delta\theta + (\ldots)\, \delta\phi + \ldots \right\} dt.$$

As the left side vanishes, when the terminal positions are unvaried, for all values of the current variations $\delta\theta, \delta\phi, \ldots,$ and these are all

* But there can be no conservation of mechanical energy unless the configuration is unchanging. For instance, an isolated earth, shrinking from loss of heat, would yet be gaining mechanical energy of rotation.

independent and arbitrary, the coordinate quantities $\theta, \phi, \ldots$ being just sufficient to determine the system, the coefficient of each of these variations must vanish separately in the integrand*. Thus we obtain a set of equations of type

Leads to Lagrangian equations of motion.

$$\frac{d}{dt}\frac{\partial L}{\partial \dot\theta} - \frac{\partial L}{\partial \theta} = 0$$

which are the Lagrangian equations of motion of any general conservative dynamical system. If there are in addition extraneous forces in action on the system, the appropriate component force F_θ, defined as that part whose work $F_\theta \delta\theta$ is confined to change of the one coordinate θ, must be added on the right-hand side. These applied forces may vary with t in any manner: they can be merged in W by addition of terms $- F_\theta \theta - \ldots$ to it: their presence will prevent the energy of the system from remaining constant.

If we restrict this comparison of paths to variation from a free path of the system *to adjacent free paths*, we have

The theory of Varying Action foreshadowed.

$$\delta \int_{t_1}^{t} L\, dt = \frac{\partial L}{\partial \dot\theta}\, \delta\theta + \frac{\partial L}{\partial \dot\phi}\, \delta\phi + \ldots$$

now as an *exact* equation, and so capable of further differentiation; and it provides the basis of the Hamiltonian theory of Varying Action.

It will be convenient at this stage to remove the restriction that the time is not to be varied: to allow for this change we must substitute in the equation in place of $\delta\theta$ the expression $\delta\theta - \dot\theta\delta t$ which deducts from the total variation of θ that part of it which arises from the motion in the interval of varied time δt. We must also add $L\delta t$ in order to include in the time of transit the new interval of time δt added on at the end by the variation. Thus now

Variation completed by extending to include the time.

$$\delta \int_{t_1}^{t} L\, \delta t = L\, \delta t + \frac{\partial L}{\partial \dot\theta}\, (\delta\theta - \dot\theta\delta t) + \frac{\partial L}{\partial \dot\phi}\, (\delta\phi - \dot\phi\delta t) + \ldots.$$

Also $L = T - W$; and T being a homogeneous quadratic function,

$$\frac{\partial L}{\partial \dot\theta}\dot\theta + \frac{\partial L}{\partial \dot\phi}\dot\phi + \ldots = 2T;$$

hence $$\delta \int_{t_1}^{t} L\, dt = \frac{\partial L}{\partial \dot\theta}\, \delta\theta + \frac{\partial L}{\partial \dot\phi}\, \delta\phi + \ldots - E\,\delta t,$$

where E is the final value of the total energy $T + W$.

* It is the gradients, whether velocities or strains, that on account of the atomic origins are physically significant. If their variations are infinitesimals of the same order, then the variations of the coordinates are also of that order: thus in the language of the modern Calculus of Variations only weak variations are concerned.

When no extraneous forces are supposed to be in action E is, from the results, constant at all times: thus

$$-E\,\delta t = t\delta E - \delta\,(Et) = t\delta E - \delta\int E\,dt.$$

Hence, transposing the last term, the alternative form arises,

The Eulerian form.

$$\delta\int_{t_1}^{t} 2T\,dt = \frac{\partial T}{\partial\theta}\,\delta\theta + \frac{\partial T}{\partial\phi}\,\delta\phi + \ldots + t\delta E,$$

for variations throughout which the energy is conserved.

This is the generalization of the previous form $\delta\int mvds = 0$ for a particle, except that now the time also is involved, and is determined as $\partial A/\partial E$, where A is the time-integral of $2T$ as expressed in terms of initial and final configurations and the conserved energy.

The groups of independent momenta as new variables.

This involves the analytical result that if Θ, Φ, ... are the momenta corresponding to the coordinates θ, ϕ, ..., then there must exist a certain function A (of form however that is usually difficult to calculate) of θ, ϕ, ... E, such that in varying from the free path to adjacent free paths of the system,

$$\delta A = \Theta\delta\theta + \Phi\delta\phi + \ldots + t\delta E.$$

A more explicit and wider form, especially for optical applications, is immediately involved in this formula, that there is a function $A|_1^2$ of the initial and final configurations of the system and the energy, such that

The original binary form for rays.

$$\delta A|_1^2 = \Theta_2\delta\theta_2 + \Phi_2\delta\phi_2 + \ldots - \Theta_1\delta\theta_1 - \Phi_1\delta\phi_1 - \ldots + (t_2 - t_1)\,\delta E.$$

There also exists a function $P|_1^2$ of the final and initial coordinates and the time, equal in value to $\int_{t_1}^{t_2}(T - W)\,dt$, such that

$$\delta P|_1^2 = \Theta_2\delta\theta_2 + \Phi_2\delta\phi_2 + \ldots - \Theta_1\delta\theta_1 - \Phi_1\delta\phi_1 - \ldots - E_2\delta t_2 + E_1\delta t_1,$$

on varying from any free orbit to adjacent free orbits; but now as there is no restriction to E remaining constant along an orbit, the forces may be in part extraneous forces whose work will impart new energy to the system.

Hamiltonian reciprocal dynamical relations.

The mere fact that such a function P or A exists involves a crowd of reciprocal differential relations connecting directly the initial and final configurations of the system or a group of systems, of type such as

$$\partial\Theta_1/\partial\phi_2 = -\,\partial\Phi_2/\partial\theta_1,$$

which are often the expression of important physical results. Moreover in the form of δP, and therefore in such resulting relations, the final set of coordinates may be different from the initial set.

Action method for general perturbed orbital system.

The influence of disturbing agencies on any dynamical system, whose undisturbed path was known, is by these principles reduced to determining by approximation (from a differential equation which it satisfies) the slight change they produce in this single function P or A which expresses the system, a method perfect in idea but amenable to further simplifications in practice.

This beautiful theory of variation of the Action from any free path to the adjacent ones was fully elaborated by Hamilton in a single memoir in two parts (*Phil. Trans.* 1834 and 1835), and soon further expanded in analytical directions by Jacobi and other investigators. It brings a set of final positions of a dynamical system into direct relations with the corresponding initial positions, independently of any knowledge whatever of the details of the paths of transition. In connection with the simplest case of orbits it has been characterized by Thomson and Tait as a theory of aim, connecting up, so to say, the deviations on a final target, arising from changes of aim at a firing point, with the corresponding quantities of the reversed motion. In geometrical optics, from which the original clue to the theory came, where the rays might be regarded as orbits of imagined Newtonian corpuscles of light, it involves the general relations of apparent distance for any two points, leading to those of image to object, that must hold for all types of instrument, as originally discovered by Huygens and put in this form by Cotes. Its scope now extends all through physical science.

Historical.

Early ray developments.

In certain cases the number of coordinate variables required for the discussion of a dynamical problem can be diminished. Thus if the kinetic potential involves one or more coordinates only through their velocities, the corresponding equations of motion merely express the constancy throughout time of the momentum that is associated with each such coordinate: this holds for instance for the case of freely spinning fly-wheels attached to any system of machinery, and for all other cases in which configuration is not affected by the changing value of the coordinate. In all such cases the velocity can be eliminated, being replaced by its momentum which is a physical constant of the motion. The kinetic potential can thus be modified (Routh, Kelvin, Helmholtz) so as to involve one or more variables the less, but still to maintain the stationary property of its time-integral. It is now no longer a homogeneous quadratic, but involves terms containing the other velocities to the first degree, multiplied of course by these constant momenta as all the terms must be of the same dimensions. Every such kinetic potential belongs to a system possessing one or more *latent* unchanging (steady) motions; and a general theory of this important physical class of systems, and of the transformation of their energies, arises.

Latent conserved momenta:

merged in structural specification of the system.

Kinetic potential thus of widened form:

interpreted:

In fact if

$$L' = L - \Psi\dot{\psi} - \dots,$$

actual form thus modified.

where ψ ... are a group of coordinates and Ψ ... the related momenta, then

$$\delta L' = \left(\frac{\partial L}{\partial \dot{\psi}} - \Psi\right)\delta\dot{\psi} - \left(\dot{\psi} - \frac{\partial L}{\partial \Psi}\right)\delta\Psi + \frac{\partial L}{\partial \psi}\delta\psi + \ldots + \delta_1 L$$

in which the first term vanishes identically, while $\delta_1 L$ is the variation of L with regard to the remaining variables. Hence* if L do not involve the coordinates ψ ..., so that Ψ ... are constant and are not made subject to variation, and $\dot{\psi}$ ... are eliminated from L' by introduction of Ψ, ... then

$$\delta \int L' dt = \left| \Theta\delta\theta + \Phi\delta\phi + \ldots - E\delta t \right|_1^2$$

depending only on the variations of the explicit coordinates at the limits, provided Ψ ... are kept unvaried, or the fly-wheels of the system are not tampered with.

Varying cyclic Action.

Although the cyclic coordinates do not appear at all in L, yet it is only in terms of L' modified as here that we can avoid their asserting themselves in the domain of varying Action†.

Criterion that a system is of dynamical type.

The ultimate aim of theoretical physical science is to reduce the laws of change in the physical world as far as possible to dynamical principles. It is not necessary to insist on the fundamental position which the kinetic potential and the stationary property of its time-integral assume in this connection. Two dynamical systems whose kinetic potentials have the same algebraic form are thoroughly correlative as regards their phenomena, however different they may be in actuality. If any range of physical phenomena can be brought under such a stationary variational form, its dynamical nature is suggested: there still remains the problem to extricate the coordinates and velocities and momenta, and to render their relations familiar by comparison with analogous systems that are more amenable to inspection and so better known‡.

To extricate the dynamics inherent in the complexity of Nature.

Note on Chapter IX, § 9.

It has appeared above, as Lagrange long ago emphasized, that the principle of Conservation of Energy can provide only one of the equations that are required to determine the motion of a dynamical system. It follows that the reasoning of this section (§ 9), which seems to deduce them all, must be insufficient. The argument there begins by supposing the system to move in any arbitrary way; that

* Compare the argument after Maxwell on the F function, *infra*, p. 69.

† The relation connecting E with L' readily arises from the dynamical equations: if, arranged by homogeneous functions of the remaining velocities, $L' = L_2' + L_1' + L_0'$ then $E = L_2' - L_1'$ as *supra*, p. 49.

‡ For Least Action in relation to the problem of space-time, see *Nature*, Supplement, April 6, 1927.

is, it assumes motions determined by the various possible types of frictionless constraint that are consistent with the constitution of the system. The equation (9) is then derived correctly from (7) and (8), as the variations δq are fully arbitrary. But the imposed constraints introduce new and unknown constraining forces which must be included in the *applied* forces F_r; and they would make the result, so far as there demonstrated, nugatory.

The equations (9) are however valid, though this deduction of them fails. As explained above, the Lagrangian equations (20) are derivable immediately from the Principle of Least Action, independently established as here: and then the equations (9) can be derived by reversing the argument.

Action is a complete formulation.

The procedure of § 12 seems to lead to a noteworthy result. It asserts that if

$$F = T_p + T_{\dot{q}} - 2T_{p\dot{q}},$$

The function F of

where p represents momentum, and involving also the coordinates q, and their velocities, all independent, then the single relation

$$\delta F = 0$$

involves all the equations connecting coordinates, velocities and momenta in the system. This will remain true when the three sets of variables, regarded still as independent, are changed to new ones by any equations of transformation, so that this threefold classification into types becomes lost. Now there are cases in which the steady motion of a system, or an instantaneous phase of a varying mode of change, can be thoroughly explored experimentally, leading to the recognition say of $3n$ physical quantities of which only $2n$ can be independent; but it is not indicated by our knowledge how we are to deduce from them a scheme of n coordinates, n corresponding velocities, and n momenta. We have arrived at the result that in every such case a function F must exist, and is capable of construction, such that $\delta F = 0$ provides a set of $3n$ equations containing all the knowledge that is needed. The relations (treated after Maxwell) of a network of mutually influencing electric coils carrying currents would form an example.

a redundant specification: how to be resolved: an example.

In cognate manner we may assert another type of equation of Variation of Action

$$\delta \int (T_p - 2T_{p\dot{q}} + W)\, dt = 0,$$

where $T_{p\dot{q}} = \frac{1}{2}\Sigma \dot{q} p$ as above, containing n coordinates q, their n velocities $\dot{q}$ and their n momenta p. For this equation is equivalent to

$$\int \Sigma \left\{ \left(\frac{\partial T_p}{\partial p} - \dot{q} \right) \delta p - p \delta \dot{q} + \frac{\partial T_p}{\partial q} \delta q + \frac{\partial W}{\partial q} \delta q \right\} dt = 0$$

leading on integration by parts as usual to two sets of relations of the types

$$\frac{\partial T_p}{\partial p} = \dot{q}, \quad \frac{dp}{dt} = -\frac{\partial T_p}{\partial q} - \frac{\partial W}{\partial q}$$

if in it the momenta and coordinates are regarded as the independent variables. As $\frac{\partial T_p}{\partial q} = -\frac{\partial T_{\dot{q}}}{\partial q}$, the second set are the Lagrangian dynamical equations.

Thus we have here a single function

$$\phi = T_p - 2T_{p\dot{q}} + W$$

involving coordinates and their velocities, linear in the latter, and an equal number of quantities p of the nature of momenta, the coordinates and momenta being taken as independent variables, such that the relation

A wider Action function,

$$\delta \int \phi \, dt = 0$$

leads both to the identification of the relations in which the momenta stand to the coordinates and to the dynamical equations of motion of the system.

after Hamilton.

This result is virtually the same as equation 12*a* in Hamilton, *Phil. Trans.* 1835, p. 247. In Helmholtz's memoir on Least Action, Crelle's *Journal*, Vol. 100 (1886), *Collected Papers*, Vol. III. p. 218, another function is introduced, apparently with less fitness, in which the velocities are regarded as independent of their coordinates, but the momenta are the gradients of L with regard to the velocities. Cf. also *Proc. Lond. Math. Soc.* 1884, *supra*, p. 46.

Cf. Helmholtz.

A main source of the great power of these dynamical relations of minimal or stationary value, as exploring agents in physical science, is that the results remain valid however the physical character of the functions involved may be disguised by transformation to new variables, given in terms of the more fundamental dynamical ones by any equations whatever. This function ϕ may thus be expressed in terms of $2n$ quantities which are in any way mixed functions of coordinates and momenta and their gradients with respect to time —remaining a linear function of the latter and subject to other limitation—and the equation $\delta \int \phi dt = 0$ will still subsist and will express all the dynamical relations of the physical system.

Problem to distil an all-sufficing Action from Nature,

The existence of a variational relation of this type may be taken as the ultimate criterion that a partially explored physical system conforms to the general laws of dynamics; while from its nature the coordinate quantities, in terms of which the configuration and motion of the system happen to be expressed, shrink to subsidiary importance.

so far as it is subject to ordered reversible dynamics.

5

ON POSSIBLE SYSTEMS OF JOINTED WICKER-WORK, AND THEIR DEGREES OF INTERNAL FREEDOM.

[*Proc. Camb. Phil. Soc.* v. (1884) pp. 161–167.]

IF the two sets of generating lines of a hyperboloid of one sheet be constructed by rods jointed where they cross one another, the system so formed will not be stiff. This statement is verified by the simplest examination of an ordinary paper-basket, or—much better—of one of the jointed frameworks of wooden rods that are sometimes placed round flower-pots.

Jointed-rod hyperboloid is not stiff:

Mr A. G. Greenhill has remarked (Math. Tripos Examination Papers, 1879) that the forms assumed by the framework on deformation are those of a confocal system of hyperboloids. This result may be proved synthetically as follows. Consider such a confocal system in position; to the points which lie on a straight line on one of them there correspond (in Ivory's manner) points on any other, which also lie on a straight line, since the correspondence is of the first order or linear. Thus to a generator corresponds a generator, and the points of intersection of pairs of generators also correspond. Again, all points corresponding to a given point lie on a curve which cuts the system of surfaces normally, being in fact the curve of intersection of two confocals of the other kinds. So that if we consider any generator and the corresponding one on a consecutive surface, the lines joining their extremities (where they meet generators of the other system) are normal to the surface, and therefore to the generators, and the generators are therefore of equal length[1]. The condition necessary for deformation is thus satisfied, and the surface, supposed made up of jointed rods, may be deformed without straining into the consecutive confocal surface, and therefore by successive steps, into any other confocal surface.

deforms into the set of confocal surfaces.

Relation to Ivory's theorem.

We propose to investigate directly the cause of this want of stiffness, and to determine the number of degrees of internal freedom possessed by other systems (which we shall prove to exist) composed of three sets of rods connected by ball-joints, there being three rods at each joint.

Generalized problems,

[1] Cf. H. J. S. Smith, *Proc. Lond. Math. Soc.* Vol. II. p. 244.

In discussing the first problem, we may confine our attention to three rods crossing three other rods: for we shall prove that every other rod that crosses one set of them meets each rod that crosses the other set, at a point *in the rod* which is unaltered by the deformation. And for similar reasons, we shall only have to consider in the second case the quasi-cubical framework formed by three sets of six placed one above the other, and tied together by nine rods passing through them, making twenty-seven rods in all.

formulated,

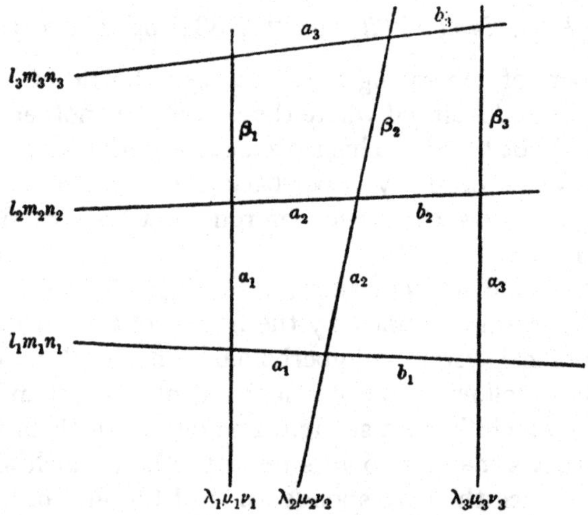

Consider then a system of six, represented on the flat by the annexed scheme, in which their direction cosines to fixed axes $l_1m_1n_1$, ..., $\lambda_1\mu_1\nu_1$, ... are indicated, and also the lengths of the three segments of each a_1, b_1, c_1 $(= a_1 + b_1)$, ... $\alpha_1, \beta_1, \gamma_1$ $(= \alpha_1 + \beta_1)$, ... By expressing that the projections of each independent circuit on the axes of coordinates are zero, we exhaust *all* the independent relations of the system. We thus obtain four sets of three equations each, of which the following is one:

of indeterminacy.

$$a_1l_1 - a_2l_2 = \alpha_1\lambda_1 - \alpha_2\lambda_2$$
$$a_1m_1 - a_2m_2 = \alpha_1\mu_1 - \alpha_2\mu_2$$
$$a_1n_1 - a_2n_2 = \alpha_1\nu_1 - \alpha_2\nu_2$$

and we have in addition six equations of the type

$$l^2 + m^2 + n^2 = 1.$$

Freedoms:

Thus we have 18 equations between 18 variables. But these variables involve 3 indeterminates, depending on the directions of the axes: and as we know that the system is not rigid, there is a fourth indeterminate. Therefore the 18 equations are equivalent to only 14 independent equations, and that can only be by reason of the existence of four relations between the coefficients, *i.e.* between the lengths of the segments of the rods. And, conversely, if we obtain these four relations independently, we can infer that the jointed system is not rigid.

We can readily obtain them as follows: Let $a_1\alpha_1$ denote the angle between the lines a_1, α_1: then by equating two expressions for the square of the diagonal of the reticulation $a_1\alpha_1a_2\alpha_2$ we obtain

$$a_1^2 + \alpha_1^2 - 2a_1\alpha_1 \cos a_1\alpha_1 = a_2^2 + \alpha_2^2 - 2a_2\alpha_2 \cos a_2\alpha_2,$$

or $$a_1\alpha_1 \cos a_1\alpha_1 - a_2\alpha_2 \cos a_2\alpha_2 = \tfrac{1}{2}\,(a_1^2 + \alpha_1^2 - a_2^2 - \alpha_2^2),$$ verified for surface-frame:

and similarly

$$b_3\beta_3 \cos b_3\beta_3 - b_2\beta_2 \cos b_2\beta_2 = \tfrac{1}{2}\,(b_3^2 + \beta_3^2 - b_2^2 - \beta_2^2),$$

$$c_1\gamma_1 \cos c_1\gamma_1 - c_3\gamma_3 \cos c_3\gamma_3 = \tfrac{1}{2}\,(c_1^2 + \gamma_1^2 - c_3^2 - \gamma_3^2),$$

three equations between the cosines of the angles $a_1\alpha_1$, $b_2\beta_2$, $c_3\gamma_3$. But we know that these angles are not determinate, therefore the result of eliminating $\cos b_2\beta_2$ between the first two equations must be equivalent to the third. That result is

$$a_1\alpha_1 . b_2\beta_2 \cos c_1\gamma_1 - a_2\alpha_2 . b_3\beta_3 \cos c_3\gamma_3$$
$$= \tfrac{1}{2}a_2\alpha_2\,(b_2^2 + \beta_2^2 - b_3^2 - \beta_3^2) - \tfrac{1}{2}b_2\beta_2\,(a_2^2 + \alpha_2^2 - a_1^2 - \alpha_1^2),$$

and we therefore have

$$\frac{a_1\alpha_1 . b_2\beta_2}{c_1\gamma_1} = \frac{a_2\alpha_2 . b_3\beta_3}{c_3\gamma_3} = \ldots,$$

therefore $$a_1b_2c_3 . \alpha_1\beta_2\gamma_3 = a_2b_3c_1 . \alpha_2\beta_3\gamma_1 .$$

And in the same way we can obtain three other similar relations, thus making up the four relations required.

Having now obtained these relations between the segments of two triads of mutually intersecting lines in space, we may easily verify their truth in other ways. We notice that they are projective for the same reason that anharmonic ratios are projective. Projecting therefore on the principal plane of the hyperboloid to which they belong, we have two triads of tangent lines to a plane ellipse. We can now project the ellipse into a parabola. But three fixed tangents to a parabola cut all variable tangents similarly, since they with the tangent line at infinity cut them in a constant anharmonic ratio: hence now also projectively.

$$a_1 : b_1 : c_1 = a_2 : b_2 : c_2 = a_3 : b_3 : c_3,$$

$$\alpha_1 : \beta_1 : \gamma_1 = \alpha_2 : \beta_2 : \gamma_2 = \alpha_3 : \beta_3 : \gamma_3,$$

and the relations are obviously true.

(We may express this argument differently by changing the hyperboloid into a hyperbolic paraboloid by a linear transformation (which we may call a projection in space of four dimensions), and noticing Confocal paraboloids.

that the theorems are true for the paraboloid because the generators of one system divide all those of the other system similarly.)

A flat jointed frame exceptional.

It is to be noticed that they are not true in general for two triads in a plane: also, inasmuch as there is only one condition necessary that six lines should touch a conic, that three *other* relations do hold in a plane.

When three lines cross three other lines in a plane the three relations between the segments formed are however still true for lines crossing in space, and are moreover clearly of a projective character. We may obtain one of them as follows. From the equations between the cosines already given, we find

$$\cos a_2\alpha_2 = \Delta'/\Delta,$$

where

$$\Delta = \begin{vmatrix} a_1\alpha_1 & -a_2\alpha_2 & 0 \\ 0 & -b_2\beta_2 & b_3\beta_3 \\ c_1\gamma_1 & 0 & -c_3\gamma_3 \end{vmatrix}$$

$$= a_1b_2c_3 . \alpha_1\beta_2\gamma_3 - a_2b_3c_1 . \alpha_2\beta_3\gamma_1,$$

$$\Delta' = b_3c_1\beta_3\gamma_1(a_1^2 + \alpha_1^2 - a_2^2 - \alpha_2^2)$$
$$+ c_3a_1\gamma_3\alpha_1(b_2^2 + \beta_2^2 - b_3^2 - \beta_3^2)$$
$$+ a_1b_3\alpha_1\beta_3(c_3^2 + \gamma_3^2 - c_1^2 - \gamma_1^2).$$

Proceeding in a similar manner we find a like expression for $\cos b_2\alpha_2$. Therefore, since $\cos b_2\alpha_2 = -\cos a_2\alpha_2$ we obtain a relation of the twelfth degree between the 12 segments: and we may find two other similar ones in the same way. It is to be noticed that the diagram is not really symmetrical, so that we cannot proceed from one expression to another by simple permutation of the symbols.

Having thus independently established the existence of these four relations, we establish at the same time the flexibility of the system of six rods. Now every line that crosses three of the rods meets every line that crosses the other three. For, if we denote the two systems of rods for an instant by 123... 1′2′3′... the planes through 1′ and 1234 cut all lines in a constant anharmonic ratio, therefore 123 are each divided in the same anharmonic ratio by 1′2′3′4′. Now consider 4, which is drawn across 1′2′3′: the plane 14′ with the lines 1′2′3′ divides it in the same anharmonic ratio as 1, 2 or 3 is divided by them: so does the plane 24′ with the lines 1′2′3′: therefore the planes 14′, 24′ are met by 4 in a common point, or, in other words, 4 meets 4′. Further, the point in which each of these lines crosses another is unaltered by deformation: for the relations already established are sufficient to determine definitely the segments of these lines in terms of the segments of the six rods: we can therefore replace the lines by jointed rods.

Synthesis for jointed surface-frame.

In the case of the paraboloidal system, in which all the rods of the same series are divided similarly, we have relations of remarkable simplicity. For the orthogonal projection on any plane consists of two series of parallel lines, and the segments of each set of rods are proportional to their projections. By considering the projections on two different planes, the above results follow immediately. **Paraboloidal framework.**

Let us consider now the quasi-cubical system of jointed rods*. In the first place, such a system is abundantly possible; for assuming the 9 rods connecting the three layers which lie the same way, and denoting them by the 9 digits, from any point on 1 draw the line which intersects 2 and 3, from the point in which it meets 3 draw the line which intersects 6 and 9, from the point in which it meets 9 draw the line which intersects 8 and 7, and from the point in which it meets 7 draw the line which intersects 4 and 1 (diagram omitted). The last line must meet 1 in the point from which we started, which gives one condition, and the three other independent circuits in the same layer give three more. Thus the three layers give twelve conditions, which can easily be satisfied by the nine lines we started with, especially as three of them may be removed by properly choosing the positions of the layers. **Spatial jointed frame.**

Having thus proved the possibility of the arrangement, we proceed as before to count *all* the independent relations of the system, and find whether they are sufficient to fix it absolutely,—or, if not, to find how many modes of deformation it possesses. We project all the independent circuits on the axes, just as before in the case of the binary system. There are 9 binary systems contained in the ternary, 3 sets of 3 each; but it will be clear on consideration that the existence of 2 of these sets determines the third set, which crosses them both. The independent circuits of the ternary system are therefore those of these two sets of binaries, and give equations 6.4.3 in number; while the metrical relations of the binaries give 6.4 conditions among the lengths of the segments, which are necessarily included in the former: so that there are 6.4.2 or 48 independent equations. There are also 27 relations between the direction cosines of the 27 lines, which are the variables. Thus there are 75 equations in all between these 81 variable direction cosines. But the arbitrary axes introduce into them 3 degrees of indeterminateness. There are therefore still 3 degrees remaining: that is, the jointed system possesses three degrees of internal freedom. ***Not* flexible:**

And now the same considerations that we employed in the case of a binary system show that we may introduce any additional number

* It has been pointed out to the author that flexibility of the triple system is not here correctly established, and is in fact not generally true.

of rods in each set, so that three rods shall meet at each joint, when the system will still possess its three degrees of internal freedom.

except in special cases. This remarkable general result is in agreement with what we can see to be true in particular cases. The simplest case of all is that of a parallelepipedal system formed of three sets of parallel jointed rods: here we can alter all the three angles between the directions of the rods. Another simple case is that of a series of equal and similar binary systems (forming paraboloids or hyperboloids) placed one over the other with corresponding joints connected by a third system of parallel rods, whose segments between two of the binary systems are therefore all equal: here the parallel rods have two degrees of freedom, and the binary systems have the third.

The sufficient specification of a jointed surface-frame. The fact that in a binary system four segments are determinable in terms of the others shows that such a system is itself determined by two rods crossing three others and jointed to them: in other words, that a system of confocal hyperboloids is so determined. So also a ternary system is determined by four rods jointed together at different parts of their lengths by three sets of four rods each.

6

ON HYDROKINETIC SYMMETRY.

[*Quart. Journ. Pure and Applied Math.* xx. (1884) pp. 261–266.]

I.

1. The foundation of the theory of the motion of a solid in an infinite [ideal] frictionless incompressible fluid without circulation is Sir W. Thomson's theorem, that the ordinary dynamical equations are at once applicable to the problem when the expression for the kinetic energy of the system has been obtained. That expression is a homogeneous quadratic function of the six velocity components of the solid, and therefore involves twenty-one coefficients, whose values depend on the shape of the solid and the distribution of its mass. Adopting Lamb's notation[1], we may write

Dynamics of free solid in infinite fluid:

$$\begin{aligned} 2T = {} & Au^2 + Bv^2 + Cw^2 + 2A'vw + 2B'wu + 2C'uv \\ & + Pp^2 + Qq^2 + Rr^2 + 2P'qr + 2Q'rp + 2R'pq \\ & + 2p\,(Lu + Mv + Nw) \\ & + 2q\,(L'u + M'v + N'w) \\ & + 2r\,(L''u + M''v + N''w), \end{aligned} \qquad (1)$$

its translatory and rotatory motions interlocked.

where u, v, w are the velocities of translation, and p, q, r the velocities of rotation of the solid referred to axes fixed in itself.

2. The simplified form which this expression assumes when the solid is one of revolution is well known, and was given in Thomson and Tait's *Nat. Phil.* 1st edn. 1867.

Conditions for an axial dynamic symmetry:

Kirchhoff pointed out the remarkable result[2] that this simple form also applies to any solid which is symmetrical with respect to two planes through its axis at right angles to one another, and also with respect to two other such planes through the same axis. Thus it applies to a right prism or pyramid standing on a square or regular hexagonal base. Such a solid may be said to have the *character* of a solid of revolution.

Again, the form which applies to a sphere also applies to a solid having two such axes of symmetry at right angles to one another, Kirchhoff's examples being a cube and a regular octahedron. These solids therefore move through the fluid in the same manner as a

for spherical symmetry, *e.g.* a cube.

[1] H. Lamb, *Motion of Fluids*, § 110.

[2] Crelle's *Journal*, t. 71, 1870; *Vorlesungen*, 19, § 3.

Motion of solid then of same type as in free space.

sphere would move, *i.e.* just as they would move in free space under the action of the same forces if their mass and their single moment of inertia about the centre were increased to a certain calculable amount.

3. We can, however, extend these conclusions to solids whose cross-sections are *any* regular figures, and to *any* regular solids, respectively; so that, for example, a right prism or pyramid on an equilateral triangular base, or even a rhombohedron, has the character of a solid of revolution, and a regular tetrahedron has the character of a sphere.

We shall first consider the case in which the only condidition imposed is, that the solid shall retain the same relations to the axes when the axes of x and y are turned round the axis of z through a definite angle α in either direction. To transform $2T$ to the new axis, we must substitute

$$\begin{aligned} u &= u' \cos\theta - v' \sin\theta, \\ v &= u' \sin\theta + v' \cos\theta, \\ w &= w', \\ p &= p' \cos\theta - q' \sin\theta, \\ q &= p' \sin\theta + q' \sin\theta, \\ r &= r', \end{aligned}$$

and the transformed expression must be of form identical with the original, whether θ is put equal to $+\alpha$ or $-\alpha$. It follows by a simple, but rather long process, that it must be of the form

$$2T = A\,(u^2 + v^2) + Cw^2 + A'\,(p^2 + q^2) + C'r^2 + 2L\,(up + vq) + 2M\,(vp - uq) + 2N''wr. \qquad (2)$$

In this expression M also can be made to vanish by moving the origin along the axis of z[1].

Dynamic helical symmetry:

The solid therefore possesses helicoidal symmetry with respect to the axis of z. When projected along that axis it executes a screw motion of pitch $-L/A$.

axial.

If, however, we are given the additional condition that there is a plane through the axis with respect to which the solid is symmetrical, the helicoidal terms necessarily disappear, and also N'' is zero; and the body possesses the character of a solid of revolution. In this case the reduction might have been much simplified by taking account of the latter condition first.

The class of geometric forms.

This latter case is that in which all the cross-sections are regular polygons (it may be with rounded faces and corners), derived from the same fundamental form, and having one common axial plane of

[1] Thomson and Tait; Lamb's *Motion of Fluids*, § 116.

symmetry. The former (helicoidal) case is that of a columnar body built up in the same way, but with no axial plane of symmetry.

4. Returning now to the form (2), let us suppose that there exists another axis situated anyhow, which possesses the same kind of symmetry as the axis of z, *i.e.* which has the character of helicoidal symmetry. Turning the axes of x and y round that of z does not affect the form of (2), as $up + vq$ and $vp - uq$ are obviously invariants for such a transformation. Let us therefore suppose them placed, so that the other axis of helicoidal symmetry lies in the plane xz; then by turning the axes of x and z round that of y through a certain angle ϕ, the new axis of x will be this axis of helicoidal symmetry, and the expression for the energy will remain of the type (2), but with the axes permuted. The process is conducted as before, and easily shows that the energy must be of the form

$$2T = A\,(u^2 + v^2 + w^2) + A'\,(p^2 + q^2 + r^2) + 2L\,(up + vq + wr), \quad (3)$$ Isotropic helical:

and therefore (since each of these three terms is an invariant for all rotations), that the solid is what Sir W. Thomson calls an isotropic helicoid[1].

It follows, therefore, that if a solid possesses the character of helicoidal symmetry about any two intersecting axes, it is an isotropic helicoid; and, as a particular case, if it possesses the character of perfect symmetry about two intersecting axes, it possesses the perfect isotropic character of a sphere. two axes give a sufficient criterion.

It may be shown similarly, as we would expect *à priori*, that if a solid have the character of helicoidal symmetry round two separate axes, they *must* intersect.

5. As examples of perfect isotropy we have all regular solids (with their edges and corners symmetrically blunted, so as to avoid discontinuous motions in the fluid), and all other bodies with the same characteristics of symmetry. Isotropic shapes:

If, however, we take a regular tetrahedron (or other regular solid), and replace the edges by skew bevel faces placed in such wise that when looked at from any corner they all slope the same way, we have an example of an isotropic helicoid; which possesses the property that, in whatever manner started, it describes a screw motion of the same pitch about an axis through its centre. This would also be the result if three plagiedral faces sloping the same way were imposed on each vertex of the tetrahedron. The first process probably gives the simplest form that a solid of this class can have. A form equivalent to the second is obtained by fixing four equal symmetrical screw- shapes with isotropic helical quality. Screw-propellers in perfect fluid.

[1] *Phil. Mag.* 1871.

propellers on the surface of a sphere at the corners of an inscribed regular tetrahedron. Sir W. Thomson's example is different from either.

II.

Extension to fluid cavities (without cyclic motion).

6. When a solid which contains cavities filled with incompressible fluid is in motion, the kinetic energy is the same as if the whole were replaced by another solid of equal mass but different moments of inertia. The criterion of kinetic symmetry is therefore concerned only with a quadratic function of p, q, r. For example, the fluid filling a regular tetrahedron, or other cavity of regular or isotropic helicoidal form, may be replaced by a sphere of equal mass and definite radius at the centre of the cavity, and rigidly attached to the solid, without altering the motion.

Sphere oscillating at centre of cavity.

7. Again, suppose a sphere to be in equilibrium, under the action of external forces, at the centre of a cavity filled with fluid, of the form of a regular solid or an isotropic helicoid. It will, when disturbed, execute vibrations which are in all respects the same as would be executed by another sphere of equal size but different density situated at the same point in free space. When the outer boundary is spherical, the expression for the ratio of the densities is well known[1].

Criteria applicable also to viscous fluids.

8. Another example of this principle of symmetry is afforded by the problem of a sphere executing torsional vibrations at the centre of a cavity filled with viscous fluid. For simple harmonic vibrations, the kinetic energy of the system, when steady motion is established, is a quadratic function of p, q, r, the component angular velocities of the sphere with respect to a system of axes fixed with reference to the boundary; and the dissipation function, which measures the rate at which the energy of motion is destroyed by friction, is also a quadratic function of the same quantities. The argument previously employed then shows, that if the cavity is of regular or isotropic helicoidal form, both these expressions take the form $C(p^2 + q^2 + r^2)$; and the character of the vibrations is therefore quite independent of the direction of the axis of torsion with reference to the boundary.

Elastic symmetries are of different type.

9. It is to be noticed that in the theory of the elasticity of crystals, it is only the *inclinations* of the edges and faces that constitute the invariable characteristics; a regular tetrahedron therefore occurs as a hemihedral form of a regular octahedron, *i.e.* as an octahedron whose alternate faces have diminished indefinitely and disappeared. The reduction of the expression for the energy in Green's theory of elastic solids is therefore not analogous to the preceding.

[1] Stokes, *Camb. Trans.* 1846.

[*Note* (1927).—In further simple ideal illustration, we can imagine a symmetrical torpedo driven by an ideal screw mounted axially without friction. Though there are here two solids, yet effectively the same configuration is always presented to the *ideal frictionless* fluid. Let M be the total effective mass, i the moment of inertia of the screw, G the driving torque and X the axial thrust it produces, so that *e.g.* the efficiency e is X/G. Then the energy of motion, there being on the ideal hypothesis no wake, would be

Torpedo driven by a screw.

$$T = \tfrac{1}{2}Mu^2 + \tfrac{1}{2}I\Omega^2 + \tfrac{1}{2}\iota\omega^2 + ju\omega:$$

so that $$X = M\dot{u} + j\dot{\omega}, \quad G = j\dot{u} + i\dot{\omega}, \quad G + I\dot{\Omega} = 0.$$

If in the ideal fluid it starts from rest and travels free without axial resistance, thus with negligible wake, u would remain equal to $-j\omega/M$. Then $G = (i - j^2/M)\,\dot{\omega}$, so that Ω/ω is constant; thus showing how the effective moment of inertia of the screw is reduced by the lengthwise freedom to move subject to inertia M, this being the interpretation of the modulus j.

Or we may consider two bodies, both with helical quality, on the same lines.]

7

ON THE THEORY OF A SYSTEM OF FORCES EQUILIBRATING AN ASTATIC SOLID.

[*Messenger of Mathematics*, Aug. 1884, pp. 61–73.]

The Complex of Central Axes.

Minding's theorem, of the congruence of single resultants in the solid.

1. In the *Aperçu Historique*, Notes, pp. 555–6, Chasles has enunciated Minding's theorem, and has suggested that a generalization of the problem would probably lead to interesting results. The theorem relates to a system of forces, each constant in magnitude and direction, applied each at a definite point fixed in a rigid body. When the body is subjected to angular displacement, there will be positions in which the system admits of a single resultant; and Minding proved that the line of action of that resultant will always intersect two conics fixed in the body, which are the focal conics of a system of confocal quadrics, and so are limiting forms of a system of quadric surfaces inscribed in the same developable. It is to be noticed that the resultant *force* of the system is always of constant magnitude; so that the system will reduce to a wrench of constant magnitude if the couple of the wrench

Chasles' suggested extension:

is constant. Chasles suggests the question, whether in this latter case the axis of the wrench will not be a common tangent to two of a system of quadrics inscribed in the same developable. He also proposes the problems of finding the locus on which lie all the axes which pass through the same point, and the envelope of all the axes which

negatived.

lie in the same plane. It will be seen, however, that only four axes pass through any point, and only a definite number lie in any plane.

Chrystal's extension: and Somoff.

2. The analysis of this system of forces has been attacked by Chrystal (*Trans. Roy. Soc. Edin.* XXIX. 1880), who finds that the axes of the resultant wrench in all possible positions form a complex of the second order in the body; he refers to Somoff's *Mechanik* (German Edition, II. Theil, 1879), where the same theorem is also proved, with the important addition that when the wrench is of constant magnitude, its axis belongs to the congruency which consists of the lines common to the above complex and another complex of the same (*second*) order.

Somoff's analysis is long, and proceeds by aid of Rodrigues' angular coordinates. Chrystal also, in discussing wrenches of finite couple, makes use of the same variables, but he has also deduced the equation of the general complex and Minding's result by a different and shorter method; while Tait has given a quaternion investigation. Tait.

The discussion immediately following is in substance the same as Somoff's, but the preliminary analysis and the use of Rodrigues' coordinates have been evaded.

3. In questions relating to systems of lines, it is natural to employ the six coordinates of a line that are used by Plücker and Cayley. For statical applications we obtain a very convenient form, by taking them to be the components along the axes, and the moments round the axes of a unit force acting along the line. If l, m, n are the direction cosines of the line, and ξ, η, ζ the coordinates of *any* point on it, these coordinates are clearly Cayley's six coordinates of an axis of a force:

$$l,\ m,\ n,\ m\zeta - n\eta,\ n\xi - l\zeta,\ l\eta - m\xi \qquad (1)$$

or, say,

$$l,\ m,\ n,\ \lambda,\ \mu,\ \nu,$$

wherein

$$l^2 + m^2 + n^2 = 1,$$

$$l\lambda + m\mu + n\nu = 0.$$

restricted, as not of a wrench.

A single homogeneous equation of the nth degree between these coordinates represents a complex of lines of the nth order. On substituting the values of the coordinates given in (1), we obtain the relation between the direction cosines of all lines which pass through any fixed point $\xi\eta\zeta$, and it is clear that these lines form a cone of the nth order. The lines of the complex which lie in any plane envelope a curve of the mth class, m being the class of the complex. The lines common to two complexes form a congruency; if the complexes are of the mth and nth orders, the congruency is of the mnth order, and mn lines of it pass through any point. The lines common to three complexes form a ruled surface. A line-complex: its degree: its class: its congruences: its ruled surfaces.

4. Returning now to the system of forces, we can replace each by its three rectangular components in directions fixed in space, thus obtaining three sets of parallel forces; and we can replace each set by a resultant force acting at its centre. Thus the whole system is equivalent to three forces at right angles to one another of constant magnitude, acting each at a fixed point in the body. Any set of forces reducible to three forces,

It is clear that we will make a further simplification by taking one of the forces in the direction of the total resultant force R, when the points of application of the other two will go off to infinity, and they will become couples. It is also clear that, by choosing proper directions from these other two forces, we can secure that the vectors to their two points of application shall be at right angles. in various ways:

We can, however, avoid the inconvenience of having the points of application at infinity. For, taking for the instant the line of action of R as axis of z, and its point of application as origin, we can apply pairs of equal and opposite forces, each equal to R for simplicity along the axes of x and y. We can now find the centre of the X-components together with R at the orgin, and the centre of the Y-components together with R at the origin, and we can, as before, turn the axes of x, y round that of z till the radii vectores to these centres are at right angles. We have thus reduced the system to the force R acting at the origin, a couple whose forces act at the origin and at another centre fixed in the body at a distance p from the origin, and another couple whose forces act at the origin and at another centre fixed in the body at a distance q from the origin, the three sets of forces, each equal to R, acting in directions at right angles fixed in space, and the lines p and q being also at right angles.

one convenient form.

5. The simplest system of axes fixed in the body must clearly consist of those lines p and q and their common perpendicular. Taking them as axes of x, y, z respectively, let the direction cosines of the resultant force and the forces of these two resultant couples be $l_1m_1n_1$, $l_2m_2n_2$, $l_3m_3n_3$ respectively. This whole system will then clearly be equivalent to

component forces $Rl_1,\quad Rm_1,\quad Rn_1,$

component couples $R\,(pn_2 - qm_3),\quad R.ql_3,\quad -R.pl_2,$

with respect to the axes of coordinates.

If now the system is equivalent to a wrench R, G, on an axis whose coordinates are

$$l,\ m,\ n,\ \lambda,\ \mu,\ \nu,$$

and if k represent the parameter G/R, which is in Professor Ball's nomenclature the pitch, we have

$$l,\ m,\ n = l_1,\ m_1,\ n_1,\quad \lambda + lk = pn_2 - qm_3, \tag{2, 3}$$

$$\mu + mk = ql_3,\quad \nu + nk = -pl_2, \tag{4, 5}$$

these latter equations simply expressing that the wrench has the same moments round the axes as the equivalent system.

6. From them we obtain

$$k = (pn_2 - qm_3)\,l_1 + ql_3m_1 - pl_2m_1$$

$$= \pm\,(pm_3 - qn_2) \tag{6}$$

and $$\lambda^2 + \mu^2 + \nu^2 + k^2 = (pn_2 - qm_3)^2 + q^2l_3^2 + p^2l_2^2;$$

therefore

The complex of astatic wrenches:

$$\lambda^2 + \mu^2 + \nu^2 = p^2\,(l_2^2 + n_2^2 - m_3^2) + q^2\,(l_3^2 + m_3^2 - n_2^2),\ \text{by (6)},$$

$$= p^2m^2 + q^2n^2. \tag{7}$$

Also, from (2) and (3),

$$\frac{(\mu + mk)^2}{q^2} + \frac{(\nu + nk)^2}{p^2} = l_2^2 + l_3^2$$
$$= m^2 + n^2. \qquad (8)$$

We have therefore obtained (7), a complex of the second order, and, as may be easily seen, of the second class, for the locus of all the axes; and (8), another complex of the second order, to which belong all those which have a given pitch. These are Somoff's general results. congruence of those of given pitch.

7. When $k = 0$, so that the wrench reduces to a single resultant force, these equations lead at once to Minding's special case.

$$\frac{\mu^2}{q^2} + \frac{\lambda^2}{q^2 - p^2} = n^2, \quad \frac{\nu^2}{p^2} + \frac{\lambda^2}{p^2 - q^2} = m^2. \qquad (9, 10)$$

To find where this congruency of lines intersects the plane of xy we put $\zeta = 0$ in (1), and so obtain from (9),

$$\frac{\xi^2}{q^2} + \frac{\eta^2}{q^2 - p^2} = 1. \qquad (11)$$

To find where it intersects the plane of xz we put $\eta = 0$, and so obtain from (10)

$$\frac{\xi^2}{p^2} + \frac{\zeta^2}{p^2 - q^2} = 1. \qquad (12)$$

The congruency therefore consists of all the lines which intersect these two conics, which are the focal conics of the system of quadrics confocal with

$$\frac{\xi^2}{p^2 + q^2} + \frac{\eta^2}{q^2} + \frac{\zeta^2}{p^2} = 1, \qquad (13)$$

and we have Minding's theorem.

When $R = 0$, so that the wrench reduces to a couple, the equations (3), (4), (5) must be modified, as the origin passes off to infinity. It is easily seen that we can then reduce the system to three couples acting round the origin with forces, say each equal to R, in fixed rectangular directions in space, and points of application at distances p, q, r along the coordinate axes. If L, M, N represent the components of the resultant G, we have therefore Case of pure torque.

$$L = (qn_2 - rm_3)\, R, \quad M = (rl_3 - pn_1)\, R, \quad N = (pm_1 - ql_2)\, R.$$

The axis of the resultant can assume any direction in the body.

8. Equation (8) shows that to any line of the complex (5) there correspond two values of k. Duplicity of wrenches.

These values may be equal; there are then two positions of the body which lead to identically the same wrench, and the axis of this wrench is a line of the congruency of the eighth order in which the complex (7) meets the complex

$$\left(\frac{\mu^2}{q^2}+\frac{\nu^2}{p^2}-m^2-n^2\right)\left(\frac{m^2}{q^2}+\frac{n^2}{p^2}\right)=\left(\frac{m\mu}{q^2}+\frac{n\nu}{p^2}\right)^2.$$

Eight such axes therefore pass through any point.

The two wrenches on any line of the complex are of equal and opposite pitch when the lines belong also to the complex

$$\frac{m\mu}{q^2}+\frac{n\nu}{p^2}=0.$$

The lines of Minding's Theorem are also axes of wrenches for which k is equal to

$$-2\left(\frac{m\mu}{q^2}+\frac{n\nu}{p^2}\right)\left(\frac{m^2}{q^2}+\frac{n^2}{p^2}\right)^{-1}.$$

9. It was proved by Sir W. R. Hamilton, that every congruency is composed of the bitangents of a certain surface; and that each line touches the surface at the two points in which it meets consecutive lines (cf. Salmon's *Solid Geometry*, 3rd edit., Appendix III).

To determine this surface we must write down the equations of the two cones which contain all the lines belonging to the two separate complexes that pass through the point $\xi\eta\zeta$, and the condition that these cones touch one another gives the equation of the surface in terms of $\xi\eta\zeta$ as running coordinates. When the complexes are both of the second order these are quadric cones, and the condition of contact is of the 12th order in the coefficients; so that as the coefficients are of the second order in $\xi\eta\zeta$, the equation of the surface will usually be of the 24th order[1].

If the lines were the common tangents to two quadrics, these cones must have double contact, and this would necessitate two conditions each of the 8th order in the coefficients. In the case of the axes of the wrenches, there is nothing to indicate that these conditions can reduce for the general problem to a single one of the second order; so that Chasles' query is apparently to be answered in the negative.

Chasles' query.

[1] The very high order of this surface requires some explanation. This is afforded by an elaborate discussion of Kummer's (*Berlin Abhandlungen*, 1866), in the course of which he shows (p. 40) that this surface (Brennfläche) for a congruency of the second order is always of necessity of the fourth order, unless nodal curves (Brenncurven) such as the conics in Minding's theorem exist. But the whole array of double tangents to a surface of the fourth order form a congruency of the twelfth order, which must therefore include the given congruency of the second order. And all possible congruencies of the second order without nodal curves are factors in the general congruency of the twelfth order which is made up of the bitangents of the general quartic surface.

A simple case, however, including that in which the forces are all parallel to a plane, is fully worked out below, § 13.

Astatic Equilibration.

10. In the above, the general system of forces has been reduced to various simpler systems of the same kind. If we reverse one of these latter, it will, together with the original, produce equilibrium *in all positions of the body*. We are thus led to the theory of *astatic* equilibrium which has been developed by Möbius, Minding, Moigno, Somoff, and Minchin (*Proc. Lond. Math. Soc.* IX).

Simplest general astatic equilibration:

Thus any system can be astatically equilibrated by three rectangular forces, or more generally by three forces inclined at any angles; for we can resolve all the forces into three sets of parallel forces inclined at those angles, and take the resultant of each set. It is easy to see that the points of application of all these resultants lie in a fixed *central plane*. As a particular case of the rectangular system, we have, as above, a single force R acting at the *centre* of this plane, and two couples, consisting of forces equal to R applied in fixed directions at the origin and at the ends of arms of lengths p, q, which are in this plane and are fixed in the body; and these arms are related to a certain ellipse, and may be its semi-axes.

by two forces:

11. Again, if the system can be astatically equilibrated by *two* forces, the components of the system in a direction perpendicular to these two must be independently in equilibrium. This direction is at right angles to the resultant force, and by taking it for one of the directions of resolution in § 10, the corresponding couple vanishes, and the system is equivalent astatically to a force and one couple. These again may be reduced to two forces, which must be in every case parallel to a plane fixed in space, obviously the plane perpendicular to the direction of independent equilibrium. The points of application of these two lie in the central plane; and if we consider any equivalent astatic rectangular system of three, their points of application must lie in the straight line joining the former, for otherwise the moments of the three round that line could not vanish however they are turned, *e.g.* when two of them are in the central plane and the other perpendicular to it. Thus the central plane reduces to a central line, and the relations of equivalent force-pairs are easily determined.

details:

12. For we can, by virtue of the astatic relation, turn the body so that the forces shall be perpendicular to the central line. The forces are then equivalent to a wrench R, G, whose axis (x) intersects the central line at a point which we shall take as the origin, the central

line being the axis of z. Let the force-pair be inclined at *any* constant angle α. This wrench is then equivalent to a force R_1 acting at z_1 at an angle θ with the axis of x, and a force R_2 acting at z_2 at an angle $\theta+\alpha$ with the axis of x. But the force R is equivalent to forces $-R\sin(\theta+\alpha)/\sin\alpha$ parallel to R_1, and $R\sin\theta/\sin\alpha$ parallel to R_2; and the couple G is equivalent to couples $G\cos(\theta+\alpha)/\sin\alpha$ in the plane (R_1z) and $-G\cos\theta/\cos\alpha$ in the plane (R_2z). These couples must therefore be just sufficient to transfer the corresponding forces to distances z_1 and z_2 along the axis; hence

$$z_1 = \frac{G}{R}\tan(\theta+\alpha), \quad z_2 = \frac{G}{R}\tan\theta. \qquad (14, 15)$$

On turning the axis of x, y through an angle α in the negative direction round that of z, the first equation gives $z_1 = \frac{G}{R}\frac{y}{x}$, which is the equation of a paraboloid on which lies the line of action of R_1.

The line of action of R_2 lies on a paraboloid which has the same equation when referred to axes turned through an angle α in the other direction.

The form of the paraboloid is independent of the value of α. Turning it through an angle π leaves it as at first, so that when the two forces are at right angles they both act along generators of the same paraboloid, which is a known theorem.

If $G/R = k$, we have from (14), (15),

$$z_1z_2 + k\cot\alpha\,(z_1 - z_2) + k_2 = 0,$$

so that the points of application are in involution.

When the body is turned into any other position, the directions of the forces being unaltered, all that happens is that the axis is now oblique to the plane of each pair; and (as follows necessarily from the astatic relations) the same equations, to the same axes now oblique, hold as before, so that the same conclusions are true. The axis of the resultant wrench does *not* now intersect this central axis; but, as already seen, we may replace the force-pair by a force acting at a fixed point on it, and a couple whose forces act at two fixed points on it, both in fixed directions in space.

its complex, 13. The discussion of the general complex for this case leads to results of interest. Since the system now reduces to a force R at the origin, and a couple whose forces, each R and perpendicular to the former, are applied at points distant by p on the central axis, we see that this axis is an axis of symmetry.

We have simply to put $q = 0$ in the general discussion of § 6.

If $$\lambda^2 + \mu^2 + \nu^2 = \varpi^2,$$

we have from (7) and (8)

$$\varpi = pm, \quad \mu = -km. \qquad (16, 17)$$

Therefore the congruency of constant pitch is of the second order.

Also, $$\frac{\mu}{\varpi} = -\frac{k}{p} = \sin\beta, \text{ say}, \qquad (18)$$

which shows that the plane containing the axis and the origin makes a constant angle β with the central axis.

Again, using the values of $\lambda\mu\nu$ in (1), and eliminating m by (17), we have

$$\begin{aligned}\varpi^2 = \quad & l^2\left\{\frac{\zeta^2}{k^2}(\xi^2+\zeta^2+k^2)+\eta^2-\frac{2}{k}\xi\eta\zeta\right\}\\ + \quad & n^2\left\{\frac{\xi^2}{k^2}(\xi^2+\zeta^2+k^2)+\eta^2-\frac{2}{k}\xi\eta\zeta\right\}\\ - \quad & 2ln\left\{\frac{\xi\zeta}{k^2}(\xi^2+\zeta^2+k^2)-\frac{\eta}{k}(\xi^2+\zeta^2)\right\}.\end{aligned}$$

Therefore $\dfrac{p^2}{k^2}(l\zeta - n\xi)^2 =$ same expression, by (16).

This equation determines the values of l/n for the two lines of the congruency, made up of axes of pitch k, which pass through $\xi\eta\zeta$. We notice that it becomes indeterminate for the points on the curve

$$\eta = 0, \quad \xi^2+\zeta^2+k^2-p^2 = 0;$$

therefore this is a nodal curve of the congruency. It is a circle in the plane of $\xi\zeta$ of radius $\sqrt{(p^2-k^2)}$, which is always real. It follows by (18) that the congruency consists of all the lines which intersect this circle, and also touch a right cone of semi-angle β with the central point as vertex and the central line as axis. **its congruences.**

We have, therefore, for this case completely determined the nodal surface and the nodal curve of the congruency.

If, however, $k = 0$, we have $\mu = 0$, and all the axes therefore lie in planes which contain the central axis, so that the congruency consists of the lines which intersect the central axis and a circle round it of radius p in the diametral plane through the origin. **Simple cases of congruences.**

When the wrench reduces to a single force the result is clear by inspection. For, that the equivalent force and couple may reduce to a single force, it is necessary that they be in one plane, and then it is evident directly that the system is balanced by a force R which meets the diametral plane of the system in a point O' at a distance p from the origin.

It is not difficult to prove the more general result in the same way.

14. When the system of forces is confined to a plane, they can always be astatically equilibrated by a single force, as, also, when

they remain parallel to a plane and are subjected to one other condition of an obvious character. But converse propositions are clearly not true, for any portion of a general system in astatic equilibrium is balanced by the remainder of it, which we may take to be a single force.

Generalized System.

Astatic linear forcive between two solids.

15. It has apparently not been observed that the same analytical methods which apply to the system of forces already considered also apply to the more general system, in which the forces act along the lines connecting pairs of points A_1B_1, A_2B_2, ..., A_nB_n in two solids, and are proportional to the lengths of those lines.

Thus let X, Y, Z be the coordinates of a point A in the first solid, and x, y, z those of the corresponding point B in the second solid, let α, β be the centres of equal masses at A_1A_2 ..., and at B_1B_2 ... respectively, and let O be the origin of coordinates. Then we may replace the force A_1B_1 by forces $-OA_1$ and OB_1, and a couple represented by twice the area of the triangle OA_1B_1; and the other forces similarly. The resultant of OA_1, OA_2, ... is $nO\alpha$, and that of OB_1, OB_2, ... is $nO\beta$; thus the resultant of all the force-components is $n\alpha\beta$ acting through O. The components of the separate couples round the axes of coordinates are of the forms

$$yZ - zY, \quad zX - xZ, \quad xY - yX,$$

For rotation round mass-centres:

and are therefore the same as the moments of a force XYZ acting at the point xyz. Now, clearly, if we keep the A-solid fixed and allow the B-solid to rotate round a centre coinciding with α, the mass-centre of the A's, we shall have an analysis agreeing exactly with that of the problem already considered. The force of the resultant wrench will be constant, and its axis will lie on the complex (7) fixed in the B-solid. Its pitch will be given by (8), and the axes of all wrenches of constant pitch will form a congruency of the fourth order; while, if the pitch is zero, and the wrench therefore reduces to a single resultant force, its axis will intersect two conics. Finally, the system of forces can be astatically equilibrated by three forces of the same kind.

the complex of astatic wrenches:

congruence for constant pitch:

In a particular case, however, which includes that in which the points A_1A_2 ... lie in the same plane, the axes of the wrenches of constant pitch will lie on a congruency of the second order, which is made up of the tangent lines to a right cone with vertex at the origin α, which intersect a coaxial circle in a plane through the vertex; while if the pitch is zero they intersect a circle and its axial line. Two forces of the same kind are sufficient to astatically equilibrate the system.

simple special case.

16. If, however, the B-solid is free to trun round a centre O which does not coincide with α, we shall have, in addition to the forces of the previous analysis, a constant force $O\alpha$. If, then, we now denote by l, m, n, the direction cosines of the axis of the resultant wrench R, G (where $G/R = k$), and by l_1, m_1, n_1, those of $O\beta$, and if R_1 represent the constant force $O\alpha$, and X_1, Y_1, Z_1 the constant components of the force $O\beta$, we have, in place of the previous equations, the following: Generalized:

$$\begin{aligned} lR &= l_1R_1 + X_1, \\ mR &= m_1R_1 + Y_1, \\ nR &= n_1R_1 + Z_1, \\ (\lambda + lk)\, R &= (pn_2 - qm_3)\, R_1, \\ (\mu + mk)\, R &= ql_3R_1, \\ (\nu + nk)\, R &= -pl_2R_1. \end{aligned}$$

We deduce

$$(lR - X_1)^2 + (mR - Y_1)^2 + (nR - Z_1)^2 = R_1^2, \qquad (19')$$

and $$\lambda^2 + \mu^2 + \nu^2 + k^2 = \{p^2m_1^2 + q^2n_1^2 + (pm_3 - qn_2)\}\frac{R_1^2}{R^2},$$

where

$$(\lambda + lk)\, l_1 + (\mu + mk)\, m_1 + (\nu + nk)\, n_1 = \pm\, (pm_3 - qn_2)\frac{R_1}{R}, \qquad (20')$$

and $$\frac{(\mu + mk)^2}{q^2} + \frac{(\nu + nk)^2}{p^2} = \frac{R_1^2 - (lR - X_1)^2}{R^2}. \qquad (21)$$

The first two of these may be written

$$2R\,(lX_1 + mY_1 + nZ_1) = R^2 - R_1^2 + X_1^2 + Y_1^2 + Z_1^2, \qquad (19)$$

$$\{kR - \lambda X_1 - \mu Y_1 - \nu Z_1 - k\,(lX_1 + mY_1 + nZ_1)\}^2$$

$$= (\lambda^2 + \mu^2 + \nu^2 + k^2)\, R_1^2 - \frac{R_1^2}{R^2}\{p^2\,(mR - Y_1)^2 + q^2\,(nR - Z_1)^2\}. \qquad (20)$$

We readily draw the following inferences: results.

(1) If $k = 0$, the axis of the resultant force lies on a congruency which is made up of the lines common to a complex of the fourth order and another of the third order.

(2) If R and k are constants, *i.e.* if the system reduces to a wrench of constant magnitude, its axis lies on a ruled surface, whose generators are all equally inclined to the fixed direction of $O\alpha$.

These results follow by elimination from the last three equations. When $k = 0$, that elimination leads to two equations of the third and fourth orders in the line coordinates, and homogeneous in them on account of the relation $l\lambda + m\mu + n\nu = 0$. In the second case we substitute from (1), and eliminate l, m, n. It is interesting to notice, that the locus of the axis becomes more general by one order when O coincides with β, as then R is always constant.

An alternative view: 17. The connection with the previous problems may also be seen (more immediately) on observing that the force AB is equivalent to a force OA, and a force through A equal and parallel to OB. The assemblage of the first components is equivalent to the constant force $n.O\alpha$, and the latter form a system such as was before discussed.

18. If Q_1, Q_2, Q_3, Q_4 be any four points in the B-solid which are not all co-planar nor any three of them collinear, the force AB can always be replaced by forces along the lines joining A to these points and having a definite relation to the lengths of these lines, by Leibnitz's theorem, which we have employed in § 15. If we thus decompose all the forces and apply the same theorem to the components through Q_1, we see that they are together equivalent to a force joining Q_1 to a point P_1 fixed in the A solid, and proportional to its length. Thus the system of forces is equivalent to four forces acting along P_1Q_1, P_2Q_2, P_3Q_3, P_4Q_4, and always in fixed proportions to these lines. It follows that, if we regard the P, Q points as multiple points in the solids, the system of forces can be astatically equilibrated by these four, for all positions of either solid. But when the one set of points lies in a plane, three forces are sufficient; and so on.

reduction of forcive: results.

If one of the bodies is fixed, and the other free to turn round a fixed centre, the addition of three more forces of the same kind to the system will reduce it to a force constant and fixed in space; while if the centre of rotation is the mass centre of the points in the fixed body, the three additional forces will equilibrate the system.

19. For simplicity we have assumed that the forces are all in the same proportion to the lengths of their lines of action. But we may have any number of coincident pairs of points in the system; and generally, we may take each force in any constant ratio whatever to the length of the corresponding line.

Extension of problem. 20. Finally, we might investigate the case of a solid which is acted on by a system of forces fixed in itself (*e.g.* the reactions of another solid rigidly connected with it), in addition to the system of forces already considered which has fixed relations in space. To do this, we would have merely to add on terms $-L_1, -M_1, -N_1$, to the left sides of the equations of moments in § 15; so that $X_1, Y_1, Z_1, L_1, M_1, N_1$, would represent the system of forces fixed in the body. The eliminations could then be effected precisely as in § 16, but the results prove to be of a more complicated character. The locus of axes of constant wrench is a ruled surface.

8

ON THE EXTENSION OF IVORY'S AND JACOBI'S DISTANCE-CORRESPONDENCES FOR QUADRIC SURFACES.

[*Proc. London Math. Soc.* Vol. XVI. (1885) pp. 189–200.]

1. The theorem of Ivory for confocal quadrics,—that, if P, Q be two points on a quadric and P', Q' the corresponding points on a confocal, then $PQ' = P'Q$,—is of a fundamental character in a general theory of distance relations. *Ivory's theorem of confocal surfaces,*

The correspondence between P and P' is determined by their lying on the same orthogonal trajectory of the system of confocals, which is of course a curve of intersection of two confocals of the other species. The correspondence is linear, each principal coordinate being proportional to the related semi-axis of its surface.

By means of the theorem, Ivory established his very general result relating to the attractions of solids bounded by quadric surfaces, which is true, as Poisson pointed out, for all laws of attraction depending only on the distance. *and of their attraction for any law of force.*

Jacobi's focal relation, which affirms that a quadric may be specified as the locus of a point whose distances from any three points on a focal conic are respectively equal to the distances of any arbitrary point in a plane from three fixed points in that plane, also flows at once from Ivory's theorem (cf. Salmon's *Geometry of Three Dimensions*). But it is important to remark that the proposition requires *equality* in the corresponding distances. If, as we sometimes find, it is only postulated that the same relations exist between the one set of distances as between the other set, *i.e.*, if only *proportionality* is required, there is no longer any locus. To obtain a locus, the comparison must then be instituted between four pairs of distances, as in the investigation given below. *Jacobi's planar correlation for a quadric.*

Confocal hyperboloids of one sheet may be considered as generated by two systems of mutually intersecting straight lines. On different confocals these lines clearly correspond to one another, as also do their points of intersection; and their corresponding segments are equal. So that in fact, if the lines were two systems of jointed rods, they could be deformed into a confocal without straining*. Now, if *Jointed ruled hyperboloid, flexible into confocals.*

* The relations of this system, and of the corresponding system of three dimensions, which proves [not] to be also flexible, are worked out in a paper in the *Proceedings of the Cambridge Philosophical Society*, 1884, Vol. v. Part 2 [*supra*, p. 71].

we take $ABCD$ a quadrilateral on the surface of a hyperboloid, and $A'B'C'D'$ the corresponding one on a confocal, we see that corresponding sides are equal, and that there are also six relations of the form $AB' = A'B$, $AC' = A'C$, and so on. Thus we have a prismoidal figure bounded by *gauche* quadrilateral faces such that the corresponding sides of two opposite faces are equal, and the diagonals of any of the six faces or diagonal sections connecting them are also equal: it is easy to trace out in this way its relations, and also to see that any further relation of equality would make it a regular prism.

Again, we may apply Ivory's theorem to the focal ellipse and hyperbola of the confocal system: we thus come directly to the result that the distances of a variable point on either of these curves from any two fixed points on the other are equal to the distances of a variable point on a straight line from any two fixed points on that line. This is Dupin's theorem that each of these conics is the locus of the foci in space of the other, and that any two points on the locus possess the focal property*.

Dupin's spatial foci of a conic.

These examples illustrate the fundamental character of the theorem. The subject is, however, still treated as a special property of confocal quadrics, rather than as a part of a theory of distances in general. The object of this paper is to discuss the question from the latter point of view[1]. It will be shown that a similar theory applies to a

* For an intuitive presentation see *infra* p. 152, from *Proc. Lond. Math. Soc.* 1887.

[1] Sections 7 (*a*) and 9 (*a*) have been added since the paper was read. It was not, however, until it had been completed, that I was able to see Darboux's Memoir "Sur les Théorèmes d'Ivory, relatifs aux surfaces homofocales du second degré," *Mémoires de Bordeaux*, T. VIII. pp. 197–280. In this paper, Jacobi's transformation is discussed and applied in an elegant and elaborate manner; so that the statement made above must be limited. Much of the results of the discussion given here is to be found in this Memoir, but the methods followed in it are usually different, and sometimes more analytical and special. On the other hand, some of the results given in this paper, such as (4), (10) and (11), lend themselves easily to an extension of Darboux's theory. Of the problem of § 8, a posthumous solution of Jacobi's for the case of conics has, it appears, been published by Hermes, and Darboux gives two solutions for the general case of quadrics. These are both different from the one indicated in § 8, which possesses the advantage that it can be at once extended to the case of cyclides.

Darboux's discussion,

M. Darboux considers also the generalization of the theory, in which the square of the distance of the point in question from a fixed point is replaced by the square of the length of the tangent drawn from it to a fixed sphere, *i.e.* by the *power* of the sphere. It may be remarked that all that is given here may be at once extended to that case by making use of the well-known relations between the lengths of such tangents, which are identical in form with (2) and (3) (Salmon's *Conic Sections*, § 132 *a*, Ex. 4), and of the other similar relation which corresponds to (4).

and extension to spheres.

system of confocal cyclides. But it will also be seen that these are the only systems of surfaces for which such relations are true.

2. Let 2, 3, 4 represent fixed points, and 1 any point in their plane, and let 23 represent the distance between 2 and 3; the relation satisfied by the distances of 1 from 2, 3, 4 is well known and easily found, and is

Distance relation of four co-planar points,

$$\Sigma_4 12^2.24^2.41^2 = \Sigma_6 34^2\,(12^2.34^2 - 13^2.24^2 - 14^2.23^2). \qquad (1)$$

We may invert this relation with respect to a point 6; this is done by writing for 12 say, the expression $\frac{12}{61.62}$; and the result is

$$\Sigma_4 12^2.24^2.41^2.63^4 = \Sigma_6 61^2.62^2.34^2(12^2.34^2 - 13^2.24^2 - 14^2.23^2). \quad (2)$$

We have here a relation connecting the ratios of the distances of any point 6 on a plane from four fixed points 1, 2, 3, 4 on the plane.

If we take the centre of inversion 6 to be out of the original plane, we see that the same relation is true also when the five points are on the surface of a sphere.

and five co-spherical:

3. This relation between the ratios of the distances of a point in a plane from four fixed points in the plane, as well as the corresponding theorem in space for five points, also flows readily from Prof. Cayley's method of multiplication of matrices (Salmon, *Higher Algebra*, Ch. III. Ex. 7, 8). Thus, if $x_r y_r z_r$ be the coordinates of the point r, and if the point 6 be the origin, and we write down the arrays

$$\begin{vmatrix} 61^2 & x_1 & y_1 & z_1 & 1 \\ 62^2 & x_2 & y_2 & z_2 & 1 \\ 63^2 & x_3 & y_3 & z_3 & 1 \\ 64^2 & x_4 & y_4 & z_4 & 1 \\ 65^2 & x_5 & y_5 & z_5 & 1 \\ 1 & 0 & 0 & 0 & 0 \end{vmatrix} . \begin{vmatrix} 1 & -2x_1 & -2y_1 & -2z_1 & 61^2 \\ 1 & -2x_2 & -2y_2 & -2z_2 & 62^2 \\ 1 & -2x_3 & -2y_3 & -2z_3 & 63^2 \\ 1 & -2x_4 & -2y_4 & -2z_4 & 64^2 \\ 1 & -2x_5 & -2y_5 & -2z_5 & 65^2 \\ 0 & 0 & 0 & 0 & 1 \end{vmatrix}$$

their product is zero, as each has one row more than the number of columns; therefore, by the ordinary rule, we have, as Prof. Cayley found,

$$\begin{vmatrix} 0 & 12^2 & 13^2 & 14^2 & 15^2 & 1 \\ 21^2 & 0 & 23^2 & 24^2 & 25^2 & 1 \\ 31^2 & 32^2 & 0 & 34^2 & 35^2 & 1 \\ 41^2 & 42^2 & 43^2 & 0 & 45^2 & 1 \\ 51^2 & 52^2 & 53^2 & 54^2 & 0 & 1 \\ 1 & 1 & 1 & 1 & 1 & 0 \end{vmatrix} = 0, \qquad (3)$$

also in addition for any five points in space.

which is Prof. Cayley's form of Carnot's relation connecting the distances of five points in space.

But if we invert the order of the last rows of the matrices, so that the units may appear at the other ends, the product gives

$$\begin{vmatrix} 0 & 12^2 & 13^2 & 14^2 & 15^2 & 16^2 \\ 21^2 & 0 & 23^2 & 24^2 & 25^2 & 26^2 \\ 31^2 & 32^2 & 0 & 34^2 & 35^2 & 36^2 \\ 41^2 & 42^2 & 43^2 & 0 & 45^2 & 46^2 \\ 51^2 & 52^2 & 53^2 & 54^2 & 0 & 56^2 \\ 61^2 & 62^2 & 63^2 & 64^2 & 65^2 & 0 \end{vmatrix} = 0, \qquad (4)$$

which is the relation connecting the ratios of the distances of a variable point, say 6, in space from the other five points,—for it is homogeneous in those distances.

We have seen that it may be deduced from the previous relation by inversion; which is also evident from the present form of expression.

The relations written out at length in § 2, being those connecting the mutual distances of four points in a plane, and connecting the ratios of the distances of a variable point in a plane from four fixed points in it, are obtained at once as above by omitting in the arrays the rows and columns which contain the third dimension z, and the fifth point 5; they are the same determinants as (3) and (4), with a row and column omitted. But to obtain the latter relation, (2), for points on a sphere, the z terms must be retained, and each array becomes an ordinary determinant, with an equal number of rows and columns; it is, however, equal to zero, as the condition that the points 1, 2, 3, 4 lie on a sphere passing through the origin 6.

Jacobi's theorem: 4. Jacobi's theorem follows, as usual, from (1). If 1′ be a point whose distances from fixed points 2′, 3′, 4′ are respectively equal to the distances of a point 1 in a plane from fixed points 2, 3, 4 in the same plane, the locus of 1′ is obtained by substituting 1′2′, 1′3′, 1′4′ for 12, 13, 14 in (1): it is a quadric surface.

generalized. Now in the same manner may be investigated the locus of a point 1′ whose distances from any four fixed points 2′, 3′, 4′, 5′ are respectively equal to the distances of some point 1 in space from four other fixed points 2, 3, 4, 5. This is obtained by substituting in (3) the distances 1′2′, 1′3′, 1′4′, 1′5′, instead of 12, 13, 14, 15 in the first row and first column of the determinant, and is

$$\begin{vmatrix} 0 & 1'2'^2 & 1'3'^2 & 1'4'^2 & 1'5'^2 & 1 \\ 2'1'^2 & 0 & 23^2 & 24^2 & 25^2 & 1 \\ 3'1'^2 & 32^2 & 0 & 34^2 & 35^2 & 1 \\ 4'1'^2 & 42^2 & 43^2 & 0 & 45^2 & 1 \\ 5'1'^2 & 52^2 & 53^2 & 54^2 & 0 & 1 \\ 1 & 1 & 1 & 1 & 1 & 0 \end{vmatrix} = 0. \qquad (5)$$

The locus is clearly a quadric surface; for, when expressed in Cartesian coordinates, $x^2 + y^2 + z^2$ is common to each member of the first row and column, and therefore its coefficient in the expanded determinant is a constant, viz. the corresponding minor with sign changed.

But the point 1, with which the correspondence is established, is also confined to a locus in space, which is the quadric

$$\begin{vmatrix} 0 & 12^2 & 13^2 & 14^2 & 15^2 & 1 \\ 21^2 & 0 & 2'3'^2 & 2'4'^2 & 2'5'^2 & 1 \\ 31^2 & 3'2'^2 & 0 & 3'4'^2 & 3'5'^2 & 1 \\ 41^2 & 4'2'^2 & 4'3'^2 & 0 & 4'5'^2 & 1 \\ 51^2 & 5'2'^2 & 5'3'^2 & 5'4'^2 & 0 & 1 \\ 1 & 1 & 1 & 1 & 1 & 0 \end{vmatrix} = 0. \qquad (6)$$

Now we may take any four points on the quadric locus of 1′, and the four related points on the quadric locus of 1, and determine the corresponding foci for them. These loci will be quadrics, and they will, by hypothesis, pass through 2′, 3′, 4′, 5′ and 2, 3, 4, 5 respectively. We have thus found two pairs of quadrics so related that the distance between any two points on the first pair (one on each) is equal to the distance between the corresponding points on the other pair.

It remains to show that these pairs of quadrics are equal, and may be superposed. We can transfer the points 2, 3, 4, 5 considered as a rigid system into such position that the six relations of the form $23' = 2'3$ are satisfied; for every rigid system possesses six degrees of freedom, which may be accommodated to these conditions. When the points are in this new position, the locus of 1 passes through 2′, 3′, 4′, 5′, and that of 1′ through 2, 3, 4, 5, and these points correspond to one another. The considerations just given then show that these two loci are so related that Ivory's theorem holds, and that they are the only class of surfaces which possess this property.

Ivory's relation is restricted to quadrics.

5. Again, the locus of a point 1′ in a plane whose distances from three fixed points 2′, 3,′ 4′ on a line in that plane are proportional to the distances of a variable point 1 on a line from fixed points 2, 3, 4 on that line, is a circular cubic with 2′, 3′, 4′ as foci. For the relation between the ratios of the distances of 1 from 2, 3, 4, is

$$23.41 + 34.21 + 42.31 = 0, \qquad (7)$$

and therefore the equation of the locus is

$$23.4'1' + 34.2'1' + 42.3'1' = 0. \qquad (8)$$

The relation (5) is unaltered by inversion; it becomes Ptolemy's

Cognate Jacobian relations: circle and bicircular quartic.

theorem for a circle. Hence the locus of a point whose distances from three points are proportional to the distances of a variable point on a circle from three fixed points on it, is a bicircular quartic with those points as foci. But this need not be pursued, as it is a case of what follows.

6. The locus of a point 1′ whose distances from four fixed points 2′, 3′, 4′, 5′ are respectively proportional to the distances of a point 1 on a plane (or sphere) from four fixed points 2, 3, 4, 5 on that plane (or sphere), may be determined similarly by aid of (2), or, what is the same, of the relation in a plane which corresponds to (4) in space. Its equation is

$$\begin{vmatrix} 0 & 1'2'^2 & 1'3'^2 & 1'4'^2 & 1'5'^2 \\ 2'1'^2 & 0 & 23^2 & 24^2 & 25^2 \\ 3'1'^2 & 32^2 & 0 & 34^2 & 35^2 \\ 4'1'^2 & 42^2 & 43^2 & 0 & 45^2 \\ 5'1'^2 & 52^2 & 53^2 & 54^2 & 0 \end{vmatrix} = 0. \qquad (9)$$

Sphere and cyclide:

When expressed in Cartesian coordinates, the only terms which contain the running coordinates are those of the first row and column, each of which is of the form $x^2 + y^2 + z^2 +$ terms of lower orders. The locus is usually a surface of the fourth order, and, as the imaginary circle at infinity is a nodal curve on the locus, it is a *cyclide*. But a reduction takes place when the points 2, 3, 4, 5 lie in a plane; for the coefficient of $(x^2 + y^2 + z^2)^2$, being the minor determinant formed by omitting the first row and column, is then zero by (3), and the locus is a cyclide of the third order.

Again, the locus of a point 1′ whose distances from five fixed points 2′, 3′, 4′, 5′, 6′ in space are respectively proportional to the distances of a variable point 1 in space from five other points 2, 3, 4, 5, 6, is determined by aid of (4), and its equation is

$$\begin{vmatrix} 0 & 1'2'^2 & 1'3'^2 & 1'4'^2 & 1'5'^2 & 1'6'^2 \\ 2'1'^2 & 0 & 23^2 & 24^2 & 25^2 & 26^2 \\ 3'1'^2 & 32^2 & 0 & 34^2 & 35^2 & 36^2 \\ 4'1'^2 & 42^2 & 43^2 & 0 & 45^2 & 46^2 \\ 5'1'^2 & 52^2 & 53^2 & 54^2 & 0 & 56^2 \\ 6'1'^2 & 62^2 & 63^2 & 64^2 & 65^2 & 0 \end{vmatrix} = 0. \qquad (10)$$

The locus is therefore a *cyclide*, and it is of the *third order*; for the coefficient of the terms of the fourth order, viz. of $(x^2 + y^2 + z^2)^2$, is the determinant which, by (3), is identically zero. And, as for quadrics in § 4, the point 1 with which the correspondence is determined must also lie on a related cylcide of the third order, whose equation is

$$\begin{vmatrix} 0 & 12^2 & 13^2 & 14^2 & 15^2 & 16^2 \\ 21^2 & 0 & 2'3'^2 & 2'4'^2 & 2'5'^2 & 2'6'^2 \\ 31^2 & 3'2'^2 & 0 & 3'4'^2 & 3'5'^2 & 3'6'^2 \\ 41^2 & 4'2'^2 & 4'3'^2 & 0 & 4'5'^2 & 4'6'^2 \\ 51^2 & 5'2'^2 & 5'3'^2 & 5'4'^2 & 0 & 5'6'^2 \\ 61^2 & 6'2'^2 & 6'3'^2 & 6'4'^2 & 6'5'^2 & 0 \end{vmatrix} = 0. \qquad (11)$$

We may say that the distances of any point P' on the first cyclide from $2', 3', 4', 5', 6'$ are to the distances of the corresponding point P on the second cyclide from 2, 3, 4, 5, 6, as $\phi(P')$ to $\phi(P)$, where $\phi(P)$ is a function of the position of P.

Now, if we take *any* five points on the first cyclide, and their related points on the other one, and find the corresponding loci for them, we arrive at a second pair of cyclides of the third order, which pass, by hypothesis, through $2', 3', 4', 5', 6'$ and 2, 3, 4, 5, 6 respectively. We thus obtain two pairs of cyclides, such that, if P, Q are points on the first pair (one on each), and P', Q' the corresponding points on the second pair,

$$\frac{PQ}{\phi(P)\,\phi(Q)} = \frac{P'Q'}{\phi(P')\,\phi(Q')}. \qquad (12)$$

generalized Ivory relation for cyclides,

This result points to a further generalization. Instead of taking distances from corresponding fixed points proportional, we may take each distance in the first diagram proportional to a definite multiple of the corresponding distance in the second diagram. This will introduce constant multipliers into the first row and column of the determinants of (10), (11). The loci will still be cyclides, but of the fourth order, and the same reasoning as above shows that the relation (12) also holds for a system of this kind.

7. Now, considering the two pairs of cyclides, we can transfer the points 2, 3, 4, 5, 6, taken as a rigid system, into such position that the distances of any one of them from $2', 3', 4', 5', 6'$ are respectively proportional to the corresponding multiples of the distances of the related points from 2, 3, 4, 5, 6; for proportionality of distances from five points involves the same number of conditions as equality of distances from four points, and these conditions can therefore be satisfied by means of six degrees of freedom. It follows then, as for quadrics in § 4, that the two pairs of cyclides coincide; and we have this generalization of Ivory's theorem, that if P, Q be points on one, and P', Q' the corresponding points on the other cyclide,

$$\frac{PQ'}{\phi(P)\,\phi(Q')} = \frac{P'Q}{\phi(P')\,\phi(Q)}.$$

(7 *a*. Starting from a given cyclide, it is evidently possible to form a whole series of related cyclides which shall possess this property *with respect to the former*; and we can draw surfaces of the series which shall be very close and consecutive to the original one, for, when the two surfaces coincide, the relation becomes an identity. Now take points 2, 3, 4, 5 ... on one surface indefinitely near one another, and their corresponding points 2′, 3′, 4′, 5′ ... on the consecutive surface: for our purpose we may suppose them to lie in two corresponding tangent planes, and we may suppose these planes parallel, for we thereby neglect only small quantities of the second order. For these points, the denominators in the relation (12) are equal to the first order of small quantities, and therefore the diagonal distances between pairs of corresponding points are equal, *i.e.* 23′ = 2′3, and so on, so that the question is reduced to Ivory's case. Further, these two systems of points are homographically situated in the two parallel tangent planes. Now it is clear that, under these circumstances, the distance-equalities can only be satisfied if the lines joining corresponding points are normal to the planes, and therefore perpendicular to the surfaces.

forming an orthogonal system,

No direct method of showing that these cyclides are related to one another as well as to the original cyclide presents itself; but it will be seen in other ways that this is true. Assuming it for the present, we can take a further step. For corresponding points lie on the curves, which, as has been seen, are the orthogonal trajectories of the system of surfaces. But these curves themselves are loci for which the relation (12) is true, for that relation may be interpreted to mean either that P and P', Q and Q' correspond, or else that P and Q, P' and Q' correspond. Now, as we have seen that all such loci are included in a system of surfaces, viz. cyclides, it follows that this congruence of orthogonal trajectories will make up a system of cyclides, which are normal to the former system. And there must clearly also be a third such system on which the curves lie; this follows by considerations of symmetry, or by considering as above the orthogonal trajectories of the second system of surfaces. The three systems of cyclides intersect everywhere at right angles, and therefore along lines of curvature on each; which is known to be a property of confocal surfaces.

which are confocal.

The truth of the relation (12) for a system of *confocal* cyclides might also be inferred from the examples in the following sections (see § 10), but I find that Darboux has already deduced this very result from the analytical theory of a confocal system, as the analogue of Ivory's theorem. (*Sur une Classe remarquable de Courbes et de Surfaces algébriques*, Note XVI.)

Darboux.

If now we show that through the points 2, 3, 4, 5 and 2′, 3′, 4′, 5′ respectively there can always be drawn two quadrics which, when properly placed, are confocal with these points in correspondence, and that through the points 2, 3, 4, 5, 6 and 2′, 3′, 4′, 5′, 6′ respectively, there can always be drawn cyclides which, when properly placed, are confocal with these points in correspondence, it will follow that all the relations here investigated are *confined to* these classes of surfaces. This can be proved by counting the disposable constants.

Each quadric can satisfy 9 conditions, while the conditions of confocality absorb 2, the conditions that four points are on one quadric absorb 4 more, and the conditions that the other four points are in corresponding positions absorb 3×4 more,—making up, in all, the 18 conditions. The problem is therefore *definite* for quadrics.

Each cyclide can satisfy 13 conditions, that being the number of constants in its equation; while the condition of confocality absorbs 4, the conditions that five points lie on one cyclide absorb 5 more, and the conditions that the other five points are in the corresponding positions absorb 3×5 more—making 24 in all. There are therefore 2 of the 26 that remain arbitrary. This is in agreement with § 6; for cubic cyclides the problem would be definite, but in quartics there are additional modes of freedom.)

8. It is interesting to notice the very compact analytical solution of the problem, to draw quadrics through these two sets of four points respectively which, when properly placed, shall be confocal with these points in correspondence, that is contained in the equations (5) and (6). **Construction of confocals through four points.**

When in the general case the points 6 and 6′ are both in the plane at infinity, so that they are equidistant from all other points not at infinity, we obtain a particular class of loci, viz. those for which, with four fixed reference-points, the one set of distances are *equal* respectively to definite multiples of the other set. The loci are quartic cyclides, and their equations might also be obtained by the method of § 4. If in the cubic cyclides represented by (10) and (11) we make this supposition, the surfaces reduce to confocal *quadrics*; but this merely shows that corresponding points at an infinite distance on confocal cubic cyclides must be regarded as at different infinities, and therefore not equidistant from finite points.

9. The results here investigated for confocal cyclides are of course true for the special case *in plano* of confocal bicircular quartics. For this case a direct verification is not difficult.

It is well known that a system of confocal conics can be represented by the equation in complex variables

$$x + \iota y = \sin^2 (\phi + \iota\psi),$$

and Greenhill has shown that a system of confocal Cartesians may be represented by the equation

Relation verified for confocal Cartesians.

$$x + \iota y = \text{sn}^2 (\phi + \iota\psi),$$

where ϕ, ψ are the parameters of the two sets of mutually orthogonal curves.

Denoting the distance of the point ϕ, ψ from the point ϕ', ψ' by $\phi\psi.\phi'\psi'$, we have to show that the above relation subsists between $\phi\psi.\phi'\psi'$ and $\phi\psi'.\phi'\psi$. Now, if x, y and x', y' are the coordinates of ϕ, ψ and ϕ', ψ', we have

$$x - x' + \iota (y - y') = \text{sn}^2 (\phi + \iota\psi) - \text{sn}^2 (\phi' + \iota\psi'),$$

and, changing the sign of ι,

$$x - x' - \iota (y - y') = \text{sn}^2 (\phi - \iota\psi) - \text{sn}^2 (\phi' - \iota\psi').$$

The product of these expressions is the square of $\phi\psi.\phi'\psi'$.

Now let

$$\begin{aligned} \Theta &\equiv \{\text{sn} (\phi + \iota\psi) - \text{sn} (\phi' + \iota\psi')\} \{\text{sn} (\phi - \iota\psi) - \text{sn} (\phi' - \iota\psi')\} \\ &= \text{sn} (\phi + \iota\psi) \text{sn} (\phi - \iota\psi) - \text{sn} (\phi' + \iota\psi') \text{sn} (\phi' - \iota\psi') \\ &\quad + \tfrac{1}{2} \{\text{sn} (\phi + \iota\psi) - \text{sn} (\phi - \iota\psi)\} \{\text{sn} (\phi' + \iota\psi') - \text{sn} (\phi' - \iota\psi')\} \\ &\quad - \tfrac{1}{2} \{\text{sn} (\phi + \iota\psi) + \text{sn} (\phi - \iota\psi)\} \{\text{sn} (\phi' + \iota\psi') + \text{sn} (\phi' - \iota\psi')\} \\ &= \frac{\text{sn}^2 \phi - \text{sn}^2 \iota\psi}{1 - k^2 \text{sn}^2 \phi \, \text{sn}^2 \iota\psi} + \frac{\text{sn}^2 \phi' - \text{sn}^2 \iota\psi'}{1 - k^2 \text{sn}^2 \phi' \, \text{sn}^2 \iota\psi'} \\ &\quad + \frac{\text{sn}\,\phi\,\text{cn}\,\iota\psi\,\text{dn}\,\iota\psi . \text{sn}\,\phi'\,\text{cn}\,\iota\psi'\,\text{dn}\,\iota\psi' + \text{cn}\,\phi\,\text{sn}\,\iota\psi\,\text{dn}\,\phi . \text{cn}\,\phi'\,\text{sn}\,\iota\psi\,\text{dn}\,\phi'}{(1 - k^2 \text{sn}^2 \phi \, \text{sn}^2 \iota\psi)(1 - k^2 \text{sn}^2 \phi' \, \text{sn}^2 \iota\psi')}. \end{aligned}$$

Hence we easily find

$$\begin{aligned} &(1 - k^2 \text{sn}^2 \phi \, \text{sn}^2 \iota\psi)(1 - k^2 \text{sn}^2 \phi' \, \text{sn}^2 \iota\psi') \Theta \\ &= (\text{sn}^2\phi + \text{sn}^2\phi')(1 + k^2 \text{sn}^2 \iota\psi \, \text{sn}^2 \iota\psi') - (\text{sn}^2 \iota\psi + \text{sn}^2 \iota\psi')(1 - k^2 \text{sn}^2\phi \, \text{sn}^2\phi') \\ &\quad + \text{sn}\,\phi\,\text{cn}\,\iota\psi\,\text{dn}\,\iota\psi . \text{sn}\,\phi'\,\text{cn}\,\iota\psi'\,\text{dn}\,\iota\psi' + \text{cn}\,\phi\,\text{sn}\,\iota\psi\,\text{dn}\,\phi . \text{cn}\,\phi'\,\text{sn}\,\iota\psi'\,\text{dn}\,\phi', \end{aligned}$$

which is not altered in value by permuting the accents on ϕ, ψ, ϕ', ψ'.

Changing the signs of ϕ', ψ', we obtain a similar expression for Θ', where also

$$\Theta\Theta' = (\phi\psi.\phi'\psi')^2.$$

It follows that

$$\begin{aligned} &(1 - k^2 \text{sn}^2 \phi \, \text{sn}^2 \iota\psi)(1 - k^2 \text{sn}^2 \phi' \, \text{sn}^2 \iota\psi')(\phi\psi.\phi'\psi') \\ &\qquad = (1 - k^2 \text{sn}^2 \phi \, \text{sn}^2 \iota\psi')(1 - k^2 \text{sn}^2 \phi' \, \text{sn}^2 \iota\psi)(\phi\psi'.\phi'\psi), \end{aligned}$$

which is the verification sought.

The relation is unaltered by inversion, and by inverting confocal Cartesians we obtain general confocal bicircular quartics.

(9 *a*. If ϕ, ψ are the parameters of two sets of curves, we can express the coordinates of any point in their plane in the form

$$x = \theta_1 (\phi, \psi), \quad y = \theta_2 (\phi, \psi),$$

where θ_1, θ_2 are functional symbols.

Now we have shown that Ivory's theorem *in plano* holds only for confocal conics, therefore the functional equation

Solution of a functional equation.

$$\{\theta_1 (\phi, \psi) - \theta_1 (\phi', \psi')\}^2 + \{\theta_2 (\phi, \psi) - \theta_2 (\phi' \psi')\}^2$$
$$= \text{an even function of } \phi - \phi', \psi - \psi',$$

has a solution, which is *unique*, and represents a system of confocal conics, and can therefore be put into the form

$$\theta_1 (\phi, \psi) + \iota\theta_2 (\phi, \psi) + A = C \sin^2 (\phi + \iota\psi),$$

where A, C are any constants, real or imaginary.

Similar analytical statements apply in the other cases.)

10. The inverses with respect to any point of a system of confocal quadrics form a special system of confocal cyclides for which the relation (12) is clearly true. Also, if the pair of cyclides of that theorem are symmetrical with respect to a plane (*i.e.* if they have focal curves in that plane), their sections by the plane are bicircular quartics for which the theorem holds, and which must therefore, as we have seen, be confocal. These considerations (see § 7 *a*) point to the conclusion that all the pairs of cyclides which satisfy the relation are confocal.

Space foci of bicircular quartic.

Assuming their confocality, and applying the relation to two focal curves, which are limiting forms of a confocal system, we arrive, by the same method as in § 1 for conics, at the result that the locus of the foci in space of a plane or spherical bicircular quartic curve is one of the other focal curves of the system of confocal cyclides of which the original is a focal curve. This theorem has been given by Darboux (*Sur une Classe remarquable de Courbes et de Surfaces algébriques*, Paris, 1873, p. 44).

9

ON THE FLOW OF ELECTRICITY IN A SYSTEM OF LINEAR CONDUCTORS.

[*Proc. London Math. Soc.* Vol. XVI (1885) pp. 262–272.]

Historical.

1. The analytical determination of the currents that are set up by given steady electromotive forces in a system of linear conducting bodies has been treated by Kirchhoff[1], who takes a separate variable to represent the current flowing in each branch of the system.

Maxwell has given a discussion[2] in which the number of variables is reduced by the equation of continuity, which requires that the total current flowing into any junction is equal to the current flowing from that junction; he takes as variables the potentials of the junctions. This method regards the system as a compound conductor, into which currents are introduced from without; and it gives symmetrical expressions for its resistance measured between any two corners.

Method of component circuits.

We may also express the phenomena in terms of a series of currents flowing in closed circuits whose number is sufficient to completely determine the system. This specification by closed currents is more fundamental in character. When the strengths of the currents are variable owing to electromagnetic action, this method is the only one very conveniently applicable. For we have, to express the electrokinetic energy, the ordinary quadratic function of the currents in the separate circuits, the coefficients of which are the coefficients of self and mutual induction of those circuits; and those coefficients are just the quantities that can be most easily determined by known methods of calculation or experiment. Nor is the facility of application disturbed when condensers are included in the system.

In the second edition of Maxwell's *Electricity and Magnetism*, there are two insertions from his lecture notes (§ 282 *b*, § 755, on the theory of Wheatstone's Bridge with steady currents, and on the same applied to the determination of coefficients of electrokinetic induction) which point to this mode of treatment; so that it is likely that the reconstruction of these two theories which (Preface, p. XVI) the author contemplated would have had some reference to it.

It is proposed to develop concisely the principles of this method of investigation.

2. Consider first the case in which the conductors form a simple

[1] *Pogg. Annal.* Bd. 72, 1847; *Gesammelte Abhandlungen*, pp. 22–23.

[2] *Elec. and Mag.* Vol. I. §§ 280–82. Also Chrystal, *Encyc. Brit.*, Art. "Electricity."

network. We may take for circuits the separate meshes of the network, and, if we suppose a barrier drawn across each mesh, we obtain a quasi-polyhedron whose faces are formed by the barriers. Let S, E, F denote the number of its summits, edges, and faces, respectively. It is obvious that only $F - 1$ of the circuits are independent, for we can represent the current round the remaining face by a system of equal currents round each of the others, just as Ampère replaced a current in a finite circuit by equal currents in a system of infinitesimal circuits forming a network bounded by the original one. Surface network:

This number $F - 1$ of independent variables is necessarily in agreement with the results of Kirchhoff's method. For there are E branch currents, subject apparently to S conditions of continuity, but really subject to only $S - 1$ independent conditions; because the sum of all the currents flowing into all the junctions is zero—without any condition, as each current flowing into one junction must flow from another. There are therefore $E - S + 1$ independent variables, which is equal to the preceding estimate $F - 1$ by virtue of Euler's relation its independent currents.

$$S + F = E + 2.$$

Euler's polyhedron relation:

This discussion throws light on the general case in which the conductors do not form a simple network, and a conductor may therefore necessarily belong to more than two independent circuits. To select a proper system of circuits, begin with any one, and suppose a barrier surface drawn across it; select a second, and suppose it also closed by a barrier; and proceed in this way, subject to the condition that the barriers do not disturb the singly continuous character of the space by forming a closed boundary round any portion of it, *i.e.* by making it periphractic. The process will teminate of itself when there is no conductor left which is not abutted on by a barrier.

But, having thus secured that the space bounded by the barriers is not periphractic, it must also be made certain that it does not possess any character of multiple continuity (cyclosis); *i.e.* any closed circuit drawn in the space must be capable of being contracted to a point without cutting through any of the barriers. The space bounded by the surface of an anchor ring is thus doubly continuous, since, to secure simple continuity, it is necessary to draw a barrier surface across the opening of the ring if the outside space is considered,—or across the section of the ring if the inside space is in question. The nature and necessity of this proviso will be made clear by considering again the simple network of this section. We may imagine the system of barriers as forming a continuous sheet; and the removal of one of them will make a hole in this sheet, through which a degree of cyclosis is established. It is clear from these considerations that every degree of cyclosis implies the absence of a necessary independent variable, generalized for cyclic solid, which can be supplied by adding a new barrier closing the corresponding circuit.

With a notation corresponding to that given above for a network, we have in this case for the number of independent circuits the value F, which must on Kirchhoff's principles be equivalent to $E - S + 1$; and we thus come upon the theorem that for a polygonal system of barriers which does not impair the singly continuous character of the space, by either periphraxy or cyclosis,

$$S + F = E + 1.$$

as by Listing. This is a particular case of Listing's generalization of Euler's theorem, now equally obvious, which asserts that, if the system of barriers divide space into R unconnected regions, each of them singly continuous, then

$$S + F = E + R;$$

for removing each superfluous barrier diminishes the number of regions by one.

The general theorem, as given by Listing[1], takes account of the corrections to be applied to this formula when the periphractic and cyclomatic numbers of the system are given constants, different from zero; but with these we are not at present concerned, though the ideas here employed would probably yield a simple method for their discussion.

The particular case of the theorem at which we have arrived verifies the correctness of the method that has been given for choosing the independent circuits of the current system; and this plan of construction by barriers has the advantage of easily showing the precise amount of liberty there is in the selection of the circuits.

3. Having thus selected the circuits, let them be denoted by the natural numbers 1, 2, 3 ... n; the currents circulating in them by $C_1, C_2 \ldots C_n$, their resistances by $R_1, R_2 \ldots R_n$, the electromotive forces *placed* in them by $E_1, E_2 \ldots E_n$ measured positive in the directions of the currents; and let also the current in the conductor pq which is common to the circuits p and q be denoted by C_{pq} and the resistance of that conductor by R_{pq} and the electromotive force *placed* in it by E_{pq}: so that we have

The independent variables.

$$R_p = \quad R_{p1} + R_{p2} + \ldots + R_{pn}, \qquad (1)$$

$$E_p = \pm E_{p1} \pm E_{p2} \ldots\ldots \pm E_{pn}. \qquad (2)$$

The signs in this last relation are determined in each special case by the directions of the component electromotive forces as compared with E_p. In the case of a simple network we can secure that the

[1] J. B. Listing, "Der Census raümlicher Complexe," *Göttingen Abhandlungen*, Band x. 1861–2, quoted by Maxwell.

positive directions of circulation of all the currents as seen from one side of the network shall be the same, and we shall then have also relations of the form

$$C_{pq} = C_q - C_p. \qquad (3)$$

But, when the system does not form a simple network, the same conductor may be common to three or more circuits, say *pqr* ...; and then the current in that conductor is denoted by C_{pq} or C_{pr} or C_{qr}, where, the signs being properly determined,

$$C_{pq} = \pm C_p \pm C_q \pm C_r. \qquad (4)$$

It will not be necessary to have an explicit notation for the case in which more than one conductor is common to two circuits.

With this notation, the expression for the heat generated per second by steady currents in the conductors can be obtained as follows:

The fundamental function: of dissipation:

$$H = \Sigma R_{pq} C_{pq}^2,$$

the summation extending over all the conductors of the system,

$$= \Sigma R_{pq} (C_p - C_q)^2$$

$$= R_1 C_1^2 + R_2 C_2^2 + \ldots \pm 2R_{pq} C_p C_q \pm \ldots, \qquad (5)$$

the latter signs being all negative for a simple network; while for any other case the rule is that, when C_p and C_q are taken to flow in the same direction along the conductor *pq*, the sign of the corresponding term is positive. Terms involving products of all pairs of contiguous closed currents are included in the expression.

4. The theorem connecting the electromotive forces with H may now be investigated by the method common to all analyses which turn upon quadratic functions of this kind. Let accented letters denote any other system of electromotive forces and corresponding currents imposed upon the same system of conductors: then we have

theory of its variation.

$$\Sigma C_p E_p' = \Sigma C_{pq} e_{pq}',$$

where e_{pq}' represents the total *gradual* fall of potential along the conductor *pq* in the direction of the current C_{pq}, aggregating for a circuit to the sum of the sudden rises,

$$= \Sigma \frac{e_{pq} e_{pq}'}{R_{pq}}, \text{ by Ohm's law,}$$

$$= \Sigma C_p' E_p, \text{ by symmetry.}$$

Therefore, if $C_p' = C_p + \delta C_p$, $E_p' = E_p + \delta E_p$, we have

$$\Sigma C_p \delta E_p = \Sigma E_p \delta C_p. \qquad (6)$$

Now $$H = \Sigma C_p E_p,$$

therefore $$\delta H = \Sigma C_p \delta E_p + \Sigma E_p \delta C_p.$$

Hence finally, by (6), $$\delta H = 2\Sigma E_p \delta C_p,$$

or, proceeding from finite increments to infinitesimals,

$$\frac{dH_c}{dC_p} = 2E_p, \quad \frac{dH_c}{dC_q} = 2E_q, \ \ldots, \tag{7}$$

where H_c is H expressed as in (5) by a function of C_p, C_q, ...

Again, when, by means of (7), C_p, C_q ... are eliminated from the expression for H_c in (5), we have for H a quadratic function H_e of the electromotive forces, and then

Equations of flow.

$$\frac{dH_e}{dE_p} = 2C_p, \quad \frac{dH_e}{dE_q} = 2C_q, \ \ldots. \tag{8}$$

5. The equations (7) are the linear system to which we are led for the determination of the currents. Writing them in full, we have

$$\left.\begin{array}{l} R_1C_1 - R_{12}C_2 - R_{13}C_3 \ldots - R_{1n}C_n = E_1 \\ -R_{21}C_1 + R_2C_2 - R_{23}C_3 \ldots - R_{2n}C_n = E_2 \\ \ldots\ldots\ldots\ldots\ldots\ldots\ldots\ldots\ldots\ldots \\ \ldots\ldots\ldots\ldots\ldots\ldots\ldots\ldots\ldots\ldots \end{array}\right\}, \tag{9}$$

a symmetrical system, in which the proper signs are here supposed to be given to the R's with double suffixes, viz. all positive in the case of a simple network, and in other cases according to the rule already given. The solution can be expressed by symmetrical determinants in the ordinary manner, and by means of (4) we can obtain at once an expression for the total current in any separate conductor.

We may thus also obtain the condition that an electromotive force in one conductor may not give rise to a current in a certain other. But for such purposes it is more convenient to employ the reciprocal expression for H in terms of the E's. To obtain this, proceed in the usual way, by joining on to (9) the equation

$$E_1C_1 + E_2C_2 + \ldots + E_nC_n = H, \tag{10}$$

and by linear elimination we find

The reciprocal dissipation function:

$$H = -\frac{1}{\Delta}\left|\begin{array}{ccccc} R_1 & -R_{12} & -R_{13} \ldots & -R_{1n} & E_1 \\ -R_{21} & R_2 & -R_{23} \ldots & -R_{2n} & E_2 \\ -R_{31} & \ldots\ldots & \ldots\ldots & \ldots\ldots & \\ \ldots\ldots & \ldots\ldots & \ldots\ldots & \ldots\ldots & \\ \ldots\ldots & \ldots\ldots & \ldots\ldots & \ldots\ldots & \\ E_1 & E_2 & E_3 \ldots & E_n & 0 \end{array}\right|, \tag{11}$$

in which Δ is the discriminant of the expression for H_c, *i.e.* is the minor of the last (zero) constituent of this determinant.

We may write this quadratic expression in the form

$$H_e = K_1E_1^2 + K_2E_2^2 + \ldots + K_nE_n^2 + 2K_{12}E_1E_2 + 2K_{23}E_2E_3 + \ldots, \tag{12}$$

in which K_1, K_{12} ... are immediately written down as minors of the determinant Δ. We have then, by (8),

$$\left.\begin{array}{l} K_1E_1 + K_{12}E_2 + K_{13}E_3 + \dots + K_{1n}E_n = C_1 \\ K_{21}E_1 + K_2E_2 + K_{23}E_3 + \dots + K_{2n}E_n = C_2 \\ \dots\dots\dots\dots\dots\dots\dots\dots\dots\dots \\ \dots\dots\dots\dots\dots\dots\dots\dots\dots\dots \end{array}\right\}. \qquad (13)$$

its equations.

The constants K_1, K_{12} ... are clearly coefficients of conductivity.

Now suppose the only electromotive force considered is E_{pq} in the conductor *pq*. If this conductor abuts on two circuits only, we have $E_p = \pm E_q = E_{pq}$ and the other *E*'s zero; if on three or more, we have $E_p = \pm E_q = \pm E_r = \dots = E_{pq}$, in which the signs have been predetermined; while, if it abuts on only one circuit, we have only one fundamental *E*. To determine the current in the conductor *lm*, we have

$$C_{lm} = C_l \pm C_m \pm C_n \pm \dots. \qquad (14)$$

We find thus, by (8) or (13),

$$C_{lm} = E_{pq}\left\{\begin{array}{l} K_{lp} \pm K_{lq} \pm K_{lr} \dots \\ \pm (K_{mp} \pm K_{mq} \pm K_{mr} \dots) \\ \pm (K_{np} \pm K_{nq} \pm K_{nr} \dots) \end{array}\right\}, \qquad (15)$$

which reduces, for the case of a simple network, to

$$C_{lm} = E_{pq}(K_{lp} - K_{lq} - K_{mp} + K_{mq}). \qquad (16)$$

We conclude that an electromotive force in *pq* will produce no current in *lm* if the coefficient of E_{pq} in this result is zero; and, as the coefficient remains unaltered when *lm* and *pq* change places, we see that equal currents are produced in either conductor by unit electromotive force in the other.

Condition for conjugate conductors.

These results are of the same general form as the ones obtained in Maxwell's investigation; and the similarity is intelligible when it is remembered that the two investigations are in a sense reciprocal to each other, the variables in the one case being related to the corners and in the other case to the faces of the diagram.

6. A simple case is that of the six conductors of Wheatstone's Bridge[1], which form a network, as in the diagram. We have, by § 3,

Functions for bridge system:

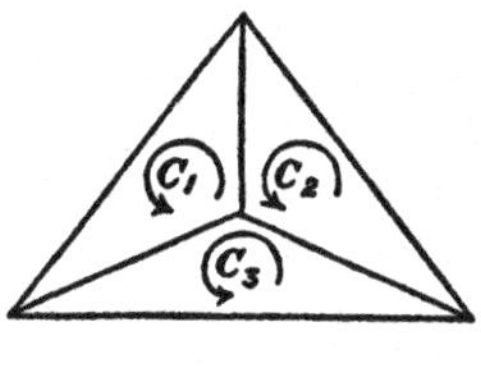

$$H_c = R_1C_1^2 + R_2C_2^2 + R_3C_3^2 - 2R_{12}C_1C_2 - 2R_{23}C_2C_3 - 2R_{31}C_3C_1. \qquad (17)$$

[1] Maxwell's *Elec. and Mag.* Vol. I. 2nd ed. §§ 282 *b*, 347; *Elementary Electricity*, p. 206.

The reciprocal function is

$$H_e = -\frac{1}{\Delta}\begin{vmatrix} R_1 & -R_{12} & -R_{13} & E_1 \\ -R_{21} & R_2 & -R_{23} & E_2 \\ -R_{31} & -R_{32} & R_3 & E_3 \\ E_1 & E_2 & E_3 & 0 \end{vmatrix} \qquad (18)$$

$$= \frac{1}{\Delta}\{(R_2R_3 - R_{23}^2)E_1^2 + \ldots + \ldots + 2(R_{12}R_{13} + R_1R_{23})E_2E_3 \ldots + \ldots\}, \qquad (19)$$

Δ being the discriminant of H_e.

interpretations. The coefficients in this expression for H_e are the conductivities of the system. The coefficient of E_1^2 represents the current in E_1 that would be produced by unit electromotive force in E_1 *alone*; the coefficient of E_2E_3 represents double the current produced in either circuit E_2 or E_3 by unit electromotive force in the other one *alone**.

Analogy to impulsive motions of solid bodies. 7. The form of this analytical theory is exactly analogous to that of the theory of initial motions in a system of bodies. We have therefore an analogue of Sir W. Thomson's theorem of least energy with certain given imposed velocities; and also one of Bertrand's theorem of greatest energy with certain given imposed impulses, subject to the condition that the only allowable variations of the motion are those caused by pure constraint, or by such other impulses as do no work in the process: viz. we have the theorems:

Minimal theorems. 1°. When given currents are introduced into a system, they distribute themselves in such a manner that the heat developed in the conductors is the least possible.

2°. When given electromotive forces are introduced into a system of conductors, the heat actually generated by the currents is greater than it would be if any of the conductors were removed[1]; for the removal of a conductor is equivalent to the introduction of such an electromotive force in it as reduces its current to zero.

8. The general theory of linear conductors, and the conjugate relations involved, clearly apply also to cases in which the conducting system is partly linear and partly continuous in other dimensions.

* An application for which this method of component circuits is essentially the proper one is (cf. § 1) the theory of the commutated divided circuits in dynamos and motors. The necessary data for the distribution of current, at each moment of time, are the electromotive forces induced *round the completed circuits*, as determined by Faraday's law. The value of the driving electric force at each point, as the theory of electrons would give it, is not required, so long as the alternations are slow enough to avoid sensible electric waves on the wires: but potential gradients can be reduced at the end from the values of the currents as here determined, on this hypothesis that radiational adjustments along the wires are made with velocity practically infinite.

[1] Lord Rayleigh, *Phil. Mag.* Vol. XLVIII. 1875.

If the system include any continuous conductor (with or without helical property) with a number of electrodes on its surface, we can replace that part of it by a system of linear conductors of the proper resistances connecting all the electrodes in pairs,—their resistances being determinable by calculation or experiment from the shape of the conductor. For, from the linearity of the law of conduction, it is permissible to superpose different current systems*.

Flow through electrodes: the equivalent linear system:

Again, we can imagine a continuous conductor such that in its ultimate structure it is composed of a thicket of interlacing conducting filaments whose cross-sections are negligible in comparison with their lengths; though this will not, of course, be a probable representation of the constitution of an ordinary conducting solid. We can show that, for a small right-angled element of such a body, the equations of conduction will be self-conjugate; *i.e.* there will be no helical coefficient. For suppose (for purposes of analysis) such an element cut out of the solid, and its faces backed up by six perfectly conducting plates, and the opposite pairs of faces connected by conductors of no resistance in which electromotive forces E_1, E_2, E_3 are placed. We shall then have a linear system, in the sense of the above theory; and H_e will be a quadratic function of these three electromotive forces; and the corresponding currents across the faces of the element will be derived from H_e by differentiation, according to (8), with respect to E_1, E_2, E_3, which proves the proposition.

transition to conducting solid.

9. It remains to indicate briefly how the method here used may be applied to the general case of currents variable owing to electrodynamic action or to the gradual charge of condensers whose terminals are connected with the system at given points.

Kinetic inductions included:

We may treat such a condenser as a branch of the system whose resistance is infinite; or, if we make allowance for the small degree of conductivity which may exist between its faces, it will be a conductor of resistance very large.

As the currents are now variable we shall want to change their notation. Let x_1, x_2, ... x_n represent the *integral* currents that have flowed round the specifying circuits from the beginning of the motion; then, with the usual fluxional notation, $\dot{x}_1$, $\dot{x}_2$, ... $\dot{x}_n$ will represent the currents in those circuits at the instant considered.

* An illustration of the general principle of superposition, useful in practical calculation, is given by F. Wenner (*Phys. Soc. Proc.* Jan. 1927): to determine *e.g.* the flux in a wire belonging to a complex inductive system, even with linkages across space. Suppose this wire is cut: the electric flux will assume a steady state developing a potential difference E between the cut ends. Now restore the continuities by inserting a generator $-E$ between the ends: this generator, supposed working by itself alone, will superpose a new flow in the system against its resistances and inertias, which gives the flux required.

Following the notation of § 3, let rs denote a condenser branch; if k_{rs} denote the capacity of the condenser, the energy stored *from the system* in its charge and in the charges of the others, if any, will be

$$V = \Sigma \tfrac{1}{2} k_{rs}^{-1} x_{rs}^2 = \Sigma \tfrac{1}{2} k_{rs}^{-1} (x_r \pm x_s \pm \ldots)^2 \qquad (20)$$

by (4).

The electrokinetic energy due to the motion of the conductors will be

$$T = \tfrac{1}{2} M_1 \dot{x}^2 + \tfrac{1}{2} M_{22} \dot{x}^2 + \ldots + M_{12} \dot{x}_1 \dot{x}_2 + \ldots, \qquad (21)$$

where M_1, M_2 ... are the coefficients of self-induction of the respective circuits, and M_{12} ... are the coefficients of mutual induction of the different pairs of circuits.

If there is any part of the electrokinetic energy due to the influence of external fixed systems, it will be represented by a function of $\dot{x}_1$, $\dot{x}_2$ ... of the first degree, which must be added on to T.

(In the general electrokinetic theory mentioned below, the coefficients of this linear function may involve constant electrokinetic momenta whose corresponding variables have been eliminated from the expression for the energy; and we shall thus have an example of the general dynamical equations with ignored coordinates[1].)

The expression for the amount of energy that runs down into heat per second, owing to the resistances of the wires, is, as in (5),

$$H = R_1 \dot{x}_1^2 + R_2 \dot{x}_2^2 + \ldots \pm 2R_{12} \dot{x}_1 \dot{x}_2 + \ldots. \qquad (22)$$

And we have also the ordinary quadratic function for the kinetic energy $\mathfrak{T}$ of the masses of the system in terms of the generalized velocities.

General dynamical synthesis.

From these four expressions equations of the currents $\dot{x}_1$, $\dot{x}_2$ ... may be formulated by aid of the principle of energy (as used originally by Helmholtz and Sir W. Thomson) combined with Ohm's law; and it is well known, as Maxwell has shown, that these equations are the same as those of a purely dynamical system, of which x_1, x_2 ... are additional coordinates, $\dot{x}_1$, $\dot{x}_2$ ... the corresponding velocities, T the corresponding part of the kinetic energy, V the corresponding part of the potential energy, and H the corresponding *dissipation function* of Lord Rayleigh[2].

The equations of the currents are, then, of types

$$\frac{d}{dt}\frac{dT}{d\dot{x}_1} + \frac{dV}{dx_1} + \tfrac{1}{2}\frac{dH}{d\dot{x}_1} = 0, \quad \frac{d}{dt}\frac{d\mathfrak{T}}{d\dot{\theta}_1} - \frac{d\,(T+\mathfrak{T})}{d\theta_1} + \frac{dU}{d\theta_1} = 0, \qquad (23)$$

where $\mathfrak{T}$ is the kinetic and U is the potential energy of the masses of the system, and θ_1, θ_2 ... are their generalized coordinates.

[1] Thomson and Tait's *Natural Philosophy*, Vol. I. Part I. p. 320.

[2] *Proc. Lond. Math. Soc.* May 1873; *Theory of Sound*, Vol. I. § 81. [Also the chapter on "Electrical Oscillations" in edition 2 for a very complete analysis, adapted to practical requirements, of the topics sketched in general terms in this section.]

10

SOME APPLICATIONS OF GENERALIZED SPACE-COORDINATES TO DIFFERENTIAL ANALYSIS: POTENTIALS AND ISOTROPIC ELASTICITY.

[*Camb. Phil. Trans.* Vol. XIV. pp. 121–137. Read April 27, 1885.]

1. There are two well-known methods of dealing with physical questions of continuous differential analysis.

In the first and more ordinary one the differential equations satisfied by the quantities involved are investigated directly from the relations and properties of the system; and their transformation from simple rectangular to curvilinear coordinates may be effected either by direct transformation of the quantities and their differential coefficients, or by an independent investigation with the new variables. This latter process is easy where only differential coefficients of the first order are concerned, for these obey the laws of all vector quantities such as forces; but when, as is usual, differential coefficients of the second order occur it involves the use of vector differentiation or some similar process, which is of a more complicated character.

Complexity of vector differentiation:

In the second method, which was invented by Lagrange, and developed and applied by Gauss, Green, and others, the equations are expressed as the conditions that a certain quadratic function of the differential coefficients of the first order, integrated over the system, shall retain a stationary value when small variations are imposed on the variables. In statical questions this function is the potential energy per unit volume at the place, and its integral is the total potential energy of the system. When its value is known, the equations of the system are obtained at once by application of the Method of Variations. Now this quadratic function, being a purely scalar quantity, will usually be expressible in a form which does not closely connect it with any special system of coordinate directions; and in any case, it may be transferred without difficulty from one system of coordinates to another by means of the vector laws obeyed by the fluxions.

evaded by the variational method.

The second method of procedure is therefore an easy and straightforward one when it is wished to express the equations in terms of special systems of coordinates. The first method involves a detailed examination of the internal properties of the system, and is therefore well suited to the clear exposition of the relations of a system whose internal structure is known, and to their expression in terms of the more simple coordinates. And the two are of course complementary to each other.

The object here proposed is to illustrate the use of the second method by its general application to some problems of common occurrence.

Coordinate frame determined by its distance-element:

2. The character of a system of space-coordinates is completely determined by the nature of the expression for the square of the distance between two neighbouring points in terms of the differentials of the coordinates of those points: this will be a quadratic function of the differentials, with coefficients which may or may not be functions of the coordinates themselves.

or by its resultant scalar gradient:

There is another function related to this, and in a manner reciprocal to it, viz. the expression for the square of the resultant force or flux at a point corresponding to a given potential function, in terms of the rates of variation of that function with respect to the coordinates.

sufficient for isotropic physics.

These two expressions can be made the basis of the whole analysis, when the relations considered are of an isotropic character.

Forms of distance-element.

3. Suppose the position of a point to be expressed in terms of ξ, η, ζ, which are given functions of the rectangular Cartesian coordinates x, y, z by which it may be originally specified. These coordinates will retain constant values over the surfaces

$$\xi = \text{constant}, \quad \eta = \text{constant}, \quad \zeta = \text{constant}, \tag{1}$$

respectively.

The square of the distance ds between two consecutive points ξ, η, ζ and $\xi + d\xi$, $\eta + d\eta$, $\zeta + d\zeta$ in the space will be given by

$$ds^2 = Ad\xi^2 + Bd\eta^2 + Cd\zeta^2 + 2Dd\eta d\zeta + 2Ed\zeta d\xi + 2Fd\xi d\eta, \tag{2}$$

where A, B, C, D, E, F are constants or functions of ξ, η, ζ, which are determined by the nature of the coordinate system.

If the system is determined by a triple series of parallel planes, the coefficients will be constants; and if α, β, γ be the angles between the directions in which these planes intersect respectively, *i.e.* between the axes of coordinates, we have

$$ds^2 = d\xi^2 + d\eta^2 + d\zeta^2 + 2d\eta d\zeta \cos\alpha + 2d\zeta d\xi \cos\beta + 2d\xi d\eta \cos\gamma.$$

If the system is determined by a triple series of orthogonal surfaces, D, E, F are zero, and, in Lamé's notation,

$$A = \frac{1}{h_1^2}, \text{ where } h_1^2 = \left(\frac{d\xi}{dx}\right)^2 + \left(\frac{d\xi}{dy}\right)^2 + \left(\frac{d\xi}{dz}\right)^2,$$

$$B = \frac{1}{h_2^2}, \text{ where } h_2^2 = \left(\frac{d\eta}{dx}\right)^2 + \left(\frac{d\eta}{dy}\right)^2 + \left(\frac{d\eta}{dz}\right)^2,$$

$$C = \frac{1}{h_3^2}, \text{ where } h_3^2 = \left(\frac{d\zeta}{dx}\right)^2 + \left(\frac{d\zeta}{dy}\right)^2 + \left(\frac{d\zeta}{dz}\right)^2, \qquad (3)$$

so that
$$ds^2 = \frac{d\xi^2}{h_1^2} + \frac{d\eta^2}{h_2^2} + \frac{d\zeta^2}{h_3^2}. \qquad (4)$$

As examples, there are the common cases of rectangular, polar, and ellipsoidal coordinates.

If the system is determined by the triple series represented by (1), whose curves of intersection drawn through the point ξ, η, ζ cut at angles α, β, γ, respectively, and if h_1, h_2, h_3 are defined as above, then

$$ds^2 = \frac{d\xi^2}{h_1^2} + \frac{d\eta^2}{h_2^2} + \frac{d\zeta^2}{h_3^2} + 2\frac{d\eta d\zeta}{h_2 h_3}\cos\alpha + 2\frac{d\zeta d\xi}{h_3 h_1}\cos\beta + 2\frac{d\xi d\eta}{h_1 h_2}\cos\gamma. \qquad (5)$$

4. The properties of the coordinate system ξ, η, ζ will in every case be completely specified if the coefficients of the expression in (2) are known.

Comparing it with (5), we find that the element of volume is

Elements of volume:

$$\begin{vmatrix} A & F & E \\ F & B & D \\ E & D & C \end{vmatrix}^{\frac{1}{2}} d\xi d\eta d\zeta, \qquad (6)$$

and the elements of area, on each of the coordinate surfaces ξ = constant, η = constant, ζ = constant, are respectively

and area.

$$\begin{vmatrix} B & D \\ D & C \end{vmatrix}^{\frac{1}{2}} d\eta d\zeta, \quad \begin{vmatrix} C & E \\ E & A \end{vmatrix}^{\frac{1}{2}} d\zeta d\xi, \quad \begin{vmatrix} A & F \\ F & B \end{vmatrix}^{\frac{1}{2}} d\xi d\eta. \qquad (7)$$

5. Suppose the potential function V to be expressed in terms of the coordinates ξ, η, ζ, and it is required to find the expression for the resultant force or flux at the point ξ, η, ζ in terms of V.

Gradient and its resultant.

Let the direction of this resultant make angles a, b, c respectively with the lines of intersection of the coordinate surfaces at the point ξ, η, ζ; these angles will be related to α, β, γ, as in the annexed scheme supposed drawn on a spherical surface.

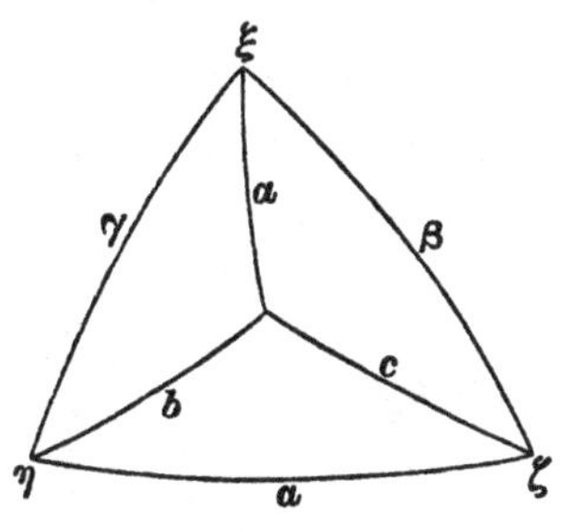

If R represent the magnitude of the required resultant, we shall have, by the fundamental property of the potential,

$$h_1\frac{dV}{d\xi} = R\cos a, \quad h_2\frac{dV}{d\eta} = R\cos b, \quad h_3\frac{dV}{d\zeta} = R\cos c. \qquad (8)$$

But the relation between the mutual distances of four points on a sphere gives

$$\begin{vmatrix} 1 & \cos\gamma & \cos\beta & \cos a \\ \cos\gamma & 1 & \cos\alpha & \cos b \\ \cos\beta & \cos\alpha & 1 & \cos c \\ \cos a & \cos b & \cos c & 1 \end{vmatrix} = 0. \tag{9}$$

Therefore

$$\begin{vmatrix} 1 & \cos\gamma & \cos\beta & h_1\dfrac{dV}{d\xi} \\ \cos\gamma & 1 & \cos\alpha & h_2\dfrac{dV}{d\eta} \\ \cos\beta & \cos\alpha & 1 & h_3\dfrac{dV}{d\zeta} \\ h_1\dfrac{dV}{d\xi} & h_2\dfrac{dV}{d\eta} & h_3\dfrac{dV}{d\zeta} & R^2 \end{vmatrix} = 0, \tag{10}$$

or, what is the same,

$$\begin{aligned}(1 - \cos^2\alpha - \cos^2\beta - \cos^2\gamma + 2\cos\alpha\cos\beta\cos\gamma)\,R^2 \\ = \sin^2\alpha.h_1^2\left(\frac{dV}{d\xi}\right)^2 + \sin^2\beta.h_2^2\left(\frac{dV}{d\eta}\right)^2 + \sin^2\gamma.h_3^2\left(\frac{dV}{d\zeta}\right)^2 \\ + 2(\cos\beta\cos\gamma - \cos\alpha)\,h_2h_3\frac{dV}{d\eta}\frac{dV}{d\zeta} \\ + 2(\cos\gamma\cos\alpha - \cos\beta)\,h_3h_1\frac{dV}{d\zeta}\frac{dV}{d\xi} \\ + 2(\cos\alpha\cos\beta - \cos\gamma)\,h_1h_2\frac{dV}{d\xi}\frac{dV}{d\eta},\end{aligned} \tag{11}$$

the expression for R^2 which was required.

In particular, if the coordinates form an orthogonal system,

$$R^2 = h_1^2\left(\frac{dV}{d\xi}\right)^2 + h_2^2\left(\frac{dV}{d\eta}\right)^2 + h_3^2\left(\frac{dV}{d\zeta}\right)^2, \tag{12}$$

as obviously should be true.

Geometrical: 6. To complete the expressions (10) and (11), we should express $\cos\alpha$, $\cos\beta$, $\cos\gamma$ in terms of ξ, η, ζ. This is easily done, and it will be found—if we represent

the array $\left\|\begin{matrix} \dfrac{d\eta}{dx} & \dfrac{d\eta}{dy} & \dfrac{d\eta}{dz} \\ \dfrac{d\zeta}{dx} & \dfrac{d\zeta}{dy} & \dfrac{d\zeta}{dz} \end{matrix}\right\|$ by $\dfrac{d(\eta, \zeta)}{d(x, y, z)}$,

and the array $\left\|\begin{matrix} \dfrac{d\xi}{dx} & \dfrac{d\xi}{dy} & \dfrac{d\xi}{dz} \end{matrix}\right\|$ by $\dfrac{d(\xi)}{d(x, y, z)}$, with similar notations for other cases (as already employed by Prof. Cayley)—that

$$\cos^2\gamma = \frac{\left\{\frac{d(\eta,\zeta)}{d(x,y,z)}\cdot\frac{d(\zeta,\xi)}{d(x,y,z)}\right\}^2}{\left[h_2{}^2h_3{}^2 - \left\{\frac{d(\eta)}{d(x,y,z)}\cdot\frac{d(\zeta)}{d(x,y,z)}\right\}^2\right]\left[h_3{}^2h_1{}^2 - \left\{\frac{d(\zeta)}{d(x,y,z)}\cdot\frac{d(\xi)}{d(x,y,z)}\right\}^2\right]}, \tag{13}$$

with corresponding expressions for $\cos^2\alpha$ and $\cos^2\beta$.

If, however, we adopt (2) as the fundamental relation, we have the simple expressions of form

$$\cos\gamma = h_1h_2F = \left(\frac{F^2}{AB}\right)^{\frac{1}{2}}, \tag{14}$$

as in the next section.

7. The reciprocal relation between these two fundamental scalar formulae (2), (10) will be clearly brought out by the following purely analytical method of deducing the second from the first:

If ds denote an element of length in any direction through the point ξ, η, ζ, we have *reciprocity of distance and gradient.*

$$\frac{dV}{ds} = \frac{d\xi}{ds}\frac{dV}{d\xi} + \frac{d\eta}{ds}\frac{dV}{d\eta} + \frac{d\zeta}{ds}\frac{dV}{d\zeta}. \tag{15}$$

The value of R is the maximum value that this expression can have subject to the necessary relation (2) which connects $d\xi$, $d\eta$, $d\zeta$ and ds, viz. to

$$\phi \equiv A\left(\frac{d\xi}{ds}\right)^2 + B\left(\frac{d\eta}{ds}\right)^2 + C\left(\frac{d\zeta}{ds}\right)^2 + 2D\frac{d\eta}{ds}\frac{d\zeta}{ds} + 2E\frac{d\zeta}{ds}\frac{d\xi}{ds} + 2F\frac{d\xi}{ds}\frac{d\eta}{ds} = 1. \tag{16}$$

We are therefore to have $\delta\dfrac{dV}{ds} = 0$ subject to the relation $\delta\phi = 0$; and, proceeding by Lagrange's method of undetermined multipliers, we obtain

$$A\frac{d\xi}{ds} + F\frac{d\eta}{ds} + E\frac{d\zeta}{ds} + \lambda\frac{dV}{d\xi} = 0,$$

$$F\frac{d\xi}{ds} + B\frac{d\eta}{ds} + D\frac{d\zeta}{ds} + \lambda\frac{dV}{d\eta} = 0,$$

$$E\frac{d\xi}{ds} + D\frac{d\eta}{ds} + C\frac{d\zeta}{ds} + \lambda\frac{dV}{d\zeta} = 0,$$

together with

$$\frac{dV}{d\xi}\frac{d\xi}{ds} + \frac{dV}{d\eta}\frac{d\eta}{ds} + \frac{dV}{d\zeta}\frac{d\zeta}{ds} - R = 0. \tag{17}$$

From the first three equations, we find immediately

$$1 + \lambda R = 0; \tag{18}$$

and then, eliminating the unknowns $\frac{d\xi}{ds}, \frac{d\eta}{ds}, \frac{d\zeta}{ds}$, we have for R^2 the equation

$$\begin{vmatrix} A & F & E & \frac{dV}{d\xi} \\ F & B & D & \frac{dV}{d\eta} \\ E & D & C & \frac{dV}{d\zeta} \\ \frac{dV}{d\xi} & \frac{dV}{d\eta} & \frac{dV}{d\zeta} & R^2 \end{vmatrix} = 0, \tag{19}$$

or

$$\begin{vmatrix} A & F & E \\ F & B & D \\ E & D & C \end{vmatrix} R^2 = (BC - D^2)\left(\frac{dV}{d\xi}\right)^2 + (CA - E^2)\left(\frac{dV}{d\eta}\right)^2 + (AB - F^2)\left(\frac{dV}{d\zeta}\right)^2 + 2(EF - AD)\frac{dV}{d\eta}\frac{dV}{d\zeta} + 2(FD - BE)\frac{dV}{d\zeta}\frac{dV}{d\xi} + 2(DE - CF)\frac{dV}{d\xi}\frac{dV}{d\eta}, \tag{20}$$

which agrees with (11).

This equation is also clearly that which determines an element of area in terms of its projections on the three coordinate surfaces at the place where it is situated.

I. 8. We now proceed to the expression of Laplace and Poisson's equation in terms of the coordinates ξ, η, ζ.

The potential energy of a system of attracting masses is well known to be

$$U = \frac{1}{8\pi}\int R^2 . d\,\text{vol.}, \tag{21}$$

extended throughout all space, where R is the expression for the resultant force. If we take Cartesian coordinates x, y, z

Quasi-energy distribution: invariant:

$$U = \frac{1}{8\pi}\iiint\left\{\left(\frac{dV}{dx}\right)^2 + \left(\frac{dV}{dy}\right)^2 + \left(\frac{dV}{dz}\right)^2\right\} dxdydz, \tag{22}$$

and the ordinary method of variation gives

its variation leads to the Laplacian.

$$\delta U = \frac{1}{4\pi}\iiint\left(\frac{dV}{dx}\frac{d\delta V}{dx} + \frac{dV}{dy}\frac{d\delta V}{dy} + \frac{dV}{dz}\frac{d\delta V}{dz}\right) dxdydz$$

$$= -\frac{1}{4\pi}\int\left(\frac{d^2V}{dx^2} + \frac{d^2V}{dy^2} + \frac{d^2V}{dz^2}\right)\delta V . d\,\text{vol.}, \tag{23}$$

on integration by parts; for the terms at the limits vanish, the attracting bodies being supposed to lie at finite distances.

If we take generalized coordinates, R^2 is given by (10) or (11). Using the former expression, and, writing for shortness, Therefore also in generalized coordinates.

$$\Theta \text{ for } 1 - \cos^2\alpha - \cos^2\beta - \cos^2\gamma + 2\cos\alpha\cos\beta\cos\gamma, \quad (24)$$

we have

$$U = \frac{1}{8\pi}\iiint \Theta^{-1}\left[\sin^2\alpha . h_1^2\left(\frac{dV}{d\xi}\right)^2 + \sin^2\beta . h_2^2\left(\frac{dV}{d\eta}\right)^2 + \sin^2\gamma . h_3^2\left(\frac{dV}{d\zeta}\right)\right.$$

$$+ 2(\cos\beta\cos\gamma - \cos\alpha)\, h_2 h_3 \frac{dV}{d\eta}\frac{dV}{d\zeta}$$

$$+ 2(\cos\gamma\cos\alpha - \cos\beta)\, h_3 h_1 \frac{dV}{d\zeta}\frac{dV}{d\xi}$$

$$\left. + 2(\cos\alpha\cos\beta - \cos\gamma)\, h_1 h_2 \frac{dV}{d\xi}\frac{dV}{d\eta}\right] . \Theta^{\frac{1}{2}} \frac{d\xi}{h_1}\frac{d\eta}{h_2}\frac{d\zeta}{h_3}, \quad (25)$$

therefore

$$\delta U = \frac{1}{4\pi}\iiint\left[\Theta^{-\frac{1}{2}}\left\{\frac{h_1}{h_2 h_3}\sin^2\alpha . \frac{dV}{d\xi} + \frac{1}{h_3}(\cos\alpha\cos\beta - \cos\gamma)\frac{dV}{d\eta}\right.\right.$$

$$\left. + \frac{1}{h_2}(\cos\gamma\cos\alpha - \cos\beta)\frac{dV}{d\zeta}\right\}\frac{d\delta V}{d\xi}$$

$$+ \Theta^{-\frac{1}{2}}\left\{\frac{1}{h_3}(\cos\alpha\cos\beta - \cos\gamma)\frac{dV}{d\xi} + \frac{h_2}{h_3 h_1}\sin^2\beta . \frac{dV}{d\eta}\right.$$

$$\left. + \frac{1}{h_1}(\cos\beta\cos\gamma - \cos\alpha)\frac{dV}{d\zeta}\right\}\frac{d\delta V}{d\eta}$$

$$+ \Theta^{-\frac{1}{2}}\left\{\frac{1}{h_2}(\cos\gamma\cos\alpha - \cos\beta)\frac{dV}{d\xi} + \frac{1}{h_1}(\cos\beta\cos\gamma - \cos\alpha)\frac{dV}{d\eta}\right.$$

$$\left.\left. + \frac{h_3}{h_1 h_2}\sin^2\gamma . \frac{dV}{d\zeta}\right\}\frac{d\delta V}{d\zeta}\right] d\xi d\eta d\zeta \quad (26)$$

$$= -\frac{1}{4\pi}\int\left[\frac{d}{d\xi}(\Theta^{-\frac{1}{2}}\{...\}) + \frac{d}{d\eta}(\Theta^{-\frac{1}{2}}\{...\})\right.$$

$$\left. + \frac{d}{d\zeta}(\Theta^{-\frac{1}{2}}\{...\})\right]\Theta^{-\frac{1}{2}} h_1 h_2 h_3 \delta V d \text{ vol.}, \quad (27)$$

on integration by parts as before.

Comparing this result with (23), with which it must be identical, we see that

$$\nabla^2 V \text{ or } \frac{d^2V}{dx^2} + \frac{d^2V}{dy^2} + \frac{d^2V}{dz^2}$$

is equivalent to

$$\Theta^{-\frac{1}{2}} h_1 h_2 h_3 \left[\frac{d}{d\xi}(\Theta^{-\frac{1}{2}}\{...\}) + \frac{d}{d\eta}(\Theta^{-\frac{1}{2}}\{...\}) + \frac{d}{d\zeta}(\Theta^{-\frac{1}{2}}\{...\})\right], \quad (28)$$

where {...} denotes the respective expressions with the { } in (26).

The equation $\nabla^2 V = -4\pi\rho$ is thus transformed to the new coordinates*; and the result agrees and may be compared with that obtained by the method of flux in Thomson and Tait's *Natural Philosophy*, 2nd edn., Appendix A_0.

If the coordinates form a rectangular system, we obtain Lamé's equation

$$h_1h_2h_3\left[\frac{d}{d\xi}\left(\frac{h_1}{h_2h_3}\frac{dV}{d\xi}\right)+\frac{d}{d\eta}\left(\frac{h_2}{h_3h_1}\frac{dV}{d\eta}\right)+\frac{d}{d\zeta}\left(\frac{h_3}{h_1h_2}\frac{dV}{d\zeta}\right)\right] = -4\pi\rho. \quad (29)$$

Potential coordinates: 9. If in the general case (28) ξ, η, ζ are potential functions, *i.e.* are such that they satisfy

$$\nabla^2\xi = 0, \quad \nabla^2\eta = 0, \quad \nabla^2\zeta = 0, \quad (30)$$

throughout the space considered, then ξ, η, or ζ substituted for V in (28) makes the expression vanish; and by virtue of this simplification the expression for $\nabla^2 V$ reduces to

$$\frac{h_1h_2h_3}{\Theta}\cdot\left[\frac{h_1}{h_2h_3}\sin^2\alpha\,.\frac{d^2V}{d\xi^2}+\frac{h_2}{h_3h_1}\sin^2\beta\,.\frac{d^2V}{d\eta^2}+\frac{h_3}{h_1h_2}\sin^2\gamma\,.\frac{d^2V}{d\zeta^2}\right.$$
$$+\frac{2}{h_1}(\cos\beta\cos\gamma-\cos\alpha)\frac{d^2V}{d\eta d\zeta}+\frac{2}{h_2}(\cos\gamma\cos\alpha-\cos\beta)\frac{d^2V}{d\zeta d\xi}$$
$$\left.+\frac{2}{h_3}(\cos\alpha\cos\beta-\cos\gamma)\frac{d^2V}{d\xi d\eta}\right],$$

i.e. to

$$\Theta^{-1}\left[h_1^2\sin^2\alpha\,.\frac{d^2V}{d\xi^2}+h_2^2\sin^2\beta\,.\frac{d^2V}{d\eta^2}+h_3^2\sin^2\gamma\,.\frac{d^2V}{d\zeta^2}\right.$$
$$+2h_2h_3(\cos\beta\cos\gamma-\cos\alpha)\frac{d^2V}{d\eta d\zeta}+2h_3h_1(\cos\gamma\cos\alpha-\cos\beta)\frac{d^2V}{d\zeta d\xi}$$
$$\left.+2h_1h_2(\cos\alpha\cos\beta-\cos\gamma)\frac{d^2V}{d\xi d\eta}\right]. \quad (31)$$

When the coordinates form a rectangular system, this reduces again to Lamé's result

$$\nabla^2V \equiv h_1^2\frac{d^2V}{d\xi^2}+h_2^2\frac{d^2V}{d\eta^2}+h_3^2\frac{d^2V}{d\zeta^2}. \quad (32)$$

The more general result (31) may be expressed by saying that if, with the conditions of this section,

simplified Laplacian invariant.

$$R^2 = f\left(\frac{dV}{d\xi}, \frac{dV}{d\eta}, \frac{dV}{d\zeta}\right),$$

then

$$\nabla^2V = f\left(\frac{d}{d\xi}, \frac{d}{d\eta}, \frac{d}{d\zeta}\right)V. \quad (33)$$

* This method of variations is the subject of a memoir by Jacobi, *Crelle*, 36 (1848), pp. 113–134; *Werke*, II. pp. 193–216, which finally branches off into consideration of solutions by Lamé's and other special functions.

It may be observed that it is always possible to choose a coordinate, say ξ, so that it shall preserve a constant value over any bounding surface, and shall satisfy the equation $\nabla^2\xi = 0$; for it has only to be taken proportional to the potential of a free electric distribution on the surface, supposed conducting, in free space,—or it may be in the presence of other charged bodies, but in that case there will be discontinuity at the places occupied by those charges, if they are in the part of the field considered.

10. If we employ the other form (2) for R^2, and denote the determinant Alternatively.

$$\begin{vmatrix} A & F & E \\ F & B & D \\ E & D & C \end{vmatrix} \text{ by } \vartheta, \tag{34}$$

we have at once

$$U = \frac{1}{8\pi}\int R^2 \,.\, d \text{ vol.}$$
$$= \frac{1}{8\pi}\iiint \vartheta^{-\frac{1}{2}}\Big[(BC - D^2)\left(\frac{dV}{d\xi}\right)^2 + \ldots + \ldots$$
$$+ 2\,(EF - AD)\frac{dV}{d\eta}\frac{dV}{d\zeta} + \ldots + \ldots\Big] d\xi d\eta d\zeta, \tag{35}$$

and it follows as before that $\nabla^2 V$ is equal to

$$\vartheta^{-\frac{1}{2}}\left[\frac{d}{d\xi}\vartheta^{-\frac{1}{2}}\begin{vmatrix} \frac{dV}{d\xi} & \frac{dV}{d\eta} & \frac{dV}{d\zeta} \\ F & B & D \\ E & D & C \end{vmatrix} + \frac{d}{d\eta}\vartheta^{-\frac{1}{2}}\begin{vmatrix} A & F & E \\ \frac{dV}{d\xi} & \frac{dV}{d\eta} & \frac{dV}{d\zeta} \\ E & D & C \end{vmatrix} + \frac{d}{d\zeta}\vartheta^{-\frac{1}{2}}\begin{vmatrix} A & F & E \\ F & B & D \\ \frac{dV}{d\xi} & \frac{dV}{d\eta} & \frac{dV}{d\zeta} \end{vmatrix}\right]; \tag{36}$$

and corresponding simplifications may occur.

II. 11. We now proceed to apply a similar analysis to the dynamical theory of an isotropic elastic solid.

We have first to determine the quadratic expression for the energy of deformation per unit volume. Following Kirchhoff [1], let us take an element of the solid, and let f, g, h denote its three principal elongations, and F, G, H the three tensions which act on it per unit area in the directions of those elongations; we may conveniently take the element to be a right solid with its edges along these directions. In any case F, G, H represent the complete system of forces to which the element is subjected from the action of the contiguous parts, for with this specification there are no shears. Assuming, as usual, the

Analysis of stress-strain relations for isotropy.

[1] Crelle's *Journal*, Bd. 40; *Gesammelte Abhandlungen*, p. 247.

truth of Hooke's law for the displacements considered, we have, from the isotropic character of the solid, equations of the form

$$F = af + bg + bh,$$
$$G = bf + ag + bh,$$
$$H = bf + bg + ah, \qquad (37)$$

where a, b are two constants which specify the elastic qualities of the body, viz.

$a + 2b$ is the modulus of compression,

$\frac{1}{2}(a - b)$ is the modulus of shear,

$a - \dfrac{2b^2}{a+b}$ is Young's modulus. (38)

The potential energy of deformation per unit volume at the point considered is

Energy of strain:

$$\tfrac{1}{2}(Ff + Gg + Hh), \qquad (39)$$

and is therefore

$$\tfrac{1}{2}a(f^2 + g^2 + h^2) + b(fg + gh + hf),$$

or, say $$V \equiv \tfrac{1}{2}a(f + g + h)^2 + (b - a)(fg + gh + hf). \qquad (40)$$

The equations of equilibrium will be obtained as the conditions that the variation of the total potential energy

$$\int V d \text{ vol.} + \text{the part due to external forces}$$

shall be zero.

transferred to generalized coordinates.

12. Suppose now the position in space of a point to be given by the generalized coordinates ϕ, χ, ψ.

Consider in the undisturbed solid a point ϕ, χ, ψ, and a consecutive point

$$\phi + \xi, \quad \chi + \eta, \quad \psi + \zeta;$$

when the solid is deformed these points will assume new positions

$$\left.\begin{matrix}\phi + u\\ \chi + v\\ \psi + w\end{matrix}\right\} \text{ and } \begin{matrix}\phi + u + \xi + \dfrac{du}{d\phi}\xi + \dfrac{du}{d\chi}\eta + \dfrac{du}{d\psi}\zeta,\\ \chi + v + \eta + \dfrac{dv}{d\phi}\xi + \dfrac{dv}{d\chi}\eta + \dfrac{dv}{d\psi}\zeta,\\ \psi + w + \zeta + \dfrac{dw}{d\phi}\xi + \dfrac{dw}{d\chi}\eta + \dfrac{dw}{d\psi}\zeta.\end{matrix} \qquad (41)$$

The square of the distance between these points will therefore, by (2), be given by

Elongation of linear element.

$$\mathfrak{L}^2 = A'\left(\xi + \frac{du}{d\phi}\xi + \frac{du}{d\chi}\eta + \frac{du}{d\psi}\zeta\right)^2$$

$$+ B'\left(\eta + \frac{dv}{d\phi}\xi + \frac{dv}{d\chi}\eta + \frac{dv}{d\psi}\zeta\right)^2 + C'\left(\zeta + \frac{dw}{d\phi}\xi + \frac{dw}{d\chi}\eta + \frac{dw}{d\psi}\zeta\right)^2$$

$$+ 2D'\left(\eta + \frac{dv}{d\phi}\xi + \frac{dv}{d\chi}\eta + \frac{dv}{d\psi}\zeta\right)\left(\zeta + \frac{dw}{d\phi}\xi + \frac{dw}{d\chi}\eta + \frac{dw}{d\psi}\zeta\right)$$

$$+ 2E'\left(\zeta + \frac{dw}{d\phi}\xi + \frac{dw}{d\chi}\eta + \frac{dw}{d\psi}\zeta\right)\left(\xi + \frac{du}{d\phi}\xi + \frac{du}{d\chi}\eta + \frac{du}{d\psi}\zeta\right)$$

$$+ 2F'\left(\xi + \frac{du}{d\phi}\xi + \frac{du}{d\chi}\eta + \frac{du}{d\psi}\zeta\right)\left(\eta + \frac{dv}{d\phi}\xi + \frac{dv}{d\chi}\eta + \frac{dv}{d\psi}\zeta\right), \quad (42)$$

where A', B', C', D', E', F' are the values of A, B, C, D, E, F at the point

$$\phi + u, \quad \chi + v, \quad \psi + w,$$

consecutive to ϕ, χ, ψ, and therefore

$$A' = A + \frac{dA}{d\phi}u + \frac{dA}{d\chi}v + \frac{dA}{d\psi}w$$

$$= A + \mathfrak{d}A, \text{ say,} \quad (43)$$

with similar expressions for B', ... F'.

Therefore, neglecting squares of small quantities $\frac{du}{d\phi}$, $\frac{du}{d\psi}$, ...

$$\mathfrak{L}^2 = \mathfrak{L}_0^2 + 2\left(A\frac{du}{d\phi} + E\frac{dw}{d\phi} + F\frac{dv}{d\phi} + \tfrac{1}{2}\mathfrak{d}A\right)\xi^2 + 2\left(\ldots\right)\eta^2 + 2\left(\ldots\right)\zeta^2$$

$$+ 2\left\{B\frac{dv}{d\psi} + C\frac{dw}{d\chi} + D\left(\frac{dw}{d\psi} + \frac{dv}{d\chi}\right) + E\frac{du}{d\chi} + F\frac{du}{d\psi} + \mathfrak{d}D\right\}\eta\zeta$$

$$+ 2\left\{\ldots\right\}\zeta\xi + 2\left\{\ldots\right\}\xi\eta, \quad (44)$$

where $\mathfrak{L}_0$ is the value of $\mathfrak{L}$ before deformation.

Since $\mathfrak{L}$ and $\mathfrak{L}_0$ differ by a quantity very small compared with either of them (*i.e.* of the second order), we have

$$\tfrac{1}{2}(\mathfrak{L}^2 - \mathfrak{L}_0^2) = \mathfrak{L}_0^2\epsilon, \quad (45)$$

where ϵ is the elongation, per unit length, of the solid, in the direction of $\mathfrak{L}$, at the point ϕ, χ, ψ.

We may therefore write (44) in the form

$$\mathfrak{L}_0^2\epsilon = \mathfrak{A}\xi^2 + \mathfrak{B}\eta^2 + \mathfrak{C}\zeta^2 + 2\mathfrak{D}\eta\zeta + 2\mathfrak{E}\zeta\xi + 2\mathfrak{F}\xi\eta. \quad (46)$$

13. The values of f, g, h are the maximum and minimum and the

stationary maximum-minimum value of this quantity ϵ; the value of $\mathfrak{L}_0$ being given by (2), or, what is the same, (5), viz.

$$\mathfrak{L}_0^2 = A\xi^2 + B\eta^2 + C\zeta^2 + 2D\eta\zeta + 2E\zeta\xi + 2F\xi\eta.$$

These values may be determined in the ordinary manner. Substituting from (2) in (46), we have

$$(\mathfrak{A} - A\epsilon)\,\xi^2 + (\mathfrak{B} - B\epsilon)\,\eta^2 + (\mathfrak{C} - C\epsilon)\,\zeta^2 + 2\,(\mathfrak{D} - D\epsilon)\,\eta\zeta + 2\,(\mathfrak{E} - E\epsilon)\,\zeta\xi + 2\,(\mathfrak{F} - F\epsilon)\,\xi\eta = 0. \tag{47}$$

Differentiating with respect to ξ, η, ζ, and remembering that for the values in question the differentials of ϵ are zero, we obtain

$$\begin{aligned}(\mathfrak{A} - A\epsilon)\,\xi + (\mathfrak{F} - F\epsilon)\,\eta + (\mathfrak{E} - E\epsilon)\,\zeta &= 0,\\ (\mathfrak{F} - F\epsilon)\,\xi + (\mathfrak{B} - B\epsilon)\,\eta + (\mathfrak{D} - D\epsilon)\,\zeta &= 0,\\ (\mathfrak{E} - E\epsilon)\,\xi + (\mathfrak{D} - D\epsilon)\,\eta + (\mathfrak{C} - C\epsilon)\,\zeta &= 0,\end{aligned} \tag{48}$$

which lead to the eliminant

$$\begin{vmatrix} \mathfrak{A} - A\epsilon & \mathfrak{F} - F\epsilon & \mathfrak{E} - E\epsilon \\ \mathfrak{F} - F\epsilon & \mathfrak{B} - B\epsilon & \mathfrak{D} - D\epsilon \\ \mathfrak{E} - E\epsilon & \mathfrak{D} - D\epsilon & \mathfrak{C} - C\epsilon \end{vmatrix} = 0. \tag{49}$$

The principal elongations.

The values of f, g, h are therefore the roots of this cubic equation in ϵ. Therefore

$$(f+g+h)\begin{vmatrix} A & F & E \\ F & B & D \\ E & D & C \end{vmatrix} = \begin{vmatrix} A & F & \mathfrak{E} \\ F & B & \mathfrak{D} \\ E & D & \mathfrak{C} \end{vmatrix} + \begin{vmatrix} A & \mathfrak{F} & E \\ F & \mathfrak{B} & D \\ E & \mathfrak{D} & C \end{vmatrix} + \begin{vmatrix} \mathfrak{A} & F & E \\ \mathfrak{F} & B & D \\ \mathfrak{E} & D & C \end{vmatrix}$$

$$= \mathfrak{A}\,(BC - D^2) + \mathfrak{B}\,(CA - E^2) + \mathfrak{C}\,(AB - F^2) + 2\mathfrak{D}\,(EF - AD) + 2\mathfrak{E}\,(FD - BE) + 2\mathfrak{F}\,(DE - CF), \tag{50}$$

$$-(fg+gh+hf)\begin{vmatrix} A & F & E \\ F & B & D \\ E & D & C \end{vmatrix} = \begin{vmatrix} \mathfrak{A} & \mathfrak{F} & E \\ \mathfrak{F} & \mathfrak{B} & D \\ \mathfrak{E} & \mathfrak{D} & C \end{vmatrix} + \begin{vmatrix} \mathfrak{A} & F & \mathfrak{E} \\ \mathfrak{F} & B & \mathfrak{D} \\ \mathfrak{E} & D & \mathfrak{C} \end{vmatrix} + \begin{vmatrix} A & \mathfrak{F} & \mathfrak{E} \\ F & \mathfrak{B} & \mathfrak{D} \\ E & \mathfrak{D} & \mathfrak{C} \end{vmatrix}$$

$$= (\mathfrak{B}\mathfrak{C} - \mathfrak{D}^2)\,A + (\mathfrak{C}\mathfrak{A} - \mathfrak{E}^2)\,B + (\mathfrak{A}\mathfrak{B} - \mathfrak{F}^2)\,C + 2\,(\mathfrak{E}\mathfrak{F} - \mathfrak{A}\mathfrak{D})\,D + 2\,(\mathfrak{F}\mathfrak{D} - \mathfrak{B}\mathfrak{E})\,E + 2\,(\mathfrak{D}\mathfrak{E} - \mathfrak{C}\mathfrak{F})\,F, \tag{51}$$

and the value of V as given by equation (40) may be written down.

14. Having thus determined the value of V, the equations of internal and surface equilibrium may be deduced by the Method of Variations in the ordinary manner.

If Φ, X, Ψ denote the components (oblique) of the surface-force, per unit area, which acts on the boundary of the solid, taken in the directions of the lines of intersection of the pairs of surfaces $\chi\psi$, $\psi\phi$, and $\phi\chi$ respectively, and so as when positive to tend to increase the values of ϕ, χ, ψ; and if W denote the potential energy per unit volume at any point in the solid, due to the action from a distance of external systems,—then the energy condition of equilibrium is that

Equation of variation of the strain-energy:

$$\delta U \equiv \delta \int V d \text{ vol.} + \delta \int W d \text{ vol.}$$
$$- \int (\Phi A^{\frac{1}{2}} \delta u + \mathrm{X} B^{\frac{1}{2}} \delta v + \Psi C^{\frac{1}{2}} \delta w)\, d \text{ surface} = 0 \quad (52)$$

for all possible consistent variations of u, v, w.

Therefore, with the notation of § 10, we have

$$\iiint \vartheta^{\frac{1}{2}} \delta (V + W)\, d\phi d\chi d\psi$$
$$- \iint (\Phi A^{\frac{1}{2}} \delta u + \mathrm{X} B^{\frac{1}{2}} \delta v + \Psi C^{\frac{1}{2}} \delta w)\, dS = 0. \quad (53)$$

Now V is a function (quadratic) of the differential coefficients of u, v, w with respect to ϕ, χ and ψ. If A, B, C, D, E, F are constants, which corresponds to the case of oblique Cartesian coordinates, V is a function of those differential coefficients alone, with constant coefficients. If $A, B, \ldots F$ are not constants, the terms $\mathfrak{d}A, \mathfrak{d}B, \ldots \mathfrak{d}F$ will introduce u, v, w themselves; so that V is now a quadratic function of u, v, w and their differential coefficients with respect to ϕ, χ, ψ.

Thus we have

$$\delta V \equiv \frac{dV}{d\dfrac{du}{d\phi}} \frac{d\delta u}{d\phi} + \frac{dV}{d\dfrac{du}{d\chi}} \frac{d\delta u}{d\chi} + \ldots + \frac{dV}{d\dfrac{dv}{d\phi}} \frac{d\delta V}{d\phi} + \ldots$$
$$+ \frac{dV}{du} \delta u + \frac{dV}{dv} \delta v + \frac{dV}{dw} \delta w. \quad (54)$$

And on substituting this value in (53), and integrating by parts in the usual manner the terms which contain differential coefficients of $\delta u, \delta v, \delta w$, we have, writing down explicitly only the terms depending on δu,

$$\iint \vartheta^{\frac{1}{2}} \frac{dV}{d\dfrac{du}{d\phi}} \delta u \,.\, d\chi d\psi + \iint \vartheta^{\frac{1}{2}} \frac{dV}{d\dfrac{du}{d\chi}} \delta u \,.\, d\psi d\phi$$
$$+ \iint \vartheta^{\frac{1}{2}} \frac{dV}{d\dfrac{du}{d\psi}} \delta u \,.\, d\phi d\chi - \iint \Phi A^{\frac{1}{2}} \delta u\, dS$$
$$- \iiint \left\{ \frac{d}{d\phi} \left(\vartheta^{\frac{1}{2}} \frac{dV}{d\dfrac{du}{d\phi}} \right) + \frac{d}{d\chi} \left(\vartheta^{\frac{1}{2}} \frac{dV}{d\dfrac{du}{d\chi}} \right) + \frac{d}{d\psi} \left(\vartheta^{\frac{1}{2}} \frac{dV}{d\dfrac{du}{d\psi}} \right) \right.$$
$$\left. - \vartheta^{\frac{1}{2}} \frac{dV}{du} - \vartheta^{\frac{1}{2}} \frac{dW}{d\phi} \right\} \delta u \,.\, d\phi d\chi d\psi$$
$$+ \text{ terms containing } \delta v \text{ and } \delta w = 0. \quad (55)$$

leads to the equations of stress and of equilibrium.

Now as the forms of δu, δv, δw are quite at our disposal, subject to conditions of continuity, we can choose them so as to make either the volume integral or the surface integral in this equation zero at pleasure; and we may also have any two of these variations of u, v, w zero at pleasure. This equation therefore cannot be satisfied unless the quantity under each separate type of volume or surface integral is zero; and we thus obtain the internal and surface equations of equilibrium.

15. To express the latter in their simplest form, it will be convenient to take the solid considered to be the element of volume $d\phi d\chi d\psi$ of a larger body. The surface element dS is now an element of area of a face of this solid, and is, by § 4,

$$\left(\frac{d\vartheta}{dA}\right)^{\frac{1}{2}} d\chi d\psi, \quad \left(\frac{d\vartheta}{dB}\right)^{\frac{1}{2}} d\psi d\phi, \quad \text{or} \quad \left(\frac{d\vartheta}{dC}\right)^{\frac{1}{2}} d\phi d\chi$$

according to its position.

The force Φ of (52) is the force exerted across this element of area by the matter on the other side of it, per unit area, in a direction such as to increase ϕ without altering χ or ψ, *i.e.* in a direction parallel to the intersection of χ, ψ. If we adopt the notation that $T_{\phi\psi}$ represent the component force per unit area acting from outside on that face of the element which lies on the surface ϕ in a direction such as to increase ψ without altering the values of ϕ, χ, *i.e.* in a direction parallel to the intersection of ϕ, χ, we have therefore from the surface integrals the relations

The stress:

$$T_{\phi\phi} = A^{-\frac{1}{2}} \left(\frac{d\vartheta}{dA}\right)^{-\frac{1}{2}} \vartheta^{\frac{1}{2}} \frac{dV}{d\dfrac{du}{d\phi}},$$

$$T_{\chi\phi} = A^{-\frac{1}{2}} \left(\frac{d\vartheta}{dB}\right)^{-\frac{1}{2}} \vartheta^{\frac{1}{2}} \frac{dV}{d\dfrac{du}{d\chi}}, \qquad (56)$$

with similar expressions for the other surface-stresses.

From the volume integral in (55) we may now obtain the equations of internal equilibrium. We may shorten their expression by making use of the values of the stresses just obtained in (56).

In the first place, we remark that, by (43) and (44),

$$\frac{dV}{du} = \frac{1}{2}\frac{dA}{d\phi}\frac{dV}{d\mathfrak{d}A} + \frac{1}{2}\frac{dB}{d\phi}\frac{dV}{d\mathfrak{d}B} + \dots + \frac{dF}{d\phi}\frac{dV}{d\mathfrak{d}F}. \qquad (57)$$

We have then

and equations of its internal equilibrium.

$$\frac{d}{d\phi}\left\{A^{\frac{1}{2}}\left(\frac{d\vartheta}{dA}\right)^{\frac{1}{2}} T_{\phi\phi}\right\} + \frac{d}{d\chi}\left\{A^{\frac{1}{2}}\left(\frac{d\vartheta}{dB}\right)^{\frac{1}{2}} T_{\chi\phi}\right\} + \frac{d}{d\psi}\left\{A^{\frac{1}{2}}\left(\frac{d\vartheta}{dC}\right)^{\frac{1}{2}} T_{\psi\phi}\right\}$$
$$- \vartheta^{\frac{1}{2}}\frac{dV}{du} - \vartheta^{\frac{1}{2}}\frac{dW}{d\phi} = 0,$$

with the similar equations

$$\frac{d}{d\phi}\left\{B^{\frac{1}{2}}\left(\frac{d\vartheta}{dA}\right)^{\frac{1}{2}} T_{\phi\chi}\right\}+\frac{d}{d\chi}\left\{B^{\frac{1}{2}}\left(\frac{d\vartheta}{dB}\right)^{\frac{1}{2}} T_{\chi\chi}\right\}+\frac{d}{d\psi}\left\{B^{\frac{1}{2}}\left(\frac{d\vartheta}{dC}\right)^{\frac{1}{2}} T_{\psi\chi}\right\}$$

$$-\vartheta^{\frac{1}{2}}\frac{dV}{dv}-\vartheta^{\frac{1}{2}}\frac{dW}{d\chi}=0,$$

$$\frac{d}{d\phi}\left\{C^{\frac{1}{2}}\left(\frac{d\vartheta}{dA}\right)^{\frac{1}{2}} T_{\phi\psi}\right\}+\frac{d}{d\chi}\left\{C^{\frac{1}{2}}\left(\frac{d\vartheta}{dB}\right)^{\frac{1}{2}} T_{\chi\psi}\right\}+\frac{d}{d\psi}\left\{C^{\frac{1}{2}}\left(\frac{d\vartheta}{dC}\right)^{\frac{1}{2}} T_{\psi\psi}\right\}$$

$$-\vartheta^{\frac{1}{2}}\frac{dV}{dw}-\vartheta^{\frac{1}{2}}\frac{dW}{d\psi}=0, \tag{58}$$

the values of $\frac{dV}{du}, \frac{dV}{dv}, \frac{dV}{dw}$ being most easily determined by means of (57).

16. In the case in which the coordinate surfaces ϕ, χ, ψ form an orthogonal system, the formulae become more simple. Simplified form in orthogonal coordinates.

We have now

$$D=0, \qquad E=0, \qquad F=0,$$

and, if we wish to revert to Lamé's notation, we have (59)

$$A=\frac{1}{h_1^2}, \qquad B=\frac{1}{h_2^2}, \qquad C=\frac{1}{h_3^2}.$$

By (44),

$$\mathfrak{L}_0^2\epsilon=\left(A\frac{du}{d\phi}+\tfrac{1}{2}\mathfrak{d}A\right)\xi^2+\left(B\frac{dv}{d\chi}+\tfrac{1}{2}\mathfrak{d}B\right)\eta^2+\left(C\frac{dw}{d\psi}+\tfrac{1}{2}\mathfrak{d}C\right)\zeta^2$$

$$+\left(B\frac{dv}{d\psi}+C\frac{dw}{d\chi}\right)\eta\zeta+\left(C\frac{dw}{d\phi}+A\frac{du}{d\psi}\right)\zeta\xi+\left(A\frac{du}{d\chi}+B\frac{dv}{d\phi}\right)\xi\eta, \tag{60}$$

where $\mathfrak{L}_0^2=A\xi^2+B\eta^2+C\zeta^2$.

The cubic equation whose roots are f, g, h is now

$$\begin{vmatrix}\mathfrak{A}-A\epsilon & \mathfrak{F} & \mathfrak{E}\\ \mathfrak{F} & \mathfrak{B}-B\epsilon & \mathfrak{D}\\ \mathfrak{E} & \mathfrak{D} & \mathfrak{C}-C\epsilon\end{vmatrix}=0, \tag{61}$$

and therefore

$$f+g+h=\frac{\mathfrak{A}}{A}+\frac{\mathfrak{B}}{B}+\frac{\mathfrak{C}}{C}, \tag{62}$$

$$fg+gh+hf=\frac{\mathfrak{B}\mathfrak{C}-\mathfrak{D}^2}{BC}+\frac{\mathfrak{C}\mathfrak{A}-\mathfrak{E}^2}{CA}+\frac{\mathfrak{A}\mathfrak{B}-\mathfrak{F}^2}{AB}. \tag{63}$$

We have now, by (40),

$$V=\tfrac{1}{2}a\left(\frac{\mathfrak{A}}{A}+\frac{\mathfrak{B}}{B}+\frac{\mathfrak{C}}{C}\right)^2-(a-b)\left(\frac{\mathfrak{B}\mathfrak{C}-\mathfrak{D}^2}{BC}+\frac{\mathfrak{C}\mathfrak{A}-\mathfrak{E}^2}{CA}+\frac{\mathfrak{A}\mathfrak{B}-\mathfrak{F}^2}{AB}\right). \tag{64}$$

Therefore, by (60),

$$V = \tfrac{1}{2}a\left(\frac{du}{d\phi}+\frac{dv}{d\chi}+\frac{dw}{d\psi}+\frac{1}{2}\frac{\mathfrak{d}A}{A}+\frac{1}{2}\frac{\mathfrak{d}B}{B}+\frac{1}{2}\frac{\mathfrak{d}C}{C}\right)^2$$

$$-(a-b)\left[\frac{dv}{d\chi}\frac{dw}{d\psi}+\frac{dw}{d\psi}\frac{du}{d\phi}+\frac{du}{d\phi}\frac{dv}{d\chi}\right.$$

$$-\tfrac{1}{2}\left(\frac{dv}{d\psi}\frac{dw}{d\chi}+\frac{dw}{d\phi}\frac{du}{d\psi}+\frac{du}{d\chi}\frac{dv}{d\phi}\right)$$

$$+\frac{1}{2}\frac{du}{d\phi}\left(\frac{\mathfrak{d}B}{B}+\frac{\mathfrak{d}C}{C}\right)+\frac{1}{2}\frac{dv}{d\chi}\left(\frac{\mathfrak{d}C}{C}+\frac{\mathfrak{d}A}{A}\right)+\frac{1}{2}\frac{dw}{d\psi}\left(\frac{\mathfrak{d}A}{A}+\frac{\mathfrak{d}B}{B}\right)$$

$$+\tfrac{1}{4}\left(\frac{\mathfrak{d}B}{B}\frac{\mathfrak{d}C}{C}+\frac{\mathfrak{d}C}{C}\frac{\mathfrak{d}A}{A}+\frac{\mathfrak{d}A}{A}\frac{\mathfrak{d}B}{B}\right)$$

$$-\tfrac{1}{4}\left\{\frac{A}{C}\left(\frac{du}{d\psi}\right)^2+\frac{A}{B}\left(\frac{du}{d\chi}\right)^2+\frac{B}{C}\left(\frac{dv}{d\psi}\right)^2+\frac{B}{A}\left(\frac{dv}{d\phi}\right)^2\right.$$

$$\left.\left.+\frac{C}{B}\left(\frac{dw}{d\chi}\right)^2+\frac{C}{A}\left(\frac{dw}{d\phi}\right)^2\right\}\right], \qquad (65)$$

wherein for brevity, as before,

$$\mathfrak{d}A \equiv \frac{dA}{d\phi}u+\frac{dA}{d\chi}v+\frac{dA}{d\psi}w,$$

$$\mathfrak{d}B = \frac{dB}{d\phi}u+\frac{dB}{d\chi}v+\frac{dB}{d\psi}w,$$

$$\mathfrak{d}C = \frac{dC}{d\phi}u+\frac{dC}{d\chi}v+\frac{dC}{d\psi}w.$$

We have thus obtained the expression in general orthogonal coordinates, for the potential energy of deformation of a strained isotropic solid.

17. By (56), we find

The stresses:

$$T_{\phi\phi} = \frac{dV}{d\dfrac{du}{d\phi}}$$

$$= a\left(\frac{du}{d\phi}+\frac{1}{2}\frac{\mathfrak{d}A}{A}\right)+b\left(\frac{dv}{d\chi}+\frac{1}{2}\frac{\mathfrak{d}B}{B}\right)+b\left(\frac{dw}{d\psi}+\frac{1}{2}\frac{\mathfrak{d}C}{C}\right),$$

$$T_{\chi\phi} = B^{\frac{1}{2}}A^{-\frac{1}{2}}\frac{dV}{d\dfrac{du}{d\chi}}$$

$$= \tfrac{1}{2}(a-b)\left\{A^{\frac{1}{2}}B^{-\frac{1}{2}}\frac{du}{d\chi}+B^{\frac{1}{2}}A^{-\frac{1}{2}}\frac{dv}{d\phi}\right\}$$

$$= T_{\phi\chi},\ \text{by symmetry;} \qquad (66)$$

with similar expressions for the other stresses.

To find the equations of internal equilibrium, we first observe that the values of $\frac{dV}{du}, \frac{dV}{dv}, \frac{dV}{dw}$ are obtained in a simple form by (57). It is clear from the way in which δA, δB, δC enter into (60) that

$$2\frac{dV}{du} = \frac{1}{A}\frac{dA}{d\phi}\frac{dV}{d\frac{du}{d\phi}} + \frac{1}{B}\frac{dB}{d\phi}\frac{dV}{d\frac{dv}{d\chi}} + \frac{1}{C}\frac{dC}{d\phi}\frac{dV}{d\frac{dw}{d\psi}}$$

$$= \frac{1}{A}\frac{dA}{d\phi}T_{\phi\phi} + \frac{1}{B}\frac{dB}{d\phi}T_{\chi\chi} + \frac{1}{C}\frac{dC}{d\phi}T_{\psi\psi}, \qquad (67)$$

with similar expressions for

$$\frac{dV}{dv} \text{ and } \frac{dV}{dw}.$$

Therefore we have the equations for this case — their internal equilibrium.

$$\frac{d}{d\phi}\{(BCA)^{\frac{1}{2}}T_{\phi\phi}\} + \frac{d}{d\chi}\{(CAA)^{\frac{1}{2}}T_{\chi\phi}\} + \frac{d}{d\psi}\{(ABA)^{\frac{1}{2}}T_{\psi\phi}\}$$
$$-\tfrac{1}{2}\left[\left(\frac{BC}{A}\right)^{\frac{1}{2}}\frac{dA}{d\phi}T_{\phi\phi} + \left(\frac{CA}{B}\right)^{\frac{1}{2}}\frac{dB}{d\phi}T_{\chi\chi}\right.$$
$$\left. + \left(\frac{AB}{C}\right)^{\frac{1}{2}}\frac{dC}{d\phi}T_{\psi\psi}\right] - (ABC)^{\frac{1}{2}}\frac{dW}{d\phi} = 0,$$

$$\frac{d}{d\phi}\{(BCB)^{\frac{1}{2}}T_{\phi\chi}\} + \frac{d}{d\chi}\{(CAB)^{\frac{1}{2}}T_{\chi\chi}\} + \frac{d}{d\psi}\{(ABB)^{\frac{1}{2}}T_{\psi\chi}\}$$
$$-\tfrac{1}{2}\left[\left(\frac{BC}{A}\right)^{\frac{1}{2}}\frac{dA}{d\chi}T_{\phi\phi} + \left(\frac{CA}{B}\right)^{\frac{1}{2}}\frac{dB}{d\chi}T_{\chi\chi}\right.$$
$$\left. + \left(\frac{AB}{C}\right)^{\frac{1}{2}}\frac{dC}{d\chi}T_{\psi\psi}\right] - (ABC)^{\frac{1}{2}}\frac{dW}{d\chi} = 0,$$

$$\frac{d}{d\phi}\{(BCC)^{\frac{1}{2}}T_{\phi\psi}\} + \frac{d}{d\chi}\{(CAC)^{\frac{1}{2}}T_{\chi\psi}\} + \frac{d}{d\psi}\{(ABC)^{\frac{1}{2}}T_{\psi\psi}\}$$
$$-\tfrac{1}{2}\left[\left(\frac{BC}{A}\right)^{\frac{1}{2}}\frac{dA}{d\psi}T_{\phi\phi} + \left(\frac{CA}{B}\right)^{\frac{1}{2}}\frac{dB}{d\psi}T_{\chi\chi}\right.$$
$$\left. + \left(\frac{AB}{C}\right)^{\frac{1}{2}}\frac{dC}{d\psi}T_{\psi\psi}\right] - (ABC)^{\frac{1}{2}}\frac{dW}{d\psi} = 0, \qquad (68)$$

in which the values of the stresses may be substituted from (66).

18. The equations now determined are those that apply to the most general cases of curvilinear coordinates. In any special system, the analysis might be very much cut down by dealing at once with the special values of A, B, C ... which will often be either constants of functions of only one variable.

Without here going back through the analysis, but merely by taking advantage of the general results already obtained, we may now apply the theory to some special cases.

Oblique Cartesian coordinates: 1st. The case of oblique Cartesian coordinates possesses some interest. The coefficients A, B, C ... are now constants, so that $\mathfrak{d}A$, $\mathfrak{d}B$ ... are all zero, and the analysis of §§ 12–15 applies with this important simplification. The potential energy V of the strain is a quadric function of the differential coefficients of u, v, w, with coefficients which are constants: it is given by (50) and (51). The stresses are given in terms of its differential coefficients by (56). The equations of internal equilibrium are given in terms of the stresses and the applied external forces *only* by equations (58); for $\frac{dV}{du}$, $\frac{dV}{dv}$, $\frac{dV}{dw}$ are now each zero, and the coefficients are all constants and can therefore be taken outside the sign of differentiation.

19. 2nd. As examples of orthogonal systems we may take the well-known cases of columnar and polar coordinates.

columnar: (α) Columnar coordinates r, θ, z. Here

$$ds^2 = dr^2 + r^2 d\theta^2 + dz^2,$$

and therefore $A = 1, \quad B = r^2, \quad C = 1$

and $\mathfrak{d}A = 0, \quad \mathfrak{d}B = 2ru, \quad \mathfrak{d}C = 0$, (69)

and the value of V is at once written down from (65).

It is however to be noticed that u, v, w are the increments of r, θ, z respectively, and that v is not here the increment of $rd\theta$ as is sometimes the case.

The values of the internal stresses may be written down by (66), viz.

$$T_{rr} = a\frac{du}{dr} + b\left(\frac{dv}{d\theta} + \frac{u}{r}\right) + b\frac{dw}{dz}, \qquad T_{r\theta} = \tfrac{1}{2}(a-b)\left(\frac{1}{r}\frac{du}{d\theta} + r\frac{dv}{dr}\right),$$

$$T_{\theta\theta} = b\frac{du}{dr} + a\left(\frac{dv}{d\theta} + \frac{u}{r}\right) + b\frac{dw}{dz}, \qquad T_{\theta z} = \tfrac{1}{2}(a-b)\left(r\frac{dv}{dz} + \frac{1}{r}\frac{dw}{d\theta}\right),$$

$$T_{zz} = b\frac{du}{dr} + b\left(\frac{dv}{d\theta} + \frac{u}{r}\right) + a\frac{dw}{dz}, \qquad T_{zr} = \tfrac{1}{2}(a-b)\left(\frac{du}{dz} + \frac{dw}{dr}\right). \quad (70)$$

The equations of internal equilibrium are, by (68),

$$\frac{d}{dr}(rT_{rr}) + \frac{d}{d\theta}(T_{r\theta}) + \frac{d}{dz}(rT_{rz}) - T_{\theta\theta} - r\frac{dW}{dr} = 0,$$

$$\frac{d}{dr}(r^2T_{r\theta}) + \frac{d}{d\theta}(rT_{\theta\theta}) + \frac{d}{dz}(r^2T_{z\theta}) - r\frac{dW}{d\theta} = 0,$$

$$\frac{d}{dr}(rT_{rz}) + \frac{d}{d\theta}(T_{\theta z}) + \frac{d}{dz}(rT_{zz}) - r\frac{dW}{dz} = 0. \quad (71)$$

polar. (β) Polar coordinates r, θ, ω. Here

$$ds^2 = dr^2 + r^2 d\theta^2 + r^2 \sin^2\theta \,.\, d\omega^2,$$

and therefore

$$\left.\begin{array}{lll} A = 1, & B = r^2, & C = r^2 \sin^2\theta, \\ \mathfrak{d}A = 0, & \mathfrak{d}B = 2ru, & \mathfrak{d}C = 2r\sin^2\theta u + 2r^2 \sin\theta\cos\theta v \end{array}\right\}, \quad (72)$$

wherein u, v, w are the displacements of r, θ, ω respectively.

By (65) the value of V may be at once written down.

The values of the internal stresses are, by (66),

$$T_{rr} = a\frac{du}{dr} + b\left(\frac{dv}{d\theta} + \frac{u}{r}\right) + b\left(\frac{dw}{d\omega} + \frac{u}{r} + \tan\theta v\right),$$

$$T_{r\theta} = \tfrac{1}{2}(a-b)\left(\frac{1}{r}\frac{du}{d\theta} + r\frac{dv}{dr}\right),$$

$$T_{\theta\theta} = b\frac{du}{dr} + a\left(\frac{dv}{d\theta} + \frac{u}{r}\right) + b\left(\frac{dw}{d\omega} + \frac{u}{r} + \tan\theta v\right),$$

$$T_{\theta\omega} = \tfrac{1}{2}(a-b)\left(\frac{1}{\sin\theta}\frac{dv}{d\omega} + \sin\theta\frac{dw}{d\theta}\right),$$

$$T_{\omega\omega} = b\frac{du}{dr} + b\left(\frac{dv}{d\theta} + \frac{u}{r}\right) + a\left(\frac{dw}{d\omega} + \frac{u}{r} + \tan\theta v\right),$$

$$T_{\omega r} = \tfrac{1}{2}(a-b)\left(r\sin\theta\frac{dw}{dr} + \frac{1}{r\sin\theta}\frac{du}{d\omega}\right). \quad (73)$$

The equations of internal equilibrium are, by (68),

$$\frac{d}{dr}(r^2\sin\theta T_{rr}) + \frac{d}{d\theta}(r\sin\theta T_{r\theta})$$
$$+ \frac{d}{d\omega}(rT_{r\omega}) - r\sin\theta T_{\theta\theta} - r\sin\theta T_{\omega\omega} - r^2\sin\theta\frac{dW}{dr} = 0,$$

$$\frac{d}{dr}(r^3\sin\theta T_{\theta r}) + \frac{d}{d\theta}(r^2\sin\theta T_{\theta\theta})$$
$$+ \frac{d}{d\omega}(r^2 T_{\theta\omega}) - r^2\cos\theta T_{\omega\omega} - r^2\sin\theta\frac{dW}{d\theta} = 0,$$

$$\frac{d}{dr}(r^3\sin^2\theta T_{\omega r}) + \frac{d}{d\theta}(r^2\sin^2\theta T_{\omega\theta})$$
$$+ \frac{d}{d\omega}(r^2\sin\theta T_{\omega\omega}) - r^2\sin\theta\frac{dW}{d\omega} = 0. \quad (74)$$

20. The equations thus obtained for these cases, (α) and (β), may be simplified considerably by performing the differentiations. They are then the same as were obtained originally by Lamé by transformation from Cartesian coordinates, and applied by him to the consideration of the elastic yielding of a sphere under given forces and other similar problems. They may also be obtained from first principles by a process involving vector differentiation (see Webb, *Messenger of Mathematics*, 1882, Vol. XI. pp. 146–155) which might be extended to the general theory of orthogonal coordinates, at the cost however of increasing the complication and the number of terms. **Historical.**

The general equations (66), (68) also lead to the equations which apply in the cases of ellipsoidal coordinates and elliptic columnar coordinates and such other orthogonal systems as have been used. These are the systems of equations which would naturally be applied to the case of elastic solids bounded by confocal quadric surfaces or by elliptic cylindrical surfaces respectively: but their degree of complication is such as to render them of not much practical value except in the more simple forms of strain, in which, for example, one or two of the three displacements are zero or very small. In these latter cases the general analysis is clearly much shortened by introducing the simplification at the beginning of the work*.

* These transformations may be completed by treating the remaining case, involving rotations or curls. The Stokes circulation theorem, in purely analytic form, is

$$\int(\lambda dp+\mu dq+\nu dr)=\iint\left\{\left(\frac{\partial\nu}{\partial q}-\frac{\partial\mu}{\partial r}\right)dq\,dr+\left(\frac{\partial\lambda}{\partial r}-\frac{\partial\nu}{\partial p}\right)dr\,dp+\left(\frac{\partial\mu}{\partial p}-\frac{\partial\lambda}{\partial q}\right)dp\,dq\right\}.$$

In it (p, q, r) may be interpreted as Cartesian coordinates of a point in an auxiliary surface and the first integral as taken round its edge. If now (p, q, r) are any orthogonal coordinates in our actual space, the theorem expresses that the circulation round a circuit is equal to the integral of twice the scalar product $|\omega\delta S|$ of differential rotation ω and surface-element δS, over any barrier surface abutting on the circuit. This circulation line integral, in *e.g.* polar coordinates, can be transformed directly by the formula: and it is easy to pick out from the resulting surface integral the expressions for the components of the differential rotation. When the coordinates are not orthogonal, the natural course is to express, as can readily be done, the energy of a strained rotational medium as $\int\frac{1}{2}\epsilon\omega^2 d\tau$, where $d\tau$ is the element of volume and ϵ the rotational elasticity, and proceed by the method of variation as above, interpreting as before by the invariant descriptive form of the result.

II

ON THE MOLECULAR THEORY OF GALVANIC POLARIZATION.

[*Philosophical Magazine*, Nov. 1885, pp. 422–435.]

1. It was first pointed out by C. F. Varley and Sir W. Thomson that the polarizing action of a galvanic cell may be explained by considering the cell to act as an electrical condenser of very large capacity. The mechanism of this action has since been examined in detail, especially by Helmholtz[1]. Polarization capacity of a cell.

In the polarization of a water-voltameter with platinum plates for electrodes, the action according to Clausius's well-known molecular theory consists in the transfer through the fluid of the temporarily dissociated hydrogen and oxygen constituents under the action of the electric force; so that in the course of time a layer of hydrogen particles with their positive charges accumulates in the immediate neighbourhood of the kathode plate, and the complementary layer of oxygen particles with their negative charges at the anode. Accumulated ions form a double sheet.

Each of these layers will form a sheet, with positive or negative charge, lying close to the metal plate. On the plate will therefore appear an equal and opposite charge by induction. There is thus a double electric layer formed at each electrode; the charged particles forming one side of it being prevented from coming up to and discharging themselves in contact with the metal, in obedience to the electrical attraction, by chemical forces of repulsion.

A double layer of this kind forms an actual condenser, whose capacity is inversely proportional to the distance between its faces. And Gauss's well-known theorem relating to magnetic shells shows, when applied to this case, that the effect of such a condenser is to cause a sudden rise or fall of potential in passing through it without producing any change in the distribution of the electric force in the neighbourhood. The notion of a condenser, therefore, gives a complete account of the principal feature of the galvanic polarization. The electromotive fall across a double electric sheet: as polarization:

Direct measures of the charge by Kohlrausch showed that on dividing this polarization-fall of potential equally between the anode

[1] See his *Wissenschaftliche Abhandlungen*, Vol. i. section "Galvanismus," and his Faraday Lecture, in the *Journal* of the Chemical Society for 1882 [reprinted posthumously in *Abhandl.* Vol. iii. pp. 52–87].

and kathode plates, the distance between the faces of the condenser comes out to be about the fifteen-millionth part of a millimetre; while more careful observations by Helmholtz on cells in which absorbed gases have been removed from the fluid, give the greater value of a ten-millionth of a millimetre. And Helmholtz makes out the very important fact that for all electromotive forces which do not exceed a certain moderate value, the capacity, and therefore the distance of the surface-layers, is very sensibly constant[1].

its dielectric interval, or capacity, is constant.

2. The most accurate and convenient method of observing the polarization at the common surface of two liquids is probably the electro-capillary method invented and applied by Lippmann.

When a surface of separation can persist between the fluids, the energy, reckoned as potential, of pairs of particles close to the surface must exceed that of the same particles when in the interior of their respective fluids. The difference may, as Gauss pointed out, be reckoned as surface-energy, and specified by its amount per unit area of surface. If T represent this amount, it follows, as is well known, that the forces which arise from it may be represented by a surface-tension equal to T across each unit of length, tending to contract the surface in all directions.

Influence on surface-tension:

Now, if the common surface is polarized with constant charges $+Q$ and $-Q$ on its two faces, there will exist an additional *electrical* energy, which is also reckoned by its amount per unit surface, and whose total value is

by method of energy:

$$E = \tfrac{1}{2}QV,$$

or, what is the same, $$E = \frac{1}{2}\frac{Q^2}{CS};$$

where S is the area of the surface and C is its electrical capacity per unit area.

The effect of this surface-energy will therefore, the system being conservative, be represented by a surface-tension T', where

when charges are insulated:

$$T' = \frac{dE}{dS}, \quad = -\frac{1}{2}\frac{Q^2}{CS^2}, \quad = -\frac{E}{S};$$

i.e. is equal to the electrical surface-energy per unit area of the surface with negative sign.

The effect of the galvanic polarization will therefore be to diminish the capillary surface-tension by this amount E/S.

In the actual case in which the polarization is maintained by a battery, the difference of potential V between the faces of the

[1] Faraday Lecture, p. 296; *Wissen. Abh.* i, p. 858.

condenser is what remains constant; and the system is no longer conservative, because the battery can be drawn upon; we have then

$$E = \tfrac{1}{2}CSV^2,$$

and

$$\frac{dE}{dS} = \tfrac{1}{2}CV^2, \quad = \frac{E}{S};$$

therefore, integrating, $dE = -T'dS;$

i.e. this force T' now acts so as to *increase* the total energy of electrification E, and is measured by its *rate of increase* per unit extension of area; for the work done by the contractile force T' in an extension of surface dS is equal to $-T'dS$, which is now the increment of the energy, but under the previous conditions was its decrement. **the opposite when potential is maintained.**

It follows that under these circumstances the battery is drawn upon for an amount of energy equal to *twice* that required to do the electrical work of extension, viz. the energy required to do this work together with the equal amount used up in increasing E, as has just been found. This is a particular case of a general theorem of Sir W. Thomson's. We have gone into the matter here to show the consistency of the propositions which make the capillary surface-tension equal to the rate of increase of the *ordinary* surface-energy per unit extension, while the electric surface-tension is equal to the rate of decrease of the electrical surface-energy per unit extension. **Results concordant.**

Once the surface-tension becomes negative, a free surface becomes unstable, and therefore practically impossible. We notice therefore that, as the polarization is made stronger and stronger, this state of affairs would finally supervene were not the polarization previously relieved by electrolytic separation of the charged layer. **Ultimate instability of surface.**

3. We have proved that the surface-tension is diminished by galvanic polarization by an amount equal to $\tfrac{1}{2}CV^2$, where C is the electric capacity of the surface per unit area.

The polarization-charge is therefore zero when T is a maximum, and the surface is then most curved. **Test of null polarization.**

The method that we have employed to determine the capillary effect of the polarization charge is different from that used by Lippmann and by Helmholtz. In their mode of procedure the variation of the energy of the system is expressed in terms of the variations of the surface-area and the surface-density, and it is claimed that this expression is an exact differential, *i.e.* that any series of operations whereby the area or density, or both, are changed so as finally to come back to the original values, will also bring back the energy of the system to its original value. This assumption seems to require justification when it is remembered how complex such a series of changes really is, and what a number of other variations **Indirect energy method.**

besides those of volume and density may enter into it. Helmholtz appeals to Lippmann's experiments on the influence of extension of the surface-film on its electrification and *vice versa*, and to his capillary engine, as pointing in a general way to the truth of the assumption.

In the method adopted above, we have proved the general theorem that the mechanical action of two layers of positive and negative electricity of equal amounts, spread over the two faces of a flexible sheet, may be represented by a negative surface-tension of amount numerically measured by the energy of the electrification per unit area. It follows, then, on this representation of the phenomenon, that no matter what other changes are taking place, the effect of the existing surface-charges is to diminish the surface-tension of the sheet by the amount just mentioned.

The case contemplated in the present application of this general proposition is that of a sheet of uniform thickness; but we can clearly extend the result to flexible condensers of variable thickness of dielectric, provided always that the thickness be small compared with any radius of curvature of the surface at the place considered.

Flexible thin condenser: In this case the mechanical effect on the condenser of a charge to potential V is to produce a negative surface-tension, numerically equal to $KV^2/8\pi t$, K being the dielectric constant and t the thickness of the sheet; this surface-tension varies from point to point of the sheet, and is at any place inversely proportional to its thickness.

the mechanical stress in it. This result may also be at once deduced from the expression for the stress transverse to the lines of force in the dielectric on Maxwell's well-known theory.

Lippmann electrometer. 4. Lippmann's original form of capillary electrometer consists of two mercury electrodes in contact with acidulated water. One of the electrodes is in an extremely fine capillary glass tube, so that the surface of contact is very small; and the other is of considerable area. It follows that when a battery is applied, all the polarization and consequent change in the surface-tension practically takes place at the fine electrode, as the corresponding charge at the other electrode is spread over so much greater area. The change in the capillary constant is measured by the column of mercury whose pressure is required to restore the meniscus to its former position.

Lippmann has given a series of observations with this instrument in his paper in the *Annales de Chimie*, Vol. v. p. 507, the electrolyte being water containing one-sixth part by volume of sulphuric acid. He finds that the maximum surface-tension is attained when the applied electromotive force is ·905 of a Daniell's cell. This value, therefore, corresponds to absence of polarization at the electrode meniscus. The following table, calculated from his results, gives δe the excess (positive or negative) of the electromotive force above this value, δp the excess of the pressure required to neutralize its effect over its value when $e = \cdot 905\ D$, and $(\delta e)^2/\delta p$, which is proportional to the capacity of the electrode, and therefore inversely to the distance between the two electrified layers,—on the supposition that the condensing arrangement remains analogous to an ordinary condenser, viz. consists of two infinitely thin layers separated by a dielectric sheet.

Tension in an unpolarized surface is the maximum.

δe	δp	$\frac{(\delta e)^2}{\delta p} \propto$
− ·89 *D*	343½	23·07
− ·88	337	23·0
− ·86	318½	23·07
− ·805	269½	23·4
− ·76	247½	23·3
− ·71	210½	24·0
− ·63	170	23·3
− ·54	123½	23·6
− ·405	70	23·26
− ·32	44½	23·01
− ·07	2	24·05
+ ·35	57½	21·3
+ ·43	79½	23 3
+ ·54	119½	24·4
+ ·81	230½	28·5
+ ·93	248½	34·8
+ ·98	254½	34·2
+1·10	264½	45·8

The pressure supported by the tension of the meniscus when $e = \cdot 905\ D$ was 1108½ millim. of mercury, which is therefore proportional to the maximum surface-tension of the film. The surface-tension, as ordinarily measured, corresponds to $e = 0$, and is therefore proportional to the pressure, which was then 750 millim.

The last column of this table is in good agreement with Helmholtz's result, that for electromotive forces from zero up to a limit of considerable magnitude the capacity of the condensing arrangement remains constant.

Constancy of capacity verified.

As δp is measured from a minimum value, it follows that in the immediate neighbourhood of that value $(\delta e)^2$ *must* vary as δp; so that the discrepancies for small values of δp in the third column are merely to be attributed to the special difficulty of the observations in that part of the series.

Taking the second line of the table to give the average value of this constant, we may calculate the thickness of the dielectric, supposed to have the properties of vacuum, and therefore to have unit specific inductive capacity, on the supposition that the arrangement acts as an ordinary condenser. When $e = 0$, Lippmann found by direct measurement that the surface-tension was $\cdot 304 \times 981$ C.G.S.

The polarization is then a single sheet of atoms:

units, which therefore corresponds to $p = 750$ millim. When $e = \cdot 024\,D$, we have $\delta e = -\cdot 88$, $\delta p = 337$; therefore the change of the surface-tension corresponding to δe is

$$\delta T \equiv \frac{337}{750} \times \cdot 304 \times 981.$$

This, as we have seen, is equal to the energy of the polarization charge per unit area. Now, taking a Daniell to be 1·1 volts, *i.e.* $1 \cdot 1 \times 10^8$ C.G.S. electromagnetic units, which is the same as

$$1 \cdot 1 \times 10^8 \div (2 \cdot 98 \times 10^{10}) \text{ C.G.S.}$$

electrostatic units of potential, we have

$$\delta T = \frac{(\delta e)^2}{8\pi \,.\, \text{thickness}};$$

therefore thickness of dielectric

$$= \frac{(\delta e)^2}{8\pi \,.\, \delta T},$$

$$= \left(\frac{1 \cdot 1 \times \cdot 88}{298}\right)^2 \div \left(8\pi \frac{337}{750} \times \cdot 304 \times 981\right),$$

$$= \frac{(\cdot 00325)^2}{3370}, \quad = \cdot 313 \times 10^{-10} \text{ metre.}$$

as indicated by atomic sizes: This calculation has already been made by Lippmann (*Comptes Rendus*, 1882, quoted in Thomson and Tait's *Natural Philosophy*, 2nd edn., Appendix, "On Size of Atoms"). It gives an estimate of a molecular distance, viz. that at which the two electrified layers are held by molecular chemical forces, which, notwithstanding the very rough suppositions on which it is founded, ought to be of the true order of magnitude; and Lippmann has pointed out that it agrees sufficiently with the estimates assigned by Sir W. Thomson and others from different considerations.

It is a satisfactory verification of the general notions involved in this discussion to find that, notwithstanding the large factors occurring in the calculation, such as the ratio of the electrostatic and electromagnetic units, it yet agrees so closely in order of magnitude with the result 1×10^{-10} metre, obtained by Helmholtz from actual measurement of the polarization capacity of platinum plates.

until the sheet becomes overcrowded. 5. But, on the principles we have been following, we may carry the analysis of the phenomenon still further. The polarization consists in the transfer of charged particles towards the electrode under the action of the electromotive force, and they are finally brought to equilibrium at a distance from the electrode, whose order of magnitude has just been determined. As these equally charged particles repel one another, they will tend to settle down in equidistant positions along the electrode surface. Instead therefore of two electrified sheets analogous to an ordinary condenser, we have really two sheets, one consisting of equidistant electrified particles, and the other of the charges brought opposite to them on the electrode by induction. Each charged particle and its corresponding induced charge will be brought by their mutual attraction so close together that this attraction will just be balanced by the chemical forces which hold them apart.

For polarizations of sufficiently small amount, the sidelong action of the neighbouring particles will be so small as to have no appreciable effect on the distance of any one particle from the electrode surface; because, in the first place, the distances of neighbouring particles must be at first large compared with the distance of two opposed charges, and, in the second place, the smaller forces exerted by these neighbouring particles must be resolved along the normal to the surface, in which direction they have no appreciable component. The radii of curvature of the surfaces are of course extremely great compared with the distance between opposed charges.

It follows that as the polarization is increased the number of charged particles over unit area of electrode increases in the same proportion, and these particles all come to rest at the same distance from the electrode surface, whatever be the amount of the polarization. And we can clearly expect this uniformity of distance to hold good until the neighbouring particles come within a distance of one another which is of the *same order* as the distance of a pair of the opposed charges.

Local effects rejected.

The pair of opposed surfaces which is thus arrived at, not uniformly charged, but each with a system of equal isolated point-charges arranged uniformly all over it, does not, of course, act as an ordinary condenser in the sense of producing a constant fall of potential in crossing it at all points, in positions whose distances from it are of the same order as the distance between neighbouring particles. But when we compare two points on opposite sides at distances from it great compared with this latter distance, it is immaterial whether the distribution is supposed to be in isolated points or uniformly spread over the surfaces. Therefore, as regards points not in the immediate molecular neighbourhood of the electrode, the effect of this polarization is still to produce simply a difference of potential on the two sides, which is just the same as if the charges were uniformly spread over the surfaces at the actual distance apart.

These considerations, then, give a reason for the fact which is brought out by the table given above, deduced from Lippmann's experiments with the capillary electrometer, and also independently by Helmholtz from direct measurement of the capacity of platinum electrodes in fluid with no dissolved gas (which would disturb the action); viz. that the polarization capacity is constant for all values of the applied electromotive force up to limits of considerable magnitude.

Capacity falls off by crowding,

6. In order to form an estimate of the nearness of the neighbouring molecules on a face of the double sheet when they begin to exert an influence on one another comparable with that exerted by the opposite charges, we must assign a limit to the interval of potentials within which the capacity remains constant*. The table in § 4 shows that we shall attain the correct order of magnitude by taking it to be, say, 1 volt in the case there considered.

We may now make the following calculation, bearing in mind that the sign = is to be interpreted as meaning that the quantities are of the same order of magnitude.

Let t be the thickness of dielectric layer;

d the distance between neighbouring atoms when their effective mutual action becomes comparable to that between opposed atoms (the important part of this action being that between any atom and the neighbours of its opposed charge);

t' the mean molecular distance in the electrolyte;

e the constant aggregate charge of a single atom or radical;

so that

$$t'^{-3}e = \text{the electro-chemical equivalent of 1 cubic centim. of water}$$

$$= \tfrac{1}{9} \cdot 10^5 \text{ coulombs, approximately,}$$

$$= \tfrac{1}{9} \cdot 10^5 \times 3 \cdot 10^9 \text{ electrostatic C.G.S. units,}$$

and $d^{-2}e$ = surface-density.

We have, then, for the condensing sheet,

$$d^{-2} = \frac{V}{4\pi t},$$

where $V = 1$ volt, $= 10^8 \div (3 \times 10^{10})$ electrostatic C.G.S. units.

Therefore $$\tfrac{1}{3} \cdot 10^{14} t'^3 d^{-2} = \frac{1}{3 \cdot 10^2 \cdot 4\pi t};$$

therefore $$t'^3 d^{-2} t = \tfrac{1}{12} \cdot 10^{-16}.$$

If now we write for t the value found above, $\cdot 3 \times 10^{-8}$, and put

* The recent investigations on capillary action of monomolecular films by Rayleigh, W. B. Hardy, Langmuir, N. K. Adam and many others, may be recalled.

t' and d equal to each other, both being molecular distances of the same kind, we obtain for either the value

giving a second estimate of atomic size:

$$\tfrac{4}{15} \times 10^{-8} \text{ centimetres,}$$

which is precisely of the same order as the value for molecular intervals obtained already from the other considerations.

On looking through this calculation it will be seen that quantities which we have designated as of the same order of magnitude do not differ nearly so much as in the ratio ten to one.

7. The two estimates of molecular distance which have thus been found on independent considerations connected with galvanic polarization therefore agree within very close limits; and they come very close to the third value determined by Helmholtz on the same theory of galvanic polarization, viz. 1×10^{-8} centim.; and they are also just below the superior limit assigned by Sir W. Thomson to molecular intervals from various considerations connected with different physical phenomena, viz. 10^{-8} centim., his inferior limit being $\tfrac{1}{20} \times 10^{-8}$ centim.

consistent:

Sir W. Thomson's different arguments lead to the following superior and inferior limits of the average distance of molecules from one another in solid and liquid substances:

résumé.

	Centim.	Centim.
Contact electricity	1×10^{-8}	$\tfrac{1}{4} \times 10^{-8}$
Surface-tension	$\tfrac{1}{2} \times 10^{-8}$	
Kinetic theory of gases		$\tfrac{1}{5} \times 10^{-8}$
Solids and liquids	$\tfrac{7}{10} \times 10^{-8}$	$\tfrac{1}{5} \times 10^{-8}$

to which we may now add

Helmholtz	1×10^{-8}	

while Lippmann's method places the mean at $\tfrac{3}{10} \times 10^{-8}$ centim.; and the other method here given places it at $\tfrac{4}{15} \times 10^{-8}$ centim., with as small limits of error as any of the methods given above.

8. The chief value of this discussion seems, however, to be not so much that it gives an estimate of molecular distance, but that its very close agreement with the other independent estimates derived from considerations connected with the same phenomenon of galvanic polarization is strong evidence of the substantial and ultimate truth of that representation of the phenomenon which has formed the basis of the discussion.

Argument confirms existence of independent ions.

This argument seems to derive very great weight from the wide variety and very different magnitudes of the physical constants employed in the three calculations, one depending on the direct measurement of the polarizing charge, another on the direct measurement of change in the capillary constant, and the third involving, in addition, the knowledge of the quantity of electricity required to decompose a gramme of water; while they all involve in different ways a constant of such large numerical magnitude as the ratio of the two electrical units of quantity.

Contact electrification.

9. The critical value of ·905 *D* in § 4 appears to have an important bearing on the much discussed question of contact electrification.

As was pointed out by Helmholtz, a discontinuous change of potential in crossing a surface can only be produced by the existence of an electrical double layer on that surface; so long as we look upon electrification or electric distribution as the cause of electrical phenomena, this is the only explanation open to us.

It has been seen that this electrification represents a distribution of purely surface-energy; and if its properties are to be investigated, it is to be expected that much light will be thrown upon them by their relations to other purely surface-distributions of energy, of which the best known is that leading to capillary phenomena.

Cause of natural voltaic potentials:

We are not required to explain the manner in which this double layer at the surface of contact of two dissimilar substances is brought about. We may illustrate it by the rather crude hypothesis that each molecule of an electrolyte consists of a positively charged cation radical and a negatively charged anion radical held together by electrical forces, but partly also by their forces of chemical affinity, so as to be analogous to a magnetic molecule with north and south poles; that along the surface of the electrode these molecules are all turned into the same direction (polarized) by reason of the greater chemical affinity of one of their constituents for the matter of the electrode; and that they thus form a double sheet analogous to a magnetic shell*. This illustration will at any rate show that it is possible to give an account of the matter which shall be in unison with the commonly received ideas of electrical and chemical action, without having to speculate on the deeper question of the relation of the material atom to its electrical charge†.

The electro-capillary observations of Lippmann quoted above, and the later ones of Koenig for various electrolytic fluids, show that, for one definite amount of polarization, each of these fluids in contact with mercury shows a maximum surface-tension. As we have seen

* This intrinsic double sheet also diminishes the capillary tension, as verified *infra*, but by an invariable amount.

† It appears (cf. R. K. Schofield, *Science Progress*, 1927, p. 396) that the idea of a partially orientated monomolecular layer, for voltaic potential differences, has been explored recently for organic liquids with large molecules of chain type by Frumkin, *Zeit. Phys. Chem.* 109 (1924) and 110 (1925).

that the existence of an electrical double layer on the surface must diminish the surface-tension, it follows that the critical value ·905 D for Lippmann's acidulated water is that difference of potential which, applied from without, just neutralizes the naturally existing double electrical layer on the surface. It would seem therefore that the natural contact-difference of potential between Lippmann's mercury and acidulated water is ·905 D, and that an absolute measure of a contact electromotive force has thus been obtained. their absolute values.

APPENDIX.

Results illustrated by direct calculation of surface-forces:

The result that the mechanical effect of the electrification on a charged condenser with thin uniform dielectric, whether flexible or not, is equivalent to a uniform negative surface-tension, has been derived in § 2 from the Principle of Energy without the use of any analysis.

The same result will of course follow from direct calculation of the mutual forces exerted by the charged elements of the surfaces on one another. As it forms a good example of the theory of surface-energy which Gauss has made the foundation of the doctrine of capillary action, in a case in which all the circumstances of the forces are known, and as it also illustrates some other points, the direct calculation is here appended.

for a plane sheet.

Consider, first, an infinite plane electrified surface, and imagine a straight line drawn across it. The mutual repulsion of the electrified parts on the two sides of this line will result in a tension tending to tear the parts of the surface asunder along the line, and whose intensity, measured across unit length, we can calculate as follows.

Imagine a unit of electricity situated at a point distant ξ from the line of division; the repulsion exerted on it by the other half of the electrification is easily expressed in polar coordinates, r, θ; θ being measured from the shortest distance to the line of separation.

It will, however, be more convenient to take this unit charge at a distance h from the plane, and to measure r, θ from its projection on the plane as pole. The repulsion exerted on it, resolved parallel to the plane, is

$$2\int_\xi^\infty dr\int_0^\phi d\theta\, r\,\frac{r}{(h^2+r^2)^{\frac{1}{2}}}\,\frac{\rho\cos\theta}{h^2+r^2},$$

where ρ is the surface-density of the electrification, and $\cos\phi = \xi/r$.

Therefore the repulsion

$$= 2\rho\int_\xi^\infty dr\,\frac{r\,(r^2-\xi^2)^{\frac{1}{2}}}{(r^2+h^2)^{\frac{3}{2}}}.$$

To integrate this, write

$$r^2 - \xi^2 = (r^2 + h^2)\, z^2;$$

therefore

$$r dr = \frac{r^2 + h^2}{1 - z^2} z dz;$$

and the integral

$$= \int_0^1 \frac{z^2}{1 - z^2} dz$$

$$= \left[\tfrac{1}{2} \log \frac{1 + z}{1 - z} - z\right]_{z=0}^{z=1}.$$

This quantity becomes infinite at the upper limit; so that for an infinite plane sheet the tearing-force due to the electrification would be infinite; a result which would also follow readily from simple consideration of the dimensions of the variable involved in the integral.

Suppose, however, we take a finite sheet bounded on the further side by the circular arc $r = r_0$; the repulsion now is

$$R \equiv 2\rho \log \{(r_0^2 + h^2)^{\frac{1}{2}} + (r_0^2 - \xi^2)^{\frac{1}{2}}\} - \rho \log (h^2 + \xi^2) - 2\rho \left(\frac{r_0^2 - \xi^2}{r_0^2 + h^2}\right)^{\frac{1}{2}}.$$

Convergence assured when sheet is double.

As we have seen, this expression increases indefinitely as r_0 increases. But if now, instead of a single electrical sheet, we had a double electrical layer with an intervening vacuum dielectric of thickness t, the repulsion exerted by it on the unit charge in the plane of the positive face will be equal to

$$t \frac{dR}{dh}.$$

But on differentiating the expression for R, it is obvious that the first and last terms give parts which become zero when r_0 is infinite; so that the repulsion of the infinite double layer on the unit charge is finite, and is equal to

$$\frac{2t\rho h}{h^2 + \xi^2}.$$

The repulsion exerted on a strip of unit breadth of density ρ and extending from $\xi = 0$ to $\xi = \infty$ therefore

$$= \rho \int_0^\infty \frac{2t\rho h}{h^2 + \xi^2} d\xi$$

$$= \pi \rho^2 t,$$

which is independent of h.

The repulsion exerted on a strip of the same *double* sheet is therefore

Result a surface-tension.

$$2\pi \rho^2 t,$$

i.e. it is the electrical energy of the distribution per unit area. And this quantity that we have thus calculated is clearly the surface-tension required.

It is clear also that the stress across any line drawn on the sheet is wholly tangential, and has no component normal to the sheet; so that this surface-tension is its complete specification.

The calculation just made has been only for the case of an infinite plane double sheet. For a single sheet the distant parts exert a finite effect; and we have seen that the stress increases indefinitely when the size of the sheet increases. But for a double sheet the parts very distant relatively to the thickness no longer contribute sensibly to the result, and the integrals converge. Thus, if the double sheet be of sensible but finite curvature, we may calculate the integrals either from the sheet itself or from the portion of the sheet which coincides sensibly with the tangent plane at the place considered, or from an infinite plane sheet coinciding with that tangent plane. This is on the assumption that the part of the sheet which coincides sensibly with the tangent plane is of large dimensions compared with the thickness of the dielectric, *i.e.* that the latter is small compared with any radius of curvature of the sheet.

Conditions for a stress satisfied, only for a double sheet.

The result obtained therefore holds for curved double sheets as well as for plane ones.

Now, if a curved sheet be under a uniform surface-tension T, it is well known that the stress experienced by any element δS of its surface is along the normal, and equal to

$$T\left(\frac{1}{R_1}+\frac{1}{R_2}\right)\delta S,$$

R_1, R_2 being the principal radii of curvature where δS is situated. When we apply this to the electrical double layer, we obtain the same result as comes from the direct expression, on Green's theory, of the force exerted by the electrical system on the two charged faces which belong to the element δS.

For a single sheet, not a uniform tension.

For a single curved electrified layer of finite dimensions, open or closed, the surface-tension is different at different points, and at the same point across different lines on the surface*; except in the case of an electrified spherical sheet, in which it is easily seen to be constant and equal to $-\pi\rho^2 a$, where a is the radius of the sphere.

* Referring to previous footnotes, it appears mathematically (*Roy. Soc. Proc.* 1921, as *infra*) that a substance with electric polarization uniform right up to its surface, thus with sharp transition, behaves superficially as if it had a uniform electric charge, and that would not give rise to a definite uniform surface-tension, representative of uniform local surface-energy, at all.

On the other hand, a unimolecular compact layer of bipolar electric molecules, with polar distance 10^{-8} cm. and diameters of the same order, would give a step of potential across the surface of the order of a volt, as in the text. The change produced by a dissolved substance on the surface step seems to be actually a small percentage of such amount.

A concise account of the remarkable recent experimental progress may be found in E. K. Rideal's book *Surface Chemistry* (1926).

12

ON THE FORM AND POSITION OF THE HOROPTER.

[*Proc. Camb. Phil. Soc.* Vol. VI. (1887) pp. 60–65.]

1. When the two eyes are kept converged upon a fixed point the images of another point will usually fall upon non-corresponding points of their retinas, and it will therefore be seen double. But there is a system of points forming a curved line in the field of view which are such that the images correspond, and they are therefore seen single*. The locus of points possessing this property was called the horopter, first by Aquilonius. It is of importance in the theory of stereoscopic vision as defining the neighbourhood in which the images formed by the two eyes are perfectly fused together; and accordingly its properties have been investigated by Helmholtz, Hering, and other physiologists.

Historical. Locus of exact single vision with two eyes.

The final investigation of Helmholtz was published in 1867, and presents the theory under an analytical form. Geometrically it is a case of the theory of linear congruences of the first order, and forms a good example of the Geometry of Rays† which has been explicitly introduced and applied chiefly by Plücker and his successors, since Helmholtz's papers were published.

It may be of advantage to give a brief account of the way in which the general results flow directly from the geometrical relations without recourse to symbolical reasoning. The direction of the horopter curve where it passes through the point of vision will then be investigated, as that would appear to be the most important matter for practical purposes on account of the complexity of the complete equations of the curve in the general case.

2. The eye being an optical instrument symmetrical round an axis, the aspect of external objects which is presented to its external nodal point is projected unchanged from its internal nodal point upon the retina.

Problem stated.

The nodal points of the eye are usually taken as coincident. This can always be done in an optical instrument, without sensible error, if their distance apart is a small fraction of the distances of the points

* Cf. the general problem of relativity-theory, which is how far the optical presentations to observers travelling at various speeds can be regarded as aspects of one world.

† Cf. a recent paper by H. Lamb for another aspect of the subject.

in the field of vision; and in the case of the eye they almost coincide. But the theory which follows applies equally well without this simplification, as is evident from the remark at the beginning of this section.

Corresponding points on the retinas are those which would coincide when they are superposed without perversion, *i.e.* right corresponds to right, and left to left.

3. In the first place, it is to be remarked that curved lines can be constructed to any extent which are seen by single vision with both eyes focussed on some given point. For draw any curve on one retina and the corresponding curve on the other; join these to the internal nodal points by cones; transfer the vertices of these cones to the external nodal points without introducing any rotation; the intersection of the two cones will then be a curve possessing the property in question.

That of curves seen as single is much wider.

For example, suppose the curves on the retinas are conics; the cones will be quadric cones; their curve of intersection may include a conic, in which case the remaining part of it is another conic. If therefore a conic curve in space be such that it is seen singly with both eyes, there is another conic in space which is seen by both eyes in coincidence with it, and undistinguishable from it so long as no accommodation of the eyes is allowed.

Ex. of two curves thus seen superposed.

4. But the simplest group of figures of this kind is that of the straight lines in space which are seen singly. They are constructed as before: draw two corresponding lines on the retinas, and join them by planes to the internal nodal points; the parallel planes drawn through the external nodal points intersect on a line of the group.

Case of straight lines:

Now all lines in space may be viewed as the intersections of planes, one passing through each of these nodal points; but the intersecting planes here considered are allied to each other by two lineo-linear (because projective) relations derived from the correspondence of their intersections with two given planes (the retinas).

Their lines of intersection therefore form a congruence of the first order, in the general sense that through any point in space one line of the congruence can be drawn.

a linear congruence of them:

This property also appears more directly in the following manner. Consider any point in space; mark its corresponding points on the two retinas; these will not usually themselves correspond, so mark the point on the other retina that corresponds to each; these points determine two lines on the retinas which correspond to each other; the line in space constructed from them passes through the point considered. Thus through any point in space passes usually one, and only one, line of the system.

the two foci along a ray:

Now it was shown by Sir W. R. Hamilton[1], that the well-known propositions of Malus respecting the shortest distances of a normal to a surface from the consecutive normals, and the locus of their points of intersection, can all be extended to the general case of a system of rays which satisfy two conditions. One of these propositions is that each ray of the system is intersected at two points on its length by a consecutive ray, and therefore that each ray of the system is a bitangent to a focal surface which is the locus of these points, and is the analogue of the surface of centres in the simpler case of the normals.

here the rays all meet same twisted cubic curve twice.

In the system under consideration, one, and only one, ray can usually be drawn through any point in space; therefore a point of intersection of two consecutive rays is a singular point, and must be the point of intersection of an infinite number of rays, forming a cone. The focal surface, which is the locus of such points, must therefore degenerate into a curve. Further, through any point in space can be drawn one line which meets this curve twice; therefore all conical projections of the curve possess only one double point; therefore the curve is a twisted cubic, and is the partial intersection of two quadric surfaces, but it may degenerate into two lines[2].

It may be here remarked that the congruence of the first order of Plücker is a more special form, corresponding to two linear relations between the six coordinates of the ray: for it the focal surface degenerates into two straight lines, each of which is met by all rays of the congruence.

Theorems of homologous ray-systems.

5. The results just proved, when modified by projection, give geometrical theorems of interest as being the extensions to space of three dimensions of well-known plane theories. Thus, if two homologous systems of planes pass through two given points, the lines of intersection of corresponding planes are the chords of a twisted cubic curve: if two homologous systems of rays pass through two fixed points, the points of intersection of those corresponding rays which intersect lie on a twisted cubic curve. Cases in which the cubic breaks up into a line and a plane conic are examined in detail by Helmholtz.

6. The curves and lines hitherto considered are seen singly because their images occupy corresponding lines on the retinas; but it is not necessary for this that the images of a definite point on one of them should occupy corresponding points. This would be a more difficult condition to fulfil, and only holds for the points of intersection of

[1] (Previously by Monge in 1781: see Prof. Cayley in *Proc. Lond. Math. Soc.* Vol. XIV. p. 139.)

[2] Salmon's *Solid Geometry*, Appendix III.: Salmon, *loc. cit.* § 364.

lines of the group, for then the image points are determined as lying on each of two lines. It is in fact clear that through such a point a singly infinite series of lines of the group can be drawn, forming a cone: for through its images on the retinas a singly infinite number of corresponding pairs of lines can be drawn, and each pair determines one of the group[1]. Thus the locus of points seen singly is the twisted cubic curve which is the nodal curve of the congruence.

This cubic nodal curve is the point-horopter:

It is worth while to point out that the pairs of points on one of the lines of the congruence whose images occupy corresponding positions on the retinas form a geometrical involution; that the line meets the horopter curve in its double points; and that its foci are therefore equidistant from these points.

relation to foci.

7. If a third condition is given between the parameters of the line, the locus of the line becomes a ruled surface; if this new condition is linear, *e.g.* if the images of the line on the retinas correspond under normal conditions to a horizontal line or to a vertical line in space, the surface is a ruled hyperboloid.

These two surfaces are the horizontal and vertical line horopters of Helmholtz, and the cubic curve is the point horopter.

Helmholtz's terms.

8. Inasmuch as the field of simultaneous vision of the eye is necessarily small, the most important part of the point horopter is that in the neighbourhood of the point on which the eyes are fixed. And it may be observed also, that the more oblique portions of the field of view do not practically come under the conditions of the geometrical problem, for the image on the retina cannot be considered as plane, except in its central portions.

The horopter may therefore be identified with the tangent to it at that point for most practical purposes. It seems therefore of importance to obtain the direction of this tangent line, especially as the results come out to be comparatively simple. At all points in the field of view in the neighbourhood of this line, the binocular vision will preserve completely the single character.

The direction defines the practical horopter:

The result obtained will apply immediately to all the simpler cases in which the horopter curve breaks up into a line and a conic, the only ones that have been completely discussed[2]. For the more general case the angle of rotation ϕ of this section is given only approximately by the law of Listing; which is an additional reason for the sufficiency of the result here obtained.

Let then a_1, a_2 be the distances of the point of vision from the

[1] Helmholtz, *Wissen. Abhandl.* II. p. 488; *Physiological Optics*, § 31.

[2] Helmholtz, *Physiological Optics*, § 31; Hermann, *Physiologie*, 7th ed. pp. 419–25.

external nodal points of the two eyes, and 2γ the angle between the axes of vision of the two eyes. Let the radius of the second retina corresponding to the radius of the first which is in the plane of the axes of vision make an angle ϕ with that plane, owing to the action of the converging muscles.

Consider two right cones of equal small angle α round the axes of vision of the two eyes as axes, and suppose the corresponding generating lines on them to be marked in such way as to identify them. These cones will intersect in a curve, and at two opposite points on this curve corresponding generators will meet one another, but at no other points. These two points are situated on the point horopter, and determine its direction in the neighbourhood of the point of vision.

determined: Taking for axis of x the bisector of the internal angle between the axes of vision drawn outwards, for axis of y the bisector of the external angle between the same lines drawn towards the first of them, and for axis of z the normal to the plane of the same lines, the coordinates of this point P of the horopter are easily determined. For, draw PM perpendicular to the plane of the axes of vision, and MN_1, MN_2 perpendicular to the axes of the two eyes. If θ denote the azimuth of P round the axis of the first cone, measured towards z from the positive direction of the axis of y, then $\theta + \phi$ will be its azimuth round the axis of the second cone. We therefore have the equations

$$PM = z = \alpha a_1 \sin\theta \qquad = \alpha a_2 \sin(\theta + \phi), \qquad (1)$$

$$MN_1 = x\sin\gamma + y\cos\gamma \qquad = \alpha a_1 \cos\theta, \qquad (2)$$

$$MN_2 = -x\sin\gamma + y\cos\gamma = \alpha a_2 \cos(\theta + \phi). \qquad (3)$$

Care has been taken to introduce only distances that are multiplied by the small quantity α, so that it is admissible to write a_1 and a_2 for the distances of P from the nodal points.

Thus
$$\frac{x\sin\gamma + y\cos\gamma}{a_1\cot\theta} = \frac{-x\sin\gamma + y\cos\gamma}{a_2(\cot\theta\cos\phi - \sin\phi)} = \frac{z}{a_1} \qquad (4)$$

where, by (1)
$$\cot\theta = \frac{-a_2\cos\phi + a_1}{a_2\sin\phi}.$$

Substituting this value in (4),

$$\frac{x\sin\gamma + y\cos\gamma}{-a_1a_2\cos\phi + a_1^2} = \frac{-x\sin\gamma + y\cos\gamma}{a_1a_2\cos\phi - a_2^2} = \frac{z}{a_1a_2\sin\phi};$$

therefore
$$\frac{x\sin\gamma}{a_1^2 + a_2^2 - 2a_1a_2\cos\phi} = \frac{y\cos\gamma}{a_1^2 - a_2^2} = \frac{z}{2a_1a_2\sin\phi}. \qquad (5)$$

the analytical result: These values of x, y, z are proportional to the direction cosines of the tangent to the horopter curve.

They show that when a_1 exceeds a_2, and ϕ is measured round in such direction that it is less than two right angles, the horopter lies in the quadrant of the field of vision which ranges from a_1 in the positive direction, and that it slopes away from the eyes in that quadrant. described.

Further approximation in equations (1), (2), (3) would lead easily to the determination of its curvature, but the results are too complicated to be of much interest.

9. When the point of vision is in the medial plane, $a_1 = a_2$, and the direction of the horopter is in the medial plane, inclined to the plane of vision at an angle whose tangent is The more important cases.

$$\sin \gamma \cot \tfrac{1}{2} \phi;$$

and it slopes away from the eyes in the upward direction.

When $\phi = 0$, the direction is in the plane of vision, and makes an angle with the axis of x whose tangent is

$$\frac{a_1 + a_2}{a_1 - a_2} \tan \gamma.$$

These are in agreement with known results: in the first case the arc in question is part of a line, in the second it is part of Müller's circle.

13

GENERAL THEORY OF DUPIN'S SPACE-EXTENSION OF THE FOCAL PROPERTIES OF CONIC SECTIONS.

[*Proc. London Math. Soc.* XVIII. (1887) pp. 363–369.]

THE following considerations may have value in reducing to direct geometrical principles a remarkable theory whose treatment has usually been of an analytical and somewhat indirect character.

I. 1. The locus of a point in a plane, the sum or difference of whose distances from two fixed points in the plane is constant, is a conic section.

The locus of a point in space which possesses this property is a quadric of revolution. Any plane section of such a surface is a conic section. It follows that there are for a conic section points outside its plane (the foci of the quadric) which possess this focal property. It is required to find their relations.

A space-locus of foci exists:

Let S_1, S_2 be the foci of the conic in its plane. The foci in space that are conjugate to S_1,—viz. that with S_1 give a focal relation,—belong to a certain locus which must include S_2. The locus will consist of two parts, say an elliptic part which corresponds to a constant sum of the distances, and a hyperbolic part which corresponds to a constant difference. Let P_1, P_2, ... be points on the first part, and Q_1, Q_2, ... points on the second part. We have then, if Π be any point on the conic,

$$S\Pi + P_1\Pi = \text{const.},$$

$$S\Pi + P_2\Pi = \text{const.},$$

$$S\Pi - Q_1\Pi = \text{const.},$$

$$S\Pi - Q_2\Pi = \text{const.}$$

Therefore

$$P_1\Pi - P_2\Pi = \text{const.},$$

$$P_1\Pi + Q_1\Pi = \text{const.}$$

any two points on it are conjugate foci:

Therefore any two points on the locus are conjugate foci; if they are on the same part of the locus their relation is hyperbolic, if on different parts it is elliptic.

locus is one curve:

It follows that the foci conjugate to *any* focus other than S also lie on the same locus.

Further, if P, Q are any two foci, and p_1, p_2 any two points on the conic,

$$Pp_1 \pm Qp_1 = \pm (Pp_2 \pm Qp_2),$$

the + or − sign being taken outside the bracket according as p_1, p_2 are on the same or different branches of the conic, and the other pairs of signs corresponding to each other; therefore

$$p_1P \mp p_2P = \mp (p_1Q \mp p_2Q)$$
$$= \text{constant}$$

for all points P, Q, ... on the locus.

Therefore the locus is a conic, of which any two points p_1, p_2 on the original conic are foci; the relation of these foci is hyperbolic if they lie on the same branch of the orginal conic, and elliptic if on different branches. **is another conic:**

The relations of the locus-conic and the original one are therefore perfectly reciprocal. Each passes through the foci of the other. They clearly lie on perpendicular planes by symmetry, and the vertices of each are therefore the ordinary foci of the other. This is Dupin's theorem. **located symmetrically.**

There is, of course, an imaginary conic related to the ordinary imaginary foci on the minor axis, as the one here discussed is related to the ordinary real foci.

2. It is easy to place this in immediate relation to the elementary geometrical theory of a conic defined as a section of a right cone.

For the focus S of a section is the point of contact with its plane of a sphere inscribed in the cone. Therefore, if the line ΠO joining any point Π on the section to the vertex O of the cone touch this sphere in π, we have $\Pi\pi = \Pi S$. Therefore, as πO is of constant length, **Relation to focal theory for a section of a right cone.**

$$O\Pi - S\Pi = \text{const.},$$

i.e. the vertex O of a right cone standing on the conic as base is a focus of the conic. Now, from consideration of the inscribed sphere, we have at once, as is well known,

$$A_1O \pm A_2O = A_1S \pm A_2S = \text{const.},$$

A_1, A_2 being the vertices of the conic; so that the locus of O is a conic in a perpendicular plane with the vertices of the given one for foci and its foci for vertices.

The discrimination of the cases then follows as before.

(*Camb. Math. Tripos*, May 18, 1887; quoted in *Educational Times Reprint*, 1887.)

II. 3. Those considerations may now be extended to bicircular quartics without much difficulty.

Extension to bicircular quartics: The locus of a point in a plane whose distances r_1, r_2, r_3 from three fixed points in the plane are connected by a linear relation

$$ar_1 + br_2 + cr_3 = 0,$$

is a bicircular quartic with these points for foci. All quartic curves which have the circular points at infinity for nodes are bicircular, and may be defined in this way (Salmon's *Higher Plane Curves*, § 296), but one of these foci may be imaginary.

The locus of a point in space which possesses this property is a species of cyclide. Now all quartic surfaces which have the circle at infinity for a double line are cyclides. Any section of a cyclide, therefore, has the circular points at infinity in its plane for double points, *i.e.* is a bicircular quartic. It follows that this bicircular quartic has also sets of foci (one set the foci of this surface) out of its plane; and we may proceed, as before, in the case of conics, to find their relations.

which also have a locus of foci:

Extended to curves on a sphere. 4. We notice that, as the inverse by reciprocal radii vectores of a plane bicircular quartic is a spherical curve which possesses the same focal property, and whose ordinary foci lie on the same sphere as itself, it follows that these more general spherical curves have also sets of foci outside their spherical surfaces.

Let us take P_1, P_2, P_3, a set of three foci of such a curve, and also Π_1, Π_2, Π_3, any three points on the curve. We have relations of the form

$$a_1P_1\Pi_1 + a_2P_2\Pi_1 + a_3P_3\Pi_1 = 0,$$
$$a_1P_1\Pi_2 + a_2P_2\Pi_2 + a_3P_3\Pi_2 = 0,$$
$$a_1P_1\Pi_3 + a_2P_2\Pi_3 + a_3P_3\Pi_3 = 0.$$

Eliminating a_1, a_2, a_3, we obtain

$$\begin{vmatrix} \Pi_1P_3, & \Pi_2P_3, & \Pi_3P_3 \\ \Pi_1P_2, & \Pi_2P_2, & \Pi_3P_2 \\ \Pi_1P_1, & \Pi_2P_1, & \Pi_3P_1 \end{vmatrix} = 0^{1}.$$

If we take P_1, P_2 fixed points, this is a homogeneous linear relation connecting the distances of a variable point P_3 from the fixed points Π_1, Π_2, Π_3. Therefore all the foci conjugate to P_1, P_2 lie on a cyclide which has *any three points* Π_1, Π_2, Π_3 on the original curve for foci. But two such cyclides cannot coincide; therefore they intersect in a curve which is the locus.

Locus a curve common to cyclides.

Further, it is easy to see, as above, that any two foci, each of them conjugate to P_1 and P_2, form a pair which is conjugate to P_1 or P_2 separately; and that any three foci, each of them conjugate to P_1

[1] This equation is, I find, given by Darboux in his memoir, *Sur une Classe remarquable...*, § 19, to establish the reciprocal relation between $P_1P_2P_3$ and $\Pi_1\Pi_2\Pi_3$, but it does not lead him to the focal curve.

and P_2, are conjugate to one another. Thus any three points on the locus of P_3 form a set of three conjugate foci; and the locus of P_3 includes the points P_1 and P_2.

The two curves are therefore reciprocally related, each being the locus of the foci of the other. Reciprocal relation as for conics.

Moreover, as the locus is a quartic, and it cuts the plane of the original quartic in three real points, it must also cut it in a fourth, *i.e.* the bicircular quartic has a fourth real focus in its plane. A fourth real coplanar focus.

It is not obvious from this method that the locus is itself a plane or spherical bicircular quartic; but, when we remember that the inverse of such a curve with respect to a focus is a Cartesian oval, for which the locus of foci must by symmetry lie in the perpendicular plane through its axis, which inverts again into a curve on a sphere, the result follows. The spheres on which the two conjugate quartics are situated intersect orthogonally. Locus a bicircular quartic.

The consideration that confines this argument to quartics on planes or spheres is the one at the beginning of Section II. The argument does not apply to space of more than three dimensions.

The curve which we have found as the locus of foci is a quartic on a sphere, and therefore the four ordinary foci in which it meets the plane or sphere of the original quartic lie on a circle. Any three of these four are conjugate to one another. Four coplanar foci concyclic.

It is to be noted that, as for conics, there are other focal curves, three in all, which intersect the plane or sphere of the quartic, each in four foci, which are however all imaginary when the above-mentioned four are real; otherwise two of them may be real. (Darboux, *loc. cit.*, or Casey.) Other focal curves, may be real:

The quartics here found turn out to be focal curves of a system of confocal cyclides, just as the conics treated of in (I) are the focal conics of a system of confocal quadrics. An investigation of their relations from that point of view, based upon the application of Ivory's theorem, will be found *Proc. Math. Soc.* Vol. XVI. p. 189, *supra*, p. 100. are the focal conics of a set of cyclides.

5. When one of the ordinary foci of the plane quartic is at an infinite distance, the circle on which the four foci lie becomes a right line, and the quartic is a Cartesian with equation of the form

$$ar_1 + br_2 = \text{const.},$$

Case of Cartesian oval.

where r_1, r_2 denote distances from two of the finite foci.

It is clear that, under the circumstances, the line at infinity is a part of the focal curve, and the remaining part is therefore a circular cubic. Conversely, the focal curve of a circular cubic is a Cartesian.

III. 6. These remarks have also an application to what is perhaps one of the most remarkable theories in Pure Geometry.

Extension to thread-loci for quadrics:

It has been shown by Chasles, in extension of Mr M. Roberts' theorem, that if a thread has its extremities attached to two fixed points A, B on a quadric (i), and is strained by a pencil-point P which is constrained to move on a confocal quadric (ii), so that the parts of the thread near A and B lie along (i), and the parts near P lie along (ii), then the point P traces out a line of curvature on (ii).

analogue of the umbilicar threads on a quadric:

Applying the reasoning of § 1, it follows that *the whole surface* of the quadric (i) may be similarly traced out by a thread whose extremities are attached to two points P, P' on a line of curvature of (ii); which at first sight seems a result of much too sweeping a character. It is, however, true in the same sense as that the whole surface of an ellipsoid may be traced out by a pencil constrained by a thread stretched over the surface between two opposite umbilics. The thread can be made to assume any azimuth on the surface, for there are an infinite number of geodesics which join two opposite umbilics, all necessarily of equal lengths.

Geodesic properties.

7. In the case in question, the parameter (pD) of each geodesic portion of the thread on either surface has a constant value determined by the property that the tangent line to that geodesic touches a fixed confocal quadric, so that as one extremity of the thread is fixed at A on the first surface, the only result of moving the pencil P is to wind more or less thread along the geodesic through A without disturbing the part already in contact with the surface. Instead of A we may therefore fix any other point A' on the geodesic; so that the restriction that A, B lie on the same line of curvature is withdrawn.

Again, the two geodesic arcs of PA, PB on the second quadric have the same parameter, which shows that either

(*a*) They are equally inclined to the lines of curvature through P, which leads to Chasles' theorem; or

(*b*) They are in one straight line, in which case through any point P can be drawn a continuous line geodesically connecting A and B; and since each of these geodesic connecting lines is the shortest distance between A and B, subject to the constraint of the surfaces, they are all of equal length, and the explanation suggested is verified.

Thread results verified.

We have therefore the two results:

(*a*) A pencil P guided by a string stretched slackly between A and B traces out a line of curvature on (ii); while

(*b*) A string stretched tightly between A and B can be made to slip over the surfaces with perfect freedom.

8. To obtain a clearer representation, we bear in mind that on an ellipsoid a line of curvature of given pD consists of two closed curves on opposite faces of the surface, and that a geodesic of that pD may be imagined as a thread wound on the surface as on a reel, so as to touch each branch in every revolution. Usually the thread will be endless, and will in a sense form the whole geodesic system of that parameter; in certain cases it will be re-entrant.

Geodesic of type of a reel of thread:

Whatever be the points A and B on (i), the portion of the thread on (ii) will be an arc of that geodesic system which touches the curve of intersection of (i) and (ii).

Let us then imagine a thread wrapped round and round the surface (ii), so as to touch alternately the two ovals of this curve of intersection. Taking any portion of it, the tangent lines along it touch the surface (i), and, when produced geodesically along it, form the corresponding arcs of the geodesic system of (i), *i.e.* those arcs which run in the same direction, in the sense that we can proceed by continuous change from the one to the other through the intermediate arcs.

case of intersecting confocals.

If, therefore, we take any two points A and B on (i), the corresponding geodesic arcs drawn from them, and carried geodesically on to (ii), finally run into one another and are continuous.

Again, if we take any point on (ii) and follow the two geodesic arcs of the system which cross each other there, they will run on to (i) so as to be non-corresponding geodesic arcs on it.

If, therefore, we take any two points A and B on (i), the non-corresponding geodesic arcs drawn from them and carried geodesically on to (ii) will meet in a point P on (ii), such that the sum or difference of AP and BP is constant, according to the relative position of A and B considered fixed, in a way that is most easily illustrated by the analogy of Roberts' theorem for a single quadric.

9. If the quadric (i) is taken to be a focal conic, and the points A and B are taken on its boundary, all lines through A and B are tangential to (i), so that the condition of geodesic contact with (i) imposes no restriction.

Case of quadric and focal conic.

Thus taking, for definiteness, (ii) to be an ellipsoid, we see that a thread drawn tight over it between two points A and B on opposite quadrants of the focal hyperbola is free to slip all over it; and also that the locus of the space-foci, in the geodesic sense, of the lines of curvature is the focal hyperbola, which, of course, contains the umbilics.

Space-foci extended to geodesic geometry on quadric.

14

ON DIRECT PRINCIPLES IN THE THEORY OF PARTIAL DIFFERENTIAL EQUATIONS.

[*Proc. Royal Society*, Vol. 43 (1887), p. 176.]

If an equation involving total differentials of any number of variables can be expressed in the form

$$\delta u + \sigma \delta v = 0,$$

where u, v are any functions of the variables, then the only *single* integral algebraic relations that are consistent with it are included under the form

$$u = \phi(v).$$

When the form of σ is assigned, the functional symbol ϕ is to be chosen, if possible, so as to agree with that form; and if this is not possible, then the equation has no integral expressible as a *single* relation. This statement holds because the equation expresses a particular case of the proposition that if $\delta u = 0$, then $\delta v = 0$, and conversely, *i.e.* that u remains constant (does not vary) when v is constant, and only then, whatever be the particular values assigned to the variables: but this is simply the definition of the algebraic idea of functionality.

The direct implications of functionality:

If, however, σ involve differentials, the alternative $\delta u = 0$ when $\sigma = 0$ may lead to integrals of a new type.

In the same way, an equation of the form

$$\delta u + \sigma_1 \delta v + \sigma_2 \delta w = 0,$$

must have all its *single* integrals included under the form

$$u = \phi(v, w),$$

where the form of ϕ is to be chosen so as to agree with the expressions for σ_1, σ_2, when these are assigned.

When no *single* integral exists*, equations of this type may be

* Cf. Clausius, in analytical elucidation of Carnot's thermal principle, *Pogg. Ann.* (1854), feeling his way towards Entropy, also in a mathematical Introduction (1858), both in Hirst's translation (1867) of the memoirs: also modern inferences from dimensional theory in physics, after Rayleigh.

But the weightiest illustration of the physical scope of these considerations, algebraically trivial though they may appear to be, is in Willard Gibbs' generalized formulation (1876–8) of thermodynamics into equations of total differentials, which opened up a new science of chemical physics. *Cf.* explanations *infra*.

satisfied by two simultaneous integral relations, one of which may be arbitrarily assumed, as orginally pointed out by Monge. This kind of exception, however, need not trouble when partial differential coefficients are concerned; for these implicitly assume the existence of a single relation connecting the dependent variable with the independent ones. regarded objectively.

Traces of this idea are to be found throughout the writings of Boole —and of Monge long previously. It can be applied to the non-analytical exposition of the differential criteria of algebraic functionality given by Jacobi, and to the discussion in a similar manner of the theory of partial differential equations of the first and second order, particularly those named after Lagrange, Monge, and Ampère.

15

THE TRANSFORMATION OF MULTIPLE SURFACE INTEGRALS INTO MULTIPLE LINE INTEGRALS.

[*Messenger of Mathematics*, June 1887, pp. 23–30.]

AN integral extended throughout a volume can in various ways be expressed as a surface integral over its boundary. Many elegant theorems of this kind have been given by Gauss[1].

I. But in order that the integral over a surface, of a vector function, meaning thereby the integral of its normal component over the surface, may be expressible by a line integral over its contour, the function must satisfy a certain condition.

In fact the integrals over any two surfaces abutting on the same contour would then be equal, and the two together would form a closed surface, such that the integral taken in the same sense over the whole of it would be equal to zero. Now if R denote the vector, X, Y, Z its components parallel to the axes, and $R\cos\epsilon$ its normal component,

Surface integrals determined by the edge alone only for a stream vector.

$$\iint R\cos\epsilon\, dS = \iiint\left(\frac{dX}{dx}+\frac{dY}{dy}+\frac{dZ}{dz}\right)d\,\text{vol}. \tag{1}$$

Therefore if this condition of zero integral is to hold for all closed surfaces, we must have identically, throughout the space considered,

$$\frac{dX}{dx}+\frac{dY}{dy}+\frac{dZ}{dz}=0, \tag{2}$$

as the condition required.

The truth of the formula (1) requires that the vector should not become discontinuous or its differential coefficients infinite anywhere in the space in question; for if that were not provided for, the integration of its right-hand side might introduce other terms: cf. Maxwell's *Electricity*, Chap. I.

Singularities.

The proposition must therefore be applied in its simple form, only when the region in question does not contain places where the vector is discontinuous or its differential coefficients infinite.

If X, Y, Z are the components of a flux R, the condition (2) is the well-known "Equation of Continuity," which secures that the flux is that of an incompressible substance. Thus in continuous motion

[1] *Theoria Attractionis...*, *Comm. Soc. Gotting.* II. 1813, or *Werke*, Band V.

of incompressible fluids the flux through any ideal aperture is expressible as a line integral round its contour; the reason for which is obvious.

To determine the form of the integral relation in question, we may first take the case of a small plane surface.

Then $$\int(\alpha dx+\beta dy)=\iint dx\,dy\left(\frac{d\beta}{dx}-\frac{d\alpha}{dy}\right) \qquad (3)$$

by immediate integration, the rule of signs being that the line integral proceeds round the contour in the direction from x to y in the first quadrant.

In the same way, for areas in the planes of yz and zx, we have

$$\int(\beta dy+\gamma dz)=\iint dy dz\left(\frac{d\gamma}{dy}-\frac{d\beta}{dz}\right), \qquad (4)$$

$$\int(\gamma dz+\alpha dx)=\iint dz dx\left(\frac{d\alpha}{dz}-\frac{d\gamma}{dx}\right). \qquad (5)$$

By what precedes, expressions to be integrated on the right-hand are to be taken as the components normal to the coordinate planes of the vector function R; and we remark that they satisfy (2).

We are entitled therefore to assert for any small plane contour, that

$$\int(\alpha dx+\beta dy+\gamma dz)=\iint dS\,.\,R\cos\epsilon, \qquad (6)$$

where the components of R are

$$X=\frac{d\gamma}{dy}-\frac{d\beta}{dz},\quad Y=\frac{d\alpha}{dz}-\frac{d\gamma}{dx},\quad Z=\frac{d\beta}{dx}-\frac{d\alpha}{dy}. \qquad (7)$$ Curl of a vector.

And by adding the results for the series of infinitesimal plane circuits into which any finite circuit may be divided, we see that the theorem, due originally to Stokes, holds for a contour of any form. Stokes' theorem of surface and edge integrals, with proper screw rule:

The rule of signs now is that the direction of integration round the contour corresponds to that of a right-handed screw along the direction of R. For this rule is in agreement with the constituent formulae (3), (4), (5), when the system of axes forms a right-handed system, as it always should in such directional investigations, viz. when the directions of rotation in the positive quadrant from y to z, z to x, and x to y correspond to right-handed screws along the positive directions of the other axes.

When there is discontinuity in X, Y, Z such that for any region included in the space between two surfaces (A) and (B) abutting on the same contour, extended,

$$\iiint\left(\frac{dX}{dx}+\frac{dY}{dy}+\frac{dZ}{dz}\right)d\,\text{vol}. \qquad (8)$$

is not zero, the surface integrals over the two surfaces are no longer equal, but differ by the value of the expression (8).

The explanation of this discontinuity is most clearly seen from the representation as a flux. The value of (8) extended over any region then denotes an emission of fluid in that region at a rate per second given by that expression. If we consider the simplest case of fluid welling out at a single point, then as the surface (A) is gradually altered into (B), when it passes over that point the direction of the velocity due to the source there situated is changed with respect to the surface, and a finite alteration is thereby produced at that stage in the value of the surface integral.

Thus if
$$R = \frac{1}{r^2},$$

modification when the surface crosses sources:

so that
$$X = \frac{x}{r^3}, \quad Y = \frac{y}{r^3}, \quad Z = \frac{z}{r^3},$$

the condition (2) is satisfied, and it is not difficult to verify that

$$\iint R \cos \epsilon \, dS = \int \frac{x\,(y\,dz - z\,dy)}{r\,(y^2 + z^2)} - A, \tag{9}$$

where A is equal to zero or 4π according to the side of the origin on which the surface bounded by the given contour lies. We may obtain other forms for the theorem by adding to the right-hand side of (9) the line integral of any exact differential, which will add nothing when taken round the circuit.

This formula (9) expresses as a line integral the flux due to a single source of fluid at the origin of coordinates, or the induction due to a single attracting particle situated there; and from it any more general case might be deduced by summation. But development in this direction simply leads to the well-known theory of the vector potential in Electrodynamics.

leading to the vector potential of a distribution.

Integrals round two contours.

II. There is another class of integrals related to Mathematical Physics in which the integrals are extended over two contours. For instance, a uniformly luminous open surface emits a quantity of radiation through a given aperture which depends only on the contours of the surface and aperture, care being taken that all parts of one contour are visible from all parts of the other. Again, the mutual energy of two closed electric currents may be expressed either as an integral extended over their circuits, or as a surface integral derived from the equivalent magnetic shells.

Example of illumination: of mutual energy of two currents.

We propose now to investigate the general forms of such relations.

If a line integral round a contour is to be expressible as a surface integral over a sheet bounded by the contour, by means of (6), it must involve the elements of the contour linearly. Therefore the most general type of double line integral in question must involve both contours linearly. The function to be integrated can only involve the distance between two elements of the contours and the mutual inclinations of the distance and these elements. If r denote the distance of the elements ds, ds', and ϑ, ϑ' the angles it makes with these elements, and ϵ the angle between the directions of the elements, the most general forms therefore involve only

Form of intrinsic double contour integral,

$$\iint ds ds' f(r) \cos \epsilon, \qquad (10)$$

reduces to two types.

and
$$\iint ds ds' \phi(r) \cos \vartheta \cos \vartheta'. \qquad (11)$$

Of these the latter [not] clearly vanishes when either circuit is complete.

One type deferred.

The former

$$= \int ds' \int f(r) (l'dx + m'dy + n'dz),$$

where l', m', n' are the direction cosines of ds';

$$= \int ds' \iint dS (X\lambda + Y\mu + Z\nu), \text{ by (6)},$$

where λ, μ, ν are the direction cosines of the normal to dS, and

$$X = f'(r) \frac{yn' - zm'}{r},$$

$$Y = f'(r) \frac{zl' - xn'}{r},$$

$$Z = f'(r) \frac{xm' - yl'}{r},$$

x, y, z being the components of r, the origin being taken temporarily at the position of ds'.

Thus changing the order of integration, and transferring the origin to the position of dS, so that we write $-x'$, $-y'$, $-z'$ for x, y, z, we have

$$\iint dS \int f'(r) \left(\frac{y'\nu - z'\mu}{r} dx' + \frac{z'\lambda - x'\nu}{r} dy' + \frac{x'\mu - y'\lambda}{r} dz'\right)$$

$$= \iint dS \int (\alpha' dx' + \beta' dy' + \gamma' dz'), \text{ say},$$

$$= \iint dS \iint dS' (X'\lambda' + Y'\mu' + z'\nu'),$$

where $$X' = \frac{d\gamma'}{dy'} - \frac{d\beta'}{dz'}, \text{ by (7)},$$

$$= \frac{d}{dr}\left\{\frac{1}{r}f'(r)\right\}\left[\frac{y'}{r}(x'\mu - y'\lambda) - \frac{z'}{r}(z'\lambda - x'\nu) + \ldots + \ldots\right]$$

$$+ \frac{1}{r}f'(r)\left[-2\lambda - \ldots - \ldots\right];$$

so that $$X'\lambda' + Y'\mu' + Z'\nu'$$

$$= \frac{1}{r}\frac{d}{dr}\left\{\frac{1}{r}f'(r)\right\}\left[-(y'^2 + z'^2)\lambda\lambda' + x'\lambda(y'\mu' + z'\nu') + \ldots + \ldots\right]$$

$$- \frac{2}{r}f'(r)\left[\lambda\lambda' + \mu\mu' + \nu\nu'\right]$$

$$= r\frac{d}{dr}\left\{\frac{1}{r}f'(r)\right\}\left[-\cos\eta + \cos\theta\cos\theta'\right] - \frac{2}{r}f'(r)\cos\eta$$

$$= -\left\{f''(r) + \frac{1}{r}f'(r)\right\}\cos\eta + \left\{f''(r) - \frac{1}{r}f'(r)\right\}\cos\theta\cos\theta',$$

$$= -\frac{1}{r}\frac{d}{dr}\{rf'(r)\}\cos\eta + r\frac{d}{dr}\left\{\frac{1}{r}f'(r)\right\}\cos\theta\cos\theta',$$

where η is the angle between the normals to dS, dS' each drawn towards the positive side of the surface, and θ, θ' are the angles between these normals and r, whose direction is the same in both cases.

First general type transformed to double surface integral.

Therefore, finally,

$$\iint dS \iint dS'\left[r\frac{d}{dr}\left\{\frac{1}{r}f'(r)\right\}\cos\theta\cos\theta' - \frac{1}{r}\frac{d}{dr}\{rf'(r)\}\cos\eta\right]$$

$$= \int ds \int ds' f(r)\cos\epsilon, \qquad (12)$$

where the positive side of the surface is determined by the rule that a right-handed screw in that direction corresponds to the direction of the line integral round it.

We have proved that this result is the most general possible of its class.

Special cases:

Particular cases may be noted as follows:

(i) Make the two circuits coincide.

(ii) Make the two open surfaces coincide, and we express the double surface integral by a double line integral round the contour. To avoid infinities, $f'(r)$ must not contain powers of r lower than the inverse first.

(iii) Make the surfaces plane, so that η is constant.

(iv) Make $$f'(r) = \frac{C}{r};$$

theorem of illumination:

then $$\iint dS \iint dS' \frac{\cos\theta\cos\theta'}{r^2} = -\tfrac{1}{2}\int ds \int ds' \log r \cos\epsilon. \qquad (13)$$

The left-hand side is the expression for the illumination from S that is intercepted by S' when the brightness of S is unity; and it follows from elementary optical principles that this quantity must be expressible as a line integral round the contours of S and S'.

When S and S' coincide, we have

case of a sheet with plane edge.

$$2\pi S = -\tfrac{1}{2}\int ds \int ds \log r \cos \epsilon, \qquad (14)$$

true only when S is plane; for when S is not plane the real optical interpretation fails, the parts of the surface not being in full view of each other.

(v) Make $\quad f(r) = \dfrac{C}{r}$, so that $f'(r) = -\dfrac{C}{r^2}$;

then
$$\iint dS \iint dS' \frac{\cos\eta - 3\cos\theta\cos\theta'}{r^3} = -\int ds \int ds' \frac{\cos\epsilon}{r}, \qquad (15)$$

Energy of two linear currents.

which is Neumann's well-known expression for the mutual energy of two simple magnetic shells, or of two linear electric currents.

(vi) Make $\quad f'(r) = Cr;$

then
$$\int dS \int dS' \cos\eta = -\tfrac{1}{4}\int ds \int ds' \, r^2 \cos\epsilon, \qquad (16)$$

thus giving a double line integral form for $\int \Pi' dS$, where Π' denotes the area of the projection of S' on the tangent plane at dS. It was clear *à priori* that such a form must exist, for this integral depends only on S and the contours of S', while the other form $\int \Pi dS'$ shows that it depends only on the contour of S; thus the form of the function of r that multiplies cos in ϵ is all that remained *à priori* to be determined, and that might have been found from the simplest particular case.

Integral of projected area:

When one surface S is plane, we have

$$S\Pi' = -\tfrac{1}{4}\int ds \int ds' \, r^2 \cos\epsilon, \qquad (17)$$

special case:

where Π' denotes the projection of S' on the plane of S.

Where S, S' coincide in one plane, we have

$$S^2 = -\tfrac{1}{4}\int ds \int ds \, r^2 \cos\epsilon. \qquad (18)$$

And comparing this with (14) we deduce

$$(4\pi S)^2 = \left\{\int ds \int ds \log r \cos\epsilon\right\}^2 = -4\pi^2 \int ds \int ds \, r^2 \cos\epsilon \qquad (19)$$

theorems for a plane sheet

for any plane circuit; a striking result.

The theorems just given may be verified by direct integration when the surfaces are plane circles, and (18) without much difficulty for

Evaluations of definite integrals.

the general surface; by applying them to surfaces bounded by other curves, we obtain evaluations of a crop of definite integrals of somewhat unusual form.

Formulation of triple contour integral:

III. If elements of three surfaces enter into a triple integral, the components of the elements of their three contours must enter, each linearly, into the corresponding line integral. The most general form of such line integral, independent of special coordinate systems, which gives a finite value when taken over complete circuits, is

$$\iiint \phi\,(r, r', r'') \begin{vmatrix} dx\,, & dy\,, & dz \\ dx', & dy', & dz' \\ dx'', & dy'', & dz'' \end{vmatrix},$$

where r, r', r'' are the mutual distances of the three elements of contour; and the determinant is equal to $3\Theta ds\,ds' ds''$, where

$$\Theta^2 = \sin\tfrac{1}{2}(a+b+c)\sin\tfrac{1}{2}(b+c-a)\sin\tfrac{1}{2}(c+a-b)\sin\tfrac{1}{2}(a+b-c),$$

a, b, c being the sides of the spherical triangle determined by the directions of ds, ds', ds''.

its transformation.

The integral may therefore by application of the method of II be expressed as a symmetrical triple surface integral; the general formulae are long, but the degenerate cases would probably be interesting.

Finally, there does not seem to be any reason why the considerations on which these theorems are founded should be restricted to the three dimensions x, y, z of ordinary space; but the more general results would probably be of only analytical interest.

16

ELECTROMAGNETIC AND OTHER IMAGES IN SPHERES AND PLANES.

[*Quarterly Journal of Pure and Applied Mathematics*, 1888, pp. 1–8.]

THE object here proposed is a brief discussion in their mutual relations of some problems in Mathematical Physics in which the idea of images is of use, either as facilitating calculation, or as leading to a clear representation of phenomena without the need of calculation. There are cases in which such questions throw light on one another, and lead to simplifications which, though obvious enough, do not seem to be much noticed, perhaps because considerations of this kind are not usually wanted in connection with special investigations.

General utility of image systems.

The results of the analysis with which we start may be considered to form a special case of the general analytical method given by Mr R. A. Herman in the *Quarterly Journal*, Vol. XXII (pp. 248 *seqq.*).

We begin with the most general problem that involves only a single point source, viz. steady conduction in a medium of permeability κ_2 which contains a sphere of radius a and permeability κ_1, when a source of strength e is situated in the medium at Q, at a distance c from the centre of the sphere.

Sphere in the field of flow of a point source:

The conditions of the problem are satisfied by the following values for the potential function outside and inside the sphere, provided $r < c$;

$$V_0 = \frac{e}{4\pi\kappa_2}\left(\frac{1}{c} + \frac{P_1 r}{c^2} + \dots \frac{P_n r^n}{c^{n+1}} + \dots\right) + \frac{A_0 P_0}{r} + \frac{A_1 a P_1}{r^2} + \dots \frac{A_n a^n P_n}{r^{n+1}} + \dots, \qquad (1)$$

$$V_i = \frac{B_0}{a} + \frac{B_1 P_1 r}{a^2} + \dots \frac{B_n P_n r^n}{a^{n+1}} + \dots. \qquad (2)$$

where P_n is the zonal harmonic of degree n.

The continuity of potential and flux at the surface give the equations between the coefficients

$$\frac{e}{4\pi\kappa_2}\frac{a^n}{c^{n+1}} + \frac{A_n}{a} = \frac{B_n}{a}$$

$$\frac{e\kappa_2}{4\pi\kappa_2}\frac{na^{n-1}}{c^{n+1}} - (n+1)\,\kappa_2\frac{A_n}{a^2} = n\kappa_1\frac{B_n}{a^2}. \qquad (3)$$

Therefore $$A_n = \frac{n(\kappa_2 - \kappa_1)}{\kappa_2 + n(\kappa_2 + \kappa_1)} \frac{e}{4\pi\kappa_2} \left(\frac{a}{c}\right)^{n+1},$$

$$B_n = \frac{(2n+1)\kappa_2}{\kappa_2 + n(\kappa_2 + \kappa_1)} \frac{e}{4\pi\kappa_2} \left(\frac{a}{c}\right)^{n+1}.$$

Therefore, writing $c' = a^2/c$, if P be the point at which the potential is estimated, and Q' the image of Q in the sphere,

solution for the disturbed flow:

$$V_0 = \frac{e}{4\pi\kappa_2}\left(\frac{1}{c} + \frac{P_1 r}{c^2} + \dots \frac{P_n r^n}{c^{n+1}} + \dots\right)$$
$$+ \frac{e(\kappa_2 - \kappa_1)}{4\pi\kappa_2}\frac{a}{c}\left\{\frac{c'}{\kappa_2 + (\kappa_2 + \kappa_1)}\frac{P_1}{r^2} + \dots + \frac{nc'^n}{\kappa_2 + n(\kappa_2 + \kappa_1)}\frac{P_n}{r^{n+1}} + \dots\right\}$$
$$= \frac{e}{4\pi\kappa_2}\frac{1}{QP} + \frac{e}{4\pi\kappa_2}\frac{\kappa_2 - \kappa_1}{\kappa_2 + \kappa_1}\frac{a}{c}\left(\frac{1}{Q'P} - \frac{1}{r}\right)$$
$$- \frac{e}{4\pi}\frac{\kappa_2 - \kappa_1}{\kappa_2 + \kappa_1}\frac{a}{c}\left\{\frac{c'}{\kappa_2 + (\kappa_2 + \kappa_1)}\frac{P_1}{r^2} + \dots + \frac{c'^n}{\kappa_2 + n(\kappa_2 + \kappa_1)}\frac{P_n}{r^{n+1}} + \dots\right\}. \quad (4)$$

The last expression in the value of V may be identified with the potential due to a line distribution along the axis, as follows. If the line density at a distance ξ from the centre be ξ^p, the potential due to it is

$$\int_0 \frac{\xi^p d\xi}{(r^2 + \xi^2 - 2r\xi\cos\theta)^{\frac{1}{2}}}$$
$$= \int_0 \xi^p \left(\frac{1}{r} + \frac{\xi P_1}{r^2} + \dots \frac{\xi^n P_n}{r^{n+1}} + \dots\right) d\xi;$$

and if we make c' the upper limit, we obtain

$$\frac{c'^{p+1}}{p+1}\frac{1}{r} + \frac{c'^{p+2}}{p+2}\frac{P_1}{r^2} + \dots + \frac{c'^{p+n+1}}{p+n+1}\frac{P_n}{r^{n+1}} + \dots. \quad (5)$$

This is equal to $$(\kappa_1 + \kappa_2)\, c'^{p+1}\,\Theta + \frac{c'^{p+1}}{p+1}\frac{1}{r}, \quad (6)$$

where Θ is the expression in brackets in (4), provided

$$p + 1 = \frac{\kappa_2}{\kappa_2 + \kappa_1}.$$

Therefore

$$\Theta = -\frac{1}{\kappa_2 r} + \frac{c'^{-p-1}}{\kappa_1 + \kappa_2}\int_0^{c'} \frac{\xi^p d\xi}{(r^2 + \xi^2 - 2r\xi\cos\theta)^{\frac{1}{2}}}, \quad (7)$$

and we have

$$V_0 = \frac{e}{4\pi\kappa_2}\frac{1}{QP} + \frac{e}{4\pi\kappa_2}\frac{\kappa_2 - \kappa_1}{\kappa_2 + \kappa_1}\frac{a}{c}\frac{1}{Q'P}$$
$$- \frac{e}{4\pi}\frac{\kappa_2 - \kappa_1}{(\kappa_2 + \kappa_1)^2}\frac{a}{c}\left(\frac{c}{a^2}\right)^{\frac{\kappa_2}{\kappa_2+\kappa_1}}\int_0^{c'} \frac{\xi^p d\xi}{(r^2 + \xi^2 - 2r\xi\cos\theta)^{\frac{1}{2}}}. \quad (8)$$

This shows that the external effect of the presence of the sphere of different permeability is the same as if the sphere were absent, while a point source of strength

the image-system for the region outside:

$$e' = e\,\frac{\kappa_2 - \kappa_1}{\kappa_2 + \kappa_1}\,\frac{a}{c}$$

were placed at Q', and a line source, of density

$$-\,e'\,\frac{\kappa_2}{\kappa_2 + \kappa_1}\left(\frac{c}{a^2}\right)^{\frac{\kappa_2}{\kappa_2+\kappa_1}}\xi^{\frac{\kappa_2}{\kappa_2+\kappa_1}-1},$$

at distance ξ from the centre, extended from Q' to the centre.

As the total flux through a distant closed surface must be unaltered by this substitution, it follows that the aggregate intensity of this new system of sources is zero, which is easy to verify.

Treating the internal potential in the same way, we find

$$\begin{aligned} V_i = \frac{e}{4\pi}\left\{\frac{1}{\kappa_2}\frac{1}{c} + \frac{3}{\kappa_2 + (\kappa_2+\kappa_1)}\frac{P_1 r}{c^2} + \dots \right.\\ \left. + \frac{2n+1}{\kappa_2 + n(\kappa_2+\kappa_1)}\frac{P_n r^n}{c^{n+1}} + \dots\right\} \\ = \frac{e}{4\pi\kappa_2}\frac{2\kappa_2}{\kappa_2+\kappa_1}\frac{1}{QP} \\ -\frac{e}{4\pi}\frac{\kappa_2-\kappa_1}{\kappa_2+\kappa_1}\left\{\frac{1}{\kappa_2+(\kappa_2+\kappa_1)}\frac{1}{c} + \dots + \frac{1}{\kappa_2+n(\kappa_2+\kappa_1)}\frac{P_n r^n}{c^{n+1}} + \dots\right\}. \end{aligned} \quad (9)$$

Now

$$\begin{aligned} \int^{\infty} \frac{\xi^{-p}\,d\xi}{(\xi^2 + r^2 - 2\xi r\cos\theta)^{\frac{1}{2}}} \\ = \int^{\infty} \xi^{-p}\left(\frac{1}{\xi} + \frac{P_1 r}{\xi^2} + \dots + \frac{P_n r^n}{\xi^{n+1}} + \dots\right\} d\xi \\ = \frac{1}{p}c^{-p} + \frac{1}{p+1}c^{-p-1}P_1 r + \dots + \frac{1}{p+n}c^{-p-n}P_n r^n + \dots, \end{aligned} \quad (10)$$

when c is taken for the lower limit; hence if we take $p = \dfrac{\kappa_2}{\kappa_2+\kappa_1}$, we have

$$\begin{aligned} V_i = \frac{e}{4\pi\kappa_2}\frac{2\kappa_2}{\kappa_2+\kappa_1}\frac{1}{QP} \\ -\frac{e}{4\pi\kappa_2}\frac{\kappa_2(\kappa_2-\kappa_1)}{(\kappa_2+\kappa_1)^2}c^{p-1}\int_c^{\infty}\frac{\xi^{-p}\,d\xi}{(\xi^2+r^2-2\xi r\cos\theta)^{\frac{1}{2}}}. \end{aligned} \quad (11)$$

This shows that the effect inside the sphere is the same as if the medium were homogeneous throughout, and instead of the source e at P there were introduced a point source $\dfrac{2e\kappa_2}{\kappa_2+\kappa_1}$ at P, together with a line source extending outwards from P to an infinite distance, whose density at distance ξ from the centre is

for the inside.

$$-\frac{e\kappa_2(\kappa_2-\kappa_1)}{(\kappa_2+\kappa_1)^2 c}\left(\frac{c}{\xi}\right)^{\frac{\kappa_2}{\kappa_2+\kappa_1}-1}$$

Special cases:

We now numerate various cases which have physical applications.

electrostatic image:

(i) When $\kappa_1=\infty$, we have the problem of the electric distribution on a conducting sphere in presence of a charge e; and the well-known image system is obtained.

image for spherical obstacle in fluid motion:

(ii) When $\kappa_1=0$, we have the problem of the velocity potential in an infinite liquid, due to the source e in the presence of the solid sphere; and there results the image system given by W. M. Hicks, viz. a source ea/c at the inverse point, and a line sink of strength e/a from that point to the centre.

magnetism of soft iron with plane face,

(iii) When $a=\infty$, we have the problem of the magnetization of a very large mass of soft iron with a plane face, in the neighbourhood of which a magnetic pole is situated, the force being everywhere so small that the permeability is practically constant. The effect of the line image vanishes, and we obtain the image system first given by

Green's image-system.

Green, viz. the potential outside the iron is due to e at Q and $e\,\frac{\kappa_2-\kappa_1}{\kappa_2+\kappa_1}$ at its image point in the face, while the potential inside the iron is due to $e\,\frac{2\kappa_2}{\kappa_2+\kappa_2}$ at Q.

In case (ii) it follows that the image of a source e at Q_1, and a source $-e$ at Q_2 on the same radius nearer the centre, consists of a source ea/c_1 at Q_1', a source $-ea/c_2$ at Q_2', and a line source of density e/a along $Q_1'Q_2'$; the total strength of the image system being as usual zero.

When Q_1Q_2 is very small, we have $Q_1'Q_2' : Q_1Q_2 = a^2 : c^2$. The external system is then the equivalent of the motion of a sphere along the axis, from the centre, its moment $e\,.\,Q_1Q_2$ being equal to $3/8\pi$ multiplied by the volume of the sphere and by its velocity; and

Small sphere reflected in large one:

similarly for the image system. The image of a small sphere moving in this manner, or of the corresponding doublet source, is therefore a like sphere, or doublet, at the inverse point, with its moment

Stokes:

reduced in the ratio $-(a/c)^3$. This is a result given by Stokes in 1847.

It may be observed that, when Stokes' result for a doublet is known, the image of a simple source may be immediately deduced, by treating the source as the nearer end of a system of doublets

reversed procedure.

extending to infinity along the axis, and such that contiguous poles cancel by superposition. For if Q_1Q_2, Q_2Q_3, ... denote the doublets, and $Q_1'Q_2'$, $Q_2'Q_3'$, ... their images, the adjacent poles at Q_r', due to the images of the two doublets which abut on Q_r, are of strength ea/c_r and $-ea/(c_r+\delta c_r)$; so that the two together amount to a pole of strength $ea\delta c_r/c_r^2$, *i.e.* $-e\delta c_r'/a$, situated on the element δc_r; which forms the element of a line distribution of density e/a, along Q_1C. This is the result already obtained.

It is however to be remarked that the image system here obtained in the general case is by no means the only one that will represent the external circumstances due to the presence of the sphere: it is not even the only one that is situated along the axis. For example, the given harmonic expansion outside the sphere may be considered as due to a multiple pole, in Maxwell's sense, situated at the centre of the sphere, or at any other point inside it. The only title to consideration of any special image system is its geometrical simplicity. Image not unique:

This simplicity exists in one important respect, for the general image system, when the problem is to find the effect of the induction in a soft iron sphere due to a bar magnet of any kind presented lengthwise to the sphere; for we then obtain the specification of a bar magnet situated in the space inside the sphere, which would produce an effect equivalent to that of the sphere. convenient:

The image inside the sphere of a magnetic molecule of moment m lying at Q along the radius is easily found to be a magnetic molecule of moment

$$-m\frac{\kappa_2-\kappa_1}{\kappa_2+\kappa_1}\left(\frac{a}{c}\right)^3$$

e.g. image of a magnetic doublet in soft iron sphere:

along the radius at Q', a distribution of free magnetism along the line from Q' to the centre of line density

$$m\frac{\kappa_1\kappa_2(\kappa_2-\kappa_1)}{(\kappa_2+\kappa_1)^3}\left(\frac{c}{a^2}\right)^{-\frac{\kappa_2}{\kappa_2+\kappa_1}}\frac{a}{c^2}\,\xi^{-\frac{\kappa_1}{\kappa_2+\kappa_1}}$$

at a point at distance ξ from the centre, and a quantity of free magnetism $-m\dfrac{\kappa_1(\kappa_2-\kappa_1)}{(\kappa_2+\kappa_1)^2}\dfrac{a}{c^2}$ collected at Q', which forms the aggregate of the complementary poles of the other part of the distribution just specified. The result of superposition then gives at once a bar magnet along the radius as the image of a bar magnet along the radius.

Taking Thalén's estimate for soft iron $\kappa_1 = 400$, which is a low one for small forces, and taking $\kappa_2 = 1$ for air, we see how little the result is affected by this line magnetization, and how closely the effect of the sphere due to an inducing molecule is represented by a molecule at the image point. simple approximate result.

We may pass from the image of a magnetic molecule to that of an electric current, or from the analogous image of a hydrodynamical doublet to that of a vortex filament, by making use of the Ampèrean method of replacing the current by a magnetic shell bounded by its circuit, or the vortex filament by a doublet sheet bounded by it.

For the case of a vortex filament which lies on a concentric sphere, we are clearly prompted to take the doublet sheet on that sphere; its image will then be a uniform doublet sheet on a concentric sphere, with moment per unit area altered in the ratio $-(a/c)^3(c/a)^2$; because the element of area is itself altered in the ratio $(c/a)^2$. The image is therefore a vortex filament, the optical image of the given one in the sphere, with strength altered in the ratio $-a/c$. As this is true for any such closed filament, we can take it to be true for any element of a filament which lies in a plane perpendicular to the radius vector drawn to the centre of the sphere, the ordinary analytical supposition being made that such action is ascribed to an element as gives the proper physical result when integrated for a complete filament. This leads to the result given by T. C. Lewis, *Quarterly Journal*, Vol. XVI. p. 338. The method just employed will also of course give an image for a filament not on a concentric sphere, but it is too complicated to be of any use.

Image for a coaxial vortex ring:

For the case of the magnetic effect of a sphere of soft iron in the presence of an electric current i, circulating on a concentric sphere, the result has to be derived from (8). The effect is the same as that of an electric current of strength

for a current in a sphere of soft iron,

$$-i\frac{\kappa_2-\kappa_1}{\kappa_2+\kappa_1}\frac{a}{c}$$

circulating round the inverse in the sphere of the given circuit, and a coil wound on the cone with this inverse circuit as base and the centre of the sphere as vertex, with current density

$$-i\frac{\kappa_2(\kappa_2-\kappa_1)}{(\kappa_2+\kappa_1)^2}\frac{1}{a}\left(\frac{c\xi}{a^2}\right)^{\frac{\kappa_2}{\kappa_2+\kappa_1}}$$

at a distance ξ from the centre.

case of current close to the surface

When c is but slightly greater than a, we might imagine that the effect of the sphere would be practically the same as that of a large mass of iron with a plane face near the inducing magnet. This assumption would be correct in the electrostatic problem, for which κ_1 is infinite; but it is noticeable, and important with regard to the practical behaviour of iron in a magnetic field, that it is not true for the magnetic problem when κ_1, κ_2 are of the same order of magnitude; though it approximates to the truth for large values of κ_1 because the magnetic problem then approximates to the electrostatic one. When κ_1 is not large, the line magnetization along the radius is of the same order

peculiar:

as the image at the inverse point, and produces an effect comparable at all points except those in the neighbourhood of the inducing molecule, where the higher inverse power of the distance tells. For points whose distance from the inducing molecule is small compared with a radius of curvature of the surface we may replace the large mass of iron by a mass bounded by an infinite plane face, but not for more distant points.

For the external medium (air) we have $\kappa_2 = 1$; the results just given apply whether κ_1, the permeability for the iron, is large or small. But for soft iron, with magnetic forces so small that κ_1 is sensibly constant, its value is very large, upwards of 300 frequently; and we can then obtain another form of approximation when the surface is not plane by considering κ_1 to be infinite. This reduces the problem to the electrostatic one of a conducting sphere, and the effect of the iron due to each elementary complete current will be that of its inverse current reduced in the ratio a/c. This result is of course in agreement with what the more general one becomes for this case. very soft iron,

For the case of a large mass of uniform soft iron bounded by a plane face of large extent compared with the dimensions of the current or other magnetic system, the result is so simple that it would appear to yield a valuable practical method of roughly estimating the effect of the presence of masses of iron on electromagnetic instruments. The effect of the iron on the inducing system is then the same as that of the image of that system in its face, with strength reduced $-\dfrac{\kappa_2 - \kappa_1}{\kappa_2 + \kappa_1}$ times. And this result applies not merely for a plane face, but to any case of a current system which is coiled on to or lies close to the surface of a mass of iron, whose surface radii of curvature are everywhere large compared with the distance of the electric currents from the surface—provided the coil is of narrow section, and it is only the magnetic force in the neighbourhood of the coil that is in question. The portion of the current which is very efficient in producing magnetic force at any point is then only that which is close to the point, and with respect to which the surface may be considered as plane. simple practical rule for uniform permeability.

The kind of error introduced by this approximation is readily estimated from the results in (9) and (11).

17

ON PROFESSOR MILLER'S OBSERVATIONS OF SUPERNUMERARY RAINBOWS.

[*Proc. Camb. Phil. Soc.* VI. (1888), pp. 280–286.]

THE theory of the supernumerary bows which accompany the primary and secondary rainbows has, it is well known, been placed on an exact mathematical basis by Airy[1].

Rainbow in a water jet measured.

A series of observations on narrow jets of water were made soon after by the late Prof. Miller[2] with the object of comparing the magnitudes involved with their theoretical values. But so far as appears the comparison was never completed, although the most difficult calculation connected with it was supplied by Prof. Stokes. Prof. Miller contented himself with giving tables of his observations and pointing out that the relative positions of the first few diffraction fringes agreed fairly with the indications of theory.

The bands expressed in terms of emergent wave-front:

The rule given by Airy to determine the absolute magnitudes of the bands which accompany the principal bow for homogeneous light of index μ is as follows. Obtain the equation of the emerging wave-front, which will be of the form

$$\mu z + bx^3 + \ldots = 0$$

in the neighbourhood of the part efficient in the formation of the bow. The dark and bright bands correspond respectively to the values of m which give the maxima and minima values of the expression

by Airy's integral,

$$\left[\int_0^\infty \cos\frac{\pi}{2}(w^3 - mw)\,dw\right]^2.$$

If m_r denote such a value, the angular separation of the corresponding band from the geometrical bow is χ, where

$$\chi = \left(\frac{\lambda}{4}\right)^{\frac{2}{3}}\left(\frac{b}{\mu}\right)^{\frac{1}{3}} m_r,$$

λ denoting the wave-length of the light.

To make a comparison, it remains therefore only to determine the value of b. In the case of Nature, when the refracting drop is a sphere and the incident beam parallel, this is easily accomplished.

and the curvature of the bow caustic.

For the geometrical caustic is the evolute of the wave-front; and its radius of curvature ρ at the bow is easily found to be given by

$$\rho = -\frac{6b}{\mu}r^3,$$

[1] *Camb. Phil. Trans.* Vol. VI. (1848) p. 79.
[2] *Camb. Phil. Trans.* Vol. VII. p. 277.

where r is the radius of curvature of the wave-front, *i.e.* the distance of the caustic from it measured along the ray. Now to calculate ρ; let ϕ denote the angle of incidence of a ray, ϕ' its angle of refraction; a the radius of the drop, and p the perpendicular from the centre of the drop on the emergent ray whose deviation is D. Let us consider the nth rainbow. We have

$$\rho = p + \frac{d^2p}{dD^2},$$

where

$$p = a \sin \phi,$$

$$D = 2(\phi - \phi') + n(\pi - 2\phi')$$

$$= n\pi + 2\phi - 2(n+1)\phi';$$

and as D is stationary at the bow,

$$\frac{dD}{d\phi} = 0, \text{ so that } (n+1)\frac{d\phi'}{d\phi} = 1.$$

Thus

$$\left.\begin{aligned} \sin\phi &= \mu \sin\phi' \\ (n+1)\cos\phi &= \mu\cos\phi' \end{aligned}\right\},$$

which determine the position of the geometrical bow.

Now

$$\frac{dp}{dD} = a\cos\phi \Big/ \frac{dD}{d\phi};$$

$$\frac{d^2p}{dD^2} = -\left(a\sin\phi\frac{dD}{d\phi} + a\cos\phi\frac{d^2D}{d\phi^2}\right)\Big/\left(\frac{dD}{d\phi}\right)^3;$$

and r is very great, so that

$$r = \frac{dp}{dD} = a\cos\phi\Big/\frac{dD}{d\phi}.$$

Hence

$$\frac{\rho}{r^3} = \frac{-1}{(a\cos\phi)^2}\frac{d^2D}{d\phi^2}$$

$$= \frac{2(n+1)}{(a\cos\phi)^2}\frac{d^2\phi'}{d\phi^2}$$

$$= \frac{2(n+1)}{(a\cos\phi)^2}\,\frac{\mu\sin\phi'(n+1)^{-2} - \sin\phi}{\mu\cos\phi}$$

$$= -\frac{2}{a^2}\frac{n(n+2)}{(n+1)^2}\frac{\sin\phi}{\cos^2\phi}$$

$$= -\frac{2n^2}{a^2}\left(\frac{n+2}{n+1}\right)^2\frac{\{(n+1)^2 - \mu^2\}^{\frac{1}{2}}}{(\mu^2-1)^{\frac{3}{2}}},\text{[1]}$$ General formula.

which gives the value of b/μ by the formula above.

[1] This result was given in a question in the Mathematical Tripos, June 2, 1888 (*Camb. Univ. Exam. Papers*, 1888, p. 560). I find that the same expression is given in the *Comptes Rendus*, May 28, 1888, by M. Boitel. See also *Philosophical Magazine*, Aug. 1888, p. 239.

The observations of Prof. Miller relate to the cases $n = 1$, $n = 2$, the primary and secondary bows.

Although the formulae here given apply strictly only to the bands whose angular deviation from the geometrical bow is not considerable, it has been thought well to make the comparison with observation through a considerable range of angle.

For the distant bands interference of rays gives better results (Stokes).

I owe to Prof. Stokes the remark that, at a sufficiently great angular distance from the principal bow, the interval between successive bands may be calculated simply from the interference of the two effective rays, as in Young's original *aperçu*.

We first examine the primary bows. Applying the formulae just obtained to Miller's series marked (C) we obtain the following results.

The index from air to water is given as 1·3346; the radius of the cylinder of water is 0·01052 inch. There is considerable uncertainty as to the value of λ which corresponds to this index, inasmuch as the temperature at the time of observation is not given. If we take it to be 12° C., it appears by interpolation from Landolt and Börnstein's Tables that the light corresponds to a place near the b lines in the spectrum, and that we may take its wave-length in air to be 5200×10^{-10} metres.

The value of m for the first bright band is 1·0845 (Airy), and the complete system of succeeding values for the other bands has been calculated by Prof. Stokes[1].

Calculating by ordinary logarithms, we obtain $\phi = 59° 19'·03$, $\phi' = 40° 7'·2$, and for the radius of the geometrical bow

W. H. Miller's measures of deviations of the bands.

$$4\phi' - 2\phi = 41° 50'·7,$$

which agrees sufficiently with Miller's value 41° 50'·4.

The deviation of the first bright band (the primary bow) from its geometrical position comes out from these data to be

$$\chi = -27'·8.$$

This series (C) is the most consistent of those given. It consists of seven sets, of which the first two extend to 28 bars. But after the 23rd bar the law of succession breaks down completely, as is confirmed for instance by the fact that observations of the 25th are entirely absent. This is conceivably owing to mixture with another series of bands due to some other caustic, which there destroys the continuity of the system under consideration.

If we exclude the bright primary bow, whose position of maximum was, it appears, difficult to fix upon, the series of 23 dark bands

[1] *Camb. Phil. Trans.* Vol. IX. Part I; *Collected Papers*, Vol. II. p. 349.

agree very perfectly in the different sets, and correspond very closely throughout their whole range to the theoretical values assigned by Prof. Stokes' table.

The value for the deviation of the primary bow from its geometrical position which best suits the observations is, however, 26′·4, though this is 3 or 4 minutes greater than the observations of the primary alone would give. Calculating from this value, the following series of numbers shows how closely the observed deviations of the first 23 dark bars from the position of the geometrical bow agree with the theory. The observations are the mean of Miller's first three series, and correspond very nearly to the second series. Comparison for set of 23 bands.

Number of bar	Deviation Calculated	Deviation Observed	Number of bar	Deviation Calculated	Deviation Observed
1	1° 0′·8	1° 0′·7	13	6° 39′	6° 42′·7
2	1 46·5	1 46·7	14	7 0	7 2·7
3	2 23·6	2 23·7	15	7 21	7 22·7
4	2 57·5	2 58·7	16	7 40·3	7 41·7
5	3 27	3 28	17	7 59	8 0·7
6	3 55	3 58	18	8 18	8 18·7
7	4 21	4 23·7	19	8 36·5	8 36·7
8	4 46·5	4 48	20	8 54·6	8 55
9	5 11	5 12·7	21	9 12	9 12
10	5 34	5 36·7	22	9 30	9 29
11	5 56·5	5 58·7	23	9 48	9 47
12	6 18·3	6 20·7			

The uncertainty in the value adopted by interpolation for the wave-length corresponding to the given index involves an error in the calculated deviations which, it has been found, cannot exceed 1/400 of their values.

An increase of ·00012 in the index of refraction of the light in the drop or cylinder, corresponding to a decrease of 1° C. in temperature, will so affect the value of λ as to diminish the calculated deviations by 1/240 of themselves. An increase of ·0001 in this index will diminish the value of θ_1 by 0′·7, and the radius of the geometrical bow by 1′·9. Uncertainties slight.

But, what is more important to notice, an unobserved decrease of 1° C. in the value assumed for the temperature of the prism of water by which the index is determined will produce a decrease in the value of λ estimated from the index, that will have the same effect on the calculated deviations as the corresponding increase of ·00012 in the true index of refraction would have, viz. a diminution of 1/240 of their amount. Temperature effect.

In Professor Miller's observations the temperatures are not recorded. He remarks that all the observations were liable to be affected by a sudden shifting of the bars, which was seen occasionally to take place through a small space to the right or left. It is possible that this may be explained as due to temperature variations in the stream of water.

The discrepancy between this value 26′·4 here adopted for the displacement of the primary bow and the value 27′·8 which is the result of the calculations corresponds to a difference of 12° C. of temperature. The observations would therefore agree exactly with theory if we supposed the hollow prism by which the index is determined to have been filled with water from a reservoir at 0° C. The sudden shiftings in the positions of the bars might then be explained (as above) as due to variations of temperature in the filament of water in which the bars are observed.

From the table of deviations which has been calculated for this series of experiments, the values which apply for any other index and temperature may now be deduced by introducing corrections according to the data just given: while for different values of the radius of the cylinder the deviation is proportional to $(\text{radius})^{-\frac{2}{3}}$.

Another set of bands. In series (A) the index was 1·3318 and the radius of the cylinder of water 0·0103 inch. The light was not so homogeneous, and as a consequence the results are not so concordant. But treating them as has been done for (C), they agree very well with theory so far as the first 10 dark bars, on the hypothesis that the displacement of the primary bow is 29′.

The value of this displacement, deduced from that for (C) by applying the corrections given above for the change of index and of radius of the cylinder, is 29′·4 for a temperature 0° C.: it would be 29′ if the temperature of the water cylinder were 3°·3 C.

A third set. In series (E) the radius of the cylinder was 0·00675 inch. The index was somewhat doubtful; the value 1·33453 leads to 41° 52′ as the radius of the geometrical bow; the value 1·3348 leads to 41° 46′·9. The second value of the index may be rejected at once, as not in agreement with the results.

The theoretical value of the displacement of the bright primary bow deduced from that in (C) by the necessary corrections is 36′·5 for 0° C. This agrees with the mean value deduced from the observations of the first 8 bars, which is 36′·3; but the succeeding bars deviate from the positions assigned by the theory.

The secondary bow, comparison for a long set of bands. Consider now the circumstances of the secondary bow ($n = 2$). In the series (D) which corresponds to (C) for the primary bow, we find $\phi = 71° 47'·35$, $\phi' = 45° 22'·75$, radius of geometrical bow

$$= \pi + 2\phi - 6\phi' = 51° 18'·2.$$

An increase of ·0001 in the index increases this radius by 1′·93.

Thus for the index 1·33464, the radius is 51° 19′·1, which agrees with Miller's result in series (D).

The theoretical displacement of the bright secondary bow from the geometrical position is 49′·61 for a temperature 12° C.

An increase of ·0001 in the index leads to a decrease of 1/220 of itself in the value of the displacement: so that for a temperature 0° C. at the time of observation the displacement would be 47′·4. The alteration is, as before, due to the different value of λ which corresponds to the given index at the altered temperature. This value is in exact agreement with the result deduced from the deviations of the first and second dark bars, and agrees very well with the succeeding ones; though as usual (in accordance with Miller's remark) the number given for the position of maximum brightness of the first band deviates considerably from it. It is to be noticed that the corresponding set of observations (C) of the primary bow required the same temperature correction.

The following table relates to (D), taking the first set of observations.

Number of bar	Deviation Calculated	Deviation Observed	Number of bar	Deviation Calculated	Deviation Observed
1	1° 49′	1° 46′	13	11° 55′	11° 50′
2	3 10·6	3 8	14	12 32·5	12 24
3	4 17·3	4 17	15	13 8	12 59
4	5 16·5	5 16	16	13 44	13 33
5	6 10·6	6 10	17	14 18	14 6
6	7 1	7 0	18	14 51·5	14 38
7	7 48	7 47	19	15 25	15 10
8	8 33·5	8 31	20	15 58	15 43
9	9 17	9 14	21	16 30	16 13
10	9 58	9 56	22	17 1	16 41
11	10 38·5	10 33	23	17 33	17 16
12	11 17·2	11 13			

The agreement is not so good as in (C), as might be expected from the greater values of the deviation.

Both (C) and (D) seem to show effects of temperature differences in the stream of water as evidenced by nearly constant differences in the readings in parallel columns persisting for a considerable time. Thus in the primary, a variation of 6° C. in the temperature of the cylinder will alter the position of the geometrical bow by 11′·4. The index was observed before and after the experiments; but the error might not thus be revealed, as the index was found by means of still water in a hollow prism which might be filled from another source of more steady temperature. Temperature uncertainties.

The series (B), which corresponds to (A) for the primary bow, is not sufficiently consistent to be of much value. The light was not Other sets.

very homogeneous. The observations of the first three dark bars give a displacement of 52′·8 for the principal bow, as compared with 51′·8 given by the theory.

The first dark bars in the series (F), which corresponds to (E) for the primary bow, yield a value 64′ for the displacement of the principal bow, which agrees very well with that given by the theory, viz. 64′·8 calculated from the first and sixth dark bars, and 65′·5 from the first and third.

[For a general theory of caustic surfaces banded by diffraction see *infra*.]

18

THE CHARACTERISTICS OF AN ASYMMETRIC OPTICAL COMBINATION.

[*Proc. London Math. Society*, xx. (1889) pp. 181–194.]

1. The general properties of optical combinations which are symmetrical round an axis, such as ordinary telescopes and microscopes, have, as is well known, been analyzed by Gauss*; and the performance of these instruments, when aberrations are left out of account, has been shown to depend simply on three constants—which may be taken geometrically as the coordinates of the two principal points and the two principal foci, between whose mutual distances one linear relation exists. Symmetrical optical systems:

It is now customary, and it conduces to clearness of view, to throw the theory into a geometrical form in the manner first completely set forth by Maxwell†. To effect this, without entering into details of the construction of the special instrument, we are confined to the use of only those properties that are *characteristic* of rays of light in general. For the purposes of the problem, when restricted to symmetry round the axis, these properties may be stated in the simple approximate form that (i) all rays proceeding from a point go to form the image of that point, (ii) for all such rays the time of passage from point to image ($\Sigma\mu\delta s$) is the same, because they belong to the same wave spreading out from the object point, and finally, after passing through the instrument, converging to the image. The application of the former of these principles requires that the rays are everywhere inclined at a small angle to the axis, as is usually the case in practice. theory immediate and general, when aberrations neglected.

2. It will be convenient to begin by briefly analyzing the parts played by these two fundamental principles in the theory‡; the following mode of procedure is simple and comprehensive.

The direct geometric correlation of object and image:

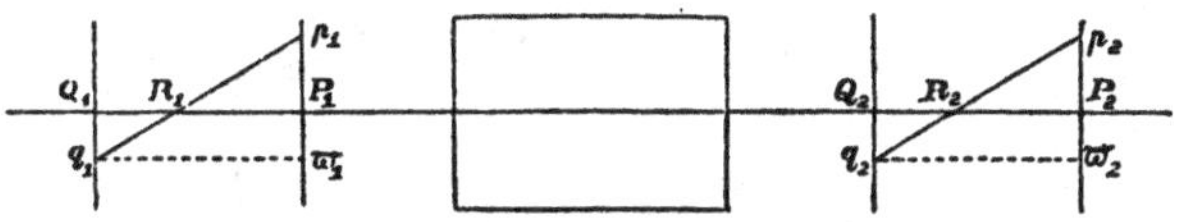

Let P_1, P_2 and Q_1, Q_2 be two pairs of conjugate foci on the axis, and let the linear magnification transverse to the axis (appropriately

* Much earlier in more geometric manner by Huygens, Cotes, Möbius.

† *Quarterly Journal of Mathematics*, II. 1858.

‡ It is implied of course that the two limited regions concerned, on the central ray which need not be straight, are regions of approximate conjugate foci.

called simply the *magnification*) of small objects at P_1, Q_1 be m_p, m_q, respectively; these are clearly the same in all azimuths round the axis. Draw any ray $q_1R_1p_1$ meeting the planes through P_1, Q_1, transverse to the axis in p_1, q_1, and meeting the axis in R_1; after passing through the instrument let its path be $q_2R_2p_2$. Then, by the principle of images, $Q_2q_2 = m_q \cdot Q_1q_1$, $P_2p_2 = m_p \cdot P_1p_1$ (if the image were inverted, the sign of m would be negative); so that

position along axis:

$$m_p \frac{P_1R_1}{P_2R_2} = m_q \frac{Q_1R_1}{Q_2R_2} = \ldots. \tag{1}$$

This law of simple proportions determines absolutely the relative positions of conjugate foci R_1, R_2. To find the corresponding magnification m_r, it is only necessary to draw a ray through Q_1, meeting the transverse planes through R_1, P_1 in ρ_1, ϖ_1, and passing out at the other side through Q_2, ρ_2, ϖ_2. Then

magnification.

$$m_r = \frac{R_2\rho_2}{R_1\rho_1} = m_p \frac{Q_2R_2}{Q_2P_2} \bigg/ \frac{Q_1R_1}{Q_1P_1}, \tag{2}$$

which is merely another form of (1). The principle of images is thus sufficient to determine completely the performance of the instrument.

Restriction, because the rays belong to waves.

3. But hitherto no notice has been taken of the nature of the optical media in which the object and image lie. We may introduce this consideration by aid of the second general principle. It can be applied by comparing two similarly situated rays, which have therefore the same value of $\Sigma\mu\delta s$ in passage through the instrument. Two such are the rays through q_1, one, q_1p_1, through R_1, the middle point of Q_1P_1, and the other $q_1\varpi_1$, thus parallel to the axis. The expression $\Sigma\mu\delta s$ has the same value S from p_1 to p_2 as from ϖ_1 to ϖ_2, because these points are symmetrically situated above and below the axis. Hence, considering the two rays from q_1 to q_2, we have

$$\mu_1 \cdot q_1p_1 + S - \mu_2 \cdot q_2p_2 = \mu_1 \cdot q_1\varpi_1 + S - \mu_2 \cdot q_2\varpi_2,$$

therefore $$\mu_1 (q_1p_1 - q_1\varpi_1) = \mu_2 (q_2p_2 - q_2\varpi_2)^*.$$

Therefore, approximately, since the transverse distances are supposed small compared with the longitudinal, and $P_1p_1 = P_1\varpi_1$, we have

$$\mu_1 \frac{4Q_1q_1^2}{2Q_1P_1} = \mu_2 \frac{(Q_2q_2 + P_2p_2)^2 - (Q_2q_2 - P_2\varpi_2)^2}{2Q_2P_2},$$

$$\mu_1 \frac{4}{Q_1P_1} = \mu_2 \frac{(m_q + m_p)^2 - (m_q - m_p)^2}{Q_2P_2},$$

$$\frac{Q_2P_2}{Q_1P_1} = \frac{\mu_2}{\mu_1} m_p m_q. \tag{3}$$

* This formula involves that, for a pencil limited by an aperture, the diffraction image at P_2 of a luminous point P_1 and that at P_1 of a luminous point P_2, are related as geometric object and image: which is the fundamental proposition limiting the possibilities of microscopic vision.

The results (1) and (2) give the position and transverse magnification of the image of a small object anywhere situated. The result (3) forms a very elegant expression (given by Maxwell) for the longitudinal magnification (termed by Maxwell the *elongation*) of any object not confined to be small, viz. it is equal to the ratio of the indices multiplied by the product of the transverse magnifications of its extremities.

Magnification in terms of convergence.

It follows from (3) that $m_v = \left(\frac{\mu_2\alpha_2}{\mu_1\alpha_1}\right)^{-1}$, where α_1, α_2 are the inclinations of the ray to the axis, and conversely.

Final pair of relations.

It is also clear that (3) includes (2), so that (1) and (3) form a complete system of fundamental formulae. The points P_1, Q_1 may be taken as the origins of measurement; and then three observations suffice to determine all the constants of the instrument.

Gaussian principal planes.

4. The scheme introduced by Gauss makes use of the *principal points* A_1, A_2, for which $m = +1$, and of the principal foci F_1, F_2.

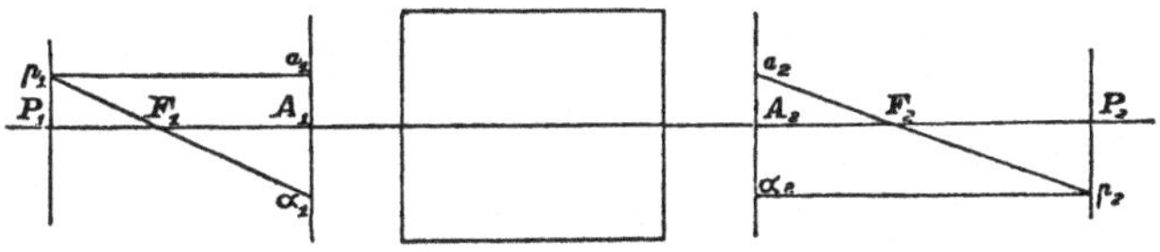

The image of a point p_1 may now be constructed by drawing rays p_1a_1 parallel to the axis, and $p_1F_1\alpha_1$ through F_1, and tracing them on from their points of emergence on the other principal plane to their point of intersection at p_2. Since

$$A_1a_1 = A_2a_2, \quad A_1\alpha_1 = A_2\alpha_2,$$

(because $m = +1$) it follows that the instrument behaves as if the principal planes through A_1, A_2 were coincident, if we leave out of account the shifting of the image system along the axis consequent on moving A_2 into coincidence with A_1. With the exception of this shifting, the instrument therefore behaves as a single thin lens whose principal foci are at distances from it equal to A_1F_1, A_2F_2. In fact it follows in a well-known manner from the diagram that, if we consider lines measured from a principal point or focus as positive when away from the instrument,

Relation of conjugate foci:

$$F_1P_1 \,.\, F_2P_2 = F_1A_1 \,.\, F_2A_2 = \text{constant}, \qquad (4)$$

the generalization of Newton's formula for a lens; which also leads to

$$\frac{A_1P_1}{A_1F_1} + \frac{A_2P_2}{A_2F_2} = 1. \qquad (5)$$

The principle $\Sigma\mu\delta s$ = constant for all rays from p_1 to p_2 requires in the same way as above that

of the two focal lengths.

$$\frac{A_1F_1}{\mu_1} = \frac{A_2F_2}{\mu_2}. \tag{6}$$

Thus when $\mu_1 = \mu_2$, $A_1F_1 = A_2F_2 = f$,

An equivalent simple lens.

where f may be called the focal length of the instrument; for it is the focal length of the *simple equivalent lens*, convex when f is positive, which, as we have seen, is equivalent to the instrument in all respects except as regards the situation of the image system along the axis. The instrument has, in this case, an *optical centre*, which is the middle point of A_1A_2*.

Its experimental determination.

5. To determine experimentally the constants of the instrument, we may proceed by any of the known methods that apply to lenses. (i) The positions of the principal foci F_1, F_2 may be marked by the aid of a parallel beam; then the positions of any pair of conjugate foci yield the value of f^2 by formula (4). To determine whether the positive or negative value of f is to be taken, we must observe whether the image is inverted or erect. (ii) The bright point may be moved along a graduated scale till its conjugate focus approaches it most nearly; the distance between them is then $A_1A_2 + 4f$, and the centre of the instrument is equidistant from them. Another observation will complete the determination. This method requires modification if $A_1A_2 + 4f$ is small or negative. (iii) Two conjugate foci at a distance c apart being selected, the instrument is shifted along its axis through a distance a till they again become conjugate; then, if $A_1A_2 = b$,

$$a^2 = (c - b)(c - b - 4f).$$

Another observation will suffice.

The special telescopic system.

6. With an instrument (telescope) so focussed that a system of rays parallel to the axis emerges parallel to the axis, the magnification is obviously constant at all distances. Therefore so also is the elongation, by (3). The value of the magnification is now the essential constant of the instrument; the exact position of the image system along the axis being, as before, a thing inessential.

Two lenses are a complete equivalent.

7. This brief sketch has shown that a symmetrical optical combination may usually be replaced by its simple equivalent lens, so far as regards the relative position and magnitudes of the images formed. It is easy to see that it may be replaced by two lenses so as to give their *exact* positions on the axis.

* Maxwell's proposition may be recalled, that if a system could be constructed to give perfect images in two planes it would give perfect images in every plane. *Q. J. Math.* II. (1858).

The principal motive of this discussion is to inquire how far similar specifications hold for instruments not symmetrical.

General system: Hamilton's analytic method.

The properties *characteristic* of a ray system now take a more complicated form, viz. that elaborated by Sir W. Rowan Hamilton, from the fundamental property that $\Sigma\mu\delta s$ is stationary along a ray, between any two points, whether in the first medium, or in the final medium, or situated in any other way, and that it has the same value, proportional to the time of propagation, for all the rays from one wave-front to another.

Reduced optical path.

The general problem before us is, given the path of the central ray of a pencil of light which traverses any system of media which may be heterogeneous and may be doubly-refracting, but passes from an initial homogeneous isotropic medium to a final one of the same character, to determine how many and what kind of observations in the initial and final media would be necessary in order to obtain a complete account of the nature of the change produced by transmission through the system in any pencil proceeding along this path.

The method of discussing the propagation of a narrow pencil on Hamilton's principles* has been set forth afresh by Maxwell†; in what follows we shall use a similar analysis.

Reduced path across the system is its characteristic function:

8. Take origins O_1, O_2 on the central ray in the initial and final media, with axes Z_1, Z_2 tangential to the ray and directed *both away from* the refracting system. Let the value of the reduced path ($\Sigma\mu\delta s$) for a ray between points $(x_1, y_1, 0)$, $(x_2, y_2, 0)$ in the transverse planes through these origins be

$$U \equiv \text{const.} + \tfrac{1}{2}a_1x_1^2 + c_1x_1y_1 + \tfrac{1}{2}b_1y_1^2 + px_1x_2 + qx_1y_2 + rx_2y_1 + sy_1y_2 + \tfrac{1}{2}a_2x_2^2 + c_2x_2y_2 + \tfrac{1}{2}b_2y_2^2. \qquad (7)$$

expressible in terms of any two initial and final coordinate frames.

As U is stationary near the axis, because the transverse planes are tangential to the wave-fronts, there can be no terms of the first degree in it.

The ten constants in its expression may be called the constants of the optical combination, when referred to these origins.

9. The first aim must therefore be to choose new origins which will reduce this expression to its simplest form. Taking then the planes $Z_1 = \gamma_1^{-1}$, $Z_2 = \gamma_2^{-1}$ as new transverse planes, we have for the reduced

* Set out especially in a brief note in *Brit. Assoc. Report* Cambridge Meeting, 1838, quoted in full by Lord Rayleigh (*Scientific Papers*, v. p. 456) in connection with the theory of aberrations at focal points, which involves higher terms than are here discussed.

† Maxwell, *Proc. Lond. Math. Soc.* VI. 1874 and 1875.

Transformation to new frames.

distance between the points $(\xi_1, \eta_1, \gamma_1^{-1})$, $(\xi_2, \eta_2, \gamma_2^{-1})$ in these planes the expression

$$V = \mu_1 s_1 + U + \mu_2 s_2, \tag{8}$$

where

$$s_1 = \{\gamma_1^{-2} + (x_1 - \xi_1)^2 + (y_1 - \eta_1)^2\}^{\frac{1}{2}}$$
$$= \gamma_1^{-1} + \tfrac{1}{2}\gamma_1 (x_1^2 + y_1^2) - \gamma_1 (x_1\xi_1 + y_1\eta_1) + \tfrac{1}{2}\gamma_1 (\xi_1^2 + \eta_1^2), \quad (9_1)$$
$$s_2 = \gamma_2^{-1} + \tfrac{1}{2}\gamma_2 (x_2^2 + y_2^2) - \gamma_2 (x_2\xi_2 + y_2\eta_2) + \tfrac{1}{2}\gamma_2 (\xi_2^2 + \eta_2^2). \quad (9_2)$$

We have to eliminate x_1, y_1 and x_2, y_2 from V. Now, since $(x_1, y_1, 0)$ and $(x_2, y_2, 0)$ are in the free path of this ray, we must have

$$\frac{\partial V}{\partial x_1} = 0, \quad \frac{\partial V}{\partial y_1} = 0, \quad \frac{\partial V}{\partial x_2} = 0, \quad \frac{\partial V}{\partial y_2} = 0;$$

therefore

$$(\mu_1\gamma_1 + a_1)\, x_1 + c_1 y_1 + p x_2 + q y_2 = \mu_1\gamma_1\xi_1,$$
$$c_1 x_1 + (\mu_1\gamma_1 + b_1)\, y_1 + r x_2 + s y_2 = \mu_1\gamma_1\eta_1,$$
$$p x_1 + r y_1 + (\mu_2\gamma_2 + a_2)\, x_2 + c_2 y_2 = \mu_2\gamma_2\xi_2,$$
$$q x_1 + s y_1 + c_2 x_2 + (\mu_2\gamma_2 + b_2)\, y_2 = \mu_2\gamma_2\eta_2. \tag{10}$$

These equations determine the paths, in the initial and final media, of the ray which goes from $(\xi_1, \eta_1, \gamma_1^{-1})$ to $(\xi_2, \eta_2, \gamma_2^{-1})$.

10. By solving these equations (10) we obtain the coordinates of the points in which the ray from $(\xi_1, \eta_1, \gamma_1^{-1})$ to $(\xi_2, \eta_2, \gamma_2^{-1})$ meets the transverse planes through the origins. If, however, the determinant Δ of their left-hand sides vanishes, the system is equivalent to only three independent equations, together with the condition

$$A_1\mu_1\gamma_1\xi_1 + C_1\mu_1\gamma_1\eta_1 + P\mu_2\gamma_2\xi_2 + Q\mu_2\gamma_2\eta_2 = 0, \tag{11}$$

in which A_1, C_1, P, Q are the minors of the first column (or any other column) of the determinant. The three independent equations do not now determine the coordinates x_1, y_1 in terms of $(\xi_1, \eta_1, \gamma_1^{-1})$ and $(\xi_2, \eta_2, \gamma_2^{-1})$, but merely lead to a linear relation between these coordinates, of the form

$$S x_1 + Q y_1 = \text{constant}, \tag{12}$$

where Q, S are the minors of the terms q, s in the determinant Δ. Also

$$Q x_2 + P y_2 = \text{constant}.$$

Plane sheaf of rays from one point to another.

The interpretation is clearly as follows. By (12) the rays from $(\xi_1, \eta_1, \gamma_1^{-1})$ to $(\xi_2, \eta_2, \gamma_2^{-1})$ now form a singly infinite system, which cut the transverse plane through O_1 along a line, and similarly cut the transverse through O_2 along a line. For different pairs of points subject to the relation (11), on the same transverse planes $Z_1 = \gamma_1^{-1}$, $Z_2 = \gamma_2^{-1}$, these lines form parallel systems. The meaning of this relation (11) itself is, that to the point $(\xi_1, \eta_1, \gamma_1^{-1})$ there correspond a system of points which form a line in the plane $Z_2 = \gamma_2^{-1}$; and that, corresponding to different points in the plane $Z_1 = \gamma_1^{-1}$, these lines are all parallel.

Focal lines of point sources:

The positions of the two transverse planes $Z_1 = \gamma_1^{-1}$, $Z_2 = \gamma_2^{-1}$ are connected by the relation $\Delta = 0$, which is quadratic in both γ_1 and γ_2. Thus for a beam proceeding from any point in one of these planes, the *focal lines* lie in the two conjugate planes determined by the equation $\Delta = 0$. For different points in the same transverse plane γ_1 the focal lines lie in the same pair of conjugate planes γ_2, and form two parallel systems, each inclined to the corresponding axis x_2 at an angle whose tangent is $(-Q/P)$. For all points in either of the planes γ_2, one of the focal lines lies in the plane γ_1, and is inclined to the axis x_1 at an angle whose tangent is $(-S/Q)$; and this line is one of a parallel system in that plane, such that all points on one of them have the same focal line in the plane γ_2. their (2, 2) correspondence.

11. The equation (11) now shows that all the rays emanating from a luminous line or slit whose equation is Conjugate slits:

$$\mu_1\gamma_1 (A_1\xi_1 + C_1\eta_1) = H, \tag{13}$$

pass through the line whose equation is

$$\mu_2\gamma_2 (P\xi_2 + Q\eta_2) = -H. \tag{14}$$

This remark has an interesting bearing on the working of refracting spectroscopes. It shows, in fact, that without any special adjustment of the refracting prismatic surfaces the slit may always be rotated into such a position that its image for any monochromatic light will be a sharp line. But it will not be an image in the ordinary sense of corresponding point for point with the slit. This position of the slit is the one inclined to the axis of x, at an angle whose tangent is $(-A_1/C_1)$, in which A_1, C_1 are expressed as above in terms of the position of the image, and the constants of the combination. realizable in spectroscopes.

12. To return to § 9, by solving (10) and substituting in the expression for V, we obtain the function characteristic of the combination when referred to the new transverse planes. The work is simplified by making immediate use of the fact that V is a minimum as above, so that

$$\frac{\partial V}{\partial x_1} = 0, \quad \frac{\partial V}{\partial x_2} = 0, \ldots .$$

By Euler's theorem of homogeneous functions

$$V = \text{const.} + \mu_1 s_1 + \mu_2 s_2 + 2\left(\frac{\partial U}{\partial x_1} x_1 + \frac{\partial U}{\partial x_2} x_2 + \frac{\partial U}{\partial y_1} y_1 + \frac{\partial U}{\partial y_2} y_2\right),$$

where $\dfrac{\partial U}{\partial x_1} = \dfrac{\partial V}{\partial x_1} - \mu_1\gamma_1 (x_1 - \xi_1)$, $\quad \dfrac{\partial U}{\partial x_2} = \dfrac{\partial V}{\partial x_2} - \mu_2\gamma_2 (x_2 - \xi_2)$,

$$\frac{\partial U}{\partial y_1} = \frac{\partial V}{\partial y_1} - \mu_1\gamma_1 (y_1 - \eta_1), \quad \frac{\partial U}{\partial y_2} = \frac{\partial V}{\partial y_2} - \mu_2\gamma_2 (y_2 - \eta_2).$$

Reduced expression for characteristic.

Therefore

$$V = \text{const.} + \mu_1 s_1 + \mu_2 s_2 - \tfrac{1}{2}\mu_1\gamma_1 x_1 (x_1 - \xi_1) - \tfrac{1}{2}\mu_1\gamma_1 y_1 (y_1 - \eta_1)$$
$$- \tfrac{1}{2}\mu_2\gamma_2 x_2 (x_2 - \xi_2) - \tfrac{1}{2}\mu_2\gamma_2 y_2 (y_2 - \eta_2)$$
$$= \text{const.} + \mu_1\gamma_1^{-1} - \tfrac{1}{2}\mu_1\gamma_1 (\xi_1 x_1 + \eta_1 y_1) + \tfrac{1}{2}\mu_1\gamma_1 (\xi_1^2 + \eta_1^2)$$
$$+ \mu_2\gamma_2^{-1} - \tfrac{1}{2}\mu_2\gamma_2 (\xi_2 x_2 + \eta_2 y_2) + \tfrac{1}{2}\mu_2\gamma_2 (\xi_2^2 + \eta_2^2). \quad (15)$$

In this expression, which is remarkable as not yet involving explicitly any of the constants, the values of $x_1, y_1, \ldots$ in terms of $\xi_1, \eta_1, \ldots$ are to be substituted.

13. Before proceeding with this substitution it will be convenient to examine how the coefficients p, q, r, s may be simplified by rotation of the axes. Changing to polar coordinates by writing

$$\left.\begin{aligned} x_1 &= \rho_1 \cos\theta_1 \\ y_1 &= \rho_1 \sin\theta_1 \end{aligned}\right\}, \quad \left.\begin{aligned} x_2 &= \rho_2 \cos\theta_2 \\ y_2 &= \rho_2 \sin\theta_2 \end{aligned}\right\},$$

we have $\qquad px_1x_2 + qx_1y_2 + rx_2y_1 + sy_1y_2$

$$= \tfrac{1}{2}\rho_1\rho_2 \{(p+s)\cos\chi - (q-r)\sin\chi + (p-s)\cos\psi + (q+r)\sin\psi\},$$

where $\qquad \psi = \theta_1 + \theta_2, \; \chi = \theta_1 - \theta_2,$

thus $\qquad = \tfrac{1}{2}\rho_1\rho_2 \{P\cos(\chi+\epsilon) + P'\cos(\psi+\epsilon')\}. \qquad (16)$

The expressions $\qquad P^2 = (p+s)^2 + (q-r)^2,$

$$P'^2 = (p-s)^2 + (q+r)^2$$

Mutual invariants for rotations.

are therefore invariant for the transformation; that is,

$$\text{(i)} \; p^2 + s^2 + q^2 + r^2, \qquad (17)$$

$$\text{(ii)} \; ps - qr \qquad (18)$$

are invariant.

The possible simplifications:

We cannot therefore make p, q, r, s all vanish. It is possible to have $q = 0$, $r = 0$; and then $p = s$ will involve a transformation of origins such that the first invariant (i) is twice the second (ii).

Conversely this latter condition by itself gives

$$(p-s)^2 + (q+r)^2 = 0,$$

case of axial symmetry.

so that, as the quantities are to remain real, it amounts to two conditions $p = s$, $q = -r$, whatever be the directions of the axes.

14. We proceed to work out the transformation to new origins. It is easy to see that

$$V = \text{const.} + \tfrac{1}{2}\mu_1\gamma_1 (\xi_1^2 + \eta_1^2) + \tfrac{1}{2}\mu_2\gamma_2 (\xi_2^2 + \eta_2^2) + \tfrac{1}{2}W, \quad (19)$$

where

$$\begin{vmatrix} \mu_1\gamma_1 + a_1 & c_1 & p & q \\ c_1 & \mu_1\gamma_1 + b_1 & r & s \\ p & r & \mu_2\gamma_2 + a_2 & c_2 \\ q & s & c_2 & \mu_2\gamma_2 + b_2 \end{vmatrix} W$$

$$= \begin{vmatrix} \mu_1\gamma_1 + a_1 & c_1 & p & q & \mu_1\gamma_1\xi_1 \\ c_1 & \mu_1\gamma_1 + b_1 & r & s & \mu_1\gamma_1\eta_1 \\ p & r & \mu_2\gamma_2 + a_2 & c_2 & \mu_2\gamma_2\xi_2 \\ q & s & c_2 & \mu_2\gamma_2 + b_2 & \mu_2\gamma_2\eta_2 \\ \mu_1\gamma_1\xi_1 & \mu_1\gamma_1\eta_1 & \mu_2\gamma_2\xi_2 & \mu_2\gamma_2\eta_2 & \end{vmatrix}. \quad (20)$$

The coefficients of the terms

$$p'\xi_1\xi_2 + q'\xi_1\eta_2 + r'\xi_2\eta_1 + s'\eta_1\eta_2$$

on the right-hand side are twice the corresponding minors in Δ, the coefficient of W, multiplied by $\mu_1\gamma_1\mu_2\gamma_2$; they involve, if common multipliers are laid aside, only γ_1, γ_2, and $\gamma_1\gamma_2$, and these linearly.

We notice that

$$(p's' - q'r')\,\Delta = (ps - qr)\,(\mu_1\gamma_1\mu_2\gamma_2)^2.$$

The conditions $p' = s'$, $q' = -r'$, lead to a quadratic equation to determine γ_1, and then a linear equation for γ_2. If the roots of the quadratic are real, there are determined in this way two pairs of principal points on the central ray. We cannot then, by turning both sets of axes through the same angle round that ray, make $q' = 0$, so that also $r' = 0$ by the relation of invariance in § 13; but this can be done by rotation through different angles.

The condition for the reality of these principal points is that a certain function of the coefficients, which it is unnecessary to write out, should be positive.

Condition for existence of principal points, as *infra.*

When this condition is not satisfied, we might attempt to make the reduction in another way. By turning the axes through different angles and moving the origins, we can make $p' = s'$, $q' = 0$, $r' = 0$ in an infinite number of ways. The considerations now to be given will show that this process leads to the same result as the above.

The case of imaginary roots may, however, be reduced to the real one by the addition of a known astigmatic lens or obliquely-placed simple lens, orientated to the proper azimuth, whose presence may be afterwards allowed for.

15. Resuming the original notation, there are therefore, with the above restriction, two pairs of origins and axes for either of which the function characteristic of the combination assumes the form

Final canonical characteristics under above condition:

$$\begin{aligned} U \equiv \text{const.} &+ \tfrac{1}{2}a_1x_1^2 + c_1x_1y_1 + \tfrac{1}{2}b_1y_1^2 \\ &+ p\,(x_1x_2 + y_1y_2) \\ &+ \tfrac{1}{2}a_2x_2^2 + c_2x_2y_2 + \tfrac{1}{2}b_2y_2^2 + \ldots . \quad (21) \end{aligned}$$

This may in a manner be taken as the canonical form of the characteristic.

We can now express very simply in Maxwell's manner the equations which determine the form of the emergent beam. Let the characteristic functions of the same beam in the neighbourhoods of the two origins O_1, O_2 be

$$V_1 = \mu_1 z_1 + \tfrac{1}{2}A_1 x_1^2 + C_1 x_1 y_1 + \tfrac{1}{2}B_1 y_1^2 + \dots, \qquad (22_1)$$

$$V_2 = \mu_2 z_2 + \tfrac{1}{2}A_2 x_2^2 + C_2 x_2 y_2 + \tfrac{1}{2}B_2 y_2^2 + \dots. \qquad (22_2)$$

The other possible quadratic terms are absent because we must have, identically satisfied, the characteristic equation

$$\left(\frac{\partial V}{\partial x}\right)^2 + \left(\frac{\partial V}{\partial y}\right)^2 + \left(\frac{\partial V}{\partial z}\right)^2 = \mu^2. \qquad (23)$$

The value of the reduced path from ξ_1, η_1, ζ_1 to ξ_2, η_2, ζ_2 is

$$V = -(\mu_1\zeta_1 + \tfrac{1}{2}A_1\xi_1^2 + \dots) + \mu_1 z_1 + \tfrac{1}{2}A_1 x_1^2 + \dots$$
$$+ U + \mu_2 z_2 + \tfrac{1}{2}A_2 x_2^2 + \dots - (\mu_2\zeta_2 + \tfrac{1}{2}A_2\xi_2^2 + \dots), \qquad (24)$$

in which, as V is stationary, we must have

$$\frac{\partial V}{\partial x_1} = 0, \quad \frac{\partial V}{\partial x_2} = 0, \quad \frac{\partial V}{\partial y_1} = 0, \quad \frac{\partial V}{\partial y_2} = 0,$$

so that

$$(a_1 + A_1)\, x_1 + (c_1 + C_1)\, y_1 + p x_2 = 0,$$
$$(c_1 + C_1)\, x_1 + (b_1 + B_1)\, y_1 + p y_2 = 0,$$
$$p x_1 + (a_2 + A_2)\, x_2 + (c_2 + C_2)\, y_2 = 0,$$
$$p y_2 + (c_2 + C_2)\, x_2 + (b_2 + B_2)\, y_2 = 0. \qquad (25)$$

These equations determine the point x_2, y_2 which corresponds to any point x_1, y_1, in two ways. On solving the first pair for x_1, y_1, and comparing with the second pair, there result the relations,

determination of emergent beam.

$$\frac{a_2 + A_2}{b_1 + B_1} = \frac{b_2 + B_2}{a_1 + A_1} = -\frac{c_2 + C_2}{c_1 + C_1} = \frac{p^2}{\Delta'}, \qquad (26)$$

where

$$\Delta' = (a_1 + A_1)(b_1 + B_1) - (c_1 + C_1)^2.$$

These equations determine the constants of the emergent beam.

When the incident pencil comes from a focus, $A_1 = B_1$, $C_1 = 0$. It is easy to verify that $(A_2 - B_2)/C_2$ is constant for all positions of that focus, only provided

Focal lines for conical incident pencil.

$$(a_2 - b_2)/c_2 = (a_1 - b_1)/c_1;$$

it is only under this condition that the focal lines corresponding to all foci are parallel.

Axial symmetry:

16. By changing U to polar coordinates, we see that the conditions that the combination is symmetrical round the axis, or acts as such a one, are $a_1 = b_1$, $c_1 = 0$, $a_2 = b_2$, $c_2 = 0$.

If we add to the instruments a thin astigmatic lens at O_1, of index μ and thickness t_1, whose effective thickness $(\mu - \mu_1)\, t_1$ is equal to

$$\tfrac{1}{2}(a_1 + \lambda_1)\, x_1^2 + c_1 x_1 y_1 + \tfrac{1}{2}(b_1 + \lambda_1)\, y_1^2,$$

and a thin astigmatic lens at O_2 whose effective thickness $(\mu - \mu_2)\, t_2$ is equal to

$$\tfrac{1}{2}(a_2 + \lambda_2)\, x_2^2 + c_2 x_2 y_2 + \tfrac{1}{2}(b_2 + \lambda_2)\, y_2^2,$$

obtained by adding lenses in front and rear.

the total resulting combination will be given by

$$U = \text{const.} - \lambda_1 (x_1^2 + y_1^2) + p\,(x_1 x_2 + y_1 y_2) - \lambda_2 (x_2^2 + y_2^2), \quad (27)$$

and therefore it will act as if it were symmetrical round the axis.

Circles of least confusion:

A pencil from O_1 as focus will not be affected by the first lens. It will therefore reach the second lens with a circular cross-section if it has started with a circular section in the first medium, and has not been so wide as to be partially stopped out somewhere in the instrument. The (so-called) circle of least confusion of the emergent pencil will therefore be at the second lens.

Now it is to be remarked that a pencil of light rays, though it always passes through two focal lines, does not in general possess a circular cross-section at any point on its course. It is evident in fact that the condition for its having a circular section is that it should diverge symmetrically from each focal line.

used when they exist to determine the principal points:

Hence to find the position of O_1, place a stop in the path of the incident beam so as to make it circular; and move along the axis the luminous point which emits it until the emergent beam is symmetrical with respect to either of its focal lines; as may be tested by focussing a telescope on the line, covering it with a cross-wire, and then putting the telescope out of focus without rotating the wire. The position of O_2 is then at the corresponding circular cross-section of the emergent beam; or it may be similarly determined by placing the luminous point on the other side of the instrument. The result incidentally appears that if a conical pencil from O_2 has a circular cross-section at O_1,[1] then a conical pencil from O_1 has a circular section at O_2.

Find now by Prof. Stokes'[2] or any other method, the astigmatic lens which, placed in the final medium at O_2, would convert a parallel beam into the actual emergent beam; this will be a lens whose principal focal lengths are equal and of opposite signs to the distances of the focal lines from the circular cross-section of the pencil, measured in the direction the light is travelling.

[1] This theorem of reciprocity may be proved directly from consideration of the general function of $x_1 y_1 z_1$ and $x_2 y_2 z_2$ characteristic of the combination. It is a case of the proposition of dynamical aim formulated and discussed in Thomson and Tait's *Natural Philosophy* (1867), §§ 334, 335.

[2] *Brit. Assoc. Report*, 1849, p. 10; *Collected Papers*, Vol. II. p. 172.

and thence the system completely. Proceed similarly with light emitted from the point O_2, and determine the corresponding lens in the medium at O_1.

The simplest equivalent. The combination is equivalent to these two lenses, together with an instrument symmetrical with respect to the axis, whose principal foci are at O_1, O_2. The principal point P_1 of the latter on the side O_1 is at once determined as the point at which a transverse plane must be placed that the beam from O_1 may mark out on it a circle equal to its circular cross-section at O_2 on emergence; and in the same way the other principal point P_2 may be determined—or else by the relation $O_1P_1/\mu_1 = O_2P_2/\mu_2$.

17. The experimental determinations here sketched amount to the specification of an instrument which is equivalent to the combination; that is, they give the optical constants of the combination.

Special cases. The special case in which O_1 or O_2 is at an infinite distance may be noticed. The effect of the lens at O_1 is then to alter a pencil coming from a focus to one coming from two focal lines in fixed directions at constant distances from that focus.

A more special case still would be that of a *quasi*-telescope for which all parallel incident pencils emerge parallel.

18. The positions of these principal points O_1, O_2 become indeterminate, in the case of an instrument with a straight axis, when the planes of the principal curvatures of the pair of astigmatic lenses are parallel; and in the more general case, when c_1 and c_2 vanish or can be made to vanish by rotating the axes on both sides through the same angle. This leads to the condition

Case of rectangular symmetry:

$$(a_1 - b_1)/c_1 = (a_2 - b_2)/c_2,$$

as at the end of § 15.

This is the single condition necessary that the combination should behave as one having a straight axis and symmetrical with respect to two perpendicular planes through the axis. In the general notation of § 9, it is therefore the condition that P/Q = constant, when $\Delta = 0$.

When the instrument is of this simple character the course of any ray may be constructed by finding those of its traces on the two pairs of corresponding principal planes in the initial and final media, passing through the axis of the combination. For it is clear, as Maxwell remarks, from the form

its character:

$$U = \text{const.} + \tfrac{1}{2}a_1x_1^2 + px_1x_2 + \tfrac{1}{2}a_2x_2^2 + \tfrac{1}{2}b_1y_1^2 + sy_1y_2 + \tfrac{1}{2}b_2y_2^2, \qquad (28)$$

that in the determination of the path of a ray by the method of § 15, the terms containing x_1, x_2 are now separated from the terms containing y_1, y_2.

Therefore the projections of any ray on the planes of x_1z_1 and x_2z_2 are now related by a construction similar to that which applies to instruments symmetrical round an axis, *i.e.* there are two pairs of *quasi*-principal foci in these planes, and two pairs of *quasi*-principal points, by means of which these projections may be constructed. And having thus obtained the traces of a given emergent ray on the planes x_2z_2, y_2z_2, in terms of those of the same ray when incident on the planes x_1z_1, y_1z_1, the problem is solved. **including oblique rays:**

This also follows more directly from the remark that the instrument is essentially the same as a straight one with the principal planes of its two terminal aplanatic lenses parallel, and therefore planes of symmetry.

It remains to show how these planes, and their cardinal points, may be identified experimentally. All rays incident in the plane x_1z_1 emerge in the plane x_2z_2. Therefore the plane x_2z_2 must be that containing the axis z_2, and a focal line corresponding to any focus on the axis z_1; there are therefore two such planes, which are the same whatever point on the axis z_1 is taken for focus, as has been already seen. In the same way the two planes x_1z_1 may be found by taking an incident focus on the axis z_2. These planes may be grouped into the two corresponding pairs in obvious ways. The positions of the cardinal points in a corresponding pair of planes may be determined by methods exactly analogous to those briefly sketched in § 5. The optical combination in question will then have been completely explored. **how recognized.**

It might at first sight be imagined that the four possible adjustments of the axes would be sufficient to make c_1, c_2, q, r vanish always, and so reduce every combination to this type. But from the above, these conditions must be inconsistent with each other, unless a special relation between the constants is satisfied.

19

A SCHEME OF THE SIMULTANEOUS MOTIONS OF A SYSTEM OF RIGIDLY CONNECTED POINTS, AND THE CURVATURES OF THEIR TRAJECTORIES.

[*Proc. Camb. Phil. Soc.* Vol. VII. (1890) pp. 36–42.]

THE following analysis is suggested by the theorems of De la Hire and Savary, whereby the determination of the curvatures of the trajectories of the different points of a solid moving in one plane is reduced to geometrical construction. In this theory the construction is based on the circle which at the instant in question is the locus of points for which the curvature is zero, the well-known circle of inflexions. See Williamson's *Differential Calculus*, Chapter XIX.[1]

Circular locus of null curvatures on the plane:

In the generalized theory, when the motion of the solid is not confined to be uniplanar, the first problem is to determine the nature of the locus of inflexions. This is easily effected by kinematical considerations; for the criterion of a point x, y, z being on the locus is that its acceleration is in the same direction as its velocity, viz. that

generalized to locus of the inflexional points in space.

$$\frac{\ddot{x}}{\dot{x}}=\frac{\ddot{y}}{\dot{y}}=\frac{\ddot{z}}{\dot{z}}. \tag{1}$$

Now we may specify the motion of the solid by u, v, w the components of the velocity of the origin, and ω_x, ω_y, ω_z the component angular velocities of the body round the axes of coordinates. Then, as usual,

$$\dot{x} = u - y\omega_z + z\omega_y, \tag{2}$$

$$\ddot{x} = \dot{u} - y\dot{\omega}_z + z\dot{\omega}_y - \omega_z\,(v - z\omega_x + x\omega_z) + \omega_y\,(w - x\omega_y + y\omega_x), \tag{3}$$

with two pairs of other similar formulae.

The equations of the curve of inflexions are now obtained by substitution in (1).

[1] [Also *supra*, p. 1.] I find that questions similar to the ones here discussed are analyzed by the method of vectors from a fixed origin in the *Comptes Rendus*, 1888, pp. 162–5, by Gilbert, who also gives references to other writers on this subject. His investigations relate chiefly to the case when a point of the system is fixed.

The principal results obtained in this note have been stated in the *Cambridge Mathematical Tripos, Part II.* June 1, 1889. (*Camb. Exam. Papers*, 1888–9, p. 569.)

The result will be simplified if we take the central axis of the motion for the axis of x, so that $u = V$, $\omega_x = \Omega$, while the other components vanish, though their fluxions remain finite. We thus obtain the equations

Referred to screw axis of the motion:

$$\frac{L}{\varpi} = \frac{M - \Omega^2 y}{z} = \frac{N - \Omega^2 z}{-y}, \qquad (4)$$

wherein

$$\left. \begin{aligned} L &= \dot{u} - y\dot{\omega}_z + z\dot{\omega}_y \\ M &= \dot{v} - z\dot{\omega}_x + x\dot{\omega}_z \\ N &= \dot{w} - x\dot{\omega}_y + y\dot{\omega}_x \end{aligned} \right\}, \qquad (5)$$

and ϖ is written for V/Ω, the pitch of the given screw motion.

The equations (4) thus obtained may readily be verified by intuition; for L, M, N represent the component accelerations due to the motion of the origin and the change of values of the angular velocities, while o, $-\Omega^2 y$, $-\Omega^2 z$ are the components of the centrifugal force round the central axis, and it is clear that these together make up the total acceleration.

equations of inflexional locus:

These equations (4) represent the curve of intersection of two paraboloids. To reduce them to the simplest possible form, first turn the axes of y, z round that of x, so as to make $\dot{\omega}_z$ zero. Then move the origin along the central axis a distance h, so that the equation referred to this new origin is obtained by writing $x + h$ in place of x, and take $h = \dot{w}/\dot{\omega}_y$. We thus have finally

$$\frac{\dot{u} + z\dot{\omega}_y}{\varpi} = \frac{\dot{v} - z\dot{\omega}_x - \Omega^2 y}{z} = \frac{-x\dot{\omega}_y + y\dot{\omega}_x - \Omega^2 x}{-y}, \qquad (6)$$

where we notice by the way that the numerators give the simplest form to which the rectangular component accelerations for a moving solid can be reduced.

The equations (6) represent the curve of intersection of a parabolic cylinder having its generators parallel to the axis of x with a rectangular-hyperbolic paraboloid having its axis in the same direction. These surfaces intersect on the plane infinity along the line where any plane z = constant meets it. The finite part of their curve of intersection, which is the proper inflexional curve, is therefore a *twisted cubic*. Its equations (6) may be put in the form

the locus a twisted cubic curve:

$$y = A + Bz(\alpha + \beta z), \quad x = y(\alpha + \beta z), \qquad (7)$$

which are unicursal in the parameter z.

We remark that a wire of this form is the most general solid that can be moved with rotation so that all its points are instantaneously describing straight paths; also that any wire whose form is given by (7) possesses this property, the movement being a screw of pitch $-\beta^{-1}(1 + B^{-1})$ round the axis of x, eased off in a way that retains one degree of indeterminateness.

its interpretation.

We proceed to investigate the trajectory of any point of the solid by the aid of this cubic.

The general problem of curvatures: Through any point, as is well known, one and only one chord of the cubic can be drawn. We may regard this chord as a line of constant length moving with its extremities on two fixed lines, which

stated: may be considered *straight* so far as the determination of accelerations and curvatures is concerned.

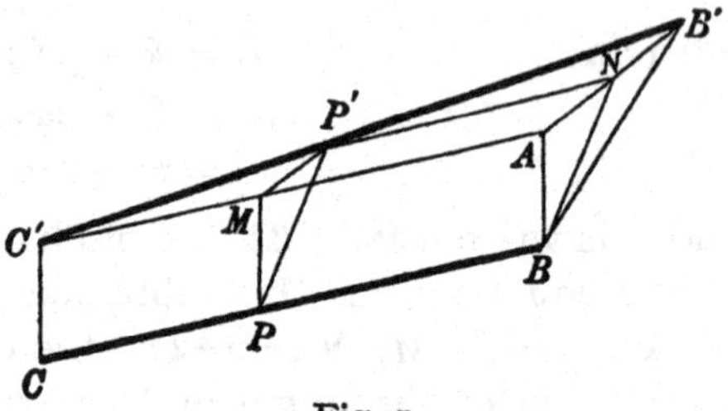

Fig. 1

Consider two consecutive positions of it, BC and $B'C'$; let

$$BP = B'P' = \rho,$$

and $$CP = C'P' = \rho',$$

and let $\rho + \rho' = a$. Complete the parallelogram $C'CBA$, and draw $P'M$, $P'N$ parallel to AB', AC', as in Fig. 1. The circumstances of the motion are given by the velocities of the extremities of BC; let then

$$BB' = bt + \tfrac{1}{2}\dot{b}t^2,\quad CC' = ct + \tfrac{1}{2}\dot{c}t^2, \tag{8}$$

so that b, c are the velocities, and $\dot{b}$, $\dot{c}$ the accelerations of B and C along their straight trajectories.

The point Q moves in a fixed plane which is parallel to both BB' and CC', being parallel to the plane ABB'. The coordinates of Q referred to axes of x and y parallel to BB' and CC' are the same as the coordinates of N referred to axes BB' and BA. They are therefore given by

$$BB' = \frac{a}{\rho}x = bt + \tfrac{1}{2}\dot{b}t^2,\quad CC' = \frac{\rho}{a}y = ct + \tfrac{1}{2}\dot{c}t^2. \tag{9}$$

These are the equations of the path of P, correct as far as the second order, and referred to the parameter t.

The determination of the radius of curvature R at the origin may now be made by the usual methods, bearing in mind that the angle between the axes is ω, the angle between BB' and CC'. It is sufficient to give the result

result:

$$R = \frac{(b^2\rho'^2 + c^2\rho^2 + 2bc\rho\rho' \cos\omega)^{\frac{3}{2}}}{a \sin\omega\,(b\dot{c} - \dot{b}c)\,\rho\rho'}, \tag{10}$$

for its interpretation suggests a purely geometrical method of arriving at it, as follows.

geometrically verified: The velocity V of P' is the same as that of N, and is therefore the resultant of velocities $\frac{\rho'}{a}b$ and $\frac{\rho}{a}c$ parallel to BB' and BA; thus

$$V^2 = \frac{1}{a^2}(b^2\rho'^2 + c^2\rho^2 + 2bc\rho\rho' \cos\omega). \tag{11}$$

In the same way, the acceleration f of P' is the resultant of accelerations $\frac{\rho'}{a}\dot{b}$ and $\frac{\rho}{a}\dot{c}$ parallel to BB' and BA; its value may therefore be written down.

Also, if θ denote the angle between V and f, we have $Vf\sin\theta$ equal to the area of the parallelogram contained by the vectors representing V and f; therefore

$$Vf\sin\theta = \left(\frac{\rho'}{a}b\frac{\rho}{a}\dot{c} - \frac{\rho}{a}c\frac{\rho'}{a}\dot{b}\right)\sin\omega$$

$$= \frac{\rho\rho'}{a^2}(b\dot{c} - \dot{b}c)\sin\omega. \qquad (12)$$

Now by Huygens' fundamental formula of centripetal acceleration in a curve,

$$f\sin\theta = \frac{V^2}{R}, \qquad (13)$$

therefore we arrive at the formula (10), which may also be written

$$R = \frac{a^2}{(b\dot{c} - \dot{b}c)\sin\omega}\,\frac{V^3}{\rho\rho'}. \qquad (14)$$

We may exhibit the result in a geometrical form.

Draw $O\beta$, $O\gamma$ to represent the velocities of B and C, both on the locus of inflexions, in magnitude and direction; divide the fixed line $\beta\gamma$ in ν so that $\beta\nu$ is to $\nu\gamma$ as BP to CP. Then $O\nu$ represents the velocity of P in magnitude and direction, and the radius of curvature of its path is

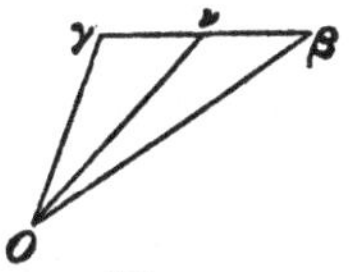

Fig. 2

$$R = \kappa\frac{O\nu^3}{\beta\nu\,.\,\gamma\nu}, \qquad (15)$$ a complete geometrical specification,

where κ represents the constant factor

$$\frac{\beta\gamma^2}{(b\dot{c} - c\dot{b})\sin\omega}.$$

It is worthy of notice that altering the accelerations of B or C only alters the value of the factor κ; so that the curvatures of the trajectories of all points on BC are altered in the same ratio.

We have thus from Fig. 2 a complete specification of the velocity and curvature for any point P on BC; for the plane of the trajectory is that parallel to the directions of motion of B and C. The acceleration of P may be constructed from those of B and C in the same way as the velocity. The value of the curvature involves the accelerations of B and C only as entering into κ. A known value of the curvature at any one point determines κ once for all, and the values of these accelerations are no longer necessary. A geometrical form for κ is given by (21).

for velocity: The construction of Fig. 2 applies to any line in the moving solids in so far as the determination of velocities only is concerned; for the determination of accelerations and curvatures it applies only to a chord of the curve of inflexions, as above.

more special for curvature: If however the planes of the curvatures at any two points on the line are parallel, it must be such a chord. This may be established by an easy extension of the method of Fig. 1. For taking B, C to represent these two points, we have now BB' and also CC' and BA curved instead of straight lines, and we obtain a quadratic equation giving two positions of N on AB' for which the curvature of the path of P is zero; this gives two points, real or imaginary, on BC, which are also on the curve of inflexions.

The formula (10) for the curvature should be reducible to a form depending only on the geometry of the diagram. In fact, if

$$BB' = h, \quad CC' = k, \quad B'BC = \beta, \quad C'CB = \gamma,$$

we have

$$B'C'^2 = a^2 + h^2 + k^2 - 2ah\cos\beta - 2ak\cos\gamma - 2hk\cos\omega,$$

so that, as $B'C' = a$, h is determined in terms of k by the equation

$$h^2 - 2h(a\cos\beta + k\cos\omega) - 2ak\cos\gamma + k^2 = 0, \tag{16}$$

which gives, to the first order,

$$h = -k\frac{\cos\gamma}{\cos\beta}, \tag{17}$$

to the second order,

$$h = \frac{-1}{2a\cos\beta}\left(1 - \frac{k\cos\omega}{a\cos\beta}\right)\left(2ak\cos\gamma - k^2 - k^2\frac{\cos^2\gamma}{\cos^2\beta}\right)$$

$$= -k\frac{\cos\gamma}{\cos\beta} + \frac{k^2}{2a\cos^3\beta}(\cos^2\beta + \cos^2\gamma + 2\cos\beta\cos\gamma\cos\omega). \tag{18}$$

Now by (8) we have to the same order

$$h = \frac{b}{c}k + \frac{1}{2}\left(\ddot{b} - \frac{b\ddot{c}}{c}\right)t^2$$

$$= \frac{b}{c}k + \frac{\ddot{b}c - b\ddot{c}}{2c^3}k^2. \tag{19}$$

Comparing these, we have

$$b\cos\beta = -c\cos\gamma, \tag{20}$$

$$b\ddot{c} - \ddot{b}c = \frac{c^3}{a\cos^3\beta}(\cos^2\beta + \cos^2\gamma + 2\cos\beta\cos\gamma\cos\omega). \tag{21}$$

Substituting in (10),

in terms of a diagram alone.

$$R = \frac{(\rho^2\cos^2\beta + \rho'^2\cos^2\gamma - 2\rho\rho'\cos\beta\cos\gamma\cos\omega)^{\frac{3}{2}}}{\rho\rho'\sin\omega\,(\cos^2\beta + \cos^2\gamma + 2\cos\beta\cos\gamma\cos\omega)}, \tag{22}$$

where β and γ are the angles which BC makes with the curve of inflexions at B and C, and ω is the angle between the tangents to the curve at those points; so that the curvature is expressed in terms of purely geometrical quantities.

It is easy to verify that this theory leads to the correct results for uniplanar motion. For in (4) we have now $\varpi = 0$, $\dot{\omega}_z = 0$, $\dot{\omega}_y = 0$; the curve of inflexions is therefore a *circle* passing through the central axis I. If now the special chord PI meet this circle again in Q, we may apply (14) if we write in it $b = 0$, $b = \omega^2 IC$; thus we obtain (and still more directly from (22)) the correct result

Uniplanar case:

$$R \,.\, PQ = IP^2. \tag{23}$$

The theory for uniplanar motion may also be reduced to simple kinematic considerations as follows. There is one point connected with the solid which has no acceleration; by taking this point for origin and reducing it to rest in the usual manner, we see that the acceleration of any other point with respect to it, *i.e.* in this case the total acceleration, is made up of a component $r\dot{\omega}$ transverse to the radius vector and a component $r\omega^2$ towards the origin. Therefore the resultant acceleration is

independent simple treatment:

$$(\omega^4 + \dot{\omega}^2)^{\frac{1}{2}}\, r,$$

and it acts at a constant inclination α to the radius vector, which is given by $\tan\alpha = \dot{\omega}/\omega^2$; a known theorem. Thus if I denote the instantaneous centre of the motion, I' this centre of accelerations, and P any point, so that

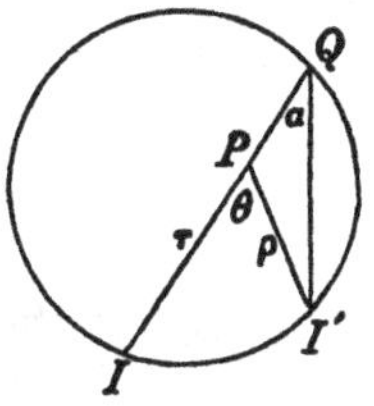

Fig. 3

$$IP = r, \quad I'P = \rho, \quad I\rho I' = \theta,$$

we have, by the theorem of centripetal acceleration,

$$(\omega^4 + \dot{\omega}^2)^{\frac{1}{2}}\, \rho \sin(\theta - \alpha) = \frac{\omega^2 r^2}{R},$$

therefore $$R = \frac{r^2 \sin\alpha}{\rho \sin(\theta - \alpha)}.$$

The points whose paths have zero curvature are given by $\theta = \alpha$, and therefore lie on a circle through I and I', the circle of inflexions.

circle of inflexions.

Let this circle cut PI in Q; then

$$\rho \sin(\theta - \alpha) = PQ \sin\alpha,$$

therefore $$R = IP^2/PQ.$$

20

THE INFLUENCE OF ELECTRIFICATION ON RIPPLES.

[*Proc. Camb. Phil. Soc.* Vol. VII. (1890) pp. 69–71.]

THE relation between the period and the wave-length of ripples on the surface of a liquid must obviously be sensibly affected by an electric charge communicated to the surface.

The problem of wave-motion on fluids.

To investigate the amount of the effect, let us take the origin of coordinates on the surface, the axis of y vertically *downwards*, and the axis of x along the direction of propagation of the ripples.

A suitable form for the electric potential V above the liquid is

$$V = - Ay + ACe^{+my} \cos mx.$$

The equation of the surface ($V = 0$) is then

$$y = C \cos mx,$$

and the surface density σ of the electric charge is equal to $- A/4\pi$.

The velocity potential of the wave-motion is a function ϕ which satisfies

$$\frac{\partial^2\phi}{\partial x^2} + \frac{\partial^2\phi}{\partial y^2} = 0,$$

and gives at the surface ($y = 0$),

$$\frac{\partial\phi}{dy} = \frac{dC}{dt} \cos mx.$$

Thus when the depth is so great that the ripples do not disturb the bottom

$$\phi = - \frac{1}{m}\frac{dC}{dt} e^{-my} \cos mx,$$

as this leads to zero velocity at a great depth.

Ripples with electric charge and surface tension.

The electric charge diminishes the fluid pressure by $- \frac{1}{2}\sigma \frac{\partial V}{\partial n}$, or $\frac{1}{8\pi}\left(\frac{\partial V}{\partial n}\right)^2$, or approximately $\frac{1}{8\pi}\left(\frac{\partial V}{\partial y}\right)^2$, which is equal to

$$\frac{A^2}{8\pi}(1 - 2Cm \cos mx).$$

The surface tension T diminishes the surface pressure by

$$- T\frac{\partial^2 y}{\partial x^2}, \text{ or } TCm^2 \cos mx.$$

Now the equation of fluid pressure is

$$\frac{p}{\rho} = \text{const.} + gy - \frac{\partial\phi}{\partial t} - \tfrac{1}{2}v^2;$$

so that we must have at the surface, approximately,

$$-\frac{A^2}{8\pi\rho}(1 - 2Cm\cos mx) - \frac{TCm^2}{\rho}\cos mx$$
$$= \text{const.} + gC\cos mx + \frac{1}{m}\frac{d^2C}{dt^2}\cos mx,$$

which requires $$\frac{d^2C}{dt^2} + C\frac{m^2}{\rho}\left(Tm + \frac{g\rho}{m} - \frac{A^2}{4\pi}\right) = 0.$$

This gives the periodic time

$$\tau = \frac{2\pi}{m}\rho^{\frac{1}{2}} \bigg/ \left(Tm + \frac{g\rho}{m} - \frac{A^2}{4\pi}\right)^{\frac{1}{2}}$$
$$= \lambda \bigg/ \left(\frac{2\pi T}{\lambda\rho} + \frac{g\lambda}{2\pi} - \frac{4\pi\sigma^2}{\rho}\right)^{\frac{1}{2}},$$

where λ is the wave-length; and the velocity of propagation is

$$\left(\frac{g\lambda}{2\pi} + \frac{2\pi T}{\rho\lambda} - \frac{4\pi\sigma^2}{\rho}\right)^{\frac{1}{2}}.$$

The result is in fact the *same as would be produced by a decrease in surface tension of amount* **Effect of charge.**

$$4\pi\sigma^2/m, \text{ or } 2\sigma^2\lambda.$$

This quantity depends on λ, as might have been expected, for the mechanical effects of an electric charge on the surface cannot be represented as a diminution of surface tension. To produce a simple reduction of tension electrically we must have the double condenser layer that has been assigned by von Helmholtz as the cause of voltaic polarization (cf. *supra*, p. 142). **Effect of surface polarization.**

When, as in the case of voltaic polarization, the ripples occur at the interface between two liquids of densities ρ_1 and ρ_2 the above formulae will clearly be applicable on substitution of $\rho_1 + \rho_2$ for ρ and $g(\rho_1 - \rho_2)/(\rho_1 + \rho_2)$ for g, and of σ/K_2 for σ if the lower one only is a conductor.

The actual values here deduced come from the form of ϕ that belongs to fluid of some depth compared with λ; but it is obvious that the surface tension effect combines with the electric effect in the same way in every case, and that the statement just made holds generally.

As the length of the ripples diminishes, the effect of the electrification is ultimately negligible compared with that of the surface tension, though it persists much longer than the influence of gravity. **Persistence of effect for short ripples.**

In the special case considered above the period becomes imaginary if λ lie between the values

Range of impossible wave-lengths.

$$\frac{4\pi^2\sigma^2}{g\rho}\left(1 \pm \sqrt{1 - \frac{g\rho T}{(2\pi\sigma^2)^2}}\right);$$

so that, if σ can be made so great that these limits are real, the wave-lengths that lie between them cannot exist. For a given period there will be a wave-length above these limits for which gravity is chiefly operative, and one below them for which surface tension is chiefly operative*.

To obtain a rough numerical estimate: On a circular plate of radius a changed to potential V the electric density at distance r from the centre is $\sigma = V/\pi^2 (a^2 - r^2)^{\frac{1}{2}}$, while on a sphere of the same radius the electric density is $\sigma = V/4\pi a$, which is rather less than that at the centre of the plate. At the centre of the plate the effective diminution of surface tension will be $4V^2\lambda/\pi^4a^2$. If we take a 10 cm. and λ 1 cm. this gives about $V^2/2500$ in C.G.S. units. The value 33 for V makes the striking distance in air between balls 2 cm. in diameter about ·3 cm., while according to Mascart the value 400 makes the striking distance between balls 2·2 cm. in diameter about 10 cm.; the former value gives an effective diminution of surface tension of $\frac{11}{60}$, the latter gives 32. For water the actual surface tension is about 80, for mercury 540. The electric effect is therefore considerable: thus Prof. C. Michie Smith (*Proc. R. S. Edin.* March 1890) has observed an effective diminution of 20 per cent. in the case of mercury owing to electrification.

The electric effect can be large,

as already observed.

[Numerical corrections have been introduced.

The problem for superposed dielectric liquids is more important, as affording explanation of the violent instability which ensues on the approach of a charged glass rod to the surface, involving projection of fine jets or threads of liquid. See *Phil. Trans.* 1897, § 80; also *Roy. Soc. Proc.* May 1898, at end: reprinted *infra*.

The presence of a conducting film on the surface of the dielectric would involve yet different conditions.]

* Unless this range of wave-lengths is inhibited, there will be instability of the surface when the electric force at it exceeds $\sqrt{8\pi g\rho T}$. For water this is as much as $5 . 10^5$ volts/cm. for the uniform field here considered, the unstable wave-lengths being around $\frac{1}{4}$ cm.

21

ON THE CURVATURE OF PRISMATIC IMAGES, AND ON AMICI'S PRISM TELESCOPE.

[*Proc. Camb. Phil. Soc.* Vol. VII. (1890) pp. 85–87.]

IT is well known that, in homogeneous light, a prism acts as a telescope in magnifying transversely the dimensions of objects in the field of view, while their longitudinal dimensions remain unchanged. In fact an incident parallel beam emerges as a parallel beam, so that the prism forms a telescopic system; and the transverse magnification along any ray is, by the general law applicable to such systems, equal to the inverse ratio of the breadths of these incident and emergent pencils.

Law of magnification by a set of prisms.

These considerations apply equally to any battery of prisms.

As a single prism has a position of minimum dispersion which is different from the position of minimum deviation, it follows that two prisms, of the same kind of glass, may be combined in opposing fashion so as to form an achromatic pair, while some deviation remains, and therefore also some magnification. By combining in perpendicular planes two such achromatic doublets, of equal magnifying power, Amici long ago succeeded in producing a telescope which magnified equally (about 4 times) in all directions, was made of the same kind of glass throughout and yet achromatic, and, according to Sir John Herschel's experience of it (*Encyc. Metrop.* "Light," § 453), gave images of remarkable perfection.

Dispersion differs from deviation: hence an achromatic pair: two such can be a telescope.

The image of a straight-edge or slit, seen through a prism without a collimator, is a curved arch or bow: and it has been pointed out by Sir Howard Grubb that when the prism is rotated the curvature of this arch is proportional to the dispersion of the spectrum produced. This law will be formally established below for any cylindrical optical system whatever which is composed throughout of the same kind of glass: it may be readily verified by a cursory examination of a pair of prisms standing on a flat plate. It follows from it that each of Amici's doublets gives images of straight lines which are free from curvature for the very reason that they are free from chromatic defect; and the remarkable absence of distortion noticed by Sir John Herschel is explained.

Grubb's law that curvature of a spectral line varies as dispersion: hence perfection of prism telescope.

We proceed to obtain a formula for the curvature of the image of a vertical slit seen a distance a through a system of prisms and

Proof of the law, after Stokes:

cylindrical lenses of the same material, standing on a horizontal plane. The horizontal projection of a ray which is travelling at an inclination θ to the horizontal plane—and is therefore refracted to an inclination θ', given by the same law $\sin\theta = \mu \sin\theta'$ as that of Snell—will be refracted according to the variable index $\mu \cos\theta'/\cos\theta$, or approximately $\mu + \frac{1}{2}(\mu - \mu^{-1})\,\theta^2$, as θ is small. This principle, as was pointed out by Stokes, will suffice for the solution of the problem.

Thus the coordinates of a point on the image are

$$x = a\theta,$$

$$y = a\frac{dD}{d\mu}\cdot\tfrac{1}{2}(\mu - \mu^{-1})\,\theta^2,$$

where D is the deviation of a horizontal ray.

The curvature is equal to $2y/x^2$, and is therefore

result.

$$\frac{\mu - \mu^{-1}}{a}\frac{dD}{d\mu},$$

wherein $\frac{dD}{d\mu}$ is clearly the angular dispersion of the spectrum produced by the combination.

The investigation is no more difficult for a slit inclined at an angle ϵ to the vertical. In this case

$$x = a\theta,$$

$$y = a\theta\tan\epsilon + a\left\{\frac{dD}{d\phi}\,\theta\tan\epsilon + \frac{dD}{d\mu}\,\tfrac{1}{2}(\mu - \mu^{-1})\,\theta^2\right\};$$

therefore $$y - \tan\epsilon\left(1 + \frac{dD}{d\phi}\right)x = \frac{1}{2a}\frac{dD}{d\mu}(\mu - \mu^{-1})\,x^2.$$

Thus the image is parabolic, of the same curvature

Result for inclined slit.

$$\frac{\mu - \mu^{-1}}{a}\frac{dD}{d\mu},$$

as above; and $$\frac{\tan\eta}{\tan\epsilon} = 1 + \frac{dD}{d\phi},$$

where η is the inclination of the image to the vertical, and ϕ is the angle of incidence of the axial ray on the first face.

22

ROTATORY POLARIZATION, ILLUSTRATED BY THE VIBRATIONS OF A GYROSTATICALLY LOADED CHAIN.

[*Proc. Lond. Math. Soc.* Vol. XXI. (1890) pp. 423–432.]

THE existence of rotation of the plane of polarization of light by quartz and certain organic bodies, led in Fresnel's hands to the conclusion that in these substances the waves that are propagated without change of type are circularly polarized, and that their velocity is different according as the direction in which the vibrating particles revolve round their circles is that of a right-handed or a left-handed screw. **The circular type of light-waves.** **Fresnel.**

The analytical modification of the ordinary dynamical equations of wave-propagation that would account for this peculiarity was first explicitly given by Airy. **Airy.**

Of strictly dynamical systems possessing the rotatory property, perhaps the only ones that have been actually realized are those founded on Sir W. Thomson's conception of *gyrostatic domination*. This idea is based on the principle that a fly-wheel, mounted without friction on an axis rigidly attached to a solid body, and initially set into rotation, modifies the dynamical properties of the solid in a peculiar and permanent manner—permanent because the fly-wheel conserves for ever the angular momentum with which it was originally endowed. A distribution of these gyrostats, scattered throughout an elastic medium, will clearly have the effect of giving rotatory properties to waves propagated in the medium. **Kelvin's gyrostatic illustrations:**

In particular, Sir W. Thomson has considered (*Proc. Lond. Math. Soc.* Vol. VI) the propagation of waves of transverse displacement on a stretched chain consisting of straight links, each of which has a fly-wheel rotating on it as axis with a certain constant angular momentum. **his chain of gyrostats:**

The object here proposed is to work out in some detail the more general case in which the axes of the fly-wheels are not perpendicular to the length of the chain. We can, in this case, specify the angular momentum attached to each link by its components with reference to three axes at right angles, of which one is parallel to the chain. And it will conduce much to simplicity to proceed at once to the **here generalized,** links short. limit when the number of links is indefinitely great; we thus gain the idea of a uniform chain, of mass ρ per unit length, stretched with tension τ, possessed of angular momentum whose components per unit length along the chain, and in two fixed directions at right angles to it, are N, L, M.

It may be remarked that the examination of the properties of a system like this may be more than a matter of mere analytical interest. For, though the general kinematic features of light-propagation in quartz and sugar-solutions, as well as in magnetized media, are well understood, yet the nature of the dynamical mechanism by which they are produced is not known; the examination of real mechanisms possessing similar properties may therefore be useful in evolving general relations common to all such systems, and in furnishing correction to ideas that may be entertained on general grounds without a sufficient dynamical basis.

Analogous to magneto-optic rotation. The gyrostatic systems are, of course, the analogue to the action on polarized light of magnetized media, but are not related to the rotation of a molecular character exhibited by quartz and certain organic substances.

Gyrostatic Chain.

To begin with, a change θ in the direction of the axis of a fly-wheel carrying a constant angular momentum G involves a change of the direction of this momentum, and therefore the imparting of new angular momentum $G\theta$ to the system round an axis parallel to the third side of a triangle whose other sides represent in magnitude and direction the initial and final momenta. Hence, when the change is continuous, and ω is its angular velocity, the system reacts with a couple $G\omega$, of which the axis bisects the obtuse angle between consecutive positions of the axis of the fly-wheel.

Gyrostatic motional torque.

If a solid possesses gyrostatic angular momentum whose components with respect to a right-handed rectangular system of axes are L, M, N, and it is made to move with angular velocity whose components are $\omega_1, \omega_2, \omega_3$, this motion is resisted by a gyrostatic couple whose components with respect to the axes are

$$M\omega_3 - N\omega_2, \quad N\omega_1 - L\omega_3, \quad L\omega_2 - M\omega_1.$$

Consider now a gyrostatically dominated stretched chain referred to an axis of z measured along its length, and axes of x and y perpendicular to its length. Let its gyrostatic angular momentum per unit length have components N, L, M with respect to these axes. Let ρ be the linear density; and let k_1, k_2, k_3 be its *effective* radii of gyration per unit length, *i.e.* suppose each link of length δs and mass $\rho\delta s$ to have the axes of coordinates for principal axes of inertia, and the *effective* principal moments of inertia with respect to the centre of mass of the link to be $\rho k_1^2\delta s$, $\rho k_2^2\delta s$, $\rho k_3^2\delta s$. Let the forces at the joints between this link and the two consecutive ones have for components X, Y, Z, and $X+\delta X$, $Y+\delta Y$, $Z+\delta Z$; the only couples at the joints are those due to the torsion, which must be specified directly by means of it. The six equations of motion of the link are

Equations of motion of chain:

$$\rho\delta s\,\frac{d^2x}{dt^2}=\delta X+\frac{d}{ds}\left(Z\frac{dx}{ds}\right)\delta s,$$

$$\rho\delta s\,\frac{d^2y}{dt^2}=\delta Y+\frac{d}{ds}\left(Z\frac{dy}{ds}\right)\delta s,$$

$$0=\delta Z,$$

$$-\rho k_1^2\delta s\,\frac{d^2}{dt^2}\frac{dy}{ds}=Y\delta s-\left(M\frac{d\phi}{dt}-N\frac{d^2x}{ds\,dt}\right)\delta s,$$

$$\rho k_2^2\delta s\,\frac{d^2}{dt^2}\frac{dx}{ds}=-X\delta s-\left(L\frac{d\phi}{dt}+N\frac{d^2y}{ds\,dt}\right)\delta s,$$

$$\rho k_3^2\delta s\,\frac{d^2\phi}{dt^2}=\frac{d}{ds}\left(\mu\frac{d\phi}{ds}\right)\delta s+\left(L\frac{d^2x}{ds\,dt}+M\frac{d^2y}{ds\,dt}\right)\delta s;$$

in which ϕ denotes the angle of twist of the chain, and μ its modulus of torsion.

When we pass to the limit, differentials are replaced by differential coefficients, and these equations become

with short links:

$$\rho\frac{d^2x}{dt^2}=T\frac{d^2x}{ds^2}+\frac{dX}{ds},$$

$$\rho\frac{d^2y}{dt^2}=T\frac{d^2y}{ds^2}+\frac{dY}{ds},$$

$$Z=T,$$

$$\rho k_1^2\,\frac{d^2}{dt^2}\frac{dy}{ds}=-Y+M\frac{d\phi}{dt}-N\frac{d^2x}{ds\,dt},$$

$$\rho k_2^2\,\frac{d^2}{dt^2}\frac{dx}{ds}=-X+L\frac{d\phi}{dt}+N\frac{d^2y}{ds\,dt},$$

$$\rho k_3^2\,\frac{d^2\phi}{dt^2}=\frac{d}{ds}\left(\mu\frac{d\phi}{ds}\right)+L\frac{d^2x}{ds\,dt}+M\frac{d^2y}{ds\,dt}.$$

The differential equations of the motion of the whole chain are now obtained by eliminating the internal stress X, Y, Z, and are

$$\rho\frac{d^2y}{dt^2}+\rho k_1^2\frac{d^2}{dt^2}\frac{d^2y}{ds^2}=T\frac{d^2y}{ds^2}+\frac{d}{ds}\left(M\frac{d\phi}{dt}-N\frac{d^2x}{ds\,dt}\right),$$

$$\rho\frac{d^2x}{dt^2}+\rho k_2^2\frac{d^2}{dt^2}\frac{d^2x}{ds^2}=T\frac{d^2x}{ds^2}+\frac{d}{ds}\left(L\frac{d\phi}{dt}+N\frac{d^2y}{ds\,dt}\right),$$

obtained.

$$\rho k_3^2\frac{d^2\phi}{dt^2}=\frac{d}{ds}\left(\mu\frac{d\phi}{ds}\right)+\left(M\frac{d^2y}{ds\,dt}+L\frac{d^2x}{ds\,dt}\right).$$

The simplest case will be when the breadth of the chain is very small throughout, in which case the terms involving k_1, k_2, which introduce the rotatory inertia of the chain, would be neglected, and we may write κ for k_3 in the torsional equation[1].

The solution of these equations will naturally represent waves travelling along the chain. To allow them to remain of permanent type, we will take L, M, N, μ constant. To obtain the circumstances of such wave-motion, we may substitute

$$(x, y, z) = (x_0, y_0, z_0) \text{ exp. } \iota c\,(s - at),$$

the wave-length being $2\pi/c$, and the velocity of propagation a. Thus

Type of undulation.

$$-\iota \frac{c}{a} N x_0 + \left(\frac{T}{a^2} + \rho k_1^2 c^2 - \rho\right) y_0 + \frac{1}{a} M \phi_0 = 0,$$

$$\left(\frac{T}{a^2} + \rho k_2^2 c^2 - \rho\right) x_0 + \iota \frac{c}{a} N y_0 + \frac{1}{a} L \phi_0 = 0,$$

$$L x_0 + M y_0 + \left(\frac{\mu}{a} - \rho k_3^2 a\right) \phi_0 = 0.$$

Hence we have for a the equation

$$\begin{vmatrix} -\iota \frac{c}{a} N, & \frac{T}{a^2} + \rho k_1^2 c^2 - \rho, & M \\ \frac{T}{a^2} + \rho k_2^2 c^2 - \rho, & \iota \frac{c}{a} N, & L \\ L, & M, & \mu - \rho k_3^2 a^2 \end{vmatrix} = 0;$$

or $$\left(\frac{\mu}{a^2} - \rho k_3^2\right)\left\{\left(\frac{T}{a^2} + \rho k_1^2 c^2 - \rho\right)\left(\frac{T}{a^2} + \rho k_2^2 c^2 - \rho\right) - \frac{c^2}{a^2} N^2\right\}$$
$$- \frac{L^2}{a^2}\left(\frac{T}{a^2} + \rho k_1^2 c^2 - \rho\right) - \frac{M^2}{a^2}\left(\frac{T}{a^2} + \rho k_2^2 c^2 - \rho\right) = 0.$$

Velocities of wave-types,

This is an equation of the third degree in $1/a^2$, so that there are three types of motion.

When there is no gyrostatic element, there are two transverse waves, and one torsional wave. If the reciprocals of their velocities are u_1, u_2, u_3, we have

$$\frac{T u_1^2}{\rho} = 1 - \left(\frac{2\pi k_1}{\lambda}\right)^2, \quad \frac{T u_2^2}{\rho} = 1 - \left(\frac{2\pi k_2}{\lambda}\right)^2, \quad \frac{\mu u_3^2}{\rho} = k_3^2.$$

as affected by the gyrostats:

If u denote $1/a$, the general velocity equation is therefore

$$\mu T^2 (u^2 - u_1^2)(u^2 - u_2^2)(u^2 - u_3^2) - L^2 T u^2 (u^2 - u_1^2)$$
$$- M^2 T u^2 (u^2 - u_2^2) - c^2 N^2 \mu u^2 (u^2 - u_3^2) = 0.$$

This equation determines the character of the motions. We notice by the signs of the left-hand side that its three roots are real and

[1] Cf. *Mathematical Tripos, Part II.* 1890. (*Camb. Exam. Papers*, 1890, p. 585.)

positive, one lower than the least of u_1, u_2, u_3, one higher than the greatest, and one intermediate between them. Thus, T being positive, a disturbance usually starts three waves, which are propagated along the chain both ways.

It may even happen that when T is negative the roots are all positive, provided the gyrostatic terms are great enough. A finite piece of the perfectly flexible chain may then be subjected to end-thrust, and yet will not collapse unless the thrust exceeds a certain limit—a noteworthy example of gyrostatically conferred rigidity. Rigidity conferred.

In the general case, the ratios $x_0 : y_0 : \phi_0$ will usually be complex quantities; hence, in any of the three types of undulation, the links of the chain will describe similar ellipses transverse to its length, and the wave will be elliptically polarized. Elliptic polarization of wave-train.

In the case which corresponds to the rotation of polarized light, the effect of the gyrostatic influence will be to modify only slightly the circumstances of the waves produced by tension and elasticity. The nature of the change can, if necessary, be examined by an obvious method of approximation.

We proceed to review some simple cases of the general result.

(i) $L = 0$, $M = 0$. Here the torsional wave is unaffected; and the velocities of the other two are given by Gyrostats mounted axially:

$$\mu T^2 (u^2 - u_1^2)(u^2 - u_2^2) - c^2 N^2 \mu u^2 = 0,$$

or

$$u^4 - (u_1^2 + u_2^2 + c^2N^2/T^2)\, u^2 + u_1^2 u_2^2 = 0.$$

The roots are both real and positive, even when T is negative. The chain will therefore remain straight under any end-thrust, however great, provided it is applied exactly along the chain; a result we are prepared for by the behaviour of an ordinary top.

The value of $x_0 : y_0$ shows that the transverse vibrations are elliptically polarized. The polarization will, of course, become circular when $k_1 = k_2$; but the velocity of propagation will be different according as the rotation is right-handed or left-handed.

(ii) $N = 0$, $L = M$, $k_1 = k_2$, so that $u_1 = u_2$. Here transversely with symmetry:

$$u = u_1,$$

or

$$\mu T (u^2 - u_1^2)(u^2 - u_3^2) - L^2 u^2 = 0,$$

that is,

$$u^4 - (u_1^2 + u_3^2 + L^2/\mu T)\, u^2 + u_1^2 u_3^2 = 0.$$

The roots of this latter are real and positive, even when T is negative, provided $(u_1 - u_3)^2$ numerically exceed $L^2 \mu T$, that is, if T exceeds $L^2/\mu\,(u_1 - u_3)^2$. So far as these vibrations are concerned, the chain could resist thrusts beyond a certain value, but would collapse under smaller ones; this anomaly will not however occur, for the vibrations of the type corresponding to $u = u_1$ give instability for all thrusts.

The root $u = u_1$ corresponds to a plane-polarized flexural vibration, unaccompanied by torsion. The other roots correspond each to a plane-polarized flexural vibration accompanied by a torsional one of the same phase.

in a transverse plane.

(iii) $M = 0$, $N = 0$. Here

$$u = u_1,$$

or

$$\mu T (u^2 - u_2^2)(u^2 - u_3^2) - L^2u^2 = 0,$$

which may be discussed as in case (ii).

The root $u = u_1$ corresponds to a plane-polarized vibration in the plane containing the gyrostats. The other roots correspond each to a plane-polarized vibration perpendicular to the plane of the gyrostats, accompanied by a torsional vibration of the same phase.

Reflexion of Circularly Polarized Waves.

The circumstances of the passage of light through a piece of magnetized heavy-glass may be illustrated by the passage of a circularly polarized wave through a piece of gyrostatic chain of length l inserted in a chain free from gyrostatic influence. The equations of vibration for this part are those of case (i) above, viz.

A segment l of chain is gyrostatically dominated:

$$\frac{d^2x}{dt^2} = a^2\frac{d^2x}{ds^2} + c\frac{d^3y}{ds^2dt},$$

$$\frac{d^2y}{dt^2} = a^2\frac{d^2y}{ds^2} - c\frac{d^3x}{ds^2dt}.$$

If we introduce the complex variable

$$w = x + \iota y,$$

these equations are both included in the form

$$\frac{d^2w}{dt^2} = a^2\frac{d^2w}{ds^2} - \iota c\frac{d^3w}{ds^2dt}.$$

incident polarized wave-train:

An incident right-handed circularly polarized series of waves is represented by

$$w = A_1 \exp \iota (nt - ex).$$

Let us trace its progress: we have in the successively traversed parts of the chain

the waves reflected and transmitted:

$$w_1 = A_1 \exp \iota (nt - ex) + B_1 \exp \iota (nt + ex),$$

$$w_2 = A' \exp \iota (nt - e'x) + B' \exp \iota (nt + e'x),$$

$$w_3 = A_3 \exp \iota (nt - ex);$$

for in the parts (1) and (2) there is a reflected wave travelling backward, as well as the direct one.

Substitution in the equation of motion gives

$$e = n/a$$

(or say n/α, if the a were not the same for both parts)

$$e' = n/(a^2 + cn)^{\frac{1}{2}}.$$

The conditions necessary for continuity are

$$\text{at } x = 0, \qquad w_1 = w_2, \quad \frac{dw_1}{dx} = \frac{dw_2}{dx};$$

$$\text{at } x = l, \qquad w_2 = w_3, \quad \frac{dw_2}{dx} = \frac{dw_3}{dx};$$

giving

$$A_1 + B_1 = A' + B',$$
$$(A_1 - B_1)\, e = (A' - B')\, e',$$

and

$$A_3 \exp(-\iota el) = A' \exp(-\iota e'l) + B' \exp(\iota e'l),$$
$$eA_3 \exp(-\iota el) = e' \{A' \exp(-\iota e'l) - B' \exp(\iota e'l)\}.$$

Thus

$$2A_1 = A'\left(1 + \frac{e'}{e}\right) + B'\left(1 - \frac{e'}{e}\right),$$

so that

$$2\frac{A_1}{A_3} = \tfrac{1}{2} \exp(-\iota el) \left\{\left(\frac{e}{e'} + \frac{e'}{e} + 2\right) \exp(\iota e'l) - \left(\frac{e}{e'} + \frac{e'}{e} - 2\right) \exp(-\iota e'l)\right\}$$

$$= C \exp(-\iota el + \iota\kappa);$$

where

$$\left.\begin{aligned} C \sin \kappa &= \left(\frac{e}{e'} + \frac{e'}{e}\right) \sin e'l \\ C \cos \kappa &= 2 \cos e'l \end{aligned}\right\}.$$

Hence

$$\frac{A_3}{A_1} = \frac{2}{C} \exp(\iota el - \iota\kappa).$$

emergent amplitude:

Therefore w_3 represents a right-handed circularly polarized system of waves, with amplitude $2A_1/C$, and acceleration of phase $el - \kappa$, where

change of phase.

$$\tan \kappa = \tfrac{1}{2}\left(\frac{e}{e'} + \frac{e'}{e}\right) \tan e'l$$

$$= \frac{a^2 + \frac{1}{2}cn}{a\,(a^2 + cn)^{\frac{1}{2}}} \tan \frac{nl}{(a^2 + cn)^{\frac{1}{2}}}.$$

Consider now an incident left-handed system of waves. Its equation is

Transition to left-handed train.

$$x - \iota y = A_1 \exp \iota\,(nt - ex),$$

so that we must employ the independent variable $u = x - \iota y$, and the equation of motion is

$$\frac{d^2u}{dt^2} = a^2 \frac{d^2u}{ds^2} + \iota c \frac{d^3u}{ds^2 dt}.$$

The results may therefore be derived from the case already worked out by changing the sign of c.

Plane-polarized train, Finally, an incident system of plane-polarized waves may be split up into two equal systems of right-handed and left-handed waves, each of which, as has been seen, will remain of the same type during transmission.

The emergent waves w_3 and u_3 will be of different amplitudes, and their accelerations of phase will be different. If they had the same amplitude, they would combine into a plane-polarized wave rotated through an angle equal to half the difference of their accelerations of phase. But, the amplitudes being different, they actually combine into an elliptically polarized wave with the major axes of its vibrations rotated through this angle.

emerges with elliptic polarization rotated: In this general case the emergent vibration is not plane-polarized, nor is the rotation proportional to the thickness of the gyrostatic medium. In the optical problem, however, the rotatory terms are very small compared with the others. And it will be found, on making c small, that to the *second* order of small quantities included, the results agree with the ordinary optical rule, which makes the approximation: emergent light plane-polarized and the rotation proportional to the thickness of the rotating medium and inversely as the square of the wave-length.

The reason of this agreement is that the values of a have been taken the same in both media, so that the only difference between them is the slight one denoted by the rotational coefficient c. The why rotation is here per unit length. intensity of the reflexion is therefore very small, and the circular waves may be treated as if they passed through in unchanged type without any reflexion, and were finally compounded by the kinematic rule. This is the mode of explanation usually employed to illustrate or account for the optical phenomena. But strictly the velocities of propagation are very different—for example, in glass and air; and if we write $e = n/a$ instead of n/a in the above analysis, we are led to a different result. In fact, the successive reflexions backward and Contrast with optics. forward at the ends of the optical plate have now a material influence, and in an exact investigation must be taken into account, either by a method analogous to the above, or by the mode of calculation usually adopted in the exact theory of interferences by reflexions at the surfaces of a thin plate.

In experimental investigations with thick plates, the influence of the reflected light on the result is usually avoided by a slight inclination of the rays, which separates the reflected from the direct part[1].

[1] Cf. Lord Rayleigh, *Phil. Trans.* II. 1885, p. 345.

The influence of the reflected light, when it is not thus eliminated, is, in fact, that we would have approximately the series

The influence of the reflexions.

$$\exp \iota \left(\frac{2\pi t}{\tau} - \frac{2\pi l}{\lambda} + \gamma l\right) + \epsilon^2 \exp \iota \left(\frac{2\pi t}{\tau} - \frac{6\pi l}{\lambda} + 3\gamma l\right) + \epsilon^4 \exp \iota \left(\frac{2\pi t}{\tau} - \frac{8\pi l}{\lambda} + 5\gamma l\right) + \dots$$

to represent the emergent undulation, where ϵ^2 is the coefficient of intensity of a single reflexion.

This sums to
$$\frac{\exp \iota \left(\frac{2\pi t}{\tau} - \frac{2\pi l}{\lambda} + \gamma l\right)}{1 - \epsilon^2 \exp \iota \left(-\frac{4\pi l}{\lambda} + 2\gamma l\right)},$$

or, when ϵ^4 is neglected,

$$\exp \iota \left\{\frac{2\pi t}{\tau} - \frac{2\pi l}{\lambda}(1 + 2\epsilon^2) + \gamma l\,(1 + 2\epsilon^2)\right\}.$$

Thus the coefficient of rotation would be increased in the ratio $1 + 2\epsilon^2$ to unity.

23

THE LAWS OF THE DIFFRACTION ALONG CAUSTIC SURFACES.

[*Proc. Camb. Phil. Soc.* Vol. VII. (1891) pp. 131–137.]

Caustic a free boundary of wave-train.

1. One of the most striking phenomena in connection with the propagation of light or other undulations is the circumstance that under certain conditions, common in optics and easily realizable in the case of superficial water waves with varying depth of water, there exists a geometrical boundary beyond which the undulations cannot penetrate at all, but in the neighbourhood of which the disturbance is very much intensified.

Rays:

In the first approximations of Geometrical Optics, where the undulations are treated as a system of rays, and the energy is considered to be propagated along them, the caustic or envelope of the rays

concentration:

appears as a surface of infinitely great concentration of energy, which is also the boundary of the space into which the energy can penetrate. The conception that must replace this in a more exact view of the phenomena is that of the theory investigated by Sir George Airy* for the case of the rainbow, on the basis of Fresnel's theory of

actually into a set of parallel bright sheets,

diffraction. It was shown by him that, outside the real caustic of maximum concentration, the energy of the undulations gradually fades away, so that with very minute wave-lengths the boundary of the caustic is quite sharp; but that inside the caustic there is presented a series of successive maxima and minima, in bands running parallel to the absolute maximum or caustic surface.

As the calculations of Sir George Airy had reference chiefly to the phenomena of supernumerary rainbows, he only cared to obtain the relative distances and illuminations of the succession of bands along the asymptote of the caustic.

of similar forms everywhere:

But the peculiarity of this case of diffraction is that there is no question of an aperture limiting the beam of light, so that the degree of closeness and other relations of the bands must depend only on the character of the caustic surface itself, along which they run. The law, connecting these elements, which is thus suggested for investigation, comes out to be very simple. It appears that for homogeneous light the system of bands is similar to itself all along the caustic, as

* *Camb. Trans.* VI. 1848; see also *supra*, p. 174.

regards relative positions and relative brightness, and that they are therefore similar to the supernumerary rainbows calculated by Airy and verified experimentally by W. H. Miller; while the absolute breadths at different parts vary inversely as the cube root of the curvature of the caustic surface along the direction of the rays. For different kinds of light the breadths vary as the wave-length raised to the power two-thirds. These laws are exact for the first few bands, usually all that are visible, owing to the extreme closeness of the subsequent ones; they form in fact the physical specification of the nature of caustic surfaces.

their law of breadth.

2. These statements will be verified in the course of the following analysis of the diffraction near the surface of centres of a wave-front, which forms the natural extension of Sir George Airy's investigation for that portion of the caustic which sensibly coincides with its asymptote.

In the first place, taking a cylindrical wave-front, and referring it to the tangent and normal as axes, we have for its equation in the neighbourhood of the origin

The cylindrical wave-front:

$$z = ax^2 + bx^3 + \dots.$$

The inclination of the tangent at the point x is $\phi = 2ax$, the radius of curvature is

$$R = \left(\frac{d^2z}{dx^2}\right)^{-1} = \frac{1}{2a}\left(1 - \frac{3b}{2a}x\right),$$

and the radius of curvature of the evolute or caustic is

$$\rho = \frac{dR}{d\phi} = -\frac{3b}{(2a)^3}.$$

We have to determine the disturbance, at a point $(\xi, 1/2a)$ in the focal plane of the origin, due to the propagation of this wave. If the amplitude of the motion in the wave-front is $\iota \sin 2\pi t/\tau$ per unit length, the value required for the point in question will be

$$\int \iota \sin\left(\frac{2\pi t}{\tau} - \frac{2\pi r}{\lambda}\right) dS,$$

resultant undulation over its focal plane:

where for the part in the neighbourhood of the origin

$$r = \left\{(\xi - x)^2 + \left(\frac{1}{2a} - z\right)^2\right\}^{\frac{1}{2}}$$

$$= \gamma^{-1}\{1 - \gamma^2\xi x + \tfrac{1}{2}\gamma^4\xi^2x^2 - \tfrac{1}{2}(\gamma^2ba^{-1} + \gamma^4\xi + \gamma^6\xi^3)\,x^3 + \dots\}$$

correct as far as terms in x^3; where $\gamma^{-2} = (4a)^{-2} + \xi^2$.

This integral is to be taken throughout the extent of the wave-front. The phenomena of optics show however that it is only the parts of the wave-front in the neighbourhood of the normal that are efficient in producing illumination along the normal, for the more remote parts may be blocked out without affecting it. The integral may therefore be confined to the immediate neighbourhood of the origin, and we may proceed by approximation. Taking ξ to be small of the same order as b, we have as far as cubes

estimated by approximation.

$$r = \gamma^{-1} - \gamma\xi x - \tfrac{1}{2}\gamma b a^{-1} x^3,$$

$$ds = dx\,(1 + 2a^2x^2 + 6abx^3),$$

and ι is of the form $\iota = \iota_0\,(1 + \alpha x + \beta x + \gamma x^3)$,

ι_0 being the amplitude at the origin.

Thus $\displaystyle\int \iota \sin\left(\frac{2\pi t}{\tau} - \frac{2\pi r}{\lambda}\right) dS$

$$= \iota_0 \int \sin\left(\frac{2\pi t}{\tau} - \frac{2\pi}{\lambda\gamma} + \frac{2\pi\gamma\xi}{\lambda}\,x + \frac{\pi\gamma b}{a\lambda}\,x^3\right)\{1 + \alpha x + (2a^2 + \beta)\,x^2 + \dots\}\,dx.$$

Writing $\qquad x' = x + \tfrac{1}{2}\alpha x^2 + \tfrac{1}{3}(2a^2 + \beta)\,x^3,$

so that $\qquad x = x' - \tfrac{1}{2}\alpha x'^2 - \tfrac{1}{3}(2a^2 + \beta)\,x'^3 + \tfrac{1}{2}\alpha^2 x'^3,$

this becomes

$$\iota_0 \int \sin\left\{\frac{2\pi t}{\tau} - \frac{2\pi}{\lambda\gamma} + \frac{2\pi\gamma\xi}{\lambda}\,x' - \frac{\pi\gamma\alpha\xi}{\lambda}\,x'^2 + \frac{1}{\lambda}\left[-\tfrac{2}{3}\pi\gamma\xi\,(2a^2 + \beta) + \pi\gamma\xi\alpha^2 + \pi\gamma b\right] x'^3\right\} dx',$$

or say $\qquad \displaystyle \iota_0 \int \sin\left\{\frac{2\pi}{\tau}\left(t - \frac{\tau}{\lambda\gamma}\right) + Ax' + Bx'^2 + Cx'^3\right\} dx',$

Airy's integral for field of illumination:

which $\qquad \displaystyle = \iota_0 D^{-1} \int \sin\left\{\frac{2\pi}{\tau}\,(t - \text{const.}) + \tfrac{1}{2}\pi\,(w^3 - mw)\right\} dw,$

where $w = D\left(x + \dfrac{B}{3C}\right)$, so that $\tfrac{1}{2}\pi D^3 = C$,

and $\qquad \displaystyle -\tfrac{1}{2}\pi m = \frac{1}{D}\left(A - \frac{B^2}{3C}\right) = (\tfrac{1}{2}\pi)^{\frac{1}{3}} A C^{-\frac{1}{3}}\left(1 - \frac{B^2}{3AC}\right);$

which gives on reduction, writing unity for $\gamma/2a$,

its parameter.

$$-m = b^{-\frac{1}{3}}\left(\frac{4}{\lambda}\right)^{\frac{2}{3}} 2a\xi\left\{1 - \frac{a}{b}\,(\tfrac{1}{3}\alpha^2 - \tfrac{5}{18}a^2 - \tfrac{2}{9}\beta)\,\xi\right\},$$

so that $\quad 2a\xi = m\,(\tfrac{1}{4}\lambda)^{\frac{2}{3}} b^{\frac{1}{3}}\,\{1 + \tfrac{1}{2}mb^{-\frac{2}{3}}\,(\tfrac{1}{4}\lambda)^{\frac{2}{3}}\,(\tfrac{1}{3}\alpha^2 - \tfrac{5}{18}a^2 - \tfrac{2}{9}\beta)\,\xi\},$

or in terms of the radius of curvature ($R = 1/2a$) of the wave-front at the origin and the radius of curvature ρ of the caustic

$$\xi = -\rho^{\frac{1}{3}} m\left(\frac{\lambda^2}{96}\right)^{\frac{1}{3}}\left[1 + \tfrac{1}{2}m\left(\frac{3\lambda}{160}\right)^{\frac{2}{3}}\{(\tfrac{1}{12}\alpha^2 - \tfrac{1}{18}\beta)\,R^2 - \tfrac{5}{18}\}\right].$$

3. Neglecting the term of the second order in this expression we have

Bright bands at the Airy-Stokes maximal values:

$$\xi = -m\left(\frac{\lambda^2\rho}{96}\right)^{\frac{1}{3}},$$

showing that the course of the ray-caustic is bordered by a series of fringes which remain similar to each other throughout, and therefore are of the same type as the asymptotic fringes of Airy's supernumerary rainbow; but they come closest together at places of greatest curvature of the caustic according to the law that their separation at any place is proportional to the cube root of the radius of curvature of the caustic at that place.

their law of succession,

The investigation shows that unless for fringes at a considerable distance from the ray-caustic their form is not sensibly affected by the varying intensity in the wave-front.

modified near a cusp.

As we proceed along a caustic, the curvature gradually increases and the fringes therefore come together when we approach a cusp. At the cusp itself $b = 0$; and very near to it b is very small, so that to determine the state of matters for an unlimited beam another term would have to be included in the equation of its front; but if the beam is limited in any way the fringes produced by this limitation will there rise in importance, and practically obliterate the ones now under discussion[1].

4. The results of this analysis indicate how we may proceed in the general case where the wave is not cylindrical, but is curved in two dimensions.

Referred to the normal as axis of z, and the tangents to the arcs of principal curvature as axes of x and y, the equation of its front is

$$z = ax^2 + by^2 + px^3 + 3qx^2y + 3rxy^2 + sy^3 + \ldots.$$

Doubly curved wave-front:

It is required to find the disturbance propagated to the point

$$\left(\xi, 0, \frac{1}{2a}\right).$$

Here
$$r_1 = \left\{(\xi - x)^2 + y^2 + \left(\frac{1}{2a} - z\right)^2\right\}^{\frac{1}{2}}$$

$$= \left(\xi^2 + \frac{1}{4a^2} - 2\xi x + x^2 + y^2 - \frac{z}{a}\right)^{\frac{1}{2}}$$

$$= \gamma^{-1} - \gamma\left\{\xi x + \left(1 - \frac{b}{a}\right)y^2 - px^3 - 3qx^2y - 3rxy^2 - sy^3\right\},$$

so that
$$\iint \iota \sin\left(\frac{2\pi t}{\tau} - \frac{2\pi r_1}{\lambda}\right) dx\,dy$$

will be complicated.

[1] See Rayleigh, "Investigations in Optics," *Phil. Mag.* Nov. 1879, pp. 408–10. [*Scientific Papers*, Vol. 1. p. 440.]

But the considerations already mentioned show that the value of the integral is practically settled by the elements in the neighbourhood of the origin, for which x and y are small. We may therefore consider only a small rectangular portion of the wave-front bounded by arcs parallel to the axes of x and y. To determine the diffraction in the plane $z = 1/2a$, we may consider only the plane problem presented by a wave of the form

reduction to cylindric case.

$$z = ax^2 + px^3;$$

for the uniform curvature in the perpendicular plane represented by the coefficient b will not affect the result at all, as is also obvious on continuing the general calculation. The variation of that curvature represented by the coefficient r will slightly displace the fringes, as it will alter the mean value of $\gamma\xi$.

The dissymmetry indicated by the coefficients q and s will on the average produce no effect on the disturbance at a point in the plane xz; these coefficients introduce odd powers of y which integrated over equal positive and negative range leave no appreciable result.

Thus the illumination in the plane $z = 1/2a$ is determined by the values of a and p only.

Two-sheeted caustic surfaces.

Every beam in a homogeneous medium therefore converges to two ray-caustic surfaces which are the two sheets of the surface of centres of curvature of the wave-fronts. Each of these surfaces is physically made up of a series of parallel bright and dark sheets, of which the first is much the brightest, whose distances and relative intensities always retain the same proportions. These distances are at any point proportional to the cube root of the radius of curvature of the normal section of the caustic surface containing the ray which touches it at that point.

Intra-ocular caustic bands;

5. It is easy enough to obtain actual examples of this general proposition. On looking at a bright lamp, sufficiently distant to be treated as a luminous point, through a plate of glass covered with fine rain-drops, the caustic surfaces after refraction through the drops are produced within the eye itself, and their sections by the plane of the retina appear as bright curves projected into the field of vision. These curves are each accompanied by the other parallel diffraction bands, which separate and become more marked as the curves recede asymptotically, while they assume a different character near the cusps which are a feature of all sections of caustic surfaces. Near these cusps in the cross-section of the caustic surface two different pencils of light come into interference.

distinct from shadow diffraction

These phenomena are quite different from the ordinary cases of entoptic diffraction, in which when the eye is put out of focus by a lens, and a bright sky is viewed through a pinhole, the pencil of light coming through the pinhole projects on the retina shadows of the *muscae volitantes* floating in the aqueous humour, and these are accompanied by the ordinary bands at the boundaries of shadows. This case of an obstacle is the exact complement of that of a similar hole in a screen as regards the position of the bands; so that when the obstacle is small, the diffraction bands round the shadow form exact circles, irrespective of its shape, which is the ordinary visual appearance.

with its circular rings.

The cusped caustic bands are easily seen when a distant street-lamp is viewed through a spectacle lens with minute rain-drops deposited on it.

The diffraction problem which has here been discussed includes diffraction at a focal line, with an unlimited beam. In practical questions such as those relating to spectroscopes and the Herschelian telescope, the exact focussing would however introduce the nature of the aperture into the discussion, and the limits of the integral would enter.

6. In the case of a cylindrical beam the bands near the caustic have been counted up to 30 or more by W. H. Miller, and the divergence of the more remote ones is too great to allow an approximate theory like the above to be applied with much certainty. For them, as Prof. Stokes has remarked, a perfectly satisfactory procedure is to simply consider the difference of path of the two pencils of light which reach a given band by different ways; these may be considered as two separate interfering rays, exactly in the manner of Thomas Young's first *aperçu* of the supernumerary rainbow. Now the difference of paths of the two rays up to the point P is clearly the excess of the two tangents from P to the geometrical caustic over the arc between their points of contact. Thus we obtain the following simple and elegant graphical construction, which applies to all the system of bands except the first two or three; imagine the caustic curve constructed as a disc, and let an endless thread be placed round it, the bands will be traced by a pencil strained by this thread in the same manner as in the ordinary construction of an ellipse by a thread passing round its foci; and successive bands will correspond to equal increments in the length of the thread.

The Miller-Stokes bands:

a thread construction for the outer bands.

24

THE MOST GENERAL TYPE OF ELECTRICAL WAVES IN DIELECTRIC MEDIA THAT IS CONSISTENT WITH ASCERTAINED LAWS.

[*Proc. Camb. Phil. Soc.* Vol. VII. (1891) pp. 165–166.]

Maxwell's theory the simplest possible: It was explained that Maxwell's hypothesis of complete circuits reduces the whole of electrodynamics to the ascertained Ampère-Faraday-Neumann laws for such circuits, as it completely defines the character of the electrostatic polarization which must be postulated as a part of the theory. The question as to how far the theory of electrodynamics may depart from this simple form when the circuits are not assumed to be complete was then examined mathematically on the lines of v. Helmholtz's investigations. It was shown that waves of transverse displacement will always be propagated in a dielectric, irrespective of what hypothesis is assumed as to the law of the mutual action of incomplete currents, whether that adopted by v. Helmholtz or the still more general one which is formally possible. **the relation of velocity to *K* is crucial.** But it was also shown that, if the velocity of these waves in a non-magnetic dielectric is equal to the inverse square root of the specific inductive capacity, the currents are necessarily complete. The experimental evidence is strongly in favour of this relation, and in so far constitutes a demonstration of Maxwell's theory of electrodynamic action and its mode of propagation in stationary media[1].

[1] See *Proc. Roy. Soc.* 1891, Vol. XLIX. p. 521 [*infra*, p. 234].

25

A MECHANICAL REPRESENTATION OF A VIBRATING ELECTRICAL SYSTEM, AND ITS RADIATION.

[*Proc. Camb. Phil. Soc.* Vol. VII. (1891) pp. 166–176.]

THE propagation of undulations of electric polarization in a dielectric is exactly similar to the propagation of elastic waves in a solid medium*, which must be absolutely incompressible if we follow Maxwell's scheme, but may transmit waves of condensation as well as shearing waves if we admit the more general scheme developed by von Helmholtz.

The propagation of electrical actions in a medium which is a conductor of ordinary type (so that the number which expresses its inductive capacity in C.G.S. electromagnetic measure is very small, while that which expresses its specific resistance is large) follows the same law as the diffusion of heat in a conducting medium, with the proviso that the thermal diffusivity is to be proportional to the reciprocal of the electric conductivity. For this reason rapidly alternating disturbances in the medium surrounding a conducting mass are only transmitted skin deep into the conductor, the depth of this skin diminishing with increasing rapidity of the alternations and increasing conductivity, in the same manner as in the corresponding question of the propagation beneath the surface of the ground of the daily and annual alternations of temperature. For this reason also a sheet of conducting metal acts as a screen against alternating electrodynamic influence.

Conductor an inelastic region:

analogy of thermal diffusion.

In development of this elastic solid analogy it has recently been explained by Sir W. Thomson[1] that the magnetic induction due to a steady electric current traversing a circuital channel in the medium is represented by twice the vorticity of the elastic strain when a longitudinal force of amount represented by the current is applied to the portion of the medium which coincides with this channel.

But the magnetic field is here correlated with vorticity:

It is also a well-known relation, and forms in many respects the simplest and most symmetrical mathematical specification of the

* The discrepancy appears only when reflexion and refraction of waves at the boundary of two media is treated: see "A dynamical theory..., Part I." *Phil. Trans.* (1892), as *infra*.

[1] *Math. and Phys. Papers*, Vol. III. p. 451.

connections of the electric field on Maxwell's scheme, that the magnetic induction (abc) is represented by the vorticity of the electric force (PQR), and the electric force by the vorticity of the magnetic induction, according to the equations

$$\frac{dQ}{dz} - \frac{dR}{dy} = \frac{da}{dt}, \ldots, \ldots,$$

$$\frac{db}{dz} - \frac{dc}{dy} = -4\pi\mu\left(K\frac{d}{dt} + \sigma\right)P, \ldots, \ldots.$$

In the dielectric, σ is zero, and the electric displacement (fgh) is proportional to the electric force according to the relation

$$(f, g, h) = \frac{K}{4\pi}(P, Q, R).$$

Thus the vorticity of the electric displacement is $\frac{K}{8\pi}\frac{d}{dt}$ of the magnetic induction; and the time integrals of (fgh) represent on Sir W. Thomson's analogy the displacements of the elastic medium multi plied by the factor $K/4\pi$.

and electric field with actual displacement: Or we may, as in the following sections, take the electric displacement to represent the actual displacement of the elastic medium, and then the magnetic induction will be equal to the time integral of its vorticity multiplied by $8\pi/K$; and the electric current will be equal to the time integral of the impressed forcive multiplied by $4\pi/K$. This forcive must on Maxwell's scheme be circuital, that is, circulating in ideal channels in the manner of the velocity system of an incompressible fluid.

giving a parallel mechanical system. These considerations thus present a mechanical view of the electric propagation in dielectrics, with the exception that the current on the wire must be imitated by an applied forcive of some kind. If the media are magnetic, differences in rigidity or of density must be introduced between them.

For an electrical vibrator we can however complete the mechanical analogy, provided the wave-length of the undulations is not smaller than a few centimetres in the case when the vibrator is made of an ordinary conducting metal. For experimental types of Hertzian vibrators the analogy will therefore be practically exact.

To do this we have to examine the conditions as regards electric displacement that must be satisfied at the surface of a conductor.

The displacement is proportional to the electric force, which consists of a kinetic and a static part. The kinetic part is

$$-\frac{d}{dt}(F, G, H),$$

where $$\nabla^2(F, G, H) = -4\pi\mu(u, v, w),$$

because *FGH* is the vector potential of the current distribution; it is therefore continuous everywhere as the potentials of all volume or surface distributions must be. The static part

$$-\left(\frac{d}{dx}, \frac{d}{dy}, \frac{d}{dz}\right) V$$

can only be due to a surface density s on the conducting surface; therefore its components along the surface must be continuous when we cross it, while the discontinuity in the component normal to the surface must be equal to $4\pi s$. The tangential components of the total electric force are therefore continuous.

Surface conditions.

Now it is known that for rapid alternations the currents are induced only in the skin of the conductor; the vibrations of the system and its free periods are in the limit quite independent of the conductivity and of the nature of the interior parts of the conductor: they are practically the same as if the conductivity were infinite, and there is no sensible decay owing to any degradation into heat.

We require the proper boundary conditions to express these facts.

The currents in the conductors may be treated as surface sheets inside which the electric force is zero.

Outside the sheet therefore the components of the electric force along the surface must be zero also; and therefore the tangential electric displacements must be zero.

The components of the electric force normal to the surface will differ on the two sides of it by an amount determined by the surface density s.

In the elastic solid analogy we may therefore consider the conductors to be cavities in the solid, provided we confer infinite rigidity on the skin of each cavity so that no point of it can have any tangential displacement. If we assume that the solid is incompressible, its vibrations are completely determined by these conditions; the component of the displacement normal to the surface must naturally adjust itself in such manner that the condensation remains always null. This will be the analogue of an electric surface density on the conductor which adjusts itself instantaneously to the equilibrium value corresponding to the actual phase of the disturbance in the dielectric, according to the equation $\nabla^2 V = 0$ which of course indicates infinite velocity of propagation or adjustment of disturbances.

Conductors as cavities with rigid skins.

The medium being incompressible, the total volume of all the cavities will remain the same. We may put this necessary property of the motion in evidence by supposing each cavity to be filled with incompressible fluid. The displacement of this fluid at the surface will then be continuous with the displacement of the solid and will represent the electric surface density. The inertia of the fluid must not sensibly interfere with the adjustment of the normal displacement; so that if the rigidity of the solid is taken to be finite, the fluid must be of negligible density. But it is to be borne in mind that, as the fluid represents a conductor, the motion of the fluid does not represent electrostatic displacement except on the surface.

Working elastic solid analogue of electric radiator:

We may thus represent the circumstances of an electric vibrator by the annexed diagram.

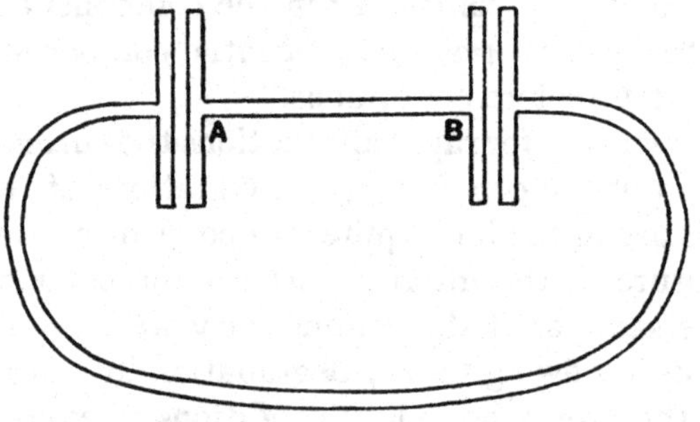

Two condensers, A, B, are represented, with their inner coatings connected by a conducting wire in which the spark gap required for the production of the initial disturbance is usually situated; and their outer coatings are connected by another wire which may be to earth. Each conducting system forms a cavity in the elastic dielectric, which may be considered to be filled with incompressible massless liquid; in this case there are two such cavities with pairs of plane faces opposed to each other.

The disturbance may be supposed to be originated by getting an excess of liquid into the inner coating of the condenser A; that involves pushing away the plate of dielectric between the two coatings of this condenser, and therefore removing an equal amount of liquid from the outer coating. For the parts of the conducting surfaces other than the opposed faces are all backed up by thick masses of dielectric, which will not yield sensibly as compared with the thin plates of dielectric which belong to the condensers. The greater mobility of the latter accounts both for the store of energy that the condenser can acquire, and for the equality of the charges of its two coatings.

When the system is disturbed the liquid will sway backwards and forwards between the condenser coatings in a way which gives a very real representation of the actual electric oscillation.

The only element in this representation that is not easily realizable is the skin, of rigidity great compared with that of the solid; if however the solid were a jelly the skin would be naturally provided by

supposing the system represented in the diagram to be constructed of very flexible sheet metal.

explored.

The greater the capacity of the condensers compared with the section of the connecting wire the longer is the period of the graver vibrations; thus illustrating the dependence of the period on the capacity and self-induction of the vibrator. There are also overtones in which the pulse of normal displacement is almost confined to the connecting wire, the greater mobility at its ends making those points approximately nodes; and their nodal character will be the more prominent the greater the capacity per unit area at the points where they are attached to the condensers. For these overtones the half wave-length is thus a sub-multiple of the length of the connecting wire.

Hertzian electric oscillator.

Owing to the small surface of the connecting wire these overtones would not have much chance of being communicated to the surrounding medium, except by reason of the general principle which requires that the shorter the period for given dimensions of the vibrator the more of the vibrational energy travels outwards into the medium, and that in a ratio which increases very rapidly with increasing frequency. In the Hertzian oscillator the condensers are replaced by large metallic plates, and it seems clear that the waves that are the chief subject of experiment issue from the large surfaces afforded by these plates, and so belong to the lower periods of the vibrator.

The waves of a Hertzian vibrator are therefore radiated from the plates of the vibrator: to obtain considerable radiation to a distance it is essential that the dimensions of a plate should be considerable compared with the length of a wave of the radiation. This condition would not be fulfilled if the plates were replaced by condensers, for though the energy of the vibration would thereby be increased, its wave-length for the fundamental periods would be lengthened, and the radiation, therefore, very much diminished. Condensers would be allowable instead of plates when the disturbance is guided along conducting wires, but as a rule they would give only slight undulations in the dielectric at a distance from conductors.

The plates are thus like two poles radiating in opposite phases: and the general equations given by Kirchhoff to represent this type of motion are those used by Hertz in his discussion of the general character of the radiation at a distance from the vibrator.

Analysis of the model.

(The different types of vibration that may theoretically exist would therefore appear to be as follows. A reciprocating flow may be set up between the plates, along the connecting wires; if the capacities are at all considerable its period will be comparatively slow, being calculable from capacity of condenser and self-induction of wire alone, because there will be involved very little disturbance in the dielectric except the static value of the displacement at each instant, and there will be no sensible amount of radiation. An oscillation of the dielectric to and fro between the two plates may be set up, corresponding to a very small period, the wave-length being twice the distance between the plates in the case in the diagram, when the earth connection is good; this will not be sensibly affected by the presence or nature of the connecting wire, and it will involve rapid decay by radiation, but it will probably be very difficult to excite to any sensible amount. And there may be superficial dielectric waves running along the connecting wire, of period about the same as the preceding when the wire is straight.

For the linear type of vibrator consisting of two equal cylinders with a spark gap between them, and no capacities at the ends, the wave-length would be the length of the vibrator simply, as from the mode of excitation its two halves would always be in opposite phases.)

Soakage into a conductor:

The nature of the approximations involved in this method of representation will be sufficiently put in evidence by the following investigation of the circumstances of the reflexion of a system of waves from a metallic plate. It will be found that for wave-lengths of a centimetre or more the reflexion of all ordinary metallic plates is sensibly perfect, and involves an acceleration of phase of half a wave-length. For smaller wave-lengths, corresponding to those of light waves, the circumstances of the reflexion are more complicated.

restriction to Maxwellian theory.

We shall proceed on Maxwell's theory, which postulates the non-existence of condensational effects: no other theory can make the velocity of propagation of waves of transverse displacement inversely proportional to the square root of the specific inductive capacity of the medium[1].

The first point is to assume precise definitions of the quantities that enter into the equations. With the usual notation

$$c = \frac{dG}{dx} - \frac{dF}{dy}, \quad a = \ldots, \quad b = \ldots,$$

so that, if J denote

$$\frac{dF}{dx} + \frac{dG}{dy} + \frac{dH}{dz},$$

we have

$$\nabla^2 H - \frac{dJ}{dx} = \frac{db}{dx} - \frac{da}{dy}$$

$$= -4\pi\mu w.$$

[1] See *Proc. Roy. Soc.* Vol. XLIX. (1891) p. 521: as *infra*, p. 264.

Hence *defining FGH* by the formula

Static potential of *total* current,

$$F, G, H = \int \frac{d\tau}{r} \mu (u, v, w),$$

we annul J, since $$\frac{du}{dx} + \frac{dv}{dy} + \frac{dw}{dz} = 0,$$

owing to the absence of condensation.

The components of electric force are

$$P = -\frac{dF}{dt} - \frac{dV}{dx},\ Q = \ldots,\ R = \ldots,$$

where V is a function whose presence is necessitated by the fact that the currents must be circuital; it must enter in this form in order that it may produce no result on integration round a circuit, and it is completely determined by the conditions of any special problem. The fact that V must be a single-valued function so as to give a null result on integrating round a circuit shows that it may be expressed as the potential of an electrostatic distribution; thus its introduction into the equations is accounted for consistently with ordinary electrical ideas.

and of charge.

In a dielectric the total current *uvw* is the displacement current in the dielectric, so that

$$(u, v, w) = \frac{K}{4\pi} \frac{d}{dt} (P, Q, R).$$

In a conductor the displacement current is evanescent compared with the current conducted according to Ohm's law, so that

$$\sigma (u, v, w) = (P, Q, R),$$

where σ is the specific resistance of the medium*.

Also, by definition above,

$$\nabla^2 (F, G, H) = -4\pi\mu (u, v, w).$$

Thus in a dielectric the equations of propagation of the vector *FGH* are of the type

Dielectric propagation.

$$\frac{1}{\mu K} \nabla^2 F = \frac{d^2F}{dt^2} + \frac{d^2V}{dx\,dt};$$

in a conductor, they are of the type

$$\frac{\sigma}{4\pi\mu} \nabla^2 F = \frac{dF}{dt} + \frac{dV}{dx}.$$

Slow conduction.

In the former case we derive at once

$$\frac{d}{dt} \nabla^2 V = 0,$$

No convection in uniform dielectric.

* For short periods comparable with light σ must be replaced throughout by σ', equal to $\sigma + \frac{K}{4\pi c^2}\frac{d}{dt}$, where $\frac{d}{dt}$ is ιp.

which merely expresses the fact that there can be no accumulation of volume-density in the dielectric, on account of the assumed non-condensational character of the phenomena.

No free density in uniform conductor.

In the latter case also

$$\nabla^2 V = 0.$$

Thus, when the dielectric has no initial charge, the charge is throughout confined to the interfaces separating media of different quality, such as the surfaces of conductors.

It will form a sufficiently general case for our purpose if we consider the reflexion of a train of plane polarized waves at a plane metallic surface. The axis of z may be taken normal to the surface and the axis of x along it in train with the waves. If the vibration of the vector potential is in the plane of incidence, we have

$$\frac{dF}{dx} + \frac{dH}{dz} = 0, \quad G = 0,$$

so that the variables are reduced to dependence on a single function χ by the substitution

$$F = \frac{d\chi}{dz}, \quad H = -\frac{d\chi}{dx}.$$

The important case is when everything is periodic, and so involves the factor exp $(-\iota pt)$. We have then

$$F = F' - \frac{1}{\iota p}\frac{dV}{dx}, \quad H = H' - \frac{1}{\iota p}\frac{dV}{dz},$$

where

$$F' = \frac{d\chi'}{dz}, \quad H' = -\frac{d\chi'}{dx};$$

and χ' is determined by the equations

$$\frac{1}{\mu_1 K_1}\nabla^2\chi' = \frac{d^2\chi'}{dt^2}$$

in the dielectric, and

$$\frac{\sigma_2}{4\pi\mu_2}\nabla^2\chi' = \frac{d\chi'}{dt}$$

in the conductor.

At the interface F is continuous, so that $\frac{d\chi}{dz}$ is continuous; and H is continuous, so that

$$-\frac{d\chi'}{dx} - \frac{1}{\iota p}\frac{dV}{dz} \text{ is continuous;}$$

these follow from the definition of F, G, H as the potentials of volume distributions.

Further, the component magnetic induction along the normal must be continuous across the interface, which being zero it is; and the magnetic force along the interface must be continuous, so that $\frac{1}{\mu}\left(\frac{dF}{dz} - \frac{dH}{dx}\right)$ is continuous across the surface, that is

$$\frac{1}{\mu}\nabla^2\chi' \text{ is continuous,}$$

or by the equations for χ',

$$K_1 \frac{d^2\chi_1'}{dt^2} = \frac{4\pi}{\sigma_2}\frac{d\chi_2'}{dt},$$

or finally, for this special case of harmonic waves

$$-\iota p K_1 \chi_1' = \frac{4\pi}{\sigma_2}\chi_2'.$$

Let $\chi_1' = A_1 \exp \iota (lx - n_1 z - pt) + B_1 \exp \iota (lx + n_1 z - pt)$, Wave-train reflected by a metal.

$\chi_2' = A_2 \exp \iota (lx - n_2 z - pt)$,

A_1, B_1 thus representing the coefficients of the incident and reflected waves, and A_2 that of the surface-wave in the conductor for which n_2 must in the result be complex.

The value of l must be the same for all these waves, as their traces on the surface must move along it with the same velocity.

The differential equations satisfied by χ' give

$$\frac{1}{\mu_1 K_1}(l^2 + n_1^2) = p^2, \quad \frac{\sigma_2}{4\pi\mu_2}(l^2 + n_2^2) = \iota p.$$

The first of these gives the velocity of the dielectric wave to be $(\mu_1 K_1)^{-\frac{1}{2}}$, as it ought to be. The second gives the penetration of the surface-wave into the conductor by the equation

$$n_2^2 = l^2 - \frac{4\pi\mu_2 p}{\sigma_2}\iota.$$

If i denote the angle of incidence of the waves, λ their length, and v their velocity, we must have

$$ln + n_1 z - pt \equiv \frac{2\pi}{\lambda}(x \sin i + z \cos i - vt).$$

Thus
$$\frac{p}{l^2} = \frac{v}{\sin i}\,\frac{\lambda}{2\pi \sin i};$$

where $v = (\mu_1 K_1)^{-\frac{1}{2}}$ is of the order $3 \cdot 10^{10}$ C.G.S. For light waves λ is of the order 10^{-4}, and for copper σ_2 is about 1600 so that σ_2' of the footnote replaces σ_2.

Hence for all realizable wave-lengths of electric wave-trains, long or short,

$$n_2 = \left(\frac{4\pi\mu_2 p}{\sigma_2}\right)^{\frac{1}{2}}\frac{1-\iota}{2} = \pi\left(\frac{2\mu_2 v}{\sigma_2 \lambda}\right)^{\frac{1}{2}}(1 - \iota),$$

which for light waves is of the order 10^5, and is still very large for all realizable electric waves, unless for media of slight conductivity compared with metals. The amplitude of the surface-wave in the conductor is reduced in the ratio e^{-1} at a depth n_2^{-1}; this wave is therefore absolutely superficial, and is fully developed even in a very thin sheet of metal.

The surface conditions give

$$n_1 (A_1 - B_1) = n_2 A_2,$$

$$l (A_1 + B_1 - A_2) = -\frac{\iota}{p} 4\pi\sigma_0$$

$$-\iota p K_1 (A_1 + B_1) = \frac{4\pi}{\sigma_2} A_2,$$

of which the first and third equations determine the amplitudes of the waves produced by the reflexion of A_1, and the second determines the surface density

$$\sigma_0 \exp \iota (lx - pt)$$

of the superficial electric charge.

Hence $$\left(\frac{n_2}{n_1} + \frac{4\pi}{p\sigma_2 K_1}\iota\right) A_2 = 2A_1,$$

i.e. $$A_2 = 2\frac{\dfrac{n_2}{n_1} - \dfrac{4\pi}{p\sigma_2 K_1}\iota}{\left(\dfrac{n_2}{n_1}\right)^2 + \left(\dfrac{4\pi}{p\sigma_2 K_1}\right)^2} A_1,$$

and $$B_1 = A_1 - \frac{n_2}{n_1} A_2.$$

The relative magnitudes of the two terms in the numerator of A_2 have to be estimated; the terms to be compared are of the orders

$$\left(\frac{v\lambda}{\sigma_2}\right)^{\frac{1}{2}}, \text{ and } \frac{\lambda}{v\sigma_2 K_1} \text{ or } \frac{\lambda v}{\sigma_2},$$

that is $10^4\lambda^{\frac{1}{2}}$ and $10^7\lambda$, roughly.

Conduction and polarization equally important for light periods: For light waves, λ is of the order 10^{-4}, so that neither of these terms can be neglected compared with the other*; and the completion of the solution will correspond to the somewhat complicated circumstances of the metallic reflexion of light.

But for electric waves comparable to a centimetre in length, or longer, the second term is negligible; and then

not for longer ones.

$$A_2 = 2\left(\frac{n_2}{n_1}\right)^{-1} A_1,$$

and $$B_1 = -A_1.$$

* For waves of visible light σ must be replaced by σ' throughout. Contrast the Rubens infra-red defect from perfect reflexion from metals, found to be proportional to $\sqrt{\sigma}$ where σ is now static resistance, K proving to be already unimportant.

The wave is therefore reflected clean but with opposite phase. And the value of n_2 given above shows that the longer the waves the slighter is their penetration into the conductor; so that even for a curved surface like that of a wire this solution has an application.

Metals actually perfectly labile to electric waves.

The meaning of this approximation is that the first surface condition, in the form it assumes when n_2 is very great, supplies all the necessary data for the motion in the dielectric. That surface condition is equivalent to the statement that F is zero in the dielectric along the surface, and therefore so is the tangential displacement.

Thus the tangential surface conditions suffice in this case to give a full account of the dielectric phenomena, the normal conditions being simply left to take care of themselves.

Static potentials instantaneous so artificial.

The function V by which the adjustment is made in the conductor to obtain the normal displacement at the surface which shall satisfy the condition of zero condensation is derived from the characteristic equation $\nabla^2 V = 0$, the same equation as that for the pressure in a homogeneous massless fluid; it of course indicates instantaneous adjustment to an equilibrium value throughout the volume.

26

ON A GENERALIZED THEORY OF ELECTRODYNAMICS.

[*Proc. Roy. Soc.* Vol. XLIX. (1891) pp. 521–536.]

THE electrical ideas of Clerk Maxwell, which were cultivated partly in relation to mechanical models of electrodynamic action, led him to the general principle that electrical currents always flow round complete circuits.

Maxwell's hypothesis of currents: effectively circuital:

To verify this principle for the case of the current which charges a condenser, it was necessary to postulate an electrodynamic action of the same type as that of a current for the electric displacement across the dielectric, in which the excitation of the dielectric may be supposed, after Faraday, to consist. The existence of such an action has subsequently been deduced *qualitatively* from the general principle of action and reaction[1], and has also been detected by various experimenters.

involving a total dielectric displacement of solenoidal character:

The principle also requires that the electric displacement shall not lead to any accumulation of charge in the interior of the dielectric, therefore that it shall be solenoidal or circuital[2], its characteristic equation being of the type

$$\frac{d}{dx}\left(K\frac{dV}{dx}\right)+\frac{d}{dy}\left(K\frac{dV}{dy}\right)+\frac{d}{dz}\left(K\frac{dV}{dz}\right)=0,$$

where V is the electric potential, and K a dielectric constant. The surface density of the electricity conducted to a face of a condenser must neutralize the electric displacement, and not leave any residual effective electrification on the surface. On taking the displacement and the surface density each equal to $KF/4\pi$, where F denotes the electric force, the value of K becomes unity for a vacuum dielectric; and K represents the specific inductive capacity as measured by electrostatic experiments.

When this principle of circuital currents is postulated, the theory of electrodynamics is reduced to the Ampère-Neumann theory of complete circuits, of which the truth has been fully established. It leads, as shown by Maxwell, to the propagation of electrical action in dielectric media by waves of transverse electric displacement,

leads to purely transverse waves like light.

[1] Cf. J. J. Thomson, *Brit. Assoc. Report*, 1885.

[2] A term recently introduced by Sir W. Thomson.

which have the intimate relations to waves of light that are now well known.

Generalized Polarization Theory.

The problem of determining how far these remarkable conclusions will still hold good when a more general view of the nature of dielectric polarization is assumed was considered by von Helmholtz[1] in a series of memoirs. Comparison with a wider theory,

The most general conception of the polarization of a medium which has been formed is the Poisson theory of magnetization. The magnetized element, whether actually produced by the orientation of polar molecules or otherwise, may be mathematically considered to be formed by the displacement of a quantity of ideal magnetic matter from its negative to its positive pole, thereby producing defect at the one end, and excess at the other end. The element is defined magnetically by its moment, which is the product of the displaced quantity and the distance through which it is displaced. The displacement per unit volume, measured by this product, is equal to the magnetic moment per unit volume, whether the magnetized molecules fill up the whole of that volume or are a system of discrete particles with unoccupied space between them.

In the electric analogue we replace ideal magnetic matter by ideal electric matter; the displacement thus measured constitutes the electric displacement, and its rate of change per unit time represents the displacement current in the dielectric. We have to consider whether a displacement current of this type suffices to make all electric currents circuital; and it will be sufficient and convenient to examine the case of a condenser which is charged through a wire connecting its two plates. In the first place this notion of electric displacement leads to the same distribution of potential between the plates as the ordinary one, adopted by Maxwell; for in the theory of induced magnetism there occurs a vector quantity of circuital character, the magnetic induction of Maxwell, of which the components are $-\mu\,(dV/dx)$, $-\mu\,(dV/dy)$, $-\mu\,(dV/dz)$, and which, therefore, leads to the characteristic equation of the potential analogous to actual magnetic polarization

$$\frac{d}{dx}\left(\mu\frac{dV}{dx}\right)+\frac{d}{dy}\left(\mu\frac{dV}{dy}\right)+\frac{d}{dz}\left(\mu\frac{dV}{dz}\right)=0,$$

corresponding to the one given above. If the displacement in the dielectric is $-\kappa\,(dV/dx)$, $-\kappa\,(dV/dy)$, $-\kappa\,(dV/dz)$, then

$$\mu = 1 + 4\pi\kappa.$$

The displacement in a unit cube may, of course, be considered as a displacement across the opposite faces of the cube.

[1] *Wissenschaftliche Abhandlungen*, I. p. 545, *et seq.*

Now, considering the case of a plane condenser, let F be the electric force in the dielectric between the plates; then the displacement is κF. Let σ be the surface density of the charge conducted to a plate; then the effective electrification along that plate will be of surface density $\sigma' = \sigma - \kappa F$; therefore, by Coulomb's principle,

$$F = 4\pi\sigma'$$
$$= 4\pi(\sigma - \kappa F);$$

so that
$$\sigma = \frac{\mu}{4\pi} F = \kappa F + \frac{1}{4\pi} F.$$

for which the current is not circuital at the interface,

Thus the current is not circuital, but there is an excess of the surface density conducted to the surface over the displacement current from the surface, which is equal to $F/4\pi$.

The specific inductive capacity, as determined by static experiments on capacity, is here measured by μ, the coefficient in the expression for σ.

nor in the dielectric,

In addition to this discontinuity at the face of a condenser plate, the induction in the mass of the dielectric will not be circuital unless the electric force is itself circuital, which it is not in the general electrodynamic theory to be presently discussed.

except in a limiting case,

The current becomes more nearly circuital the greater the value of μ. If μ, and therefore κ, were infinite we should attain the limit when the currents are circuital. If the values of μ for all dielectrics were multiplied by the same infinite constant, so as to keep their ratios unchanged, the distribution of electric potential would not be altered, provided the charges on all conducting surfaces were also increased in that ratio; the displacement or induction, which is now the essential quantity in the theory, thus maintaining its original value. This comes to the same thing as measuring the actual charges in a unit which is diminished in that ratio.

which represents Maxwell's scheme,

In this way the Maxwell scheme of circuital currents reveals itself as a limiting case of the more general polarization theory. The infinite dielectric constant makes the excited polarization of very great amount in comparison with the exciting cause; so that in the limit we may, in a sense, imagine the system as one of self-excited circuital polarization, a point of view which approaches somewhat to that of Maxwell himself.

in Helmholtz's transition.

This mode of connecting the two theories was pointed out by von Helmholtz. But his scheme takes for the new unit of charge the electrostatic unit corresponding to vacuum with its new infinitely great dielectric constant, so that this unit is reduced proportionally to the square root of the infinite ratio; the displacement is then infinitely great, and the potential infinitely small, according to the square root of this ratio.

(This, however, should be expressed more precisely as follows: The *absolute* dimensions of electric charge and electric displacement in K are both $K_2^{\frac{1}{2}}$, those of electric force (static) $K_2^{-\frac{1}{2}}$. These dimensional relations must persist when the transition is made from von Helmholtz's system to Maxwell's, so that the changes in the units are as von Helmholtz indicates; and the ratio of the electrostatic to the electrokinetic unit of quantity in an ideal absolute medium with K_2 unity will now be the ascertained value of this constant for air or vacuum multiplied by the square root of the value of K_2 for air. The electric pressure in a fluid dielectric, however, depends, in this limiting form of the theory, on the square of the value of the electric displacement, as may be proved: thus the circumstances of ordinary cases of statical electrification are those of finite numerical value of the displacement, notwithstanding the smallness of this absolute unit of charge.)

Generalized Electrodynamic Theory.

To obtain the general type of the modification of which the theory of electrodynamics is susceptible owing to the existence of non-circuital currents, we start, following von Helmholtz, from the ascertained laws for circuital currents, which may be developed in the manner of Neumann and Maxwell from the electrodynamic potential* — **The Neumann electrodynamic energy function,**

$$T = u' \iint \frac{\cos \epsilon}{r} \, ds \, ds'.$$

The value of T with the sign here given to it is to be reckoned as kinetic energy; the mechanical forces are to be derived by its

* In the fourfold relativity-group of optical frames of reference, conditioned by invariance of

$$\delta x^2 + \delta y^2 + \delta z^2 - c^2 \delta t^2,$$

it is

$$\Sigma\Sigma \frac{1}{r_{12}} (e_1 dx_1 e_2 dx_2 + \ldots + \ldots - c^2 e_1 dt_1 e_2 dt_2)$$

that is invariant as regards the differential factor, and thus far must be adopted as the fourfold mutual Action. As $\Sigma e dx$ averaged over the element of volume $d\tau$ becomes $d\tau . u dt$, this becomes for a pair of elements of volume

$$d\tau_1 d\tau_2 \frac{1}{r_{12}} (u_1 u_2 + v_1 v_2 + w_1 w_2 - c^2 \rho_1 \rho_2) \, dt_1 dt_2,$$

where (u, v, w) is current and ρ is density of charge, thus integrating for the complete system to

$$\iint d\tau_1 d\tau_2 \iint \frac{1}{r_{12}} (\iota_1 \iota_2 \cos \epsilon - c^2 \rho_1 \rho_2) \, dt_1 dt_2, \quad \epsilon = (\iota_1 \iota_2).$$

This is identical with the Neumann electrodynamic form, now completed however by the electrostatic term. It is optically invariant (except as regards $1/r_{12}$) while the Helmholtz modification is not. — **Optically invariant.**

It has to be varied subject to the condition of persistence of charge

$$\frac{du}{dx} + \frac{dv}{dy} + \frac{dw}{dz} + \frac{d\rho}{dt} = 0,$$

variation due to any virtual displacement of the system, a force acting in the direction of the displacement producing an increment of T; the electric forces are derived according to Lenz's law or Maxwell's kinetic theory. The equations of the field are thus all expressible in terms of this function T. When non-circuital currents are contemplated, the currents ι, ι', now varying with s, s', must be placed inside the integral signs; and to T must be added the most general type of expression that will vanish when either current is circuital. Thus we must write

now generalized, after Helmholtz:

$$T = \iint \iota ds\, \iota' ds' \left(\frac{\cos\epsilon}{r} + \frac{d^2\Psi}{ds\,ds'}\right),$$

where Ψ is a function such that

$$\int \frac{d\Psi}{ds} ds = 0, \qquad \int \frac{d\Psi}{ds'} ds' = 0,$$

i.e. it is, so far, any function which has no cyclic constant round either circuit. The distribution of the energy between the pairs of elements is now supposed to be specified by the elements of this integral.

The form of Ψ is limited by the fact that it must be a function of the geometrical conformation of the pair of elements. The elements of this conformation are given by the equations

$$\cos\theta = -\frac{dr}{ds},$$

$$\cos\theta' = \frac{dr}{ds'},$$

$$\cos\epsilon = -\frac{d}{ds'}\left(r\frac{dr}{ds}\right)$$

$$= -\frac{dr}{ds}\frac{dr}{ds'} - r\frac{d^2r}{ds\,ds'},$$

which also is optically invariant, being the fourfold convergence of $(u, v, w, \iota c\rho)$ which is a vector because its ultimate element $(edx, edy, edz, e\iota cdt)$ is a vector in the fourfold.

Modified Action-theory. The development in *Aether and Matter*, Chapter VI, but introducing the fourfold potential (F, G, H, V) by its curl instead of separate vector and scalar forms, leads to the usual equations of the field now with retarded potentials, thus evading direct intervention of the aether.

The contrast to Maxwell's tentative discussion, *Treatise*, §§ 598–600, is instructive. But this is not yet satisfactory. In the fourfold the potential factor $1/r_{12}$ is not invariant: it must be replaced by $1/R_{12}^2$, where

$$R_{12}^2 = (x_1 - x_2)^2 + \ldots + \ldots - (ct_1 - ct_2)^2.$$

How propagated potentials arise. It will appear, however, in another connection that the infinities in the integration conduct straight to a three-dimensional theory in ordinary time, with $1/|r_{12}|$ now as a propagated potential, but unfortunately requiring incoming as well as outgoing propagation.

where r is their distance apart, $\theta, \theta', \epsilon$ represent the angles $r \,.\, ds, r \,.\, ds'$, $ds \,.\, ds'$, r being measured positive from ds to ds'.

The only function of the type $d^2\Psi/ds\,ds'$ which can be specified in terms of these quantities is $d^2\phi(r)/ds\,ds'$, which is equal to

$$r^{-1}\phi'(r)(\cos\theta\cos\theta' - \cos\epsilon) + \phi''(r)\cos\theta\cos\theta'.$$

On substitution we have

expressed analytically.

$$T = \iint \iota ds\, \iota' ds' \left\{ -\frac{1}{r}\frac{dr}{ds}\frac{dr}{ds'} - \frac{d^2(r - \phi r)}{ds\,ds'} \right\},$$

in which the elements of the energy are supposed to be correctly localized.

To obtain the mutual mechanical forces between the conductors we have to determine the variation in T produced by the most general virtual displacements of the separate elements which do not alter these elements, nor break the continuity of either circuit. Thus ds, ds', ι, ι' are not to be varied.

Mode of conducting its dynamical variation.

The shortest way to take account of currents which are not of the same strength all along the circuit is to consider two uniform currents ι, ι' flowing in interrupted circuits, and examine the terms of the variation involving the terminal points at which electric charges are being accumulated by the currents flowing into them. Of course the same general results would flow from taking ι, ι' functions of s, s' respectively and neglecting the ends. Thus, employing electromagnetic units and so avoiding a numerical coefficient, we have, after F. E. Neumann and von Helmholtz,

$$T = \iint \iota ds\, \iota' ds' \left\{ -\frac{1}{r}\frac{dr}{ds}\frac{dr}{ds'} - \frac{d^2(r - \phi r)}{ds\,ds'} \right\};$$

$$\delta T = \iint \iota ds\, \iota' ds' \left\{ \frac{\delta r}{r^2}\frac{dr}{ds}\frac{dr}{ds'} - \frac{1}{r}\frac{d\delta r}{ds}\frac{dr}{ds'} - \frac{1}{r}\frac{dr}{ds}\frac{d\delta r}{ds'} - \frac{d^2\delta(r - \phi r)}{ds\,ds'} \right\}$$

$$= -\int \left| \delta r \right|_{s_1}^{s_2} \iota' ds'\, \iota \frac{1}{r}\frac{dr}{ds'} - \int \left| \delta r \right|_{s_1'}^{s_2'} \iota ds\, \iota' \frac{1}{r}\frac{dr}{ds}$$

$$+ \iint \iota ds\, \iota' ds'\, \delta r \left\{ \frac{1}{r^2}\frac{dr}{ds}\frac{dr}{ds'} + \frac{d}{ds}\left(\frac{1}{r}\frac{dr}{ds'}\right) + \frac{d}{ds'}\left(\frac{1}{r}\frac{dr}{ds}\right) \right\}$$

$$- \left\| \iota\iota' (1 - \phi' r)\, \delta r \right|_{s_2}^{s_1} \Big|_{s_1'}^{s_2'}.$$

This variation is accounted for by the following forces of repulsion, tending to increase r.

Helmholtz's interpretation of result as mutual forces

(i) Between the elements ιds and $\iota' ds'$, equal to

$$\iota ds\, \iota' ds' \left(-\frac{1}{r^2}\frac{dr}{ds}\frac{dr}{ds'} + \frac{2}{r}\frac{d^2 r}{ds\,ds'} \right),$$

or

$$-2\iota ds\, \iota' ds' \frac{1}{r^2}\left(\cos\epsilon - \tfrac{3}{2}\cos\theta\cos\theta'\right),$$

which is Ampère's law.

(ii) Between the element ιds and the positive end of the conductor ds',

$$\iota ds\, \iota' \frac{1}{r}\frac{dr}{ds'},$$

or
$$-\iota ds \frac{de'}{dt}\frac{1}{r}\cos(r\,.\,ds),$$

where de'/dt is the rate at which the charge at that end is increasing.

(iii) Between the element $\iota' ds'$ and the end of the conductor ds,

$$\iota' ds'\, \iota \frac{1}{r}\frac{dr}{ds'},$$

or
$$\iota' ds' \frac{de}{dt}\frac{1}{r}\cos(r\,.\,ds'),$$

r being here measured away from ds'.

(iv) Between an end of one conductor and an end of another conductor,

$$-\iota\iota'(1-\phi' r),$$

or
$$-\frac{de}{dt}\frac{de'}{dt}(1-\phi' r).$$

It is to be observed that the form of $\phi(r)$ affects only the forces (iv) in this scheme of attraction, as one would expect from the fact that $\phi(r)$ disappears if either current flows round a complete circuit.

Not to refer to (ii) and (iii), we notice from (iv) that two changing electrifications attract each other with a force involving a term which is constant at all distances, unless a special form of $\phi(r)$ be assigned differing from any of the values which occur in the sequel. It is difficult to imagine the mechanical basis of such an action; the remarks of von Helmholtz in justification (against Bertrand) may, however, be referred to[1].

An anomaly.

Present form more general.

This investigation of the mechanical forces is equivalent to von Helmholtz's with the exception that he takes at the beginning $\phi(r)$ to be proportional to r, on the general ground that the potential energy of two elements in all natural actions involves only the inverse first power of the distance. The validity of this consideration seems to be weakened by the fact noticed above that $\phi(r)$ occurs only in the force (iv). For what follows it will not be necessary to restrict the form of $\phi(r)$.

Translation from linear into spatial current analysis.

To discuss the propagation of electrical action in continuous media, we have to translate T from the form suitable to linear distributions to the form suitable to volume distributions. Following the method first developed by Kirchhoff, and for this case the analysis of von Helmholtz, the energy function for any field of currents is to be

[1] *Wissen. Abhandl.* I. p. 708.

obtained by summation of the energy functions of all the pairs of elementary filaments of currents that compose it, care being taken that no pair is counted twice over. The proper form will be a volume integral; instead of ds, ds', the elements of the filament, it will involve $d\tau$, $d\tau'$, the elements of volume, and instead of ι, ι', the resultant currents, it will involve their components per unit sectional area uvw and $u'v'w'$.

Thus $$T = \iint \frac{\iota ds\, \iota' ds' \cos\epsilon}{r} + \iint \iota \frac{d}{ds}\, \iota' \frac{d}{ds'}\, \phi(r)\, ds\, ds'$$

$$= \tfrac{1}{2}\iint \frac{1}{r}(uu' + vv' + ww')\, d\tau\, d\tau'$$

$$+ \tfrac{1}{2}\int d\tau \left(u\frac{d}{dx} + v\frac{d}{dy} + w\frac{d}{dz}\right)\int d\tau' \left(u'\frac{d}{dx'} + v'\frac{d}{dy'} + w'\frac{d}{dz'}\right)\phi(r),$$

the factors $\frac{1}{2}$ being inserted because the volume integrals, being extended all over the system, take each pair of elements twice over.

Hence
$$T = \tfrac{1}{2}\int (Fu + Gv + Hw)\, d\tau + \tfrac{1}{2}\int \left(u\frac{d\chi}{dx} + v\frac{d\chi}{dy} + w\frac{d\chi}{dz}\right) d\tau,$$

where $$F = \int \frac{u'}{r}\, d\tau', \quad G = \int \frac{v'}{r}\, d\tau', \quad \cdot H = \int \frac{w'}{r}\, d\tau',$$

In addition to the vector potential an undetermined function appears:

$$\chi = \int \left(u'\frac{d\phi}{dx'} + v'\frac{d\phi}{dy'} + w'\frac{d\phi}{dz'}\right) d\tau';$$

in these formulae the accents may now be dropped, as the integrals are extended over the whole system.

It is through this function χ that the indeterminateness enters into the equations of electrodynamics. In a certain class of cases the function may be expressed in another form, which is useful in the subsequent analysis. By integration by parts throughout space, we obtain

its form rearranged by integration by parts:

$$\chi = \int dS\phi\, (lu + mv + nw) - \int d\tau\phi \left(\frac{du}{dx} + \frac{dv}{dy} + \frac{dw}{dz}\right)$$

$$= \int d\tau\phi \frac{d\rho}{dt},$$

provided we can neglect the surface integral over the infinite sphere; and this we can do, if the system is confined to a finite region and ϕ contains only inverse powers of r, or it may be direct powers of r when there is no total current flow to infinity. Thus

$$\chi = -\frac{1}{4\pi}\int d\tau\phi\nabla^2\frac{dV}{dt}$$

$$= -\frac{1}{4\pi}\int d\tau\frac{dV}{dt}\nabla^2\phi,$$

by Green's theorem, provided the surface integrals vanish as before.

In this equation,

$$\nabla^2\phi = \left(\frac{d^2}{dr^2} + \frac{2}{r}\frac{d}{dr}\right)\phi;$$

simplified by Helmholtz's hypothesis,

so that, on von Helmholtz's assumption

$$\phi(r) = Cr,$$

we have

$$\nabla^2\phi = 2C/r;$$

and therefore

$$\chi = -\frac{1}{4\pi}\int d\tau \frac{2C}{r}\frac{dV}{dt}$$

so that

$$\nabla^2\chi = 2C\frac{dV}{dt},$$

a result to be used immediately.

The final result is

$$T = \tfrac{1}{2}\int d\tau\,(F_1 u + G_1 v + H_1 w),$$

where

$$F_1 = F + \frac{d\chi}{dx}, \quad G_1 = G + \frac{d\chi}{dy}, \quad H_1 = H + \frac{d\chi}{dz},$$

to his final formulation.

$$\chi = \int d\tau \left(u\frac{d\phi}{dx} + v\frac{d\phi}{dy} + w\frac{d\phi}{dz}\right)$$

$$= \frac{1}{4\pi}\int d\tau \frac{dV}{dt}\left(\frac{d^2\phi}{dr^2} + \frac{2}{r}\frac{d\phi}{dr}\right).$$

To obtain the components *PQR* of the electric force we assume, following F. E. Neumann and von Helmholtz, that the principle involved in Lenz's law is applicable to the element as well as to the circuit as a whole. This is the same principle as flows from Maxwell's dynamical theory, and is justified, if we assume that *T* is the energy function of an actual dynamical system. To the kinetic part of the electric force so determined the electrostatic part must be added, giving in all the components

Results of dynamical variation: localized by a scalar potential:

$$P = -\frac{dF_1}{dt} - \frac{dV}{dx}, \quad Q = -\frac{dG_1}{dt} - \frac{dV}{dy}, \quad R = -\frac{dH_1}{dt} - \frac{dV}{dz}.$$

The conduction current is given by

completed scheme.

$$\sigma(u_1, v_1, w_1) = (P, Q, R),$$

where σ is the specific resistance. The total current is

$$(u, v, w) = \left(u_1 + \frac{df}{dt}, \quad v_1 + \frac{dg}{dt}, \quad w_1 + \frac{dh}{dt}\right),$$

where

$$(f, g, h) = \frac{K_1}{4\pi}(P, Q, R).$$

The vector potential FGH is connected by definition above with uvw by the equations of potential

$$\nabla^2 (F, G, H) = -4\pi (u, v, w);$$

while the characteristic equation of V is

$$\nabla^2 V = -4\pi\rho$$
$$= 4\pi \left(\frac{df}{dx} + \frac{dg}{dy} + \frac{dh}{dz}\right).$$

The equations of electric propagation are involved in these results. The value of K_1 in electromagnetic units is very small, the square of the reciprocal of the velocity of light in the medium; so that there are, broadly, two classes of media, (i) conductors in which K_1 is neglected, (ii) insulators in which $u_1 v_1 w_1$ are zero. The equations of propagation for each case are involved in the above equations.

Propagation in Dielectric Media.

The simplest and most important case of this generalized theory, as displacement currents in conductors are negligible, is that of dielectrics.

In the first place, we may consider the propagation of V. We have **Propagation of the scalar potential.**

$$\nabla^2 V = 4\pi \left(\frac{df}{dx} + \frac{dg}{dy} + \frac{dh}{dz}\right)$$
$$= -K_1 \frac{d}{dt}\left(\frac{dF_1}{dx} + \frac{dG_1}{dy} + \frac{dH_1}{dz}\right) - K_1 \nabla^2 V.$$

Now
$$\frac{dF}{dx} + \frac{dG}{dy} + \frac{dH}{dz} = \int \frac{1}{r}\left(\frac{du}{dx} + \frac{dv}{dy} + \frac{dw}{dz}\right) d\tau$$
$$= -\int \frac{1}{r}\frac{d\rho}{dt}\, d\tau$$
$$= -\frac{dV}{dt}.$$

Therefore
$$\frac{1 + K_1}{K_1} \nabla^2 V = \frac{d^2 V}{dt^2} - \frac{d}{dt}\nabla^2 \chi,$$

and, finally,
$$\frac{1 + K_1}{K_1} \nabla^2 V = \frac{d^2 V}{dt^2} - \frac{1}{4\pi}\nabla^2 \int \nabla^2 \phi \frac{d^2 V}{dt^2}\, d\tau,$$

in which ϕ is a function of r.

This equation determines the mode of propagation of V. It represents wave-motion of a complicated character which may be analysed most easily by applying the equation to the case of a plane wave with the displacement at right angles to its front. There are two comparatively simple cases.

A simplifying assumption:

(i) If $\nabla^2\phi = 0$, *i.e.* $\phi = A + Br^{-1}$, the equation becomes

$$\frac{1+K_1}{K_1}\nabla^2 V = \frac{d^2V}{dt^2} + B\frac{d^2}{dt^2}\nabla^2 V,$$

which represents wave-propagation with velocity depending on the wave-length, and therefore involving dispersion.

For the plane wave $V \propto \exp\iota\,(mx - nt)$,
it leads to the condition

$$\frac{1+K_1}{K_1}m^2 = n^2 - Bm^2n^2,$$

gives for it propagation dispersively in waves:

and the velocity of propagation is

$$v = \left(\frac{1+K_1}{K_1}\right)^{\frac{1}{2}}\left(1 - \frac{4\pi^2 B}{\lambda^2}\right)^{-\frac{1}{2}},$$

where λ is the wave-length.

unless ϕ is constant:

The special case of $\phi\,(r)$ equal to zero is worth notice, as that would represent a theory in which the element of Neumann's integral, viz. $\iota ds\,\iota' ds' \cos\epsilon/r$, is the mutual energy of two current elements. When the currents are not circuital, this leads to a condensational wave of the type here given.

or equal to Cr.

(ii) If $\phi\,(r) = Cr$ (von Helmholtz's hypothesis) the equation becomes

$$\frac{1+K_1}{K_1}\nabla^2 V = (1 + 2C)\frac{d^2V}{dt^2},$$

denoting undulatory propagation with constant velocity

$$\left\{\frac{1+K_1}{(1+2C)K_1}\right\}^{\frac{1}{2}},$$

which agrees with von Helmholtz's result, when his notation $\frac{1}{2}(k-1)$ is written for C.

There is apparently nothing self-contradictory in the more general value of $\phi\,(r)$. The form $\phi\,(r) = Cr + A + Br^{-1}$, here considered, is notable for the case $C = 1$; as then the law of electrodynamic action ((iv), *supra*) between two changing charges would depend on B and be simply that of the inverse square.

Propagation of dielectric polarization:

Next we shall consider the propagation of the electric displacement fgh. We have

$$\nabla^2 f = \frac{K_1}{4\pi}\left(-\frac{d}{dt}\nabla^2 F - \frac{d^2}{dx\,dt}\nabla^2\chi - \frac{d}{dx}\nabla^2 V\right)$$

$$= K_1\frac{d^2f}{dt^2} - \frac{K_1}{4\pi}\nabla^2\frac{d^2\chi}{dx\,dt} + K_1\frac{d\rho}{dx},$$

with two similar equations in g and h.

From these equations χ may be eliminated by means of the equations found for V

$$\frac{1+K_1}{K_1}\nabla^2 V = \frac{d^2V}{dt^2} - \frac{d}{dt}\nabla^2\chi,$$

$$\nabla^2 V = -4\pi\rho,$$

where
$$\rho = \frac{df}{dx} + \frac{dg}{dy} + \frac{dh}{dz}.$$

There result equations of the type

$$\left(\nabla^2 - K_1\frac{d^2}{dt^2}\right)f = \frac{K_1}{4\pi}\frac{d}{dx}\left(\frac{1+K_1}{K_1}\nabla^2 V - \frac{d^2V}{dt^2}\right) + K_1\frac{d\rho}{dx}.$$

Thus, finally, $\left(\nabla^2 - K_1\frac{d^2}{dt^2}\right)\left(\nabla^2 f + \frac{d\rho}{dx}\right) = 0,$

or
$$\left(\nabla^2 - K_1\frac{d^2}{dt^2}\right)\left\{\frac{d}{dz}\left(\frac{df}{dz} - \frac{dh}{dx}\right) - \frac{d}{dy}\left(\frac{dg}{dx} - \frac{df}{dy}\right)\right\} = 0.$$ its equations:

The three equations of this type are equivalent to only two independent equations.

They show that all displacements fgh for which the condensation ρ is zero are propagated with the constant velocity $K_1^{-\frac{1}{2}}$, whatever be the form assigned to $\phi(r)$. For, write independent propagation.

$$(f, g, h) = \left(\frac{dS}{dx} + f', \frac{dS}{dy} + g', \frac{dS}{dz} + h'\right),$$

so that
$$\frac{df'}{dx} + \frac{dg'}{dy} + \frac{dh'}{dz} = 0;$$

this is possible, for to determine S we have simply

$$-\rho = \nabla^2 S,$$

so that
$$S = V/4\pi.$$

These equations will then determine the mode of propagation of $f'g'h'$ subject to this condition of no condensation, because S disappears from them. The propagation of S or $V/4\pi$ has already been considered.

For a system of non-condensational waves of this kind, propagated along the axis of x, all the quantities must be functions of x; therefore f must vanish; that is, the displacement must be perpendicular to the direction of propagation. These waves are therefore waves of transverse displacement.

We conclude that the propagation of waves of transverse displacement with this velocity $K_1^{-\frac{1}{2}}$ is not a characteristic of any special theory, but forms a part of any conceivable theory which admits some sort of polarization in the dielectric, and leads to the correct results for Ampère's case of circuital currents. The transverse wave-systems not special to Maxwell's theory:

This cardinal result will still follow, even if χ is any function whatever. The degree of (mathematical) generality which this remark imparts may be expressed as follows. In a complete circuit the one thing essential to the established theory is that the electric force integrated round the circuit should be equal to the time rate of change of the magnetic induction through it, and, therefore, have an ascertainable value, though its distribution round the circuit is a subject of hypothesis. The conclusion that waves of transverse displacement will be propagated in a dielectric with velocity $K_1^{-\frac{1}{2}}$ will hold good if we assume any form whatever for the electric force which does not violate this one relation, and also assume an electrostatic polarization of the medium, equal at each point to the electric force multiplied by a constant $K_1/4\pi$. For the indeterminateness that may exist in the vector potential (or electric momentum) FGH is of the same type as that which may exist in the electric force PQR, and, therefore, as the equations show, may be merged in the latter. It would, perhaps, be difficult to conceive any more general hypothesis than this.

but appear under conditions of very wide generality, here specified:

The increased generality which can be imparted to the theory merely leads to various modes of propagation of a condensational wave.

though waves of condensation also arise.

Comparison with Experimental Knowledge in 1891.

In the general theory of polarization sketched at the beginning of this paper,

$$(f, g, h) = \kappa\,(P, Q, R);$$

therefore $$K_1 = 4\pi\kappa.$$

The specific inductive capacity of the medium is

$$K_2 = \mu = 1 + 4\pi\kappa.$$

Thus $$K_2 = 1 + K_1,$$

The dielectric constant.

the units being here electrostatic.

Now, the results of various experimental investigations seem to place it beyond doubt that for dielectrics of simple chemical constitution the velocity of propagation varies as $K_2^{-\frac{1}{2}}$. Thus, in the recent experiments of Arons and Rubens[1], the velocity of waves, 6 metres long, guided by a pair of parallel wires, was measured by interference experiments when a part of the circuit was surrounded by various liquid dielectrics. The great length of the wave compared with the section of the conductor ensures that it travels with its front sensibly in the direction of propagation, and, therefore, that its velocity is normal; while the presence of the return wire limits

Dielectric waves measured recently along wire system:

[1] Wiedemann's *Annalen*, Vol. XLII. 1891, p. 581.

its divergence into space. Their results are expressed in the following table which gives $K_2^{\frac{1}{2}}$, the index of refraction m for light waves of length $6 \cdot 10^{-7}$ metres, and the index of refraction m' for the observed waves of about 6 metres long:

	$K_2^{\frac{1}{2}}$	m	m'
Castor oil	2·16	1·48	2·05
Olive oil	1·75	1·47	1·71
Xylene	1·53	1·49	1·50
Petroleum	1·44	1·45	1·40

Thus the greatest deviation from correspondence for the longer waves is about 5 per cent. The correspondence of these numbers requires that the values of K_1 and K_2 should be sensibly equal for the substances tested, which can only be the case in the limiting form of the polarization theory which constitutes Maxwell's displacement theory. In that case, as has been seen, the currents are all circuital; the Ampère-Neumann theory of electrodynamics suffices for all purposes, and there is no condensational wave.

their velocities require Maxwell's theory.

The standpoint from which the theory of dielectric polarization has been generalized in the theory here expounded is that of polar elements attracting according to the law of inverse squares in the manner of small magnets. In the results, however, this conception disappears and the phenomena are all expressed in the continuous manner by means of partial differential equations.

Expression as a formal scheme, with attractions eliminated:

It is also possible, in Maxwell's manner, to ignore the attractions of the elements from the beginning, and simply to define the displacement as proportional to the electric force. The statical theory of condensers shows that in the dielectric the displacement must be circuital, for the characteristic equation of the potential must hold good. The displacement constant assumed by Maxwell is equal to the specific inductive capacity, in order to ensure that the charging current shall be continuous across the faces of a condenser. It might be proposed to take a less restricted form for this constant, with the result, of course, that the currents would be non-circuital. The investigation of this paper, however, proves that in all cases the velocity of the waves of transverse displacement is specified in terms of this displacement constant; and the experimental fact that in the simpler media it is determined in the same manner by the specific inductive capacity confines us to that value of the constant which is assumed by Maxwell[1]. It is necessary to emphasize that it is of the very essence of a theory of this kind that the current in the dielectric is not circuital, and, therefore, that the electric volume density

[1] Cf. J. J. Thomson, *Brit. Assoc. Report*, 1885, p. 140.

must be Maxwell's,

produced by the electric displacement varies with the time. This is so because the electrodynamic part of the electric force is not derived from a potential. Any investigation which restricts the current to be circuital is necessarily inconsistent with itself, except for the limiting case which forms Maxwell's theory*.

within specified limits of uncertainty:

A discrepancy of n per cent. (n a small number less than 5) between the observed velocity and $K_2^{-\frac{1}{2}}$ would involve, by the formulae at the beginning of this section, a difference of about $2n$ per cent. between K_2 and $K_2 - 1$, so that K_2 would be of numerical magnitude about $100/2n$; which determines the ratio in which the ordinary values of the inductive capacities of all media, including vacuum, would have to be multiplied, to make the polarization theory not discordant with the observations.

The amount of discontinuity in the current at the surface of a conductor is the fraction K_2^{-1} of the total current across the surface. At the interface between two dielectric media, denoted by the values K_2 and K_2', the normal components of the displacement on the two sides are

$$(K_2 - 1)\, N/4\pi \quad \text{and} \quad (K_2' - 1)\, N'/4\pi,$$

where N, N' are the normal components of the electric force, so that

$$K_2 N = K_2' N'.$$

Thus the discontinuity in the displacement is $(N' - N)/4\pi$ or $(K_2/K_2' - 1)\, N/4\pi$ compared with a total displacement $(K_2 - 1)\, N/4\pi$; the ratio of these is $(K_2 - K_2')/K_2'\,(K_2 - 1)$, which is less than the fraction $K_2'^{-1}$, which corresponds to the surface of a conductor.

thus involving possible slight condensational waves.

Thus, under the assumed circumstances, the ratio of the amplitudes of the condensational waves to those of the transverse waves would have a superior limit of the order $2n/100$; in the observations quoted this limit is at 5 per cent.

The complete transition not unnatural.

It is worth while to emphasize that if the polarization theory were to take K_2 equal to unity for a vacuum, K_1 would be zero, and in a vacuum there would be nothing but action at a distance. It is thus an essential part of a theory like this that a vacuum has an absolute inductive capacity greater than unity, so that the ordinary value unity is merely a relative unit. Thus the transition to Maxwell's scheme, where the absolute coefficients are all assumed infinite, does not involve any undue stretch of the original hypothesis.

In the above, the relative velocities in different media of the transverse waves have been considered. The absolute velocity in a vacuum must take account of the fact that the ratio of the electro-

* The existence of electric radiation, which involved ultimately the certain decision of these questions in favour of Maxwell's scheme of purely aethereal transmission, had been detected by Hertz in 1888.

static and electromagnetic units of quantity has been altered by the factor $K_2'^{\frac{1}{2}}$ in the transition to Maxwell's theory, where K_2' now represents the assumed absolute inductive capacity of the vacuum: thus the velocity for vacuum is $(1 - K_2'^{-1})^{-\frac{1}{2}}$ multiplied by the ratio of the electric units in vacuum, agreeing with von Helmholtz's result[1], on writing this inductive capacity K_2' for his constant $1 + 4\pi\epsilon_0$, and exceeding the velocity of light unless K_2' is very great. **Wave-speed in vacuum.**

The theory of electrodynamics would thus appear to be, on all sides, limited to Maxwell's scheme, which has also so much to recommend it on the score of intrinsic simplicity. **Conclusion.**

Note (1927).—Boltzmann was led into similar discussions on K as early as 1872, when Maxwell's *Treatise* was published. They were the source of his classical researches on the inductive capacities of dielectrics, especially of gases (*Abhandl.* I. pp. 403–615), which were so important for the early reception abroad of the views of Maxwell. It appears from a letter of his of date 1902 to Koenigsberger, written in relation to the "Life of Helmholtz" (English trans., p. 288), in whose laboratory he had measured the dielectric constants (K) of various substances, that they were disappointed in not finding K equal to the index of refraction μ, as they both imagined was demanded by Maxwell's theory, and so concluded that it must be "entirely wrong." He observed however on returning to the subject, some time after leaving Berlin, that his results gave $\sqrt{K}$, instead of K, equal to μ: and on consulting Maxwell's *Treatise* he found to his delight, and reported in a letter to his master, there quoted, that all was well.

[1] *Brit. Assoc. Report*, 1885, p. 627.

27

THE EQUATIONS OF PROPAGATION OF DISTURBANCES IN GYROSTATICALLY LOADED MEDIA, AND THE CIRCULAR POLARIZATION OF LIGHT.

[*Proc. Lond. Math. Soc.* Vol. XXIII. (1891) pp. 127–135.]

Influence of latent angular momentum on an elastic medium.

THE object of the following analysis is to investigate the modification produced, in the characteristics of the propagation of undulations in a solid medium, by a distribution of angular momentum throughout the medium. We may, to make the problem physically realizable, imagine an elastic solid permeated by small spherical cavities, in each of which a flywheel is mounted, and is given rotating, with angular velocity which is not itself affected by any motions of the medium, but which reacts against angular motions.

The chief interest of such an investigation is as an illustration of the effect of a magnetic field in rotating the plane of polarization of light. The conclusion drawn by Sir W. Thomson, that this phenomenon reveals the presence of rotation, or rather angular momentum*, round the lines of magnetic force in the medium, is of deep significance, and has been one of the keystones in the development of electrical theory by Clerk Maxwell. It may also be of interest to compare a purely dynamical influence with the various hypotheses that have been in use to form a theory of the rotation of the plane of polarization of electrical waves, based on assumptions wholly electrical.

Kelvin.

Clerk Maxwell.

The hypothesis of gyrostatic cells interspersed throughout the medium, though at first sight artificial, is a correct realization of the current views of the influence of ponderable matter on the undulations of the aether. Any exhaustive optical investigation must take cognizance of the mutual influence of the two interpenetrating media, the aether and the ordinary matter. The treatment of the ordinary problem of undulatory propagation under these circumstances is rendered obscure and hypothetical. But when once the results of this interaction are ascertained as experimental laws, the superposition of an angular momentum in the material medium does not require the introduction of any further hypothetical element in the analysis, and its effects may be unfolded by ordinary dynamics.

Aether and matter:

including spin.

From the point of view of pure dynamical analysis, the problem is interesting, as forming an extension, which can be realized, of the

* Now, however, recognized to be, in the spirit of this discussion, that of the orbital electrons in the orientated molecules, which have been detected in experiment.

dynamics of a continuous material system. Ordinary matter reacts against force simply by its mass or inertia, represented in the equations by a scalar coefficient; matter gyrostatically endowed possesses, in addition, another coefficient of inertia, of the vector or directed type, representing the gyrostatic momentum per unit volume, which may be referred to as its rotary inertia.

Idea of rotational inertia.

In the formation of the differential equations of an elastic solid in this way endowed, we shall follow the method of analysis which has been already illustrated in a discussion of the vibrations of a gyrostatically loaded chain[1].

As a preliminary, it is interesting to remark, though not essential, that the disturbance in a strained medium, due to a small cell or cavity in it, is of a very local character, and falls off rapidly as the distance from the cavity increases; thus the presence of cavities, interspersed so thickly that their distances apart are even small multiples of their diameters, will not very much affect the elastic quality of the medium as a whole*. But, if the distribution of angular momentum connected with these cells were so intense that the forces produced by its rotatory displacements were of the same order as the stresses in the medium, these local disturbances would then affect the gyrostatic forces, and at any rate alter the coefficients in the results of the analysis which follows. In the optical analogue, however, the gyrostatic forces which produce rotation are extremely small, compared with the stresses in the aether; we may, therefore, consider that the components of the rotation of the imbedded gyrostats are the same as the average components of rotation for the elements of the medium in which they lie.

Cellular structures permissible.

We shall find it convenient, in the first instance, to consider the medium divided up into right solid elements, of the type $\delta x \delta y \delta z$. We take uvw to represent the components of the elastic displacement at the point xyz, and LMN to represent the components of the angular momentum per unit volume near that point. The cubic dilatation δ and the components of the angular velocity ω_x, ω_y, ω_z of the medium† are given by the equations

The local rotation of the medium.

$$\delta = \frac{du}{dx} + \frac{dv}{dy} + \frac{dw}{dz},$$

$$\omega_x = \tfrac{1}{2}\frac{d}{dt}\left(\frac{dv}{dz} - \frac{dw}{dy}\right), \quad \omega_y = \tfrac{1}{2}\frac{d}{dt}\left(\frac{dw}{dx} - \frac{du}{dz}\right), \quad \omega_z = \tfrac{1}{2}\frac{d}{dt}\left(\frac{du}{dy} - \frac{dv}{dx}\right).$$

[1] *Proc. Lond. Math. Soc.* Vol. XXI. 1891, p. 423: as *supra*, p. 205.

* Though they will affect its strength: cf. *infra*, p. 256.

† This rotation is superposed on a motion of pure strain, which on the average is of no effect when the axes of the gyrostats are distributed sporadically.

The components of the couples produced by these rotations, applied to this constant angular momentum, are, per unit volume,

$$2\mathfrak{L} = -M\omega_z + N\omega_y, \quad 2\mathfrak{M} = -N\omega_x + L\omega_z, \quad 2\mathfrak{N} = -L\omega_y + M\omega_x.$$

These will make themselves felt by additional shearing stresses acting over the surface of the element $\delta x \delta y \delta z$. In forming the equations of elastic displacement for the medium we are justified in replacing them by these shearing stresses. Thus, we may, so far as these rectangular elements and the resulting equations for the medium are concerned, replace the component couple $2\mathfrak{L}$ by either of the stresses

$$(YZ) = 2\mathfrak{L} \text{ or } (ZY) = -2\mathfrak{L}$$

(or partly by one and partly by the other), where (YZ) denotes the tangential stress in the plane perpendicular to Y which acts parallel to Z, and similarly for (ZY).

Gyrostatic torque transmitted to the medium,

To eliminate this indeterminateness, and to find the actual distribution of this stress, we have to consider volume-elements of other forms; the consideration of an element of circular cross-section shows at once, by symmetry, that the component of this tangential stress round the circular section must be uniform, and therefore that

$$(YZ) = \mathfrak{L}, \; (ZY) = -\mathfrak{L}.$$

involving non-conjugate stress.

We observe that when the gyrostatic quality is thus merged, and included in the elastic specification of the medium, the stress is no longer self-conjugate, in the sense that $(XY) = (YX)$, when the medium is in rotatory motion.

Denoting the elastic constants of an isotropic medium by λ and μ, so that $\lambda + \frac{2}{3}\mu$ is the compressibility, and μ is the rigidity, we have therefore the stress-components of the types

Stress-strain relations.

$$\left.\begin{aligned} (XX) &= \lambda\delta + 2\mu\frac{du}{dx}, & (YY) &= \ldots, \; (ZZ) = \ldots, \\ (YZ) &= \mu\left(\frac{dw}{dy} + \frac{dv}{dz}\right) + \mathfrak{L}, & (ZX) &= \ldots, \; (XY) = \ldots, \\ (ZY) &= \mu\left(\frac{dw}{dy} + \frac{dv}{dz}\right) - \mathfrak{L}, & (XZ) &= \ldots, \; (YX) = \ldots, \end{aligned}\right\}$$

wherein the shearing forces $2\mathfrak{L}$, $2\mathfrak{M}$, $2\mathfrak{N}$ are equally distributed between the tangential stresses, for the reason given above.

The dynamical equations are of the type

$$\frac{d\,(XX)}{dx} + \frac{d\,(YX)}{dy} + \frac{d\,(ZX)}{dz} = \rho\,\frac{d^2u}{dt^2},$$

which lead to

Dynamical equations.

$$\left.\begin{aligned}(\lambda+\mu)\frac{d\delta}{dx}+\mu\nabla^2 u-\frac{d\mathfrak{N}}{dy}+\frac{d\mathfrak{M}}{dz}&=\rho\frac{d^2u}{dt^2},\\(\lambda+\mu)\frac{d\delta}{dy}+\mu\nabla^2 v-\frac{d\mathfrak{L}}{dz}+\frac{d\mathfrak{N}}{dx}&=\rho\frac{d^2v}{dt^2},\\(\lambda+\mu)\frac{d\delta}{dz}+\mu\nabla^2 w-\frac{d\mathfrak{M}}{dx}+\frac{d\mathfrak{L}}{dy}&=\rho\frac{d^2w}{dt^2},\end{aligned}\right\}$$

in which $\mathfrak{L}$, $\mathfrak{M}$, $\mathfrak{N}$ have already been expressed in terms of u, v, w.

Compressional waves unaffected.

It is clear at once that compressional waves are unaffected by the gyrostatic momentum.

Wider problems.

It will be sufficient for our present purpose to examine the case of LMN constant; though interesting mathematical problems are presented by other distributions of angular momentum—for example, radiating from a point or radiating from an axis.

When L, M, N are constant, we have

$$\frac{d\mathfrak{M}}{dz}-\frac{d\mathfrak{N}}{dy}=\tfrac{1}{2}L\vartheta-\tfrac{1}{2}G\frac{d\omega_x}{d\sigma},$$

where
$$\vartheta=\frac{d\omega_x}{dx}+\frac{d\omega_y}{dy}+\frac{d\omega_z}{dz}=0,$$

and
$$G\frac{d}{d\sigma}=L\frac{d}{dx}+M\frac{d}{dy}+N\frac{d}{dz}.$$

Expressed in full, the equations of motion are therefore

Equations with uniform gyrostatic distribution.

$$\left.\begin{aligned}(\lambda+\mu)\frac{d\delta}{dx}+\mu\nabla^2 u-\tfrac{1}{4}G\frac{d}{d\sigma}\frac{d}{dt}\left(\frac{dv}{dz}-\frac{dw}{dy}\right)&=\rho\frac{d^2u}{dt^2},\\(\lambda+\mu)\frac{d\delta}{dy}+\mu\nabla^2 v-\tfrac{1}{4}G\frac{d}{d\sigma}\frac{d}{dt}\left(\frac{dw}{dx}-\frac{du}{dz}\right)&=\rho\frac{d^2v}{dt^2},\\(\lambda+\mu)\frac{d\delta}{dz}+\mu\nabla^2 w-\tfrac{1}{4}G\frac{d}{d\sigma}\frac{d}{dt}\left(\frac{du}{dy}-\frac{dv}{dx}\right)&=\rho\frac{d^2w}{dt^2}.\end{aligned}\right\}$$

The modification of the equations of a crystalline medium, produced by rotary inertia, is represented by the same new terms.

The analysis may now be shortened, by taking the axis of the momentum LMN as the axis of x; so that M and N will be zero, and

Plane wave-train.

$$G\frac{d}{d\sigma}=L\frac{d}{dx}.$$

For a given wave travelling in the medium, we may take the plane of xy to contain the direction of propagation, so that d/dz is zero, and then

$$\begin{aligned}(\lambda+\mu)\frac{d\delta}{dx}+\mu\nabla^2 u-\tfrac{1}{4}L\frac{d^3w}{dx\,dy\,dt}&=\rho\frac{d^2u}{dt^2},\\(\lambda+\mu)\frac{d\delta}{dy}+\mu\nabla^2 v+\tfrac{1}{4}L\frac{d^3w}{dx^2dt}&=\rho\frac{d^2v}{dt^2},\\\mu\nabla^2 w-\tfrac{1}{4}L\frac{d^2}{dx\,dt}\left(\frac{du}{dy}-\frac{dv}{dx}\right)&=\rho\frac{d^2w}{dt^2}.\end{aligned}$$

Or, still shorter, we may retain the general equations, and consider a wave travelling with its front at right angles to the axis of x, so that d/dy and d/dz are both zero; then

$$(\lambda + 2\mu)\frac{d^2u}{dx^2} = \rho\frac{d^2u}{dt^2},$$

$$\left.\begin{aligned} \mu\frac{d^2v}{dx^2} - \tfrac{1}{4}L\frac{d^3w}{dx^2dt} &= \rho\frac{d^2v}{dt^2}, \\ \mu\frac{d^2w}{dx^2} + \tfrac{1}{4}L\frac{d^3v}{dx^2dt} &= \rho\frac{d^2w}{dt^2}. \end{aligned}\right\}$$

Types of the possible plane wave-trains.

Of these, the first equation exhibits the wave of normal displacement propagated with unaltered velocity. The other two represent the wave of tangential displacement propagated with a rotating plane of polarization[1], the coefficient of rotation being proportional simply to the component angular momentum at right angles to its front[2].

Problem of reflexion of waves.

In the problem of reflexion, whether for isotropic or crystalline media, the surface conditions are that the three components of displacement are continuous, and the two tangential stresses on the plane of reflexion, as well as the normal stress at right angles to it, are continuous on both sides of the interface. Of these it is only the two tangential stresses that are affected by the gyratory terms; one of them is increased by $\mathfrak{L}$, and the other diminished by $\mathfrak{L}$, when the plane of reflexion is that of yz, where

$$4\mathfrak{L} = \frac{d}{dt}\left\{-M\left(\frac{du}{dy} - \frac{dv}{dx}\right) + N\left(\frac{dw}{dx} - \frac{du}{dz}\right)\right\}.$$

Thus, if the axis of resultant momentum is at right angles to the interface, the surface conditions are not affected by it.

The optical analogue is more complex.

The difficulties in the way of the explanation of magnetic reflexion of light, on cognate principles, are discussed in Sir W. Thomson's *Baltimore Lectures*, pp. 317–18. In that work a compound gyrostatic molecule is described, which operates by change of internal configuration, and is susceptible to influence from translation instead of rotation; it does not require so great an angular velocity to produce in a small body the momentum necessary for finite effects. As this molecule has free periods of its own, it introduces intrinsic dispersion, with respect both to refraction and rotation.

There also arises the problem of the reflexion, at the surface of a magnet, of the electric waves discovered by Hertz. This is a real question of actual phenomena, of which the solution may or may not

[1] *Proc. Lond. Math. Soc.* Vol. XXI. 1891, p. 429.

[2] The corresponding law, for magnetic reflexion of light from metals, has been verified by Du Bois, *Phil. Mag.* 1890 (1), p. 259.

correspond to the optical phenomena when the wave-length is made very small; though, if light waves run on all fours with the electric waves, we should expect a complete correspondence in this respect, owing to the smallness of the rotatory coefficient. The theory of electric propagation—whether as deduced by Faraday and Maxwell from the fundamental notion of an electric medium and the simplest formal type-equations that are consistent with ascertained phenomena, or, as arrived at in the method of F. E. Neumann and Helmholtz, by assuming the most general conceivable type of action at a distance between two electric systems and reducing to definiteness the formulae so obtained by the aid of crucial experiments[1]—is, in its present form, not an ultimate dynamical theory. The dynamical conception of rotary inertia must therefore be replaced in it by some formal quality of rotational type, associated with the electric displacement; the verification of the existence of such a quality is a necessary part of the electric theory of light. As is well known, a rotational quality of this kind relating to electric currents in a conducting medium has been detected by Hall; and its hypothetical extension to electric displacements in an insulating medium has been applied to the theory of electric waves by Rowland. The equations of propagation are of the same formal type as those given above. But when we come to consider the question of reflexion, there is room for further variety of hypothesis. This problem has in fact been recently treated by Mr Basset[2], on the hypothesis that the total electric force consists of the electrostatic and electrokinetic parts which fit in with Maxwell's scheme, together with an additional though small part depending for its origin on an external magnetic force, which cannot be said to have a natural place in an ideal simple scheme of electrodynamic relations, but must rather be classed as a residual phenomenon of a higher order, as is the rotational property of quartz in the elastic solid theory of optics. It will be seen, on comparison, that the boundary conditions found in that way are not the same as those here obtained by referring the phenomena to an ultimate dynamical theory, in which it is essential that the stress shall not be discontinuous in crossing an interface. Whether the gyratory coefficient is due to a single rotator, as taken above, or to a more complicated rotating system, the equations of propagation are the same in form—at any rate, so far as they depend only on the angular momentum of the rotator, and are not sensibly influenced by its free periods. The gyratory terms in them, as they involve the time to a lower negative degree than the terms due to ordinary inertia, must,

Electric theory not however mechanical.

An electric analogue in the Hall effect.

Various schemes.

[1] Cf. *Proc. Roy. Soc.* Vol. XLIX. 1891, p. 521.

[2] *Phil. Trans.* 1891.

for purposes of boundary conditions, be considered as derived from a bodily stress, and not classed with the inertia terms; for the latter procedure would involve infinite accelerations at the interface. If the relations were deduced, in Green's manner, from an expression for the energy of the medium, this latter procedure would involve a distribution of energy over the interface, of finite surface density (that is, infinite volume density at the surface), which would take part in the propagation of the waves; it may, however, be possible to express the surface-integral representing this distribution as a volume-integral, and so refer it to the medium as a whole. There is thus room for different types of boundary conditions, depending on the structure of the gyrostatic elements. In the case of compound elements there would also occur the anomalous rotatory dispersion corresponding to the free periods of the elements; and this would affect the equations of propagation as well as the boundary conditions.

Interfacial perplexities.

This leads to, and is proved by, another remark, with reference to the three types of term originally suggested by Sir G. B. Airy as competent to render an account of the optical magnetic rotation[1]. The case considered by him being that of symmetry round the axis of z, the equation of propagation may be expressed in the form

Airy's rotational types:

$$\rho \frac{d^2\theta}{dt^2} = k \frac{d^2\theta}{dz^2} + P,$$

where θ represents $u + v\sqrt{-1}$, and P may have any of the forms

$$\kappa_1 \frac{d^3\theta}{dz^2 dt}, \quad \kappa_2 \frac{d^3\theta}{dt^3}, \quad \kappa_3 \frac{d\theta}{dt},$$

or may include them all.

their dimensional significance:

Now the dimensions of κ_1/ρ, κ_2/ρ, κ_3/ρ, considered as physical quantities involved in this equation, are $[L^2T^{-1}]$, $[T]$, and $[T^{-1}]$. Thus the first of these forms must involve a distribution of angular momentum throughout the volume; while the second and third could only be derived from theories which involve the introduction of an absolute time-constant, such, for example, as a period of free molecular vibration, thus verifying the statement of the last paragraph. The somewhat exact correspondence between Verdet's experiments[2], on the relation of magnetic rotation to dispersion for a substance of simple chemical structure like bisulphide of carbon, and the law deducible from the first of the three types of equations, is in accord with these considerations.

the magnetic type:

In the same way the *reversible* rotatory property of quartz, and

[1] Maxwell, *Electricity and Magnetism*, Vol. II. § 830.

[2] Maxwell, *loc. cit.*

sugar solutions, requires an additional term in the equation of motion of a wave parallel to z, of one of the types

$$\frac{d^3\theta}{dz\,dt^2} \text{ and } \frac{d^3\theta}{dz^3},$$

where $\theta = u + v\sqrt{-1}$. The first of these represents a coefficient of inertia of the type $\rho + \sigma\frac{d}{dz}$, *i.e.* one which, in the wave-motion considered, has a varying term which is a harmonic function of z. The second connotes a law of elasticity involving differential coefficients of odd order of the displacement. Of these terms the latter is thus the only one that can be derived from an elastic or other statical theory based on a symmetry in the medium; while the former would require some sort of motional structure (*i.e.* the existence of steady forces arising from the inertia opposing unrecognized steady motions) to justify its adoption. the structural types.

28

THE INFLUENCE OF FLAWS AND AIR-CAVITIES ON THE STRENGTH OF MATERIALS.

[*Philosophical Magazine*, January 1892, pp. 70–78.]

Necessity for a factor of elastic safety, IN practical estimates of the strength of materials it is usual to take the greatest compressive or tensile stress which the material is found in experiment to sustain, and divide it by a factor of safety to ensure against sudden applications and reversals of the load, and against flaws or sources of weakness that cannot be foreseen. Among the latter, cavities or air bubbles in the material hold a place; and these may also be taken in a general way for purposes of calculation as the type of flaws consisting of a defect or weakening which is confined to a limited volume of the substance.

owing to concealed flaws, Thus, in the case of a shaft transmitting a torque or couple, the shearing-stress is annulled over the volume of the cavity, and this may lead to greater than average shearing-stress in some part of its immediate neighbourhood. In the case of a column supporting a load, the supporting thrust is absent over the part of the cross-section occupied by the cavity, and this defect of support must be compensated by a greater thrust elsewhere.

which weaken by their form rather than size, When the cavity or flaw is at a great distance from the surface of the casting compared with its linear dimensions, the changes produced by it in the intensity of the stress are the same at corresponding points, whatever be the dimensions of the cavity. For when the latter is altered in linear dimensions but not in form, and the displacement of the material at corresponding points is altered in the same ratio, the components of the strain will maintain their intensities unaltered at corresponding points, and so will the components of the stress. Thus the traction over the surface of the cavity will be unaltered, and therefore remain zero; while the displacements over the surface will be changed in the above ratio. The practical statement of this principle of similarity is that the effect which is produced by a cavity on the strength of a piece under uniform stress is dependent on the form but not on the size of the cavity, provided the distance of the nearest part of the surface from it is at least two or three times its greatest diameter.

The amount of this increase of internal stress determines the theoretical factor of safety which the possibility of a flaw of the type in question would necessitate; and it is possible to arrive at an estimate for the case of spherical or cylindrical cavities, which may be of use as a general indication of the order of magnitude involved in other similar cases also. Even for the actual cases worked out the result is not, however, to be interpreted exactly. For, in the first place, to make calculation possible the proportionality of stress to strain (Hooke's law) is assumed, and this ceases to hold, the material sometimes even begins to flow, before the critical condition is attained; and, secondly, the conditions that produce a breakdown of the material are but vaguely understood.

unless the stress is relieved by local plastic flow.

A spherical portion of the mass becomes, when strained, an ellipsoid of which the principal axes determine the three principal elongations which constitute the strain. Now it is sometimes assumed that a simple change of volume by compression or expansion cannot produce or affect rupture, and therefore this ellipsoid need only be compared with the sphere of equal volume from which it is derived by three simple shears in mutually rectangular planes. The value of the greatest of these shears may then perhaps be taken to be the circumstance determining the limiting strength of the material. It may, however, be remarked that, as the forces of cohesion between the elements of the material are not infinite, it must be possible to break it down or pull it asunder by a tension uniform in all directions (say a negative hydrostatic pressure); and it is quite conceivable that a pressure equal in all directions may by the opposite displacement loosen the bonds of cohesion and so produce a plastic condition which will give other forces play to act. The experiments of W. Spring, in which an intimate mixture of two solid substances which do not combine chemically under ordinary circumstances is caused to combine by the application of great pressure, may have a bearing on this question. The fact that cast iron supports compression much better than tension is also in point. If it is, however, the case with any material that the range of tension uniform in all directions which it can stand is very much greater than the range of stresses involving shear, the rupture would depend for that material on the shears only, and the greatest of them might be taken to be its determining cause. Thus rupture would be determined by the difference between the greatest and least axes of the ellipsoid into which a sphere of unit radius is strained. When this supposition is not valid the greatest elongation would be a more likely criterion; but in any case the assumed law will be a sufficient indication for our purpose, because any more precise specification would be vitiated in its application by the causes above mentioned, which render elastic calculations illustrative rather than exact when pushed towards the limit of strength of the materials.

Shearing criterion for limits of elastic strength:

but uniform pressure may change molecular configuration.

Other conceivable criteria.

In the most important examples we shall be concerned only with shears.

Small cavity in a column is innocuous:

It will appear on consideration that a small spherical cavity in a column or other mass under tension or compression cannot seriously affect its strength. For its strength could be reduced only by an increase of shear in the neighbourhood of the cavity. Now this shear must act all round it in the planes containing that diameter which lies along the direction of the stress, and at the free surface of the cavity it must be zero, in the absence of surface-tractions there; hence the shear is diminished in the neighbourhood of the cavity. The compression may be slightly increased in a ratio depending on that of the area of the section of the cavity to the area of the section of the shaft. The same argument applies of course to any symmetrical form of cavity, and generally to any cavity of regular shape.

not so in a shaft transmitting torque.

The hydrodynamic analogy applied for cylindrical cavities:

The case is different, however, when the cavity exists in a shaft which transmits a couple. If we suppose the cavity to consist of a narrow tunnel bored down the length of the shaft, we may make use of the result of St Venant's torsion problem. The distribution of the shear across the section of the shaft is simply and succinctly expressed by hydrodynamical analogy[1]. If a cylindrical shell of the same form of cross-section as the shaft is filled with frictionless fluid and is set in rotation, the velocity of the fluid relative to the shell will at each point represent the shear, in direction and magnitude; and the momentum of the fluid relative to the shell, which must necessarily have no linear component, will be proportional to the torsional rigidity of the shaft. For the present purpose it is convenient to state the proposition in a form less practically realizable: suppose the shell fixed and the fluid circulating inside it with uniform vorticity, the velocity at each point will represent the shear, and its resultant momentum (angular) will be proportional to the rigidity of the shaft.

the resulting diminution of strength.

Now the result of boring a small tunnel will be to modify the velocity system in the neighbourhood in the same way as a solid cylinder changes the velocities in a stream flowing past it. The velocities in front and rear are reduced to zero, while those at the sides are doubled. A tunnel of this kind therefore halves the strength of the portion of the shaft in which it is situated; and the same statement practically applies to any cavity of elongated form and circular section which lies parallel to the axis of the shaft. The possibility of a flaw of this kind near the part of the cross-section where the shear is greatest will therefore necessitate the use of a factor of safety equal to two. As the cavity is taken shorter in proportion to its diameter, its effect might at first sight be taken to diminish till we come to the

[1] Thomson and Tait's *Natural Philosophy*, § 705.

spherical form, which is again amenable to calculation, though with considerable intricacy: we might perhaps expect for it a factor considerably less than two. The result of the mathematical investigation for a spherical cavity which follows, for which I am indebted to Mr A. E. H. Love, gives, however, a factor which is never very far from the value two, unless the material is but slightly compressible, like a jelly. If we now suppose the spherical cavity to elongate in a direction perpendicular to the shear the factor may be expected still to diminish; and when it is so long as to be sensibly cylindrical the shear is itself diminished in its neighbourhood, for reasons specified above. But if it elongates in the direction perpendicular to the axis of the shaft, and in the plane of the shear, the factor two is recovered. Result for a spherical cavity.

If the cylindrical cavity is of flat cross-section the hydrodynamical analogy shows that its action is intensified. If it were absolutely flat with a sharp edge the strain would be infinite there and rupture would take place, unless in the test there is a chance of smoothing the edge of the flaw by a local flow or adjustment of the material. Re-entrant edges are deleterious:

A semicircular groove, running along the surface of a shaft, would (in the absence of local flow*) nearly halve its torsional strength. even in the form of surface scratches.

Adaptation of St Venant's *Solution for a Shaft.*

The displacement in St Venant's solution is Shear in a shaft:

$$u = \omega yz, \quad v = -\omega xz, \quad w = f(x, y);$$

where u, v represent a simple torsion round the axis of z, and w represents the warping of the cross-section which is necessary to annul the shear in a plane normal to the free boundary. The value of this shear is $\frac{dw}{dn} - \omega p$, where p is the perpendicular from the axis on the tangent plane to the boundary, and dn is an element of the normal. Thus the boundary condition is requires an equilibrating warping of the section:

$$\frac{dw}{dn} = \omega p;$$

and these displacements maintain internal equilibrium provided

$$\nabla^2 w = 0.$$

These equations show that w is the velocity-potential of the absolute motion in space of liquid contained in a box rotating with angular velocity ω. its character, in terms of hydrodynamic analogy:

The tractions exerted across the section of the shaft are (with unit rigidity)

$$X = \omega y + \frac{dw}{dy}, \quad Y = -\omega x + \frac{dw}{dx}.$$

* That scratches on the surface are actually deleterious has more recently come into prominence.

These must vanish when integrated over the area of the section; therefore the box containing liquid must be supposed to rotate round an axis through the centre of gravity of the section.

The couple transmitted across the section is

$$G = -\int (Xy - Yx)\, dS$$
$$= \int \omega\, (x^2 + y^2)\, dS - \int \left(y \frac{dw}{dx} - x \frac{dw}{dy}\right) dS;$$

it is therefore less than the couple due to simple torsion by the absolute angular momentum of the liquid.

Also X, Y are the component velocities of liquid circulating in a fixed box with vorticity ω; its resultant velocity represents the shear at each point, and its angular momentum represents the couple transmitted. Its linear momentum is null.

The analysis of this well-known result has been here indicated in full, partly in order to point out that in the first form of the analogy in which the box is made to rotate, the velocity of the liquid relative to the box represents the shear whatever be the axis of rotation, but the angular momentum of the liquid represents the correction to the rigidity only when taken about that axis for which its value is least, viz. the axis through the centre of gravity of the section. If the motion is referred to any other axis, as in the case of a rectangle bounded by two concentric arcs and two radii (Thomson and Tait, § 707), then from the angular momentum round that axis must be subtracted the moment of the linear momentum of the whole mass of fluid supposed collected at its centre of gravity.

correction to rigidity of shaft: precaution in its application.

Suppose, now, a cylindrical tunnel of small circular section bored down the shaft at a place where the velocity of the rotational fluid motion is V; the stream function near it will be changed from the form $\psi_1 = Vy$ to the form $\psi_2 = Vy - V\frac{a^2}{r^2}y$, because the boundary of the tunnel, $r = a$, must become a stream line, and therefore give a constant value to ψ_2. The velocity along the tunnel is $-d\psi_2/dn$, and is therefore $2V$ at the sides, as stated above.

The angular momentum of the fluid is altered, owing to the tunnel, by

Circular tunnel:

$$\iint \frac{d\,(\psi_2 - \psi_1)}{dy}\, dx dy,$$

that is $\int (\psi_2 - \psi_1) \frac{dx}{ds}\, ds$ round the boundary. For a circular boundary this is equal to $Va \int \cos^2 \theta\, ds$, or $\pi a^2 V$. The rigidity of the surrounding parts is therefore diminished by the presence of the cavity, just as if the shearing over the material which originally occupied its place

compound effect on rigidity of shaft.

were reversed in direction; the loss of rigidity is due in equal proportion to the removal of the matter and the release on the constraint of the surrounding parts.

The case when the section of the cavity is an elliptic cylinder is of interest, as it illustrates the effect of making it more and more flat until it is finally a mere crack for which the strain is theoretically infinite at the edge. The corresponding hydrodynamical problem has been solved by Prof. Lamb[1]: his value of the stream function ψ, which may easily be verified, is

$$\psi = -V\left(\frac{a+b}{a-b}\right)^{\frac{1}{2}} e^{-\eta}\sin\xi - Vx;$$

where a, b are the semiaxes of the ellipse, V is the velocity of the stream past it parallel to the axis b, and ξ, η are the conjugate functions given by

$$x + \iota y = c\sin(\xi + \iota\eta).$$

The velocity at the end of the longer axis is the value of $d\psi/dx$ when $y = 0$, that is, when $\xi = \frac{1}{2}\pi$, and is found to be

$$V\left(1 + \frac{a}{b}\right).$$

The weakness due to a flat elliptic cavity or crack.

Thus in the elastic problem the increase of shear produced by the cavity is the original shear multiplied by a/b.

Analysis for Strained Spherical Cavity. (By Mr A. E. H. Love.)

To investigate the strain in an infinite solid containing a spherical cavity, the displacements at an infinite distance being

$$u = \alpha y,\quad v = 0,\quad w = 0.$$

Analysis for spherical cavity:

From the spherical harmonic solutions of Thomson and Tait it can be shown that the forms of the displacements at a point (x, y, z) at a distance r from the centre of the cavity are

$$u = A\frac{y}{r^3} + (B + Cr^2)\frac{d}{dx}\frac{xy}{r^5} + \alpha y,$$

$$v = A\frac{x}{r^3} + (B + Cr^2)\frac{d}{dy}\frac{xy}{r^5},$$

$$w = \qquad (B + Cr^2)\frac{d}{dz}\frac{xy}{r^5}, \qquad (1)$$

where A, B, C are constants to be determined.

The cubical dilatation δ is

$$\delta = -6(A + C)\frac{xy}{r^5}. \qquad (2)$$

[1] *Quart. Journ. of Math.* 1875; Lamb, "Fluid Motion," p. 90.

The equations of equilibrium are three such as

$$(\lambda + \mu)\frac{d\delta}{dx} + \mu\nabla^2 u = 0, \tag{3}$$

which gives $\quad -(\lambda + \mu)\,6\,(A + C) - 10\mu C = 0,$

or

$$3\,(\lambda + \mu)\,A + (3\lambda + 8\mu)\,C = 0. \tag{4}$$

The remaining equations to determine the constants are to be found from the condition that the surface $r = a$ of the cavity is free from stress. It is shown in Thomson and Tait's *Nat. Phil.* Part II. art. 737, that if F, G, H be the component surface-tractions parallel to the axes across a spherical surface whose centre is the origin and radius is r, then

$$Fr = \lambda x\delta + \mu\left(\frac{d\xi}{dx} + r\frac{du}{dr} - u\right),$$

where $\xi = ux + vy + wz$ and similar equations hold for G and H.

Now

$$x\delta = -6\,(A + C)\frac{x^2y}{r^5},$$

$$\xi = (2A - 3C)\frac{xy}{r^3} - 3B\frac{xy}{r^5} + \alpha xy,$$

$$\frac{d\xi}{dx} = (2A - 3C)\frac{y}{r^3} - 3B\frac{y}{r^5} - (6A - 9C)\frac{x^2y}{r^5} + 15B\frac{x^2y}{r^7} + \alpha y,$$

$$r\frac{du}{dr} - u = -3\,(A + C)\frac{y}{r^3} - 5B\frac{y}{r^5} + 15C\frac{x^2y}{r^5} + 25B\frac{x^2y}{r^7}.$$

Hence the equations

$$-\left(A + 6C + \frac{8B}{r^2}\right)\frac{y}{r^3} + \alpha y - \left[\frac{\lambda}{\mu}6\,(A + C) + 6A - 24C - \frac{40B}{r^2}\right]\frac{x^2y}{r^5} = 0,$$

$$-\left(A + 6C + \frac{8B}{r^2}\right)\frac{x}{r^3} + \alpha x - \left[\frac{\lambda}{\mu}6\,(A + C) + 6A - 24C - \frac{40B}{r^2}\right]\frac{xy^2}{r^5} = 0;$$

and similarly from the z equation

$$-\left[\frac{\lambda}{\mu}6\,(A + C) + 6A - 24C - \frac{40B}{r^2}\right]\frac{xyz}{r^5} = 0.$$

These three equations hold when $r = a$.

Hence
$$\frac{40B}{a^2} - 6\,(A + C)\frac{\lambda + \mu}{\mu} + 30C = 0,$$

and
$$\frac{8B}{a^2} + (A + 6C) = \alpha a^3.$$

From which and (4) we find

$$A = \frac{3\lambda + 8\mu}{9\lambda + 14\mu}\alpha a^3,\quad B = \frac{3\,(\lambda + \mu)}{9\lambda + 14\mu}\alpha a^5,\quad C = -\frac{3\,(\lambda + \mu)}{9\lambda + 14\mu}\alpha a^3. \tag{5}$$

The shear is given by

$$\frac{du}{dy}+\frac{dv}{dx}=\alpha+\frac{2}{r^3}\left(A+C+\frac{B}{r^2}\right)-\frac{x^2+y^2}{r^5}\left(8C+3A+\frac{10B}{r^2}\right)$$
$$+\frac{x^2y^2}{r^7}\left(50C+\frac{70B}{r^2}\right).$$

To see the greatest magnitude of this take $x=0, y=0, r=a$, then

$$\frac{du}{dy}+\frac{dv}{dx}=\alpha\left[1+\frac{10\mu+6(\lambda+\mu)}{9\lambda+14\mu}\right]=\alpha\frac{15\lambda+30\mu}{9\lambda+14\mu},$$

which depends on the value of λ/μ, but for all known isotropic materials differs little from 2α, where α is the uniform impressed shear. resulting factor of safety.

When $\lambda/\mu=\infty$, or the material is incompressible, the maximum local shear is $\frac{5}{3}\alpha$: when $\mu/\lambda=\infty$, or the stretch-squeeze ratio vanishes, it is $\frac{15}{7}\alpha$: when $\lambda=\mu$, Poisson's condition, it is $\frac{45}{23}\alpha$.

29

THE SIMPLEST SPECIFICATION OF A GIVEN OPTICAL PATH, AND THE OBSERVATIONS REQUIRED TO DETERMINE IT.

[*Proc. Lond. Math. Soc.* Vol. XXIII. (1892) pp. 165–172.]

1. The complete specification of an optical path through a heterogeneous medium like the atmosphere*, or through a combination of transparent substances which may form an optical instrument, requires not merely the form of the curve which a narrow beam or filament of light traverses, but also a statement of the character of the modification impressed upon the filament by traversing that path. The mode of division of a beam of light into filaments, or linear elements of waves, in the way thus suggested, corresponds much more closely to the physical reality than the more ordinary analysis which splits these filaments up into the rays of which they are supposed to be constituted; and it is fortunate that, by making use of the dioptrical methods introduced by Sir W. R. Hamilton, and further developed more recently by Maxwell, the consideration of filaments leads to a deeper and more coordinated analysis than that of rays. That this is the case follows from the fact that the effects impressed upon all filaments by traversing a given path may be determined in a simple manner by observation or calculation relating to a few cases, and may be considered as expressed in terms of certain optical properties of the path. In addition to the geometrical gain which results from recognizing that the treatment of a group of rays is nearly as simple as that of a single ray, there is the fact that all phenomena of diffraction, arising from limitation of the beam, are most simply connected with the form of the wave-fronts which determine the filaments.

The ray-filament as the element of radiation: its Hamiltonian analysis.

The physical unit.

2. In the *Proceedings*, Vol. XX. (1889) p. 185, *supra*, p. 181, it has been explained, after Maxwell[1], that the characteristic function for an optical path, from a point $(x_1, y_1, 0)$ referred to a frame at one end of it to a point $(x_2, y_2, 0)$ referred to a frame at the other end, is of the form

The characteristic function between two regions:

$$U = \text{const.} + \tfrac{1}{2}a_1x_1^2 + c_1x_1y_1 + \tfrac{1}{2}b_1y_1^2 + px_1x_2 + qx_1y_2 + rx_2y_1 + sy_1y_2 + \tfrac{1}{2}a_2x_2^2 + c_2x_2y_2 + \tfrac{1}{2}b_2y_2^2, \qquad (1)$$

* We may now add, of a wireless electric ray.

[1] *Proc. Math. Soc.* Vol. VI. (1874); *Collected Papers*, Vol. II. p. 381. On Hamiltonian theory see the historical Appendix at end of this volume.

the terms of the first degree being omitted when the axes of z_1 and z_2 are tangential to the path.

The course of a ray, from a point $(\xi_1, \eta_1, \gamma_1^{-1})$ at one end to a point $(\xi_2, \eta_2, \gamma_2^{-1})$ at the other end, was then determined by the fact that the ray must strike the planes $z_1 = 0$ and $z_2 = 0$ at points $(x_1, y_1, 0)$ and $(x_2, y_2, 0)$, which satisfy the condition of making the characteristic function stationary for all variations of these points. When the media beyond the two ends of the path are homogeneous and isotropic, this leads to the determination of these points by the equations

its stationary property,

$$\begin{aligned}(\mu_1\gamma_1 + a_1)\, x_1 + c_1 y_1 + p x_2 + q y_2 &= \mu_1\gamma_1\xi_1,\\ c_1 x_1 + (\mu_1\gamma_1 + b_1)\, y_1 + r x_2 + s y_2 &= \mu_1\gamma_1\eta_1,\\ p x_1 + r y_1 + (\mu_2\gamma_2 + a_2)\, x_2 + c_2 y_2 &= \mu_2\gamma_2\xi_2,\\ q x_1 + s y_1 + c_2 x_2 + (\mu_2\gamma_2 + b_2)\, y_2 &= \mu_2\gamma_2\eta_2. \qquad (2)\end{aligned}$$

correlates the two ends of the filament,

When, however, the media beyond the two ends of the path are isotropic but not homogeneous, these equations determine the inclinations of the ray, as it enters the path at the point $(x_1, y_1, 0)$, and emerges from it at the point $(x_2, y_2, 0)$; for its direction cosines

$$(l_1, m_1, n_1) \text{ and } (l_2, m_2, n_2)$$

at these two points are equal to

$$\{\gamma_1(\xi_1 - x_1), \gamma_1(\eta_1 - y_1), 1\} \text{ and } \{\gamma_2(\xi_2 - x_2), \gamma_2(\eta_2 - y_2), 1\},$$

which are $\left\{\dfrac{1}{\mu_1}\dfrac{\partial U}{\partial x_1}, \dfrac{1}{\mu_1}\dfrac{\partial U}{\partial y_1}, 1\right\}$ and $\left\{\dfrac{1}{\mu_2}\dfrac{\partial U}{\partial x_2}, \dfrac{1}{\mu_2}\dfrac{\partial U}{\partial y_2}, 1\right\}$, (3)

and the terminal directions of its rays.

as they ought to be from the fundamental properties of the characteristic function. This remark in fact suggests an alternative method of stating the proof of the formulae (2).

When the determinantal equation

$$\Delta \equiv \begin{vmatrix} \mu_1\gamma_1 + a_1 & c_1 & p & q \\ c_1 & \mu_1\gamma_1 + b_1 & r & s \\ p & r & \mu_2\gamma_2 + a_2 & c_2 \\ q & s & c_2 & \mu_2\gamma_2 + b_2 \end{vmatrix} = 0 \qquad (4)$$

is satisfied, there is a plane sheaf of rays from the point $(\xi_1, \eta_1, \gamma_1^{-1})$ to the point $(\xi_2, \eta_2, \gamma_2^{-1})$, which intersect the plane $z_1 = 0$ along the line (of constant direction)

Plane sheaf between terminal points.

$$S x_1 + Q y_1 = \text{constant}, \qquad (5_1)$$

Focal lines:

and also the plane $z_2 = 0$ along the line

$$Q x_2 + P y_2 = \text{constant}; \qquad (5_2)$$

while we have the identical relation

$$A_1\mu_1\gamma_1\xi_1 + C_1\mu_1\gamma_1\eta_1 + P\mu_2\gamma_2\xi_2 + Q\mu_2\gamma_2\eta_2 = 0, \qquad (6)$$

the capital letters representing the minors of the corresponding small-letter constituents in the determinant Δ.

The interpretation of these formulae is that, under the condition

their complex mutual relations. (4), a plane sheaf of rays, instead of a single one, passes from any point in the plane $z_1 = \gamma_1^{-1}$ to any point in the plane $z_2 = \gamma_2^{-1}$; that the planes of all these sheaves are parallel at each end of the path, being given by the equations (5_1) and (5_2); that all the rays issuing from a point in the plane $z_1 = \gamma_1^{-1}$ converge to a line in the plane $z_2 = \gamma_2^{-1}$; or, more generally, all rays issuing from the line

$$A_1\mu_1\gamma_1\xi_1 + C_1\mu_1\gamma_1\eta_1 = H \qquad (7_1)$$

converge to the line

$$P\mu_2\gamma_2\xi_2 + Q\mu_2\gamma_2\eta_2 = -H, \qquad (7_2)$$

and *vice versa*.

Their (2, 2) correspondence: 3. The equation (4) establishes a 2-to-2 correspondence between the planes $z_1 = \gamma_1^{-1}$ and $z_2 = \gamma_2^{-1}$. Any point-focus on one side of the path has, for its conjugate, two focal lines on the other side of the path, whose distances from the origin are determined by this equation, and whose actual positions are determined by the equations (7).

roots always real: The roots of (4), treated as an equation in γ_1 or γ_2, are necessarily always real; a result which admits of extension to any determinantal equation of this type, and so leads to a generalization of the proposition which asserts the reality of the roots of the period equation for the vibrations of a frictionless dynamical system with a positive energy function. in extension of the dynamical theorem. The roots of any determinantal equation of this type are in fact separated by those of the equation obtained by equating to zero the leading minor determinant; and in this way a system of equations of orders diminishing by units may be obtained, each of which separates the roots of the one before it, thus showing, on Sturm's principles, that the roots of each are all real.

When the two values of γ_1 in (4) are equal, their common value might at first sight be taken to be the root of either first minor; this would lead to

$$-\begin{vmatrix} \mu_2\gamma_2+a_2 & c_2 \\ c_2 & \mu_2\gamma_2+b_2 \end{vmatrix}\mu_1\gamma_1 = \begin{vmatrix} a_1 & p & q \\ p & \mu_2\gamma_2+a_2 & c_2 \\ q & c_2 & \mu_2\gamma_2+b_2 \end{vmatrix} = \begin{vmatrix} b_1 & r & s \\ r & \mu_2\gamma_2+a_2 & c_2 \\ s & c_2 & \mu_2\gamma_2+b_2 \end{vmatrix},$$

so that there would be two values of γ_2 which lead to equal roots for γ_1. But this argument is vitiated by the fact (noted below) that the values of γ_2 thus obtained are imaginary, while Sturm's theory relates only to equations with real coefficients. A direct procedure shows that there are four such values of γ_2.

Precautions of interpretation: The interpretation of these results requires some care. It might appear at first sight that any focus in the plane $z_2 = \gamma_2^{-1}$ has two coincident focal lines, and therefore a conjugate focus, in the plane $z_1 = \gamma_1^{-1}$; and thus that there are always four pairs of planes of conjugate foci. But we shall see presently, on other grounds, that this cannot generally be true. The explanation of the apparent paradox is that the equation (4) only ensures that a certain plane sheaf of rays from the one point converges as a plane sheaf to the other point; and that the equality of the roots further ensures that two plane sheaves from the one point converge as plane sheaves to the other point. But any two plane sheaves from a point to another do not make up a wave-front; that they should do so another condition is required, viz. that the time of passage should be the same for each—in other words, that they should have the same characteristic function; and this has not been secured.

The state of matters may be illustrated geometrically by the Poncelet 2-to-2 correspondence of points on two conics. Each point on a conic is given by its parameter; to a point on one, say the outer, there correspond, on the inner, the two points of contact of tangents from it; to a point on the inner, there correspond, on the outer, the two intersections of the tangent at it. There are four points of intersection of two conics, and four common tangents, which are the correlatives of the two sets of four points referred to in the last paragraph. It is at once evident geometrically that the correspondence is not symmetrical; where a coincidence occurs on one conic it does not occur on the other. To make a coincidence at the same point on both conics, the conics must touch at that point, instead of intersecting. Similarly, to have conjugate foci, not only must the values of γ_1 obtained from (4) be equal, but also the values of γ_2; to make the conditions for both these equalities consistent, a relation must hold between the coefficients, and therefore the optical path must have a special character. enforced by geometric analogy:

That the geometrical may agree with the optical correspondence, in referring only to real points, the conic which we have called the inner one must lie wholly inside the other; otherwise the tangents will become imaginary over a certain range of the correspondence. The conics cannot therefore cross each other in real points. Thus the values of γ_2 which give equal roots for γ_1 in (4) must be imaginary, a fact which we have already used in the analytical theory. restricted optical form.

4. If we revert for a moment to the more general question in which no part of the medium is homogeneous, the planes $z_1 = 0$, $z_2 = 0$ will be in correspondence if

$$\Delta' \equiv \begin{vmatrix} a_1 & c_1 & p & q \\ c_1 & b_1 & r & s \\ p & r & a_2 & c_2 \\ q & s & c_2 & b_2 \end{vmatrix} = 0. \qquad (4')$$

General plane sheaf.

The rays from a point in the plane $z_1 = 0$, which emerge from the plane $z_2 = 0$ in a constant direction, there form a plane sheaf parallel to the plane

$$S'x_1 + Q'y_1 = 0, \qquad (5')$$

and *vice versa*. The identical relation

$$A_1'\mu_1 l_1 + C_1'\mu_1 m_1 + P'\mu_2 l_2 + Q'\mu_2 m_2 = 0 \qquad (6')$$

connects the direction cosines $(l_1, m_1, 1)$, $(l_2, m_2, 1)$ of the same ray at incidence and emergence. In this statement, the capital letters represent the minors of the determinant Δ'.

Filament referred to its cardinal points. 5. These considerations will in their analytical form be much simplified if we refer the path to its two cardinal points, so that in the characteristic function $p = s$, while q and r vanish. The determination of the emergent beam

$$V_2 = \mu_2 z_2 + \tfrac{1}{2}A_2 x_2^2 + C_2 x_2 y_2 + \tfrac{1}{2}B_2 y_2^2,$$

which corresponds to an incident beam

$$V_1 = \mu_1 z_1 + \tfrac{1}{2}A_1 x_1^2 + C_1 x_1 y_1 + \tfrac{1}{2}B_1 y_1^2,$$

is completed in § 15 of the previous paper already quoted (*Proceedings*, Vol. XX. p. 190, *supra*). The result is

$$\frac{a_2 + A_2}{b_1 + A_1} = \frac{b_2 + B_2}{a_1 + B_1} = -\frac{c_2 + C_2}{c_1 + C_1} = \frac{p^2}{\Delta_1},$$

where $$\Delta_1 = (a_1 + A_1)(b_1 + B_1) - (c_1 + C_1)^2. \qquad (8)$$

A pair of point-foci exist, To obtain a pair of conjugate foci, we must have together

$$A_1 = B_1, \quad C_1 = 0,$$
$$A_2 = B_2, \quad C_2 = 0,$$

which leads, on subtraction of numerators and of denominators in (8), to a relation between the constants

$$\frac{a_2 - b_2}{a_1 - b_1} = \frac{c_2}{c_1}. \qquad (9)$$

only when the filament has rectangular symmetry: This is exactly the condition that, by a rotation of both systems of axes through the same angle, the characteristic may assume the form

$$U = \text{const.} + \tfrac{1}{2}a_1 x_1^2 + \tfrac{1}{2}b_1 y_1^2 + p(x_1 x_2 + y_1 y_2) + \tfrac{1}{2}a_2 x_2^2 + \tfrac{1}{2}b_2 y_2^2. \quad (10)$$

The path is then such that the equations connecting x_1, x_2 are independent of the equations connecting y_1, y_2, each set being the same as those for an ordinary two-dimensional Gaussian system, except as regards an absolute rotation round the tangent to the path at either end. The focal lines corresponding to all foci will then be parallel, and will determine the directions of these principal sections at the two ends of the path. the practical test thereof.

To pass to the case of symmetry round the axis, two further conditions are necessary, $a_1 = b_1$, and $a_2 = b_2$.

6. It will be observed that there are these two pairs of planes of conjugate foci, and no more; the conics of the analogy have double contact, and cannot touch again. The images formed in these planes will have a linear correspondence, and by means of them the path of any ray may be constructed. For the ray which connects any two points on one side must connect their image points on the other side, when such exist. The point-foci then unique: they determine the system:

The fact that it is only under a special condition that two sets of image points exist, and that there cannot be more than two sets, may be more directly demonstrated, as in the following sketch. Let the points (ax, by, z) and $(ax, by, z_1 + R_1)$ on one side correspond to the points (x, y, z_2) and $(ax \cos \epsilon - \beta y \sin \epsilon,\ ax \sin \epsilon + \beta y \cos \epsilon,\ z_2 + R_2)$ on the other side, the correspondence of the second pair thus involving a rotation through an angle ϵ. As the incident rays form a parallel system, the emergent rays must be normal to a surface alternative view.

$$z = z_2 + \tfrac{1}{2}Ax^2 + Cxy + \tfrac{1}{2}By^2.$$

Therefore, for all values of x, y,

$$\frac{(1 - \alpha \cos \epsilon)\, x + \beta \sin \epsilon\, y}{Ax + Cy} = \frac{-\alpha \sin \epsilon\, x + (1 - \beta \cos \epsilon)\, y}{Cx + By} = \frac{-R_2}{1};$$

$$\text{therefore}\quad \left.\begin{aligned} 1 - \alpha \cos \epsilon &= -R_2 A \\ \beta \sin \epsilon &= -R_2 C \end{aligned}\right\}, \qquad \left.\begin{aligned} \alpha \sin \epsilon &= R_2 C \\ 1 - \beta \cos \epsilon &= -R_2 B \end{aligned}\right\},$$

so that $\epsilon = 0$, or else $\alpha + \beta = 0$.

The first alternative gives principal sections at each end of the beam, the case discussed above. The second alternative requires also, by consideration of a parallel beam incident on the opposite side of the path, that $a + b = 0$; thus the two image planes must be similar to the two object planes, but turned through different angles; but even then it will be found that any other beam incident normally to one surface will not emerge normally to another.

7. The theorem of *Proceedings* [*supra*, p. 192], which gives the simplest equivalent of a general optical path, may be expressed in a somewhat simpler form, which is worth notice. The modification follows on observing that an optical instrument symmetrical round its axis may be replaced by two thin lenses at any two specified points on its axis.

Consider any unsymmetrical optical system, separating two homogeneous media; and let the path of a central ray through it be considered as its axis. Then the system is optically equivalent to two thin astigmatic lenses placed at definite points O_1 and O_2 on a straight axis; and that in two ways. It may, however, be necessary, in order to make these points real, to add on a simple lens to the original Equivalent lens-pair:

system. The positions of these cardinal points may be found experimentally, without any previous knowledge of the constitution of the instrument, on making the incident pencil symmetrical round the axis by the use of a circular stop, and shifting the luminous point from which it comes, until, on emergence into the second medium, it diverges symmetrically from either of its focal lines, as may be tested by focussing that line in an observing telescope and examining the appearance when it is put slightly out of focus. The focus of the incident beam is then the point O_1, and the circular section of the emergent beam is the point O_2. The determination may be checked by passing a beam over the path in the opposite direction. The astigmatic lens to be placed at O_2 is now determined as that lens which will refract an incident pencil from O_1 to the same focal lines as the actual instrument does; these focal lines are parallel to the principal sections of the lens, and its focal lengths in these sections are at once determined. In a similar manner the astigmatic lens to be placed at O_1 is determined. The instrument is then optically equivalent in all respects to this pair of lenses, with the exception that all emergent beams may in addition have to be rotated round the axis through a constant angle, which, however, is optically of no consequence. When the axis of the emergent beam is in a different direction from that of the incident beam, these points O_1, O_2 must be laid off on a subsidiary straight axis. The distance between them is to be adjusted by the fact that the section on the plane O_2 of a given beam from the side of O_1 must be the same on the subsidiary axis as on the actual path. The astigmatic lenses are then determined, with reference to the distance O_1O_2, and in the result the incident and emergent beams will correspond exactly in both systems.

how their positions can be recognized:

and the lenses determined by observations.

The system of observations here sketched therefore suffices theoretically to determine the optical effect of any transparent system, however complicated, on all beams of light passing across it in a given path, straight or curved.

The filament thus experimentally determined.

30

THE APPLICATION OF THE SPHEROMETER TO SURFACES WHICH ARE NOT SPHERICAL.

[*Proc. Camb. Phil. Soc.* Vol. VII. (1892) pp. 327–329.]

Equilateral tripod spherometer.

THE ordinary form of spherometer, which is used for measuring the curvatures of lenses, rests on the surface to be measured by three legs which are at the corners of an equilateral triangle; and the mode of using it consists in finding the length of the ordinate drawn up to the surface from the centre of the triangle formed by the points of support, by means of a micrometer screw moving along the axis of the instrument.

In the actual use of the instrument the surface to be measured is assumed to be spherical; and the question has apparently not occurred to examine the character of the results which may be derived from its application to a surface of double curvature.

Reading independent of orientation on a surface:

On actual trial with such a surface, for example the cylindrical surface of an iron pipe, it appears at once that when the centre is set at a given point the instrument may be rotated anyhow on its axis without affecting its reading. It therefore measures some definite quality of the double curvature of the surface at the point. There is a temptation to hastily assume that the plane of support is parallel to the tangent plane at the centre of the instrument, that it is in fact the indicatrix plane of that point, and to deduce that the reading gives the mean of the principal curvatures of the surface; this result is correct, but the assumption just mentioned is erroneous.

To obtain a rigorous investigation, let us assume that the points of support form an isosceles triangle, let the base subtend an angle 2α at the centre of the circumscribing circle, and let c be the radius of this circle and h the ordinate drawn from its centre up to the surface. If this ordinate is taken as axis of z, the equation of the surface will be

$$z = px + qy + \frac{x^2}{2R_1} + \frac{y^2}{2R_2},$$

where $(p, q, -1)$ is the direction of the tangent plane at the origin, and R_1, R_2 are the radii of principal curvature. As the three legs rest on the surface, we have

$$h = cp\cos(\theta+\alpha) + cq\sin(\theta+\alpha) + \tfrac{1}{2}c^2\left\{\frac{\cos^2(\theta+\alpha)}{R_1} + \frac{\sin^2(\theta+\alpha)}{R_2}\right\},$$

$$h = cp\cos(\theta-\alpha) + cq\sin(\theta-\alpha) + \tfrac{1}{2}c^2\left\{\frac{\cos^2(\theta-\alpha)}{R_1} + \frac{\sin^2(\theta-\alpha)}{R_2}\right\},$$

$$h = -cp\cos\theta \quad - cq\sin\theta \quad + \tfrac{1}{2}c^2\left\{\frac{\cos^2\theta}{R_1} + \frac{\sin^2\theta}{R_2}\right\},$$

where $\pi + \theta$ is the azimuth of the vertex of the triangle of support. We are to eliminate p, q, and so connect h with R_1, R_2 and θ. By addition of the first pair of relations

$$2h = 2cp\cos\theta\cos\alpha + 2cq\sin\theta\cos\alpha + \tfrac{1}{2}c^2\left\{\left(\frac{1}{R_1}+\frac{1}{R_2}\right) + \left(\frac{1}{R_1}-\frac{1}{R_2}\right)\cos 2\theta\cos 2\alpha\right\};$$

therefore by use of the third

$$2h(1+\cos\alpha) = \tfrac{1}{2}c^2\left\{\left(\frac{1}{R_1}+\frac{1}{R_2}\right)(1+\cos\alpha) + \left(\frac{1}{R_1}-\frac{1}{R_2}\right)\cos 2\theta(\cos 2\alpha + \cos\alpha)\right\},$$

or on rejecting the factor $1 + \cos\alpha$,

$$\frac{4h}{c^2} = \left(\frac{1}{R_1}+\frac{1}{R_2}\right) + \left(\frac{1}{R_1}-\frac{1}{R_2}\right)\cos 2\theta\,(2\cos\alpha - 1).$$

only when it is equilateral: The value of h therefore depends on the azimuth θ except in one case, when α is $\frac{1}{3}\pi$ so that the triangle of support is equilateral, which is the case referred to above. The quantity involved in the formula is then $\frac{1}{R_1}+\frac{1}{R_2}$; and by referring back to the original case of a spherical surface we see that the instrument measures the arithmetic mean of the principal curvatures.

it then measures the mean of the principal curvatures:

Thus for example the equilateral form of the instrument may be conveniently used to measure the curvature of a cylindrical lens or a cylindrical pipe, but for that purpose its indication must be doubled.

it cannot test deviation from local sphericity. The equilateral form will be of no use for testing deviation from sphericity at a given point of a surface. The isosceles form may however be so used, the difference of the extreme curvature indications given by it for any point being by the above formula

The isosceles form needed.

$$\left(\frac{1}{R_1}-\frac{1}{R_2}\right)(2\cos\alpha - 1),$$

that is directly proportional to the difference of the principal curvatures. The curvature may thus be completely explored[1].

In all these formulae the usual assumption is made that the span of the instrument is small compared with the radii of curvature of the surface.

If the instrument had four legs at the corners of a rectangle, there would be only two positions in azimuth, corresponding to the sections

[1] Mr H. F. Newall informs me that an isosceles spherometer is used by Dr Common for exploring the curvatures of his large specula.

of greatest and least curvature, in which it would rest firmly at a given point on a surface, with all its legs in contact; and the plane of contact would in this case be parallel to the tangent plane at the summit of the surface. The readings for these positions would give

$$\frac{\cos^2 \alpha}{R_1} + \frac{\sin^2 \alpha}{R_2}$$

and

$$\frac{\sin^2 \alpha}{R_1} + \frac{\cos^2 \alpha}{R_2},$$

where α is an angle made by a diagonal of the rectangle with a side; so that the values of both principal curvatures might thus be determined. **Rectangular form.**

31

ON THE THEORY OF ELECTRODYNAMICS, AS AFFECTED BY THE NATURE OF THE MECHANICAL STRESSES IN EXCITED DIELECTRICS.

[*Proc. Roy. Soc.* Vol. LII. (1892) pp. 55–66.]

1. A theory of electrodynamics was first precisely developed by Maxwell, which based the phenomena on Faraday's view of the play of elasticity in a medium, instead of the conception of action at a distance, by means of which the mathematical laws had been primarily evolved. The electromotive equations of Maxwell however involve nothing directly of the elastic structure of this medium, which remains wholly in the background. They involve simply the assumption of a displacement across dielectrics with such properties as to make all electric currents circuital; all the equations of Ampère and Neumann for closed or circuital currents have then a universal validity, and no further hypothesis is required for the full development of the subject.

Maxwell's theory formally the simplest.

The theory was next discussed by Helmholtz in his memoirs on electrodynamics, in a way which took direct advantage of the picture of a polarized dielectric supplied by Mossotti's adaptation of the Poisson theory of induced magnetization. Stated absolutely, this simply builds upon the assumption that at each point in the excited dielectric there is something which has the properties of a current element (electric transfer or displacement), which is represented both in direction and magnitude by the electric force at the point multiplied by a constant factor; no more general starting point seems possible for an isotropic dielectric. The development of this hypothesis, exactly on the analogy of a similar discussion with the Poisson-Mossotti phraseology in a previous paper[1], leads to the necessity of recognizing the existence of absolute electric charges on the faces of an excited condenser; so that the exciting current causes the accumulation of these charges, and therefore is not circuital or solenoidal. This defect of circuital character however practically disappears in the limiting case when the constant ratio of the polarization to the electric force is extremely great; and then the theory becomes a concrete illustration of the general statements of Maxwell with respect to electric displacement.

The Helmholtz extension:

[1] "The Theory of Electrodynamics," *Proc. Roy. Soc.* 1891: *supra*, p. 232.

It was shown in the paper above referred to that this hypothesis, adopted by Helmholtz, led by itself—without any necessity for further assumptions which its author introduced on various grounds—to the undulatory propagation of electromotive disturbances across dielectric media, with the same transverse type of waves as constitute light. There will usually be, in addition, a disturbance of a *quasi*-compressional character, which, on the more special hypothesis of Helmholtz, is also propagated as a wave of permanent type, but with different velocity. The electric undulations of transverse type have been detected by Hertz; and the balance of evidence, from the experiments of different authors, seems to point to the conclusion that their velocities in different media are inversely as the square roots of the specific inductive capacities. Should this be fully verified, it would follow demonstratively that the Helmholtz hypothesis must be restricted to the special form which represents the Maxwell displacement theory; and the general equations of electrodynamics, or rather the electromotive part of them, will be definitely established.

any such must lead to transverse waves such as light: also other types of waves. Discrimination by velocity.

2. The object here proposed is to pass on from the electromotive to the ponderomotive properties of the electric field, and examine whether the latter lend any strength* to the conclusions derived from the former. Instead of a kinetic phenomenon like undulatory propagation, we shall now consider the static phenomena of the stress produced in the material of a dielectric by its excitation; and, to avoid the complexity, both optical and mechanical, introduced by the elasticity of solids, we shall consider solely liquid dielectrics, on which a very valuable series of experiments has been made by Quincke[1]. The mechanical stress in a fluid depends on one variable, the intensity of the hydrostatic pressure, and therefore may be connected immediately with the distribution of the energy in the medium, by means of the principle of work.

Static stresses in dielectrics: fluids.

The arguments for the actual existence of a stress of the Maxwell type may be exhibited in a synthetical manner as follows. Consider a condenser formed by two closed conducting sheets, one inside the other; and imagine the equipotential surfaces to be traced in the excited fluid dielectric between them. It is a matter of experimental knowledge that there is a traction on each face, acting inwards, and equal, at any rate approximately, to $KF^2/8\pi$ per unit surface, where F is the electric force. Now the electric potential, and therefore the state of the dielectric fluid, will be in no wise altered if we imagine a very thin stratum along one of the equipotential closed surfaces to

Physical geometric synthesis of the stress:

* As now corrected and elucidated the result is mainly negative, as these forces arise through the mediation of the electrons.

[1] Wiedemann's *Annalen*, Vol. XIX. 1883.

the tension: become conducting. There will therefore be a normal traction given by the same formula, on each element of area of this surface. If this traction is an affair transmitted across the medium, the transmitting stress must be a tension $KF^2/8\pi$ along the lines of force. To form an opinion as to whether a medium transmitting stress in this way could be imagined, let us suppose the dielectric divided into thin layers, like those of an onion, by much thinner conducting sheets, which coincide with the equipotential surfaces. The potential will not thereby be altered; if we run a tube of force across the dielectric, equal and opposite charges will reside on the portions of the two faces of each sheet intercepted by it. The layers of dielectric will be electrically independent of each other, being separated by conducting layers. Each dielectric layer will, therefore, form a condenser, and the energy of its electrification per unit surface will be $K(\delta V)^2/8\pi t$, or $KF^2 t/8\pi$, where t is the thickness at the point, and δV the difference of potential between the faces; that is, there will be a distribution of energy $KF^2/8\pi$ per unit volume. The resultant traction on both the equal and opposite charges, each σ per unit area, on the two faces of a layer of dielectric, will be normal to the layer, and equal to $\frac{1}{2}\sigma\,(dF/dn)\,\delta n$ per unit surface; now, by Green's form of Laplace's equation, $\frac{dF}{dn} = F\left(\frac{1}{R_1}+\frac{1}{R_2}\right)$, where R_1, R_2 are the radii of principal curvature of the sheet; thus the traction is $\frac{KF^2\delta n}{8\pi}\left(\frac{1}{R_1}+\frac{1}{R_2}\right)$. By the theorem of surface tension, this normal traction will produce and be balanced by a uniform surface tension along the sheet, of intensity $KF^2\delta n/8\pi$, or $KF^2/8\pi$ per unit thickness. In this laminated medium, owing to the attraction across the layers of very small thickness, we have thus set up a tension $KF^2/8\pi$ along the lines of force, which by reaction on the medium produces a pressure uniform in all directions round the lines of force, of the same numerical value. Or, again, we might, following Maxwell, postulate that the stress system in the medium must be symmetrical round the lines of force, and deduce, by the condition of internal equilibrium, that the tension and pressure of which it must thus consist are equal. A spherical system will form a simple illustration, capable of elementary treatment.

its energy distribution: *the balancing transverse thrust.* *Alternative argument.*

Elevation of a fluid between condenser plates: The fact that the surface of a dielectric liquid like petroleum is raised up by attraction, towards an electrified body brought near it, also affords evidence that this tension must exist. Consider two horizontal condenser plates, one inside the petroleum and the other over its surface in air. When the condenser is charged, the surface of the fluid rises between the two plates. There must, therefore, be some traction acting on it upwards to sustain it against gravity.

The intensity of this traction is, in fact, according to Maxwell's law, $\frac{F^2}{8\pi} - \frac{K}{8\pi}\left(\frac{F}{K}\right)^2$, that is, $\frac{F^2}{8\pi}\left(1 - \frac{1}{K}\right)$, where F is the electric force in the air; being positive, it acts upwards, in accordance with the actual phenomenon. Without the assistance of a traction of this kind, the fact would be unexplained, unless by assuming, with Helmholtz, the existence of a *quasi*-magnetic polarization of the elements of the medium; that would lead, on the interface between two media, to an uncompensated sheet of poles of density $\frac{K_1 - 1}{4\pi} F_1 - \frac{K_2 - 1}{4\pi} F_2$, subject to a mean force $\frac{1}{2}(F_1 + F_2)$; so that, as $K_1F_1 = K_2F_2$, the traction would be $(F_1^2 - F_2^2)/8\pi$, or, in the example chosen above, $\frac{F^2}{8\pi}\left(1 - \frac{1}{K^2}\right)$. The discrepancy between these values might, perhaps, be amenable to experiment; but I find, on trial, that the difficulty of obtaining a clean unelectrified surface is not easily overcome*.

Maxwell's value:

discrepant values of polarization theory.

The observation of Faraday, that short filaments of silk or other dielectric material suspended in a fluid dielectric set themselves along the lines of force when it is excited, is also evidence of actual internal polarization related to the lines of force.

For the case of a fluid, the Faraday-Maxwell stress is made up of a hydrostatic pressure, $KF^2/8\pi$, which is consistent with simple fluidity, together with a tension $KF^2/4\pi$ along the lines of force, which requires for its maintenance qualities other than those of isotropic mechanical fluidity.

The stress not consistent with fluidity.

3. The polarization theory, in the form of Mossotti and Helmholtz, which locates part of the electrification in a displacement existing in the elements of the dielectric, and part of it in an absolute electric charge situated on the plates of the condenser the cause of that displacement, is the representation of a wider theory which supposes the electrostatic energy to be in part distributed through the dielectric as a volume-density of energy, and in part over the plates as a surface-density. If experiment show that the latter part is null, we are precluded from imagining any superficial change on the plates which has a separate existence, and is not merely the aspect at one end of the displacement across the volume of the dielectric. We shall find reason to conclude that there is no superficial part in the distribution of energy; this would carry the result that the excitation of a condenser consists in producing a displacement across

Energy distribution all in dielectric:

or else aethereal part and polarization part.

* As the matter is represented by the polarized electrons, the polarization result is the right one for a local surface traction: but the two should lead to the same result for the resultant forcive on any portion of the body.

the dielectric which just neutralizes the charge conducted to the plates; it would also carry the result that all currents, whether in conductors or in dielectrics, must flow in complete circuits, and would therefore confirm the Maxwell theory of electrodynamics.

Consequences:

The conclusion that the location of all the electrostatic energy in the dielectrics involves that all currents flow in complete circuits seems of importance sufficient to justify a few remarks on the nature of the evidence on which it is based. The only precise notion or illustration of the nature of the dielectric polarization which has yet been advanced is that of Poisson, which has been at various times used and developed by Mossotti, Faraday, Thomson, and Helmholtz. It might be held merely on general grounds that it gives a correct formal view of the phenomenon, though the dynamical machinery of which it represents the action is quite unknown. But this presumption is very much strengthened by the fact that the displacement or polarization is known to present qualitatively the properties of a true current, and also that the theory of dielectric propagation developed from this basis presents all the general analogies to light propagation that have been experimentally confirmed by Hertz and others. Taking it then that dielectric polarization is formally of this type, the absence of a sensible absolute exciting charge on the bounding plates will show that it must be, so to speak, self-excited, that it is of the formal character of a displacement, of something pushed across from one plate towards the other like an incompressible substance*.

suggests a homogeneous Maxwellian total current.

4. Let us confine our attention for definiteness to the case of two metallic plates immersed horizontally near together in an extended mass of a fluid dielectric, so as to form a condenser. The traction T per unit area on the upper plate may by this arrangement be directly weighed. Suppose that there is a small aperture in the centre of the upper plate through which a volume of a different dielectric, say, a bubble of air, may be introduced between the plates, so as to form a flat cylinder coaxial with the plates and bounded above and below by them. The extra pressure P in this air bubble, when the condenser

Quincke's methods for fluid stress.

* The synthesis in § 2, on which this conclusion is based, already implies that all is polarization. That synthesis would, however, have to be applicable to a pure vacuum as dielectric. On the other hand, the theory of electrons (*infra, Phil. Trans.* 1894, §§ 114 *seq.*) discriminates between polarization of the atoms with their electron distributions, and aethereal disturbance: the Maxwell stress would then be a compound and purely formal concatenation of the actual forces on the material and electronic system. The argument here belongs to the limiting case of the Helmholtz theory (cf. *supra, On the Theory of Electrodynamics* (1891)), where K is very large so that the aethereal part can be neglected.

is excited, may be measured by a manometer in connection with it, and it will give the means of determining the pressure in the surrounding liquid dielectric. This arrangement describes in fact the plan of Quincke's experiments.

At a point in the dielectric where the electric force is F, the electric pressure will be proportional to F^2, say, A_2F^2 for the liquid, and A_1F^2 for the air. The air column in the manometer tube would thus be in internal equilibrium with an electric pressure A_1F^2 next the liquid and null at the manometer end. We might at first glance infer that under these circumstances the pressure A_1F^2 is not indicated by the manometer, which would thus record simply the electric pressure in the liquid. But this air pressure is an internal stress; the equilibrium of any section of the air column requires, in order to maintain it, the electric pressure against it of the air on the other side of that section; therefore, the indication of the manometer really gives the differential effect $A_2F_2{}^2 - A_1F_1{}^2$.

Let now the volume energy of the electrification be C_2F^2 in the liquid, and C_1F^2 in the air, each per unit volume; and let the energies of such real surface electric distributions as might exist on the plates in contact with these dielectrics be Σ_2 and Σ_1 respectively, each per unit surface of both plates. These surface energies would involve in their expression the electric potential as well as the force. We may apply the principle of virtual work to determine the relations between the quantities thus defined.

Suppose that the distance between the plates, with no air bubble introduced, is slowly increased from c to $c + \delta c$ against the tension T_2 which tends to draw them together. The work done against T_2 in raising the upper plate is the only source of the additional energy of the system which appears when that separation is effected. Now the value of F is not thereby altered, as the electrification remains constant. The volume energy contained in a cylinder of the liquid of unit sectional area is therefore changed from C_2F^2c to $C_2F^2c\,(1 + \delta c/c)$. The corresponding surface energy is changed from Σ_2 to $\Sigma_2 + \dfrac{d\Sigma_2}{dc}\delta c$.

Analysis of results of internal distribution of energy,

Therefore, by the principle of work, $T_2 = C_2F^2 + \dfrac{d\Sigma_2}{dc}$.

Now suppose a large cylindrical air bubble introduced between the plates; and suppose its volume to be increased by δv owing to a virtual displacement produced by pressing in more air. The virtual work $P\delta v$ so done must be equal to the increase of the internal energy of the system due to the displacement. This increase may be calculated from the energy function of the actual conformation, not of the displaced position, as internal equilibrium subsists; and all

by method of virtual work.

considerations as to change of intrinsic energy may thus be evaded[1]. For the flat cylindrical bubble its surface would, in fact, be increased by the supposed displacement, and so there would be an increase of intrinsic capillary energy; with surface tension τ, radius a, and a semicircular meniscus πc, there would be an increase of amount $2\pi^2 c\tau\delta a$, that is $\pi\tau c\delta v/a$. Again, the electric forces F_1 and F_2 in the two media, measured not close up to the meniscus, are equal, because the plates are each at uniform potential. Close to the centre of the meniscus they are both tangential to it, and must also be equal on the two sides. But along the slope of the meniscus they are oblique to it and are unequal, their relation being determined on the ordinary theory by the continuity of the normal component induction and of the tangential component force. Thus the difference of electric pressure across the meniscus will vary from point to point of it, and its form will, therefore, be slightly altered by the electrification. It follows from the observations of Quincke and others that its capillary constant will not thereby be altered, as was to be expected, because there is no extra supply of molecular surface energy. There is also the alteration of intrinsic energy due to the fact that the expansion of the air bubble alters the electric force. But, according to the principles just stated, these changes of intrinsic energy balance each other, because all the parts of the medium are in internal equilibrium. We may therefore consider the annular mass between two ideal coaxial cylindric surfaces at a distance from the meniscus, one in the bubble and one in the liquid, and reckon the change in the energy contained in the space originally occupied by this annulus, when it receives a small displacement outwards from the axis under the action of the manometer pressure P. That change is $-(C_2 - C_1) F^2\delta v$, where F is the electric force at a distance from the meniscus. Therefore, by the principle of virtual work, $P\delta v - (C_2 - C_1) F^2\delta v = 0$, so that $P = (C_2 - C_1) F^2$.

Surface charge has no capillary effect.

Transverse thrust:

tension along lines of force

The value of the traction between the two plates in air is given by this formula as $T_1 = C_1 F^2 + \frac{d\Sigma_1}{dc}$; this must, therefore, be the same as the well-ascertained experimental value $F^2/8\pi$. Now the experiments of Quincke and others on liquid dielectrics have given reason to believe that, within the limits of experimental uncertainty due to want of purity of the materials and other causes, $T_2 = P + F^2/8\pi$. It follows that we must have $d\Sigma_2/dc = d\Sigma_1/dc$; that is, $d\Sigma/dc$ must be the same for all media, which is physically consistent only with the non-existence of this surface energy, unless we can suppose it to be

[1] For applications of this principle, cf. Helmholtz, Wied. *Ann.* Vol. XIII. p. 388; and Kirchhoff, Wied. *Ann.* Vol. XXIV. p. 57.

the energy of an action at a distance or in the aether which is independent of the intervening medium altogether.

The argument may also be expressed somewhat differently as follows. The plates of the condenser being supported independently, the existence of an extra pressure on the dielectric when the condenser is excited shows that part, at any rate, of the electric energy resides in the dielectric. That part must, on any view, either of action by contact or of *quasi*-magnetic polarization, be proportional to the square of the electric force at the point, which is in fact confirmed by the experiments of Silow on a quadrant electrometer with its needle immersed in a dielectric liquid filling the quadrants. If this were the whole of the electric energy, the traction between the plates would be equal to the hydrostatic pressure in the dielectric, or at most differ from it by an amount which would be the same for all media[1]. If this were only part of the electric energy, the difference would depend on the other superficial part. The experiments show that for a large number of liquids the difference is very nearly the same, being in fact an aether contribution, as in the next sentences, so that if, after Quincke, we suppose it to be null for air or vacuum, it is null for all the others[2]. Hence either the superficial energy must be absolutely independent of the nature of the dielectric or else it must be non-existent. The precise logical statement of Quincke's results is, in fact, that the difference between the electric stress in a ponderable fluid dielectric K and the electric stress in a vacuum, in a field of force F, consists of a tension $\frac{K-1}{8\pi}F^2$ along the lines of force, combined with a pressure of equal amount in all directions at right angles to them; and this is consistent with a distribution of

Conclusions, that the stress is merely a formal aggregate.

[1] Cf. J. Hopkinson, *Proc. Roy. Soc.* 1886, p. 453, for the use of a similar argument in the converse manner, to show that the tension and pressure must be equal; but in it the energy of the polarization of the medium is apparently not sufficiently traced.

[2] The results of Quincke are calculated so as to give values for K_p, the inductive capacity deduced from experiments on fluid pressure, and K_s, the inductive capacity deduced from experiments on the traction between the plates, on the assumption that the stress is of the Faraday-Maxwell type. The following examples show the order of magnitude of the discrepancies:

Confirmation of formulae.

	K_p	K_s
Ether	4·62	4·66
Carbon disulphide	2·69	2·75
Benzol	2·32	2·37
Turpentine	2·26	2·35
Petroleum	2·14	2·15

The chief difficulty seemed to be to avoid conduction, owing to want of purity of the dielectric fluid.

polarization energy in the fluid, added to the electric energy for a vacuum with the same intensity of force, but not entering into combination with it.

Now the propagation of electrical waves across air or vacuum shows that even then, when there is no ponderable dielectric present, there must be a store of statical energy in the dielectric; and this fact appears to remove the only explanation which seems assignable for the division of the energy into two parts, one located in the dielectric, and the other located on the plates and absolutely independent of the dielectric, viz. that the latter might be the energy of a direct action across space which is not affected by the dielectric. The experimental facts, therefore, so far tend to the conclusion that at any rate the basis of electrical theory is to be laid on Maxwell's lines, with a reservation for possible modification in the form of residual corrections, but not for change of principle.

Yet the aethereal energy is also in the dielectric volume.

A theory has been developed by Helmholtz for fluids, and by Kirchhoff, following him, for solid material dielectrics, in which slight residual differences between the intensities of the tension and pressure may be accounted for on the supposition that the inductive capacity, instead of being constant, is a function of the material stress due to the electric force. This theory is primarily expounded in terms of a polarization scheme, and in so far is subject to the remarks of the next section; but it may in the end be based, as Helmholtz suggested, on the principle of energy applied with the aid of the ascertained form of the characteristic equation of the potential treated as a condition of internal equilibrium. If we adopt the view that the difference to be explained has not certainly been detected[1], this theory need not here be considered.

Helmholtz's energy method.

Some of the points in the general treatment given above will also be illustrated by the following brief discussion, which has special reference to the Mossotti-Helmholtz polarization theory. In the course of it reasons will appear that even the special limit of that theory which coincides with Maxwell's as to form must be abandoned as inconsistent with the dynamical phenomena, in favour of a theory of pure contiguous action or strain of an incompressible aether.

General polarization theory.

Without entering here into detail as to the general characteristics of this kind of polarization, it will suffice to point out some of its principal relations with regard to which misconception is easy, and also to point out the modifications which are necessitated in its usual form by the recognition of the discrete or molecular character of

[1] Cf. Bos, "Inaugural Dissertation," abstracted in *Philosophical Magazine*, February 1891. Helmholtz in *Abhandl.* under year 1881, Kirchhoff under year 1884.

the polarized elements. In the Poisson theory of induced magnetism the magnetic potential is the potential not of the actual magnetism, but of the continuous volume and surface distributions of ideal magnetic matter which Poisson substitutes for it. The forces on a magnetic molecule are therefore not to be derived from it[1]. But if we imagine a very elongated cavity to be scooped out in the medium along the direction of magnetization, and the molecule to be placed in the middle of the cavity, the forces of the remaining magnetized matter will be correctly derived from this potential. This part of the forcive will thus be derivable from a potential energy $MF \cos \epsilon$, where M is the moment of the molecule, F the resultant force derived from the magnetic potential, and ϵ the angle between their directions; we may thus consider a potential energy function $IF \cos \epsilon$ per unit volume. We have to add to these forces the ones due to the rejected magnetic molecules which lay in the elongated cavity. Now the mutual action of contiguous magnetic molecules will be of the nature of a tension along the lines of magnetization and a pressure at right angles to them, as Helmholtz remarked[2]; but these stresses will not necessarily be equal in intensity; nor will they represent the Faraday-Maxwell stress, since each component is proportional to the square of the coefficient of magnetization, not to its first power. In a fluid medium these forces also must be derivable from an energy function, for otherwise the medium could not be in equilibrium; and the total potential energy per unit volume with its sign changed is equal to the fluid pressure*. Thus in the polarized fluid the pressure is

Theory of induced magnetism: energy distribution.

Local molecular stress in a magnet:

$$\tfrac{1}{2}FI + \tfrac{1}{2}\lambda I^2,$$

that is, $$\tfrac{1}{2}(\kappa + \lambda\kappa^2) F^2.$$

magneto-striction.

An actual illustration in which the term involving λ is of predominant importance is afforded by a bunch of iron nails hanging end to end from a pole of a magnet; the adjacent nails hang on to each other lengthways and repel each other sideways, while the action of non-adjacent ones is but slight.

Stress in hanging bunch of iron nails.

In the electric polarization theory the specific inductive capacity is $K = 1 + 4\pi\kappa$. The results of Quincke, above mentioned, after they had been corrected for an experimental oversight in the direct

[1] In estimating these forces it is not allowable to replace the molecule by its three components parallel to the axes in the usual manner. This procedure would lead to error if there are electric currents in the field. Cf. Maxwell, *Electricity*, ed. 2, Vol. II. Ch. XI., appendix 2, p. 262.

[2] Wiedemann's *Annalen*, Vol. XIII. 1881, p. 388.

* But these forces, being purely local, are locally compensated by deformation of local structure of the medium, which is in fact the essence of magnetic striction, and are not transmitted as fluid pressure involving a λ term as above.

Stress not transmitted.

determinations of the values of K by experiments on capacity, in accordance with a suggestion made by Hopkinson[1], made the electric pressure to be $KF^2/8\pi$, consistently within the limits of experimental error for fifteen different substances. Thus, even in the limiting Maxwell form of the theory, which takes the absolute numerical value of K to be very great, this theory would not fit with the experiments unless λ is zero. Even by the purely mathematical device of taking the polarized elements to be right solids closely packed together, it does not seem possible to evade this argument.

Local strictions not transmitted.

In an actual fluid polarized in the above manner each element might on the average be considered as lying at the centre of a cavity, a sort of sphere of action within which the other molecules in their motions do not approach it further. On averaging the positions of these surrounding molecules during their motions with respect to the one under consideration, we arrive at the conception of a continuous polarized medium with a cavity in it of the form of this sphere of action. If this cavity were an actual sphere, the value of λ would be $\frac{4}{3}\pi$; and for cavities not very greatly different from the spherical form, the alteration in this value would be insensible. Under no likely circumstances could the value of λ come to be zero*.

Maxwell's not the limit of a theory of pure polarization.

Thus the limiting Helmholtz polarization representation of an excited dielectric, though complete as regards electromotive properties, would appear to fail to include the static ponderomotive phenomena of electrification, and requires to be modified into some more continuous mechanism, such as an elastic displacement in an aether loaded with the molecules of the dielectric.

It may be well to remark that, on account of the extreme smallness of the magnetic coefficient κ for all fluids, its square is of no account in comparison, and therefore magnetic pressures are sufficiently represented by the simpler formula $\frac{1}{2}\kappa F^2$, by means of which Quincke has measured the magnetic constants of various fluid media.

Conclusions (modified).

5. The principal conclusions which have been arrived at are here enumerated†.

[1] *Proc. Roy. Soc.* 1886.

* But this contemplates a single pole within the cavity: the forces on the two conjugate poles will cancel. The striction, illustrated by magnetized nails hanging together, is more intimate and not amenable to this averaging.

† The difficulties encountered in the sections *supra*, such as are focussed in (vi), arise from want of some method of bridging the gap between electromotive and ponderomotive phenomena in the field, which must both arise from its single energy specification.

The genesis of a formal transmitting quadratic stress-tensor of the type above discussed, exhibiting the emergence of a mechanical field of force-momentum-energy from virtual variation of the purely electrodynamic field of Action, can

(i) It is shown from experimental results that the equivalent, or representative stress in an excited fluid dielectric between two condenser plates consists, at any rate to a first approximation, of a tension along the lines of force and an equal pressure in all directions at right angles to them, superposed upon such stress as would exist in a vacuum with the same value of the electric force.

(ii) It is shown from experiments that the numerical value of these additional equal tensions and pressures is, at any rate to a first approximation, $(K-1)\,F^2/8\pi$, where F is the electric force, and K the inductive capacity.

(iii) Such a distribution of equal tension and pressure is the result of (might possibly be correlated with) a uniform volume distribution of energy in the dielectric, irrespective of what theory is adopted as to its mode of excitation.

(iv) If we consider the mode of excitation to be a *quasi*-magnetic polarization of its molecules, the numerical magnitude of these stresses should be

$$\frac{K-1}{8\pi}F^2\left(1+\lambda\frac{K-1}{4\pi}\right),$$

where λ is a coefficient which depends on the molecular discreteness of the medium, and expresses the indirect effect [*not transmitted*] of local striction or change of form and quality.

(v) The stress which would exist in a vacuum dielectric is certainly due in part to a volume distribution of energy, as is shown by the propagation of electric waves across a vacuum. There is thus no reason left for assuming any part of it to be due to a distribution of energy on its two surfaces, acting directly at a distance on each other*. There is therefore ground for assuming somehow a purely volume distribution of energy in the vacuous space, leading to a tension $F^2/8\pi$ along the lines of force, and a pressure $F^2/8\pi$ at right angles to them.

(vi) The *quasi*-magnetic polarization theory rests on the notion of a dielectric excited by a surface charge on the plates, and therefore involves a surface or else aethereal distribution of energy, except in the extreme case when the absolute value of K is very great; in that case a slight surface charge produces a great polarization effect, and in the limit the polarization may be taken as self-excited. Thus the

now be directly visualized from first principles: see Appendix added *infra* to Part III. (1897) § 39 of the memoir on Dynamical Theory.

* The energy of the surface electrons is located in the medium between them.

absence of a surface distribution of energy leads to Maxwell's displacement theory, in which all electric currents are circuital, and the equations of electrodynamics are therefore ascertained: but it does not lead to correct mechanical forces on the matter.

(vii) It appears that even this limiting polarization theory must be replaced, on account of the stress-formula in (iv), by some dynamical theory of displacement of a more continuous character.

6. We may perhaps attempt to form a more vivid picture of the interaction between aether and matter by following out the ideas of Lord Rayleigh's version of Young's theory of capillarity. We may conceive the compound medium, aether and matter, to consist of a very refined aethereal substratum, in which the molecular web of matter is imbedded. The range of direct action between contiguous parts of the aether would be very small, and that between contiguous elements of matter large in comparison. There exist disturbances in which the matter-web is unaffected, its free periods being too slow to follow them: these are propagated with great velocity as light, or electrical radiations. There are other disturbances in which the matter-web is alone active; these are so slow that the aether can adjust itself to an equilibrium condition at each instant; they are propagated as waves of material vibration or sound waves.

Radiation waves and sound waves.

When a dielectric is excited, we find ourselves in the presence of a strain of an aethereal origin somehow produced; it would relax on discharge of the system with the velocity of light. At an interface where one dielectric joins another, the aethereal conditions will somehow, owing to the nature of the connection with the matter [*i.e.* the electrons], only admit of a portion of the stress being transmitted across the interface; and there will thus be a residual traction on the interface which must, if equilibrium subsist, be supported by the matter-web, and be the origin of the stress which has been verified experimentally. Inside a conductor, the aether cannot sustain stress at all [electrons being there free], so that the whole aethereal stress in the dielectric is supported by the surface of the matter-web of the conductor. At such interfaces the aethereal part of the distribution of energy in the medium will be discontinuous.

Pressure of electric waves:

A formula has been given by Maxwell[1] for the intensity of the pressural force produced by electric undulations in the aether striking against a plate of conducting matter, a force which has apparently not been detected for the case of light-waves. If the notions here suggested have any basis, this force may likely be non-existent*. For

[1] *Electricity*, § 793.

* The pressure of radiation is now abundantly verified, and has become a keystone of theory. Cf. *loc. cit.* § 5 footnote.

the pulsations of the aether at this surface may be so rapid as to prevent their energy being communicated to the matter-web of the conductor; and the energy will then be scattered and lost instead of appearing as energy of material stress. We may take as an illustration a stretched cord with equidistant equal masses strung on it, for which Lagrange showed that if the period of a disturbance imparted at one end exceeds a certain limit, the disturbance will not be transmitted into the cord, but will be eased off within a short distance of the point of application. And also in a manner which forms a more exact analogy, Sir G. Stokes has shown that the higher harmonics of a telegraph wire vibrating in the wind have their pulsations too rapid to get a grip on the air around them, and their note is therefore not transmitted.

might have been problematical.

The grip on matter.

This view would place the electrostatic and electrodynamic forces on matter on a lower plane, and in the case of rapid or sudden disturbance a more uncertain one, than the electromotive phenomena.

32

THE DIOPTRICS OF GRATINGS.

[*Proc. Lond. Math. Soc.* Vol. XXIV. (1893) pp. 266–272.]

Sifting of complex waves by uniform striation: WHEN a beam of light falls upon a ruled or striated surface, a considerable portion of it is inevitably scattered and lost by the inequalities of the surface; and the residue is reflected or refracted in the ordinary manner. But when the striation varies from point to point in a continuous and fairly uniform way, there is sifted out from the incident beam, in addition to the debris of scattered light, a series of regular secondary beams, which are propagated onwards in directions inclined to that of the principal one.

elucidated by Young: The origin of such a diffracted beam, by the union of the diffracted parts from the different striae which arrive at its front in the same phase, was fully explained by Thomas Young, as also was the very perfect separation of the different chromatic constituents of a regular compound beam by a good grating of this kind. In the few pregnant sentences in which Young pointed out the reason of these phenomena[1], he, in fact, made the way perfectly clear for that extension of their range which was afterwards worked out experimentally by Fraunhofer, and which has more recently led to the development of the optical grating as the chief instrument of spectral analysis. **developed by Fraunhofer.**

A general dioptrical theory of gratings, The discussion of the action of such gratings, so far as it is usually required for practical purposes, is a simple and well-known matter. But there are questions of some importance, such as the effect of want of perpendicularity of the lines of the grating to the plane of incidence of the light, which are more readily attacked by means of a general theory; while it may also be of interest formally to include general diffracted beams within the domain of dioptrical analysis, and exhibit the rules by which the position of their focal lines, when narrow, and the determination of their caustic surfaces in other cases, is to be accomplished.

by Hamiltonian characteristics. The fundamental physical principle is that the existence of a continuous wave-front requires either (i) that the optical path measured up to it of the rays which come from all the striae shall be the same, or (ii) that for successive striae it shall differ by the index multiplied by a multiple of a wave-length of the diffracted beam, say by $n\mu_2\lambda_2$ ($= n\mu_1\lambda_1$) for the diffracted beam of the nth order. Thus, if m be the number of striations between a selected point on the grating and any origin of reference, the difference of paths for the corresponding rays will be $mn\mu\lambda$. This expression will be a function of the co-

[1] *Phil. Trans.* 1801.

ordinates of the point on the grating; and to obtain the Hamiltonian characteristic function of the diffracted beam we have simply to add this function to the characteristic of the unbroken incident beam.

Let us take the equation of the surface of the grating to be, up to the second order,

$$\zeta = \tfrac{1}{2}\alpha\xi^2 + \tfrac{1}{2}\beta\eta^2 + v\xi\eta + \dots;$$

The ruled surface of the grating:

and let the lines of the grating be parallel to the axes of η, so that

$$nm\mu\lambda = L\xi + \tfrac{1}{2}\alpha'\xi^2 + \tfrac{1}{2}\beta'\eta^2 + v'\xi\eta + \dots,$$

the forms of the rulings:

where the coefficients α', β', v' represent the effect of any continuous change in the breadths of the striae that may exist. Suppose the characteristic function of an incident beam in the medium of index μ to be

$$V_1 = \mu_1 \{l_1\xi_1 + m_1\eta_1 + n_1\zeta_1 + \tfrac{1}{2}A_1\xi_1^2 + \tfrac{1}{2}B_1\eta_1^2 + \tfrac{1}{2}C_1\zeta_1^2 + F_1\eta_1\zeta_1 + G_1\zeta_1\xi_1 + H_1\xi_1\eta_1 + \dots\};$$

characteristic of the incident beam,

while the characteristic function of the nth diffracted beam in the medium of index μ_2 is given by the similar expression with suffix 2, the value of λ_2 above also belonging to this medium; the case of reflexion is obtained by making $\mu_2 = -\mu_1$.

of the diffracted beam.

At the surface of the grating we must have

Direction of diffraction:

$$V_2 - V_1 = nm\mu\lambda.$$

Hence, considering first the terms of the first degree, we have

$$\mu_2 l_2 - \mu_1 l_1 = L,$$
$$\mu_2 m_2 - \mu_1 m_1 = 0.$$

As these relations are linear, they express that the projection of the incident ray on a normal plane parallel to the lines of striation is bent according to the ordinary law of refraction; while its projection on the normal plane at right angles to these lines is bent in the same manner as an actual ray in this direction would be diffracted, the angles of incidence and diffraction being connected by the relation

$$\mu_2 \sin\phi_2 - \mu_1 \sin\phi_1 = L,$$

where L/n is the value of $\mu\lambda$ divided by the width of a striation at the origin.

The direction of the diffracted beam being thus determined, it remains to find its focal lines. This is done by equating the terms of the second order at the diffracting surface; the equation of the surface must be used to eliminate ζ, and then the two sides of the equation of condition must agree identically. There results

focal lines of the diffracted beam.

$$(\mu_2 A_2 + n_2\alpha) - (\mu_1 A_1 + n_1\alpha) = \alpha',$$
$$(\mu_2 B_2 + n_2\beta) - (\mu_1 B_1 + n_1\beta) = \beta',$$
$$(\mu_2 H_2 + n_2 v) - (\mu_1 H_1 + n_1 v) = v';$$

the other coefficients only entering in the third order.

These remaining coefficients are, however, determined by the characteristic equation

$$\left(\frac{dV}{dx}\right)^2 + \left(\frac{dV}{dy}\right)^2 + \left(\frac{dV}{dz}\right)^2 = \mu^2,$$

which requires

$$(l + A\xi + H\eta + G\zeta)^2 + (\ldots)^2 + (\ldots)^2 = 1;$$

and this is satisfied up to the first degree of small quantities, provided

$$Al + Hm + Gn = 0,$$
$$Hl + Bm + Fn = 0,$$
$$Gl + Fm + Cn = 0.$$

These equations determine G, F, C in terms of A, B, H.

Now the distances of the focal lines of the beam are the radii of principal curvature, at the origin, of the surface $V = 0$. These radii are equal to the squares of the semi-axes of the central section of the surface

$$\tfrac{1}{2}A\xi^2 + \ldots + F\eta\zeta + \ldots = 1$$

by the plane l, m, n; therefore they are determined by making $R^2 = \xi^2 + \eta^2 + \zeta^2$ a maximum or minimum, subject to the condition

$$l\xi + m\eta + n\zeta = 0.$$

And the analysis may be completed for any special case by means of well-known formulae in Geometry of Three Dimensions.

Case of incidence in plane transverse to the rulings

A manageable case arises when the incident and diffracted rays are in the same plane, which is therefore normal to the striations. We may now refer each of the beams to its own principal axes. Thus

$$V_1 = \mu_1 \{z_1 + \tfrac{1}{2}A_1x_1^2 + \tfrac{1}{2}B_1y_1^2 + H_1x_1y_1 + \ldots\},$$
$$V_2 = \mu_2 \{z_2 + \tfrac{1}{2}A_2x_2^2 + \tfrac{1}{2}B_2y_2^2 + H_2x_2y_2 + \ldots\};$$

and we will gain symmetry by altering the equation of the diffracting surface to

on the surface.

$$0 = \zeta + \tfrac{1}{2}\alpha\xi^2 + \tfrac{1}{2}\beta\eta^2 + v\xi\eta.$$

Change of coordinates is effected by equations of the type

$$\left.\begin{aligned} x &= \xi\cos\phi - \zeta\sin\phi, \\ z &= \xi\sin\phi + \zeta\cos\phi, \\ y &= \eta. \end{aligned}\right\}$$

On eliminating ζ as before, and so identifying at the surface the two sides of the equation of condition, we have

Modified law of refraction, but

$$\mu_2 \sin\phi_2 - \mu_1 \sin\phi_1 = L_2,$$
$$\mu_2 A_2 \cos^2\phi_2 - \mu_1 A_1 \cos^2\phi_1 = \alpha_2 + (\mu_2\cos\phi_2 - \mu_1\cos\phi_1)\,\alpha,$$
$$\mu_2 B_2 - \mu_1 B_1 = \beta_2 + (\mu_2\cos\phi_2 - \mu_1\cos\phi_1)\,\beta,$$
$$\mu_2 H_2\cos\phi_2 - \mu_1 H_1\cos\phi_1 = v_2 + (\mu_2\cos\phi_2 - \mu_1\cos\phi_1)\,v.$$

In both the general problem and this more special case, it is to be observed that, if α', β', v' are null, *i.e.* if the striations are symmetrical with respect to the origin, the focal lines are determined by exactly the same formulae as would apply to simple refraction at the surface, the different direction of the diffracted ray being allowed for. In the case of Rowland's spherical gratings, this result is well known, and is made use of in the instrumental arrangements. The aberration would be expressed by terms of the third degree. focal lines determined by usual dioptric formulae,

When the incidence is direct, the circumstances of the diffracted beam will be correctly represented by imagining it to be refracted at an ideal surface situated at each point a distance $mn\mu\lambda/(\mu_2 - \mu_1)$ in front of the real one. But this rule must be modified when the incidence is oblique. The ideal surface would then vary with the angle of incidence, the distance being now $mn\mu\lambda/(\mu_2 \cos\phi_2 - \mu_1 \cos\phi_1)$; for the interposition of a thickness t of medium of index μ_2 retards the ray by an amount that corresponds to a length as from an ideal refracting surface.

$$(\mu_2 \cos\phi_2 - \mu_1 \cos\phi_1)\, t/\mu_1$$

in the medium of index μ_1. The direction of the diffracted ray will be determined by the rule given above; and, once that direction is found, the form of the diffracted beam will be given, when the striation is symmetrical, by the formulae which belong to ordinary refraction at the surface of the grating.

Another case of some theoretical interest arises when the lines of the grating are closed curves drawn round its centre. If we take A system of ruled oval curves form a diffracted image:

$$nm\mu\lambda = \tfrac{1}{2}\alpha'\xi^2 + \tfrac{1}{2}\beta'\eta^2 + v'\xi\eta + \ldots,$$

that is, if we make L null in the above analysis, these lines will be, in the neighbourhood of the vertex, the system of similar concentric conics $nm\mu\lambda =$ constant, the successive rings enclosed between them being now of equal area. The result indicated by the formulae is that the diffracted beam follows the same direction as the principal refracted beam, but the equations which give its elements differ by the terms α', β', v' on their right-hand sides. If the incidence is direct, the grating by itself acts in the same manner as a thin astigmatic lens, whose thickness t is given by equivalent astigmatic lens.

$$(\mu_2 - \mu_1)\, t = nm\mu\lambda;$$

if it is oblique at an angle ϕ_1, the law of thickness of the equivalent lens is

$$(\mu_2 \cos\phi_2 - \mu_1 \cos\phi_1)\, t = nm\mu\lambda.$$

33

THE SINGULARITIES OF THE OPTICAL WAVE-SURFACE, ELECTRIC STABILITY, AND MAGNETIC ROTATORY POLARIZATION.

[*Proc. Lond. Math. Soc.* Vol. XXIV. (1893) pp. 272–290.]

Fresnel's surface: ALTHOUGH Fresnel recognized that the two sheets of his optical wave-surface meet each other, he did not explicitly realize that they meet at conical points. This fact, and the striking phenomena of conical refraction which are dependent on it, was left for Sir W. R. Hamilton to discover. Hamilton's points of meeting of its two sheets, Yet it hardly needed very much discovery; for the roughest observation of the colours of crystalline plates in polarized light had shown that there are only two directions of single-ray velocity in a crystal, and therefore that the two sheets of the wave-surface meet each other (or at any rate come very close together) only along these directions—therefore not along any curve of intersection, but at definite points, which must be conical points on the surface. not peculiar to any special theory. Thus, as Sir G. G. Stokes remarked long ago, the prediction of conical refraction does not constitute any demonstration of the exactness of Fresnel's wave-surface; though the two experiments of Lloyd, if they could be made with an infinitely narrow beam, and with exactly parallel rays, would reveal the existence of actual singular tangent planes*, and their normals the optic axes, as well as the existence of actual conical points.

May the sheets cross? The question suggests itself whether there is anything to prevent the different sheets of the wave-surface corresponding to any crystalline elastic medium from crossing each other. The hypothesis of their intersecting along a curve would not lead to any striking anomaly like conical refraction†; but yet dynamical reasons can be readily assigned which forbid its occurrence.

Geometrical indications against. We can realize in a general way that if in an optical wave-surface the velocities of the two types of transverse undulations corresponding to any plane front coincide, these types become merged together, and undulations can be propagated which correspond to any direction of vibration in the front; and we might hence infer that the front is a singular one, and not one of the infinite number of double tangent

* *I.e.* which have contact along the thin dark curves that are observed inside the luminous band.

† A certain range of rays, or rather filaments of light, would be split into two on refraction.

planes which could be drawn to two intersecting sheets of a wave-surface. But more precise and also more general reasons can be given.

In any homogeneous material structure in three dimensions of space, there are three velocities with which plane undulations can travel with a given direction of front. These velocities must be all real, for an imaginary value would imply dynamically a real exponential, with arbitrary sign, in the expression for the vibrations of the medium, and would therefore show that the medium could not subsist, owing to instability. These three velocities will usually be given by a cubic equation, in which the variable is the velocity squared.

Three wave velocities, in crystalline medium, all real:

If the medium is incompressible, one of the velocities is infinite*, and the other pair are determined by a quadratic equation; this also applies in other cases which have been imagined in connection with physical optics. If these two velocities are equal, the quadratic equation has equal roots, and therefore the expression under the radical sign in its solution must vanish. But the reality of the roots requires that this expression can never be negative, in whatever direction the wave-front may lie. It therefore only vanishes for a direction which makes it a minimum. Thus the wave-fronts of coincident velocities are confined to a definite number of planes which correspond to minimum values of this function, and of these only the ones are to be selected which make it equal to zero. In general then there are no such values, *i.e.* there are no double tangent planes to the wave-surface to be anticipated; though it may happen exceptionally that there are certain isolated planes of this kind, which therefore touch the surface along a curve, like the singular planes of Fresnel's surface.

can reduce to two, as in optics:

condition for equality,

must apply only to a definite direction,

so the sheets can meet only at points, if at all.

It is not possible at all for the two sheets of the wave-surface to have a continuous series of tangent planes; therefore its two sheets cannot intersect along a curve. The exceptional case of one singular tangent plane touching along a plane curve implies a depression in the sheet of which that curve is the rim; it is possible that at the bottom of this depression the two sheets may come together, thus forming a conical point, but for the reason given above they must not intersect along a curve[1].

Geometrical form near the singularity.

* But a static stress, instantly adjusted, persists usually in the equations.

[1] It is interesting to compare this method of ascertaining the nature of the singular planes with the argument employed by Sir G. G. Stokes, "Report on Double Refraction," *Brit. Assoc. Report*, 1862, where he shows that if a depression exist on the wave-surface its rim must be a plane curve, from the consideration that, if it were not plane, then in certain directions four parallel tangent planes could be drawn to the surface, whereas physical reasons preclude the existence of four separate velocities of propagation.

Stokes' mode of argument.

It is easy to conceive that a slight change in the constitution of the medium would just introduce imaginary terms into that velocity which corresponds to the direction of a singular tangent plane with a curve of contact round a nodal point; so that in this sense the constitution of the medium is labile. The abnormality of conical refraction thus indicates the immediate approach of instability*.

Extension to the wider problem.

These conclusions with respect to singularities on the wave-surface also hold good in the more general case, when all three velocities of propagation are finite. For the equality of a pair of roots in the cubic equation which determines the velocities implies equality of roots in the reducing quadratic which is introduced in the algebraical solution of the cubic. Now for the roots of the cubic to be real, the roots of this quadratic must be imaginary; therefore the expression under the radical sign in them must be negative. That is, the directions of equal wave-velocity are determined by the maximum values of this expression, subject to the condition that the maximum is in each case zero. The three sheets of the wave-surface therefore in general have no double tangent planes, and therefore no curve of intersection; but in special cases they may have isolated singular planes or conical points. Also, as before, in tracing the gradual change in the form of the wave-surface which corresponds to gradual change in the constitution of the medium, the final stage is arrived at when two sheets of the surface come into contact; any further change in that direction lands us in instability.

The crystalline medium is discrete:

It is, perhaps, not fanciful to see in the singularities of the optical wave-surface an indication that it is not derived from a purely elastic medium, but is modified or dominated by the inertia and the free periods of the molecules with which the pervading aether is loaded[1].

* Provided the phenomena are really governed by perfectly uniform linear equations, so that molecular considerations do not intervene, *e.g.* by influencing vibrations, as in what follows.

Young's earlier *aperçu* towards the theory of dispersion.

[1] This view of the dynamics of refraction, which has been formulated in recent years with general approval, is, strange to say, the very first explanation that ever was offered on the Undulatory Theory. In the memoir by Thomas Young, "On the Theory of Light and Colours," *Phil. Trans.* 1801, Prop. VII., the manner in which dispersion can be thus explained is very clearly expounded; but the author has gone wrong in the calculation of the mutual influence of the free periods of the aether and the matter molecules, partly owing to his usual manner of attempting to see through the phenomena without the aid of analysis. On this point a recollection of the analytical dynamics of Lagrange and Laplace, with which he was well acquainted, or even further reflexion in his own manner, would have kept him right; his explanation would then have been in principle complete and unexceptionable, and might have directed mathematicians to more fruitful ground than the partial theories, based on simple heterogeneity of the medium, which have been in the meantime developed at great length by Cauchy and others.

For although the influence of temperature and pressure produce the most marked alterations in the positions of the optic axes of a crystal, yet they do not destroy it as an optical medium[1]. but aethereally stable.

In the light of these considerations, the famous dynamical proposition of Green[2], that it is possible so to choose the elastic constants of a crystalline medium that the displacement in two of its vibration-types shall be exactly transverse to the wave-front, and that this condition leads to Fresnel's expression for the velocities, but with opposite polarization, involves the rider that a medium so modified has been brought to the verge of instability; and this is so whatever be the velocity of its wave of normal displacement. Green's type of simple medium peculiar;

Electric Stability in Crystalline Media.

These general considerations may be illustrated and applied in a brief discussion of the question as to what relations must be assumed, between the electric and magnetic constants of a crystalline medium, in order that it may be stable as regards those electric vibrations in it which have been experimentally realized by Hertz, and which are supposed on strong grounds to involve the same mechanism as that by which light is propagated. The simpler circumstances of long waves.

The fundamental equations of electrodynamics, on Maxwell's scheme, are the two circuital relations. The first of these relations (Ampère's) expresses that the circulation of the magnetic force round any circuit is equal to the electric flow through the aperture of that circuit, multiplied by 4π; the magnetic force is therefore derived from a potential only in those parts of the field which convey no current. The second relation (Faraday's) expresses that the circulation of the electric force round any circuit is equal to the time-rate of decrease of the flux of magnetic induction through the aperture of that circuit; the electric force is therefore derived from a potential only in regions in which the magnetic field is constant or null. The electric field relations:

To obtain a complete scheme of equations for any given medium, there must be conjoined with these principles the experimental laws which connect in that medium the electric force and the magnetic force with their correlative fluxes, the electric current and the magnetic induction. When we utilize the facts that the convergence of each flux is null, while the circulation of each force is as above formulated, the specification of the electric relations of the medium will be complete. the constitutive material connections.

[1] Thus, according to Kerr (*Phil. Mag.* 1888), the effect of a tensile or compressile strain on glass is simply to convert it optically into a uniaxial crystal with its optic axis along the line of strain.

[2] Green, *Trans. Camb. Phil. Soc.* 1839; *Collected Papers*, p. 292.

The equations: To obtain analytical expressions, let (PQR) be the electric force, $(\alpha\beta\gamma)$ the magnetic force, (uvw) the electric current, and (abc) the magnetic induction, each specified by its components referred to rectangular axes; then the fundamental circuital relations of Maxwell's theory become

dynamical:

$$4\pi u = \frac{d\gamma}{dy} - \frac{d\beta}{dz}, \qquad -\frac{da}{dt} = \frac{dR}{dy} - \frac{dQ}{dz},$$

$$4\pi v = \frac{d\alpha}{dz} - \frac{d\gamma}{dx}, \qquad -\frac{db}{dt} = \frac{dP}{dz} - \frac{dR}{dx},$$

$$4\pi w = \frac{d\beta}{dx} - \frac{d\alpha}{dy}, \qquad -\frac{dc}{dt} = \frac{dQ}{dx} - \frac{dP}{dy}.$$

If the medium is non-magnetic, or only so feebly magnetic that its magnetization may be neglected, we have

$$(a, b, c) = (\alpha, \beta, \gamma)$$

exactly. But if it is magnetic, we must, in order to proceed further, assume a coefficient of magnetic permeability μ, which will in the case of crystalline structure involve the six coefficients belonging to a self-conjugate linear system of equations.

If we confine ourselves throughout to dielectrics, the electric current is the time-rate of change of the electric induction (XYZ) divided by 4π.[1] For media of no specific electric inductive capacity, or in which this capacity may be neglected, we have

$$(X, Y, Z) = (P, Q, R).$$

But for ponderable media we must assume a coefficient of electric permittivity K, which also will in the case of crystalline structure involve the six coefficients of a self-conjugate linear system of equations.

This restriction to self-conjugate equations is (under ordinary conditions) required in both the electric and magnetic relations, on the principles first applied by Lord Kelvin, to avoid the possibility of perpetual motions[2].

constitutive, as restricted. Thus we have the two systems of equations

$$X = K_{11}P + K_{12}Q + K_{13}R, \qquad a = \mu_{11}\alpha + \mu_{12}\beta + \mu_{13}\gamma,$$

$$Y = K_{21}P + K_{22}Q + K_{23}R, \qquad b = \mu_{21}\alpha + \mu_{22}\beta + \mu_{23}\gamma,$$

$$Z = K_{31}P + K_{32}Q + K_{33}R, \qquad c = \mu_{31}\alpha + \mu_{32}\beta + \mu_{33}\gamma,$$

[1] It has been remarked by Heaviside that the equations would gain in symmetry by the adoption of a new unit of current, which would suppress this factor 4π. In Maxwell's notation, usually employed in these papers, (fgh) would replace (XYZ).

[2] Maxwell, *Electricity and Magnetism*, § 297.

in each of which the conjugate coefficients, such as K_{12} and K_{21}, are equal; while also

$$4\pi\,(u, v, w) = \frac{d}{dt}\,(X, Y, Z).$$

In treating of the dielectric relations of crystals, it is usual to neglect their coefficient of magnetization in comparison with their coefficient of electric polarization. The obviously simplest course then is to choose axes of coordinates along the directions of principal electric polarization, so that the coefficients $K_{12}, K_{13}, \ldots$ do not appear, and this choice of axes will still be suitable for a more detailed analysis, in which the effect of the feeble magnetization of the medium will have to be considered; so that then $\mu_{11}, \mu_{22}, \mu_{33}$ are each nearly unity, while $\mu_{12}, \mu_{13}, \ldots$ are small.

It will be convenient to invert the form of the magnetic and electric equations, and write

Referred to principal dielectric axes:

$$\alpha = \mu_1' a + \mu_{12}' b + \mu_{13}' c, \qquad P = K_1' X,$$
$$\beta = \mu_{21}' a + \mu_2' b + \mu_{23}' c, \qquad Q = K_2' Y,$$
$$\gamma = \mu_{31}' a + \mu_{32}' b + \mu_3' c, \qquad R = K_3' Z,$$

where $\mu_{12}' = \mu_{21}'$.

Then, expressing the equations of the electric induction, we have

result.

$$\frac{d^2X}{dt^2} = \frac{d}{dt}\left(\frac{d\gamma}{dy} - \frac{d\beta}{dz}\right) = \left(\mu_{21}'\frac{d}{dz} - \mu_{31}'\frac{d}{dy}\right)\left(K_3'\frac{dZ}{dy} - K_2'\frac{dY}{dz}\right)$$
$$+ \left(\mu_{22}'\frac{d}{dz} - \mu_{32}'\frac{d}{dy}\right)\left(K_1'\frac{dX}{dz} - K_3'\frac{dZ}{dx}\right)$$
$$+ \left(\mu_{23}'\frac{d}{dz} - \mu_{33}'\frac{d}{dy}\right)\left(K_2'\frac{dY}{dx} - K_1'\frac{dX}{dy}\right),$$

with two similar equations.

Form when electric and magnetic axes coincide:

In particular, if the directions of the principal electric and magnetic axes coincide, $\mu_{23}' \ldots$ are null, and

$$\frac{d^2X}{dt^2} = \mu_{22}'\frac{d}{dz}\left(K_1'\frac{dX}{dz} - K_3'\frac{dZ}{dx}\right) + \mu_{33}'\frac{d}{dy}\left(K_1'\frac{dX}{dy} - K_2'\frac{dY}{dx}\right).$$

These relations are reduced to a simpler form by writing (P, Q, R) for $(K_1'X, K_2'Y, K_3'Z)$, that is by retaining the electric force as the variable.

In the special case of $\mu_{23}', \ldots$ null,

$$\frac{1}{K_1'}\frac{d^2P}{dt^2} = \mu_2'\frac{d}{dz}\left(\frac{dP}{dz} - \frac{dR}{dx}\right) + \mu_3'\frac{d}{dy}\left(\frac{dP}{dy} - \frac{dQ}{dx}\right)$$
$$= \mu_2'\frac{d^2P}{dz^2} + \mu_3'\frac{d^2P}{dy^2} - \frac{d}{dx}\left(\mu_2'\frac{dR}{dz} + \mu_3'\frac{dQ}{dy}\right).$$

If μ_1', μ_2', μ_3' are each unity, the type becomes

non-magnetic is Fresnel's case:

$$\frac{1}{K_1'}\frac{d^2P}{dt^2} = \nabla^2 P - \frac{d}{dx}\left(\frac{dP}{dx} + \frac{dQ}{dy} + \frac{dR}{dz}\right),$$

which leads in the well-known manner to Fresnel's laws of double refraction.

When the magnetic coefficients are different from unity, the same form of the equations is reached by making

but always reducible to it by transformation.

$$(P', Q', R') = (\mu_1'^{-\frac{1}{2}}P,\ \mu_2'^{-\frac{1}{2}}Q,\ \mu_3'^{-\frac{1}{2}}R),$$

$$(x', y', z') = (\mu_1'^{-\frac{1}{2}}x,\ \mu_2'^{-\frac{1}{2}}y,\ \mu_3'^{-\frac{1}{2}}z);$$

so that

$$(K_1'\mu_2'\mu_3')^{-1}\frac{d^2P'}{dt^2} = \left(\frac{d^2}{dx'^2} + \frac{d^2}{dy'^2} + \frac{d^2}{dz'^2}\right)P' - \frac{d}{dx'}\left(\frac{dP'}{dx'} + \frac{dQ'}{dy'} + \frac{dR'}{dz'}\right).$$

Thus the wave-surface in this case is of a form which may be derived from Fresnel's by application of homogeneous strain, according to the above specification[1]. Furthermore, the geometrical relations of the electric force to the new wave-surface are correctly represented by imposing the same strain on its relations to the Fresnel surface from which the new one has been derived. The electric induction is in every case in the plane of the wave-front, on account of its solenoidal character.

Results for general problem:

For the most general case, Heaviside finds*, by the aid of a powerful vector analysis, that the equation for the velocities of the two waves whose fronts are in the direction (lmn) is

$$V^4 - [(K_2'\mu_{33}' + K_3'\mu_{22}')\,l^2 + \dots - 2K_3'\mu_{12}'lm - \dots]\,V^2$$
$$+ (\mu_{11}l^2 + \dots + 2\mu_{12}lm + \dots)(K_1l^2 + \dots) = 0;$$

[1] This extension is due to Heaviside, *Phil. Mag.* 1885; *Electrical Papers*, Vol. II. p. 16.

* That the general problem also is reducible to the Fresnel form by simple homogeneous strain of the field is readily seen by applying the special transformation just formulated with regard to the principal magnetic axes (μ_1, μ_2, μ_3) of the medium as follows:

$$(x, y, z) = (\mu_1^{-1}x', \mu_2^{-1}y', \mu_3^{-1}z'),\quad (P, Q, R) = \mu_1\mu_2\mu_3\,(\mu_1^{-1}P', \mu_2^{-1}Q', \mu_3^{-1}R').$$

For the circuital relations are then maintained unchanged in the accented fields, with (a', b', c') equal to $(\alpha', \beta', \gamma')$; while (u', v', w'), the same as (u, u, w), are expressed in terms of (P', Q', R') by a self-conjugate set of linear equations, just as previously, but slightly modified on account of the change to the accented vector. This accented scheme is precisely the Fresnel formulation for a non-magnetic medium, but referred now to a set of axes not in its principal electric directions. Thus the intrinsic, or invariant, circumstances, form of wave-surface and constitutive relations, for the general problem are derived very directly from those of an auxiliary Fresnel case by a pure strain, but one whose principal axes are now not the principal axes of the Fresnel system. In particular, the most general wave-surface is a Fresnel surface deformed by pure strain.

and that the equation of the wave-surface is

$$1 + [\mu]\,[K]\,(\mu_{11}'x^2 + \ldots + 2\mu_{12}'xy + \ldots)\,(K_1'x^2 + \ldots)$$
$$- [(K_2\mu_{33} + K_3\mu_{22})\,x^2 + \ldots - 2K_3xy - \ldots]^2 = 0,$$

where $[\mu]$, $[K]$ are the determinants of the μ, K matrices.

The condition for the electric stability of the medium is that the two roots of this equation for V^2 should be real and positive for all values of the direction cosines l, m, n. This involves that the quartic cone stability:

$$[(K_2'\mu_{33}' + K_3'\mu_{22}')\,x^2 + \ldots]^2 = 4\,(\mu_{11}x^2 + \ldots)\,(K_1x^2 + \ldots)$$

should be wholly imaginary, or at most reduce to two rays. The directions of these two rays will in that special case be optic axes of the wave-surface, and will correspond to singular tangent planes which touch it all along a curve. special limiting case.

In the theory of crystalline refraction, based upon electric ideas, the difficulty is to some extent the opposite of that which occurs in purely dynamical theories; in the latter, the problem is so to combine the few independent relations allowed by the laws of pure dynamics as to represent all the phenomena; in the former each pair of related vectors are at the outset assumed to be connected in the most general linear manner, and we have to decide what is to be done with all the array of constants so introduced. The conditions of stability here indicated require relations between them. Contrast of electrical and dynamical procedures.

It may, however, possibly be objected to this statement of the conditions of stability of the medium, that it is in discrepancy with the ordinary theory of opacity, which makes that quality depend on the existence of an imaginary value for the velocity of propagation. But it is to be borne in mind that the terms which explain opacity, on the electromagnetic theory, involve electric conduction, and are therefore of a frictional character, and so irreversible; while, on the other hand, terms which depend on the structure of the medium are reversible, so that if a real exponential occur in the solution of the velocity equation with a negative sign, a reversal of the motion will make it appear with a positive sign, which would lead to a breaking-up of the medium. Absorption excluded. Friction can restore stability.

The consideration of an imaginary index without accompanying instability also occurs in the theory of anomalous dispersion elaborated by Sellmeier, Helmholtz, and Kelvin; but there it enters from the consideration of the sympathetic vibrations of molecular structures which are considered to be *outside* the system which transmits the undulations, and which therefore extract its energy in much the same sort of way as frictional resistances. In fact a strong argument in favour of supposing that double refraction is to be explained in this Molecular view of refraction favoured.

manner, is that we thereby avoid the incipient instability which would be a characteristic of a simple elastic medium possessing that property.

Comparison of Theories of Disperion and Rotatory Polarization.

A general formal development of the equations of the electromagnetic theory, which is necessarily wide enough to take account of all possible secondary phenomena, such as dispersion and circular polarization, was first given by Prof. Willard Gibbs[1], under the title of "An Investigation of the Velocity of Plane Waves of Light," in which they are regarded as consisting of solenoidal electrical fluxes in an indefinitely extended medium of uniform and very fine-grained structure.

Gibbs' formal generalizations:

The principle on which his investigation is based is the very general idea that the regular simple harmonic light-waves traversing the medium excite secondary vibrations in its molecular electrical structure, which is supposed very fine compared with the length of a wave. When there is absorption, the phases of these excited vibrations will differ from that of the exciting wave; but even in this most general case the simple harmonic electric flux with which we are alone concerned is at each point completely specified by six quantities, the three components of the flux itself, and the three components of its rate of change with the time. In the same way, the electric force may be similarly specified by six quantities representing itself and its time-quadrant. Now the electric elasticity of the medium, as regards its power of transmitting waves, is specified by the relation connecting average force and average flux, this average referring to a region large compared with molecular structures, but small compared with a wave-length. The most general relation of this kind, that can result from the elimination of the molecular vibrations, must be of the form of six linear equations connecting the quantities specifying the flux with the quantities specifying the force, the coefficients being functions of the wave-length. If E denote the flux and U the force, "we may therefore write in vector notation

in terms of a fine secondary structure:

averaged form:

$$[E]_{\mathrm{Ave}} = \Phi\,[U]_{\mathrm{Ave}} + \Psi\,[\dot{U}]_{\mathrm{Ave}},$$

where Φ and Ψ denote linear functions[2].

"The optical properties of the media are determined by the forms of these functions. But all forms of linear functions would not be consistent with the principle of the conservation of energy.

[1] In its final form this is contained in a paper, "On the General Equations of Monochromatic Light in Media of every Degree of Transparency," *American Journal of Science*, February 1883 (reprinted in *Scientific Papers*, Vol. 1).

[2] This requires that the coefficients K_{11}, K_{12}, ... above are complex quantities, depending in part on the period.

"In media which are more or less opaque, and which therefore absorb energy, Ψ' must be of such a form that the function always makes an acute angle (or none) with the independent variable. In perfectly transparent media Ψ' must vanish, unless the function is at right angles to the independent variable. So far as is known, the last occurs only when the medium is subject to magnetic influence. In perfectly transparent media, the principle of the conservation of energy requires that Φ should be self-conjugate, *i.e.* that for three directions at right angles to one another, the function and independent variable should coincide in direction. restricted by conservation of energy.

"In all isotropic media not subject to magnetic influence, it is probable that Φ and Ψ' reduce to numerical coefficients, as is certainly the case with Φ for transparent isotropic media."[1] Magnetic influence exceptional:

The subject of the rotatory polarization produced by a magnetic field has recently been resumed by various writers, with a view to the elucidation of the experimental results of Kerr, on the reflexion of light from magnets*. The following investigation of the degree of variety that it is permissible to import into the theoretical treatment may be of service in indicating the alternatives out of which a theory must be finally chosen; it will also exhibit the relation to each other of the various theories which have been developed. possibilities of formulation:

To begin with, we recall the fundamental circuital relations of the types

$$4\pi u = \frac{d\gamma}{dy} - \frac{d\beta}{dz}, \quad \mu_1 \frac{d\alpha}{dt} = \frac{dR}{dy} - \frac{dQ}{dz},$$

in which the axes are those of principal magnetic permeability (μ_1, μ_2, μ_3). The relation between the current and the electric induction is, in a dielectric,

$$(u, v, w) = \frac{1}{4\pi} \frac{d}{dt} (X, Y, Z);$$

in a metallic or other conducting medium this would be replaced by

$$(u, v, w) = \frac{1}{4\pi} \frac{d}{dt} (X, Y, Z) + (\sigma) (P, Q, R).$$

The hypothesis adopted by Basset[2] and by Drude[3], following FitzGerald and Rowland, is that the electric force contains, in addition to the part of it derived in Maxwell's manner from the kinetic (magnetic) energy of the medium, a part derived from a new term in historical.

[1] J. Willard Gibbs, *loc. cit.* p. 133.

* It has been carried through very completely and clearly, in comparison with the extensive experimental records, by J. G. Leathem, *Phil. Trans.* 1897, pp. 89–137, on the lines of the present train of ideas, in terms of variation of a single Action formula.

[2] *Phil. Trans.* 1891; *Physical Optics*, Ch. XX.

[3] Wiedemann's *Ann.* Vol. XLVI. 1892, p. 377.

the kinetic energy due to the indirect action of the imposed permanent magnetic field, giving the components

$$\begin{vmatrix} p_1 & p_2 & p_3 \\ \dot{X} & \dot{Y} & \dot{Z} \end{vmatrix},$$

and that this is the only change introduced. The effect is simply that the magnetic force is to be derived not from the total electric force (P, Q, R), but from the part

$$(P', Q', R') \equiv (P + p_3\dot{Y} - p_2\dot{Z},\ Q + p_1\dot{Z} - p_3\dot{X},\ R + p_2\dot{X} - p_1\dot{Y});$$

and if we take (P', Q', R') as independent variables, the relation between $4\pi(u, v, w)$ [that is $(\dot{X}, \dot{Y}, \dot{Z})$] and (P', Q', R') will be expressed by a set of linear equations with rotational coefficients.

On the other hand, the hypothesis of Willard Gibbs[1] and Goldhammer[2] takes the circuital relations as the fundamental and unmodifiable principles of electrodynamics, and assumes, after Maxwell, that the effect of the imposed magnetic field is to produce a rotational aeolotropy in the constitution of the medium, so that the system of linear equations connecting electric induction with electric force becomes rotational. This is clearly formally equivalent to the other process, if (P', Q', R') is there taken to be the electric force, *except* as to the mode in which the rotational coefficients involve the operator d/dt as a factor. Now, according to the considerations taken from Willard Gibbs, which have been sketched above, the most general possible type of his rotational coefficients, for simple harmonic vibrational disturbance, is $p + q\,d/dt$, where p and q are constants (vectorial) for the particular period involved; and Goldhammer makes the point that for slow periods, and for steady fields, this coefficient practically reduces to p, and includes the Hall effect, while for very rapid periods such as those of light-waves it reduces to $q\,d/dt$, which is the form required, as above, to make the two types of theory formally agree as to mode of propagation[3]. If the medium is non-conducting, the negation of a perpetual motion requires that the coefficients of type p shall form a self-conjugate system.

The usual interfacial conditions

We now compare the boundary conditions that must be satisfied, on these two hypotheses, in a problem of reflexion. In ordinary electrodynamics, taking for the moment the axis of z normal to the interface, all the following quantities must be continuous across it,

$$P,\ Q,\ Z,$$

$$\alpha,\ \beta,\ c.$$

[1] *American Journal of Science*, 1883. [2] Wied. *Ann.* Vol. XLVI. 1892, p. 75.

[3] A discussion is also given by J. J. Thomson, *Recent Researches in Electricity and Magnetism*, 1893, p. 509.

Of these the continuity of α, β involves that of Z, by the first circuital relation, which is taken by Maxwell to be kinematic and not kinetic; and under ordinary circumstances the continuity of P, Q would involve that of c. But under the conditions of the first hypothesis, the continuity of P, Q violates that of c, which would rather require instead that P', Q' should be continuous. The solenoidal character of the electric and magnetic induction could hardly be subject to modification; we are driven therefore on the first hypothesis to admit that the electric force parallel to the interface is discontinuous in crossing it*. The question whether both P' and Q' are continuous, or only $\frac{dP'}{dy} - \frac{dQ'}{dx}$ is so, is usually settled by considering that the flow of energy across the interface is continuous, that is that there is no *quasi*-Peltier effect at the surface; taking Maxwell's expressions for the energy in terms of the electric and the magnetic induction, this necessitates the continuity of both P' and Q'.

involve perplexities in problem of magnetic reflexion.

On the hypothesis developed by Gibbs and Goldhammer, no difficulty of this kind arises. The boundary conditions are simply the ordinary ones; and they formally agree with the conditions finally assumed on the other hypothesis, for the reasons given above.

The general dynamical considerations which verify the restriction of the rotatory coefficient to Goldhammer's form have been briefly indicated at the end of a previous paper[1], and may be recapitulated as follows. The equation of propagation of a plane wave is of type

$$\rho \frac{d^2\theta}{dt^2} = k \frac{d^2\theta}{dz^2}.$$

The introduction of small additional terms of odd order, and therefore of the types

$$\kappa_1 \frac{d^3\theta}{dz^2 dt}, \quad \kappa_2 \frac{d^3\theta}{dt^3}, \quad \kappa_3 \frac{d\theta}{dt}, \quad \kappa_4 \frac{d^3\theta}{dz\, dt^2}, \quad \kappa_5 \frac{d^3\theta}{dz^3},$$

will produce rotation of the plane of polarization. In the case of the first three types, change of sign of z does not affect the phenomenon; thus the rotation is in the same direction whether the wave travels forward or backward; it is of the magnetic kind. In the case of the fourth and fifth types, change of sign of z produces the same effect as change of sign of the rotatory coefficient; the rotation is of the kind exhibited by quartz and sugar and other active chemical compounds.

Magnetic and structural types of rotation:

* For the definite dynamical settlement of all these questions see Leathem, *loc. cit.*

[1] "The Equations of Propagation of Disturbances in Gyrostatically Loaded Media," *Proc. Math. Soc.* 1891, p. 134: *supra*, p. 248.

dimensional indications. On an ultimate dynamical theory, $\rho\, d^2\theta/dt^2$ will represent kinetic reaction; and the principle of dimensions shows that κ_1/ρ, κ_2/ρ, κ_3/ρ are respectively of dimensions $[L^2T^{-1}]$, $[T]$, $[T^{-1}]$. Thus the coefficient κ_1 will produce rotation owing to some influence of a distribution of angular momentum pervading the medium; while the coefficients κ_2 and κ_3 would produce selective rotation owing to the influence of the free periods of the fine-grained structure of the imbedded atoms of matter. The latter kind of rotation is to be expected only in the rare cases in which selective absorption is prominent; consequently we are entitled, as a first approximation, to ascribe magnetic rotation to a coefficient of type κ_1; a conclusion which is abundantly verified by the discussion of the results of the experimental measures of Verdet and others[1].

Review of theories. This review gives the conclusion that all effective theories of magnetic rotation, whether electro-optic, or gyrostatic and so purely dynamical, lead to the same modification of the equations of light-propagation, in order to take account of magnetic rotation *to a first approximation*; the consideration of the possible influence of ordinary and selective dispersion must in any case be conducted in a tentative manner. The different ways of formulating the electric theory also lead practically to the same boundary conditions in problems of reflexion at a magnet; while the consideration of what should be the boundary conditions on a gyrostatic theory[2] places in a prominent view what is the great trouble in the definite formulation of all problems of reflexion of light, viz. that a gradual transition at the boundary, taking place over a sensible portion of a wave-length, may completely alter the circumstances.

The result of a comparison of his formulae with an elaborate series of measurements of Kerr's phenomenon, made by Sissingh[3], leads Drude to the conclusion that all the features are fairly accounted for by a real coefficient of the type q; while a similar discussion by Goldhammer, a few months earlier, led to a more exact correspondence on taking q to be a complex quantity, as it is customary to assume the optical constants of metals to be. If, however, we are to found Goldhammer's equations on the physical basis afforded by the remarks of Willard Gibbs, quoted above, all the constants that occur must be real; and they must be able to account for metallic reflexion and all the other phenomena, so far as they are independent of dispersion.

[1] Maxwell, *Elec. and Mag.* § 830.

[2] *Loc. cit.*, *Proc. Math. Soc.* 1891: *supra*.

[3] Wiedemann's *Ann.* 1891; trans. in *Phil. Mag.* 1891. [But cf. Leathem, *loc. cit.*, as regards the whole subject.]

The question whether a gradual transition at the reflecting surface will sensibly influence the modification imposed by magnetization on the reflexion has apparently not yet been examined[1].

Magnetic Field and Optical Stability.

If the medium exhibits rotation of the plane of polarization in the Faraday manner, when it is placed in a powerful magnetic field parallel to the axis of z, the exact equations connecting electric induction with the electric force may thus be taken on Gibbs' theory to be

The equations lead to six periods for assigned wave-length:

$$X = K_1 P - \epsilon \frac{d}{dt} Q,$$
$$Y = K_2 Q + \epsilon \frac{d}{dt} P,$$
$$Z = K_3 R,$$

where ϵ is a rotatory coefficient which is proportional to the strength of the magnetic field. It may here be noticed again that in a transparent medium these equations involve no absorption of energy.

To examine the effect of this rotational coefficient on the periods, it will suffice to take the medium as usual magnetically isotropic, or more simply non-magnetic, so that

$$(a, b, c) = (\alpha, \beta, \gamma).$$

The equations are in that case of the type

$$\frac{d^2X}{dt^2} = \nabla^2 P - \frac{d}{dx}\left(\frac{dP}{dx} + \frac{dQ}{dy} + \frac{dR}{dz}\right).$$

From them, on substituting for (X, Y, Z) in terms of (P, Q, R), and writing

$$(P, Q, R) = (P_0, Q_0, R_0) \exp \iota \frac{2\pi}{\lambda} (lx + my + nz),$$

[1] The making the coefficients q complex, by Goldhammer, involves a virtual reintroduction of the coefficients p, but with the essential difference that the new coefficients are combined with the frequency of vibration as a factor, and so have a preponderating influence when the frequency is very great.

In Wiedemann's *Ann.* 1893, Drude returns to this subject, and discusses some new observations of Zeeman on reflexion from cobalt, which were supposed to disagree with his formulae. His conclusion seems to be that, although two coefficients naturally give a somewhat better account of the observations than one, yet the account they give is not complete, and suggests the residual influence of surface contamination or some other cause—that, in fact, the necessity for a complex constant cannot be allowed in the absence of more complete experimental data. [But cf. Leathem, *loc. cit.*]

where P_0, Q_0, R_0 are functions of the time, there arises

$$\left(\frac{\lambda}{2\pi}\right)^2 \frac{d^2}{dt^2}\left(K_1 P_0 - \epsilon \frac{d}{dt} Q_0\right) = -P_0 + l\ (lP_0 + mQ_0 + nR_0),$$

$$\left(\frac{\lambda}{2\pi}\right)^2 \frac{d^2}{dt^2}\left(K_2 Q_0 + \epsilon \frac{d}{dt} P_0\right) = -Q_0 + m\ (lP_0 + mQ_0 + nR_0),$$

$$\left(\frac{\lambda}{2\pi}\right)^2 \frac{d^2}{dt^2} K_3 R_0 = -R_0 + n\ (lP_0 + mQ_0 + nR_0).$$

The elimination of P_0, Q_0, R_0 yields an equation for the operator d/dt; caling this operator D, the equation is

$$\frac{K_1 + \epsilon \frac{m}{l} D}{K_1 + \frac{4\pi^2}{\lambda^2 D^2} - \epsilon \frac{l}{m} D} + \frac{K_2 - \epsilon \frac{l}{m} D}{K_2 + \frac{4\pi^2}{\lambda^2 D^2} + \epsilon \frac{m}{l} D} + \frac{K_3}{K_3 + \frac{4\pi^2}{\lambda^2 D^2}}$$

$$= \frac{lmn^2 K_3 \epsilon D \left\{\frac{K_1^2}{l^2} + \frac{K_2^2}{m^2} + \left(\frac{1}{l} + \frac{1}{m}\right)\frac{4\pi^2}{\lambda^2 D^2}\right\}}{\{\ldots\ldots\}\{\ldots\ldots\}\{\ldots\ldots\}}.$$

This is an equation of the sixth degree, of which, when ϵ is small, the roots are nearly equal in pairs. One pair of them disappear when ϵ vanishes, so that they are very small, and correspond to waves which are propagated with extreme slowness. Thus, as Gibbs remarks, their wave-length would be much smaller than molecular magnitudes, and they therefore cannot have an actual existence.

two of them spurious,

yet indicating incompleteness.

Yet it would appear that, in strictness, without something which exactly fulfils the *rôle* of this wave, this theory of magnetic reflexion will come to grief through inability to run parallel to the actual physical conditions; though in the first approximation which coincides with the other theory these waves do not present themselves at all.

Labile gyrostatic model.

This electric theory runs in close correspondence with Lord Kelvin's theory of a labile mechanical aether, in which the velocity of compressional disturbances is null; the introduction of rotatory coefficients makes that velocity finite but small, exactly in the manner here exemplified. If we suppose this mechanical medium to be an adynamic gyrostatic structure, an aeolotropic arrangement of the axes of the gyrostats will represent crystalline quality, but there must be as many gyrostats with their axes pointing in one direction as there are with their axes pointing in the opposite direction; any cause which in addition slightly slews round the axes towards a certain direction will introduce rotation of the magnetic type with respect to that direction.

We may now examine how far the rotatory terms or other small terms of the same order disturb the stability of a medium of biaxial character. The general analytical problem is as follows: along the direction of the optic axes the equation for V^2 has two equal roots; the coefficients in this equation are altered by the introduction of a series of new quantities of small magnitude typified by ϵ; it is required to find the condition that the originally equal roots of the equation, thus altered, shall preserve real values. Let the value of a root of the equation

$$\phi(x) = 0,$$

where x stands for V^2, be thus altered from x to $x + \delta x$; then

$$\phi(x) + \phi'(x)\,\delta x + \tfrac{1}{2}\phi''(x)\,\delta x^2 + \ldots + \Sigma\left\{\frac{d\phi}{d\epsilon}\,\delta\epsilon + \tfrac{1}{2}\frac{d^2\phi}{d\epsilon^2}\,\delta\epsilon^2 + \ldots\right\} = 0,$$

wherein $$\phi(x) = 0,$$

and also $$\phi'(x) = 0,$$

owing to the equality of roots. Thus

$$\tfrac{1}{2}\phi''(x)\,\delta x^2 = -\,\delta_\epsilon\phi,$$

where $\delta_\epsilon\phi$ represents the change in $\phi(x)$ produced by the new terms. In order that the roots may remain real, the alteration must be in such direction as to make the sign of $\delta_\epsilon\phi \Big/ \dfrac{d^2\phi}{dx^2}$ negative.

Now, as $\delta_\epsilon\phi$ is of the first degree in the increments $\delta\epsilon, \ldots$, it may have any sign at will; hence the medium thus modified is necessarily unstable. It must be concluded, if we adhere to a theory of this kind, that an imposed magnetic field will alter the electro-optic constants, not only by introducing rotatory coefficients, but also by modifying (very slightly) the double refraction of the medium so as to undo their tendency to instability. Apparent optical instability also in evidence:

The character of the circular polarization imposed magnetically on doubly-refracting media has been worked out by approximate methods by Prof. Willard Gibbs[1]; he has not, however, noticed that along the optic axes his approximation becomes nugatory, and that he has really to deal with conditions involving instability*. magnetic crystals:

The theorem that the stability depends on the sign of $\delta_\epsilon\phi$ may be applied to obtain explicitly the conditions of stability of a slightly magnetic biaxial medium from Heaviside's formula (quoted *supra*) for the velocities; but the result is too long to merit reproduction.

[1] *American Journal of Science*, June 1882.

* This discussion implies a uniform medium. In the actual medium, aether modified by the distribution of electrons in the molecules, the result would be that certain types of waves could not travel.

dynamical reason suggested for type (Airy) of wave-surface of quartz.

In the case of the natural asymmetry of quartz, a uniaxial crystal, we can make a distinction between the ordinary and the extraordinary wave; and the consideration of stability gives a necessary reason for the relation derived by Airy from an experimental examination, that the wave-surface is so modified by rotational quality that one sheet lies wholly inside the other, instead of intersecting it along a curve as would be formally possible. A similar statement of course applies to the temporary modification produced by a strong magnetic field.

Uniaxial crystals:

To examine further the form these anomalies assume in the more simple media, let us suppose $K_1 = K_2$ ($= K$, say), so that the medium is uniaxial, with its axis along the direction of the magnetic field. The result will now be pure circular polarization, so that we are prompted to replace in the analysis

$$X, Y \text{ by } \Xi = X + \iota Y, \ \mathrm{H} = X - \iota Y,$$

$$P, Q \text{ by } \Pi = P + \iota Q, \ \Sigma = P - \iota Q,$$

and to take as coordinates

$$\xi = x + \iota y, \ \eta = x - \iota y.$$

This leads to

$$\Xi = \left(K + \iota\epsilon \frac{d}{dt}\right)\Pi,$$

$$\mathrm{H} = \left(K - \iota\epsilon \frac{d}{dt}\right)\Sigma,$$

and

$$\frac{d^2\Xi}{dt^2} = \nabla^2\Pi - \frac{d}{d\xi}\left(\frac{d\Pi}{d\xi} + \frac{d\Sigma}{d\eta} + \frac{dR}{dz}\right),$$

$$\frac{d^2\mathrm{H}}{dt^2} = \nabla^2\Sigma - \frac{d}{d\eta}\left(\frac{d\Pi}{d\xi} + \frac{d\Sigma}{d\eta} + \frac{dR}{dz}\right),$$

$$\frac{d^2Z}{dt^2} = \nabla^2R - \frac{d}{dz}\left(\frac{d\Pi}{d\xi} + \frac{d\Sigma}{d\eta} + \frac{dR}{dz}\right).$$

Thus, on writing for waves, taking V as their speed,

$$(\Xi, \mathrm{H}, R) = (\Xi_0, \mathrm{H}_0, R_0) \exp \iota \frac{2\pi}{\lambda}(l\xi + m\eta + nz - Vt),$$

equations of Fresnel type but complex:

these equations are still formally the same as Fresnel's equations for biaxial media, when the reciprocals of the squares of the principal velocities are taken to be

$$K + \frac{2\pi\epsilon}{\lambda}V, \quad K - \frac{2\pi\epsilon}{\lambda}V, \quad K_3;$$

so that the velocities corresponding to the direction l, m, n are given by the equation

$$\frac{l^2}{\left(K+\frac{2\pi\epsilon}{\lambda}V\right)^{-1}-V^2}+\frac{m^2}{\left(K-\frac{2\pi\epsilon}{\lambda}V\right)^{-1}-V^2}+\frac{n^2}{K_3^{-1}-V^2}=0,$$

wave-velocities.

which has still three pairs of nearly equal roots, one pair of them small, when ϵ is small.

The equation for a medium originally isotropic is obtained on writing K instead of K_3; when expanded, it assumes the form

$$(1-KV^2)\left\{K+\frac{2\pi\epsilon}{\lambda}(l^2-m^2)KV^3-\left(\frac{2\pi\epsilon}{\lambda}\right)^2(l^2+m^2)V^4\right\}$$
$$+\left(\frac{2\pi\epsilon}{\lambda}\right)^2 n^2KV^6=0.$$

For isotropic medium the instability nugatory.

There are still three pairs of roots, of which two pairs are now very small when ϵ is very small; and, inasmuch as when ϵ is null these pairs of roots are equal, being zero, they may in the actual case become imaginary for certain values of l, m, n. Thus the instability here attaches to the very slow waves which are outside the limits of physical reality. With a rotatory coefficient of the p type, the trouble would not occur at all.

34

THE ACTION OF MAGNETISM ON LIGHT; WITH A CRITICAL CORRELATION OF THE VARIOUS THEORIES OF LIGHT-PROPAGATION.

[*Report of the British Association* (1893), pp. 335–372.]

PART I. MAGNETIC ACTION ON LIGHT.

Discovery of Magnetic Rotation.

1. The reduction of light and heat, and of electrical phenomena, to a common cause has been a cardinal subject of physical speculation from the earliest times. More recently Oersted[1] fully persuaded himself, on somewhat wider knowledge of fact, "that heat and light are the result of the electric conflict," and saw in his great discovery of the gyratory action of an electric current on a magnet the explanation of the phenomena classed under the name of polarization of light. But it was reserved for Faraday to make the first effective entrance into this domain of knowledge.

Faraday's discovery: his train of thought.

2. After failure in 1834 to discover any direct relation between light and static electrification, and after repeated attempts in other directions, he at length discovered[2] the fact that when plane polarized light is passed through a transparent body along the direction of lines of magnetic force, its plane of polarization undergoes rotation by a specific amount characteristic of the medium traversed. He thus succeeded "in magnetizing and electrifying a ray of light, and in illuminating a magnetic line of force." After observing that when the ray is oblique to the lines of magnetic force it is the component of the force in the direction of the ray which appears to be effective in producing the rotation (a law which has since been exactly verified by Verdet and more recently by Du Bois), he proceeds to inquire into the condition of the active medium with the following results:

"2171. I cannot as yet find that the heavy glass when in this state, *i.e.* with magnetic lines of force passing through it, exhibits any increased degree, or has any specific magneto-inductive action of the recognized kind. I have placed it in large quantities, and in different positions, between magnets and magnetic needles, having at the time

[1] Hans Christian Oersted, *Experimenta circa Effectum conflictus Electrici in Acum Magneticam*, Hafniae, 1820.

[2] M. Faraday, *Experimental Researches*, 19th series; *Phil. Trans.* 1845.

very delicate methods of appreciating any difference between it and air, but can find none."

"2172. Using water, alcohol, mercury, and other fluids contained in very large delicate thermometer-shaped vessels, I could not discover that any difference in volume occurred when the magnetic curves passed through them."

The rotation was in general right-handed with respect to the magnetic force; and in the case of (2165) "bodies which have a rotative power of their own, as is the case with oil of turpentine, sugar, tartaric acid, tartrates, etc., the effect of the magnetic force is to add to, or subtract from, their specific force, according as the natural rotation and that induced by the magnetism is right or left-handed." (2187) "In all these cases the superinduced magnetic rotation was according to the general law, and without reference to the previous power of the body."

Further on, after describing the diversity of the effect in different media, and its usually small amount in crystals, he adds:

"2182. With some degree of curiosity and hope, I put gold-leaf into the magnetic lines, but could perceive no effect. Considering the extremely small dimensions of the length of the path of the polarized ray in it, any positive result was hardly to be expected."

The powerful rotation discovered long after by Kundt, with films of iron, will here be called to mind.

Repeated trials with various transparent media gave no effect of lines of electrostatic force on a ray of polarized light, propagated either along them or at right angles to them. An effect in this case has been detected by Kerr long after, but presented itself as a change in the elasticity, producing double refraction, and entirely devoid of rotational character.

"2224. The magnetic forces do not act on the ray of light directly and without the intervention of matter, but through the mediation of the substance in which they and the ray have a simultaneous existence."

Any such changes of internal constitution of media must of necessity (2226) "belong also to opaque bodies: for as diamagnetics there is no distinction between them and those which are transparent. The degree of transparency can, at the utmost, in this respect only make a distinction between the individuals of a class."

After pointing out (2230) that this is "the first time that the molecular condition of a body, required to produce the circular polarization of light, has been artificially given," and is, on that account also, worthy of minute study, Faraday proceeds to draw out in very clear and striking contrast the distinction between the natural undirected rotatory property of liquids like turpentine, and the magnetic property which is related to the direction of the lines of force, as well as the distinction between the latter and the axial but undirected rotatory power of quartz.

This brief *résumé* of the topics treated in Faraday's memoir will be of interest as indicating how thoroughly he probed the problem, and how much his ideas were on the lines of the subsequent development of the subject.

Mathematical Representations of the Phenomena.

Fresnel's chiral analysis: 3. The rotation of the plane of polarization in quartz and other substances had already been explained by Fresnel as depending on the two principles, (i) that the vibrations which can be propagated without change of form as they proceed are for these substances circular (or it may be elliptical), and (ii) that the velocity of propagation is different according as the vibration runs along in the manner of a right-handed or a left-handed screw-motion. It had also been shown by MacCullagh[1] how such properties might be deduced from equations of vibration modified by the insertion of small terms involving $(d/dz)^3$, where z is the direction of propagation.

Airy's extension to magnetic effect. Soon after Faraday's discovery of magnetic rotation, Airy[2] pointed out the different modifications of the equations of vibration that would similarly account for the existence of the magnetic rotation.

We may in fact develop a complete and compact account of the matter, as follows. The equations for the displacements in a circular transverse vibration, propagated along the axis of z, are

$$u = A\cos(nt - ez), \quad v = A\sin(nt - ez);$$

for a given value of z these equations represent a circular vibration in the plane of xy, and this is propagated in spiral fashion as a wave. We may very conveniently combine the two equations into one by use of the vector $\vartheta = u + \iota v$ to represent the displacement, thus obtaining the form

$$\vartheta = Ae^{\iota(nt-ez)}.$$

As this vibration is propagated without change, the equation of **Vector vibrational equation.** propagation must be linear in ϑ, therefore of the form

$$\frac{d^2\vartheta}{dt^2} = a^2\frac{d^2\vartheta}{dz^2} + P.$$

The terms in P involve higher differential coefficients, and are necessary in order that the two values of e corresponding to a given value of n may not be equal except as to sign, in other words in order that

[1] J. MacCullagh, *Trans. R.I.A.* Vol. XVII. 1836; *Collected Works*, pp. 63, 186.

[2] G. B. Airy, *Phil. Mag.* June 1846.

right-handed and left-handed waves of the same period may be propagated at different speeds. To ensure this result, P must contain terms of odd order in the differential coefficients; if there were only terms of even order, it would still lead to an equation for the square of e, and so would represent ordinary dispersion without the rotational property.

If we confine our attention to terms of the first and third orders we can tabulate possible rotational terms as follows[1]: **Rotational types:**

$$\kappa_1 \frac{d^3\vartheta}{dz^2 dt}, \quad \kappa_2 \frac{d^3\vartheta}{dt^3}, \quad \kappa_3 \frac{d\vartheta}{dt}, \quad \kappa_4 \frac{d^3\vartheta}{dz\,dt^2}, \quad \kappa_5 \frac{d^3\vartheta}{dz^3}, \quad \kappa_6 \frac{d\vartheta}{dz}.$$

Now in the case of the first three types, change of sign of z does not affect the phenomenon; thus the rotation is in the same direction whether the wave travels forward or backward; it is of the magnetic kind. In the case of the last three types, change of sign of z produces the same effect as change of sign of the rotatory coefficient; the rotation is of the kind exhibited by quartz and sugar and other active chemical compounds.

On an ultimate dynamical theory, if ϑ denote displacement in a medium of density ρ, $\rho \frac{d^2\vartheta}{dt^2}$ will represent force per unit volume; and the principle of dimensions shows that κ_1/ρ, κ_2/ρ, κ_3/ρ, are respectively of dimensions $[L^2T^{-1}]$, $[T]$, $[T^{-1}]$, in length and time. Thus the coefficient κ_1 will produce rotation owing to some influence of a distribution of angular momentum pervading the medium; while the coefficients κ_2 and κ_3 would produce selective rotation owing to the influence of the free periods of the fine-grained structure of the imbedded atoms of matter. The latter kind of rotation is to be expected to a sensible amount only in the rare cases in which selective absorption of the light is prominent; consequently we are guided, as a first approximation, to ascribe magnetic rotation to a coefficient of the type κ_1. **interpreted.**

The last three types of term will be appropriate to represent the rotation of naturally active media. The dimensions of κ_4/ρ, κ_5/ρ, κ_6/ρ, are respectively $[L]$, $[L^3T^{-2}]$, $[LT^{-2}]$. The term actually employed by MacCullagh to illustrate that action was the statical one with κ_5 for coefficient.

Dynamical Illustrations.

4. The first direct dynamical investigation bearing on the subject is by Lord Kelvin[2]. He points out that the elastic reaction of a **Kelvin's dynamical analogies:**

[1] J. Larmor, *Proc. Lond. Math. Soc.* Vol. XXI. 1890.

[2] W. Thomson, "Dynamical Illustrations of the Magnetic and the Helicoidal Rotatory Effects of Transparent Bodies on Polarized Light," *Proc. Roy. Soc.* 1856.

homogeneously strained solid has a character essentially devoid of all helicoidal and of all dipolar asymmetry. It therefore follows that the helicoidal rotation of the plane of polarization by quartz, turpentine, etc., must be due to elastic reactions dependent on the heterogeneity of the strain through the space of a wave.

Then with regard to the magnetic or unipolar rotation the well-known paragraph occurs, quoted by Maxwell (*Treatise*, § 831) as "an exceedingly important remark," of which his own theory of molecular vortices, and also its outcome, the conception of the working model which led to the electric theory of light, is an expansion. On reversing the light the magnetic rotation is not reversed: therefore it depends on some outside influence of a vector character, exerted on the system which transmits the light. This influence makes the free period of a circular motion differ, according as it rotates in one direction or the opposite one. If the purely elastic forces maintaining the motion are supposed similar in the two cases, it will follow that "the luminiferous circular motions are only components of the whole motion." There must be another dynamical system present, linked with the one which transmits the light, and possessing motion of rotation round the lines of magnetic force, or some other motion directed with respect to those lines; and the kinetic reaction between these two systems will account for the magnetic rotation.

a compound medium is implied.

The influence which is exerted on the free periods of a vibrating system by linking it on to another system which is in rotation may be illustrated by some dynamical problems. If the angular velocity of the rotating system is supposed to be maintained constant, such illustrations admit of comparatively simple analytical treatment. We can determine the change produced by the rotation in the free period of the original system. If that system is one member of a chain or solid continuum, we can deduce the velocity of propagation of waves of given length from a knowledge of this change of period; for it is the velocity which would carry the undulation over a wave-length in the free period. A typical example of this kind, which is treated in the paper, is the motion of a Blackburn's pendulum, suspended from a horizontal bar which is made to spin round a vertical axis with angular velocity ω. The equations of motion are

Pendulum with two periods

$$\frac{d^2x}{dt^2} - \omega^2 x - 2\omega\frac{dy}{dt} = -\frac{g}{l}x;$$

$$\frac{d^2y}{dt^2} - \omega^2 y + 2\omega\frac{dx}{dt} = -\frac{g}{m}y.$$

Writing $n^2 = \frac{1}{2}\left(\frac{g}{l}+\frac{g}{m}\right)$, and $\lambda^2 = \frac{1}{2}\left(\frac{g}{l}-\frac{g}{m}\right)$, the motion for the case when ω is very great compared with n loses its original character, and reduces to a form which, [within the stable amplitudes] neglecting slight tremors, is derived approximately from the superposition of two circular motions, one in the same direction as the angular velocity ω, of period $2\pi/\sigma$, the other in the opposite direction, of period $2\pi/\rho$, where

in rapid imposed rotation.

$$\rho = n + \frac{1}{8}\frac{\lambda^4}{\omega^2 n} - \frac{1}{8}\frac{\lambda^4}{\omega^3}, \quad \sigma = n + \frac{1}{8}\frac{\lambda^4}{\omega^2 n} + \frac{1}{8}\frac{\lambda^4}{\omega^3}.$$

The rotation in such a case as this becomes dominant; a plane oscillation now subsists, but will rotate steadily round the axis with angular velocity $\frac{1}{2}(\sigma - \rho)$, which is the slower the greater the velocity ω with which the horizontal arm is carried round.

These results may be extended to any rotating system with two transverse principal periods. Thus for the case of a long stretched cord, or a long rod, rotating round its own length with an angular velocity ω which is very great compared with either of its natural transverse frequencies $(2\pi/l)^{-\frac{1}{2}}$ and $(2\pi/m)^{-\frac{1}{2}}$, the period of vibration of a wave of given type on the rotating cord will be changed to

$$\frac{2\pi}{n}\left(1 + \frac{1}{8}\frac{\lambda^4}{\omega^2 n^2}\right)^{-\frac{1}{2}};$$

Rotated vibrating string:

and the angle of rotation of its plane of polarization, during propagation through a wave-length, will be

$$\frac{\pi\lambda^4}{4n\omega^3},$$

this rotation being in the same direction as the angular velocity ω.

Again, this aeolotropic cord may have imposed on it such a (slight) rate of twist that a very long plane wave, made helicoidal by the rotation ω, will just be straightened out again by this twist, as it progresses along the cord, the natural period being still practically unaltered. From this remark it follows that "the effect of a twist amounting to one turn in a length s, a small fraction of the wave-length, is to cause the plane of vibration of a wave to turn round with the forward propagation of the wave, at the rate of one turn in $8n^4/\lambda s^3$ wave-lengths," in the same direction as the imposed twist.

contrast of twisted anisotropic string.

The first of these results illustrates magnetic rotation, the second the axial rotation of quartz; while a medium filled with spiral arrangements, like the second but devoid of special orientation, represents the rotation of turpentine. The mode of passing directly in this illustration from the effect of spin to the effect of helical structure produced by twist is noteworthy.

The subject of a vibrating chain loaded with gyrostats, having

Gyrostatic chain.

their axes all along it, is considered by Lord Kelvin in a later paper[1]: and the general behaviour, as to propagation of waves, of a chain loaded with gyrostats which are orientated in any orderly manner with respect to it, has also been developed[2].

Mathematical Representations tested by Verdet's Experiments on Magnetic Dispersion.

Maxwell.

5. The use of the term of type κ_1 to explain magnetic rotation was arrived at by Maxwell[3] by the help of a provisional theory of molecular vortices, in which it occurs as standing for the reaction of a vortical motion of the medium representing its magnetization, when that motion is disturbed by the light-vibrations passing through it.

Verdet's laws.

A very full examination has been made by Verdet[4] of the manner in which the constant of magnetic rotation (hence called Verdet's constant) depends on the direction of the ray with regard to the magnetic force, on the refractive power of the medium, on the dispersive power of the medium, and in the same medium on the wave-length of the light. The rotation comes out, as has since been verified in detail by Du Bois, to be proportional simply to the component of the magnetic force along the ray. Media of great refractive power have in general high magnetic rotatory power. For the same medium the product of the rotatory power and the square of the wave-length is nearly constant, but always increases slightly with the index of refraction; media of great dispersive power have in general also high rotatory dispersion.

Verdet's most important piece of work is, however, a precise comparison of his experimental numbers for different wave-lengths with the results of a mathematical formula adapted to express both ordinary dispersion, and the magnetic rotation according to Maxwell's theory. He assumes Cauchy's form of the ordinary dispersion terms, and so obtains equations equivalent to

Types of term tested by dispersion.

$$A_0 \frac{d^2\vartheta}{dz^2} + A_1 \frac{d^4\vartheta}{dz^4} + \dots = \rho \frac{d^2\vartheta}{dt^2} + 2C\gamma\iota \frac{d^3\vartheta}{dz^2 dt},$$

from which is derived (Maxwell, *Treatise*, §§ 828–830) the formula connecting θ, the rotation, with m, a specific constant for the medium; γ, the magnetic force resolved along the ray; c, the length of path of

[1] W. Thomson, *Proc. Lond. Math. Soc.* Vol. VI. 1875.

[2] J. Larmor, *Proc. Lond. Math. Soc.* Vol. XXI. 1890.

[3] J. C. Maxwell, *Phil. Mag.* 1861; *Treatise*, § 822 *seq.*

[4] E. Verdet, *Comptes Rendus*, 1863; *Annales de Chimie* (3), Vol. LXIX.; in *Œuvres*, Vol. I. p. 265.

the ray on the medium; λ, the wave-length of the light in air; and i, the index of refraction of the medium. This formula is

$$\theta = mc\gamma \frac{\iota^2}{\lambda^2}\left(i - \lambda \frac{di}{d\lambda}\right).$$

The comparison with experiment leads to agreement within the possible errors of observation (Maxwell, *loc. cit.*) for the case of bisulphide of carbon, but for the ordinary creosote of commerce the agreement is not so good. The fact that creosote is a chemically complex substance, or rather a mixture of different substances, may be of influence here.

A coefficient of the type κ_2 leads also to the general law of proportionality to the inverse square of the wave-length, but does not correspond nearly so well in detail as κ_1; a coefficient of the type κ_3 (C. Neumann's) must be rejected altogether.

It is to be borne in mind that it is only in substances with regular dispersion that Cauchy's dispersive terms can be taken to represent the facts; whereas the κ_2 rotatory term is, as we have seen, related to a free period of some kind in the system, and therefore to abnormal dispersion.

6. The considerations just given bring together evidence of various kinds, that for ordinary media the κ_2 rotatory term is to be taken as very subordinate to the κ_1 term. Using the κ_1 term alone, the equations of propagation of a wave travelling along the lines of magnetic force (now leaving out dispersion) will be of the form

Maxwell's tentative term:

$$\rho \frac{d^2u}{dt^2} = a^2 \frac{d^2u}{dz^2} + \kappa_1 \frac{d^3v}{dz^2dt};$$

$$\rho \frac{d^2v}{dt^2} = a^2 \frac{d^2v}{dz^2} - \kappa_1 \frac{d^3u}{dz^2dt}.$$

Let us now attempt to deduce general equations of propagation along any direction. These clearly must involve three constants, κ_x, κ_y, κ_z proportional to the components of the magnetic field along the axes of coordinates. For they must lead to the experimental law that the rotation for any direction of the wave is proportional to the component in that direction of the intensity of the magnetic field; in particular this law must be satisfied for the directions of the axes of coordinates. Further, the vibrations may be assumed to remain purely transverse, so that we must have no compression of the medium; and therefore the condition

$$\frac{du}{dx} + \frac{dv}{dy} + \frac{dw}{dz} = 0$$

is to remain satisfied after the rotational terms are added to the equations.

most general type of spatial equations:

The equations, then, must for an isotropic medium conform to the general type

$$\rho \frac{d^2u}{dt^2} = A\nabla^2 u + B \frac{d}{dx}\left(\frac{du}{dx} + \frac{dv}{dy} + \frac{dw}{dz}\right) + \frac{d}{dt} P_z,$$

in which P_x, P_y, P_z are linear functions of the second spatial differential coefficients of the displacements; and transversality of the unmodified wave requires $A + B = 0$. Further when w, d/dx and d/dy are all null, these functions must reduce to the forms

$$\kappa_z \frac{d^2v}{dz^2}, \quad -\kappa_z \frac{d^2u}{dz^2}, \quad 0.$$

Hence
$$P_x = \kappa_z \frac{d^2v}{dz^2} - \kappa_y \frac{d^2w}{dy^2} + Q_x,$$

in which Q_x involves only products of d/dx, d/dy, d/dz.

transversality

Transversality of the disturbed wave requires

$$\frac{dP_x}{dx} + \frac{dP_y}{dy} + \frac{dP_z}{dz} = 0;$$

hence changing the expression for P_x to

$$P_x = \left(\kappa_z \frac{d}{dx} + \kappa_y \frac{d}{dy} + \kappa_z \frac{d}{dz}\right)\left(\frac{dv}{dz} - \frac{dw}{dy}\right) + R_x,$$

R_x can involve only products of the operators d/dx, d/dy, d/dz, and we must have identically

$$\frac{dR_x}{dx} + \frac{dR_y}{dy} + \frac{dR_z}{dz} = 0.$$

These conditions necessitate that R_x, R_y, R_z shall be each null.

determines their form.

The equations of the magnetically modified medium are therefore restricted to a definite form by the hypothesis that the wave remains strictly transversal. The equations so obtained, of the type

$$\rho \frac{d^2u}{dt^2} = A\nabla^2 u + B \frac{d}{dx}\left(\frac{du}{dx} + \frac{dv}{dy} + \frac{dw}{dz}\right)$$
$$+ \frac{d}{dt}\left(\kappa_x \frac{d}{dx} + \kappa_y \frac{d}{dy} + \kappa_z \frac{d}{dz}\right)\left(\frac{dv}{dz} - \frac{dw}{dy}\right),$$

contain only terms that are invariantive for transformation of the coordinates; they thus retain the same form when referred to new axes. They therefore satisfy Verdet's law, that the rotatory coefficient for any other direction, which may be taken as the new axes of z, is proportional to the component of the magnetic field in that direction, as they ought to do.

Not only so, but Verdet's law *requires* that the equations shall be expressible in terms of invariants of the three vectors **Argument from spatial invariance:**

$$(\kappa_x, \kappa_y, \kappa_z), \ \left(\frac{d}{dx}, \frac{d}{dy}, \frac{d}{dz}\right), \text{ and } (u, v, w),$$

independently of particular axes of coordinates. Hence P_x, P_y, P_z must be so expressible; and they must be the components of a vector, of the first degree in the first and third of the above vectors, of the second degree in the remaining one. The invariants which can enter are simply the geometrical relations of the figure formed by the above three vectors drawn as rays from an origin. The only possible forms are the scalars

$$\begin{vmatrix} \kappa_x & \kappa_y & \kappa_z \\ \frac{d}{dx} & \frac{d}{dy} & \frac{d}{dz} \\ u & v & w \end{vmatrix},$$

$$\frac{d^2}{dx^2}+\frac{d^2}{dy^2}+\frac{d^2}{dz^2}, \quad \kappa_x\frac{d}{dx}+\kappa_y\frac{d}{dy}+\kappa_z\frac{d}{dz}, \quad \frac{du}{dx}+\frac{dv}{dy}+\frac{dw}{dz},$$

and the vectors of which the x components are

$$\kappa_y w - \kappa_z v, \quad \frac{dv}{dz}-\frac{dw}{dy}.$$

These combine to give the most general form for P_x, represented by the equation

$$\begin{aligned} P_x = {} & L\left(\frac{d^2}{dx^2}+\frac{d^2}{dy^2}+\frac{d^2}{dz^2}\right)(\kappa_y w - \kappa_z v) \\ & + M\left(\kappa\frac{d}{dx}+\kappa_y\frac{d}{dy}+\kappa_z\frac{d}{dz}\right)\left(\frac{dv}{dz}-\frac{dw}{dy}\right) \\ & + N\left(\kappa_y\frac{d}{dz}-\kappa_z\frac{d}{dy}\right)\left(\frac{du}{dx}+\frac{dv}{dy}+\frac{dw}{dz}\right) \\ & + G\left\{\left(\kappa_x\frac{d}{dz}-\kappa_z\frac{d}{dx}\right)\left(\frac{dv}{dx}-\frac{du}{dy}\right)-\left(\kappa_y\frac{d}{dx}-\kappa_x\frac{d}{dy}\right)\left(\frac{du}{dz}-\frac{dw}{dx}\right)\right\}. \end{aligned}$$

Now

when $w, \frac{d}{dx}, \frac{d}{dy}$ are null, $\quad P = -\kappa_z\frac{d^2v}{dz^2}$;

" $v, \frac{d}{dx}, \frac{d}{dz}$ " $\quad P = \kappa_y\frac{d^2w}{dy^2}$;

" $u, \frac{d}{dy}, \frac{d}{dz}$ " $\quad P = 0$;

hence $\quad L - M = 1, \ G = 0.$

This expressión for P_x with the correlative ones for P_y and P_z is the most general form which the magnetic terms can assume in an isotropic medium, independently of any condition of exact trans- **general result.**

versality of the vibrations: transversality requires in addition that L and N shall be null.

The equations of vibration of an elastic medium loaded with spinning molecular gyrostats, whose axes follow the rotations of the elements of the medium, have been formed[1], and it is of interest to observe that the rotatory terms come under this special type for which

Gyrostatic type.

L, N are null as well as G. The reason is clear: the action of the gyrostats depends solely on the rotations of the elements of the medium, while the terms involving L and N have no rotational character.

7. Before application of these magneto-optic terms to problems of reflexion at a magnet the type of the unmodified equations of propagation of light, to which they are to be added, must first be settled. The form of these equations which gives most satisfactory results for reflexion at the interface separating transparent media is (equivalent partially to Lord Kelvin's labile aether theory) expressed most simply both as to bodily equations and as to boundary conditions by the principles of the electromagnetic theory of light; and it has been shown that the introduction of electric conducting quality into these equations gives a tolerable account of the phenomena of metallic reflexion[2]. It is therefore natural to add on these rotatory terms to the equations of the electromagnetic theory, and to try to explain the phenomena of magnetic reflexion by their aid, with the boundary conditions appropriate to that theory. This is what has been done in all attempts that have had any success; though there is room for diversity in the electrical basis which has to be supplied for the rotational terms.

Dynamical Theories based on the Form of the Energy-function.

FitzGerald: his electro-dynamic Action theory:

8. The subject of magnetic rotation has been treated by G. F. FitzGerald[3] from the point of view of an additional magneto-optic term in the energy-function of the electromagnetic medium. According to theory, the energy of this medium is made up of the kinetic or electromagnetic part T, and the static part W, where in Maxwell's notation, $d\tau$ being an element of volume,

$$T = \frac{1}{8\pi}\int (a\alpha + b\beta + c\gamma)\, d\tau,$$

$$W = \frac{1}{2}\int (Pf + Qg + Rh)\, d\tau;$$

[1] J. Larmor, *Proc. Lond. Math. Soc.* Vol. XXII. 1891, *supra*, p. 248.

[2] Cf. J. J. Thomson, *Recent Researches in Electricity and Magnetism*, 1893, § 352 *seq.*

[3] G. F. FitzGerald, "On the Electromagnetic Theory of the Reflection and Refraction of Light," *Phil. Trans.* 1880.

and there is also in our problem to be added on another small term, Maxwell's hypothetical magneto-optic part T'.

Now the dynamical equations of any medium or system are most fundamentally expressed as the conditions that the characteristic function of Lagrange and Hamilton

$$\int (T + T' - W)\, dt$$

should be stationary for a given time of motion from any one definite configuration to another, subject to whatever restraints the coordinates have to obey. This form is the most fundamental, because the processes of the Calculus of Variations are purely analytical, and quite independent of whatever specifying quantities we may choose in order to represent the state of the system, the only condition being that the function $T + T' - W$ is to be expressed in terms of a sufficient number of measures of configuration and their first differential coefficients with respect to the time, and is to be of the second degree as regards these differential coefficients.

In order to obtain such an expression FitzGerald proposes to treat (α, β, γ) as velocities corresponding to coordinates (ξ, η, ζ), so that **magnetic force visualized as a velocity:**

$$(\alpha, \beta, \gamma) = \frac{d}{dt}(\xi, \eta, \zeta),$$

and then

$$T = \frac{\mu}{8\pi}\int\left(\frac{d\xi^2}{dt^2} + \frac{d\eta^2}{dt^2} + \frac{d\zeta^2}{dt^2}\right) d\tau;$$

and this will be successful if W can be represented in terms of (ξ, η, ζ) only. Now in a dielectric

$$4\pi\frac{df}{dt} = \frac{d\gamma}{dy} - \frac{d\beta}{dz}, \ \ldots, \ \ldots,$$

hence

$$4\pi f = \frac{d\zeta}{dy} - \frac{d\eta}{dz}, \ \ldots, \ \ldots;$$

also (P, Q, R) is, from the constitution of the medium, expressed in terms of (f, g, h) by the linear equations of electrostatic induction, so that the thing required is done. If, in fact, **with rotational elasticity:**

$$W = \int U d\tau,$$

where U is a quadratic function of (f, g, h), the equations of motion in non-rotational media are involved in the variational equation

$$\int dt \left\{\frac{\mu}{8\pi}\,\delta\int\left(\frac{d\xi^2}{dt^2} + \frac{d\eta^2}{dt^2} + \frac{d\zeta^2}{dt^2}\right) d\tau - \delta\int U d\tau\right\} = 0,$$

for variations of (ξ, η, ζ) subject to the *imposed* condition **and incompressibility.**

$$\frac{d\xi}{dx} + \frac{d\eta}{dy} + \frac{d\zeta}{dz} = 0,$$

which expresses that the magnetic flux is constrained to be circuital. This condition is included, in the Lagrangian manner, by adding on to the above variation, which is equated to zero, a term

$$\int dt\, \delta \int \lambda \left(\frac{d\xi}{dx} + \frac{d\eta}{dy} + \frac{d\zeta}{dz}\right) d\tau,$$

and determining λ afterwards as a function of position, so that the imposed condition shall be satisfied. Thus we have

$$\int dt \left\{\frac{\mu}{4\pi}\int\left(\frac{d\xi}{dt}\frac{d\delta\xi}{dt} + \ldots\right) d\tau - \int\left(\frac{dU}{df}\delta f + \ldots\right) d\tau \right.$$
$$\left. + \int \lambda \left(\frac{d\delta\xi}{dx} + \ldots\right) d\tau\right\} = 0.$$

Changing from the differential coefficients of $\delta\xi$ to $\delta\xi$ itself by integration by parts, and similarly for $\delta\eta$ and $\delta\zeta$ in the usual manner, we obtain finally

$$\frac{1}{4\pi}\int dt \left\{\int\left(\mu \frac{d^2\xi}{dt^2} + \frac{d}{dy}\frac{dU}{dh} - \frac{d}{dz}\frac{dU}{dg} + 4\pi \frac{d\lambda}{dx}\right) \delta\xi\, d\tau + \ldots + \ldots \right.$$
$$\left. + \int\left(\frac{dU}{dh}\delta\eta - \frac{dU}{dg}d\zeta - 4\pi\lambda\delta\xi\right) l dS + \ldots + \ldots\right\} = 0.$$

This is to be true for all forms of $\delta\xi$, $\delta\eta$, $\delta\zeta$, which necessitates equations of the type

$$\mu \frac{d^2\xi}{dt^2} + \frac{d}{dy}\frac{dU}{dh} - \frac{d}{dz}\frac{dU}{dg} + 4\pi \frac{d\lambda}{dx} = 0$$

throughout the medium; while at an interface, supposed for an instant to be normal to the axis of x, so that $(l, m, n) = (1, 0, 0)$, we must have

$$\frac{dU}{dh}\delta\eta - \frac{dU}{dg}\delta\zeta - 4\pi\lambda\delta\xi$$

continuous.

Now from the bodily equations we deduce at once

$$\nabla^2\lambda = 0;$$

therefore λ is mathematically the potential function of a mass-distribution on the interfaces only, and so is continuous across them. It follows that the only way of securing the required continuity at an interface is (i) to postulate that η and ζ are continuous across it, owing to the continuous structure of the system, and that therefore dU/dg and dU/dh are also continuous across it; and (ii) either to postulate that $\delta\xi$ is constrained to be null or else that λ is null all over it. The alternative taken in Maxwell's electrodynamics is that λ is null everywhere in the field, as in fact representing no observed physical phenomenon; then the boundary conditions are four, that the tangential components of the magnetic force are continuous, as also the tangential components of the electric force.

No pressural transmission:

Since $(\dot{\xi}, \dot{\eta}, \dot{\zeta})$ is the magnetic force, the displacement of the medium, as represented by the vector (ξ, η, ζ), is in the plane of polarization of plane-polarized light: by representing it by some function of (ξ, η, ζ), *e.g.* its curl, we could have it at right angles to this plane, as in Fresnel's work.

This quantity λ is not introduced in FitzGerald's analysis of the problem of ordinary crystalline refraction. As the results of the discussion of propagation and reflexion show, any motion propagated in a medium, homogeneous or heterogeneous, whose dynamical properties are determined by the above characteristic function, is effectively of a compressionless character, and there is no necessity to introduce a restriction to that type. But the case becomes different when magneto-optic terms are added to the energy-function.

even in crystals:

9. FitzGerald goes on to assume, after Maxwell's theory of molecular vortices, that the magneto-optic part of the energy is of the form

$$T' = 4\pi C \int \left(\frac{d\xi}{d\theta}\frac{df}{dt} + \frac{d\eta}{d\theta}\frac{dg}{dt} + \frac{d\zeta}{d\theta}\frac{dh}{dt} \right) d\tau,$$

where $d/d\theta$ denotes differentiation along the lines of imposed magnetic force; but on working out the variation of the characteristic function, he finds a difficulty about satisfying all the equations of condition at an interface. This may, I think, be got over by introducing the undetermined multiplier λ of the above analysis into the variation, and so taking into account a certain condensational tendency which is originated at the interface, and propagated throughout the medium with very great velocity. There also remains for settlement the question whether the energy represented by T' is correctly localized by its formula, or whether it involves superficial components, in addition to the bodily distribution; in Maxwell's vortex theory, from which it is taken, it has been transformed by integration by parts. So long as this doubt remains we shall not be in a position to demonstrate boundary conditions by this method.

unless possibly in a magnetic field.

A chief interest, at the present date, of FitzGerald's paper, lies in the application of the method of Least Action to deduce the equations of a dielectric medium from the expression for its energy alone. This method would not be available for a medium which is the seat of viscous forces[1]; consequently the equations for a conducting medium would have to be derived from those of a dielectric by the empirical introduction of appropriate terms to represent the viscosity. It is, in fact, clear that the scientific method, in forming a dynamical

Treatment of viscosity.

[1] It is possible, however, to introduce Lord Rayleigh's dissipation function into the formal development from the general equation of Action. [*Phil. Trans.* 1894, § 88, as *infra*, p. 488.]

theory, is to restrict it in the first instance to systems in which the interaction of stress and motion has free play, without the interference with its results that is produced by frictional agencies. The subject of the reduction of the equations of electrodynamics into the domain of the general principle of Least Action has recently been treated by von Helmholtz.

Localization of magneto-optic Action term.

10. The second of the questions raised above will now be examined. The magneto-optic energy must in reality be localized in space and not on surfaces; and it is of interest to inquire what is the most general formula that can be given for it which will lead to terms of the accepted type in the equations of motion. If we take a term in the variational equation of motion of the form $\int \frac{d^3\phi}{ds^2dt}\delta\psi d\tau$,—where ϕ and ψ each stand for one of the symbols ξ, η, ζ, and s stands for one of the symbols x, y, z,—and if we trace backwards the operation of integration by parts by which it was derived from the characteristic function, we obtain the following types under the sign of volume-integration which may exist in the direct variation of that function:

$$\frac{d^2\phi}{dsdt}\frac{d\delta\psi}{ds},\quad \frac{d\phi}{dt}\frac{d^2\delta\psi}{ds^2},\quad \frac{d\phi}{ds}\frac{d^2\delta\psi}{dsdt},\quad \phi\frac{d^3\delta\psi}{ds^2dt}.$$

These expressions may combine into complete variations of terms of any of the types

$$\frac{d^2\phi}{dsdt}\frac{d\psi}{ds},\quad \frac{d\phi}{dt}\frac{d^2\psi}{ds^2},\quad \phi\frac{d^3\psi}{ds^2dt}.$$

Now the term in the energy which comes from the linking of the optical with the magnetic motion should be of the first degree as regards the velocities of each of them; and it may involve linear and angular displacements, but not their differential coefficients, *i.e.* it should involve only second differential coefficients with respect to space. The first of these types is thus the only one available, and the term in the energy must therefore be a scalar constructed from the combination of d/dt with

$$\frac{d}{d\theta}(\xi, \eta, \zeta), \text{ and } (f, g, h) \text{ or } \left(\frac{d\eta}{dz}-\frac{d\zeta}{dy}, \ldots, \ldots\right),$$

if we exclude the scalar $v = \frac{d\xi}{dx}+\frac{d\eta}{dy}+\frac{d\zeta}{dz}$,

which would introduce compression. The term under investigation may therefore have the form either

Possible invariant forms:

$$f\frac{d^2\xi}{d\theta dt}+g\frac{d^2\eta}{d\theta dt}+h\frac{d^2\zeta}{d\theta dt}$$

or

$$\frac{d\xi}{d\theta}\frac{df}{dt}+\frac{d\eta}{d\theta}\frac{dg}{dt}+\frac{d\zeta}{d\theta}\frac{dh}{dt},$$

excluding, for the reason already given, forms such as

$$v \frac{d}{dt} (\alpha f + \beta g + \gamma h).$$

Of these the first combines together the angular *distortion* of the medium, the velocity representing the motion in the magnetic field, and the rate of change of the velocity of the medium in the direction of that motion; while the second combines the *spin* of the medium with the velocity in the magnetic field. It would be difficult to assign a physical basis to the former on either a dynamical or an electric theory; and thus we are confined on our premises to Maxwell's form as giving correctly the localization of the magneto-optic part of the energy. This dynamical conclusion if granted will restrict the purely formal results of § 6, in the same way as has been already done by the use of the hypothesis of absolutely perfect incompressibility in the resulting equations of propagation.

the second preferred: restricts formal possibilities.

11. To discuss the first question it will be convenient to reproduce the main lines of FitzGerald's analysis, with however the introduction of the new terms involving λ, the origin of which has been already explained. The variational equation of the motion is

Tentative magneto-optic pressure, for a medium not compound:

$$\int lt \left\{\frac{\mu}{8\pi} \delta \int \left(\frac{d\xi^2}{dt^2} + \frac{d\eta^2}{dt^2} + \frac{d\zeta^2}{dt^2}\right) d\tau - \delta \int U d\tau \right.$$
$$\left. + 4\pi C \delta \int \left(\frac{d\xi}{d\theta}\frac{df}{dt} + \frac{d\eta}{d\theta}\frac{dg}{dt} + \frac{d\zeta}{d\theta}\frac{dh}{dt}\right) d\tau + \delta \int \lambda \left(\frac{d\xi}{dx} + \frac{d\eta}{dy} + \frac{d\zeta}{dz}\right) d\tau \right\} = 0,$$

in which $\frac{d}{d\theta} = \alpha \frac{d}{dx} + \beta \frac{d}{dy} + \gamma \frac{d}{dz}$, where (α, β, γ) is the imposed uniform magnetic field, 4π (f, g, h) is the curl of (ξ, η, ζ) as defined above, and λ is a function of (x, y, z), analogous to a hydrostatic pressure in a dynamical theory, and to be determined afterwards as circumstances dictate. The variation is conducted in the ordinary manner; and of the final result the term involving $\delta\xi$ is here explicitly set down, for the special case of an isotropic medium for which

$$U = \frac{2\pi}{K} (f^2 + g^2 + h^2),$$

as follows, l, m, n being the direction cosines of the normal to the element of surface dS:

$$\int dt \left[\frac{1}{4\pi K} \int \left\{n\left(\frac{d\xi}{dz} - \frac{d\zeta}{dx}\right) - m\left(\frac{d\eta}{dx} - \frac{d\xi}{dy}\right)\right\} \delta\xi dS + C \int \frac{d}{d\theta}\left(n\frac{d\eta}{dt} - m\frac{d\zeta}{dt}\right) \delta\xi dS \right.$$
$$- C \int (l\alpha + m\beta + n\gamma) \frac{d}{dt}\left(\frac{d\zeta}{dy} - \frac{d\eta}{dz}\right) \delta\xi dS - \int l\lambda\delta\xi dS$$
$$+ \frac{1}{4\pi} \int \left\{\mu \frac{d^2\xi}{dt^2} + 8\pi C \frac{d^2}{d\theta\, dt}\left(\frac{d\zeta}{dy} - \frac{d\eta}{dz}\right)\right.$$
$$\left.\left. + \frac{1}{K}\left[\frac{d}{dy}\left(\frac{d\eta}{dx} - \frac{d\xi}{dy}\right) - \frac{d}{dz}\left(\frac{d\xi}{dz} - \frac{d\zeta}{dx}\right)\right] + 4\pi \frac{d\lambda}{dx}\right\} \delta\xi d\tau\right].$$

Thus the bodily equations of propagation are of type

$$\mu \frac{d^2\xi}{dt^2} = -\frac{1}{K}\left[\frac{d}{dy}\left(\frac{d\eta}{dx}-\frac{d\xi}{dy}\right)-\frac{d}{dz}\left(\frac{d\xi}{dz}-\frac{d\zeta}{dx}\right)\right]$$
$$- 8\pi C\frac{d^2}{d\theta\, dt}\left(\frac{d\zeta}{dy}-\frac{d\eta}{dz}\right) - 4\pi\frac{d\lambda}{dx}.$$

From them comes $\nabla^2\lambda = 0$, showing that λ is mathematically the potential function of a mass-distribution on the interfaces only, and so does not appear at all in an infinite homogeneous medium.

The interfacial conditions are most easily expressed by taking for the instant the axis of z at right angles to the element of surface considered. The following expression must then be continuous across the interface in order that the surface integral may be null:

$$\left\{\frac{1}{K}\left(\frac{d\xi}{dz}-\frac{d\zeta}{dx}\right) + 4\pi C\frac{d^2\eta}{d\theta\, dt} - 4\pi C\gamma\frac{d}{dt}\left(\frac{d\zeta}{dy}-\frac{d\eta}{dz}\right)\right\}\delta\xi$$
$$-\left\{\frac{1}{K}\left(\frac{d\zeta}{dy}-\frac{d\eta}{dz}\right) + 4\pi C\frac{d^2\xi}{d\theta\, dt} + 4\pi C\gamma\frac{d}{dt}\left(\frac{d\xi}{dz}-\frac{d\zeta}{dx}\right)\right\}\delta\eta$$
$$-\left\{\lambda + 4\pi C\gamma\frac{d}{dt}\left(\frac{d\eta}{dx}-\frac{d\xi}{dy}\right)\right\}\delta\zeta.$$

Now we must have $\delta\xi$ and $\delta\eta$ continuous to avoid a breach in the medium at the interface, therefore the coefficients of these quantities must also be continuous across the interface; and as regards the third term either $\delta\zeta$ is continuous,or else its coefficient must vanish[1]. The other conditions of continuity do not allow $\delta\zeta$ to be continuous; therefore the third term gives simply the surface condition as to λ in the form

can adjust the interfacial conditions in reflexion.

$$\lambda = -4\pi C\gamma\frac{d}{dt}\left(\frac{d\eta}{dx}-\frac{d\xi}{dy}\right).$$

This very slight pressure λ is by the previous analysis continuous across the interface; it is important because it appears in a rotationally active form in the equations; the formula shows that, at the

A discontinuity remains.

[1] The difficulty has been raised that this procedure leaves $\delta\zeta$ discontinuous, and so apparently leads to rupture of the media at the interface. The reply to this point would be that if the necessity of the continuity of $\delta\zeta$ is admitted, the very formulation of the problem will involve innate inconsistency, as no other equation of condition can be introduced into the variational equation; while, on the other hand, the vanishing of the coefficient of $\delta\zeta$, as above, shows that there is no resistance offered to stretching along the normal of the layer of the medium at the interface, and therefore the continuity of $\delta\zeta$ will be actually adjusted by a stretching of the interfacial layer which involves no dynamical consequences. The part of $\delta\zeta$ to be thus adjusted is very small, depending on C; the mode of adjustment would probably be more fully in evidence, if we passed to the limit through a medium of slight compressibility. Precisely the converse mode of adjustment is in fact required in Lord Kelvin's labile aether. In any case we can hold to the axiom (§ 24) that the variation of the Action introduces all the conditions that are really essential.

interface, it is proportional to the normal component of the magnetic force.

It appears therefore that we have here a consistent scheme of equations of reflexion and refraction, without the necessity of condoning any dynamical difficulties in the process, the result being in all respects implicitly involved in the expression for the energy function of the medium.

The introduction of the circumstance of conduction, or absorption of the energy of vibration, can hardly affect the analytical form of the boundary conditions as to displacement and traction across an interface. If this be allowed, the problem of reflexion at a magnet will involve the same equations of propagation in the magnet as the above, with the exception that the velocity constant is complex, both the magneto-optic terms and the boundary conditions being otherwise unaltered. How far this theory can compete with others in giving a full explanation of our experimental knowledge would take too long time at present to inquire; but the considerations to be explained in the latter part of this paper will, I think, give it strong claims to being a correct formulation of the phenomena.

Recent Electrical Theories.*

12. A recent very comprehensive memoir[1] by Drude, on this subject, begins by alluding to the enormous rotatory power of magnetized bodies discovered by Kundt, which places in strong light the direct magnetic origin of the phenomenon of rotation, and to the observation of Kundt that a film of non-magnetic metal deposited on a magnet destroys Kerr's phenomena, so that they cannot be due to magnetic rotation in the air. He also remarks on the insufficiency

Influence on reflexion is purely interfacial.

* One important criterion remains over. The rotational terms ought to be such as conform to optical relativity up to the second power of the convection-ratio v/c. The writer had concluded that agreement existed up to the first order: *Aether and Matter*, 1900, Ch. XIII. But Prof. Lorentz pointed out a mistake in the algebra: so that, now in accordance with his own previous result, relativity is not secured in any *continuous* formulation hitherto, even to the first order, and the discrepancy remains.

As regards the magnetic rotation, however, one feature has been overlooked. There is no isotropic relativity in respect to a disturbed aether. If the optical rotation is due to a magnetic field, the magnets producing the field must themselves be convected as part of the optical system. This is the condition of experiment, and contributes to the null result: but then the field of the convected magnets involves a small electric force as well as the magnetic one, and it must come into the account.

For natural rotation, as due to helical structure, whether intrinsic in the molecule or in the arrangement of the crystal, it is purely a problem of gaining a suitable specification.

[1] P. Drude, "Ueber magneto-optische Erscheinungen," Wied. *Ann.* XLVI. 1892. [The whole subject is treated exhaustively, in relation to the copious experimental data. by J. G. Leathem, *loc. cit. ante*, *Phil. Trans.* 1897.]

of the notion that before reflexion the light penetrates slightly into the magnet and so undergoes rotation in its substance; this notion is in the first place not precise or quantitative at all, and further it assigns to the surface layer of transition an influence in reflexion which is much too great in view of other optical phenomena.

He then takes up one type of the formal equations of propagation in an isotropic medium, viz. (u, v, w) being a certain vector (the rotation in an isotropic elastic medium)

$$\frac{d^2}{dt^2}(u, v, w) = \epsilon\nabla^2 (u, v, w);$$

and he works out as follows the results of adding on to the right-hand side terms of the various kinds originally suggested by Airy. On adding terms of the form which represents the theory of C. Neumann, viz.

Theories tried,

$$\begin{vmatrix} b_1 & b_2 & b_3 \\ \dfrac{du}{dt} & \dfrac{dv}{dt} & \dfrac{dw}{dt} \end{vmatrix}$$

there come equations of the type

$$\frac{d^2u}{dt^2} = \epsilon\nabla^2 u + b_3\frac{dv}{dt} - b_2\frac{dw}{dt} + \frac{d\lambda}{dx},$$

to which the term $d\lambda/dx$ is conjoined in order to allow us to have

$$\frac{du}{dx} + \frac{dv}{dy} + \frac{dw}{dz} = 0,$$

i.e. in order that the light-waves shall remain purely transversal.

The form of this new term might be derived from the general variational equation of motion of the system, worked out subject to this limitation of the displacement (u, v, w).

On writing

$$(\xi, \eta, \zeta) = \left(\frac{dw}{dy} - \frac{dv}{dz}, \frac{du}{dz} - \frac{dw}{dx}, \frac{dv}{dx} - \frac{du}{dy}\right)$$

there follows $$\nabla^2\lambda = -\left(b_1\frac{d\xi}{dt} + b_2\frac{d\eta}{dt} + b_3\frac{d\zeta}{dt}\right),$$

and also the equations of propagation in the form

$$\frac{d^2}{dt^2}\nabla^2 u = \epsilon\nabla^2\nabla^2 u - \frac{d}{dt}\left(b_1\frac{d\xi}{dx} + b_2\frac{d\xi}{dy} + b_3\frac{d\xi}{dz}\right).$$

Since (b_1, b_2, b_3) are small we may employ in the terms containing them the approximate values of (ξ, η, ζ) which neglect the rotatory action, viz. which satisfy $\left(\frac{d^2}{dt^2} - \epsilon\nabla^2\right)(\xi, \eta, \zeta) = 0$, and so obtain finally, for disturbances of period $2\pi/\tau$, the equations

$$\frac{d^2u}{dt^2} = \epsilon\nabla^2 u + \frac{\epsilon}{\tau^2}\frac{d}{dt}\left(b_1\frac{d\xi}{dx} + b_2\frac{d\xi}{dy} + b_3\frac{d\xi}{dz}\right).$$

Thus for a wave travelling along the axis of z

$$\frac{d^2u}{dt^2} = \epsilon\frac{d^2u}{dz^2} + \frac{\epsilon b_3}{\tau^2}\frac{d^3v}{dz^2dt},$$

$$\frac{d^2v}{dt^2} = \epsilon\frac{d^2v}{dz^2} - \frac{\epsilon b_3}{\tau^2}\frac{d^3u}{dz^2dt};$$

but the rotatory effect given by these equations, if sensible for waves of moderate length, would be quite insensible for light-waves. **and rejected.**

On the other hand, if we add rotational terms of the type employed by Maxwell we should have similarly

$$\frac{d^2u}{dt^2} = \epsilon\nabla^2u + b_3\frac{d^3v}{dz^2dt} - b_2\frac{d^3w}{dy^2dt} + \frac{d\lambda}{dx},$$

where the condition for transverse undulations determines λ by the equation

$$\nabla^2\lambda = -\frac{d}{dt}\left(b_1\frac{d^2\xi}{dx^2} + b_2\frac{d^2\eta}{dy^2} + b_3\frac{d^2\zeta}{dz^2}\right),$$

so that

$$\frac{d^2}{dt^2}\nabla^2u = \epsilon\nabla^2\nabla^2u + \frac{d}{dt}\left\{b_3\frac{d^2}{dz^2}\left(\frac{d\zeta}{dx} - \frac{d\xi}{dz}\right) - b_2\frac{d^2}{dy^2}\left(\frac{d\xi}{dy} - \frac{d\eta}{dx}\right)\right\}$$

$$- \frac{d^2}{dxdt}\left(b_1\frac{d^2\xi}{dx^2} + b_2\frac{d^2\eta}{dy^2} + b_3\frac{d^2\zeta}{dz^2}\right),$$

that is
$$\frac{d^2}{dt^2}\nabla^2u = \epsilon\nabla^2\nabla^2u - \frac{d}{dt}\left(b_1\frac{d^3\xi}{dx^3} + b_2\frac{d^3\xi}{dy^3} + b_3\frac{d^3\xi}{dz^3}\right).$$

Thus the equations arrived at in these two ways are not, as Drude seems to hastily assume without examination, of the same type.

It is however difficult to see why equations arrived at in this manner are worthy of the detailed discussion and refutation to which Drude subjects them: though it is to be said that they agree formally with the equations of the earliest attempt to explain magnetic reflexion, that of Lorentz based on the Hall effect. The form of the rotational terms in the first of them (leaving out of account the character of the coefficient $\epsilon b_3/\tau^2$) is the same as the one to which we have been already guided as the correct type, by various lines of argument; and in fact the equations adopted by Drude himself are obtained by adding on these terms, somewhat empirically, to the ordinary electromagnetic equations of type

$$\frac{d^2u}{dt^2} + \epsilon\left(\frac{d\zeta}{dy} - \frac{d\eta}{dz}\right) = 0.$$

It may be observed that the analysis here given would apply equally if the equations just written were substituted for the fundamental equations from which it started.

Drude's scheme of equations:

13. The electrical views by means of which Drude accounts for the addition of terms of this kind to the electromagnetic equations are as follows. He starts with the two circuital relations to which the equations of electrodynamics have been reduced by [Maxwell,] Heaviside, Hertz, and other expositors, of the types

$$4\pi u = \frac{d\gamma}{dy} - \frac{d\beta}{dz}, \qquad -\frac{da}{dt} = \frac{dR}{dy} - \frac{dQ}{dz},$$

in which as usual (u, v, w) is total electric current, (α, β, γ) is magnetic force, (a, b, c) is magnetic induction, and (P, Q, R) is electric force. To these equations we would, under ordinary circumstances, add relations depending on the structure of the medium, in the form for isotropic media,

$$(a, b, c) = \mu\,(\alpha, \beta, \gamma),$$

$$(u, v, w) = \left(\frac{K}{4\pi}\frac{d}{dt} + \sigma\right)(P, Q, R),$$

where K is specific inductive capacity and σ is specific conductivity. To introduce the magnetic rotatory property, Drude proposes to modify the second set of circuital relations "on Maxwell's analytical basis, that to the kinetic energy of the medium which is expressed in simple form by means of the components of the magnetic force certain subsidiary terms are appended; as according to Maxwell the magnetization is to be considered as a kind of molecular vortex or concealed motion (*verborgene Bewegung*)." The modification which he assumes on this ground is a replacement of the second circuital relation by one of type

magneto-optic term.

$$-\frac{da}{dt} = \frac{dR}{dy} - \frac{dQ}{dz} + \frac{d^2}{dy\,dt}(b_2P - b_1Q) - \frac{d^2}{dz\,dy}(b_1R - b_3P),$$

keeping the other equations unaltered.

On forming the expression for the transfer of energy per unit volume of the medium, there is obtained (*neglecting*, however, the magneto-optic energy) the equation

$$\frac{d}{dt}\left\{\frac{1}{2}\mu\int(\alpha^2+\beta^2+\gamma^2)\,d\tau + \frac{K}{8\pi}\int(P^2+Q^2+R^2)\,d\tau\right\}$$

$$= -4\pi\sigma\int(P^2+Q^2+R^2)\,d\tau$$

$$+\int\left\{b_1\left(Q\frac{dR}{dt} - R\frac{dQ}{dt}\right) + \ldots + \ldots\right\}d\tau$$

$$+\int\left\{\left(P + b_3\frac{dQ}{dt} - b_2\frac{dR}{dt}\right)\beta - \left(Q + b_1\frac{dR}{dt} - b_3\frac{dP}{dt}\right)\alpha\right\}n\,dS$$

$$+ \ldots + \ldots,$$

in which $d\tau$ is an element of volume, and the three integrals at the end are extended over the boundary of the medium, of which (l, m, n) are the direction cosines.

For periodic vibrations there is thus no dissipation of energy except that due to conduction. But at the interface between two media the transmission of energy without accumulation on the surface requires that, the axis of x being assumed normal to the interface for the moment, in addition to the continuity of (β, γ) the tangential magnetic force, we must have continuity in

$$\left(Q + b_1 \frac{dR}{dt} - b_3 \frac{dP}{dt}, \quad R + b_2 \frac{dP}{dt} - b_1 \frac{dQ}{dt}\right).$$

Interfacial transfer of energy must be complete:

The tangential electrical force is therefore to be taken discontinuous; and the author enters into explanations to minimize the repugnance which may be felt to such a hypothesis, their gist being that the part of the electric force derived from the relations of the system itself must be continuous, but the part imposed from without need not be so.

difficulty.

The weak point in this determination of the boundary conditions is the fact that, as the extra terms are supposed to have their origin in a new term in the energy, this term ought to have been included in the reckoning before we can draw any conclusions from the flux of energy across the interface.

The equations of propagation are, for periodic motions in which $\frac{d}{dt} = \frac{\iota}{\tau} = -\iota\tau \frac{d^2}{dt^2}$, of the type

$$\mu K' \frac{d^2\alpha}{dt^2} = \frac{dQ}{dz} - \frac{dR}{dy} - \frac{d}{dt}\left(b_1 \frac{dP}{dx} + b_2 \frac{dP}{dy} + b_3 \frac{dP}{dz}\right),$$

where K' is the complex quantity $\frac{K}{4\pi}\frac{d}{dt} + \sigma$, and where, when the axis of x is normal to an interface, the quantities above mentioned are to be continuous across it. The vector (α, β, γ) is in the wave-front of the undulations for, the magnetic permeability being constant,

$$\frac{d\alpha}{dx} + \frac{d\beta}{dy} + \frac{d\gamma}{dz} = 0;$$

and it is in the plane of polarization.

These equations may also be variously expressed in terms of other vectors, *e.g.* of (P, Q, R) which is in the plane of the wave-front and transverse to the plane of polarization, or of

$$\left(P - b_3 \frac{dQ}{dt} + b_2 \frac{dR}{dt}, \ldots, \ldots\right).$$

The magnetic rotation is here to be explained by the single real vector coefficient (b_1, b_2, b_3); and the same value of this coefficient is in fact found to give a fairly good account of the various circumstances attending the rotation of the plane of polarization in magnetic reflexion by iron and nickel, while it is also of the same order of magnitude as would correspond to Kundt's measures of the rotation produced by transmission through a thin film of iron.

Gold-hammer's scheme.

14. A theory published a few months before by Goldhammer[1] goes, on the other hand, on the assumption that the effect of the magnetic field is to produce a temporary structural change in the medium. In the ordinary case of an isotropic medium

$$(u, v, w) = \left(\frac{K}{4\pi}\frac{d}{dt} + \sigma\right)(P, Q, R);$$

so that in periodic motion for which $\frac{d}{dt} = \frac{\iota}{\tau}$ we have

$$\frac{d}{dt}(P, Q, R) = \left(\frac{K}{4\pi} + \iota\sigma\tau\right)^{-1}(u, v, w) = \frac{4\pi}{K'}(u, v, w), \text{ say,}$$

in which the complex part of the coefficient, involving σ, represents the effect of conductivity. This is now replaced by a wider relation: in the general crystalline medium he proposes the form

$$\frac{dP}{dt} = \frac{4\pi}{K'_1}u + \lambda_3 v - \lambda_2 w + \mu_x \frac{du}{dt} + \mu_3 \frac{dv}{dt} - \mu_2 \frac{dw}{dt},$$

$$\frac{dQ}{dt} = \frac{4\pi}{K'_2}v + \lambda_1 w - \lambda_3 u + \mu_y \frac{dv}{dt} + \mu_1 \frac{dw}{dt} - \mu_3 \frac{du}{dt},$$

$$\frac{dR}{dt} = \frac{4\pi}{K'_3}w + \lambda_2 u - \lambda_1 v + \mu_z \frac{dw}{dt} + \mu_2 \frac{du}{dt} - \mu_1 \frac{dv}{dt},$$

in which μ_1, μ_2, μ_3 are assumed to be complex.

When the period $\tau/2\pi$ is very great the last terms, involving $d/dt\,(u, v, w)$, exert no appreciable effect, and so may be left out of account: the vector coefficient $(\lambda_1, \lambda_2, \lambda_3)$ which is left is the representative of the Hall effect. When the period $\tau/2\pi$ is very small, as in the case of light-waves, the rotational coefficient (μ_1, μ_2, μ_3) is preponderant, and the other one $(\lambda_1, \lambda_2, \lambda_3)$ may be neglected. We may also leave out of account the slight double refraction represented by the coefficients μ_x, μ_y, μ_z, as these are in no wise rotational. Thus, for an isotropic optical medium, the structural relation which connects electric force with electric current would reduce to the form

$$\frac{dP}{dt} = \frac{4\pi}{K'}u + \mu_3 \frac{dv}{dt} - \mu_2 \frac{dw}{dt},$$

$$\frac{dQ}{dt} = \frac{4\pi}{K'}v + \mu_1 \frac{dw}{dt} - \mu_3 \frac{du}{dt},$$

$$\frac{dR}{dt} = \frac{4\pi}{K'}w + \mu_2 \frac{du}{dt} - \mu_1 \frac{dv}{dt};$$

[1] D. A. Goldhammer, "Das Kerr'sche...Phänomen," Wied. *Ann.* XLVI. 1892, p. 71.

in which Goldhammer takes (μ_1, μ_2, μ_3) to be complex (unless the medium is transparent) and proportional to the intensity of the imposed magnetic field. This structural relation between magnetic induction and magnetic force is supposed to remain unmodifiable in form by magnetic or other disturbance. But it is not easy to understand the manner in which the relation is introduced into the equations of electrodynamics, and the analysis to be given presently leads to a different result. The bodily equations are expressed in terms of Maxwell's vector-potential; they are the same in form as Drude's equations expressed in terms of magnetic force; and the boundary conditions assumed are continuity of the vector-potential and its first differential coefficients, and continuity of the electrostatic potential.

15. The equations of Drude may be subjected to an important transformation which will bring them into line with another class of electrical theories. If in them we write

$$P' = P + b_3 \frac{dQ}{dt} - b_2 \frac{dR}{dt},$$

$$Q' = Q + b_1 \frac{dR}{dt} - b_3 \frac{dP}{dt},$$

$$R' = R + b_2 \frac{dP}{dt} - b_1 \frac{dQ}{dt},$$

and take (P', Q', R') as the electric force instead of (P, Q, R), we may preserve unaltered both of the fundamental circuital relations. The first one is clearly preserved; and so will be the second one if the relation between electric current and electric force is taken to be that derived by substitution from

$$(u, v, w) = \left(\frac{K}{4\pi}\frac{d}{dt} + \sigma\right)(P, Q, R).$$

This leads to a relation of type

$$P' = \left(\frac{K}{4\pi}\frac{d}{dt} + \sigma\right)^{-1}\left(u + b_3 \frac{dv}{dt} - b_2 \frac{dw}{dt}\right),$$

which differs from the structural relation assumed by Goldhammer, but is of the same class. When this transformation is made, Drude's boundary conditions become simply the ordinary ones which express that the tangential components of the electric force and the magnetic force are continuous in crossing the interface; the difficulty as to discontinuity in the tangential electric force does not now occur. **Conjugate form of Drude's scheme.**

The special type of this relation which is assumed by Goldhammer, is

$$\frac{dP}{dt} = \left(\frac{K}{4\pi}\frac{d}{dt} + \sigma\right)^{-1} u + \mu_3 \frac{dv}{dt} - \mu_2 \frac{dw}{dt},$$

in which he asserts that μ_1, μ_2, μ_3 are each, owing in some way to their origin, of the form $\frac{K'}{4\pi}\frac{d}{dt} + \sigma'$, so that they are complex constants of which the real and imaginary parts are, for the case of light-waves, of the same order of magnitude. He had previously rejected the coefficients (λ_1, λ_2, λ_3) of the Hall effect as being for light-waves negligible in comparison with those retained, although the purely imaginary part of a coefficient of type μ must have the same character as they have, irrespective of magnitude. It is, perhaps, difficult to see any reason which would give probability to this assumption that the coefficients of type μ are complex quantities whose real and imaginary parts come to be precisely of the same order of magnitude.

Willard Gibbs' generalized outlook.

16. A general formal development of the equations of the electromagnetic theory, which is necessarily wide enough to take account of all possible secondary phenomena, such as dispersion and circular polarization, has been given in 1883 by Prof. Willard Gibbs[1], under the title of "An Investigation of the Velocity of Plane Waves of Light, in which they are regarded as consisting of solenoidal electrical fluxes in an indefinitely extended medium of uniform and very fine-grained structure."

The principle on which his investigation is based is the very general idea that the regular simple harmonic light-waves traversing the medium excite secondary vibrations in its molecular electrical structure, which is supposed very fine compared with the length of a wave. When there is absorption the phases of these excited vibrations will differ from that of the exciting wave; but even in this most general case the simple harmonic electric flux with which we are alone concerned is at each point completely specified by six quantities, the three components of the flux itself, and the three components of its rate of change with the time. In the same way, the electric force may be similarly specified by six coordinates. Now the electric elasticity of the medium, as regards its power of transmitting waves, is specified by the relation connecting average force and average flux, this average referring to a region large compared with molecular structures, but small compared with a wave-length. The most general relation of this kind that can result from the elimination of the molecular vibrations must be of the form of six linear equations connecting the quantities specifying the flux with the quantities

[1] J. Willard Gibbs, "On the General Equations of Monochromatic Light in Media of every Degree of Transparency," *American Journal of Science*, February 1883.

specifying the force, the coefficients being functions of the wave-length. If E denote the force and U the displacement, "we may therefore write in vector notation

$$[E]_{Ave} = \Phi\,[U]_{Ave} + \Psi\,[\dot{U}]_{Ave},$$

where Φ and Ψ denote linear functions.

"The optical properties of the media are determined by the forms of these functions. But all forms of linear functions would not be consistent with the principle of the conservation of energy.

"In media which are more or less opaque, and which therefore absorb energy, Ψ must be of such a form that the function always makes an acute angle (or none) with the independent variable. In perfectly transparent media Ψ must vanish, unless the function is at right angles to the independent variable. So far as is known, the last occurs only when the medium is subject to magnetic influence. In perfectly transparent media, the principle of the conservation of energy requires that Φ should be self-conjugate, *i.e.* that for three directions at right angles to one another, the function and independent variable should coincide in direction.

"In all isotropic media not subject to magnetic influence, it is probable that Φ and Ψ reduce to numerical coefficients, as is certainly the case with Φ for transparent isotropic media."[1]

For the further examination of the content of this relation connecting the two electric vectors we may express it in the symbolical form

$$[\text{flux}] = [p]\,[\text{force}] + \left[q\,\frac{d}{dt}\right][\text{force}],$$

where $[p]$ and $\left[q\,\frac{d}{dt}\right]$ represent vectorial coefficients. For the simple harmonic oscillations of period τ that are here contemplated

$$\frac{d}{dt} = \frac{2\pi}{\tau}\,\iota,\quad \frac{d^2}{dt^2} = -\left(\frac{2\pi}{\tau}\right)^2;$$

so that for the very small periods of light-vibrations multiplication of a coefficient by d/dt increases its importance enormously. When the oscillations are very slow the coefficient $[p]$ has still in a magnetic field a rotational part which reveals itself as the Hall effect; the presence of a coefficient $\left[q\,\frac{d}{dt}\right]$ could hardly be detected. On the other hand, with greater rapidity of vibrations, the importance of the rotational part in $\left[q\,\frac{d}{dt}\right]$ increases steadily, and finally absolutely overshadows any possible effect of the $[p]$ terms, unless the latter should contain a part whose origin was of the form $\left[r\,\frac{d^2}{dt^2}\right]$. If that

[1] J. Willard Gibbs, *loc. cit.* p. 133; J. Larmor, *Proc. Lond. Math. Soc.* Vol. XXIV. 1893, where, however, some of the statements need correction.

were so, at a still higher rapidity of vibrations the [p] terms would again become the important ones: but the wave-lengths would then be too small for such vibrations to have any physical reality.

Magnetization not directly effective in optics.

17. The question occurs whether to secure complete generality a corresponding rotational quality should be imparted to the linear relation connecting the magnetic flux (*i.e.* magnetic induction) with the magnetic force. It is however usual to assume, on various grounds, that the vibrations of light are too rapid to allow of their being accompanied by an oscillating magnetization of the material medium. The phenomena of magnetization of iron leave possibly no room for doubt that the magnetic movement is an affair of loosely associated groups of molecules, not of individual molecules themselves, the free periods corresponding to these groups being much too slow to follow the light-vibrations. These groups are broken up at the temperature of recalescence without the occurrence of any very striking effect: nor is there any striking difference in kind between the behaviour of iron to light and the behaviour of non-magnetic metals.

The effect of strong magnetization on light-waves would be on this view a secondary effect due to a change of structure of the medium. Soon after the experimental discovery of the Hall effect, and the attention which was concentrated on it owing chiefly to the influence of Lord Kelvin, it was pointed out by J. Hopkinson that the existence of an effect of that character had been anticipated by Maxwell in his *Treatise*, Vol. I. § 303, where in discussing the possibility of the occurrence of a rotational term in the equations expressing the general form of Ohm's law of conduction, he remarks that such a coefficient ι "we have reason to believe does not exist in any known substance. It should be found if anywhere in magnets, which have a polarization in one direction, probably due to a rotational phenomenon in the substance." The theory of such rotatory coefficients had also been worked out long before by Lord Kelvin[1], in a thermo-electric connection.

Kelvin and Maxwell on the Hall effect.

18. It seems worth while to examine how much in the way of magnetic rotation can be got out of this relation by assuming the functions Φ and Ψ to have rotational quality; though Gibbs himself later on in his memoir qualifies its use by a statement that "the equation would not hold in case of molecular vibrations excited by magnetic force. Such vibrations would constitute an oscillating magnetization of the medium, which has already been excluded from the discussion."

Willard Gibbs.

[1] Cf. Lord Kelvin (Sir W. Thomson), *Collected Papers*, Vol. II.

If the rotational quality is simply due to a magnetic field, we may take for brevity the direction of its lines of force to be along the axes of z, and the equations will be

$$u = \epsilon P - vQ, \quad v = \epsilon Q + vP, \quad w = \epsilon R,$$

where $\epsilon = \frac{K}{4\pi}\frac{d}{dt} + \sigma$, and v is of the form $\lambda \frac{d}{dt} + \nu$. The circuital relations of types

$$4\pi u = \frac{d\gamma}{dy} - \frac{d\beta}{dz}, \quad -\frac{d\alpha}{dt} = \frac{dR}{dy} - \frac{dQ}{dz}$$

lead to

$$\nabla^2 P - \frac{d}{dx}\left(\frac{dP}{dx} + \frac{dQ}{dy} + \frac{dR}{dz}\right) = 4\pi\frac{du}{dt} = 4\pi\epsilon\frac{dP}{dt} - 4\pi v\frac{dQ}{dt}.$$

Thus the rotational operator, instead of being of Maxwell's type d^3/dx^2dt, comes out of type

$$\left(\lambda\frac{d}{dt} + \nu\right)\frac{d}{dt}.$$

Though a rotational term of this latter type, entering into the relation between current and force, and conjoined with the ordinary equations of electrodynamics, leads, as we have just seen, to precisely the same scheme as Drude's to explain magnetic reflexion of waves of any single period; yet in order to take into account Verdet's laws of magnetic dispersion (in transparent media) the coefficients λ and ν have to be taken functions of the wave-lengths, whereas the coefficients in Drude's form of theory remain constant for all wave-lengths. The relation of Gibbs is competent to give an account of the laws of reflexion and of crystalline propagation for any one wave-length, by altering so to speak the electric inertia of the medium; and it fails for dispersion simply because the only method it possesses of rendering an account of dispersion is by accepting the observed facts, and making the coefficients functions of the wave-length. Thus we ought not to allow its failure to agree with magnetic dispersion to tell too much against the mode of explaining magnetic reflexion now under discussion. Yet the fact remains that the scheme embodied in Drude's equations has an advantage in comprehending a wider group of phenomena, and to that extent corresponds more fundamentally with the mechanism of the action; while on the other hand it exhibits, especially with regard to the boundary conditions, a more empirical character.

19. These equations we have named after Drude because his memoir contains by far the most detailed comparison with observation that has yet been made. The same equations, however, had been used by a number of other writers. For transparent media they had been obtained by Rowland[1] as equations of propagation, and they

Other schemes.

[1] H. A. Rowland, *Phil. Mag.* 1881.

had been used by FitzGerald and by Basset[1] to calculate the circumstances of magnetic reflexion, without, however, entering into the case of metallic media. While recently J. J. Thomson[2] has employed them in an independent discussion of the laws of magnetic reflexion, which corroborates the main conclusions of Drude without going so much into detail. In dielectric media Rowland and Basset proceed simply by assuming rotational terms in the expression for the electric force on the analogy of the Hall effect in metals; and FitzGerald, as we have seen, deduces his equations from a new term in the energy which represents the linking on of the magnetic system. It is shown by J. J. Thomson, in his discussion of Kerr's results on reflexion, that in metals as well as dielectrics it is the time-rate of change of the induction or electric displacement, and not the total electric current, that combines with the magnetic field in the formation of this new term.

The boundary conditions are determined by FitzGerald and Basset from the hypotheses that the tangential magnetic force shall be continuous, and there shall be no concentration of energy, or *quasi*-Peltier effect, at the interface, subject however in the case of the latter to the same objection as has been applied above to Drude's use of this principle; while J. J. Thomson arrives at the same boundary conditions by postulating that the part of the electric force which is derived from the system itself must be continuous tangentially, whatever may happen to the part imposed from without.

Alternative principles.

20. There are thus two ways in which the magnetic field may affect the phenomena of light-propagation. The imposed magnetization is an independent kinetic system of a vortical character which is linked on to the vibrational system which transmits the light-waves; the kinetic reaction between the two systems will add on new terms to the electric force: these terms are naturally continuous so long as the medium is continuous, but owing to their foreign origin they need not be continuous at an interface where the magnetized medium suddenly changes. At such an interface the other part of the electric force, which is derived from the vibrating system itself, has been assumed to be continuous in the ordinary manner, viz. its tangential components continuous; the total induction through the interface must of course always maintain continuity. This seems to be the type of theory developed by Maxwell in his hypothesis of molecular vortices (*Treatise*, § 822), and the conditions to which it leads have been applied to magnetic reflexion by the majority of writers on the subject, including Basset, Drude, J. J. Thomson. But against this

[1] A. B. Basset, *Phil. Trans.* 1891.

[2] J. J. Thomson, *Recent Researches*..., § 408 *seq.*

procedure there stands the pure assumption as regards discontinuity of electric force at an interface. The correct boundary conditions would be derived from the modification of FitzGerald's procedure, which has been explained above.

The other point of view is the purely formal one contemplated by Lord Kelvin and Maxwell in their discussions of possible rotational coefficients introduced into the properties of the medium by magnetization. The magnetization is supposed to slightly alter the structure of the medium which conveys the light-vibrations, but not to exert a direct dynamical effect on these vibrations.

It would appear from the analysis of Drude, and more particularly of J. J. Thomson[1], that there is some ground for assuming the correctness of the equations to which the former method leads; and those equations may be expressed in the terms of the second method somewhat as follows. The electric current is in a dielectric the rate of change of the electric displacement, which is of an elastic character; in a conducting medium part of the current is due to the continual damping of electric displacement in frictional modes: it may thus fairly be argued that the fundamental relation is primarily not between current and electric force, but between current and displacement, while the current is indirectly expressed in terms of electric force through the elastic relation between displacement and force. The equations would then run as follows, (ξ, η, ζ) being the electric displacement:

Final type of theory.

$$(u, v, w) = \left(\frac{d}{dt} + \frac{4\pi\sigma}{K}\right)(\xi, \eta, \zeta),$$

where

$$\xi = P - b_3Q + b_2R;$$
$$\eta = Q - b_1R + b_3P;$$
$$\zeta = R - b_2P + b_1Q.$$

This would make the relation between electric displacement and electric force of a rotational character, owing to the magnetization. If the medium were not magnetized, Lord Kelvin's argument might be employed for the negation of such a rotational character on the ground that a sphere rotating in an electric field would generate a perpetual motion; but as it is the rotation in the magnetic field would generate other electric forces. The frictional breaking down of displacement, viz. conduction, is known to assume a slightly rotational character, as manifested in the Hall effect*.

[1] J. J. Thomson, *Recent Researches in Electricity and Magnetism*, 1893, § 412.

* The subject of magneto-optic reflexion has been worked out very completely on cognate lines, and in comparison with copious recorded experimental data, by J. G. Leathem, *Phil. Trans.* 1897.

PART II. CORRELATION OF GENERAL OPTICAL THEORIES.

MacCullagh's Dynamical Theory of Light.

21. It has been remarked in this discussion of magneto-optic phenomena that a perfectly straightforward mechanical theory of magneto-optic reflexion would be obtained by adding on a uniaxial gyratory part to the energy-function of Lord Kelvin's labile aether[1].

Least Action. The development of such a theory as this, after the manner already indicated, from the single basis of the principle of Least Action, would compare very favourably, by the absence of subsequent adjustment and assumption, with any of the foregoing explanations.

It has possibly been observed that the energy-function of Fitz-Gerald's electrodynamic analysis considered above is identical except as to surface terms with the energy-function of the labile aether theory, when (ξ, η, ζ) is taken to denote actual displacement of the medium. The difference that, for plane-polarized light, (ξ, η, ζ) is in the former case in the plane of polarization, while in the latter case it is at right angles to that plane, is due, as we shall see, not to the fact that in the electric medium the compression $\frac{d\xi}{dx}+\frac{d\eta}{dy}+\frac{d\zeta}{dz}$ is taken to be absolutely null, while in the labile aether the pressure is taken to be absolutely null or the medium is supposed devoid of consistence to compression, but it is the result of the neglected surface terms on the energy-function. The correlation between an electric theory and a mechanical theory which follows from this comparison has already been alluded to by Willard Gibbs[2]. It will be found below that there is a similar correlation between two mechanical theories.

Correlations of theories.

FitzGerald on MacCullagh. The vector (ξ, η, ζ) of FitzGerald's equations is, as he points out, exactly the displacement in MacCullagh's[3] *quasi*-mechanical theory of optical phenomena; and his analysis is for non-rotational media very much a translation of MacCullagh's work into electric terminology. The method followed in MacCullagh's extremely powerful investigation, which was independent of and nearly contemporary with those of Green[4], and, I think, of at least equal importance, was

Green.

[1] Lord Kelvin (Sir W. Thomson), "On the Reflexion and Refraction of Light," *Phil. Mag.* 1888.

[2] J. Willard Gibbs, "A Comparison...," *Phil. Mag.* 1889.

[3] James MacCullagh, "An Essay towards a Dynamical Theory of Crystalline Reflexion and Refraction," *Trans. R.I.A.* December 1839.

[4] George Green, "On the Laws of the Reflexion and Refraction of Light at the Common Surface of two Non-crystallized Media," *Cambridge Phil. Trans.* December 1837, with Supplement, May 1839; George Green, "On the Propagation of Light in Crystallized Media," *Cambridge Phil. Trans.* May 1839.

to discover some form of the energy-function of the optical medium which shall lead by pure dynamical analysis in Lagrange's manner, without further hypothesis, to the various optical laws of Fresnel. In this he was completely successful, though Stokes[1] gives reason to doubt whether he has obtained the most general solution of his problem. His optical work has, however, to a great extent failed to receive due recognition from various causes; in particular the objection has been emphasized by Stokes (*loc. cit.*), and generally accepted, that the vector (ξ, η, ζ) which represents the light-disturbance in his analysis could not possibly be the displacement in a medium which transmits vibrations by elasticity in the manner of an ordinary elastic solid. "Indeed MacCullagh himself expressly disclaimed to have given a mechanical theory of double refraction. (It would seem, however, that he rather felt the want of a mechanical theory, from which to deduce the form of the function Q or V, than doubted the correctness of that form itself.) His methods have been characterized as a sort of mathematical induction, and led him to the discovery of the mathematical laws of certain highly important optical phenomena. The discovery of such laws can hardly fail to be a great assistance towards the future establishment of a complete dynamical theory."[2]

Stokes.

MacCullagh's theory purely formal:

Since the date of these remarks the mechanical theory sought for has, I think, been supplied by Lord Kelvin's notion[3] of a medium dominated by some form of molecular angular momentum such as may be typified by spinning gyrostats imbedded in it. The gyrostatic part of the energy of strain of such a medium can be a quadratic function of its elementary twists or rotations, precisely after MacCullagh's form. The conjugate tangential tractions on the faces of a rectangular element of volume, instead of being equal and of the same sign as in the elasticity of solid bodies, are equal and of opposite sign[4], just as Stokes pointed out they would be on MacCullagh's theory. Consequently a framework free of elasticity of its own, and carrying a system of such gyrostatic cells, would be a mechanical representation of an aether which corresponds with MacCullagh's

Kelvin's gyrostatic medium is its realization:

[1] Sir G. G. Stokes, "Report on Double Refraction," *Brit. Assoc.* 1862, p. 227.

[2] Sir G. G. Stokes, *loc. cit.* p. 279.

[3] Lord Kelvin, *Comptes Rendus*, Sept. 1889; *Collected Papers*, Vol. III. 1890, p. 467.

[4] Cf. J. Larmor, "On the Equations of Propagation of Disturbances in gyrostatically-loaded Media," *Proc. Lond. Math. Soc.* Vol. XXIII. 1891. The medium considered in this paper is dominated by simple rotators imbedded in its structure, and the forcive is proportional to angular velocity. Lord Kelvin's new rotational medium is dominated by complex gyrostatic cells, containing arrangements of Foucault gyrostats, of which only the outer cases are firmly imbedded in the medium; and the forcive is proportional to the angular displacement.

Gyro-compass type of element.

expression for the energy-function, and so would afford an explanation of optical phenomena on the lines of his analysis. The axes of the gyrostats will, in crystalline media, be concentrated in certain directions; but in any one direction as many must point backwards as forwards. Any very slight violation of the latter condition will introduce into the medium directed rotational property with respect to the resultant axes of angular momentum; such we may imagine to be the effect of an imposed magnetic field. Non-directed rotational property will be a structural effect, due to mode of aggregation. If the light-disturbance is represented by the displacement of the medium, it will be in the plane of polarization; while if it is represented by the rotation, it will be at right angles to that plane. According to this theory of light, the density of the aether will be the same in all media; but in different media the distribution of angular momentum will vary.

adapted to MacCullagh's analysis,

22. The bodily equations of MacCullagh, when formulated in connection with the boundary conditions appropriate to the theory of the elasticity of solids, which it is, I think, fair to say that their author never intended, and with which, in fact, Stokes pointed out that his whole scheme is inconsistent, have been shown[1] by various writers to lead to a wholly untenable account of reflexion.

which mastered the problem of reflexion.

The investigation of MacCullagh himself, based purely on dynamical analysis, leads him to the boundary conditions which alone are consistent with his scheme, much in the manner of FitzGerald's correlative electrodynamic theory sketched above. These conditions are quite different from the ones appropriate for an elastic solid medium.

The energy of MacCullagh's medium depends only on rotation, and not sensibly on compression. The compressional term can in general be absent only because either (i) there is no resistance offered to pressure, so that no work is done by it, or (ii) the medium is incompressible so that pressure *can* do no work. The tangential tractions on either side of an interface are expressed in terms of rotation, not of distortion as in the elastic solid theory.

Tentative gyro-optic reflexion:

The surface conditions are, however, theoretically too numerous, as MacCullagh knew but did not suffer from in the problem of crystalline reflexion, and as FitzGerald found irremediably in the magneto-optic problem. The way to remove this difficulty is to recognize, according to which of the above views we adopt, either (i) a local play of compression close to the interface which is not propagated away from it, which involves no sensible energy, but which renders it unnecessary to suppose the displacement normal to the

[1] Cf. Lord Rayleigh, "On the Reflexion of Light from Transparent Matter," *Phil. Mag.* 1871.

interface to be continuous, or (ii) a play of pressure which is propagated from the interface with infinite velocity (*i.e.* attains instantly an equilibrium distribution throughout the medium), and which therefore necessitates the modification of the equations of propagation as FitzGerald's equations are modified (*supra*, § 11), λ in that analysis being clearly a hydrostatic pressure when (ξ, η, ζ) represents linear displacement of the medium.

Any actual refracting system is of finite extent, so that the equilibrium state contemplated by (ii) is easily established throughout it: it is only for the simplification of analysis that it is customary to take the interface to be an unlimited plane.

The discussion of crystalline reflexion which is given by MacCullagh takes no account of this pressure λ, but makes an argument in favour of his theory out of the remarkable fact that although there are too many surface conditions compared with the number of variables, yet in no case is the introduction of such a pressure required by the analysis or the optical phenomena, provided the densities of both media are assumed to be the same; while FitzGerald's further application to magneto-optic reflexion simply leaves the continuity normal to the interface unsatisfied, and so far tacitly adopts the first of the above alternatives, that the medium, considered as a mechanical one, offers no resistance to compression—a hypothesis which turns out to be untenable. its difficulties absent for ordinary crystalline media.

23. If these considerations are sound, we have the following conclusions. Summary.

The phenomena of light are explained on MacCullagh's mathematical equations by a theory of pure rotational elasticity, without any accompaniment of the character of the elasticity due to change of volume or change of shape of an ordinary solid body; for linear vibrations the direction of the displacement of the medium is in the plane of polarization of the light, while the axis of its rotation is at right angles to that plane. There is, however, no occasion to take the medium devoid of resistance to compression: it may transmit longitudinal waves with finite velocity, and still no such wave will be produced by the refraction of a transverse wave.

The electric theory of light is formally the same as MacCullagh's theory, magnetic force corresponding to velocity, provided his medium is taken to be incompressible. Electric theory of light:

The labile aether theory of Lord Kelvin is one that contemplates elastic quality depending on compression and distortion, *i.e.* the ordinary elasticity of solid bodies, but the resistance of the medium to laminar compression is taken to be infinitesimal.

The difference between MacCullagh's theory and the electric theory does not, as has been just remarked, affect the problem of propagation in crystalline media, nor does it enter into the question of reflexion at an interface between either isotropic or crystalline media, the boundary conditions being all satisfied without any condensational disturbance; it is not necessary to introduce either (i) interfacial compression or (ii) hydrostatic pressure, according to the two cases above, to preserve the continuity at the interface. But we have already seen that the difference between these hypotheses makes itself felt in the problem of magneto-optic reflexion.

without propagation of electric pressure.

Kelvin's labile aether.

The labile aether theory stands, according to the remark of Willard Gibbs, already quoted, in a relation of precise duality to the electric theory, and therefore also to the other limiting interpretation of MacCullagh's theory, which postulates absence of volume elasticity; the linear displacement in the labile aether corresponds to the rotation in the rotational aether. And here there is a point which demands explanation. The energy-function is the same in both the labile aether and this rotational aether; but the boundary conditions are different, being in the one case those of the elasticity of solids and in the other those of pure rotational elasticity. Yet, in the treatment of the subject proposed here, emphasis is laid, after MacCullagh, on the fact that the energy-function implicitly involves in itself the boundary conditions. This difficulty is elucidated by observing that the expression

For reflexion problem, the energy must be correctly localized.

$$\frac{1}{2}\int B\left\{\left(\frac{d\zeta}{dy}-\frac{d\eta}{dz}\right)^2+\left(\frac{d\xi}{dz}-\frac{d\zeta}{dx}\right)^2+\left(\frac{d\eta}{dx}-\frac{d\xi}{dy}\right)^2\right\}d\tau,$$

given by Lord Kelvin[1] for the potential energy of the labile aether does not represent the localization of the energy, considered as that of an elastic solid. It is in fact derived from the appropriate expression for an elastic solid

$$\frac{1}{2}\int B\left[\left(\frac{d\zeta}{dy}+\frac{d\eta}{dz}\right)^2+\left(\frac{d\xi}{dz}+\frac{d\zeta}{dx}\right)^2+\left(\frac{d\eta}{dx}+\frac{d\xi}{dy}\right)^2 - 4B\left(\frac{d\eta}{dy}\frac{d\zeta}{dz}+\frac{d\zeta}{dz}\frac{d\xi}{dx}+\frac{d\xi}{dx}\frac{d\eta}{dy}\right)\right]d\tau$$

by integration of the second term by parts; and at an interface between different media, a surface term which will be found to be the difference for the two media of the values of the expression

$$\int B\left\{l\left(\frac{d\xi\eta}{dy}-\frac{d\xi\zeta}{dz}\right)+m\left(\frac{d\eta\zeta}{dz}-\frac{d\eta\xi}{dx}\right)+n\left(\frac{d\zeta\xi}{dx}-\frac{d\zeta\eta}{dy}\right)\right\}dS$$

Illustration: *quasi*-tension in surface.

is thus thrown away. Now a superficial distribution of energy is represented mechanically by a surface tension of equal intensity; so

[1] Lord Kelvin (Sir W. Thomson), *Phil. Mag.* 1888.

that a surface tension of this amount, varying from point to point, assists in keeping up the equilibrium of the interfacial layer, in addition to the surface forces indicated by MacCullagh's analysis.

Elucidation of a General Dynamical Principle.

24. A cardinal point in this correlation of different theories is the insistence on the validity of the proper application of MacCullagh's doctrine that the energy-function of a medium, provided it is correctly localized, contains implicitly in it the aggregate of the boundary conditions at an interface between two different media; and that, notwithstanding any apparent discrepancy in continuity that may still be outstanding after the conditions so obtained have been applied to the problem. The same principle had previously been formulated by Green[1], in similar terms; "one of the advantages of this method, of great importance, is, that we are necessarily led by the mere process of the calculation, and with little care on our part, to all the equations and conditions which are *requisite* and *sufficient* for the complete solution of any problem to which it may be applied." On the practical application of this procedure some fresh light may be thrown by the consideration of a quite similar difficulty in the dynamics of actual elastic systems, which has recently occupied the attention of several mathematicians. The vibrations of a curved elastic plate, in fact of a bell supposed of small thickness, have been worked out by Lord Rayleigh[2], simply from the energy-function of the plate. The plate being thin, it can easily be deformed by bending; on the other hand to stretch it sensibly would be very difficult. For this reason the energy-function is formed by Lord Rayleigh on the assumption that the plate is perfectly inextensible, so that terms depending on extension do not occur in its expression. Some years subsequently it was pointed out by Love[3] that this treatment does not allow of all the elastic conditions at the boundary of the plate being satisfied. Now on the principles here expounded the adjustment of these terminal conditions would be made by tensions in the plate, which, owing to the very rapid velocity of propagation of extensional disturbances, practically obey at each instant an equilibrium theory of their own, and at the same time involve the play of only a negligible amount of energy owing to the magnitude of their elastic modulus. If the plate were quite inextensible these tensions would be absolutely

The method of Action:

its power and weight: illustrated in vibrations of a thin bell,

by isolation of the essential features.

[1] George Green, "On the Laws of the Reflexion and Refraction of Light," *Trans. Camb. Phil. Soc.* December 11, 1837; *Math. Papers*, p. 246.

[2] Lord Rayleigh, "On the Infinitesimal Bending of Surfaces of Revolution," *Proc. Lond. Math. Soc.* Vol. XIII. 1882.

[3] A. E. H. Love, *Phil. Trans.* 1888.

in equilibrium at each instant, and the energy-changes involved in them would be null. And this view is, I believe, in agreement with the mode of explanation now generally accepted for that problem[1]. The solution of the problem of vibration of a bell may thus be derived, as regards all things essential, from the energy-function of the bending alone, combined explicitly or implicitly with the geometrical condition of absence of extension.

Critique of Kirchhoff's Theory.

The Kirchhoff interfacial adjustment.

25. The principle implied in MacCullagh's analysis is claimed to be identical, in its results if not in theory, with a hypothesis adopted by Kirchhoff in his discussion of crystalline reflexion[2], which is commonly quoted by German authors under the title of Kirchhoff's principle. Its author employs it avowedly as a formal mathematical representation of assumptions made explicitly by F. Neumann, and tacitly he says by MacCullagh, in their theories, which it is the object of his memoir to reproduce and amplify. He attempts no dynamical justification of its use; on the other hand, he rather formulates it as an additional hypothesis. At any rate it has been treated as a hypothesis by Kirchhoff's followers in Germany, while its validity is suspected by some other writers who have considered the subject. The explanation of Kirchhoff himself in the introductory paragraph of his memoir, in comparing Neumann's and MacCullagh's theories, is here reproduced in a free translation. "Yet at the first glance the points of departure of the two theories would appear to be different, even diametrically opposed to each other. For Neumann starts from the view that the aether in respect of light-vibrations comports itself as an elastic solid, on whose elements no forces act except such as

Neumann and MacCullagh.

are called forth by their relative displacements; while MacCullagh takes for the potential of the forces in operation on the elements of the aether an expression which does not agree with the potential of the forces called into play by the relative displacements of the parts of an elastic solid. Thus in the theory of MacCullagh, if we are to treat the aether as an elastic solid, we must treat it as one which is acted on by forces in addition to those called into play by its elasticity.

Whether both require extraneous tractions on interfaces,

Yet of these other forces it may be proved from the energy-function adopted by MacCullagh that, taken throughout a portion of the aether in a homogeneous body, they reduce to tractions which operate on its surface. We can therefore assert, that the theory of MacCullagh

[1] Cf. A. E. H. Love, *Treatise on Elasticity*, Vol. II. 1893, § 349.

[2] G. Kirchhoff, "Ueber die Reflexion und Brechung des Lichts an der Grenze krystallinischer Mittel," *Abh. der Berlin. Akad.* 1876; *Gesammelte Abhandl.* p. 352.

rests on the hypothesis that on the elements of the aether no forces act except such as are derived from its elasticity; but on the surfaces which form the boundaries of heterogeneous media tractions are imposed which have some other origin. And such tractions must also be contemplated by Neumann's theory; their function is that we are by their aid empowered to leave the compressional wave out of consideration, just as happens in the former theory: they must exist, in order that compressional waves may not be set up in the reflexion and refraction of light-waves. The two theories compared can thus be seen to be in complete accord. I propose to myself to lay before the Academy a treatment of the question from the standpoint of these theories which, I think, is more general and more comprehensive than those that have been given hitherto." These imposed interfacial forces are restricted merely to satisfy the condition that they shall do no work on any element during the actual displacements of the media; they are considered by Kirchhoff to be "tractions from without (*fremden Druckkräfte*) which act on an element of the interface, tractions which, we are accustomed to assert, arise from the forces which the ponderable parts of the two media sustain from the aether."[1] The forces contemplated by Kirchhoff's principle, in order to allow of the condition of incompressibility being satisfied, are thus only interfacial tractions, which form an equilibrating system in so far as they do no work in any displacement actually contemplated. According to the elucidation and extension of MacCullagh's principle which is here proposed, they should be taken to be a system of pressures distributed throughout the media, which do no work for the displacements actually contemplated, and which are in so far equilibrating. These pressures will be discontinuous at an interface; and will hence modify the boundary conditions in the same manner as Kirchhoff's extraneous forces.

when waves of compression are not set up.

A suggested origin:

or alternatively.

26. In the account of Kirchhoff's principle given by Volkmann[2], the view is propounded that such a principle is necessary because the equations of an elastic solid medium, with the addition of a pressure introduced in the manner indicated above (§ 11), will not lead to an account of reflexion which is in accordance with experiment. Quoting from Kirchhoff's lectures on Optics (p. 143), "We have to recognize that the elasticity of the aether is different in the various transparent media, different in glass, for example, from what it is in empty space. We are not in a position to form for ourselves a clear representation as to how the alteration of the elasticity of the aether in glass is

[1] G. Kirchhoff, *loc. cit.*; *Gesammelte Abhandl.* p. 367.

[2] P. Volkmann, *Theorie des Lichtes*, 1891, § 76.

brought about; but still we can say that it is a consequence of *forces* which the elements of the ponderable matter exert on the elements of the aether. As therefore such forces are present they must exert a *direct* influence on the motion of the elements of the aether at the boundary of the glass, though in the interior of the glass they have only an indirect influence in altering the elasticity of the aether. The relations of the direct action of these forces at the surface and in the interior are similar to those which hold with capillary forces, which also are only of influence at the surfaces of fluids, and are not felt in the interior." This quotation has been given at length, as it puts the case precisely. The reply is that it is only a confession of total ignorance as to the distribution of the energy throughout the mass of the media which would permit us to prop up the boundary conditions by extraneous forces in this manner. In the theory of capillarity the surface-tractions are derived from the distribution of energy throughout the mass of the liquid; and if they could not be deduced rationally from some possible volume distribution of energy, it would have to be held that they were erroneous. So here, if Kirchhoff's extraneous surface-tractions cannot be deduced from some energy-function of the complex medium (aether and matter) which is the seat of the undulations, there is absolutely no basis left for them. It will not suffice to say that at the boundary there is interaction between the aether and the matter, and a gradual transition in density caused by the equilibration of such action: if the depth of this layer of transition is a small fraction of the wave-length, the introduction of the energy-function appropriate to it would have but a small influence on the variation of the total energy, and so would not sensibly affect the results. In so far as the introduction of the pressure arising mathematically from the condition of incompressibility will not make an elastic theory work, that theory has simply not been sustained; in various theories above mentioned the introduction of the pressure is efficacious, and they are in so far verified and in a position to be further tested by application to more complicated phenomena.

Illustration from capillary forces:

but not a physical justification:

yet formally adequate.

Although it would seem that Kirchhoff's method cannot be maintained, yet, as he remarks, his formal equations come out the same as those of the rotational theory represented by MacCullagh's equations; so that his detailed development of the problem of crystalline reflexion will be in agreement with MacCullagh's, and holds good so far as it goes.

Neumann's procedure,

27. In the theory of Neumann, which contains one of the first attempts at a rational dynamical treatment of reflexion and refraction, he starts with equations for the strain of an elastic crystalline medium,

of the imperfect type, however, which the then current elastic theory of Navier and Poisson supplied. By assumption of special relations between the constant coefficients of these equations, he obtained a form which led approximately to Fresnel's laws of double refraction[1]. He then applied this form to the problem of crystalline reflexion[2], but found, I suppose, that the six conditions which he recognized as necessary to ensure continuity of displacement and stress at the interface could not all be satisfied. To satisfy them in a case of an ordinary compressible medium would require the introduction of a wave of longitudinal displacement in each medium, set up in the act of refraction; Neumann's medium being incompressible, he did not take account of such waves, and so was in difficulty with his boundary equations. He cut the knot by assuming that the displacement is continuous across the interface, in other words that there can be no rupture of material continuity; and by omitting altogether all conditions of continuity of stress, replacing them by the principle that there is no loss of energy in the act of refraction and reflexion. This, as Kirchhoff remarks, is equivalent to an admission that the equilibrium (or vibrational motion) of an indefinitely thin layer, including in it the interface, is maintained by the aid of forces introduced somehow from outside the vibrating system; but that, as the energy of the incident light is accounted for exactly by that of the reflected and refracted light, these forces must be subject to the condition that they do no work on any element of this surface-layer in the displacements to which the medium is *actually* subjected during the motion. On this basis Neumann obtains Fresnel's equations of reflexion, by aid of the hypotheses that the displacement of a linear wave is in the plane of polarization, and that media differ optically in elasticity but not in density.

and its difficulties:

As we have seen, Kirchhoff adopts and expounds the method initiated by Neumann for getting over the boundary difficulty. But his main argument is that if we do not assume surface forces from without we are helpless, that such forces exist, as is inferred from molecular theory, but that all we know about them is that in their play they cannot absorb any of the energy of the light. His method of procedure would therefore be to assume the most general possible type of such forces subject to this one condition, and then try by special assumption to adjust them to the final result he desires. There is clearly no dynamical validity in this, it is purely empirical;

adopted by Kirchhoff as the best that can be done on elastic-solid theory.

[1] F. E. Neumann, *Pogg. Ann.* Vol. XXV.

[2] F. E. Neumann, *Abhandlungen der Berliner Akademie*, 1835. This memoir proceeds throughout on the method of rays, without explicit consideration of the elasticity of the medium.

the surface forces may really be subject (as we shall see, are subject) to other unknown laws as well, which will not, with the assumed energy-function of the medium, allow of the desired solution. The process would then only prove that the assumed energy-function is untenable.

Energy must be located in the medium:

28. The correct method is the one indicated above. The energy of the medium is associated with the medium in bulk, is located in its elements of volume. In Gauss' theory of capillarity it is true that interfacial energy is contemplated, but that is only the actual excess or defect of the energy in the very thin layer of transition over what its amount would be if the transition was supposed sharp and the density of the energy in the elements of each medium near the surface were unaltered by the neighbourhood of the other medium. It is this portion of the energy that produces superficial effects such as surface tension, though owing to the thinness of the interfacial layer it forms only a very minute fraction of the whole energy, the distribution of the other part being uniform. Now the propagation of vibrations across the interface is an affair of the redistribution of the energy of the medium *en masse*; if we make the ordinary optical hypothesis that the layer of transition is very thin compared with the length of a wave, we may be certain that there is no superficial term of sensible importance in the vibrational energy of the system. The only superficial forces which can come in are, then, those which enter logically in the dynamical analysis of the motion, on the basis of a volume distribution of energy in the medium, the determination of whose form is part of the problem. Until the possibilities of this statement of the problem are exhausted, it would appear to be gratuitous and unscientific to assume the existence of unknown surface forces; and moreover, as these forces could only arise from the existence of a finite layer of transition, so not only would their assumption be purely empirical, but the present method of investigation of the problem of reflexion would actually no longer apply: if there is to be a finite layer of transition, the postulation of material continuity of the media across it by means of a single set of surface conditions would be meaningless.

even for Gaussian capillarity.

No room for interfacial forces.

Crystalline analysis based on Green's form of energy:

29. In the light of these remarks it will be of interest to follow somewhat in detail Kirchhoff's discussion of the general problem of crystalline reflexion and refraction, to find out how far his imposed surface forces satisfy the conditions that we here demand of them, namely, of being deducible from a bodily energy-function. Kirchhoff restricts himself to an elastic solid aether; three sets of waves will thus be possible with a given front; the restriction that the displace-

ment for two of these waves shall be in the plane of the front confines the energy-function to Green's well-known form[1].

He then neglects the first term involving the compression, in Green's formula, on the ground that in the transverse waves the density of the medium remains unaltered[2], so that such a term can have no influence on the equations. If he had definitely omitted this term from the energy, the analysis, as carried out by him without an introduced pressure, would have shown that the function so modified belongs to a medium in which a compressional wave is propagated with null velocity, in fact a medium which (like Lord Kelvin's foam) opposes no resistance to *laminar* compression, though it does resist *uniform* compression with a finite volume-elasticity. Green was not able to do away in this manner with the terms producing a normal wave, because he thought his medium would be unstable; and possibly the same idea suggested Kirchhoff's cautious procedure.

labile instead of incompressible:

This energy-function F, with the compression omitted, is easily expressed, in the notation of § 11, in the form

$$F = U - 2a_{11}\frac{d(\eta, \zeta)}{d(y, z)} + \ldots + \ldots$$

$$- 2a_{23}\left\{\frac{d(\xi, \eta)}{d(x, z)} + \frac{d(\xi, \zeta)}{d(x, y)}\right\} - \ldots - \ldots$$

where

$$2U = a_{11}f^2 + a_{22}g^2 + a_{33}h^2 + 2a_{23}gh + 2a_{31}hf + 2a_{12}fg,$$

(f, g, h) being the curl of the displacement (ξ, η, ζ) of the medium. By integration by parts, all the terms of the volume integral $\int F d\tau$ except U are clearly expressible as surface integrals; while U, the remaining volume distribution, is identical with the *complete* energy-function of MacCullagh's medium. The interfacial part of the energy F, when thus expressed, is, (l, m, n) being direction cosines, the difference in value on the two sides of the interface of the expression

transformed to MacCullagh's form of energy,

$$- 2a_{11}\left(m\eta\frac{d\zeta}{dz} - n\zeta\frac{d\eta}{dy}\right) - 2a_{22}\left(n\zeta\frac{d\xi}{dx} - l\xi\frac{d\zeta}{dz}\right) - 2a_{33}\left(l\xi\frac{d\eta}{dy} - m\eta\frac{d\xi}{dx}\right)$$

$$- 2\left(a_{23}\frac{d\xi}{dx} + a_{31}\frac{d\eta}{dy} + a_{12}\frac{d\zeta}{dz}\right)(l\xi + m\eta + n\zeta)$$

$$+ 2(a_{23}l\xi + a_{31}m\eta + a_{12}n\zeta)\left(\frac{d\xi}{dx} + \frac{d\eta}{dy} + \frac{d\zeta}{dz}\right).$$

by a redistribution with surface terms.

If we take for an instant the plane of (xy) to be the interface, so that $(l, m, n) = (0, 0, 1)$, this expression becomes

$$2\xi\left\{(a_{22} + a_{23} - a_{21})\frac{d\zeta}{dz} - (a_{33} + a_{32} - a_{31})\frac{d\eta}{dy}\right\}.$$

[1] G. Green, *Cambridge Phil. Trans.* 1839.

[2] As explicitly recognized by MacCullagh. See Sir G. G. Stokes' *Report*.

Now on any form of interpretation of MacCullagh's theory, no extraneous interfacial forces at all are required to satisfy the boundary conditions; if the present theory is to agree with it, we might expect that there will be required only interfacial forces such that their activity will for the actual motion just undo the variations of this surface energy. But the boundary conditions of MacCullagh are (§ 9)

$$\xi, \eta, \zeta, \frac{dU}{dg} \text{ and } \frac{dU}{dh}$$

all continuous, where $\int U d\tau$ is the statical energy; and these do not suffice to make this surface energy constant, *i.e.* the time variations of the above expression continuous across the interface. As already remarked, the theories of Kirchhoff and MacCullagh are formally identical; therefore there must be some discrepancy here. It is in fact the circumstance that this surface integral part of the energy has lost its correct location, and does not really belong to the place with which it is now analytically associated.

The discrepancy of Kirchhoff's from MacCullagh's.

Again, Kirchhoff's actual procedure is to take the tractions (X, Y, Z) and (X', Y', Z') on the two sides of the interface that are derived in Lagrange's manner from the energy-function, and to equate to nothing their activity

$$(X - X')\frac{d\xi}{dt} + (Y - Y')\frac{d\eta}{dt} + (Z - Z')\frac{d\zeta}{dt}.$$

Further difficulty:

If ξ, η, ζ are quite independent this will give three boundary conditions just as before, and will be no help. But in the motion to which he restricts himself, ξ, η, ζ are the displacements in a plane-wave, and so are functions of the same linear function of x, y, z and t; he finds that the introduction of this restriction reduces the conditions to two, and so allows further progress.

The reason which Kirchhoff assigns for the two theories of himself and MacCullagh being analytically in agreement is that they can only differ as to boundary conditions, that he gets to a definite theory by his principle of extraneous forces, and that MacCullagh's definite theory also satisfies this principle from the simple fact that there are no extraneous forces. But then the energy-functions are not the same in the two theories. The Fresnel laws of reflexion are obtained by Neumann really by the hypothesis that for rays, *i.e.* for *simple wave-trains*, no loss of energy occurs in the reflexion. This is a much narrower principle than its generalization by Kirchhoff; and, as we have seen, to make his generalization work, the latter has to return practically to Neumann's form in which it is restricted to plane-waves.

returns to Neumann's rays.

These considerations are set forth as showing the artificial character of Kirchhoff's principle, and illustrating the various mistakes and misconceptions which may arise in connection with a subtle point of analytical dynamics, of which the physical bearing has not, I think, been realized by many of the writers on this subject.

In contrast with these explanations, the real reason why the theories of MacCullagh and Kirchhoff agree in their results will now be stated. It is simply that, when ξ, η, ζ are functions of a linear function of x, y, z and t, and therefore are the displacements in a plane-wave of some form, the unmodified expression for Kirchhoff's energy-function F reduces to MacCullagh's energy-function U, the various Jacobian expressions $d(\eta, \zeta)/d(y, z)$, etc., contained in it being all null. For a wave with a spherical or other curved form of front, these terms would not thus disappear; and the boundary conditions could not, I think, be reduced to the proper number by Kirchhoff's process. The conclusion to be drawn from this would be as before mentioned, not that reflexion cannot be explained, but that Green's expression for the energy, as employed by Kirchhoff, is untenable.

We have seen that a labile aether gives results conjugate to, but not the same as, those of the rotational aether corresponding to MacCullagh's equations. It is also known that Neumann's simple theory which can be expressed by means of rays, without technical considerations of elasticity, leads to the same results as MacCullagh's; and we now see that Kirchhoff's method would lead to the same result. Now the elastic solid theory of Kirchhoff is in its elements just the same as the labile aether elastic solid theory; and yet Kirchhoff gets a different result out of it. This demonstrates still further the faultiness of his procedure: he is not entitled to throw away the Jacobian terms in the energy because they happen to be null for the plane-wave kind of motion which he assumes to be the only one to which the reflexion will give rise; though he happens to be led to the correct result by equilibrating them, as he can clearly do *for this particular case*, by extraneous surface tractions of null activity. Further, it thus appears that, according to the form he takes for his extraneous forces, he can arrive from the same *data* at *either* of two conjugate theories of reflexion. **Process indeterminate.**

Mechanical Illustrations of MacCullagh's Theory.

30. The conclusions here arrived at naturally tempt one to pursue the invention of mechanical illustrations of the aether. Lord Kelvin proposes to realize and illustrate his labile contractile aether by a homogeneous mass of foam free from air. Such a medium, when distorted, will have its equilibrium disturbed, and will tend to recover **Elastic qualities of foam:**

itself; when uniformly compressed it will exhibit volume-elasticity. But when it is compressed in one direction only in plane layers, there will be no tendency to recover: its Young's modulus will be null, and so there will exist a fixed ratio between its compressibility and its rigidity, an interesting result which it would be rather difficult to investigate directly. Longitudinal waves will thus not be propagated in the medium.

We have also two types of Lord Kelvin's gyrostatic aethers, one of them with pure rotational elasticity and no compressional or distortional elasticity, the other incompressible but with no distortional elasticity; either of them will represent MacCullagh's equations. A mechanical realization of an aether of the second kind has been proposed by FitzGerald as consisting of a web of long vortex filaments, interlaced together in homogeneous frictionless incompressible liquid, with any desired isotropic or crystalline quality: but even if we could be assured that such a system could subsist, and not be at once hopelessly entangled and destroyed owing to instability, as seems likely, its elasticity would appear at first sight to depend on angular velocity and not on angular displacement, so that it could not have the properties of MacCullagh's aether[1]. Lord Kelvin has recently occupied himself[2] with the dynamics of media composed of gyrostats mounted on framework having various degrees of mechanical freedom. It is possible to imagine frames devoid of distortional elasticity and either incompressible or devoid of compressional elasticity, one of the former class being simply composed of rectangular parallelepipedal webs hinged together, each web consisting of three systems of parallel rods freely jointed at their points of meeting.

of a complex of vortex filaments.

Other structural media.

But we ought not to lose sight of the fact that a gyrostatic aether will be effective, whatever be its modulus of compressibility, provided it has no purely distortional elasticity. Thus FitzGerald's fluid need not be incompressible; an oblique parallelepipedal frame on which to mount the gyrostats will do equally as well as a rectangular frame; and we may also have more complicated forms.

Compressional waves would not disturb light-waves.

The wide field of physical theory which is opened up by this remark that in a rotational aether, however heterogeneous it may be, compressional waves are propagated in perfect independence of rotational waves, must be reserved for future consideration. A generalization of Maxwell's electrodynamic equations has been already proposed and discussed by von Helmholtz, which introduces the

[1] See, however, Lord Kelvin (Sir W. Thomson), "On the Propagation of Laminar Motion through a turbulently-moving inviscid Fluid," *Phil. Mag.* 1887.

[2] Lord Kelvin (Sir W. Thomson), *Collected Papers*, Vol. III. 1890, pp. 466–472.

possibility of compressional disturbances; but that theory is on quite a different footing from the one here suggested, in that Helmholtz's compressional wave interacts with the rotational one, getting mixed up with it at each refraction into a different medium.

The only optical phenomena which the compression can affect, on MacCullagh's theory, appear to be magneto-optic reflexion and possibly other such secondary disturbances, depending on the introduction of terms of higher orders into the energy-function.

35

ELECTRIC VIBRATIONS IN CONDENSING DIELECTRIC SYSTEMS.

[*Proc. Lond. Math. Soc.* Vol. XXVI. (1894) pp. 119–144*.]

Formulation of the problem

1. In forming a theory of rapid electric vibrations, the first point to settle is as to the conditions that may be taken to hold at the boundary of the dielectric medium, where it abuts on a good conductor like a metal. It is well known that when the vibrations are of such short period as free vibrations usually are, the currents in the conductor are confined to mere sheets on the surface. Inside these surface sheets the electric force is null and the magnetic force is null; for if any such forces, of alternating character, existed, there would be currents induced by them, contrary to the fact. The circumstances are thus practically the same as if the conductors were of perfect conducting quality, that is, as if they formed simple cavities devoid of elasticity in the active dielectric medium, with proper boundary conditions over their surfaces. It is now easy to infer what these boundary conditions must be. On the surface there exists an electric current sheet, and also such a free electric charge and such a layer of magnetic poles as are required to satisfy the necessary conditions as to continuity of the fluxes and forces involved in the problem; but there cannot be any electric double sheet on the surface, except the permanent one of chemical origin which enters into the explanation of voltaic potential difference; nor can there ever be a magnetic double sheet. Now by means of these surface layers the actual electric force in the dielectric is to be made to correspond with the null electric force in the conductor. A density of free charge on the surface can always adjust the normal components of electric displacement into agreement; but, in the absence of a special double electric sheet, the tangential components of the electric force must be continuous across the surface, and therefore the tangential force at this boundary in the dielectric must be null. Again, as regards magnetic force, the normal component of magnetic induction must be continuous, by its fundamental character as a flux, therefore the normal component of magnetic force at this

of free vibrations in a partially enclosed dielectric region.

* It has been convenient to put this paper and the next in front of the longer memoir on Electrodynamic Theory, in three parts, with the rotational aether-models of which however they are involved.

boundary of the dielectric must be null; on the other hand, the discrepancy in the tangential components of the magnetic force on the two sides of the interface merely determines the intensity of the current sheet that flows there. This condition of continuity of normal magnetic induction across the surface is not of course an additional one, but is derived at once from the previous condition of continuity of tangential electric force by application of Ampère's circuital relation.

Boundary conditions.

In the equations which follow, dissipation of the skin currents into heat does not therefore appear. That is not because there is no such dissipation*, but because their being confined to the outer skin arises from inductance being much more influential than dissipation, owing to the high period. The vibrations will thus be prolonged for a very large number of periods, but will not, even theoretically, go on for ever, although there will be no radiation when the dielectric is completely surrounded by a conducting medium.

2. Suppose now we attempt to form an analogy by taking the magnetic force to represent the velocity in an elastic medium; as regards its bodily elastic equations this medium will have the properties of an incompressible elastic solid, or of a rotationally elastic incompressible perfect fluid[1], for there is no difference between the two except in the formulae for the tractions on an interface, and therefore in the boundary conditions. The actual boundary conditions in the present problem are that the normal component of the velocity of the medium, and therefore of the displacement, must vanish, while the tangential components are unrestricted.

Elastic solid analogy.

Thus, keeping to the elastic solid analogy as the most vivid for the moment, though as we shall see presently not the real representation, the circumstances of electric vibration in the active medium are of similar type to those of elastic vibrations in an incompressible solid whose boundaries (where the dielectric abuts on conductors) are free to move tangentially but not normally. There are of course no free boundaries because the dielectric extends to infinity all round; but its effective elastic properties may change at an interface where its material constitution changes; at such an interface the elastic solid analogy breaks down, and we must fall back upon the rotationally elastic incompressible fluid which completely represents all the electrical conditions under the most general circumstances.

* The mode of decay of surface density with increasing conductance may readily be worked out as a current sheet soaking into the conductor.

[1] *Phil. Trans.* 1894; or *Proc. Roy. Soc.* 1893–4, "On a Dynamical Theory of the Electric and Luminiferous Medium": or *infra*, p. 432.

Plate condenser:

3. Suppose now we examine the character of the electric vibrations in a simple condenser, consisting of a thin plate of dielectric material, plane or curved, of thickness uniform or varying, separating two conducting bodies*. The case is somewhat analogous to that of the vibrations of an elastic plate of the same form, which is constrained at the edge by the thick compact mass of the surrounding elastic medium; but at its faces, and therefore practically throughout its breadth when that is small compared with its radii of curvature, the movement is confined to be tangential; so that the vibrations are of extensional, but not at all of flexural, character.

As the elastic solid of the analogy is incompressible, on account of the circuital character of the magnetic induction, we may represent the purely tangential displacements of our problem by means of a stream function ψ, which we may consider as belonging to the mean surface of the dielectric plate; in the general problem, in which the plate is not of uniform thickness, it is displacement multiplied by its thickness τ that is derived from the stream function.

with curved faces.

For the general case of a curved sheet, it is clearly proper to employ Gaussian orthogonal coordinates p, q, so that the elementary length on the surface is given by the formula

$$\delta s^2 = h^2\delta p^2 + k^2\delta q^2,$$

where the parameters h, k are functions of position on the surface. The elements of length along the coordinate curves q = constant, p = constant, are

$$\delta x = h\delta p, \quad \delta y = k\delta q,$$

as in the diagram.

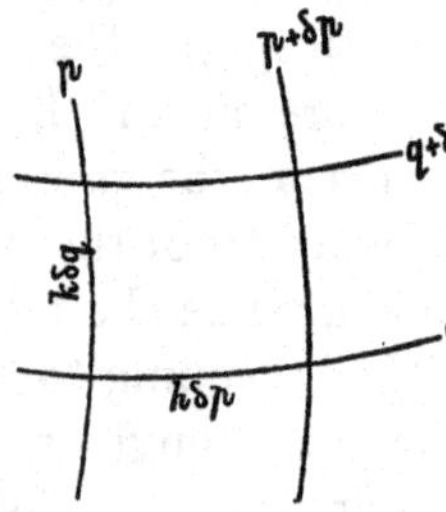

Curvilinear equations of the field.

The magnetic force in the sheet is tangential, and its components are thus

$$\frac{1}{\mu\tau}\frac{d}{dy}\frac{d\psi}{dt}, \quad -\frac{1}{\mu\tau}\frac{d}{dx}\frac{d\psi}{dt}, \quad 0;$$

that is,

$$\frac{1}{\mu\tau k}\frac{d}{dq}\frac{d\psi}{dt}, \quad -\frac{1}{\mu\tau p}\frac{d}{dp}\frac{d\psi}{dt}, \quad 0.$$

On the other hand, the electric force (P, Q, R) is purely normal at the two faces of the plate, and therefore practically throughout its thickness; thus P and Q are each null. By Faraday's circuital law (law of induced electric force) the time-rate of decrease of the magnetic induction is equal to the curl of the electric force; so

* In modern wireless transmitters, the wave-length is usually great compared with the diameter of the condenser, so that it acts as a static accumulator, internal surgings not being sensibly excited.

that the components of the magnetic force are also expressible in the form

$$-\frac{1}{\mu}\left(\frac{d}{dt}\right)^{-1}\frac{dR}{dy},\quad \frac{1}{\mu}\left(\frac{d}{dt}\right)^{-1}\frac{dR}{dx},\quad 0;$$

that is

$$-\frac{1}{\mu k}\left(\frac{d}{dt}\right)^{-1}\frac{dR}{dq},\quad \frac{1}{\mu h}\left(\frac{d}{dt}\right)^{-1}\frac{dR}{dp},\quad 0.$$

While by the other circuital law, that of Ampère, we have the current multiplied by 4π equal to the curl of the magnetic force, so that

$$K\frac{dR}{dt}\delta x\delta y = \frac{d}{dx}\left(-\frac{1}{\mu\tau}\frac{d}{dx}\frac{d\psi}{dt}\delta y\right)\delta x - \frac{d}{dy}\left(\frac{1}{\mu\tau}\frac{d}{dy}\frac{d\psi}{dt}\delta x\right)\delta y;$$

that is

$$R = -\frac{1}{Khk}\left(\frac{d}{dp}\frac{k}{\mu\tau h}\frac{d}{dp} + \frac{d}{dq}\frac{h}{\mu\tau k}\frac{d}{dq}\right)\psi.$$

Equating the two expressions thus obtained for the magnetic force, we have

$$R = -\frac{1}{\tau}\frac{d^2\psi}{dt^2};$$

thus, working with the two circuital electrodynamic laws, there was no occasion to introduce the function ψ,—these laws showing at once that R serves as a stream function for the magnetic force, so that the magnetic equipotential lines are the curves along which R is constant. The function ψ will however be essential presently, when we examine the purely dynamical aspect of the problem.

Eliminating R, the differential equation for ψ is

$$\frac{d^2\psi}{dt^2} = \frac{\tau}{Khk}\left(\frac{d}{dp}\frac{k}{\mu\tau h}\frac{d}{dp} + \frac{d}{dq}\frac{h}{\mu\tau k}\frac{d}{dq}\right)\psi,$$

representing vibratory motion in the condenser layer, unaccompanied by dissipation, as it ought under the circumstances to do.

Before proceeding to a discussion of the types and periods of the vibrations in condenser layers of simple forms, we have still to formulate the mathematical conditions which obtain round the edge of such a layer where it merges in the mass of the surrounding dielectric medium. If we held to the imperfect elastic solid analogy, we should infer that the edge is maintained fixed by the mass of dielectric beyond; and that would give ψ constant and $d\psi/dn$ null along the edge, which are more conditions than can be satisfied by the solution of a vibrational equation of the second order. As however we have seen already, the elastic solid analogy does not extend to the formulae for the tractions in the medium, so that we cannot apply it in this way. Similar difficulties attach[1] to the complete representation by means of a rotationally elastic fluid aether, as well

[1] Cf. *infra*, § 11.

as those associated with the unusual character of the elasticity; at the present stage it is simpler and safer to employ immediate electrical considerations. We therefore direct our attention to the current sheets on the two opposed faces of the condenser, and consider the magnetic field as that due to these currents. At all parts of the plate the currents on its two faces are equal and opposite, and so neutralize each others' effects except in the contiguous part of the plate, when the plate is very thin; at the edge the currents are tangential[1], therefore the magnetic force near the edge of the plate has only a very small, practically vanishing, component in the tangential direction. Thus we may take the condition at the edge to be that the tangential magnetic force is null; that is, $d\psi/dn$ is to be null along the edge. When the plate is not very thin, there will be a correction to be made to solutions thus obtained, which might be calculated according to the same principles as the well-known correction for the open ends of organ pipes in the theory of acoustical vibrations.

Conditions at the open edge, determined.

The general scheme of vibration at which we have arrived is of course in keeping with the characteristic of electric undulations in general, that the electric force and the magnetic force are in the plane of the wave-front and at right angles to each other. Thus here the undulations advance along the dielectric plate, and the electric force is across it; therefore the magnetic force is tangential, as we have seen. We do not consider the other type of vibration in which the wave sways across the plate from one face to the other, and for which the periods would of course be extremely high when the plate is thin.

Type of the electric vibrations:

A very striking result of the theory is that the types and periods of the electric vibrations in a condenser layer are unaffected by any possible deformation of the layer by bending, which does not involve stretching, and does not interfere with the condition of freedom at the edges or the distribution of thickness in the layer. This proposition forms a vivid illustration of Maxwell's fundamental position, that in the analytical formulation of electric phenomena no considerations of action across a distance need enter.

unaffected by bending of the plate without stretching.

[1] Any tendency to flow over, across the edge, to the other face of the coating, would be effectually resisted, on account of the very great increase of electric energy that such a flow would produce. In illustration of the case where the dielectric plate does not come abruptly to a sharp boundary, but gradually widens out into the surrounding medium, the experiments of Righi (*Rend. dei Lincei*, 1893) with two equal spheres close together and sparking into each other may be noticed. The observed values of the wave-length indicate that the adjacent points of the spheres are antinodes and the remote points nodes in the principal electric vibration which surges over their surfaces. The closeness of the spheres, here also, implies a fair amount of capacity and consequent potential energy, and therefore tolerable persistence of the vibrations.

4. The equations of vibration at which we have arrived are the same as those for the vibrations of a sheet of air or gas of the same form and law of thickness as the condenser plate, bounded on each side by rigid walls and with a rigid boundary round its edge, ψ now representing the velocity potential of the air. We can accordingly at once utilize in electric theory the results obtained in the discussion of this problem for spherical sheets by Lord Rayleigh, *Theory of Sound*, Vol. II. Chap. XVIII[1].

Types the same as for an enclosed thin sheet of gas.

In actual problems it is always possible and usually easy to employ a system of conformal coordinates, so that

$$h = k, \text{ and } ds^2 = h^2(dp^2 + dq^2);$$

the equation of vibration becomes

$$\frac{d^2\psi}{dt^2} + \frac{\tau}{Kh^2}\left(\frac{d}{dp}\frac{1}{\mu\tau}\frac{d}{dp} + \frac{d}{dq}\frac{1}{\mu\tau}\frac{d}{dq}\right)\psi;$$

it reduces to its simplest form when $\mu\tau$ is constant over the sheet, and then represents waves travelling with velocity $(K\mu)^{-\frac{1}{2}}$, which is of course the velocity of radiation in the medium of which the plate is composed.

Simplest case.

We proceed to some examples of this type, $\mu\tau$ constant, which includes the case of sheets of uniform thickness and uniform magnetic quality; if we take the electric coefficient K also uniform, the velocity of the waves will be uniform all over the sheet. The equation is now

$$\frac{d^2\psi}{dt^2} = \frac{c^2}{h^2}\left(\frac{d^2\psi}{dp^2} + \frac{d^2\psi}{dq^2}\right),$$

with $d\psi/dn$ null along the edge; where $c^{-2} = K\mu$.

(i) For a flat condenser

Flat condenser plate, rectangular.

$$\frac{d^2\psi}{dt^2} = c^2\left(\frac{d^2\psi}{dx^2} + \frac{d^2\psi}{dy^2}\right).$$

If it is of rectangular form with the origin of the rectangular coordinates (x, y) at one corner, and its sides of lengths a and b, the vibration is of type given by

$$\psi = A \cos px \cos qy \cos(rt + \gamma),$$

where $$r^2 = c^2(p^2 + q^2),$$

[1] In a similar manner, the magnetic transverse vibrations of a cylindrical system correspond to the acoustical vibrations of a uniform plate of air of the same form of section, but with an open edge at which the pressure remains constant. As in the above, the electric force is everywhere in a constant direction parallel to the axis of the cylinder, so that it satisfies the same equation as the velocity potential in the motion of the air; while at a conducting boundary its value is null.

Gas analogues for other types.

There is another type of electric vibration in cylindrical systems in which the electric force is transverse; the magnetic force is now longitudinal and satisfies the equation of a velocity potential in the acoustical problem, where the edge must now be a fixed boundary, as in the text.

and the remaining part of the condition at the edge, $d\psi/dx$ null when $x = a$, $d\psi/dy$ null when $y = b$, gives

$$pa = m\pi, \quad qb = n\pi,$$

when m and n are integers.

Thus $$\psi = A \cos \frac{m\pi}{a} x \cos \frac{n\pi}{b} y \cos \left\{\pi c \left(\frac{m^2}{a^2} + \frac{n^2}{b^2}\right)^{\frac{1}{2}} t + \gamma\right\};$$

so that the period for the type of vibration in which there are $m - 1$ nodal lines in the plate parallel to the side b, and $n - 1$ parallel to the side a, excluding the edges themselves, is

$$\frac{2}{c}\left(\frac{m^2}{a^2} + \frac{n^2}{b^2}\right)^{-\frac{1}{2}}.$$

Cylindrical sheet.

(ii) For a cylindrical condenser of any form of section, which we may always bend into a circular section without altering the problem, we have

$$ds^2 = a^2 d\theta^2 + dz^2,$$

so that $$p = a\theta, \quad q = z, \quad h = 1;$$

thus $$\frac{d^2\psi}{dt^2} = c^2 \left(\frac{1}{a^2}\frac{d^2\psi}{d\theta^2} + \frac{d^2\psi}{dz^2}\right),$$

with $d\psi/dz$ null at the two ends of the cylinder.

Hence we can have standing vibrations round the perimeter of the section, of wave-length a sub-multiple of this perimeter; and standing vibrations along the length of the cylinder, with the two ends both nodes in the electric vibration, so that the wave-length is a sub-multiple of half the length l of the cylinder. For the general type, we may take

$$\psi = A \cos pa\theta \cos qz \cos (rt + \gamma),$$

where $$r^2 = c^2 (p^2 + q^2),$$

and $$pa = m, \quad ql = n\pi,$$

where m and n are integers. Thus

$$\psi = A \cos m\theta \cos \frac{n\pi}{l} z \cos \left\{c \left(m^2 + \frac{n^2\pi^2}{l^2}\right)^{\frac{1}{2}} t + \gamma\right\},$$

so that the period for the type in which there are $2m$ longitudinal nodal lines, and $n - 1$ transverse nodal circles, is

$$\frac{2}{c}\left(\frac{m^2}{\pi^2 a^2} + \frac{n^2}{l^2}\right)^{-\frac{1}{2}}.$$

5. When the dielectric plate is of uniform material but varying thickness,

$$\frac{d^2\psi}{dt^2} = c^2 \frac{\tau}{h^2}\left(\frac{d}{dp}\frac{1}{\tau}\frac{d\psi}{dp} + \frac{d}{dq}\frac{1}{\tau}\frac{d\psi}{dq}\right).$$

For example, if τ is a function of p only, we may write

$$\psi = \chi e^{\iota nq + \iota rt},$$

and χ is to be determined by the equation

$$\tau \frac{d}{dp} \frac{1}{\tau} \frac{d\psi}{dp} + \left(\frac{r^2}{c^2} h^2 - n^2\right) \chi = 0.$$

(i) Thus if the flat coatings of a plate condenser are slightly inclined to each other, we may take the line of intersection of their planes for the axis of y, and (Wedge-shaped plate.)

$$x \frac{d}{dx} \frac{1}{x} \frac{d\psi}{dx} + \left(\frac{r^2}{c^2} - n^2\right) \psi = 0.$$

This equation is the same as

$$\frac{d^2\psi}{dx^2} - \frac{1}{x} \frac{d\psi}{dx} + \left(\frac{r^2}{c^2} - n^2\right) \psi = 0,$$

which reduces to Bessel's form by the substitution $\psi = x^\kappa \chi$. It turns out that $\kappa = 1$, and then

$$\frac{d^2\chi}{dx^2} + \frac{1}{r} \frac{d\chi}{dx} + \left(\frac{r^2}{c^2} - n^2 - \frac{2}{x^2}\right) \chi = 0,$$

so that

$$\chi = A J_{\surd 2} \left(\frac{r^2}{c^2} - n^2\right)^{\frac{1}{2}} x + B J_{-\surd 2} \left(\frac{r^2}{c^2} - n^2\right)^{\frac{1}{2}} x.$$

It would be, however, simpler to adopt independent treatment, which is quite straightforward, on the analogy of the Bessel equation of zero order.

(ii) In the case of circular plates, the appropriate coordinates are polar, and (Circular plate:)

$$ds^2 = dr^2 + r^2 d\theta^2$$
$$= r^2 (r^{-2} dr^2 + d\theta^2),$$

so that

$$p = \log r, \quad q = \theta, \quad h^2 = r^2;$$

hence

$$\frac{d^2\psi}{dt^2} = c^2 \frac{\tau}{r^2} \left(r \frac{d}{dr} \frac{r}{\tau} \frac{d}{dr} + \frac{d}{d\theta} \frac{1}{\tau} \frac{d}{d\theta}\right) \psi,$$

which may be discussed after the same manner as the cases next following.

(iii) In a cylindrical condenser with two coatings of circular section but slightly eccentric, (eccentric:)

$$\tau = \alpha (\beta + \cos\theta),$$

so that

$$(\beta + \cos\theta) \frac{d}{d\theta} \frac{1}{\beta + \cos\theta} \frac{d\psi}{d\theta} + a^2 \left(\frac{r^2}{c^2} - n^2\right) \psi = 0,$$

that is,

$$\frac{d^2\psi}{d\theta^2} + \frac{\sin\theta}{\beta + \cos\theta} \frac{d\psi}{d\theta} + a^2 \left(\frac{r^2}{c^2} - n^2\right) \psi = 0.$$

spherical surface, (iv) If the condenser forms a portion of a spherical surface, on which θ is co-latitude and ω longitude,

$$ds^2 = a^2 d\theta^2 + a^2 \sin^2\theta\, d\omega^2$$
$$= a^2 \sin^2\theta \left(\frac{d\theta^2}{\sin^2\theta} + d\omega^2\right),$$

so that $$p = \log\tan\tfrac{1}{2}\theta,\quad q = \omega,\quad h^2 = a^2\sin^2\theta;$$

hence $$\frac{d^2\psi}{dt^2} = \frac{c^2\tau}{a^2\sin^2\theta}\left(\sin\theta\frac{d}{d\theta}\frac{\sin\theta}{\tau}\frac{d}{d\theta} + \frac{d}{d\omega}\frac{1}{\tau}\frac{d}{d\omega}\right)\psi.$$

special law of thickness, If in this case the law of thickness is

$$\tau = \tau_0 \sin\theta,$$

then $$\frac{d^2\psi}{dt^2} = \frac{c^2}{a^2}\left(\frac{d^2\psi}{d\theta^2} + \frac{1}{\sin^2\theta}\frac{d^2\psi}{\omega^2}\right).$$

The vibration type is clearly

$$\psi = \chi e^{\iota s\omega + \iota nt},$$

where $$\frac{d^2\chi}{d\theta^2} + \left(\frac{a^2 n^2}{c^2} - \frac{s^2}{\sin^2\theta}\right)\chi = 0.$$

results. In particular, for the meridianal type of vibrations $s = 0$, and they take place just as in a flat condensing strip, the wave-length measured along the curved plate being uniform. The two coatings would come into contact at the poles of the spherical surface; but the conditions there, if such a point were included in the system, would be indeterminate. If the sheet form a zone of breadth l measured along the surface, the period when there are $k - 1$ nodal parallels in addition to the free edges, is $2l/kc$, where as above

$$c^{-2} = K\mu.$$

Spherical boundaries in contact: When the coatings form portions of two spherical surfaces, which touch at a pole, $\tau = \tau_0(1 - \cos\theta)$, and the equation becomes

$$\frac{1-\eta^2}{\eta}\frac{d}{d\eta}\eta\frac{d\psi}{d\eta} + \frac{1}{\eta^2(1-\eta^2)}\frac{d^2\psi}{d\omega^2} = \frac{a^2}{c^2}\frac{d^2\psi}{dt^2},$$

where $\eta = \cos\tfrac{1}{2}\theta$. For the purely radial types of period $2\pi/n$, we have, writing κ^2 for a^2n^2/c^2,

$$(\eta^{-2} - 1)\left(\eta\frac{d}{d\eta}\right)^2\psi + \kappa^2\psi = 0;$$

so that, attempting a serial solution

$$\psi = \ldots + A_r\eta^r + A_{r+2}\eta^{r+2} + \ldots,$$

we have $$(r+2)^2 A_{r+2} - r^2 A_r + \kappa^2 A_r = 0,$$

with A_0 arbitrary. Thus we obtain a solution

$$\psi = 1 - \frac{\kappa^2}{2^2}\eta^2 + \frac{\kappa^2 . \kappa^2 - 2^2}{2^2 . 4^2}\eta^4 - \frac{\kappa^2 . \kappa^2 - 2^2 . \kappa^2 - 4^2}{2^2 . 4^2 . 6^2}\eta^6 + \ldots,$$

which converges everywhere except at the point of contact of the spherical surfaces ($\eta = 1$), where the divergence may be considered to express that the contact is equivalent to an undefined sink of electric motions.

The condition along the edge is $d\psi/d\eta$ null, which will give the values of κ, and therefore the periods, by successive approximation; the other pole of the sheet is clearly a node, as it ought to be. For complete spherical surfaces, there are definite vibrations with free periods only when the divergence is avoided by the series terminating, that is when κ is an even integer, say $2m$; the periods are therefore the times required by radiation to traverse an even sub-multiple ($1/2m$) of the circumference of a great circle. An open shell bounded periods.

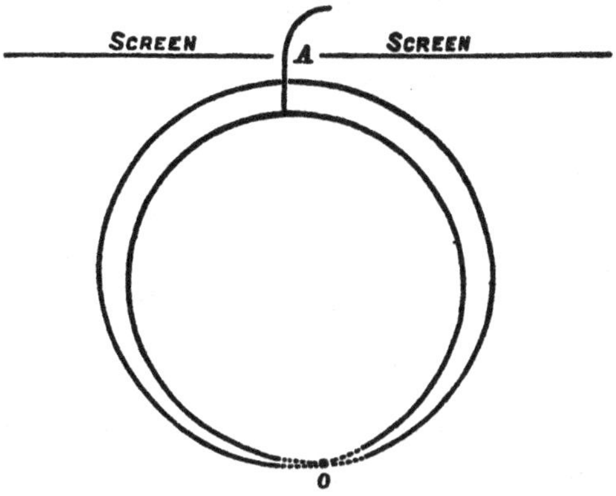

by a nodal line of a free electric vibration in the complete spherical sheet will have the same periods. For $m = 2$, there are no nodal lines except the pole; for $m = 4$, the circle $\cos \frac{1}{2}\theta = \cdot 25$ is nodal; for $m = 6$, the circles $\cos \frac{1}{2}\theta = \cdot 92$ or $\cdot 60$ are nodal.

The diagram above represents a condenser with spherical surfaces which, when completed, would come into contact at the pole O. At the opposite point A a wire is connected with the inner coating but insulated from the outer one. If this wire is connected, after passing through a hole in a metal screen, with one of the knobs of a sparking apparatus, and the outer coating of the condenser is connected similarly with the other knob, free electric oscillations will be set up in the condenser by each spark, and they will be persistent because they can be radiated away but slowly at the edge of the dielectric plate. An arrangement of this kind is not therefore, like the ordinary Hertzian vibrators, a rapidly damped system, but is rather analogous to a pipe or cord in acoustics which maintains its vibrations, once they get started, for a long series of periods without important damping. It would perhaps be better to connect the Persistent oscillations.

coatings with induction plates influenced by the sparking knobs, instead of the knobs themselves. If the free period of the exciter is nearly the same as that of the condenser, it is conceivable that a steady permanent state of vibration might be established[1], at any rate if the exciting sparks could be made to follow each other with sufficient rapidity.

Uniform spherical condenser.

6. There is also the usual class of spherical condensers, which is amenable to more complete treatment, viz. those in which the thickness of the dielectric is uniform and the aperture, if any, is very small. The differential equation is, writing μ for $\cos\theta$,

$$\frac{d}{d\mu}(1-\mu^2)\frac{d\psi}{d\mu}+\frac{1}{1-\mu^2}\frac{d^2\psi}{d\omega^2}=\frac{a^2}{c^2}\frac{d^2\psi}{dt^2};$$

or on substituting as before

$$\psi = \chi e^{\iota s\omega + \iota nt},$$

$$\frac{d}{d\mu}(1-\mu^2)\frac{d\chi}{d\mu}+\left(\frac{a^2n^2}{c^2}-\frac{s^2}{1-\mu^2}\right)\chi = 0,$$

which is the equation of tesseral spherical harmonics. It is known[2] that there is no solution which remains finite all over the complete sphere unless

$$a^2n^2/c^2 = \alpha(\alpha+1),$$

where α is an integer, in which case one of the solutions in series reduces to a finite number of terms. Whether α is integral or not, the solution is

$$\chi = (1-\mu^2)^{\frac{1}{2}s}\left(\frac{d}{d\mu}\right)^s P,$$

where P is the solution of a binomial equation, and thus easily expressible by series.

[1] In this mode of excitation by spark discharge in an influencing system, the exciter, consisting of the condenser *plus* connecting wires *plus* sparker and other accessories, has free, though transitory, periods of its own, which are not to be confounded with those of the dielectric plate, or what is the same thing, those of the swayings of the electric charges on its coatings. For these free swayings the pole A is nodal as well as the edge; and the vibrations are not transitory, for it is only at the edge that they can radiate away. The analogy is with an organ pipe excited by an air blast across a wooden lip, rather than with a pipe excited by a vibrating reed. The condenser might also be excited by waves of proper period travelling across the surrounding medium, like Helmholtz's acoustical resonators, provided the waves were very steady. The intensity of the charge of the condenser affects the strength but not the period of the vibrations.

Analogy to sound-pipes:

The Röntgen rays might conceivably be derived from such persistent oscillations in confined aether spaces within the atom.

and to X-rays.

[2] Lord Rayleigh, *Theory of Sound*, Vol. II. Ch. XVIII.

When α is an integer, one of the two forms of P is the finite series representing the zonal harmonic or Legendre's function of order α.

Thus for the case of a complete sphere, the periods are — Free periods.

$$2\pi a/c\sqrt{(1\,.\,2)},\quad 2\pi a/c\sqrt{(2\,.\,3)},\ \ldots\ 2\pi a/c\sqrt{(\alpha\,.\,\alpha+1)},\ \ldots,$$

whether the vibrations are of zonal or tesseral type. But the nodal lines of the electric vibration ($d\psi/dn$ null) are different in the two cases. For the zonal vibrations involving P_α there are $\alpha - 1$ nodal parallels, including the poles, determined by $(d/d\mu)\,P_\alpha = 0$; for the tesseral vibrations with s nodal meridian circles, there are only $\alpha - s - 1$ nodal parallels determined by

$$(d/d\mu)^{s+1} P_\alpha = 0.$$

We might consider the condenser as terminated by any of these nodal series, so that solutions are thus obtained immediately for a great variety of cases.

For example in the case $P_2 = \frac{1}{2}(3\mu^2 - 1)$, the single nodal line is the diametral circle, and the period of the lowest radial vibration in a hemispherical condenser is thus $2\pi a/c\sqrt{6}$. The other radial periods are $2\pi a/c\sqrt{(4\,.\,5)}$, $2\pi a/c\,(6\,.\,7)$, ..., viz. the alternate ones of the complete set for a sphere. Again, we have types suitable to a hemisphere which involve an even number of nodal meridian circles, and whose periods are $2\pi a/c\sqrt{(2\alpha\,.\,2\alpha+1)}$, where $2\alpha - 1$ is the number of nodal lines on the complete sphere corresponding to the type. And we have also vibration types with an odd number of nodal meridian circles of which the periods are $2\pi a/c\sqrt{(2\alpha+1\,.\,2\alpha+2)}$, where 2α is the total number of nodal lines corresponding to the type on the complete sphere; the largest period of this kind is $\pi a/\sqrt{3}$, corresponding to one nodal meridian circle, and no nodal parallel except the edge of the hemisphere, and this is the lowest free period belonging to the hemispherical condenser. — Hemispherical condenser.

Now suppose that our spherical condenser is not quite complete, but that there is a small aperture at the opposite end of the diameter from A, as in the diagram of the previous case. The effect of this aperture is merely to make its boundary a node instead of the point O, so that the periods are not perceptibly altered if the aperture is small.

7. The solutions hitherto obtained for incomplete spherical surfaces have been special ones derived from the nodal lines of the complete sphere. To attack the general problem we must transform the independent variable in the equation of vibration so as to obtain general solutions, finite at that pole of the sphere which belongs to the condenser; these will necessarily be infinite at the other pole, — General case.

which is outside the condenser, unless in the special case above considered where the series terminates.

To derive series which will be convergent over the whole spherical surface up to the other pole, we write $z = \frac{1}{2}(1 - \mu)$, thus obtaining

$$\psi = (1 - \mu^2)^{\frac{1}{2}s} \left(\frac{d}{d\mu}\right)^s P,$$

where $$z(1 - z)\frac{d^2P}{dz^2} + (1 - 2z)\frac{dP}{dz} + a(a + 1)P = 0,$$

$a(a + 1)$ representing as before n^2a^2/c^2, but a not now being integral.

The solution which remains finite when $z = 0$ is Murphy's form[1], analogous to the series in § 5 (iv),

$$P = 1 - \frac{a \,.\, a + 1}{1 \,.\, 1}\frac{1 - \mu}{2} + \frac{a - 1 \,.\, a \,.\, a + 1 \,.\, a + 2}{2!\,2!}\left(\frac{1 - \mu}{2}\right)^2 + \dots$$

$$\dots + (-)^r \frac{a - r + 1 \,.\, a - r + 2 \dots a + r}{r!\,r!}\left(\frac{1 - \mu}{2}\right)^r + \dots.$$

The equation for the period is obtained by making $d\psi/dn$ null at the edge of the sheet. Unless the sheet is more than a hemisphere, a rapid approximation to the periods may be made, as the series that occurs in the equation converges fairly. But when the sheet is nearly a complete sphere the convergence is very slow, and we must either attempt to transform P into a semi-convergent series by the method of Kummer and Stokes, or else we may work with a definite integral type of solution. The former method seems to be inapplicable, or at any rate it is not easy to transform the equation to a binomial type, after the intrinsic singularity has been reduced by introducing a logarithmic factor multiplying the dependent variable. For attempting the latter procedure, we have available C. Neumann's solution

$$P = \int_0^\infty \frac{\cosh\,(a + \frac{1}{2})\,\phi}{(\mu + \cosh\phi)^{\frac{1}{2}}}\,d\phi,$$

as this is finite when $\mu = +1$ and infinite when $\mu = -1$.

Results quoted for circular boundaries.

8. For the important case of flat circular plates, including implicitly semicircular plates and various sectors, and also conical sheets bounded by circles, the analysis is given with full numerical results by Lord Rayleigh in his treatment of the identical problems of the vibrations of a circular sheet of air, and the oscillations under gravity of water in a cylindrical vessel[2]. The wave-length of the vibrations in the circular dielectric plate is its circumference divided by x, where x is a root of $J_n'(z) = 0$; and a numerical table of these roots is given. The longest wave-length corresponds to $x = 1{\cdot}841$, and represents a

[1] Murphy, *On Electricity*, 1833, Preliminary Propositions.

[2] *Theory of Sound*, Vol. II. § 339; "Water-waves in Cylinders," *Phil. Mag.* April 1876; also *Nature*, July 29th, 1875.

swaying backwards and forwards along the direction of a diameter; the next, $x = 3{\cdot}054$, is a vibration of the same kind, with an antinode in the middle instead of a node; the next, $x = 3{\cdot}832$, is a radial vibration. In all cases, in passing from the dielectric plate to the surrounding atmosphere, the wave-length is increased in the ratio of the velocity of propagation in air to that in the plate, that is of unity to $K^{\frac{1}{2}}$ when the plate is non-magnetic.

The plates of a condenser may be divided across, along nodal lines, without interfering with the types or periods. A condenser with a guard ring will thus vibrate as two separate condensers, without sensible interference between them.

9. One application of the present formulae might be to the determination of the effective value of K for vibrations of the period under consideration, by tuning condensers with various dielectrics so as to be in unison.

From this point of view the case in which the dielectric in a horizontal plate condenser is a composite one, consisting say of a layer of alcohol or water below and a layer of air above, merits consideration, as on it might perhaps be based a method of measurement of the dielectric constants of badly conducting electrolytes. If we try to proceed as in § 3, the electric force R will not be constant across the plate, so we must employ instead the electric displacement P $(= KR/4\pi)$ which is constant. It will be found however that the conditions cannot be all satisfied, so that the magnetic force cannot now be tangential right across the plate. Composite flat plate.

We must therefore introduce a coordinate to represent depth in the plate, and employ the ordinary three-dimensional equations. But we may still escape from the analytical intricacies of the general problem if we confine ourselves to plane waves in a flat rectangular condenser, or indeed to any other case of symmetry in which the magnetic induction may be specified by means of a stream function.

Let us consider these plane waves in a rectangular condenser; let τ_1 be the thickness of the part of the dielectric which is of inductive capacity K_1, τ_2 that of K_2, so that the thickness of the plate is $\tau = \tau_1 + \tau_2$; we may take the value of μ to be uniform throughout, the materials being non-magnetic. Let the two velocities of propagation be c_1, c_2, so that $c_1^{-2} = K_1\mu$, $c_2^{-2} = K_2\mu$; we have, in each medium, equations of the type

$$\frac{d^2\alpha}{dt^2} = c^2\nabla^2\alpha, \quad \frac{d^2\gamma}{dt^2} = c^2\nabla^2\gamma,$$

where
$$\alpha = \frac{d\psi}{dz}, \quad \gamma = -\frac{d\psi}{dx};$$

the tangential magnetic force and normal magnetic induction are continuous at the interface, that is $d\psi/dx$ and $d\psi/dz$ are so; while the normal magnetic force is null at each boundary, that is $d\psi/dx$ is so. Thus measuring z from the interface

$$\frac{d^2\psi}{dt^2} = c^2 \left(\frac{d^2\psi}{dx^2} + \frac{d^2\psi}{dz^2}\right);$$

giving $$\psi = \chi e^{-\iota nt + \iota mx}, \quad \frac{d^2\chi}{dz^2} - \left(m^2 - \frac{n^2}{c^2}\right)\chi = 0;$$

so that for a plane wave we may take

$$\psi = \sin(mx - nt)(Ae^{\kappa z} + Be^{-\kappa z}), \quad m^2 - n^2/c^2 = \kappa^2.$$

At the interface,

$$A_1 + B_1 = A_2 + B_2, \quad (A_1 - B_1)\kappa_1 = (A_2 - B_2)\kappa_2.$$

At the faces of the plate,

$$A_1 e^{\kappa_1 \tau_1} - B_1 e^{-\kappa_1 \tau_1} = 0, \quad A_2 e^{-\kappa_2 \tau_2} - B_2 e^{\kappa_2 \tau_2} = 0.$$

Thus from the latter pair of equations,

$$A_1 = \lambda_1 e^{-\kappa_1 \tau_1}, \quad B_1 = \lambda_1 e^{-\kappa_1 \tau_1}, \quad A_2 = \lambda_2 e^{\kappa_2 \tau_2}, \quad B_2 = \lambda_2 e^{-\kappa_2 \tau_2};$$

where, from the former pair,

$$\lambda_1 \cosh \kappa_1 \tau_1 = \lambda_2 \cosh \kappa_2 \tau_2, \quad \kappa_1 \lambda_1 \sinh \kappa_1 \tau_1 = -\kappa_2 \lambda_2 \sinh \kappa_2 \tau_2,$$

so that $$\kappa_1 \tanh \kappa_1 \tau_1 + \kappa_2 \tanh \kappa_2 \tau_2 = 0,$$

or, written at length,

$$(m^2 - n^2 c_1^{-2})^{\frac{1}{2}} \tanh (m^2 - n^2 c_1^{-2})^{\frac{1}{2}} \tau_1 + (m^2 - n^2 c_2^{-2})^{\frac{1}{2}} \tanh (m^2 - n^2 c_2^{-2})^{\frac{1}{2}} \tau_2 = 0,$$

an equation to determine the period $2\pi/n$ of plane waves of length $2\pi/m$, travelling along a composite dielectric plate.

If either of the quantities represented by κ_1 and κ_2 is imaginary, there will be nodal planes parallel to the plate; the lowest period of a given plate has of course no such nodes.

10. The scope of the present method may also be illustrated by applying it to the case of an ordinary Leyden, with flat base of thickness τ_1 and radius a, and cylindrical sides of thickness τ_2 and length l. If the transition between the thicknesses τ_1 and τ_2 is not very abrupt compared with either of them, there will be no important disturbance by reflexion or otherwise at that place. In the circular base

Leyden jar.

$$\frac{d^2\psi}{dt^2} = \frac{c^2}{r^2}\left(r\frac{d}{dr}r\frac{d}{dr} + \frac{d^2}{d\theta^2}\right)\psi;$$

in the cylindrical sides

$$\frac{d^2\psi}{dt^2} = \left(\frac{d^2}{dr^2} + \frac{1}{a^2}\frac{d^2}{d\theta^2}\right)\psi;$$

at the junction between them $d\psi/dr$ and $d\psi/rd\theta$ must be continuous; and at the free edge $d\psi/dr$ must be null.

Let $$\psi = \chi e^{\iota nt+\iota m\theta};$$

then in the base $$\frac{d^2\chi_1}{dr^2} + \frac{1}{r}\frac{d\chi_1}{dr} + \left(\frac{n^2}{c^2} - \frac{m^2}{r^2}\right)\chi_1 = 0,$$

in the sides $$\frac{d^2\chi_2}{dr^2} + \frac{n^2}{c^2} - \frac{m^2}{a^2}\chi_2 = 0.$$

Thus $$\chi_1 = AJ_m\left(\frac{nr}{c}\right),$$

there being no source at the origin; and

$$\chi_2 = B\cos\left(\frac{n^2}{c^2} - \frac{m^2}{a^2}\right)^{\frac{1}{2}}(r - a - l)$$

with the conditions at the junction,

$$AJ_m\left(\frac{na}{c}\right) = B\cos\left(\frac{n^2}{c^2} - \frac{m^2}{a^2}\right)^{\frac{1}{2}} l,$$

$$A\frac{n}{c}J_m'\left(\frac{na}{c}\right) = -B\left(\frac{n^2}{c^2} - \frac{m^2}{a^2}\right)^{\frac{1}{2}}\sin\left(\frac{n^2}{c^2} - \frac{m^2}{a^2}\right)^{\frac{1}{2}} l.$$

Hence the free periods are $2\pi/n$, where

$$J_m\left(\frac{na}{c}\right)\left(\frac{n^2}{c^2} - \frac{m^2}{a^2}\right)^{\frac{1}{2}}\tan\left(\frac{n^2}{c^2} - \frac{m^2}{a^2}\right)^{\frac{1}{2}} l + \frac{n}{c}J_m'\left(\frac{na}{c}\right) = 0,$$

$2m$ being the number of radial nodal lines.

11. The investigation (§ 3) of the differential equations of the problem of a condenser plate above considered can be put into a purely dynamical form, which will conveniently illustrate the dynamical aspect of the electromotive equations in the theory of electromagnetism. We assume as *data*, as in fact kinematic relations of the system, that the magnetic induction is circuital, and so is derived in the present problem from a stream function ψ, and that the electric current is derived from the magnetic induction by Ampère's law. The other circuital law of Faraday should now follow as a dynamical consequence of Maxwell's expressions for the kinetic and potential energies, (Analysis by method of Action.)

$$T = \iint \frac{1}{8\pi\mu\tau}\left\{\frac{1}{k^2}\left(\frac{d^2\psi}{dq\,dt}\right)^2 + \frac{1}{h^2}\left(\frac{d^2\psi}{dp\,dt}\right)^2\right\} hk\,dp\,dq,$$

$$W = \iint \frac{\tau}{8\pi K}\frac{1}{h^2k^2}\left\{\frac{d}{dp}\left(\frac{k}{\mu\tau h}\frac{d\psi}{dp}\right) + \frac{d}{dq}\left(\frac{h}{\mu\tau k}\frac{d\psi}{dq}\right)\right\}^2 hk\,dp\,dq,$$

and should form a confirmation of the validity of these expressions.

The vibrations of the dielectric layer are to be derived from these expressions by the dynamical principle of least action

$$\delta \int (T - W)\, dt = 0,$$

in which the symbol of variation δ refers to ψ alone. On making the variation, integrating by parts in the usual manner to get rid of differential coefficients of $\delta\psi$ except at the limits of the integral, and equating to zero the coefficient of this purely arbitrary variation $\delta\psi$ in the resulting form, we have the equation of vibration

$$\frac{d^2}{dt^2}\left(\frac{d}{dp}\frac{k}{\mu\tau h}\frac{d}{dp}+\frac{d}{dq}\frac{h}{\mu\tau k}\frac{d}{dq}\right)\psi$$

$$= \left(\frac{d}{dp}\frac{k}{\mu\tau h}\frac{d}{dp}+\frac{d}{dq}\frac{h}{\mu\tau k}\frac{d}{dq}\right)\frac{\tau}{Khk}\left(\frac{d}{dp}\frac{k}{\mu\tau h}\frac{d}{dp}+\frac{d}{dq}\frac{h}{\mu\tau k}\frac{d}{dq}\right)\psi,$$

wherein K, μ, τ may all be any functions of position on the mean surface of the condenser. This equation may now be split into two, and simplified by the introduction of a subsidiary independent variable, such as R in § 3.

We thus obtain

$$\chi = -\frac{\tau}{Khk}\left(\frac{d}{dp}\frac{k}{\mu\tau h}\frac{d}{dp}+\frac{d}{dq}\frac{h}{\mu\tau k}\frac{d}{dq}\right)\psi,$$

and

$$\frac{d^2\chi}{dt^2} = \frac{\tau}{Khk}\left(\frac{d}{dp}\frac{k}{\mu\tau h}\frac{d}{dp}+\frac{d}{dq}\frac{h}{\mu\tau k}\frac{d}{dq}\right)\chi.$$

On substitution for χ from the first of these equations in the left side of the second, and integration, there follows

$$\psi = -\left(\frac{d}{dt}\right)^{-2}\chi + F(p, q),$$

where $F(p, q)$ denotes some function of the coordinates that would be a stream function, for steady electric flow without internal sources, on either coating of the condenser. Thus, the value of χ being derived as above from the solution of the second equation, the value of ψ deduced from it will involve in addition this function $F(p, q)$.

The rotationally elastic analogy is wider than electric:

The dynamical problem is therefore more general than the present electrical one[1]. For instance, if we take T and W to represent the

[1] (It is not implied here that the representation by means of a rotationally elastic fluid aether is in any way defective. If it were possible practically to prescribe the velocity of the aether round the edge of the plate, this analysis would just be sufficiently wide to determine the resulting motions; and these would consist of an elastic vibrational part and an irrotational fluid motion. But it is not possible to impose a velocity on the aether at the edge except by introducing an extraneous magnetic field; and the analysis simply states that throughout the plate this field is superposed on the electromagnetic vibrations if it is steady, and modifies them in the ordinary manner if it is variable.)

energy of a rotationally elastic fluid aether, the additional term $F(p, q)$ will represent an irrotational flow in the aether, which as we know will excite no elastic reactions and so will not interfere with the equation of vibrations. If, on the other hand, we imagine the aether to be an elastic solid, its potential energy W will include other terms in addition to those above expressed; but as they can be integrated into an expression relating only to the boundary, they will not affect the equation of propagation; in this case $F(p, q)$ will represent an irrotational strain. We might by aid of this function satisfy in either case the conditions necessary to make the edge of the plate fixed, viz. $d\psi/dn$ null and also ψ constant all along it; for the case of a cylindrical condenser of length l the periods of the vibrations which have $2m$ longitudinal nodal lines would then come out to be

$$\frac{2}{c}\left(\frac{m^2}{\pi^2 a^2}+\frac{n^2}{l^2}\right)^{-\frac{1}{2}},$$

where n is a root of the equation

$$\tanh ml \tan nl = \frac{2mn}{m^2 - n^2},$$

instead of being an integer, thus differing from the result in § 4 (ii) to which the direct electric equations led. Thus the ordinary electric equations imply that in a rotational fluid aether no irrotational flow is associated with electric vibrations relating to fixed conductors; they also imply, on this and many other grounds, that the aether is not of elastic solid constitution. On the rotational theory, flow of the aether is of course associated with a steady magnetic field; the question as to whether it is actually influenced by the movements of matter must be decided, at the present time, on grounds of optical theory and observation[1].

unless inertia of aether is very great.

The problem of electric vibrations is thus only a special problem in the dynamical theory of the electrical and optical medium; the latter is wider, because, not to mention the electromagnetic forcive on material bodies, it must include the theory of the electrics and optics of moving media, and possibly the theory of the atoms of matter themselves considered as intrinsic singularities, of motion or strain or both together, existing in the fundamental structureless medium.

[1] The term $F(p, q)$ enters in the analysis as representing a part of the magnetic force which is independent of, and not excited by, the electric force,—in other words a part of the motion of the aether which is not directly traceable to electric strain. It implies, and is derived from, activity of a hydrostatic pressure in the aether, which is probably operative only in molecular problems, the aether in bulk being on this theory stationary except in a magnetic field. Cf. *Phil. Trans.* 1894, A, p. 790: *infra*, p. 450.

Analogies with vibrating elastic solids.

It may be noticed that the elastic solid problem, whose solution has just been stated for the case of a cylinder, is identical with the problem of the extensional, wholly non-flexural, vibrations of a plane or curved elastic plate, which has been treated by Mr A. E. H. Love[1] under the much more difficult circumstances which obtain when the material is not incompressible. Such vibrations are too high in pitch and too difficult of excitation to be of acoustical importance; but if the aether were like an elastic incompressible solid they would be the ones we should have to deal with here, except that the more usual case would involve as above the simpler conditions appertaining to a fixed edge in place of those of a free one.

This mode of forming a concrete representation of electric vibrations is explained at some length, in *Proc. Camb. Phil. Soc.* Vol. VI. 1890, "On a Mechanical Representation of a Vibrating Electrical System and its Radiation."* The analogy there developed is however the *conjugate* one in which electric force represents the velocity in the elastic-solid medium; the surface conditions at a conductor are then clearly that the horizontal velocity vanishes while the normal movement is unrestricted. Though that scheme is more remote from the actual electrical conditions, it has the advantage for intuitional purposes that the surface relations are completely represented by the presence of a very thin skin of much more powerful elasticity, supposed to exist on the elastic solid of the analogy. The fact that we can always obtain conjugate representations, of the same scope but of totally different character, by replacing magnetic force by electric force as the independent variable, is very fundamental in this kind of theory; and use has been made of it in various ways by Willard Gibbs, Drude, and other writers. It was clearly indicated and utilized in Maxwell's original paper "On the Electromagnetic Field," *Phil. Trans.* 1864.

Waves along cable condensers:

12. The propagation of electrical waves in the dielectric of a long cylindrical condenser like a submarine electric cable may be discussed in a similar manner. The lines of electric force are transverse, passing radially from the core to the outer sheath across the dielectric. If the frequency of the vibrations is sufficiently great, the currents will not penetrate far into the conductors, and there will also be little viscous loss of energy; so that the dielectric layer will act like a speaking tube in acoustics, maintaining the vibrations in force, as they travel along it.

This propagation is of course very different from the ordinary working of submarine cables, when the alternations are so slow that

[1] *Phil. Trans.* 1888.

* *Supra*, *p.* 221.

electric inertia hardly counts, and the viscous forces of conduction, which are the only ones in action, make the propagation of diffusive type like the conduction of heat, instead of undulatory type as here.

It might be supposed that long submarine cables would possibly be adaptable for telephonic purposes by making the incident sound waves act, in the manner of a relay, on a spark gap, and so set loose electrical vibrations which would be propagated along the cable and received at the other end of it; but the discussion which follows negatives such an idea*.

In examining this point we shall also incidentally observe the order of frequency which is necessary to restrict the importance of the dissipative terms, so that the conclusions of the above discussion, in which they do not appear, may be valid. When as here the penetration into the conductor is to a small depth compared with its radius of curvature, its surface may be treated as plane.

The equations of propagation, for conductivity κ', are

$$\left(K\frac{d^2}{dt^2} + 4\pi\kappa'\frac{d}{dt}\right)(P, Q, R) = \mu^{-1}\nabla^2(P, Q, R).$$

It will suffice to consider the case in which Q vanishes; there is then a current function ψ, so that, x being parallel and z perpendicular to the interface, and the dielectric sheet being thin,

$$(P, R) = \left(\frac{d}{dz}, \quad -\frac{d}{dx}\right)\psi;$$

and the characteristic equation of the problem is

$$\left(c^{-2}\frac{d^2}{dt^2} + 4\pi\kappa\frac{d}{dt}\right)\psi = \left(\frac{d^2}{dx^2} + \frac{d^2}{dz^2}\right)\psi,$$

where c^{-2} stands for $K\mu$ and κ for $\kappa'\mu$. In non-magnetic matter such as the dielectric μ is unity; and c is very great, being of the order of the velocity of radiation.

For a wave of period $2\pi/n$ travelling along the dielectric layer,

$$\psi = e^{\iota m'x - \iota nt}\chi,$$

where m' is now complex, as we suppose the wave to be damped by conduction in the metals. We have therefore

$$\frac{d^2\chi}{dz^2} - (m^2 - 4\pi\kappa n\iota)\chi = 0,$$

where $$m^2 = m'^2 - n^2/c^2.$$

If the plane of xy be taken along the middle of the dielectric layer, of small breadth $2a$, then by symmetry we have in the dielectric **soakage into conducting coatings:**

$$\chi_1 = A(e^{mz} + e^{-mz}),$$

* As Quincke found later by experiment.

and in the upper conductor

$$\chi_2 = B'e^{-rz},$$

where $$r^2 = m_2^2 - 4\pi\kappa n\iota;$$

so that very approximately

$$r = (2\pi\kappa n)^{\frac{1}{2}} \left\{1 - \iota + \frac{m_2^2}{8\pi\kappa n}(1+\iota) + \ldots\right\},$$

in which the last term may also be neglected because n/m_2 is of the order of the velocity of radiation.

Thus in the dielectric plate

$$\psi_1 = A\,(e^{mz} + e^{-mz})\, e^{\iota m'x - \iota nt};$$

in the upper conductor

$$\psi_2 = Be^{-(2\pi\kappa n)^{\frac{1}{2}}(1-\iota)(z-a)}\, e^{\iota m'x - \iota nt}.$$

The constants are to be determined by the conditions that at the interface $\left(\frac{K}{4\pi}\frac{d}{dt} + \kappa'\right)\psi$ and also $d\psi/dz$ are to be continuous. Thus we have exactly, κ' being null in the dielectric,

$$-\iota nKA\,(e^{ma} + e^{-ma}) = 4\pi\kappa' B, \quad mA\,(e^{ma} - e^{-ma}) = -rB;$$

and the equation for m is therefore

$$ma\,\frac{e^{ma} - e^{-ma}}{e^{ma} + e^{-ma}} = \frac{nKra}{4\pi\kappa'}\,\iota.$$

If there were no dissipation m would vanish; so that to our order of approximation m is small and c is the velocity of propagation. Thus

$$(ma)^2 = \pi a\left(\frac{\mu}{c\kappa'\lambda^3}\right)^{\frac{1}{2}}(1+\iota),$$

so that $$\left(\frac{m'\lambda}{\pi}\right)^2 = 4 + \frac{1}{\pi a}\left(\frac{\mu\lambda}{c\kappa'}\right)^{\frac{1}{2}}(1+\iota),$$

or $$m' = \frac{2\pi}{\lambda}\left\{1 + \frac{1}{8\pi a}\left(\frac{\mu\lambda}{c\kappa'}\right)^{\frac{1}{2}}(1+\iota)\right\}.$$

decay of the train of short waves: The exponential coefficient of decay along the wave-train is therefore

$$\frac{1}{4a}\left(\frac{\mu}{c\kappa'\lambda}\right)^{\frac{1}{2}};$$

it is diminished equally by increase of conductivity κ' or diminution of magnetic permeability μ in the metal, or by increase of the wave-length, which however must not be too small if this analysis is to apply.

very rapid. For copper, $\kappa' = 1600^{-1}$, and if λ is one metre and the thickness $2a$ of the dielectric sheet one centimetre, the amplitude would thus be reduced in the ratio of 1 to e after travelling about 100 metres, that is after 100 vibrations. Thus oscillations of this kind, though enormously persistent compared with ordinary Hertzian waves, are nothing like so persistent as ordinary sound waves; nor can they be transmitted very far along a dielectric cylinder without sensible loss.

The exponential coefficient of penetration into the metal is the real part of r, which is $2\pi\,(c\mu\kappa'/\lambda)^{\frac{1}{2}}$, being independent of the thickness of the dielectric plate; this is about 3×10^3 under the above circumstances, so that the current is reduced in the ratio of 1 to e at a depth of about 3×10^{-4} centimetres, which of course amply justifies the procedure of the previous part of this paper.

13. It may be convenient to briefly treat on these lines the problem of electric vibrations in other systems than dielectric shells, a subject already referred to in the footnote to § 4.

The periods of a circular cylinder of dielectric, of radius a, with a conducting boundary, are easily expressed; when the electric force is longitudinal and the magnetic force transverse, they are given by the roots of $J_0\,(na/c) = 0$; when the magnetic force is longitudinal and the electric force transverse, by the roots of $J_0'\,(na/c) = 0$. The wave-lengths in free aether of the two types are the diameter of the cylinder divided by ·765, 1·757, 2·754, ... and 1·220, 2·233, 3·238, ... respectively (Stokes, *Camb. Trans.* Vol. IX.; Rayleigh, *Sound*, § 206). Both classes of vibrations should be excited by sparking between two knobs in the middle of a hollow metal cylinder whose length is several times its diameter; and in analogy (*supra*) with the air vibrations excited in a pipe by tapping it, they will last a considerable number of periods before disappearing by radiation from the open ends. On the same analogy also, shortening the cylinder should somewhat increase the periods; for it diminishes the constraint. **Free periods for solid dielectric cylinder with conducting boundary:**

The more general problem of the vibrations excited in a dielectric region bounded by a conducting surface of revolution is also amenable to similar treatment. There are two types of vibration symmetrical with respect to the axis; in one the lines of magnetic force are circles round the axis, and the lines of electric force are curves in the meridian planes; in the other *vice versa*. **for surface of revolution.**

In the first case if H is the intensity of the magnetic force, the component, parallel to the axis, of the electric displacement is $dH/4\pi d\rho$, where ρ denotes distance from the axis. Thus the equation satisfied by H is*

$$\left[\left(\frac{d^2}{dt^2} - c^2\nabla^2\right)\frac{dH}{d\rho} = 0, \text{ not}\right] \frac{d^2H}{dt^2} = c^2\nabla^2 H,$$

and the condition of continuity of the tangential electric force at the

* This discrepancy and its results were pointed out to the writer by the late Lord Rayleigh.

boundary makes dH/dn null at the boundary. The vibration types and periods are thus [not] precisely the same as those of a gas in a rigid envelope of the same form, H corresponding to the velocity potential of the gas. The periods are well known (Rayleigh, *Sound*, § 331) for the case of a spherical envelope, including a hemispherical envelope and various other types as sub-cases; and by combining symmetrical vibrations relating to different axes the most general class of vibrations in spheres is arrived at, the magnetic lines of force always lying on concentric spherical surfaces, and the periods being of course unaltered. If the exciting spark gap lies along a radius, the vibrations excited in the dielectric sphere will be of this type, symmetrical round that radius; and the boundary condition will not be essentially modified if an aperture is made in the conducting boundary at either pole; thus the period will be practically unaltered while the vibrations are slowly radiated away through this aperture.

Analogy to sound in gas.

In the second type of vibrations the intensity E of the electric force will satisfy the equation

$$\left[\frac{d^2}{dt^2}\frac{dE}{d\rho} = c^2\nabla^2\frac{dE}{d\rho}, \text{ not}\right] \frac{d^2E}{dt^2} = c^2\nabla^2E,$$

while over the boundary E will be constant. In a spherical boundary the periods are thus easily determined by the same analysis as applies to the previous case; they will [not] be intermediate between those of the previous set, the lowest corresponding to a wave-length 1·4 times the radius, instead of 3·02 times the radius which gives the lowest period of the previous set. These vibrations will not be sensibly excited when the spark gap lies along a radius; and their periods will not be sensibly altered if an aperture is made in the conducting boundary at an antinode of the electric force so as to admit of radiation into the outside space.

36

THE SIGNIFICANCE OF WIENER'S LOCALIZATION OF THE PHOTOGRAPHIC ACTION OF STATIONARY LIGHT-WAVES.

[*Philosophical Magazine* (Jan. 1895), pp. 97–106*.]

THE experiments by which Wiener demonstrated[1] that, when stationary plane-polarized optical undulations are produced in a photographic film, by reflexion of a stream of incident plane-polarized light at a metallic or other backing, the photographic action occurs at the antinodes of Fresnel's vibration-vector and not at the nodes, have been employed by its author and others to decide between the various theories of light. If for purposes of precise description we utilize the terminology of the electric theory of light, which formally includes all the other theories by proper choice of the vibration-vector, we may say that [for such stationary wave-trains] the photographic action takes place at the antinodes of the electric vector which corresponds to Fresnel's vibration, and not at the intermediate antinodes of the magnetic vector which corresponds to MacCullagh's and to Neumann's vibration.

It is the electric vibration that affects the photographic film:

The crucial experiment of Wiener relates to the case when the angle of incidence is half a right angle, so that the direct and reflected waves which interfere are at right angles to each other. If the vibration take place along the direction of intersection of the two wave-planes, it will present a series of nodes and antinodes; but if in the perpendicular direction there will not be such alternations of intensity. The experiment showed that when the light is polarized in the plane of incidence, the photographic plate develops a series of bands; but when it is polarized in the perpendicular plane these bands are absent.

for the parallel bands are not produced with polarization at right angles to plane of incidence.

The argument employed is that the photographic effect will be greatest at those places in the stationary wave-train where the vibration is most intense; and the conclusion is drawn from it that the actual vibration is represented by Fresnel's vector and not by MacCullagh's; in other words, that the vibrations of polarized light are at right angles to the plane of polarization. The force of this argument, as against MacCullagh's theory, would, however, be evaded if the vector of that theory were taken to represent something

But not relevant as a test of the elastic theories of light.

* See footnote, p. 356.

[1] Wiedemann's *Annalen*, 1890.

different from the linear displacement of the aether, or if vibrations were excited in the molecule by rotation instead of translation, or by stress, as Poincaré has pointed out[1].

But as a matter of fact it seems difficult to assign any reason of the above simple kind, on either theory, in favour of the photographic disturbance occurring at the antinodes rather than at the nodes of the optical vibration. The remarkable suggestion thrown out by Lord Rayleigh some time previous to Wiener's experiments, and afterwards verified by Lippmann, that certain effects in colour photography produced by Fox Talbot and Becquerel were really due to this kind of localization of the photographic effect, is not in opposition to such a view; for the consideration adduced was simply that a localization, periodic with the waves, would, if it happened to exist, produce effects like the observed ones. At any rate, the observed localization demonstrates the important result that the effect is due to a specific dynamical action of the waves, and not to mere general absorption of the radiation.

Colour photography: Rayleigh's prediction.

Let us consider the actual circumstances of the case. There are about 10^3 molecules of the sensitive medium in the length of a single wave of light: thus in the stationary wave-train all the parts of a single molecule would at any instant be moving with a sensibly uniform velocity, which increases and diminishes periodically. The vibration of the molecule would thus be, were it not for the influence of differences of inertia or elasticity between its parts and the surrounding aether, very nearly a swaying to and fro of it as a whole: if it were exactly this, it could not be expected to produce any breaking up of the molecule at all. Moreover, as at the antinodes of the vibration there is movement but no stress in the medium, so at the nodes there is stress but no movement; and it does not seem at all clear that alternating stress might not be as potent a factor in disintegration as alternating motion. A representation has been constructed by Lord Kelvin[2] of a system in which internal vibrations can be excited by simple translation, by means of the device of an outer shell imbedded in the aether and containing inside it masses with spring connections; and such a system might also be adjusted so as to respond to simple rotation, and therefore be excited at the nodes of the wave-train instead of the antinodes.

Various analogical modes of Action.

Slight inertia and high rigidity of a material aether:

A theory based in this manner on difference of inertia must take the density of the aether to be very minute compared with that of matter; therefore if the molecule is to have free periods of the same order of

[1] See a discussion on this subject in *Comptes Rendus*, CXII. 1891, in which MM. Cornu, Poincaré, and Potier took part.

[2] *Lectures on Molecular Dynamics*, Baltimore, 1884.

magnitude as the periods of the incident light-waves, the elastic forces acting between the atoms and concerned in these periods must be very intense. But Lord Kelvin's well-known estimate of the rigidity of the aether on this hypothesis makes it very small compared with the ordinary rigidity of material bodies[1]. In fact on Pouillet's data, which imply a considerable underestimate, the energy of the solar radiation near the sun's surface is about 4×10^{-5} ergs per cubic centimetre; it easily follows that if the amplitude of the aethereal disturbance is, say, ϵ times the wave-length, the density of the aether must be about $1/10^{22}\,\epsilon$, and its rigidity, which is equal to the density multiplied by the square of the velocity of propagation, therefore $1/10\,\epsilon^2$. On an elastic solid theory it is desirable to have the density very small: thus if we adopt 10^{-2} as the maximum likely value of ϵ, the density of the aether comes out 10^{-18} of that of water, and its rigidity about 10^3, whereas the rigidity of steel or glass is of the order 10^{11}. **Kelvin's estimates:**

Now at first sight it would appear that the elastic tractions exerted by an aether of such small rigidity on an imbedded molecule swaying backwards and forwards in it, would be vanishingly small compared with the elastic forces between its constituent atoms which are concerned with free vibrations of the kind of period under consideration; and that therefore they would be quite incompetent to produce violent disturbance in the molecule. But on a closer examination this difficulty may to a considerable extent be evaded.

Let us imagine an imbedded rigid nodule of linear dimension L, and let the force necessary to displace it in the aether in any manner through a distance x be [as] Lx. Let us compare with it a similar nodule of linear dimensions κL displaced through a distance κx. There is complete dynamical similarity between the two cases; the strains at corresponding points in the aether are equal, and therefore so are the tractions per unit area. Thus the forcive necessary in the latter case to produce the displacement κx is [as] $\kappa^2 Lx$, and therefore to produce the same displacement x as in the previous case a forcive κLx is required. If now instead of comparing the total forcives in the two cases we compare the forcives per unit volume, an increase of linear dimensions in the ratio of κ to one diminishes this forcive in the ratio of κ^2 to one. Thus, if only the atoms are taken small enough, an aether of very slight rigidity can exert a forcive on them which, estimated per unit volume, is of any order of magnitude we please. The features of the case are in fact analogous to those of the suspension of small bodies, such as motes, in a viscous fluid medium like the atmosphere: if only the particles are small enough they will **yet sufficient to excite vibrations in an atom:** **if it is small enough in bulk:**

[1] Cf. Maxwell, *Encyc. Brit.* article "Aether."

float for an indefinitely great time against the force of gravity, even be they as dense as platinum,—the only limit being in that case the one imposed by the molecular discreteness of the air itself.

It would appear that the application of this principle does much to vivify the notion of an elastic solid aether. A medium of this kind, which is excessively rare and, as a consequence, of very feeble elasticity, would exert practically negligible tractions on the surfaces of a mass of matter in bulk, while it may exert relatively very powerful ones on the individual atoms of which the mass is composed if only they are sufficiently small, it being of course supposed that the structure of the medium itself is absolutely continuous. And it would even appear that a medium of very small density and rigidity may be competent to excite powerful vibrations in the molecules notwithstanding the strength of the forcives which hold them together.

a very rare elastic-aether analogy thus adequate in these aspects,

We may thus imagine a working illustration of a ponderable transparent medium of elastic solid type as made up of very small spherical nodules of great density and rigidity dispersed through the aether and imbedded in it. We may even imagine these nodules to be collected into more or less independent groups, each of which will have free periods of relative vibrations of its own nearly independent of other groups, in the manner now well known in connection with Prof. Ewing's model of a magnetic medium. A wave running across such a medium may excite these groups, and thus illustrate the theory of selective absorption by means of a system in which only the elasticity of the ambient medium is operative, but no other internal forcive.

by its intense elastic tractions.

The explanation of a very weak medium exciting such powerful tractions implies of course strains of enormous intensity, so that its limits of perfect elasticity must be taken enormously wide compared with anything we know in ordinary matter. The magnitude of the strains also requires that the displacement of an atom relative to the aether must be a considerable fraction of its diameter; and this is sufficiently secured by the large value of ϵ above that which is required to keep down the density of the aether, combined with the great relative density of the atom. It would thus seem to be possible to account for sufficiently large differential tractions between the component atoms of a molecule, especially if some of them lie well under the lee of others, to produce brisk internal vibration.

The atomic picture:

In this way we could imagine the construction of a sort of model illustrative of an elastic solid theory of refraction, including selective absorption and other such phenomena, in the form in which it is presented by von Helmholtz and others. In the simpler case, in which the atoms are not grouped into systems capable of synchronous

and its vibrational scheme.

free internal vibrations, let (ξ, η, ζ) denote the mean displacement of the free aether, and (ξ_1, η_1, ζ_1) that of the atoms. Then the equations of vibration assume the forms

$$\rho \frac{d^2}{dt^2}(\xi, \eta, \zeta) = c\nabla^2(\xi, \eta, \zeta) - \left(\frac{d}{dx}, \frac{d}{dy}, \frac{d}{dz}\right)p + a(\xi_1 - \xi, \eta_1 - \eta, \zeta_1 - \zeta),$$

$$\rho_1 \frac{d^2}{dt^2}(\xi_1, \eta_1, \zeta_1) = -a(\xi_1 - \xi, \eta_1 - \eta, \zeta_1 - \zeta),^1$$

in which $p = c'\left(\frac{d\xi}{dx} + \frac{d\eta}{dy} + \frac{d\zeta}{dz}\right)$; and the phenomena of crystalline media could be included by assuming a vector coefficient instead of the scalar a.

The conclusion, then, is that in this limited range an elastic-solid theory of a very rare aether is not so much at fault as would at first sight appear.

The rotational aether model.

A theory based on difference of rigidity without difference of inertia, after MacCullagh's manner, would have to be realized by ascribing to the atom an atmosphere of intrinsic aethereal strain, instead of endowing it with great inertia; and this could only be possible in a rotational aether, and would in fact form a mechanical representation of the electric theory. As such it must be expected to give an account of the phenomena of electricity as well as those of light[2], and in such an account is founded one of its chief claims.

Bearing on Helmholtz's mobile aether.

A development of the electric theory has recently been essayed by von Helmholtz[3], on the basis of the formal equations of Heaviside and Hertz, in which the free aether is still supposed to be an elastic medium of excessively small density in which the dense atoms are imbedded. If such a view should turn out to be the basis of a consistent body of theory, the considerations given above with respect to the intensities of molecular tractions would have a bearing on it also.

Let us now consider more particularly the explanation that would be offered by the electric theory of light. The difference between a material medium and a vacuum consists in an altered effective

Helmholtz's first dispersion theory.

[1] There are introduced by von Helmholtz (*Wiss. Abh.* II. p. 216) in addition, a forcive proportional to the absolute displacement of the atom, and a frictional one proportional to its absolute velocity. The former is derived from the idea that the heavy central masses of the atoms are unmoved by the aether, and only outlying satellites are affected by its motion. On our present view this restriction might be dispensed with, except in so far as it renders possible an illustrative theory of absorption of an analytically simple character. The consequences of the above equations are set out by various writers, *e.g.* Carvallo (*Comptes Rendus*, CXII. p. 522).

[2] Cf. "A Dynamical Theory...," *Phil. Trans.* (1884), §§ 122–124: *infra*, p. 513.

[3] Wiedemann's *Ann.* 1894 [*Collected Papers*, Vol. III.].

Electric atoms vibrating in radiation field: the electric force acts with full intensity: not so the magnetic: photographic action thus explained.

dielectric coefficient. This difference is simply and naturally explained by the hypothesis that the material molecules are polar owing to their associated atoms having atomic charges equal in amount but opposite in sign, and that they therefore possess electric moments just as the molecules of a magnet possess magnetic moments. An electric force thus tends to pull the two constituents of a molecule asunder; and its full intensity is exerted in this manner, not merely its differential intensity over the range of the molecular volume. But a magnetic force has no such tendency even when we take the molecule to be magnetically polarized, because the two poles of a magnetic element cannot be dissociated from each other; the magnetic moment is thus directly associated with the atom, not with the molecule. In the case of the stationary light-waves the antinodes of the electric force are therefore places where alternating disturbances of a kind suitable to produce decomposition of the molecules are maintained, and may produce strong effects through sympathetic molecular vibration or otherwise; but at the intermediate antinodes of the magnetic force the individual ultimate atoms may be disturbed by the alternating magnetic force, but there is no tendency to separation of the constituents of the molecule. On the electric theory, therefore, there is abundant justification both for the magnitude of the effect produced, and for its localization as determined by Wiener's experimental investigation.

Considered for the orbital electric atom.

The theory, noticed first it seems by Weber, which ascribes molecular magnetism to the orbital rotation round each other of ionic charges, and which has very strong recommendations from the point of view of the dynamics of the aether, may form a partial exception to this statement. It leaves the question open as to whether the principal part of the magnetic moment is due to orbital motions in the atoms or to the motions of the constituent atoms in the molecules; though it suggests strongly the latter alternative. In that case there will usually be a differential magnetic action of the field as between these moving atoms; but the magnetic actions on positive and negative ions will be by no means equal and opposite, as is true of the electric actions. Thus, for example, in the limiting case of two equal and opposite ions revolving round each other, the elements of the equivalent ionic convection currents will be at each instant parallel, and there will be no differential magnetic forcive at all; there will also be no magnetic moment; but the electric differential action will retain its full force.

Chemical energy mainly electrostatic.

It is well understood, and in accordance with this explanation, that the energy of chemical combination of atoms into molecules is almost entirely that of electrostatic attraction of their atomic charges.

In fact the electric attraction between them diminishes according to the law of inverse square with increasing distance, their magnetic attraction according to the law of the inverse fourth power: if these forces are of the same order of magnitude in the actual configuration of the atoms in the molecule, the work done by the former during their combination must be almost indefinitely greater than the work done by the latter.

Material loading by atoms excluded,

If we contemplate the purely dynamical basis which must underlie the descriptive explanations of the electric theory of light, it is difficult to see how there can be any place for a theory of the aether loaded by the material molecules, which dynamical views usually associated with Fresnel's theory demand. There could be no polarity in the inertia of a mere load, such as the present considerations require. On the other hand, the presence of electrically polarized molecules is effectively a diminution of the elasticity of the luminiferous medium; and I have tried to show elsewhere[1] that the principles of MacCullagh's theory of optics are in substantial agreement with all the general features of our electrical and optical knowledge.

in favour of electric.

It is definitely implied in the electromotive, as distinguished from an electrodynamic, character of the electric theory of light, that the atomic charges vibrate in unison with the light-waves, quite unimpeded by any material inertia of their atoms*. This hypothesis is conceivable and natural, independently of any particular explanation, on the theory that the atoms are themselves intrinsic mobile configurations of stress or motion, or both together, in the ultimate medium.

The elastic molecular solid analogy pursued:

It is not without interest to consider how far the conception mentioned above of an isotropic solid medium of very small density, with very massive minute nodules imbedded in it but exerting no direct forcives on each other, will carry us in forming a representation of optical phenomena. The theory is of the Young-Sellmeier type, because each nodule has one or more free periods conditioned by its form and by the surrounding elasticity.

On eliminating (ξ_1, η_1, ζ_1) from the equations expressed above, we obtain the vibrational equations of the aether, supposed thus loaded. Its elastic properties are found to be conserved intact, but the effective density as regards vibrations of period τ is increased by $a\rho_1\Big/\left(a - \dfrac{4\pi^2}{\tau_2}\rho_1\right)$. When the coefficient a is of aeolotropic type, by reason either of the form or the distribution of the nodules, we have

favourably into details:

[1] *Loc. cit.*, *Phil. Trans.* 1894: as *infra*, p. 447.

* But with the relevant electric inertia: cf. "Theory of Electrons," *loc. cit.*, *infra*, p. 514, § 114, of date Aug. 13, 1894.

effectively an isotropically elastic medium with aeolotropic inertia; this leads to Fresnel's wave-surface, provided the elasticity is labile in Lord Kelvin's sense. The theory also leads to a formula for ordinary dispersion, of the usually admitted type (Ketteler's) for isotropic media; but, on the other hand, it is in default by assigning a dispersional origin to double refraction. If we wish to include the minute effect known as the dispersion of the optic axes in crystals, it will be necessary to assume for the elastic stress between aether and matter a somewhat more general form, involving (after von Helmholtz) absolute as well as relative displacement, but always of course remaining linear.

The assumption of elasticity of labile type also allows an escape from the usual difficulties of a solid aether in the matter of reflexion. In that problem the elasticity would naturally be taken continuous across the interface, the volume occupied by the molecules being on this hypothesis extremely small compared with that occupied by the aether.

We may further amend the theory by getting rid of the difficulties associated with lability, at the same time avoiding the difficulty as to how a body can move through a perfect solid medium, if we take the aether to be a rotationally elastic fluid, and retain the material load as before.

but it could not include strong electric fields.

But an essential and fundamental difficulty will still remain. It is the extremely small volume-density of the energy involved in radiation which permits a very small inertia, and consequently a small elasticity, to be assigned to the aether, and so prevents it from acting as an appreciable drag or exerting an appreciable force on finite bodies moving through it. But these very properties would incapacitate it for acquiring the very large volume-densities of energy that would have to be associated with it in order to explain electrodynamic phenomena.

Kelvin vortical type of material atoms.

Any representation which would make the aether consist of molecules of ordinary matter is open to the objection that the thermal kinetic energy of gases and other material systems must then, in accordance with Maxwell's law of distribution of energy, largely reside in it. But, on the other hand, if we hold to the view of matter which was first rendered precise by Lord Kelvin's theory of vortex atoms, namely, that the aether is the single existing medium and that atoms of matter are intrinsic singularities of motion or strain which belong to it, then there is no inducement to assume for the aether a molecular structure at all, or to make its inertia anything comparable with the inertia of the atoms on whose play the thermal energy of the movements of the matter consists. On such a theory the inertia, and the resulting kinetic energy, of the matter may be hard to explain*, but it is certainly something different from the inertia of the underlying medium in which the atom is merely a form of strain or motion. On such a theory refraction, and also double refraction, will be caused by the atmosphere of intrinsic strain which represents the electric charge on the atom; and only dispersion will be assigned to the influence of sympathetic vibrations in the atoms or molecules, thus doing away with any difficulty of the kind mentioned above.

In the theory of gases the ordinary kinetic energy of the molecules represents sensible heat, and as such may be derived for example from the dissipation by friction or otherwise of the mechanical energy of ordinary masses: it is of the nature of kinetic energy of the masses of the atoms. But the store of energy which keeps up radiation is of electromotive kind, is concerned with displacing electricity, not with moving matter except indirectly†; at least no consistent scheme has yet been forthcoming which includes both. It is quite conceivable that the disturbances which occur in the ordinary encounters of molecules are of far too gentle a character to excite the very powerful elasticity which on a certain form of the electric theory binds together the continuous medium taking part in optical propagation, any more than a system of solid balls rushing about in an enclosure bounded by a heavy continuous rigid solid can excite sensibly the elastic qualities of that body. The opinion has been widely supported, both on theoretical and experimental grounds, that a gas will not emit its definite radiations however high the temperature to which it is raised, unless there is chemical decomposition‡ of the molecules going on. If that be so, the aether does not act as an equalizer of the kinetic energy between the different modes of vibration of the molecules, and the ordinary theory of gases need make no reference to the aether.

Radiation and gas-theory.

Radiation of disruptive origin.

If I have understood aright, a similar view has been expressed as at any rate a possible explanation of the difficulty as to the application of Maxwell's distribution theorem in the theory of gases, by Prof. Boltzmann himself §. The law of distribution of energy is perhaps unassailable for the case of molecules like small spheres, with three

Maxwell-Boltzmann equipartition.

* Cf. however the mass of an electron, as definitely located in the aether: *Phil. Trans.* 1884, § 114; or *infra*, p. 522.

† The effect of the Maxwell radiation pressure is slight in this connection.

‡ Now replaced by the wider Bohr view of radiation as arising by lapses toward different static levels of energy, which are perhaps the motional configurations of minimal energy, in the atom.

§ The ascertained falling away of thermal capacities at very low temperatures has of course now revolutionized this subject.

degrees of freedom, all translational. By including the rotational modes of freedom, which may be none at all for a monad gas, only two for a diad, and three for other types, and these possibly not complete, a sufficient number of freedoms is obtained to cover the known range of values of the ratio of the specific heats. The introduction of any vibrational types would make too many; so on this ground also it is not likely that such types can enter into those among which the thermal energy is divided.

37

A DYNAMICAL THEORY OF THE ELECTRIC AND LUMINIFEROUS MEDIUM.

[*Proceedings of the Royal Society*, Vol. LIV. (1893) pp. 438–461.]

(*Abstract of memoir following: and general discussion.*)

EVER since the causes of natural phenomena began to attract attention, the interaction of the different classes of physical agencies has been taken to suggest that they are all manifestations in different ways of the energy of some fundamental medium; and the efforts of the more sanguine class of naturalists have always been in some measure directed towards the discovery of the properties of this medium. It is only at the end of the last century that the somewhat vague principle of the economy of action or effort in physical actions —which, like all other general principles in the scientific explanation of Nature, is ultimately traceable to a kind of metaphysical origin— has culminated in the hands of Lagrange in his magnificent mathematical generalization of the dynamical laws of material systems. Before the date of this concise and all-embracing formulation of the laws of dynamics there was not available any engine of sufficient power and generality to allow of a thorough and exact exploration of the properties of an ultimate medium, of which the mechanism and mode of action are almost wholly concealed from view. The precise force of Lagrange's method, in its physical application, consists in its allowing us to ignore or leave out of account altogether the details of the mechanism, whatever it is, that is in operation in the phenomena under discussion; it makes everything depend on a single analytical function representing the distribution of energy in the medium in terms of suitable coordinates of position and of their velocities; from the location of this energy, its subsequent play and the dynamical phenomena involved in it are all deducible by straightforward mathematical analysis.

The aether.

Lagrangian foundation essential,

and suitable:

The problem of the correlation of the physical forces is thus divisible into two parts, (i) the determination of the analytical function which represents the distribution of energy in the primordial medium which is assumed to be the ultimate seat of all phenomena, and (ii) the discussion of what properties may be most conveniently and simply assigned to this medium, in order to describe the play of

its formal statement

and developments. energy in it most vividly, in terms of the stock of notions which we have derived from the observation of that part of the interaction of natural forces which presents itself directly to our senses, and is formulated under the name of natural law. It may be held that the first part really involves in itself the solution of the whole problem; that the second part is rather of the nature of illustration and explanation, by comparison of the intangible primordial medium with other dynamical systems of which we can directly observe the phenomena.

The chief representative of exact physical speculation of the second of these types has been Lord Kelvin. In the older attempts of this kind the dynamical basis of theories of the constitution of the aether consisted usually in a play of forces, acting at a distance, between ultimate elements of molecules of the medium; from this we must, however, except the speculations of Greek philosophy and the continuous vortical theories of the school of Descartes, which were of necessity purely descriptive and imaginative, not built in a connected manner on any rational foundation. It has been in particular the aim of Lord Kelvin to deduce material phenomena from the play of inertia involved in the motion of a structureless primordial fluid; if this were achieved it would reduce the duality, rather the many-sidedness, of physical phenomena to a simple unity of scheme; it would be the ultimate conceivable simplification. The celebrated vortex theory of matter makes the indestructible material atoms consist in vortex rings in a primordial fluid medium, structureless homogeneous, and frictionless*, and makes the forces between the atoms which form the groundwork of less fundamental theories consist in the actions excited by these vortices on one another through the inertia of the fluid which is their basis—actions which are instantaneously transmitted if the fluid is supposed to be absolutely incompressible.

Kelvin's vortical atomic type.

In case this foundation proves insufficient, there is another idea of Lord Kelvin's by which it may be supplemented. The characteristic properties of radiation, which forms so prominent an element in actual phenomena, can be explained by the existence of an elastic medium for its transmission at a finite, though very great, speed; such a medium renders an excellent account of all its relations, if we assume it to possess inertia and to be endowed with some elastic quality of resistance to disturbance roughly analogous to what we can observe and study in ordinary elastic solids of the relatively incompressible kind, such as india-rubber and jellies. Lord Kelvin has been

Rotational elasticity superposed on fluid:

* Ultimately of course on a scheme of differential relations in space and time interpretable as expressing an ideal fluid.

the promoter and developer of a view by which the elastic forces between parts of such a medium may be to some extent got rid of as ultimate elements, and be explained by the inertia of a spinning motion of a dynamically permanent kind, which is distributed throughout its volume. If we imagine very minute rapidly-spinning flywheels or gyrostats spread through the medium, they will retain their motion for ever, in the absence of friction on their axles, and they will thus form a concrete dynamical illustration of a type of elasticity which arises solely from inertia; and this illustration will be of great use in realizing some of the peculiarities of a related type, which I believe can be thoroughly established as the actual type of elasticity transmitting all radiations, whether luminous and thermal or electrical—for they are all one and the same—through the ultimate medium of fluid character of which the vortices constitute matter.

e.g. gyrostatic.

It has always been the great puzzle of theories of radiation how the medium which conveys it by transverse vibrations, such as we know directly only in media of the elastic-solid type, could yet be so yielding as to admit of the motion of the heavenly bodies through it absolutely without resistance. According to the view of the constitution of the aether which is developed in this paper, not only are these different properties absolutely consistent with each other, but it is, in fact, their absolute and rigorous coexistence which endows the medium with the qualities necessary for the explanation of a further very wide class of phenomena. The remark which is the key to this matter has been already thrown out by Lord Kelvin, in connection with Sir George Stokes' suggested explanation of the astronomical aberration of light. The motion of the ultimate homogeneous frictionless fluid medium, conditioned by the motion of the vortices existing in it, is, outside these vortices, of an absolutely irrotational character. Now, suppose the medium is endowed with elasticity of a purely rotational type, so that its elastic quality can be called into play only by absolute rotational *displacement* of the elements of the medium; just as motion of translation of a spinning gyrostat calls into play no reaction, while any alteration of the absolute position of its axis in space is resisted by an opposing couple. As regards the motion of the medium involved in the movements of its vortices, this rotational elasticity remains completely latent, as if it did not exist; and we can at once set down the whole theory of the vortical hydrodynamical constitution of matter as a part of the manifestations of an ultimate medium of this kind.

This aether permeable to matter in bulk.

We have now to indicate some of the consequences of the assumed constitution of the aether as regards the phenomena of radiation,

Radiation.

which depend on this elasticity: to do this it will be convenient to make a fresh start, dealing more particularly with the first part of the general question.

Young. The true nature of the phenomena of light had been brought to view at the beginning of the present century by the intuition of Thomas Young; and the secret of the exact quantitative mathematical laws which govern the behaviour of light in all the various circumstances attending its propagation, reflexion, and refraction had been fathomed in a marvellous manner by the genius of Fresnel. (Fresnel's scheme.) The nature of the mathematical reasoning by which Fresnel was led to his results has for the most part never been understood; and, as presented by him in his writings, it certainly seems devoid of dynamical coherence and formal logical validity. Yet, the more the phenomena of light were afterwards experimentally examined, the stronger was the confirmation of the whole scheme of formulae at which he had arrived.

The explanation of the laws of physical optics advanced by Fresnel, and verified by comparison with the phenomena, which was possible in several very exact ways, chiefly by himself and Brewster, was, about the year 1835, engaging the attention of several of the chief mathematicians of that time—Augustin Cauchy in France, Franz Neumann in Germany, George Green in England, and James MacCullagh in Ireland. (Elastic theory of physical optics: historical.) The prevalent mode of attacking the problem was through the analogy with the propagation of elastic waves in solid bodies; and the comparison of Fresnel's laws of propagation in crystalline media with the results of the mathematical theory of the elasticity of crystalline bodies gave abundance of crucial tests for the verification, modification, or disproof of the principles assumed in these investigations. The treatment of Cauchy is earliest in date, but somewhat empirical and unsatisfactory in its logical aspects in the light of subsequent more precise knowledge of the conditions of the problem of the elasticity of solids. The treatment of Neumann is also a sound and original piece of investigation, if we except the limited view of the elasticity of solids, that of Navier and Poisson, on which he based it. The treatment by Green had the great distinction of incidentally laying, with all the generality and simplicity which we expect in an ultimate theory, the foundations on which every theory of elastic action in ordinary material bodies must in future be constructed; it proceeded, in fact, on the basis of one of those great generalizations, of which the aggregate constitutes the all-embracing modern doctrine of energy. These three authors all treated the question of reflexion and refraction of waves. Cauchy could not make much of Fresnel's formulae in any logical manner. Neumann had the merit of seeing clearly that the thing was impossible on his elastic solid theory; so he dropped it altogether, assumed a sufficient number of principles which might be taken, with fair probability, in accordance with general reasoning, to be satisfied in the reflexion and refraction of light rays, viz. complete continuity of the media and continuity of energy in crossing the boundary at which the reflexion and refraction take place, and had the satisfaction of evolving a solution which agreed with Fresnel's laws, and easily extended them to the much more complicated circumstances of crystalline media. But to obtain this solution he assumed, from what he found necessary to make his very imperfect theory of propagation in crystals agree with Fresnel's laws, that the density of the luminiferous medium is the same in all bodies, and that the displacement of plane-polarized light is in the plane of polarization. It may be shown, as is now indeed to be expected, that this is a totally wrong foundation to work upon, that Neumann's general principles for the solution of the problem of reflexion are inconsistent with his elastic theory. If he had adopted a converse procedure, and worked out the problem of the reflexion of a ray on his general principles, and then deduced, by comparison with Fresnel's formulae, the law of density of the luminiferous medium and the direction of the vibration in plane-polarized light, he would have been entitled to the credit of a joint discoverer in the domain of the dynamics of reflexion. But, for the reasons here indicated, the credit of that discovery must, I think, be assigned to MacCullagh.

Neumann's tentative theory.

The achievements by which the memory of MacCullagh is now to a great extent preserved are his very elegant investigations in the domain of pure Euclidean geometry. He may be claimed to be an instance of the numerous cases from Archimedes down through Descartes, Newton, and, we may add, Thomas Young, in which keen geometrical insight has formed a key for unlocking the formal laws of physical actions. He was first attracted to Fresnel's laws of optics by the very simple and elegant geometrical relations to which they lead. At a later period he proposed to himself the problem to hit off the extension of Fresnel's laws of reflexion which would apply to crystalline media, in the light of the crucial conditions afforded by the delicate experiments of Brewster and, at a later stage, Seebeck, to which such a theory must conform. He had thus to cast about for geometrical principles on which Fresnel's laws might be founded, such as would admit of easy extension to the more general problem. He early came upon the principle of continuity of the media, which he put in the geometrical form that the resultant of the displacements in the refracted waves is equal to the resultant of the

MacCullagh's new departure:

early formal scheme:

displacement in the incident and reflected waves. As regards the other necessary condition, he was not at first successful. The density of the medium he took to be the same in all bodies, because he could not imagine it to be aeolotropic, or different in different directions, in crystalline media. He assumed the vibrations to be in the plane of polarization, from considerations of geometrical symmetry and necessity, confirmed in the earlier stage by one of the theories of Cauchy. The other condition above referred to he took to be equality of certain pressures in the media, as imagined by Cauchy; and by this means he arrived at a satisfactory explanation of Brewster's observations on the polarizing angle in reflexion from crystals. But Seebeck pointed out that this solution would not account for the values of the deviation of the plane of polarization from the plane of reflexion, by means of which he had himself tested it. Owing to this criticism MacCullagh was finally led to abolish Cauchy's notion of pressure, and assume simply the continuity of energy in its place. This principle of energy, which gives a quadratic equation between the displacements at the interface, he succeeded to his satisfaction, as regards the confirmation of his views, in replacing by linear relations. And then he gave his two magnificent geometrical theorems —that of *transversals* and that of the *polar plane*, which contain each in a sentence the complete specification of the laws of reflexion for the most general case of a transparent medium, and which form the culmination of the geometrical relations by which he was guided throughout this whole process of synthetical discovery. His laws of reflexion are the same as Neumann's; of them, as formal laws, these two authors must be regarded as the independent discoverers—Neumann by a happy assumption suggested by reasoning at bottom illogical in the light of subsequent knowledge, MacCullagh by a resolute attack on the observed facts with a view to reducing them to simple formulae.

But the greatest achievement of MacCullagh is that contained in his memoir of 1839, two years later, entitled an "Essay towards a Dynamical Theory of Crystalline Reflexion and Refraction." He is in quest of a dynamical foundation for the whole scheme of optical laws, which had been notably extended and confirmed by himself already. He recognizes, I think for the first time in a capital physical problem, that what is required is the discovery of the potential-energy function of Lagrange on which the action of the medium depends, and that the explanation of the form of that function is another question which can be treated separately. His memoir is subsequent to, but apparently quite independent of, that of Green, in which Green restricted the medium to a constitution like an elastic

dynamical foundations,

solid, laid down the general laws of such constitution for the first time, and made a magnificent failure of his attempt to explain optical phenomena on that basis. If this thing was to be done, the power, simplicity, and logical rigour of Green's analysis might have been expected to do it; and nothing further has come of the matter until the recent new departure of Lord Kelvin in his speculation as to a labile elastic-solid aether. To return to MacCullagh, he is easily able to hit off a simple form of the potential-energy function, which—on the basis of Lagrange's general dynamics, or more compactly on the basis of the law of Least Action—absolutely sweeps the whole field of optical theory so far as all phenomena are concerned in which absorption of the light does not play a prominent part. He is confident, as anyone who follows him in detail must be, that he is on the right track. He tries hard to obtain a dynamical basis for his energy-function, that is, to imagine some material medium that shall serve as a model for it and illustrate its possibility and its mode of action; he records his failure in this respect, but at the same time he protests against the limited view which would tie down the unknown and in several ways mysterious and paradoxical properties of the luminiferous medium to be the same as those of an ordinary elastic solid. **his Action formula,**

The form of MacCullagh's energy-function was derived by him very easily from the consideration of the fact that it is required of it that it shall produce, in crystalline media, plane-polarized waves propagated by displacements in the plane of the wave-front. Though he seems to put his reasoning as demonstrative on this point, it has been pointed out by Sir George Stokes, and is indeed obvious at once from Green's results, that other forms of the energy-function besides MacCullagh's would satisfy this condition. But the important point as regards MacCullagh's function is that it makes the energy in the medium depend solely on the absolute rotational displacements of its elements from their equilibrium orientations, not at all on its distortion or compression, which are the quantities on which the elasticity of a solid would depend according to Green. **rotational, not deformational.**

Starting from this conception of rotational elasticity, it can be shown that, if we neglect for the moment optical dispersion, every crystalline optical medium has three principal elastic axes, and its wave-surface is precisely that of Fresnel, while the laws of reflexion and refraction agree precisely with experiment. Further, it follows from the observed fact of transparency in combination with dispersion, that the dispersion of a wave of permanent type is properly accounted for [formally] by the addition to the equations, therefore to the energy-function, of subsidiary terms involving spacial **Crystalline media:**

differentiations of higher order*. To preserve the medium hydrodynamically a perfect fluid, these terms also must satisfy the condition that the elasticity of the medium is thoroughly independent of compression and distortion of its elements, and wholly dependent on absolute rotation. It can be shown, I believe, that this restriction limits the terms to two kinds, one of which retains Fresnel's wave-surface unaltered, while the other modifies it in a definite manner stated without proof by MacCullagh—[but the first terms depend on an interaction between the dispersive property and the wave-motion itself, while the second terms involve the square of the dispersive quality: it seems clear that the second type involves only phenomena of a higher order of small quantities than we are here considering]—thus an account of dispersion remains which retains Fresnel's wave-surface unaltered for each homogeneous constituent of the light, while it includes the dispersion of the axes of optical symmetry in crystals as regards both their magnitudes and directions—results quite unapproached by any other theory ever entertained.

his static dispersional terms in the Action:

In this analysis of dispersions, all terms have been omitted which possess a unilateral character, such as would be indicated in actuality by rotatory polarization and other such phenomena. The laws of crystalline material structures seem to prohibit the occurrence of such asymmetry as these terms would indicate, except to the very small extent evidenced by the hemihedral faces of quartz crystals. The influence on the optical medium of this asymmetric arrangement of the molecules must be very much smaller still, for the rotatory terms are in all media exceedingly minute compared with the ordinary dispersional terms. The form of these rotatory terms in the energy-function is at once definitely assigned by our condition of perfect fluidity of the medium, both for crystals and for rotational liquids such as turpentine, and this form is the one usually accepted, on MacCullagh's suggestion, as yielding a correct account of the phenomena†.

and his formulation for chiral quality.

When dispersional terms [of this type involving higher gradients] are included in the energy-function, our continuous analysis is not any longer applicable to the problem of reflexion; the conditions at the interface are altogether too numerous to be satisfied by the available variables. There is in fact discontinuity at the interface in the discrete molecular structure, such as could not be representable by a continuous analysis. But if we proceed by the method of rays,

Conditions for regular reflexion.

* Dispersion is due actually to interaction of the atomic vibrations excited in the compound medium aether matter: as this depends on the period, the consequences in the text follow naturally.

† Cf. *Aether and Matter* (1900), Ch. XII.

and assume that there is a play of surface forces which do not absorb any energy, while they adjust the dispersional part of the stress, it appears that reflexion is independent of dispersion.

Test of theories by laws of reflexion.

The treatment of the problem of reflexion by Fresnel involves a different direction of vibration of the light, and different surface conditions, from MacCullagh's. It is of interest to remark that this theory may be stated in a dynamically rigorous form, provided the medium to which it refers possesses the properties of the labile elastic-solid aether of Lord Kelvin; and Fresnel's own account of his analysis of the problem becomes more intelligible from such a standpoint.

Kelvin's labile solid aether.

The mention of the phenomena of magnetic rotational quality will introduce us to the next division of the subject, that of the inclusion of electric and magnetic phenomena in the domain of the activity of this primordial medium.

Maxwell's electric scheme of an aether:

The problem of the aether has been first determinedly attacked from the side of electrical phenomena by Clerk Maxwell in quite recent times; his great memoir on a *Dynamical Theory of the Electromagnetic Field* is of date 1864. It is in fact only comparatively recently that the observation of Oersted, and the discoveries and deductions of Ampère, Faraday, and Thomson had accumulated sufficient material to allow the question to be profitably attacked from this side. Even as it is, our notions of what constitute electric and magnetic phenomena are of the vaguest as compared with our ideas of what constitutes radiation, so that Maxwell's views involve difficulties, not to say contradictions, and in places present obstacles which are to be surmounted, not by logical argument or any clear representation, but by the physical intuition of a mind saturated with this aspect of the phenomena. Many of these obstacles may, I think, be removed by beginning at the other end, by explaining electric actions on the basis of a mechanical theory of radiation, instead of radiation on the basis of electric actions. The strong point of Maxwell's theory is the electromotive part, which gives an account of electric radiation and of the phenomena of electromagnetic induction in fixed conductors; and this is in keeping with the remark just made. The nature of electric displacement, of electric and magnetic forces on matter, of what Maxwell calls the electrostatic and the magnetic stress in the medium, of electrochemical phenomena, are all left obscure.

its unsettled features.

The rotational aether:

We shall plunge into the subject at once from the optical side, if we assume that dielectric polarization consists in a strain in the aether, of the rotational character contemplated above. The conditions of internal equilibrium of a medium so strained are easily worked out from MacCullagh's expression for W, its potential energy. If the vector (f, g, h) denote the curl or vorticity of the actual linear displacement of the medium, or *twice* the absolute rotation of the portion of the medium at the point considered, and the medium is supposed of crystalline quality and referred to its principal axes, so that

its energy formulation for a dielectric,

$$W = \tfrac{1}{2}\int (a^2f^2 + b^2g^2 + c^2h^2)\, d\tau,$$

where $d\tau$ is an element of volume, it follows easily that for internal equilibrium we must have

gives the law of polarization:

$$a^2f\, dx + b^2g\, dy + c^2h\, dz = -\, dV,$$

a complete differential, and that over any boundary enclosing a region devoid of elasticity the value of V must be constant. Such a boundary is the surface of a conductor; V is the electric potential in the field due to charges on the conductors; (f, g, h) is the electric displacement in the field, circuital by its very nature as a rotation, and (a^2f, b^2g, c^2h) is the electric force derived from the electric potential V.

the nature of a charge.

The charge on a conductor is the integral of (f, g, h) over any surface enclosing it, and cannot be altered except by opening up a channel devoid of elasticity, in the medium, between this conductor and some other one; in other words, electric discharge can take place only by rupture of the elastic quality of the aethereal medium.

The interfacial conditions for crystals.

(At the interface between two dielectric media, taken to be crystalline as above, the condition comes out to be that the tangential electric force is continuous. When the circumstances are those of equilibrium, and therefore an electric potential may be introduced, this condition allows discontinuity in the value of the potential in crossing the interface, but demands that the amount of this discontinuity shall be the same all along the interface; these are precisely the circumstances of the observed phenomena of voltaic potential differences. The component, normal to the interface, of the electric displacement is of course always continuous, from the nature of that vector as a flux.

It may present itself as a difficulty in this theory that, as the electric displacement is the rotational displacement of the medium, its surface integral over any sheet should be equal to the line integral of the linear displacement of the medium round the edge of the sheet; therefore that for a closed sheet surrounding a conductor this integral should be null, which would involve the consequence that the electric charge on a conductor cannot be different from null. This line of argument, however, implies that the linear displacement is a perfectly continuous one, which is concomitant with and required by the electric displacement. The legitimate inference is that the electric displacement in the medium which corresponds to an actual charge cannot be set up without some kind of discontinuity* or slip in the linear displacement of the medium; in other words, that a conductor cannot receive an electric charge without rupture of the surrounding medium; nor can it lose a charge once received without a similar rupture. The part of the linear displacement that remains, after this slip or rupture has been deducted from it, is of elastic origin, and must satisfy the equations of equilibrium of the medium. —*Dec.* 7, 1893.)

Anticipation of electrons and their intrinsic fields.

We can produce in imagination a steady electric current, without introducing the complication of galvanic batteries, in the following manner, and thus examine in detail all that is involved, on the present theory, in the notion of a current. Suppose we have two charged condensers, with one pair of coatings connected by a narrow conducting channel, and the other pair connected by another such channel, as in the diagram, where the field is dielectric all except the dark paths which are conducting. If we steadily move towards each other the two plates of the condenser *A*, a current will flow round the circuit, in the form of a conduction current in the conductors and a displacement current across the dielectric plates of the condensers. Let us suppose the thicknesses of these dielectric plates to be excessively small, so as to minimize the importance of the displacement part of the current. There is then practically no electric force, and therefore no electric displacement, in the surrounding dielectric field, except between the plates of the condensers and close to the conducting wires. Consider a closed surface passing between the faces of the condenser *A*, and intersecting the wire at a place *P*. A movement of the faces of this condenser alters the electric force between them, and therefore alters the electric displacement across the portion of this closed surface which lies in that part of the field; as we have seen there is practically no displacement anywhere else in the field except at the conducting wire; therefore to preserve the law of the circuital character of displacement throughout the whole space, we

Elastic analogy of field of a current and its excitation: circuit must be open.

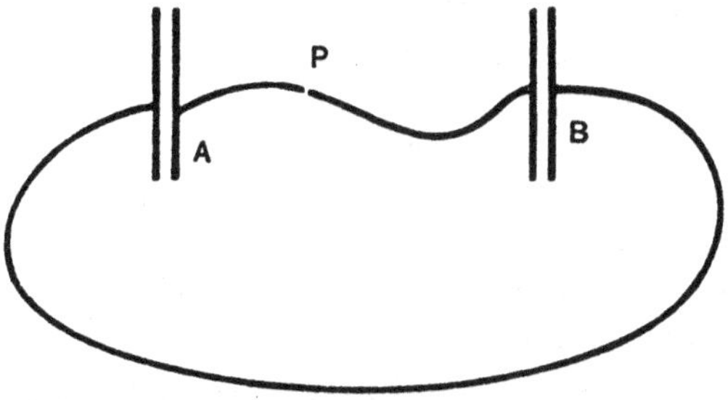

Concrete illustration of its relation to the medium.

* Such as transfer of electrons.

must suppose that this alteration is compensated by a very intense change of displacement at the conducting wire. So long as the movement of the plates continues, as long does this flow of displacement along the wire go on; it constitutes the electric current in the wire. Now, in calculating the magnetic force in the field, which is the velocity of the aethereal medium, from the change of electric displacement, we must include in our integration the effect of this sheet of electric displacement flowing along the surface of the perfectly conducting wires, for exactly the same reason as in the correlative problem in hydrodynamics, of calculating the velocity of the fluid from the distribution of vorticity in it, Helmholtz had to consider a vortex sheet as existing over each surface across which the motion is discontinuous.

Permanent current: its kinetic energy determined.

The next stage in this mode of elucidation of electrical phenomena is to suppose, once the current is started in our non-dissipative circuit, that both the condensers are instantaneously removed, and replaced by continuity of the wire. We are now left with a current circulating round a complete perfectly conducting channel, which in the absence of viscous forces will flow round permanently. The expression for the kinetic energy in the field is easily transformed from a volume integral of the magnetic force, which is represented by the velocity of the medium $d/dt\ (\xi, \eta, \zeta)$, to an integral involving the current $d/dt\ (f, g, h)$, which is in the present case a line integral round the electric circuit. The result is F. E. Neumann's celebrated formula for the electromagnetic energy of a linear electric current,

Neumann.

$$T = \tfrac{1}{2}\iota^2 \iint r^{-1} \cos \epsilon \, ds ds;$$

or we may take the case of several linear circuits in the field, and obtain the formula

$$T = \tfrac{1}{2}\Sigma\iota^2 \iint r^{-1} \cos \epsilon \, ds ds + \Sigma\iota_1\iota_2 \iint r^{-1} \cos \epsilon \, ds_1 ds_2,$$

which is sufficiently general to cover the whole ground of electrodynamics.

Our result is in fact that linear current is a vortex ring in the fluid aether, that electric current is represented by vorticity in the medium, and magnetic force by the velocity of the medium. The current being carried by a perfect conductor, the corresponding vortex is (as yet) without a core, *i.e.* it circulates round a vacuous space.

(The strength of a vortex ring is, however, permanently constant; therefore, owing to the mechanical connections and continuity of the medium, a current flowing round a complete perfectly conducting circuit would be unaffected in value by electric forces induced in the circuit, and would remain constant throughout all time. Ordinary electric currents must therefore be held to flow in incomplete conducting circuits, and to be completed either by convection across an electrolyte or by electric displacement or discharge across the intervals between the molecules, after the manner of the illustration given above.—*Dec.* 7, 1893.)

Current a non-continuous vortex.

Now we are here driven upon Ampère's theory of magnetism. Each vortex-atom in the medium is a permanent non-dissipative electric current of this kind, and we are in a position to appreciate the importance which Faraday attached to his discovery that all matter is magnetic. Indeed, on consideration, no other view than this seems tenable; for we can hardly suppose that so prominent a quality of iron as its magnetism completely disappears above the temperature of recalescence, to reappear again immediately the iron comes below that temperature; much the more reasonable view is that the molecular re-arrangement that takes place at that temperature simply masks the permanent magnetic quality. In all substances other than the magnetic metals, the vortex atoms pair into molecules and molecular aggregates in such way as to a large extent cancel each other's magnetic fields; why in iron at ordinary temperatures the molecular aggregates form so striking an exception to the general rule is for some reason peculiar to the substance, which, considering the complex character of molecular aggregation in solids, need not excite surprise.

Does the same apply to a molecular current?

Para-magnetism.

We have now to consider the cause of the pairing together of atoms into molecules. It cannot be on account of the magnetic, *i.e.* hydrodynamical, forces they exert on one another, for two electric currents would then come together so as always to reinforce each other's magnetic action, and all substances would be strongly magnetic. The ionic electric charge, which the phenomena of electrolysis show to exist on the atom, supplies the attracting agency. Furthermore, the law of attraction between these charges is that of the inverse square of the distance, and between the atomic currents is that of the inverse cube; so that, as in the equilibrium state of the molecule these forces are of the same order of intensity and counteract each other, the first force must have much the longer range, and the energy of chemical combination must therefore be very largely electrostatic, due to the attraction of the ions, as von Helmholtz has clearly made out from the phenomena of electrolysis and electrolytic polarization.

Electric forces paramount in chemistry.

All our conclusions, hitherto, relate to the aether, and are therefore about electromotive forces. We have not yet made out why two sets of molecular aggregates, such as constitute material bodies, should attract or repel each other when they are charged, or when electric currents circulate in them; we have, in other words, now to explain

Mechanical forces.

the electrostatic and electrodynamic forces which act between material systems.

Consider two charged conductors in the field; for simplicity, let their conducting quality be perfect as regards the very slow displacements of them which are contemplated in this argument. The charges will then always reside on their surfaces, and the state of the electric field will, at each instant, be one of equilibrium. The magnitude of the charge on either conductor cannot alter by any action short of a rupture in the elastic quality in the aether; but the result of movement of the conductors is to cause a re-arrangement of the charge on each conductor, and of the electric displacement (f, g, h) in the field. Now the electric energy W of the system is altered by the movement of the conductors, and no viscous forces are in action; therefore the energy that is lost to the electric field must have been somehow spent in doing mechanical work on the conductors; the loss of potential energy of the electric field reappears as a gain of potential energy of the conductors. We have to consider how this transformation is brought about. The movement of the conductors involves, while it lasts, a very intense ideal flow of electric displacement along their surfaces, and also a real change of displacement of ordinary intensity throughout the dielectric. The intense surface flow is in close proximity with the electric flows round the vortex atoms which lie at the surface; their interaction produces a very intense elastic disturbance in the medium, close at the surface of the conductor, which is distributed by radiation through the dielectric as fast as it is produced; the elastic condition of the dielectric, on account of its extreme rapidity of propagation of disturbances compared with its finite extent, being always extremely nearly one of equilibrium. It is, I believe, the reaction on the conductor of these [incipient] wavelets which are continually shooting out from its surface [during disturbance], carrying energy into the dielectric, that constitutes the mechanical forcive acting on it. But we can go further than this; the locality of this transformation of energy, so far at any rate as regards the material forcive, is the surface of the conductor; and the gain of mechanical energy by the conductor is therefore correctly located as an absorption of energy at its surface; therefore the forcive acting on the conductor is correctly determined as a surface traction, and not a bodily forcive throughout its volume. One mode of representing the distribution of this surface traction, which, as we know, gives the correct amount of work for every possible kind of virtual displacement of the surface, is to consider it in the ordinary electrostatic manner as a normal traction due to the action of the electric force on the electric density at the surface; we conclude that this distribution

Field energy spent in mechanical work.

Equilibrium due to rapid elastic transmission:

tentative for motional change:

of traction is the actual one. To recapitulate: if the dielectric did not transmit disturbance so rapidly, the result of the commotion at the surface produced by the motion of the conductor would be to continually start wavelets which would travel into the dielectric, carrying energy with them. But the very great velocity of propagation effectually prevents the elastic quality of the medium from getting hold; no sensible wave is produced and no flow of energy occurs into the dielectric. The distribution of pressure in the medium which would be the accompaniment of the wave-motion still persists, though it now does no work in the dielectric; it is this pressure of the medium against the conductor that is the cause of the mechanical forcive.

establishing definite adjusting stress.

The matter is precisely illustrated by the fundamental *aperçu* of Sir George Stokes with regard to the communication of vibrations to the air or other gas. The rapid vibrations of a tuning-fork are communicated as sound waves, but much less completely to a mobile medium like hydrogen than to air. The slow vibrations of a pendulum are not communicated as sound waves at all; the vibrating body cannot get a hold on the elasticity of the medium, which retreats before it, preserving the equilibrium condition appropriate to the configuration at the instant; there is a pressure between them, but this is instantaneously equalized [around the pendulum] as it is produced, without leading to any flow of vibrational energy*.

Familiar analogy of sound waves.

Now let us formally consider the dynamical system consisting of the dielectric media alone, and having a boundary just inside the surface of each conductor; and let us contemplate motions of the conductors so slow that the medium is always indefinitely near the state of internal equilibrium or steady motion, that is conditioned at each instant by the position and motion of the boundaries. The kinetic energy T of the medium is the electrodynamic energy of the currents, as given by Neumann's formula; and the potential energy W is the energy of the electrostatic distribution corresponding to the conformation at the instant; in addition to these energies we shall have to take into account surface tractions exerted by the enclosed conductors on the medium, at its boundaries aforesaid. The form of the general dynamical variational equation that is suitable to this problem is

$$\delta\int(T-W)\,dt+\int dt\int \delta w\,dS=0,$$

Slow electrodynamic phenomena: Action theory:

where $\delta w\,dS$ represents the work done by the tractions acting on the

* The ideas struggling towards expression in these three paragraphs find their concise and beautiful consummation in the derivation of the Maxwellian quadratic field-stress, belonging to the material sources, by a special variation of the Action that is equivalent to displacement of the sources—the equilibrium of the medium ensuring that the result of this, as of all other internal variations, vanishes except for forces coming from the sources. See Appendix *infra*.

element dS of the boundary, in the virtual displacement contemplated. If there are electromotive sources in certain circuits of the system, which are considered to introduce energy into it from outside itself, the right-hand side of this equation must also contain an expression for the work done by them in the virtual displacement contemplated of the electric coordinates. Now this variational equation can be expressed in terms of any generalized coordinates whatever, that are sufficient to determine the configuration in accordance with what we know of its properties. If we suppose such a mode of expression adopted, then, on conducting the variation in the usual manner and equating the coefficients of each arbitrary variation of a coordinate, we obtain the formulae

$$\Phi = \frac{d}{dt}\frac{dT}{d\dot{\phi}} - \frac{dT}{d\phi} + \frac{dW}{d\phi}, \quad E = \frac{d}{dt}\frac{dT}{d\dot{e}}.$$

formal results, mechanical and electromotive: In these equations Φ is a component of the mechanical forcive exerted on our dielectric system by the conductors, as specified by the rule that the work done by it in a displacement of the system represented by $\delta\phi$, a variation of a single coordinate, is $\Phi\delta\phi$: the corresponding component of the forcive exerted by the dielectric system on the conductor is of course $-\Phi$. Also E is the electromotive force which acts from outside the system in a circuit in which the electric displacement is e, so that the current in it is $\dot{e}$; the electromotive force induced in this circuit by the dielectric system is $-E$.

These equations involve the whole of the phenomena of ordinary electrodynamic actions, whether ponderomotive or electromotive, whether the conductors are fixed or in motion through the medium: in fact, in the latter respect no distinction appears between the cases. They will be completed presently by taking account of the dissipation which occurs in ordinary conductors.

static applied forcive. These equations also involve the expressions for the electrostatic ponderomotive forces, the genesis of which we have already attempted to trace in detail. The generalized component, corresponding to the coordinate ϕ, of the electrostatic traction of the conductors on the dielectric system, is $dW/d\phi$; therefore the component of the traction, somehow produced, of the dielectric system on the conductors is $-dW/d\phi$.

Aether stress is linear. The stress *in the aether* between two electrified bodies consists of a tangential traction on each element of area, equal in magnitude to the tangential component of the electric force at that place, and at right angles to its direction. Quadratic and material stress. The stress in the *material* of the dielectric is such as is produced in the ordinary manner by the surface tractions exerted on the material by the conductors that are imbedded in it. The stress in the dielectric of Faraday and Maxwell has no real existence; it is in fact such a stress as would be felt by the surface of a conductor used to explore the field, when the conductor is so formed and placed as not to disturb the electric force in the dielectric. The magnetic stress of Maxwell is simply a mathematical mode of expression of the kinetic reaction of the medium.

Dielectric currents due to convection of a charge.

The transfer of a charged body across the field with velocity not large compared with the velocity of electric propagation carries with it the whole system of electric displacement belonging to the body, and therefore produces while it lasts a system of displacement currents in the medium, of which the circuits are completed by the actual flow of charge along the lines of motion of the different charged elements of the body.

Conducting molecules foreign to present view.

The phenomena of the electrostatic polarization of dielectrics were at one time provisionally represented by Faraday as due to the orientation of electric polar elements of the medium by the electric force, just as magnetization is actually due to the orientation of the magnetic polar elements by the magnetic force of the field; and this theory was developed at length by Mossotti. At a later period Maxwell himself (*Dynamical Theory*, § 11) compared the electric displacement in a dielectric medium to an actual displacement of the electric charge on conducting molecules imbedded in it—a conception mathematically equivalent to the above. In a previous paper[1] I have explained by simple reasoning that this view is inconsistent with the circuital character of the electric current, a conclusion in agreement with that of von Helmholtz, who adopted this idea in his generalized theory of electrodynamics....

The Faraday model by spherical conductors inadequate for polarization,

As regards the second of these hypotheses, it is to be observed that the moment of electric induction in a conducting atom depends only on its size, and not on the intensity of its free electrification; for the case of conducting spherules the electric moment produced by the action of an electric force F is $3F/4\pi$ multiplied by the total volume of the atoms, and this would give a dielectric inductive coefficient equal to three times the ratio of the aggregate volume of the atoms to the whole volume of the region, a result which is, in any case, far too small to represent the facts, and may easily be so small as to be quite negligible, so as to leave the current practically circuital.

But if we add on to this the first assumption, no room will be left for the explanation of the pyro-electricity and piezo-electricity of crystalline media, by changes of orientation of polar molecules due to changes of temperature or to applied pressure. If this very rational explanation is to be retained, we are driven to assume that the electric

[1] "On a Generalized Theory of Electrodynamics," *Roy. Soc. Proc.* Vol. XLIX. (1891) p. 522; *supra*, p. 232.

force of the field does not sensibly alter the orientation of a molecule, which would then be wholly controlled by the internal electrical, chemical, and cohesive forces of the medium. The state of matters thus required is, in fact, precisely realized by a symmetrical arrangement of positive and negative atoms in the molecule, such as the hexagonal molecule recently imagined by Lord Kelvin[1], and earlier by J. and P. Curie, to account for the piezo-electric quality of quartz; the symmetry of the electric charges makes null the aggregate electric moment and therefore the turning couple in an electric field, while a differential polarity can still be developed under strain of the crystal.

but electron structure suffices.

According to the present theory of electrification, a discharge of electricity from one conductor to another can only occur by the breaking down of the elasticity of the dielectric aether along some channel connecting them; and a similar rupture is required to explain the transfer of an atomic charge to the electrode in the phenomenon of electrolysis. We can conceive the polarization increasing by the accumulation of dissociated ions at the two electrodes of a voltameter, until the stress in the portion of the medium between the ions and the conducting plate breaks down, and a path of discharge is opened from some ion to the plate. While this ion retained its charge, it repelled its neighbours; but now electric attraction will ensue, and the one that gets into chemical contact with it first will be paired with it by the chemical forces; while if the conducting path to the electrode remains open until this union is complete, the ion will receive an opposite atomic charge from the electrode, which very conceivably may have to be also of equal amount, in order to equalize the potentials of the molecule and the plate. This is on the hypothesis that the distance between the two ions of a molecule is very small compared with the distance between two neighbouring molecules. A view of this kind, if thoroughly established, would lead to the ultimate averaging of atomic charges of all atoms that have been in combination with each other, even if those charges had been originally of different magnitudes*.

Discharge either by rupture or by mobile electrons.

(The assignment of free electric charges to vortex atoms tends markedly in the direction of instability; though instability under certain circumstances is essential to electric discharge, yet it must not be allowed to become dominant.—*Dec.* 7, 1893.)

Charged vortex may be unstable.

The presence of vortex atoms, forming faults so to speak in the aether, will clearly diminish its effective rotational elasticity; and

[1] Lord Kelvin, *Phil. Mag.* October 1893.

* This became precise by the introduction of free electrons in the complete memoir (p. 514, § 114 *infra*) as the orbital elements of structure of the atom.

thus it is to be expected that the specific inductive capacities of material dielectrics should be greater than the inductive capacity of a vacuum. The readiness with which electrolytic media break down under electric stress may be connected with the extremely high values of their inductive capacities, indicating very great yielding to even a small electric force.

Dielectric capacity is elastic weakness.

The radiation of a body into the surrounding medium is wholly electrical, and is due to the electric vibrations of the atomic charges; some of these types of vibration may correspond to the single atom by itself, while others will be considerably affected by the presence of the neighbouring atoms of the molecule. The most striking fact to be explained is the total independence of temperature that is exhibited by the periodic times corresponding to the various spectral lines. The extreme smallness of an atom implies correspondingly intense electrification, and therefore independence of the external field. If it is assumed that the dimensions and configuration of the atom are determined by the very intense actions between it and its partners in the molecule, and are not sensibly affected by the comparatively feeble influence of the velocity of translation of the molecule through the medium, this fact will be accounted for; irregularities can then only occur during an encounter with another molecule.

Spectra independent of temperature: may be influenced by density.

In the hydrodynamics of ordinary liquids, when the energy of an isolated vortex ring is increased, the ring expands in radius, and therefore moves onward more slowly. But in the case of an isolated charged atom, an expansion in radius diminishes the potential energy of the electric charge. These two agencies counteract each other; if the latter one is the greater, increase in energy will involve increase in velocity, as would be required in the ordinary form of the kinetic theory of gases. But the more natural supposition is, perhaps, to consider a molecule as composed of atoms paired so that the velocity of translation does not depend intrinsically on the amount of energy associated with the molecule, but is determined by the circumstances of the encounters with other molecules. The distribution of energy between the various vibration types of the molecule, according to the law of Maxwell and Boltzmann, will not affect its configuration, while there is also perfect independence between the hydrodynamical motion of the medium due to the molecule and the radiation produced by it.

Difficulty of vortex atom gas theory.

As regards the rotational elasticity of this hydrodynamical aether, on which we have made all radiative and electrical phenomena depend, it was objected, in 1862, by Sir George Stokes[1] to Mac-

[1] (I am informed by Sir George Stokes that in the above criticism he contemplated only media of which the elements are self-contained, and devoid of internal motions.)

Cullagh's aether, that a medium of that kind would leave unbalanced the tangential surface tractions on an element of volume, and therefore could not be in internal equilibrium; and this objection has usually been recognized, and has practically led to MacCullagh's theory of light being put aside, at any rate in this country. Now, it has been already mentioned that a precisely equivalent objection will apply to the elasticity actually produced by a gyrostatic distribution of momentum in an ordinary solid medium, the only difference in the circumstances being that in the latter case the rotational elasticity is proportional to the angular velocity and not to the angular displacement[1]; and this remark suggests that there must be some way out of the difficulty. If we consider the laws of motion, stated in Newton's manner with reference to absolute space and absolute time, as fundamental principles, then it is also a fundamental principle that the energy of a spinning gyrostat has reference to absolute space, and is not relative to the material system which contains it. The gyrostat may be considered as a kind of connection binding that system to absolute immovable space by means of the forcive which it opposes to rotation; and this is the reason why the element of mass in a gyrostatic medium remains in equilibrium with its translational kinetic reactions, although the tractions of the surrounding parts on its surface are unbalanced and result in a couple. If this mode of viewing the subject is regarded as incongruous, then we must discard from dynamics the notion of absolute space, and we must set out in quest of some transcendental explanation of the directional forcives in rotational systems. In any case the general Lagrangian dynamical procedure applies precisely to the gyrostatic medium we have here taken as an illustration: nor, probably, would its application to MacCullagh's aether be questioned, once the preliminary objection was removed.

Objection to rotational elasticity:

parried by an illustration.

Aether as involving absolute space.

Action.

This question may also be instructively illustrated from another side, by the consideration of an actual medium which possesses precisely the rotational elasticity of MacCullagh's aether. I allude to a solid medium with small magnets interspersed through it in any arbitrary manner, but so that in any single element of volume there is some regularity in their orientation. If this medium when un strained is in equilibrium in a magnetic field, then when an element of it is displaced rotationally it will be acted on by a bodily couple arising from this external field; and therefore the surface tractions on the element would, in the presence of this couple, be unbalanced. Here the disturbing cause is a magnetic forcive arising either from

Magnet in a field is stressed rotationally.

[1] For a detailed discussion of equilibrium and wave-propagation in such a medium, see *Proc. Lond. Math. Soc.* 1891: *supra*, p. 248.

the medium as a whole or from some external system; it has to be considered as of a statical character, that is, the velocity of propagation of the magnetic action is supposed to be indefinitely great compared with the velocity of propagation of any disturbances that are under discussion; the magnetic influence of the whole system is supposed to be instantaneously brought to bear on the element, and not merely the influence of the surrounding parts. On this saving hypothesis, the magnetic energy is here also correctly localized, for dynamical purposes, in the element of volume of the medium, and the Lagrangian method has perfect application to the mathematical analysis of its phenomena.

Now, in the case of the aether we have at hand a *vera causa* precisely of this kind. The cause of the phenomena of gravitation has hitherto remained perfectly inscrutable. Though the present order of ideas forbids us to consider it otherwise than as propagated in time, yet all we know of its velocity of propagation is the demonstration by Laplace that it must*, at the very least, exceed the velocity of propagation of light in the same kind of proportion as the latter velocity exceeds that of ordinary motions of matter. It is not unphilosophical to assume that an explanation of gravitation might carry along with it the explanation of the fact that the tangential tractions on an element of the strained aether are unbalanced. The dynamical phenomena of mass in matter would appear to be analytically explicable by the addition of a rotational part to the kinetic energy of the element of the medium; such a term is of course practically null except in the vortex rings.

An instantaneous gravitation invoked.

In all that has been hitherto said we have kept clear of the complication of viscous forces; but in order to extend our account to the phenomena of opacity in the theory of radiation and of electric currents in ordinary conductors, it is necessary to introduce such forces and make what we can of them on general principles. It is shown that the introduction of the dissipation function into dynamics by Lord Rayleigh enables us to amend the statement of the fundamental dynamical principle, the law of Least Action, so as to include in it the very extensive class of viscous forces which are proportional to absolute or relative velocities of parts of the system. This class is the more important because it is the only one that will allow a simple wave to be propagated through a medium with period independent of its amplitude; if the viscous forces that act in light propagation were not of this kind, then on passing a beam of homogeneous light through a metallic film it should emerge as a mixture of lights of different colours. The viscous forces being thus proved

Dissipation function introduced into Action formula:

exists for all cases of pure vibration,

* If subject to aberration in case of moving masses, like that of radiation.

by the phenomena of radiation to be derived from a dissipation function, it is natural to extend the same conclusion to the elastic motions of slower periods than radiations, which constitute ordinary electric disturbances. We thus arrive, by way of an optical path, at Joule's law of dissipation of electric energy, and Ohm's linear law of electric conduction, and the whole theory of the electrodynamics of currents flowing in ordinary conductors; though the presumption is that the coefficients which apply to motions of long period are not the same as those which apply to very rapid oscillations, the characters of the matter-vibrations that are comparable in the two cases being quite different[1]. If it is assumed that the form of the dissipation function is the same for high frequencies as for low ones, we obtain the ordinary theory of metallic reflexion, which differs from the theory of reflexion at a transparent medium simply by taking the refractive index to be a complex quantity, as was done originally by Cauchy, and later for the most general case by MacCullagh. And, in fact, we could not make a more general supposition than this for the case of isotropic media; while for crystalline media the utmost generality would arise merely from assuming the principal axes of the dissipation function to be different from those of the rotational elasticity, a hypothesis which is not likely to be required.

therefore for currents: simple but adequate, for propagation in crystals.

It has been pointed out, originally by Lord Rayleigh, that to fit this theory to the facts of metallic reflexion it is necessary to take the real part of the index of refraction of the metals to be a negative quantity, which can hardly be allowed on other grounds, as it would imply instability of the medium. We might indeed, following the view of Willard Gibbs and others, imagine an interaction between the light-wave and the free vibrations of the atomic electric charges, and through them the chemical vibrations of the atoms, owing to proximity of their periods; and we might possibly conceive the electric medium to be, so to speak, held together by this kind of support. But I think there is another and simpler alternative that merits examination; we might conceive the opacity at the surface to be so great that a sensible part of the light is lost before it has penetrated more than a very small fraction of a wave-length. In the extreme case of electric waves of finite length reflected from metals, the absorption is complete in a very small fraction of the wave-length, and the result is total reflexion*, as from a vacuum; on the

Apparent optical instability in metals: the electron explanation foreshadowed. Penetration possibly of molecular order of depth. Aether not coarse grained.

[1] It is interesting to notice that, already in his memoir of 1864, Maxwell is struck by the identity of the coefficients of the free aether for all periods, which "shows how perfect and regular the elastic properties of the medium must be when not encumbered with any matter denser than air."

* Corrected however for the ordinary resistance of the metal, as Rubens actually found precisely for infra-red radiation.

other hand, if the opacity is but slight, the phenomena ought to agree approximately with those of transparent media. It seems worth while to examine the consequences of assuming that the optical phenomena of metallic reflexion are nearer the first of these limiting cases than the second. It seems worth while also to compare the facts for some medium not so opaque as metals with the formulae of Cauchy and MacCullagh; the examples of tourmaline crystal, and some of the aniline dyes which exhibit selective absorption, suggest themselves as affording crucial tests[1].

The considerations which have here been explained amount to an attempt to extend the regions of contact between three ultimate theories which have all been already widely developed, but in such a way as not to have much connection with one another. These theories are Maxwell's theory of electric phenomena, including Ampère's theory of magnetism and involving an electric theory of light, Lord Kelvin's vortex atom theory of matter, and the purely dynamical theories of light and radiation that have been proposed by Green, MacCullagh, and other authors. It is hoped that a sufficient basis of connection between them has been made out, to justify a re-statement of the whole theory of the kind here attempted, notwithstanding such errors or misconceptions on points of detail as will unavoidably be involved in it.

Three converging ranges of physical theory.

(While writing this summary it had escaped my memory that Lord Kelvin has proposed a gyrostatic adynamic medium which forms an exact representation of a rotationally elastic medium such as has been here described[2]. If the spinning bodies are imbedded in the aether so as to partake fully in its motion, the rotational forcive due to them is proportional jointly to the angular momentum of a gyrostat and the angular velocity of the element of the medium, in accordance with what is stated above. But if we consider the rotators to be free gyrostats of the Foucault type, mounted on gymbals of which the outer frame is carried by the medium, there will also come into play a steady rotatory forcive, proportional jointly to the square of the angular momentum of the gyrostat and to the absolute angular displacement of the medium. An ideal gyrostatic cell has been imagined by Lord Kelvin in which the coexistence of pairs of gyrostats

Kelvin's gyrostatic model,

1 (An alternative view, in many respects preferable, is supplied by the assumption, with Sir George Stokes, of the existence in metals of an *adamantine* property, such as was discovered by Airy for the diamond [perhaps still awaiting full elucidation]. Cf. Sir G. G. Stokes, *Proc. Roy. Soc.* February 1883.)

2 Lord Kelvin (Sir W. Thomson), *Comptes Rendus*, Sept. 16, 1889; *Collected Papers*, Vol. III. 1890, p. 467. [See *infra*, Part III (1897), or *Aether and Matter* (1900), Appendix.]

spinning on parallel axles in opposite directions cancels the first of these forcives, thus leaving only a static forcive of a purely elastic rotational type. The conception of an aether which is sketched by him on this basis[1] is essentially the same as the one we have here employed, with the exception that the elemental angular velocity of the medium is taken to represent magnetic force, and in consequence the medium fails to give an account of electric force and its static and kinetic manifestations*. A gyrostatic cell of this kind has internal freedom, and therefore free vibration periods of its own; it is necessary to imagine that these periods are very small compared with the periods of the light waves transmitted through the medium, in order to avoid partial absorption. The propagation of waves in this aether, having periods of the same order as the periods of these free vibrations, would of course be a phenomenon of an altogether different kind, involving diffusion through the medium of energy of disturbed motion of the gyrostats within the cells.

of the rotational aether.

Kelvin's vortex sponge is rotationally elastic, if stable.

Lord Kelvin has shown that a fluid medium, in turbulent motion owing to vorticity distributed throughout it, would also possess rotational elasticity provided we could be assured of its permanence. Prof. G. F. FitzGerald proposes to realize such a medium by means of a distribution of continuous vortex filaments, interlacing in all directions; if the vorticities of the filaments in an element of volume are directed indifferently in all directions, the motional part of the kinetic forcive on the element, which depends on the first power of the vorticity, will be null, while the positional part depending on the square of the vorticity will remain, just as in the gyrostatic medium above considered. The atoms may now be imagined to consist of vortex rings making their way among these vortex filaments, and thus a very graphic and suggestive scheme is obtained; the question of stability is however here all-important. No ultimate theory can be final; and schemes of the kind discussed in this paper may not inaptly be compared to structural formulae in modern chemistry; they bind together phenomena that would otherwise have to be taken as disconnected, though they are themselves provisional and may in time be replaced by more perfect representations.

Models are necessarily provisional.

FitzGerald on MacCullagh's scheme.

The electric interpretation of MacCullagh's optical equations, which forms the basis of this paper, was first stated so far as I know by Prof. G. F. FitzGerald, *Phil. Trans.* 1880. I have recently learned, from a reference in Mr Glazebrook's Address, British Association, 1893, that an electric development of Lord Kelvin's rotational aether

[1] Lord Kelvin, *Collected Papers*, Vol. III. 1890, pp. 436–472.

* Until the electrons are introduced (p. 514, *infra*) into it as sources of intrinsic strain.

has been essayed by Mr Heaviside, who found it to be unworkable as regards conduction current, and not sufficiently comprehensive (*Phil. Trans.* 1892, § 16; *Electrical Papers*, Vol. II. p. 543). A method of representing the phenomena of the electric field by the motion of tubes of electric displacement has been developed by Prof. J. J. Thomson, who draws attention to their strong analogies to tubes of vortex motion (*Recent Researches...*, 1893, p. 52).

Lodge finds no influence of a magnetic field on speed of light.

Prof. Oliver Lodge has kindly looked for an effect of a magnetic field on the velocity of light, but has not been able to detect any, though the means he employed were extremely searching; the inference would follow, on this theory, that the motion in a magnetic field is very slow, and the density of the medium correspondingly great.—*Dec.* 18, 1893.)

38

A DYNAMICAL THEORY OF THE ELECTRIC AND LUMINIFEROUS MEDIUM*. PART I.

[*Philosophical Transactions of the Royal Society*, Vol. CLXXXV. pp. 719–822. Received Nov. 15,—read Dec. 7, 1893: revised June 14, 1894. With addition formulating the Theory of Free Electrons and the Electron Theory of Matter, Aug. 13, 1894.]

The distribution of energy in the physical field is fundamental.

1. The object of this paper is to attempt to develop a method of evolving the dynamical properties of the aether from a single analytical basis. One advantage of such a procedure is that by building up everything *ab initio* from a consistent and definite foundation, we are certain of the congruity of the different parts of the structure, and are not liable to arrive at mutually contradictory conclusions. The data for such a treatment lie of course in the properties of the mathematical function which represents the distribution of energy [so far as it determines that of Action] in the medium, when it is disturbed. The consequences which should result from the disturbance are all deducible by dynamical analysis from the expression for this function; and it is the province of physical interpretation to endeavour to identify in them the various actual phenomena, and in so far to establish or disprove the explanation offered. A method of this kind has been employed by Clerk Maxwell with most brilliant results in the discovery and elucidation of the laws of electricity; he has also been led by its development into the domain of optics, and has thus arrived at the electric theory of light. His expression for the energy of the active medium has been constructed from reasoning on the phenomena of electrification and electric currents; this

Maxwell's procedure.

* The main idea of these three memoirs was to express and extend the electric theory by reference, under direct inspiration mainly from Lord Kelvin's writings, to a guiding working model framed as far as possible directly on dynamical principles. Such a model was found in MacCullagh's formal scheme expressing a rotationally elastic aether: it was realized, at any rate over a large range of phenomena, by Kelvin's medium under elastic gyrostatic domination. In the electric development the primordial electron, as for the purpose in hand a source purely electric, but actually an intrinsic structure or knot of unknown type in the medium, soon asserted itself (§ 114). Afterwards in a more formal exposition, *Aether and Matter* (1900), essentially the same theory was developed and further extended, in abstract dynamical terms directly from the formulation of density of Action in the medium, without reference to any special type of model except in explanatory appendices.

procedure offers perhaps difficulties greater than might be, owing to the intangible character of the electric coordinates, and their totally undefined connection with the coordinates of the material system which is the seat of the electric manifestations. In the following discussion, the order of development began with the optical problem, and was found to lead on naturally to the electric one. We shall show that an energy-function can be assigned for the aether which will give a complete account of what the aether has to do in order to satisfy the ordinary demands of Physical Optics; and it will then be our aim to examine how far the phenomena of electricity can be explained as non-vibrational manifestations of the activity of the same medium. The credit of applying with success the pure analytical method of energy to the elucidation of optical phenomena belongs to MacCullagh; he was, however, unable to discover a mechanical illustration such as would bring home to the mind by analogy the properties of his medium, and so his theory has fallen rather into neglect from supposed incompatibility with the ordinary manifestations of energy as exemplified in material structures*. We shall find that such difficulties are now removed by aid of the mechanical example of a gyratory aether, which has been imagined by Lord Kelvin to illustrate the properties of the luminiferous and electric medium. The aether whose properties are here to be examined is not a simple gyrostatic one[1]; it is rather the analogue of a medium filled with magnetic molecules which are under the action, from a distance, of a magnetic system. But the same peculiarities that were supposed to fatally beset MacCullagh's medium and render it inconceivable are present in an actual mechanical medium dominated by gyrostatic momentum.

MacCullagh's procedure: analytic Optical Dynamics.

2. The general dynamical principle which determines the motion of every material system is the Law of Least Action, expressible in

The fundamental Dynamical formulation of a material system:

* When these papers were written the work of MacCullagh had fallen into almost complete discredit. In this country it had suffered under the destructive criticism of Stokes: while abroad it was assumed to be merely a belated version of the theory of crystalline optics of elastic-solid type, developed by F. E. Neumann. At present it has been restored in some degree to its proper position in the historical development of physical optics: see, for example, Rayleigh's obituary notice of Stokes, *Roy. Soc. Proc.* 1903, or F. Klein, *Entwickelung der Mathematik im* 19 *Jahrhundert* (1926), *passim*. Thus the interest of much of what follows is now historical and critical, but based on direct examination of the sources.

Historical.

[1] A medium has however been invented by Lord Kelvin, containing gyrostatic cells composed of arrangements of Foucault gyrostats whose cases are imbedded in it, such as give precisely the rotational elasticity of the aether. [See Part III *infra*, §§ 3–4; or *Aether and Matter* (1900), Appendix.]

the form that $\delta\int(T - W)\,dt = 0$, where T denotes the kinetic energy and W the potential energy of the system, each formulated in terms of any coordinates that are sufficient to specify the configuration and motion in accordance with its known properties and connections; and where the variation refers to a fixed time of passage of the system from the initial to the final configuration considered. The power of this formula lies in the fact that once the energy-function is expressed in terms of any measurements of the system that are convenient and sufficient for the purpose in view, the remainder of the investigation involves only the exact processes of mathematical analysis. It is to be observed that forces which can do no work by reason of constraints of the system tacitly assumed in this specification, but which nevertheless may exist, do not enter at all into the analysis. Thus in the dynamics of an incompressible medium, the pressure in the medium will not appear in the equations, unless the absence of compression is explicitly recognized in the form of an equation of condition between coordinates otherwise redundant, which is combined into the variation in Lagrange's manner; in certain cases (*e.g.* magnetic reflexion of light, *infra*) we are in fact driven to the explicit recognition of such a pressure in order that it may be possible to satisfy all the necessary stress-conditions of the problem, while in other cases (*e.g.* ordinary reflexion of light) the pressure is not operative in the phenomena. There is also a class of cases at the other extreme—typified by a medium such as Lord Kelvin's labile aether which opposes no resistance to laminar compression,—where a certain coordinate does not enter into the energy-function because its alteration is not opposed and so involves no work; in these cases there is solution of a constraint which reduces by one the number of kinematic conditions to be satisfied. In intermediate cases the energy corresponding to the coordinate will enter into the function in the ordinary manner.

[Margin notes: forces of constraint, latent only in equation of energy. — Labile systems.]

3. It is to be assumed as a general principle that all the conditions necessary to be satisfied in any dynamical problem are those which arise from the variation of the Action of the system in the manner of Lagrange. If these conditions appear to be too numerous, the reason must be either that the forcive which compels the observance of some constraint has not been explicitly included in the analysis, or else that the number of the constraints has been over-estimated. In each problem in which the mathematical analysis proceeds without contradiction or ambiguity to a definite result, that result is to be taken as representing the course of the dynamical phenomena in so

[Margin note: Mode of entry of forces of constraint.]

far as they are determined by the energy [Action] as specified; a further more minute specification of the energy* may however lead to the inclusion of small residual phenomena which had previously not revealed themselves.

4. The object of these remarks is to justify the division of the problem of the determination of the constitution of a partly concealed dynamical system, such as the aether, into two independent parts. The first part is the determination of some form of energy-function which will explain the recognized dynamical properties of the system, and which may be further tested by its application to the discovery of new properties. The second part is the building up in actuality or in imagination of some mechanical system which will serve as a model or illustration of a medium possessing such an energy-function. There have been cases in which, after the first part of the problem has been solved, all efforts towards the realization of the other part have resulted in failure; but it may be fairly claimed that this inability to directly construct the properties assigned to the system should not be allowed to discredit the part of the solution already achieved, but should rather be taken as indicating some unauthorized restriction of our ideas on the subject. Of course where more than one solution of the question is possible on the ascertained data, that one should be preferred which lends itself most easily to interpretation, unless some of the others should prove distinctly more fertile in the prediction of new results, or in the inclusion of other known types of phenomena within the system.

General analytic schemes contrasted with special working models:

the latter not essential to a solution.

Practical question of Determinacy.

5. In illustration of some of these principles, and as a help towards the realization of the validity of some parts of the subsequent analysis, a dynamical question of sufficient complexity, which has recently occupied the attention of several mathematicians, may be briefly referred to. The problem of the deformation and vibrations of a thin open shell of elastic material has been reduced to mathematical analysis by Lord Rayleigh[1] on the assumption that, as the shell can be easily bent but can be stretched only with great difficulty, the potential energy of stretching would not appear in the energy-function from which its vibrations in which bending plays a prominent

Powerful elastic resistance can be treated for slow motions as effectively a constraint.

* If it is the form of the energy that is given in terms of coordinates and their velocities, the Lagrangian function, which is the integrand of the Action, is to be obtained from it (after Helmholtz) as the solution of a partial differential equation, analogous to Hamilton's except that its variables are the velocities, and that being linear its general solution is available explicitly.

[1] Lord Rayleigh, "On the Infinitesimal Bending of Surfaces of Revolution," *Proc. Lond. Math. Soc.* 1882.

part are to be determined,—that in fact the shell might be treated as inextensible. But a subsequent direct analysis of the problem, of a more minute character[1], led to the result that the conditions at the boundary of the shell could not all be satisfied unless stretching is taken into account. The reason of the discrepancy is that, if the question is simplified by taking the shell to be inextensible, a static extensional stress ought at the same time to be recognized as distributed all along the surface of the shell, and as assisting in the satisfaction of the necessary conditions at its free edge; the stress condition that can be adjusted in this manner may thus be left out of consideration, as taking care of itself. If we suppose the shell to be not absolutely inextensible, this tension will be propagated over the shell by extensional waves with finite but very great velocity; it will therefore still be almost instantaneously adjusted at each moment over a shell of moderate extent of surface, and the extensional waves will thus be extremely minute; such waves would have a very high period of their own, but in ordinary circumstances of vibration they would be practically unexcited*. These remarks appear to be in keeping with the explanation of this matter which is now generally accepted.

Systems changing structurally with the time.

6. The dynamical method as hitherto explained applies only to cases in which the forces are all derived from a potential-energy function, or are considered as explicitly applied from outside the system; in the latter case they may be, as von Helmholtz remarks, any arbitrary functions of the time. By means of the Dissipation Function introduced by Lord Rayleigh, the equation of Varying Action will be so modified as to include probably all the types of frictional internal forces that are of much importance in physical applications.

Systems involving linear friction.

Rudimentary vector notation.

7. A few words may be said with respect to notation. In order to reduce as much as possible the length to which formulae involving vector quantities extend themselves in ordinary Cartesian analysis, a vector will usually be specified by its three Cartesian components enclosed in brackets, in front of which may be placed such operators as act on the vector. Of particularly frequent occurrence is the operator which deduces the doubled rotation of an element of volume from the vector which represents the translation; this will, after Maxwell, receive a special designation, and will here be called the

[1] A. E. H. Love, "On the...Vibrations of a Thin Elastic Shell," *Phil. Trans.* 1888.

* Cf. Stokes' fundamental memoir on the communication of sound-vibrations to the air, *Phil. Trans.* 1868; *Math. and Phys. Papers*, Vol. IV. pp. 299–324.

vorticity or curl of that vector. If the vector represent the displacement in an incompressible medium, *i.e.* if it has no convergence, we have $(\text{curl})^2 = -\nabla^2$, where ∇^2 is Laplace's well-known scalar operator. The introduction of still more vector analysis would further shorten the formulae, and probably in practised minds lead to clearer views; but the saving would not be very great, while as yet facility in vector methods is not a common accomplishment. In the various transformations by means of integration by parts that occur, after the manner of Green's analytical theorem, it is not considered necessary to express at length the course of the analysis; so as there is no further object in indicating explicitly by a triple sign the successive steps by which a volume integration is usually effected, it will be sufficient to take the symbol $d\tau$ to represent an element of volume and cover it by a single sign of integration. In the notation of surface integrals, the ordinary usage is somewhat of this kind[1].

PART I. PHYSICAL OPTICS.

Preliminary and Historical.

8. The development of the analytical theory of the aether which will be set forth in this paper originated in an examination of Professor G. F. FitzGerald's Memoir "On the Electro-magnetic Theory of the Reflection and Refraction of Light[2]," of which the earlier part is put forward by the author as being a translation of MacCullagh's analysis of the problem of reflexion into the language of the electromagnetic theory. Later on in the Memoir the author discusses the rotation of the plane of polarization of the light, which is produced by reflexion from the surface of a magnetized medium, assumed in the analysis to be transparent; but the application of MacCullagh's method to this case leads him to more surface conditions than can be satisfied by the available variables, and the rigorous solution of the problem is not attained. After satisfying myself that this contradiction is really due to the omission from consideration of the *quasi*-hydrostatic pressure which must exist in the medium and assist in satisfying the stress conditions at an interface, though on account of the incompressible character of the medium this pressure takes no part in the play of energy on which the kinetic phenomena depend, it was natural to turn to MacCullagh's optical writings[3], in

MacCullagh: FitzGerald.

Latent non-working stresses reveal themselves at optical interfaces.

[1] Various matters have been treated from rather different points of view in the abstract of this paper. *Roy. Soc. Proc.* Vol. LIV. pp. 438–461, *supra*, p. 389.

[2] G. F. FitzGerald, *Phil. Trans.* 1880. [*Collected Scientific Writings*, pp. 41–73.]

[3] *The Collected Works of James MacCullagh*, ed. Jellett and Haughton, 1880.

order to ascertain whether a similar idea had already presented itself. An examination, particularly of "An Essay towards a Dynamical Theory of Crystalline Reflexion and Refraction[1]," led in another direction, and showed that to MacCullagh must be assigned the credit of one of the very first notable applications to physical problems of that dynamical method which in the hands of Maxwell, Lord Kelvin, von Helmholtz, and others, has since been so productive, namely, the complete realization of Lagrange's theory that all the phenomena of any purely dynamical system free from viscous forces are deducible from the single analytical function of its configuration and motion which expresses the value of its energy. The problem proposed to himself by MacCullagh was to determine the form of this function for a continuous medium[2], such as would lead to all the various laws of the propagation and reflexion of light that had been ascertained by Fresnel, supplemented by the exact and crucial observations on the polarization produced by reflexion at the surfaces of crystals and of metallic media, which had been made by Brewster and Seebeck. He arrived at a complete solution of this problem, and one characterized by that straightforward simplicity which is the mark of all theories that are true to Nature; but he was not able to imagine any mechanical model by which the properties of his energy-function could be realized. In another connection, in vindicating his equations for the rotatory polarization of quartz[3] against a theory of Cauchy's leading to different results, he however expresses himself on such a question as follows[4]. "For though, in my Paper, I have said nothing of any mechanical investigation, yet as a matter of course, before it was read to the Academy, I made every effort to connect my equations in some way with mechanical principles; and it was because I had failed in doing so to my own satisfaction, that I chose to publish the equations without comment, as bare geometrical assumptions, and contented myself with stating orally... that a mechanical account of the phenomena remained a *desideratum* which no efforts of mine had been able to supply." And again, "though for my own part I never was satisfied with that theory [of Cauchy], which seemed to

Method of Action.

Type of dynamical scheme required by crystalline optics.

A mechanical model of the medium useful, not essential to the dynamics.

[1] MacCullagh, *loc. cit.* p. 145; *Trans. Roy. Irish Acad.* Vol. XXI. Dec. 9, 1839.

Green.

[2] The problem had already been fully analyzed by Green, shortly before, and unknown to MacCullagh, precisely on these principles, but without success owing to his restriction to elasticity of the type of an ordinary solid body; cf. Green's "Memoir on Ordinary Refraction," *Trans. Camb. Phil. Soc.* Dec. 11, 1837, introduction, and his "Memoir on Crystalline Propagation," *Trans. Camb. Phil. Soc.* May 20, 1839.

[3] MacCullagh, "On the Laws of the Double Refraction of Quartz," *Trans. Roy. Irish Acad.* 1836; *Collected Works*, p. 63.

[4] MacCullagh, *Proc. Roy. Irish Acad.* 1841; *Collected Works*, pp. 198, 200.

me to possess no other merit than that of following out in detail the extremely curious, but (as I thought) very imperfect analogy which had been perceived to exist between the vibrations of the luminiferous medium and those of a common elastic solid,...still I should have been glad, in the absence of anything better, to find my equations supported by a similar theory, and their form at least countenanced by a like mechanical analogy."

9. After trying an empirical alteration of Cauchy's equations for the stress in his medium[1], which sufficed to satisfy Brewster's observations on reflexion from crystals, but did not agree with subsequent observations of a different kind by Seebeck, MacCullagh was finally led to results which were in keeping with all the experiments by means of the principles[2] that (i) the displacements in the incident and reflected waves, compounded as vectors, are geometrically equivalent at the interface to the displacements in the refracted waves, compounded in the same manner, and (ii) there is no loss of energy involved in the act of reflexion and refraction. This agreement was obtained, provided he took the displacement to be in the plane of polarization of the light, and the density of the aether to be the same in all media.

Preliminary empirical solutions of the problem of optical reflexion:

Shortly before, and unknown to MacCullagh, F. E. Neumann[3] had based the solution of the problem of reflexion on the very same principles; and he had as early as 1833 ascertained that his results agreed with Seebeck's experiments, though MacCullagh had priority in publication. He began by applying to the problem of reflexion the equations of motion of an elastic solid, as then imperfectly understood in accordance with the prevalent theory of Navier and Poisson; he recognized that there were six interfacial conditions to be satisfied, three of displacement and three of stress, while in the absence of compressional waves there were enough variables to satisfy only four of them; he cut the knot of this difficulty by assuming that the displacement must be continuous, to avoid rupture of the medium at the interface, and assuming that there is no loss of energy in the act of reflexion and refraction of the light, thus asserting the absence of waves of compression, and at the same time leaving the conditions as to continuity of stress altogether out of his account. As his displacement is in the plane of polarization, the solution arrived at by Neumann is formally the same as MacCullagh's; but it can be shown

MacCullagh's compared with F. E. Neumann's.

[1] MacCullagh, "On the Laws of Reflexion from Crystallized Surfaces," *Phil. Mag.* Vol. VIII. 1835.

[2] MacCullagh, "On the Laws of Crystalline Reflexion," Dec. 13, 1836; *Phil. Mag.* Vol. X. 1837.

[3] F. E. Neumann, *Abhandl. der Berliner Akad.* 1835, pp. 1–116.

that the reasoning by which Neumann arrived at it, from the basis of an elastic solid aether, is invalid, so that the solution as stated by him must be considered to be the result of a fortunate accident, the correctness of which he would have had no real ground, in the absence of comparison with observations, for anticipating; while MacCullagh afterwards (in 1839) placed his own empirical theory on a real dynamical foundation.

10. The hypothesis on which Neumann's surface conditions are virtually based has been expounded and amplified in more recent times by Kirchhoff[1]; and in this form it is often quoted as Kirchhoff's principle. The analysis of Kirchhoff also amends Neumann's defective energy-function by the substitution for it of the one determined by Green, by the condition that the displacements in two of the three types of waves that can travel unchanged in the medium are in the plane of the wave-front. About the rate of propagation of the third wave, involving compression in the medium, Kirchhoff makes no hypothesis, but he avails himself of the remark (originally due to MacCullagh) that the transverse waves involve no compression, and therefore are independent, as regards their propagation, of the term in the energy which involves compression. He assumes that in the act of reflexion and refraction no compressional waves are produced; and that this is so because extraneous forces act on the interface just in such manner as to establish the continuity of stress across it, while on account of the conservation of the energy they can do no work in the *actual motion* of the medium at the interface. The explicit recognition of such forces constitutes Kirchhoff's principle; as to their origin he says that it lies in traction exerted by the matter on the aether which is unbalanced at the surface of discontinuity, and that they are somehow of the same nature as the capillary force at the interface between two liquids; as to their happening to be precisely such as will extinguish the compressional waves, he merely says that it must be so, because as a matter of fact no compressional waves are produced by the reflexion, the energy being assumed to be all in the reflected and refracted light-waves. On the other hand, the pure elastic theory has been worked out on Neumann's hypothesis, for the simple case of an isotropic medium, without the assumption of these extraneous forces, by Lorenz, Lord Rayleigh, and others, and has been shown to lead to loss of light owing to the formation of compressional waves which carry away some of the energy, and to laws of reflexion quite irreconcilable with observation.

Kirchhoff's assumed extraneous non-working interfacial forces,

distinct from bodily elastic stress:

[1] G. Kirchhoff, "Ueber die Reflexion und Brechung des Lichtes an der Grenze krystallinischer Mittel," *Abh. der Berl. Akad.* 1876; *Ges. Abh.* p. 367.

11. Can then any justification be offered of Kirchhoff's doctrine of extraneous surface forces? The parallel case which is appealed to for its support is that of capillary forces at an interface between two fluids. Now on Gauss' theory of capillarity these forces are derived simply from the principle of energy; each fluid being in equilibrium, its intrinsic energy is distributed throughout its interior with so to speak uniform volume-density; if we imagine the surface of transition to be sharp, and each fluid to retain its properties unaltered right up to it, the total energy will be simply the sum of the two volume-energies and will not depend on the surface at all; as a matter of necessity, however, there is a gradual transition from one fluid to the other across a thin surface layer, and the energy per unit volume in this layer alters with the change of properties; so that to the energy estimated as if the transition were sharp, there is to be made a correction which takes the form of a surface distribution of energy; and this latter term must reveal itself, according to Gauss' well-known reasoning, in the phenomena of capillary surface tension. The relation between the volume-densities of the energy in the two fluids is determined by the proper balance of intrinsic hydrostatic pressure across the interface. Now if we adhere at all to the principle that the play of energy, as distributed throughout the masses in the field, is the proper basis for the interpretation of physical phenomena, the extraneous surface forces of Kirchhoff must also be accounted for in some such way as the above; they must arise out of the influence of a layer of gradual transition between the media. But superior limits have been obtained to the thickness of such a layer in various ways, by actual measurement; such limits are found in the thickness of the thinnest possible soap-film, as measured by Reinold and Rücker, or in the thickness of the film of silvering which in Quincke's experiments just suffices to extinguish the influence of the glass, on which it is deposited, on the phenomena of surface tension. The former limit is about one-fortieth of the wave-length of green light, the latter limit is well within one-tenth of the same wave-length[1]. The quantity with which to compare the surface energy due to this transition is the energy contained in a wave-length of the light whose reflexion is under consideration. It is plain that such an amount of surface energy as is here possible will not suffice to totally transform the circumstances of the reflexion, and therefore will not account for Kirchhoff's extraneous forces. Furthermore, a layer of transition, of thickness of the same order of magnitude as the wave-length, would

Interfacial (*quasi*-capillary) energy.

[1] Reinold and Rücker, *Roy. Soc. Proc.* 1877; *Phil. Trans.* 1883. Quincke, *Pogg. Ann.* Vol. CXXXVII. 1869. Cf. Lord Kelvin, *Popular Lectures and Addresses*, Vol. I. p. 8.

introduce a change of phase into the reflexion, such as we know, from Lord Rayleigh's and Drude's experiments on reflexion from absolutely clean surfaces of transparent media, does not exist, and such as even Kirchhoff's own theory does not allow for. It is for these reasons that it is here considered that Neumann's theory of light is, on his own dynamical basis, untenable, and leads to the correct result only by accident,—and that the credit of the solution of the fundamental dynamical problem of Physical Optics belongs essentially to MacCullagh.

An elastic solid theory untenable for optical reflexion.

12. To return now to the course of the development of optical doctrine in MacCullagh's hands, he recounts in straightforward fashion[1], somewhat after the custom usual with Faraday, the way in which after successive trials he was at last guided to the formal laws which govern the phenomena of reflexion. To his success two main elements contributed; the bent of his genius led him to apply the methods of the ancient Pure Geometry, of which he was one of the great masters, to the question, and this resulted in simple conceptions, such as the principle of equivalent vibrations already explained, which are applicable to the most general aspect of the problem; while the variety and exactness of the experiments of Brewster and Seebeck on the polarization of the light reflected from a crystal gave him plenty of material by which to mould his geometrical views. The simple theorems[2] of the *polar plane* and of *transversals*, by which he expressed without symbols in the compass of a single sentence, and in two different ways, the complete solution of the most general problem of crystalline reflexion, contrast with the very great complexity of the analytical solutions of Neumann and Kirchhoff. Thus at the end of this paper he remarks that "several other questions might be discussed, such as the reflexion of common light at the first surface, and the internal reflexion at the second surface of a crystal[3]; but these must be reserved for a future communication. It would be easy indeed to write down the algebraical solutions resulting from our theory; but this we are not content to do, because the expressions are rather complicated, and when rightly treated will probably contract themselves into a simpler form. It is the character of all true theories that the more they are studied the

MacCullagh's geometrical analysis of crystalline reflexion:

[1] MacCullagh, "On the Laws of Crystalline Reflexion and Refraction," *Trans. R.I.A.* Vol. XVIII. Jan. 9, 1837.

[2] MacCullagh, *Collected Works*, pp. 97 and 176.

[3] It is interesting to observe that, in the notes appended to the paper, MacCullagh has actually obtained the geometrical solution of this seemingly most complicated question by means of a very powerful and refined application of the principle of reversibility of the motion, which was afterwards employed to such good purpose by Sir G. G. Stokes.

more simple they appear to be." "We are obliged to confess that, with the exception of the law of *vis viva*, the hypotheses" on which the solution is founded "are nothing more than fortunate conjectures. These conjectures are very probably right, since they lead to elegant laws which are fully borne out by experiments; but that is all that we can assert respecting them. We cannot attempt to deduce them from first principles; because, in the theory of light, such principles are still to be sought for. It is certain, indeed, that light is produced by undulations, propagated, with transversal vibrations, through a highly elastic aether; but the constitution of this aether, and the laws of its connexion (if it has any connexion) with the particles of bodies, are utterly unknown. The peculiar mechanism of light is a secret which we have not yet been able to penetrate...but perhaps something might be done by pursuing a contrary course; by taking these laws for granted, and endeavouring to proceed upwards from them to higher principles...." He then allows himself to give a pure mechanical interpretation to his formal results, taking his displacement to be linear, and he derives the conclusion that the effective density of the aether is the same in all bodies.

at first tentative:

13. In the notes appended to this purely formal paper MacCullagh "afterwards proved that the laws of reflexion at the surface of a crystal are connected, in a very singular way, with the laws of double refraction, or of propagation in its interior"; he was led to infer that "all these laws and hypotheses have a common source in other and more intimate laws that remain to be discovered; and that the next step in physical optics would probably lead to those higher and more elementary principles by which the laws of reflexion and the laws of propagation are linked together as parts of the same system." And in the following memoir[1] he takes this step by developing his dynamical theory. His analysis is based on the hypothesis of constant density of the aether, and on the principle of rectilinear vibrations in crystalline media, substances like quartz being excepted. "Concerning the peculiar constitution of the ether we know nothing, and shall assume nothing, except what is involved in the foregoing assumptions," and that it may be taken as homogeneous for the problem in hand.

then consolidated into a dynamical formulation by Action.

In Section III of this paper MacCullagh proceeds to determine the potential-energy function on which the transverse rectilinear vibrations propagated through the aether must depend. He observes that such vibrations involve no condensation; and as in a plane wave all

[1] MacCullagh, "An Essay towards a Dynamical Theory of Crystalline Reflexion and Refraction," *Trans. R.I.A.* Vol. XXI. Dec. 9, 1839.

the points in the medium move in parallel directions, the effective strain produced in it may be taken to be specified by the rotation of the element, which is round a line in the plane of the wave-front and at right angles to the line of the displacement, this rotation being proportional to the rate of change of the displacement in the direction of propagation. Having previously shown, probably for the first time, that the expression now interpreted as representing the elementary rotation in the displacement of a medium by strain, enjoys the invariant properties of a vector, he at once seizes upon it as the very thing he wants, as it has a meaning independent of any particular system of axes to which the motion is referred; and he makes the potential energy of the medium a quadratic function of the components of this elementary rotation. As pointed out by Stokes[1], the possible forms of the effective strain and therefore of the energy-function are by no means thus restricted: in fact Green had a short time previously established another form, in which the energy depends on the components of the strain of the medium, as it would do if the medium possessed the properties of an elastic solid.

Invariance of form of the Action a necessary criterion.

At any rate, MacCullagh assumes a purely rotational quadratic expression for the energy, which he reduces to its principal axes in the ordinary manner; and then he deduces from it in natural and easy sequence, without a hitch, or any forcing of constants, all the known laws of propagation and reflexion for transparent isotropic and crystalline media. In common with Neumann, he cannot understand how with Fresnel the inertia in a crystal could be different in different directions, or its elasticity isotropic; so he assumes the density of the aether to be the same in all media, but its elasticity to be variable. The laws of crystalline reflexion are then established as below, and shown to be embraced in a single theorem relating either to his transversals or to his polar plane; and the memoir ends with a remark "which may be necessary to prevent any misconception as to the nature of the foundation on which" the theory stands. "Everything depends on the form of the function V; and we have seen that, when that form is properly assigned, the laws by which crystals act upon light are included in the general equations of dynamics. This fact is fully proved by the foregoing investigations. But the reasoning which has been used to account for the form of the function is indirect, and cannot be regarded as sufficient, in a mechanical point of view. It is, however, the only kind of reasoning that we are able to employ, as the constitution of the luminiferous medium is entirely unknown."

MacCullagh's scheme a formally complete dynamical system.

[1] Sir G. G. Stokes, "Report on Double Refraction," *Brit. Assoc.* 1862. MacCullagh possibly perceived this afterwards himself; cf. note at the end of his memoir.

MacCullagh's Optical Equations.

14. Let the components of the linear displacement of the primordial medium be represented by (ξ, η, ζ), and let (f, g, h) represent the curl or vorticity of this displacement, *i.e.*

His analysis by Action expressed in terms of a displacement vector:

$$(f, g, h) = \left(\frac{d\zeta}{dy} - \frac{d\eta}{dz},\ \frac{d\xi}{dz} - \frac{d\zeta}{dx},\ \frac{d\eta}{dx} - \frac{d\xi}{dy}\right),$$

so that this vector is equal to twice the absolute rotation of the element of volume. The elasticity being purely rotational, the potential energy per unit volume of the strained medium is represented by a quadratic function U of (f, g, h), so that

$$W = \int U\, d\tau,$$

where $d\tau$ denotes an element of volume. The kinetic energy is

$$T = \tfrac{1}{2}\rho\int\left(\frac{d\xi^2}{dt^2} + \frac{d\eta^2}{dt^2} + \frac{d\zeta^2}{dt^2}\right)\delta\tau.$$

The general variational equation of motion is

$$\delta\int (T - W)\, dt = 0,$$

for integration through any fixed period of time. Thus[1]

$$\int dt \left[\rho\int\left(\frac{d\xi}{dt}\frac{d\delta\xi}{dt} + \frac{d\eta}{dt}\frac{d\delta\eta}{dt} + \frac{d\zeta}{dt}\frac{d\delta\zeta}{dt}\right) d\tau \right.$$
$$\left. - \int\left\{\frac{dU}{df}\left(\frac{d\delta\zeta}{dy} - \frac{d\delta\eta}{dz}\right) + \frac{dU}{dg}\left(\frac{d\delta\xi}{dz} - \frac{d\delta\zeta}{dx}\right) + \frac{dU}{dh}\left(\frac{d\delta\eta}{dx} - \frac{d\delta\xi}{dy}\right)\right\} d\tau\right] = 0.$$

On integration by parts in order to replace the differential coefficients of $\delta\,(\xi, \eta, \zeta)$ by these variations themselves, we obtain, leaving out terms relating to the beginning and end of the time,

$$\int dt \left[-\rho\int\left(\frac{d^2\xi}{dt^2}\delta\xi + \frac{d^2\eta}{dt^2}\delta\eta + \frac{d^2\zeta}{dt^2}\delta\zeta\right) d\tau \right.$$
$$- \int\left\{\left(\frac{d}{dy}\frac{dU}{dh} - \frac{d}{dz}\frac{dU}{dg}\right)\delta\xi + \left(\frac{d}{dz}\frac{dU}{df} - \frac{d}{dx}\frac{dU}{dh}\right)\delta\eta\right.$$
$$\left. + \left(\frac{d}{dx}\frac{dU}{dg} - \frac{d}{dy}\frac{dU}{df}\right)\delta\zeta\right\} d\tau$$
$$+ \int\left\{\left(m\frac{dU}{dh} - n\frac{dU}{dg}\right)\delta\xi + \left(n\frac{dU}{df} - l\frac{dU}{dh}\right)\delta\eta\right.$$
$$\left.\left. + \left(l\frac{dU}{dg} - m\frac{dU}{df}\right)\delta\zeta\right\} dS\right] = 0,$$

[1] Cf. G. F. FitzGerald, "On the Electromagnetic Theory...," *Phil. Trans.* 1880. In that memoir the rotation is represented by $4\pi\,(f, g, h)$, instead of simply (f, g, h) as above, in order to be in line with Maxwell's electrodynamic equations.

where (l, m, n) are the direction cosines of the element of surface dS. As the displacements $\delta\,(\xi, \eta, \zeta)$ are as yet quite arbitrary, the equations of elastic vibration of the medium are therefore

$$\rho\frac{d^2\xi}{dt^2}+\frac{d}{dy}\frac{dU}{dh}-\frac{d}{dz}\frac{dU}{dg}=0,$$

$$\rho\frac{d^2\eta}{dt^2}+\frac{d}{dz}\frac{dU}{df}-\frac{d}{dx}\frac{dU}{dh}=0,$$

$$\rho\frac{d^2\zeta}{dt^2}+\frac{d}{dx}\frac{dU}{dg}-\frac{d}{dy}\frac{dU}{df}=0.$$

From them it follows that

$$\frac{d\xi}{dx}+\frac{d\eta}{dy}+\frac{d\zeta}{dz}=0,$$

giving transverse waves independent of any condensation that may exist.

in other words, that there is no compression of the medium involved in this motion, whether we assume that it has the property of incompressibility or not.

15. In accordance with the general dynamical principle, all the conditions which it is essential to explicitly satisfy at an interface between two media are those which secure that the variation of the energy [Action] shall not involve a surface integral over this interface. To express these conditions most concisely, let us take for the moment the element of the interface to be parallel to the plane of yz, so that $(l, m, n) = (1, 0, 0)$; the surface integral term corresponding to one side of the interface is now

$$\int\left(-\frac{dU}{dg}\,\delta\eta+\frac{dU}{dh}\,\delta\zeta\right)dS,$$

where $\delta\eta$, $\delta\zeta$ are perfectly arbitrary, subject only to being continuous across the interface. Thus to make the surface integral part of the variation vanish, we must have dU/dg and dU/dh, the tangential components of the traction, continuous across the interface; it follows from the first of the equations of motion that the continuity of ξ is also thereby secured, provided the density is the same on both sides; and the normal traction on the interface is null. The continuity in the flow of energy across the interface is of course also necessarily involved. Of the complete set of six conditions only four are thus independent, which is the precise number required for the problem of optical reflexion between crystalline media.

Condensational waves not excited in reflexion.

It has not been necessary to assume incompressibility of the medium in order to avoid waves of longitudinal disturbance. A medium of this type, however heterogeneous in elastic quality from part to part, whether compressible or not, will transmit waves of transverse displacement in absolute independence of waves of compression, provided its density is everywhere the same; the one type of wave cannot possibly change into the other.

16. If $$(\xi, \eta, \zeta) = \text{curl}\,(\xi_1, \eta_1, \zeta_1),$$
so that $$(f, g, h) = -\nabla^2 (\xi_1, \eta_1, \zeta_1),$$
[as there is no compression,] and if the equations of propagation are referred to the principal axes of the medium so that now

Relation to Fresnel's displacement vector.

$$U = \tfrac{1}{2}(a^2f^2 + b^2g^2 + c^2h^2),$$

they assume the form

$$\rho \frac{d^2}{dt^2}(\xi_1, \eta_1, \zeta_1) = \nabla^2 (a^2\xi_1, b^2\eta_1, c^2\zeta_1),$$

which are precisely Fresnel's equations of crystalline propagation[1]. The vector (ξ_1, η_1, ζ_1) of Fresnel is at right angles to the plane of polarization; therefore its curl (ξ, η, ζ), which is the displacement of the medium on MacCullagh's theory, is in the plane of polarization.

17. In the theory of reflexion the tangential components of the displacement are continuous, and the tangential components of the stress are continuous; these conditions, or the more direct conditions of continuity of displacement and continuity of energy, taken in conjunction with the hypothesis of effective density constant throughout space, lead immediately to Fresnel's equations of reflexion for isotropic media, and in MacCullagh's hands give a compact geometrical solution when the media are of the most general character. A medium of this kind, however heterogeneous and aeolotropic as regards elasticity, is still adapted to transmit transverse undulations without any change into the longitudinal type; and the conditions of propagation are all satisfied without setting up any normal tractions in the medium, which might if unbalanced produce motion of translation of its parts. Thus the incidence of light-waves on a body will not give rise to any mechanical forces*.

Alternative expression of interfacial conditions.

Pressure of radiation is outside this theory.

Alternative Optical Theories.

18. The equations of propagation of Fresnel above mentioned obviously agree with those which are derivable from the variational equation

Alternative models for the same Action formulation, thus dynamically equivalent.

$$\delta \int dt \left[\tfrac{1}{2}\kappa \int \left(\rho a^{-2} \frac{d\xi_1{}^2}{dt^2} + \rho b^{-2} \frac{d\eta_1{}^2}{dt^2} + \rho c^{-2} \frac{d\zeta_1{}^2}{dt^2} \right) d\tau \right.$$
$$\left. - \tfrac{1}{2}\kappa \int (f_1{}^2 + g_1{}^2 + h_1{}^2)\, d\tau \right] = 0,$$

[1] MacCullagh, *Proc. R.I.A.* Vol. II. 1841; *Collected Works*, p. 188.

* The interfacial conditions would be independent of any existing mechanical forcive of radiation if (ξ, η, ζ) were a rotational displacement instead of a linear one: then the reflexion would not give rise to any distribution of torque over the interface, but as to distribution of pressural force nothing is involved.

which belongs to a medium having aeolotropic inertia of the kind first imagined by Rankine, and having isotropic purely rotational elasticity. The coefficient of elasticity κ may be in the first instance assumed to be different in different substances. The surface conditions for the problem of reflexion which are derived from this equation are clearly, in the light of the above analysis, continuity of tangential displacement and of tangential stress. A compression of the medium now takes part in the propagation of transverse undulations, yet the compression does not appear in this isotropic potential energy-function; hence the resistance to laminar compression must be null, the other alternative infinity being on the latter account inadmissible. The surface condition as to continuity of normal displacement need not therefore be explicitly satisfied; and the remaining surface condition of continuity of normal traction is non-existent, there being no normal traction owing to the purely rotational quality of the elasticity. Whether a medium of this type could be made to lead to the correct equations of reflexion we need not inquire. [See, however, § 21.]

Wider alternatives, illustrated by Kelvin's labile mechanical aether: its modified Action.

19. It has been shown by Lord Kelvin[1] that a medium of elastic-solid type is possible which shall oppose no resistance to laminar compression, viz. to compression in any direction without change of dimensions sideways, and that its potential energy if elastically isotropic is of the same form as the above, with the addition of some terms which, integrated over the volume, are equivalent to a surface integral. The remaining coefficient of elasticity, that is the rigidity, must then be the same in all media, to avoid static instability; that condition is in fact required as below, in order that waves may be transmissible at all through a heterogeneous medium of this type.

As an illustration of this somewhat abstract discussion, let us conduct the variation of the Action in this labile elastic-solid medium. The equation takes the form

$$\delta \int dt \left[\tfrac{1}{2} \int \left(\alpha^2 \frac{d\xi^2}{dt^2} + \beta^2 \frac{d\eta^2}{dt^2} + \gamma^2 \frac{d\zeta^2}{dt^2} \right) d\tau - \tfrac{1}{2}\kappa \int \left\{ \left(\frac{d\zeta}{dy} + \frac{d\eta}{dz} \right)^2 + \left(\frac{d\xi}{dz} + \frac{d\zeta}{dx} \right)^2 \right.\right.$$
$$\left.\left. + \left(\frac{d\eta}{dx} + \frac{d\xi}{dy} \right)^2 - 4 \left(\frac{d\eta}{dy}\frac{d\zeta}{dz} + \frac{d\zeta}{dz}\frac{d\xi}{dx} + \frac{d\xi}{dx}\frac{d\eta}{dy} \right) \right\} d\tau \right] = 0;$$

it would be illegitimate for the present purpose to replace the potential energy by a surface part and a volume part, because then it would

[1] Lord Kelvin (Sir W. Thomson) "On the reflexion and refraction of light," *Phil. Mag.* 1882 (2), p. 414; Glazebrook, *Ibid.* p. 521.

not be correctly located in the medium. We obtain on the left-hand side the time-integral of the expression

$$-\int\left(\alpha^2 \frac{d^2\xi}{dt^2}\delta\xi + \beta^2\frac{d^2\eta}{dt^2}\delta\eta + \gamma^2\frac{d^2\zeta}{dt^2}\delta\zeta\right)d\tau$$

$$-\kappa\int\left\{\left(\frac{d\zeta}{dy}+\frac{d\eta}{dz}\right)(m\delta\zeta + n\delta\eta) + \left(\frac{d\xi}{dz}+\frac{d\zeta}{dx}\right)(n\delta\xi + l\delta\zeta)\right.$$

$$+\left(\frac{d\eta}{dx}+\frac{d\xi}{dy}\right)(ld\eta + md\xi) - 2\left(\frac{d\eta}{dy}+\frac{d\zeta}{dz}\right)ld\xi - 2\left(\frac{d\zeta}{dz}+\frac{d\xi}{dx}\right)m\delta\eta$$

$$\left.-2\left(\frac{d\xi}{dx}+\frac{d\eta}{dy}\right)n\delta\zeta\right\}dS$$

$$+\kappa\int\left\{\left(\delta\zeta\frac{d}{dy}+\delta\eta\frac{d}{dz}\right)\left(\frac{d\zeta}{dy}+\frac{d\eta}{dz}\right)+\left(\delta\xi\frac{d}{dz}+\delta\zeta\frac{d}{dx}\right)\left(\frac{d\xi}{dz}+\frac{d\zeta}{dx}\right)\right.$$

$$+\left(\delta\eta\frac{d}{dx}+\delta\xi\frac{d}{dy}\right)\left(\frac{d\eta}{dx}+\frac{d\xi}{dy}\right) - 2\delta\xi\frac{d}{dx}\left(\frac{d\eta}{dy}+\frac{d\zeta}{dz}\right)$$

$$\left.-2\delta\eta\frac{d}{dy}\left(\frac{d\zeta}{dz}+\frac{d\xi}{dx}\right) - 2\delta\zeta\frac{d}{dz}\left(\frac{d\xi}{dx}+\frac{d\eta}{dy}\right)\right\}d\tau,$$

or collecting and exhibiting specimen terms only,

$$-\int\left(\alpha^2\frac{d^2\xi}{dx^2}\delta\xi + \ldots\right)d\tau$$

$$-\kappa\int\left[l\left\{\left(\frac{d\eta}{dy}+\frac{d\zeta}{dz}\right)\delta\xi + \left(\frac{d\eta}{dx}+\frac{d\xi}{dy}\right)\delta\eta + \left(\frac{d\zeta}{dx}+\frac{d\xi}{dz}\right)\delta\zeta\right\} + \ldots\right]dS$$

$$-\kappa\int\left[\delta\xi\left\{\frac{d}{dy}\left(\frac{d\eta}{dx}-\frac{d\xi}{dy}\right)-\frac{d}{dz}\left(\frac{d\xi}{dz}-\frac{d\zeta}{dx}\right)\right\}+\ldots\right]d\tau.$$

The equations of motion are thus

$$\alpha^2\frac{d^2\xi}{dt^2}=\frac{dh}{dy}-\frac{dg}{dz},\quad \beta^2\frac{d^2\eta}{dt^2}=\frac{df}{dz}-\frac{dh}{dx},\quad \gamma^2\frac{d^2\zeta}{dt^2}=\frac{dg}{dx}-\frac{df}{dy},$$

reducible to MacCullagh's by changing (ξ, η, ζ) into (a^2f, b^2g, c^2h), making the corresponding change for (f, g, h), and taking

$$(\alpha^2, \beta^2, \gamma^2) = \rho\,(a^{-2}, b^{-2}, c^{-2});$$

while the surface conditions are easily seen by taking $(l, m, n) = (1, 0, 0)$ to be continuity of tangential elastic-solid tractions, and continuity of tangential displacement; both these results might of course have been foreseen from the formulae for the tractions in an elastic solid, without special analysis. The surface condition involving normal displacement can be adjusted by the lability of the medium as regards simple elongation; and the continuity of its coefficient, that is, of the normal forcive as determined by the lateral contraction, is already secured by the other surface conditions, provided the elasticity is continuous. The mode in which lability thus affects the surface conditions in the method of variations, is the chief point that required illustration; the addition to the energy of § 18 of terms which form a perfect differential is seen to be immaterial, provided they show no discontinuity at the interface.

Correct location of the energy is essential only in the problem of reflexion.

20. It is of interest to observe that a geometrical transformation, specified by the equations[1]

$$(x, y, z) = pqr\left(\frac{x'}{p}, \frac{y'}{q}, \frac{z'}{r}\right), \text{ and } (\xi, \eta, \zeta) = pqr\,(p\xi', q\eta', r\zeta'),$$

leads to $\quad d\tau = d\tau', \text{ and } (f, g, h) = pqr\left(\frac{f'}{p}, \frac{g'}{q}, \frac{h'}{r}\right),$

and so leaves the elastic quality of a purely rotational medium unaltered.

Also, the variational equation of MacCullagh

$$\delta\int dt\left[\tfrac{1}{2}\rho\int\left(\frac{d\xi^2}{dt^2}+\frac{d\eta^2}{dt^2}+\frac{d\zeta^2}{dt^2}\right)d\tau - \tfrac{1}{2}\int(a^2f^2+b^2g^2+c^2h^2)\,d\tau\right]=0$$

may be expressed, so far as regards vibrations of period $2\pi/n$, in the form

$$\delta\int dt\left[\tfrac{1}{2}\int\rho n^2(\xi^2+\eta^2+\zeta^2)\,d\tau - \tfrac{1}{2}\int(a^2f^2+b^2g^2+c^2h^2)\,d\tau\right]=0,$$

in which the distinction between coordinates and velocities, between potential and kinetic energy, has been obliterated, if we regard n as simply a numerical coefficient.

If in the above transformation (p, q, r) is taken equal to (a, b, c), this variational equation of MacCullagh is changed into the one appropriate to an aether of isotropic rotational elasticity and aeolotropic effective density, as discussed above; and the wave-surface is changed into its polar reciprocal, which is also a Fresnel's surface in which a, b, c are replaced by their reciprocals; and the geometrical relations between the two schemes may be correlated on this basis. This mode of transformation does not however extend to surface integral terms, and so cannot be applied to the problem of reflexion.

Other partial models.

The same end might have been attained by taking (f, g, h) to denote displacement and (ξ, η, ζ) proportional to rotation in the variational equation; for $\nabla^2(\xi, \eta, \zeta) = -\operatorname{curl}(f, g, h)$, and the operator ∇^2 may be replaced by a constant so far as regards light propagation in a single medium. This interchange, which has already been indicated in § 18, does not affect the development of the

[1] Cf. *Proc. Lond. Math. Soc.* 1893, p. 278, *supra*, p. 298. Afterwards extended by Heaviside to general linear transformation, such as can include magnetic aeolotropy; giving a form of wave-surface which is Fresnel's form subjected to uniform linear deformation. (See footnote, p. 298 *supra*.)

variational equation except as regards surface-integral terms; and the character of the modification of the geometrical relations of the wave-surface, on passing from the one theory to the other, is now open to inspection[1].

All such models are classifiable under the standard electric scheme of Maxwell.

(21. Added June 14. The formal relations between these various mechanical theories may be very simply traced by comparing them with the electromagnetic scheme of Maxwell. In that theory the electric and magnetic inductions, being circuital, are necessarily in the plane of the wave-front; while the electric and magnetic forces need not be in that plane. On taking the electric or the magnetic induction to represent the mechanical displacement of the medium, the electric theory coincides formally with that of Fresnel or that of MacCullagh respectively; while on taking the electric or the magnetic force to represent the mechanical displacement, we obtain the equations of the correlative theories of Boussinesq, Lord Kelvin, and other authors[2]. Thus, for example, it follows at once from this correlation that the combination of aeolotropic inertia with labile isotropic elasticity will lead, not only to Fresnel's wave-surface as Glazebrook has shown, but also to MacCullagh's theory of crystalline reflexion and refraction. If we suppose the magnetic quality of the medium to take part in the vibrations, as would probably be the case to some extent with very slow electric waves, the equations of propagation would possess features analogous to those due to an alteration of density in passing from one medium to another, on the mechanical theory here adopted. But the continuity of normal displacement of the medium could not now be satisfied in the problem of reflexion, the appropriate magnetic condition being instead continuity of induction. A homogeneous mechanical medium representing or illustrating such a case would thus have to possess suitable labile properties; in the ordinary optical circumstances in which magnetic quality is not effective, the degree of compressibility is on the other hand immaterial, and no normal wave will be started in reflexion.)

Treatment of the Problem of Reflexion by the Method of Rays.

The earlier discussions by rays:

22. We are now in a position to compare the various investigations of the problem of reflexion, by means of rays, that have been given by Fresnel, Neumann, MacCullagh and others. It is a cardinal principle in all theories of transparent media that there is no loss of energy in the act of reflexion and refraction. Consequently there is no energy

[1] Cf. J. Willard Gibbs, "A comparison of the electric theory of light and Sir W. Thomson's theory of a quasi-labile aether," *Phil. Mag.* 1889 [or *Collected Papers*].

[2] Cf. Drude, *Göttinger Nachrichten*, 1892.

carried away by longitudinal waves in the aether; and this must usually be either because the medium offers no resistance to laminar compression, or because it is incompressible, the case of rotational elasticity being however not thus restricted. The rays are most simply defined as the paths of the energy.

their definition.

23. Let us consider the first of these hypotheses, that of null velocity of longitudinal waves. At the interface the tangential components of the displacement must be continuous, otherwise there would be very intense tangential tractions acting in the thin interfacial layer of transition, such as could not be equilibrated by the tractions outside that layer. The normal components of the displacement need not be made continuous, for the neighbourhood of this thin interfacial layer will stretch without effort as much as may be required. The tangential stresses must be continuous across the layer of transition, otherwise they would produce very great acceleration of this layer which could not be continuous with the moderate accelerations outside it. As we have thus already obtained the sufficient number of conditions the normal pressure need not also be explicitly made continuous, for the continuity of tangential displacements should secure its continuity as well; if the medium is constituted so as regularly to reflect waves at all, this must be the case, and it is clear on a moment's consideration of the formula for the pressure that it is so in a labile medium of isotropic elastic-solid type. We have thus the four conditions, continuity of tangential displacement and of tangential stress; and the one sufficient condition which will secure that they also make the normal stress continuous, *i.e.* that the medium is a possible one, is that there shall be no loss of energy in the operation of reflexion and refraction. The four conditions here specified are mathematically equivalent to those of Fresnel's theory of reflexion; and the satisfaction of the fifth condition carried with it the justification of that theory for the type of medium which it implies. For the case worked out by Fresnel, that of isotropic media, the constitution of his medium is thus limited to be precisely that of the labile aether of Lord Kelvin; in order to satisfy also the fifth condition, that of continuity of energy, we are constrained to take the displacement perpendicular to the plane of polarization, which gives a reason independent of experiment for Fresnel's choice.

The Kelvin unloaded labile aether is also Fresnel's.

24. Let us consider the second form of hypothesis, that of incompressibility. At the interface all three components of the displacement must now be continuous; and to obtain a solution, there is needed only one other condition, which may be taken to be the preservation of the energy of the motion. Here, as Neumann remarks, there is

absolutely nothing assumed about the elastic condition of the media, which may in fact remain wholly unknown except as to their assumed incompressibility and as to the law of density, and the problem of reflexion will nevertheless be completely solved. But if we go further than this, and attempt to speculate about the elasticity of the optical medium, it must be limited to be of such nature as also to satisfy two other conditions which are involved in the continuity of the tangential stress at the interface*.

For an incompressible optical aether the elastic character can remain latent.

Thus on the principles that the energy is propagated along the rays, that it is at any instant half potential and half kinetic, and that there is no loss of energy of the light in the act of reflexion, and on the hypothesis that the medium is incompressible, the solution of the problem of reflexion as distinct from that of the elastic constitution of the medium is immediately derived, for all media which polarize the light linearly, without the aid of further knowledge except the law of density and the form of the wave-surface. If the density is uniform and the same in all media, the solution is that of MacCullagh and Neumann, which is known to be correct in form for isotropic (and also for crystalline) media. There is nothing so far to indicate whether the vibrations are in the plane of polarization or at right angles to it, but that point is soon settled by the most cursory comparison with observation of the resulting formulae for the two kinds of polarized light; the vibrations must be in the plane of polarization of the light. It remains in this order of procedure, to discover a form of the potential-energy function which will lead to the correct form of wave-surface in crystalline media, at the same time making the vibrations in the plane of polarization, and which also will conform to the additional surface conditions not utilized in order to obtain merely the solution of the problem of reflexion; the discovery of such a function, as a result of a precise estimation of what was really required, is MacCullagh's special achievement†.

The Neumann-MacCullagh tentative scheme.

MacCullagh's dynamical synthesis:

* Kirchhoff proposed to adjust the stress by an extraneous surface traction: as *supra*, § 11. In the problem of an incompressible elastic medium a distribution of isotropic pressure is set up by the reflexion, which adjusts itself with infinite velocity and so ensures continuity of stress across the surface, but at the cost of departure from Fresnel's laws.

† There proved to be in fact more clues towards general theory for an intuitive and profoundly attentive geometrical intellect, like that of MacCullagh, in the connections of the complex crystalline phenomena than in the simpler analysis for isotropic substance. Yet one could also imagine the original Fresnel formulas as built up tentatively from the simpler results for direct reflexion, along with other features. The exposition in Airy's *Tract* goes back to 1831, nearly to Fresnel's intuitions, so is much earlier than Green or MacCullagh: its perusal makes the magic of their analytical elucidation of total reflexion very conspicuous. As Green's dynamic precedes that of MacCullagh

which would allow independent compressional waves.

25. If the aether in crystalline media is of aeolotropic rotational elastic quality, and of isotropic effective inertia the same in all media, all the conditions of the problem of actual optical reflexion are satisfied whatever be the degree of its compressibility. While, on the other hand, if it is of isotropic elastic-solid quality and aeolotropic effective inertia, and there is no elastic discontinuity in passing from one medium to another, *i.e.* if the elasticity is the same in all media, all the conditions are satisfied when there is no resistance to laminar compression. It is somewhat remarkable that the condition of continuity of the energy assumes the same form in both these cases.

Extension into electrodynamics.

What happens under more general conditions, or in circumstances of mixed elastic-solid and rotational elasticity, or possibly yet more general types of elasticity, we shall not stop at present to inquire. [See, however, § 21.] For the explanation of electrical phenomena, MacCullagh's energy-function possesses fundamental advantages for which none of these other possible optical schemes appear to be able to offer any equivalent; it is therefore not necessary to examine whether they can survive the searching ordeal of crystalline reflexion.

Total Reflexion.

26. So long as there actually exist the full number of refracted waves, this simple mode of solution of the problem by means of rays is perfectly rigorous, and puts the matter in as clear a light as a more detailed analysis of what is going on in the media; it is not necessary to make any assumption about the character of the incident wave, except that it is propagated without change. But the case is different when the incidence on a rarer medium is so oblique that one or both the refracted waves disappear; if we simply treat these waves as non-existent, the four surface conditions cannot all be satisfied. The natural inference is that the solution of the problem now depends on the particular form of the wave; the fundamental simple-harmonic form is the obvious one to choose, so let the vibration be represented by

$$A \exp \iota\, 2\pi\lambda^{-1} (lx + my + nz - vt),$$

real parts only being in the end retained. The satisfaction of the interfacial conditions,—which must now be chosen all linear [thus not

by two years and is mentioned in the *Abstract* of MacCullagh's *Memoir*, it is conceivable that it was the source of his stimulation to search for an Action function which would translate into analysis his own formal geometrical constructs: but the internal evidence of MacCullagh's *Collected Papers* is that the ideas came independently to him.

In striking contrast to these ancient explorations are the strenuous modern endeavours to modify the free radiation theory so as to make it amenable to discrete *quanta* of molecular Action.

introducting, *e.g.* that of continuity of the energy] as we are running a real and an imaginary part concurrently, and they must not get mixed up,—leads to a complex value of n for one or both of the refracted waves and of A for both of them. The interpretation is, of course, in the first case purely surface-waves, in the second a change of phase [remarkable because sensibly abrupt] in the act of reflexion or refraction. With this modification the celebrated interpretation of the imaginary expression in his formulae, by Fresnel, becomes quite explicit, and the general problem of total or partial crystalline reflexion is solved for the type of medium virtually assumed by him, without any detailed consideration of the nature of the elasticity. The hypothesis is implied, and may be verified, that the surface-waves penetrate into the medium to a depth either great, or else small, compared with the thickness of the layer of transition between the media,—a point which has not always been sufficiently noticed.

The imaginary velocity within the reflecting medium.

Reflexion at the Surfaces of Absorbing Media.

27. The fact that homogeneous light in passing through a film of metal does not come out a mixture of various colours, or more crucially the fact that the use of a metallic speculum in a telescope does not interfere with spectrum observations, shows that the equation of vibration of light in a metallic medium is linear, and therefore that to represent the motion of the light in the metal requires simply the introduction of an ordinary exponential coefficient of absorption. The interface being the plane of xy, the light propagated in the absorbing medium will be represented by the real part of an expression of the form $A \exp \iota\, 2\pi\lambda^{-1} (lx + my + nz - vt)$, where n is now complex with its real part negative if the axis of z is towards the direction of propagation. If the opacity of the medium is so slight that the light gets down some way beyond the interfacial layer of transition without very sensible weakening, we may therefore solve the problem of reflexion by an application of the ordinary surface conditions stated in a linear form, but with a complex coefficient of elasticity; for we may treat the layer of transition as practically indefinitely thin. This comes to the same thing as the method used first by Cauchy, of simply treating the index of refraction as a complex quantity in the ordinary formulae for transparent media; and it should give a satisfactory solution of the problem, provided the opacity is not excessive.

Equations of metallic reflexion must be linear.

The results obtained for metallic reflexion are however found to suffer, when compared with observation, from several serious defects; the real part of the *quasi*-index of refraction becomes negative, which is sufficient to prevent any stable self-subsisting medium from acting

Objection of instability.

in this manner*; while on transmission through certain metallic films there is a gain of phase of the light compared with vacuum, when there ought, according to the equations, to be a loss.

Optical Dispersion in Isotropic and Crystalline Media†.

28. In order to make our luminiferous medium afford an explanation of electric and magnetic phenomena, it will be necessary to assume its potential energy to be wholly rotational, therefore quite independent of compression or distortion. When bodies are displaced through it, its motion will then be precisely that of a continuous frictionless incompressible fluid, and therefore no rotational stress will be thereby produced in it.

Mobility secured.

Dispersion, if assumed of static elastic type:

The phenomena of optical dispersion require us to recognize a dependence of the effective elasticity of the medium on the wave-length of the light; for we are bound on this theory, in the absence of sympathetic rotational vibrations of the atoms, to take the effective density of the primordial medium to be the same throughout all space. The dependence of the elasticity on the length of the wave can only arise from the presence of a structure of some sort in the medium, representing the molecular arrangement of the matter, whose linear dimensions are comparable with the wave-length of the disturbance that is propagated through it. The actual motion will now be of a very complicated character; but the fact that a wave is propagated through without change, in certain media (those which are at all transparent), shows that for the present purpose it is formally sufficient to average the disturbance into a continuous differential analysis, and thus take it to be a simple one as if there were no molecular discreteness, but with an effective elastic modulus proper to its wave-length. The expression for the potential energy of the medium will thus have to be of a form that will vary with the wave-length, while it is still a quadratic function of differential coefficients of the displacements; therefore we must now assume it to involve differential coefficients of higher order than the first. This mode of formulating the problem is what is led up to by the transparency of dispersive media, *i.e.* by the permanence of type of simple waves

a generalized uniform rotational energy-function could include it:

* The objection is nugatory for a complex medium such as aether and its electrons. See *infra*, § 114.

† Sections 28 to 36 extend the theory of a uniform medium purely rotational, after MacCullagh, to include dispersional effects, *elastic*, not *inertial*. The energy-function involves gradients higher than the first, but confined to purely rotational type. The interest of this problem, other than purely analytical, is superseded by the introduction of discrete electrons into the medium, in § 114 *seq.*, and especially in the next paper, Part II, § 12.

travelling through them, and by the rotational character of the optical elasticity which is quite distinct from that of the molecular web, and, we may assume, of a different order of magnitude. It need excite no surprise if in extreme circumstances, involving near approach to equality with free periods of vibration, it is insufficient.

29. Now if the medium is to be thoroughly and absolutely fluid as regards non-rotational motions, *i.e.* if a vortex atom theory of matter is to be part of the theory of the aether, this potential energy-function must be such that no work is done by any displacement which does not involve rotation, therefore such that the work done by any displacement whatever is of the form

$$\int (L\delta f + M\delta g + N\delta h)\, d\tau,$$

or
$$\int \left\{ L\left(\frac{d\delta\zeta}{dy} - \frac{d\delta\eta}{dz}\right) + M\left(\frac{d\delta\xi}{dz} - \frac{d\delta\zeta}{dx}\right) + N\left(\frac{d\delta\eta}{dx} - \frac{d\delta\xi}{dy}\right)\right\} d\tau,$$

together with possible surface integral terms. Integration by parts leads to the expression

$$\int\left\{\left(\frac{dN}{dy} - \frac{dM}{dz}\right)\delta\xi + \left(\frac{dL}{dz} - \frac{dN}{dx}\right)\delta\eta + \left(\frac{dM}{dx} - \frac{dL}{dy}\right)\delta\zeta\right\} d\tau.$$

This expression must be the same as the one derived by integration by parts in the usual manner from the variation of the potential energy $\delta\int W d\tau$, where W is now of the second degree in spacial differential coefficients, of various orders, of (ξ, η, ζ). The result, as far as the volume integral is concerned, will be the same as if the symbols of differentiation d/dx, d/dy, d/dz were dissociated from ξ, η, ζ and treated like symbols of quantity, after the sign of each has been changed, so that for example $d\xi/dy\, d^2\eta/dx^2$ is to be taken the same as $-\, d/dy\, d^2/dx^2\, \xi\eta$; the function W may thus be replaced for this purpose by

$$W' = A\xi^2 + B\eta^2 + C\zeta^2 + 2D\eta\zeta + 2E\zeta\xi + 2F\xi\eta,$$

where A, B, C, D, E, F are functions of d/dx, d/dy, d/dz.

We shall then have

$$\delta\int W d\tau = \int\{\ldots\}\, dS + \int\left(\frac{dW'}{d\xi}\delta\xi + \frac{dW'}{d\eta}\delta\eta + \frac{dW'}{d\zeta}\delta\zeta\right) d\tau.$$

On comparing these expressions there results

$$\left(\frac{dN}{dy} - \frac{dM}{dz},\ \frac{dL}{dz} - \frac{dN}{dx},\ \frac{dM}{dx} - \frac{dL}{dy}\right) = \left(\frac{d}{d\xi},\ \frac{d}{d\eta},\ \frac{d}{d\zeta}\right) W'.$$

Hence
$$\left(\frac{d}{dx}\right)\frac{dW'}{d\xi} + \left(\frac{d}{dy}\right)\frac{dW'}{d\eta} + \left(\frac{d}{dz}\right)\frac{dW'}{d\zeta} = 0$$

identically, where the differential operators in brackets are to be

still giving waves purely distortional or rotational.

treated as if they were symbols of quantity. The vanishing of this expression, for all values of ξ, η, ζ, involves three conditions between $A, B, \ldots$, one of which may be stated in the form that the quadratic expression W' is the product of two linear factors; these are in fact the general analytical conditions that a medium shall not propagate waves of compression involving sensible amounts of energy*.

Theories indistinguishable as regards propagation.

30. But these conditions are not sufficient to insure that the elasticity shall be purely rotational, and in no wise distortional. For example, as may be seen from the above, the elasticities of Lord Kelvin's labile elastic-solid aether and of Green's incompressible aether satisfy them. What is required is that for any displacement of a given portion of the medium, the total work done by both the bodily forcive and the surface tractions shall be expressible in terms of the rotations of its elementary parts alone. In the particular case in which the medium is in internal equilibrium in a state of strain, the part of this work which is due to bodily forcive is of course null; so that the surface tractions are then all-important.

A type of crystalline wave-surface uninfluenced by dispersion.

31. Now let us examine a form of W_2, the dispersional part of the energy, which has been put forward by MacCullagh solely in order to explain the fact that the character of the crystalline wave-surface is not altered by the dispersional energy. He assumes that W_2 is a function of (f, g, h) and of its vorticity or curl, and of the curl of that curl, say its curl squared, and so on; and he observes that if this quadratic function only involve squares and products of the respective components of odd powers of the curl, Fresnel's wave-surface is unaltered, while if even powers come in, the surface is modified in a simple and definite manner[1]; it will be clear on consideration that if an odd power of the operator is combined with an even power, in any term, rotational quality of the medium must be introduced. It will be sufficient for practical applications to attend to the dispersional terms of lowest order. Since in an incompressible medium $(\text{curl})^2 = -\nabla^2$, these terms yield two possible forms for the dispersional part of the energy,

MacCullagh's formal scheme for rotatory polarization.

$$f\nabla^2 f + g\nabla^2 g + h\nabla^2 h$$

and

$$(\nabla^2\xi)^2 + (\nabla^2\eta)^2 + (\nabla^2\zeta)^2;$$

* Rather shall only propagate waves of differential rotation, or, what comes to the same thing within the uniform medium but not at interfaces, of distortion.

[1] MacCullagh, "On the dispersion of the Optic Axes and of the Axes of Elasticity in Biaxal Crystals," *Phil. Mag.* October 1842, *Collected Works*, pp. 221–226; "On the law of Double Refraction," *Phil. Mag.* 1842, *Collected Works*, pp. 227–229.

or in a crystalline medium we might take the corresponding forms

$$\alpha^2 f\nabla^2 f + \beta^2 g\nabla^2 g + \gamma^2 h\nabla^2 h$$

and $$\alpha'^2 (\nabla^2\xi)^2 + \beta'^2 (\nabla^2\eta)^2 + \gamma'^2 (\nabla^2\zeta)^2;$$

or we could have more generally the lineo-linear function of (f, g, h) and $\nabla^2 (f, g, h)$ and the general quadratic function of $\nabla^2 (\xi, \eta, \zeta)$, respectively, which would not be symmetrical with respect to the principal optical axes of the medium.

The first of these forms, the intermediate case being taken for brevity, yields a bodily forcive

$$\nabla^2 \left(\frac{d\gamma^2 h}{dy} - \frac{d\beta^2 g}{dz}, \ \frac{d\alpha^2 f}{dz} - \frac{d\gamma^2 h}{dx}, \ \frac{d\beta^2 g}{dx} - \frac{d\alpha^2 f}{dy}\right),$$

and the second one yields a bodily forcive

$$(\alpha'^2\nabla^2\nabla^2\xi, \ \beta'^2\nabla^2\nabla^2\eta, \ \gamma'^2\nabla^2\nabla^2\zeta).$$

Both of these forcives satisfy the condition of being null when the medium is devoid of rotation. But, as in the motion of a train of plane waves of length λ the operator ∇^2 is replaceable by the constant $-(2\pi/\lambda)^2$, we see that the first forcive merges in the ordinary rotational forces of the medium, only altering its effective crystalline constants in a manner dependent on the wave-length; while the second forcive alters the character of the equations by adding to the right-hand sides terms proportional to ξ, η, ζ, and so modifies the wave-surface. If with MacCullagh we had taken the last and most general type of terms, which are not symmetrical with respect to the principal axes of optical elasticity, the observed dispersion of the optic axes of crystals would clearly have been involved in the equations. The nature of the proof of MacCullagh's general proposition is easily made out from the examination here given of this particular case.

Dispersion of optic axes in crystals.

The types of elasticity that are independent of distortion:

32. The question has still to be settled, whether the postulate of complete fluidity as regards irrotational motion limits the form of W_2 to the one assumed by MacCullagh. It will I think be found that it does. For the final form of the variation of the potential energy is

$$\delta \int W\, d\tau = \int \{\ldots\}\, dS + \int (P\,\delta f + Q\,\delta g + R\,\delta h)\, d\tau,$$

where (P, Q, R) involve (f, g, h) linearly, but with differential operators of any orders. We may change it to

$$\delta \int W\, d\tau = \int \{\ldots\}\, dS - \int \mathrm{curl}\,(P, Q, R)\, \delta\,(\xi, \eta, \zeta)\, d\tau,$$

the expression in the integral representing a scalar product; and this form shows that the bodily forcive in the medium is curl (P, Q, R).

It also shows that the curl operator persists on integration by parts. Now this forcive is linear in (ξ, η, ζ), and taking for a moment the case of an isotropic medium, it must be built up of invariant differential operators. The complete list of such operators consists of curl, convergence, and shear operators, and their powers and products; and these operators are mathematically convertible with each other. Any combination of them, operating on (ξ, η, ζ), which involves curl as a factor, will limit the medium, as has been already seen, to the propagation of waves only rotational; but in order to secure perfect fluidity as regards irrotational motions it is necessary also that the surface tractions, involved in the surface integral part of the variation of the energy, shall not depend on the shear or convergence of the medium. Now in arriving at the final form of the variational equations, by successive integrations by parts, if a convergence or shear occur in either factor of a term in W, it will emerge at some stage as an actual convergence or shear of the medium in a surface integral term, indicating a surface traction which violates the condition of fluidity. But the only forms of W_2 for an isotropic medium, which maintain an invariantive character independent of axes of coordinates, and in which each factor involves only (f, g, h), appear to be made up of MacCullagh's forms and the form

invariance:

fluidity:

$$\left(\frac{dh}{dy}+\frac{dg}{dz}\right)^2+\left(\frac{df}{dz}+\frac{dh}{dx}\right)^2+\left(\frac{dg}{dx}+\frac{df}{dy}\right)^2;$$

and if the medium is incompressible this new form is identical with the second type of MacCullagh. The conclusion thus follows that for isotropic media, the form of the potential energy, when we include dispersion and other secondary effects in it, is that of MacCullagh, the two forms given by him being in this case identical.

33. The question now presents itself, whether there is any distinction between the two types into which MacCullagh divides possible energy-functions of this kind, which will enable us to reject the one that modifies the form of the wave-surface*. It seems fair to lay stress on the circumstance that the first of MacCullagh's types of dispersional energy may represent an interaction between the average strain of the medium (f, g, h) and the average disturbance of the strain due to molecular discreteness, while the other form represents the energy of some type of disturbance of the strain which combines only with itself, and is not directly operative on the average strain. It would seem natural to infer that a term of the second type would have its

modified Fresnel wave-surface.

* MacCullagh explored this latter possibility, in relation to uncertainties of the time, in *Phil. Mag.* (1842), *Collected Papers*, pp. 227–229.

coefficient of a higher order of small quantities than the ones we are now investigating.

For the most general case of aeolotropy, the dispersional energy W_2 must be either a quadratic function of first differential coefficients of (f, g, h), or else a lineo-linear function of (f, g, h) and its second differential coefficients. If the first alternative be rejected for the reason just given, there remains a form of which MacCullagh's is the special case in which the second differential coefficients group themselves into the operator ∇^2. A reason for this restriction is not obvious, unless we may take the form already determined for an isotropic medium as showing that the dispersion arises from the interaction of (f, g, h) on $\nabla^2 (f, g, h)$; such a restriction is in fact demonstrable when we bear in mind the scalar character of the energy-function.

The Influence of Dispersion on Reflexion.

34. It has been explained that on this theory the mode of formal representation of dispersion without sensible absorption is, by the inclusion of differential coefficients of the displacement, higher than the first, in the energy-function. This makes the dispersion depend on change of elasticity, and not on any effective change of inertia of the primordial medium; in the neighbourhood of a dark band in the absorption spectrum of the medium, absorption plays an important part, rendering the phenomena anomalous, and we must then have recourse to some theory of the Young-Sellmeier type, involving perhaps change of effective inertia, which will take a more complete account of the sympathetic interaction which occurs between the electric vibrations of the molecules and the vibrations of the medium, when their periods are very nearly alike. Elastic dispersion, and inertial.

The sum of the orders of the differential coefficients in any term of the energy must usually be even; a term in which it is odd would introduce unilateral quality into the medium, typified by such phenomena as rotatory polarization; and it is known from the facts and principles of crystalline structure that such terms can be, when existent at all, only of a very minute residual kind. Rotational.

When we come to discuss the problem of reflexion, the surface terms derived from the variation of the energy-function must be retained, and they should be adjusted so as to maintain the continuity of the manifestations of energy in crossing the interface. But the dispersional terms will introduce into the variational equation surface integrals involving not only $\delta\xi$, $\delta\eta$, $\delta\zeta$, but also $\delta\,(d\xi/dx)$, $\delta\,(d^2\xi/dx^2)$, ...; and we cannot even attempt to make all these independent terms continuous across the interface. We therefore cannot Reflexion not sharp if dispersion were of elastic type.

follow in our analysis the complete circumstances of the problem of reflexion*. This is not cause for surprise, because the essence of the method of continuous analysis consists of averaging the molecular discreteness of the medium; and we are now trying to fit this analysis on to conditions at an interface where the law of the discreteness changes abruptly or rather very rapidly.

Treatment by rays and energy: and

35. In a problem of this kind the procedure by the method of rays asserts a marked superiority. The interfacial layer being assumed for other reasons to be very thin compared with a wave-length, the displacement of the medium must be continuous across it. And it may be fairly assumed that there is no sensible amount of degradation of energy in this very thin superficial layer; so that the principle of continuity of energy gives the remaining interfacial condition. The result of these hypotheses will be that, so far, the law of reflexion of each homogeneous portion of the light depends on its own index, and not on the amount of the dispersion in its neighbourhood. The assumption of continuity of energy is the same thing as recognizing that the continuity of the dispersional part of the stress at the interface is maintained by surface forces of molecular character, which absorb no energy, and which need not be further specified for the present purpose,—thus forming an instance of a perfectly valid application of a surface traction principle of the same kind as that of Neumann and Kirchhoff (§ 10).

Kirchhoff's interacting intrinsic surface stress.

This explanation is based on MacCullagh's theory of reflexion. If, merely for further illustration, we take Fresnel's analysis of that problem, the medium is thereby assumed to be labile, and we must employ a stress condition at the interface as well as the energy condition. Now it is exactly in the insufficient specification of the stress near the surface that the trouble with respect to the dispersional terms came in; thus, if Fresnel's theory were the tenable one, it would be a matter of some difficulty to get from it a clear view of reflexion in its relation to dispersion.

The Structural Rotational, or Helical, Quality of Certain Substances.

36. The quality of rotatory polarization, exhibited by quartz and turpentine, depends on the structure of the optical medium, and therefore must be expressed by a term in the potential energy W. When symbols of differentiation are imagined for the moment as separable from their operands, this term must be of the third degree

* On the theory of an aether pervaded by electrons, as introduced in § 114 *infra*, the problem is definite: for magnetic reflexion it has been worked out completely, on the lines of the present papers, by J. G. Leathem, *Phil. Trans.* 1897.

in $(d/dx, d/dy, d/dz)$; and it must be quadratic in (ξ, η, ζ). It can therefore only involve the rotation (f, g, h) and its curl, each of them linearly[1]; therefore, being a scalar, the only form it can have is that of their scalar product; thus the term we are in quest of must be* **Helical terms in energy: isotropic:**

$$C\left\{f\left(\frac{dh}{dy}-\frac{dg}{dz}\right)+g\left(\frac{df}{dz}-\frac{dh}{dx}\right)+h\left(\frac{dg}{dx}-\frac{df}{dy}\right)\right\},$$

or what is the same

$$-C\{f\nabla^2\xi+g\nabla^2\eta+h\nabla^2\zeta\}.$$

This is in fact the term invented by MacCullagh for the purpose of explaining the rotational phenomena of liquids, and of quartz in the direction of its optic axis, and shown by him and subsequent investigators to account for the facts. In the case of a crystalline medium, we might have for this term the general function of (f, g, h) and its curl, that is linear in both; but probably in all uniaxial crystals, **crystalline.** certainly in quartz, the principal axes of this term are the same as the principal axes of optical elasticity of the medium.

On the Elasticity of the Primordial Medium.

37. The objection raised by Sir G. G. Stokes[2] in 1862 against the possibility of a medium of the kind contemplated by MacCullagh's energy-function, and since that time generally admitted, is that an element of volume of such a medium when strained could not be in equilibrium under the elastic tractions on its boundaries, but would require the application of an extraneous couple, of amount proportional to its surface and [to an arm leading to] proportion to its mass, in order to keep it balanced. Such a state of matters is of course in flagrant contradiction to the character of the elasticity of solid bodies, and can only occur if there is some concealed rotational phenomenon going on in the element, the kinetic reaction of which can give rise to the requisite couple. If the medium had acquired its rotational **Nature of rotational elasticity:** elasticity by means of a distribution of rotating simple gyrostats, such a kinetic couple would be afforded by it so long as rotational

[1] (June 14. The rotatory term in the energy-function cannot involve differential coefficients with respect to the time; for to obtain the structural type of rotation these would have to appear in the second degree, which would make the term, as it involves only (f, g, h), of the fourth order in differential operators; cf. *Brit. Assoc. Report*, 1893, "Magnetic Action on Light," § 3, *supra*, p. 312. Thus MacCullagh's term involves on the present theory only the one hypothesis that the medium is self-contained, and not effectively under the influence of another interpenetrating medium.)

* For developments see *Aether and Matter* (1900), § 133 *seq*.

[2] Sir George Stokes corroborates my impression that his criticism is expressly limited to media the elements of which are at rest and self-contained, and that it is not to be regarded as effective against a medium of gyrostatic quality or of the *quasi*-magnetic quality described below.

motion of the element is going on[1], and Stokes' criticism would not apply in this case. If again we imagine an ordinary elastic medium full of elementary magnets with orientations distributed according to some law or even at random, and in internal equilibrium either in its own magnetic field or in the field of some external magnetic system, then on rotational distortion a couple will be required to hold each element in equilibrium; so that the conjugate tangential tractions on the surface of the element cannot be equal and opposite in this case either. The couple depends here on the absolute rotation of the element of volume, not on its angular velocity as in the previous illustration. The potential energy of such a medium as this will contain rotational terms of MacCullagh's type, and its condition of internal equilibrium will be correctly deduced from an energy-function containing such terms by the application of the Lagrangian analysis. The origin of the elasticity purely rotational of MacCullagh's medium is we may say unknown; the first example here given shows that it cannot be simply gyrostatic, though Lord Kelvin has invented a complex gyrostatic structure that would produce it[2]; and either example shows that we are not warranted in denying the possibility of such a medium because the equilibration of an element of it requires an extraneous couple. The explanation of gravitation is still outstanding, and necessitates some structure or property quite different from, and probably more fundamental than, simple rotational elasticity of the aether and simple molar elasticity of material aggregations in it; and this property may very well be also operative in the manner here required.

not simple gyrostatic.

Kelvin's complex gyrostatic.

A static rotational type equally ultimate:

38. It becomes indeed clear when attention is drawn to the matter, that there is something not self-contained and therefore not fundamental in the notion of even a gyrostatic medium and the resistance to absolute motion of rotation which it involves. For we want some fixed frame of reference outside the medium itself, with respect to which the absolute rotation may be specified; and we also encounter the question why it is that rotatory motion reveals absolute directions in this manner. Another aspect of the question appears when we consider the statical model with its rotational property produced by small magnets interspersed throughout it, the medium being in internal equilibrium in a magnetic field when unstrained; the unbalanced tractions on the element of volume are here supplemented by a couple due, as to sense, to magnetic action at a distance, and it

partial illustration by magnetized medium.

[1] Cf. *Proc. Lond. Math. Soc.* 1890: as *supra*, p. 205.

[2] Lord Kelvin (Sir W. Thomson), *Comptes Rendus*, Sept. 16, 1889; *Collected Papers*, Vol. III. p. 467.

is the energy of this action at a distance which constitutes the rotational part of the energy of the model. We may if we please suppose some analogous action at a distance to exist in the case of the actual aether, the ultimate explanation of which will be involved in the explanation of gravitation. Now in this magnetic analogue to our medium the equations of equilibrium and motion are clearly quite correctly determined by the analytical method of Lagrange. So long as the potential energy is derived from a forcive emanating and transmitted nearly instantaneously from all parts of the medium and not merely from the contiguous elements, its location is expressed, quite sufficiently for dynamical purposes which are concerned with a finite volume of the medium and finite velocity of propagation by attaching it to the element on which the forcive acts. The medium of MacCullagh therefore, on a saving hypothesis of this kind, appears to escape the kind of objection above mentioned.

Lagrangian method valid also for structure involving distance-action.

PART II. ELECTRICAL THEORY.

39. The next stage in the development of the present theory is the application of the properties of non-vibrational types of motion of the primordial medium to the explanation of the phenomena of electricity. In accordance with the interpretation of MacCullagh's equations, on the ideas of the electromagnetic theory of light, the electric displacement in the medium is its absolute rotation (f, g, h) at the place, and the magnetic force is the velocity of its movement $d/dt\,(\xi, \eta, \zeta)$. At the beginning, our view will be confined to rotational movements unaccompanied by translation, such namely as call into play only the elastic forces which are taken to be the cause of optical and electromotive phenomena; but later on we shall attempt to include the electrical and optical phenomena of moving bodies.

MacCullagh's type of optical Action function extended to electric fields.

In the ordinary electromagnetic system of electric units we should have $4\pi\,(f, g, h) = \text{curl}\,(\xi, \eta, \zeta)$; but in purely theoretical discussions it is a great simplification to adopt a new unit of electric quantity such as will suppress the factor 4π, as Mr Heaviside has advocated. Except in this respect, the quantities are all supposed to be specified in electromagnetic units.

Rational elastic units.

It may be mentioned that a scheme for expressing the equations of electrodynamics by a minimal theorem analogous to the principle of Least Action has recently been constructed by von Helmholtz[1].

Helmholtz's essay towards Least Action.

[1] H. von Helmholtz, "Das Princip der kleinsten Wirkung in der Electrodynamik," Wied. *Ann.* Vol. XLVII. 1892.

Conditions of Dielectric Equilibrium.

40. The conditions of electromotive equilibrium in a general aeolotropic dielectric medium are to be derived from the variation of the potential energy function [referred to its principal axes]

General dielectric energy-function.

$$W = \tfrac{1}{2}\int\left\{a^2\left(\frac{d\zeta}{dy}-\frac{d\eta}{dz}\right)^2 + b^2\left(\frac{d\xi}{dz}-\frac{d\zeta}{dx}\right)^2 + c^2\left(\frac{d\eta}{dx}-\frac{d\xi}{dy}\right)^2\right\}d\tau.$$

On conducting this variation, we have

Its variation:

$$\delta W = \int\left\{a^2 f\left(\frac{d\delta\zeta}{dy}-\frac{d\delta\eta}{dz}\right) + b^2 g\left(\frac{d\delta\xi}{dz}-\frac{d\delta\zeta}{dx}\right) + c^2 h\left(\frac{d\delta\eta}{dx}-\frac{d\delta\xi}{dy}\right)\right\}d\tau$$

$$= \int\{a^2 f(m\delta\zeta - n\delta\eta) + b^2 g(n\delta\xi - l\delta\zeta) + c^2 h(l\delta\eta - m\delta\xi)\}\,dS$$

$$-\int\left\{a^2\left(\frac{df}{dy}\delta\zeta - \frac{df}{dz}\delta\eta\right) + b^2\left(\frac{dg}{dz}\delta\xi - \frac{dg}{dx}\delta\zeta\right) + c^2\left(\frac{dh}{dx}\delta\eta - \frac{dh}{dy}\delta\xi\right)\right\}d\tau$$

$$= \int\{(nb^2g - mc^2h)\,\delta\xi + (lc^2h - na^2f)\,\delta\eta + (ma^2f - lb^2g)\,\delta\zeta\}\,dS$$

$$-\int\left\{\left(\frac{dc^2h}{dy}-\frac{db^2g}{dz}\right)\delta\xi + \left(\frac{da^2f}{dz}-\frac{dc^2h}{dx}\right)\delta\eta + \left(\frac{db^2g}{dx}-\frac{da^2f}{dy}\right)\delta\zeta\right\}d\tau,$$

where (l, m, n) represents the direction of the normal to the element dS.

The vanishing of the volume integral in this expression for all possible types of variation of (ξ, η, ζ) requires that

leads for equilibrium to a potential:

$$a^2 f dx + b^2 g dy + c^2 h dz = -dV,$$

where V is some function of position, in other words that

and to displacement expressed in terms of its gradient, namely the electric force: it is normal to every boundary at which elasticity fails.

$$(f, g, h) = -\left(\frac{1}{a^2}\frac{d}{dx},\ \frac{1}{b^2}\frac{d}{dy},\ \frac{1}{c^2}\frac{d}{dz}\right)V.$$

The vanishing of the surface integral requires that the vector (a^2f, b^2g, c^2h) shall be at each point at right angles to the surface.

It is hardly necessary to observe that in this solution V is the electric potential, from which the electric displacement (f, g, h) is here derived by the ordinary electrostatic formulae for the general type of crystalline medium, and that the surface condition is that the electric force is at right angles to the surface, or in other words that the electric potential is constant all over it.

In deducing these conditions it has been assumed that the electrostatic energy is null inside a conductor; thus in statical questions the conductors may be considered to be regions in the medium devoid of elasticity, over the surfaces of which there is no extraneous constraint or forcive applied.

41. In this analysis it has not been explicitly assumed that the electric displacement is circuital, *i.e.* that

$$\frac{df}{dx}+\frac{dg}{dy}+\frac{dh}{dz}=0.$$

If we were to introduce explicitly this equation of constraint, we must by Lagrange's method add a term

$$\tfrac{1}{2}\lambda\left(\frac{df}{dx}+\frac{dg}{dy}+\frac{dh}{dz}\right)^2$$

to the energy-function, before conducting the variation; and we must subsequently determine the function of position λ so as to satisfy the conditions of the problem. The result would now come out

$$(a^2f+\vartheta,\ b^2g+\vartheta,\ c^2h+\vartheta)=-\left(\frac{d}{dx},\frac{d}{dy},\frac{d}{dz}\right)V,$$

with the condition that V is constant over the surface of the conductor; where

$$\vartheta=\lambda\left(\frac{df}{dx}+\frac{dg}{dy}+\frac{dh}{dz}\right),$$

and would represent so to speak an electromotive pressure uniform in all directions. The introduction of such a quantity would make the equations too general for the facts of electrostatics; on this ground alone we might assume ϑ to be null, and therefore V to be subject to a characteristic equation

Inclusion of an electromotive pressure is formally possible:

$$\frac{d}{dx}\left(\frac{1}{a^2}\frac{dV}{dx}\right)+\frac{d}{dy}\left(\frac{1}{b^2}\frac{dV}{dy}\right)+\frac{d}{dz}\left(\frac{1}{c^2}\frac{dV}{dz}\right)=0.$$

but it is absent in fact:

This investigation may remain as an illustration of method; but it is not required, when we bear in mind the constitution of the medium. Since

$$(f,g,h)=\text{curl}\,(\xi,\eta,\zeta)$$

we *must* have (f, g, h) circuital; so that the characteristic equation for V is involved in the data, without the necessity of any appeal to observation; while the introduction of the quantity ϑ would be illicit, and would have to be annulled later on.

which would agree with rotational quality of the displacement.

42. If we assumed that the energy-function contained a term

The equations with elastic compression included:

$$\tfrac{1}{2}A\left(\frac{d\xi}{dx}+\frac{d\eta}{dy}+\frac{d\zeta}{dz}\right)^2,$$

the conditions of electromotive equilibrium would come out

$$\left(\frac{dc^2h}{dy}-\frac{db^2g}{dz},\ \frac{da^2f}{dz}-\frac{dc^2h}{dx},\ \frac{db^2g}{dx}-\frac{da^2f}{dy}\right)=-\left(\frac{d}{dx},\frac{d}{dy},\frac{d}{dz}\right)\vartheta'$$

and $(mc^2h - nb^2g,\ na^2f - lc^2h,\ lb^2g - ma^2f) = -(l, m, n)\,\vartheta'$,

where
$$\vartheta' = A\left(\frac{d\xi}{dx} + \frac{d\eta}{dy} + \frac{d\zeta}{dz}\right).$$

Throughout a region devoid of elasticity this electromotive pressure ϑ' must be constant, and the electric force just outside its boundary must be along the normal; in the dielectric ϑ' must satisfy Laplace's equation, and so be the potential of an ideal superficial distribution of matter; but the electric force is not now derived from a potential, although its curl is derived from the potential ϑ' just specified.

it destroys the potential:

The phenomena of electrostatics require that this term does not occur in the energy; and that may be either (i) because

$$\frac{d\xi}{dx} + \frac{d\eta}{dy} + \frac{d\zeta}{dz}$$

is null, and the medium so to speak incompressible, or (ii) because A is null, so that the medium offers no resistance to laminar compression. But there is, apparently, nothing as yet to negative a constitution of the medium approximating extremely close to either of these two limiting states for both of which the equations of electrostatics would be exact. It has been shown already that there is absolutely nothing against such a supposition in the theory of light. But the experiments of Cavendish in proof of the electrostatic law of inverse squares, as repeated by Maxwell, may be taken as showing that the ratio of any compressional effect to the rotational part of the phenomenon is at any rate excessively minute. A very small compressional term like this might possibly be of advantage in an attempt to include gravitation among the manifestations of aethereal activity, a point to be examined later on. It differs fundamentally from the compressional term introduced by von Helmholtz into the equations of electrodynamics [*supra*, p. 232].

and must be in fact null or extremely minute.

Helmholtz's electro-kinetic theory.

43. We may also apply the variational equation of equilibrium to a volume in the interior of the dielectric medium, and therefore subject to surface tractions from the surrounding parts. It thus appears that the component surface tractions in the aether in the directions of the axes of coordinates are, per unit area lying in the direction (l, m, n),

Aethereal interfacial traction always tangential:

$$nb^2g - mc^2h,\quad lc^2h - na^2f,\quad ma^2f - lb^2g;$$

their resultant is tangential, *i.e.* in the plane of the element; it is equal to the component of the electric force in that plane, and is at right angles to that component. This is the specification of the aethereal stress by which static electromotive disturbance is transmitted across a dielectric medium. This stress does not at all interfere

its specification, linear:

with any irrotational fluid motion which may be going on in the medium, or with the normal hydrostatic pressure which regulates such motion.

is not excited by irrotational displacement.

Electrostatic Attraction between Material Bodies.

44. When two charged bodies are moved relative to each other the total electrical energy of strain in the aether is altered; on the other hand, since the electrical displacement (rotation of the aether) is circuital, the charges of the bodies are maintained constant. In the absence of viscosity, this loss or gain of energy must be due to transference to some other system linked with the electric system; it reappears in fact as mechanical energy of the charged conductors, which determines the mechanical forcive between them. It is desirable to attempt a closer examination of the nature of the action by which this transfer of energy takes place between the aether and the material of the conductors, and by which the similar transfer takes place at a transition between one dielectric substance and another.

Electron system interlinked with the elastic medium, is foreshadowed.

In the displacement of a conductor through an excited dielectric there is thus an overflow of electromotive energy, and in the absence of viscous agencies and radiation it simply displays itself in ordinary mechanical forces acting on the surface of the conductor. The magnitude of these forces has been examined experimentally in different media, and has been found to correspond precisely with this account of their origin; good reason can be assigned to show that their intensity changes from point to point of the surface according to a law[1] ($KF^2/8\pi$, where F is electric force) which suggests that the energy is [abstracted by a moving] conductor at its surface. In a similar way, when a dielectric body is moved through the electric field the transformation of energy takes place at the interface between the two dielectrics [and may thence be in part radiated away].

The statical distribution of electromotive stress in the excited aethereal medium is definite and has just been determined: it involves on each element of interface in the dielectric aether a purely tangential traction at right angles to the tangential component of the electric force and equal to it. This is the denomination of stress that corresponds to the displacement (ξ, η, ζ), just as an ordinary force corresponds to a translation of matter or a couple to a rotation. If we have no direct knowledge of the aethereal displacement (ξ, η, ζ) we cannot actually recognize this stress; but when (ξ, η, ζ) is taken as here to be a linear displacement, this electromotive stress must be

The aethereal stress linear and reversible.

[1] Cf. "On the theory of Electrodynamics, as affected by the nature of the mechanical stresses in excited dielectrics," *Roy. Soc. Proc.* 1892: *supra*, p. 274.

a mechanical stress in the aether such as does work in making a linear displacement.

45. The mechanical traction along the normal, which is distributed over the surfaces of two conductors separated by an excited dielectric, as for example the coatings of a charged Leyden jar, may be balanced by supports applied to the conductors; or if there is a dielectric body between them, it may be mechanically balanced by a stress in the *material* of this dielectric. This is the only kind of mechanical stress in a dielectric of which we have direct cognizance: its amount has been calculated by Kirchhoff[1] and others for some cases, and compared with experimental measures of change of volume of dielectrics under electrification. The stress in the aether itself has been here deduced by a wholly different path.

Mechanical electro-striction.

It will possibly be a true illustration of what occurs to imagine each element of surface dS of the conductor to encroach by forward movement into the excited dielectric. As it proceeds, its superficial molecules somehow dissolve or loosen the strain of each little piece of the dielectric aether as they pass over it. Each fragmentary easing of strain sends a shiver through the dielectric aether, which however practically instantaneously readjusts itself into an equilibrium state. Thus the process goes on, the gradual molecular dissolution of the strain by the advance of the conductor shooting out minute wavelets of re-arrangement of strain into the dielectric, which are confined to the immediate neighbourhood and are quite undiscernible directly, because on account of their great velocity of propagation the aether is always excessively near an equilibrium condition[2]. The pressural reaction (§ 97) of these disturbances on the conductor may be taken to be the source of the mechanical forcive experienced by it, which does work in impelling its movement and to an equal extent exhausts the energy of the dielectric.

Instantaneous adjustment to displacements.

Imagine a very thin element dS on the surface of the conductor, thick enough, however, to include this layer of intense disturbance of the aether; it will be subject to this electric reaction of the excited

[1] G. Kirchhoff, "Ueber die Formänderung, die ein fester elastischer Körper erfährt, wenn er magnetisch oder diëlectrisch polarisirt wird," Wied. *Ann.* Vol. XXIV. 1885, p. 52; Vol. XXV. 1885, p. 601. Such a stress, involving the square of the electric intensity instead of its first power, must of necessity be of secondary character, and cannot take direct part in wave-propagation in the electric medium. [On elastic deformation of material dielectrics (electro-striction) see *infra*, Part III. The Maxwell quadratic stress tensor arises naturally, on the other hand, in very striking manner, in the Action theory, as a formal expression of the force transmitted across any aethereal boundary and so acting on the matter contained within it.]

[2] Cf. Sir G. G. Stokes, "On the Communication of Vibrations from a vibrating body to the surrounding gas," *Phil. Trans.* 1868, p. 448; or in Lord Rayleigh, *Theory of Sound*, Vol. II.

dielectric acting on it on the one side, and the elastic traction of the material of the solid conductor acting on it on the other side; and as its mass is very small compared with its surface, these forcives must equilibrate. For if this superficial element is displaced outwards through a very minute distance ds, the following changes of energy result; the energy of the dielectric is altered by the subtraction of that contained in a volume $dSds$ of it, while the elastic normal traction P of the conductor does work $PdSds$. These changes must compensate each other by the energy principle of equilibrium (compare § 58); hence the normal elastic traction P is equal to the energy in the dielectric per unit volume. The consideration of a tangential displacement of the element leads in the same way to the conclusion that the tangential elastic traction, required to be exerted by its material backing in order to maintain its equilibrium, is null.

Traction as deduced from energy.

Electrodynamic Actions between Material Bodies: [ideal permanent Currents].

46. In order to examine how far our energy-function of an aethereal medium involves an explanation of electrodynamic phenomena, we must begin with a simple case of electric currents that will avoid the introduction into the field of all complications like galvanic batteries, which could not easily be included in the energy-function. Let us therefore consider two charged condensers with their two pairs of coatings connected by thin wires as in the annexed diagram; and let us suppose the two plates of one of the condensers to be steadily moved towards each other when both pairs of coatings are thus in connection. This will produce a steady current in the conducting wires, which will flow completely round the circuit; the only breaches of linearity of the current are at the condensers themselves, and these may be made negligible by taking the dielectric plates very thin. In this way a steady current can be realized in a conductor devoid of resistance, without the aid of any complicated electromotive source[1].

Reversible electric current fed by a charged condenser.

47. Now we have to inquire what account the dynamical theory gives of this steady current. In the first place, the motion is very slow in comparison with the velocity of electric propagation; therefore the interior of the dielectric is at each instant sensibly in an equilibrium condition, for the same kind of reason that moving a body slowly to and fro does not start any appreciable sound waves in the atmosphere. Thus at each instant the vector (f, g, h) is derived as above from a potential function V; and at the surface of any of the conductors (supposed here of insensible resistance) it is directed

Radiation not excited by the continuous current:

[1] Cf. "A mechanical representation of a vibrating electrical system and its radiation," *Proc. Camb. Phil. Soc.* 1891: *supra*, p. 221.

along the normal, if the medium is isotropic. It is, in fact, in the more familiar electric language, at each instant the electric displacement determined by the charges which exist in a state of equilibrium on the faces of the condensers and on the connecting wires. This electric displacement in the dielectric field is, owing to the condensing action, very small compared with the charges involved, except between the plates of the condensers and close to the thin conducting wire. Imagine a closed surface which passes between the plates of one of the condensers, and intersects the conducting wire at a place *P*. As the vector (f, g, h) is by its nature as a rotation circuital, its total flux through any surface must be null, if we imagine the elastic continuity of the medium inside the conductors to be restored, and such an electric displacement at the same time imparted along the wire as will leave the state of the field unaltered and thus no disturbance

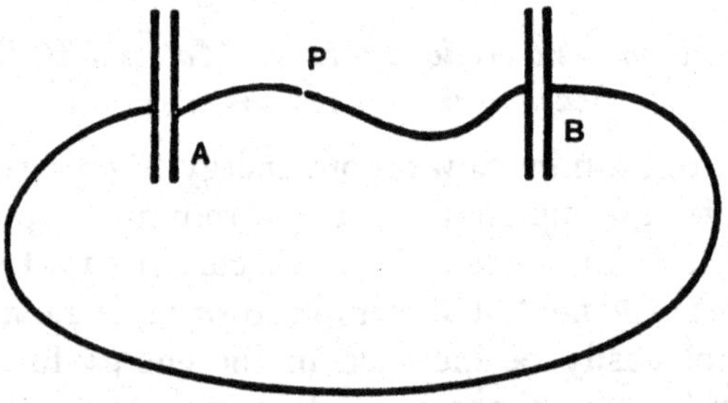

inside the conductors. And this flux must remain null when the plates of the condenser are slightly brought together; or rather we have to contemplate such a flow of displacement along the wire as will make it remain null. The movement of the plates will, however, very considerably alter the large flux across that portion of the surface which lies between them; and the total flux for the other part of the surface not near the wire is as we have seen of trifling amount; therefore the alteration just mentioned must be considered to be balanced by an intense alteration of the above ideal flux in the immediate neighbourhood of the surface of the wire, in fact along its very surface if it is a perfect conductor. Immediately this change of the capacity of the condenser is over, the vector (f, g, h) will be back in its equilibrium condition in which it is, at each point of the surface of the wire, directed along the normal. As (f, g, h) represents the electric displacement in the field, the intense flux here contemplated, close to or on the surface of the wire, when the capacity is undergoing change, is the current in the wire. But all these circumstances concerning it have been made out from the dynamics alone, electric phraseology being employed only to facilitate the quotation of known analytical theorems about potential functions, and about how their distribution through space is connected with the forms of surfaces to which their

tracing out its dielectric displacement:

fluxes are at right angles, and over which they therefore have themselves constant values.

If now while a current is flowing round the circuit, the two condensers are imagined to be instantaneously removed, and the wire made continuous, we shall be left with an ordinary circuital current, which in the absence of dissipative resistance will flow on for ever.

a permanent cyclic current, how initiated.

48. The argument in the above rests on the fact that there is circuital change of an elastic displacement $d/dt\,(f, g, h)$ distributed throughout the dielectric, while the medium is discontinuous at the surface of the perfectly conducting wire because displacement cannot be sustained inside the wire. When we for purposes of calculation imagine the elastic quality to extend across the section of the wire, and so avoid consideration of the discontinuity in the medium, we must imagine as above a flow of rotational displacement along the wire so long as the capacity of one of the condensers is being altered; and the velocity in the field will be deducible, by the ordinary formulae for a continuous medium, from this ideal flow together with the actual changes of displacement throughout the dielectric. For a perfect conductor the circumstances will be exactly represented by confining this flow to its surface; what is required to make the analytical formulae applicable, without modification on account of discontinuity in the medium, is simply the addition of such an ideal flow at the places of discontinuity as shall render the displacement (f, g, h) circuital throughout the field, without disturbing its actual distribution in the volume of the media.

The flux of electrons adumbrated:

The kinetic and potential energies of the medium may in fact either be calculated for the actual configuration, when they will involve surface integral terms extended over the surfaces of discontinuity, or they may be calculated as for a continuous medium if we take into account a flow of displacement along these surfaces, such as we would require to introduce by some agency if the medium were perfectly continuous, in order to establish the actually existing state of motion throughout it; in estimating the energy of the medium in terms of the flow of displacement these surface sheets must be included, after the manner of vortex sheets in hydrodynamics.

and sheets of displaced electrons.

In the same way, when the electric charge on a conductor is executing oscillations, a vortex sheet of changing electric displacement, such as will make the displacement in the field everywhere circuital, must be supposed to exist on the surface of the conductor.

49. There is this difference* between actual electric current systems and the permanently circulating currents, or vortex rings,

Cyclic currents cannot be hydrodynamic vortex rings:

* The chapter of general dynamical analysis in Maxwell's *Treatise*, §§ 568–578, implies this distinction, namely, that the current in each coil is of the type of a velocity, and not a conserved momentum, as adjusted to uniformity

in this aethereal medium, that the latter move in the medium so that their strengths remain constant throughout all time, while alteration of the strength of an electric current is produced by electrodynamic induction. In our condenser circuit, however, the strength of the current depends on the rate of movement of the plates of a condenser, that is, it is affected by changes in the rotational strain-energy of the portions of the medium which are situated in the gaps across the conducting circuit. Motion of the condenser-plates produces a flow of displacement across any closed surface which passes between them, and therefore is to be taken as producing an equal and opposite flow where this surface intersects the connecting circuit. That ideal flow, or current, the representation of the action of the channel of discontinuity on the elastic transmission in the medium, implies, on the other hand, a hydrodynamical circulation of the medium round the conducting circuit, which provides the kinetic energy of the electric current. A current in a conductor has practically no elastic potential energy, because for movements of ordinary velocity the medium is always sensibly in an equilibrium condition, any beginning of an electromotive disturbance of the steady motion being instantly equalized before it has time to grow. A complete current, consisting of a flexible vortex ring, or even circulating in a rigid core in the free aether, will thus maintain its strength unaltered, that is, the surrounding aether will move so that the electrodynamic induction in the circuit is always null; but if the current circuits are completed across the dielectric or through an electrolytic medium, this constraint to nullity of induction will be thereby removed, and constancy of circulation will no longer be a characteristic of such a broken vortex ring, so to speak, in the medium.

for they are discontinuous: their energy kinetic. Completion across ionized medium:

50. The above mode of representing the surface terms in the kinetic energy of course supposes that the intensities of the vortex sheets have been somehow already determined, or else that they are to be included in the scheme of variables of the problem. When the conductors are of narrow section, then as regards their action at a distance all that is wanted is the aggregate amount of flow across the section, that is, the electric current in the wire in the ordinary sense; and the introduction into the energy of terms calculated with reference only to these aggregates of flow is sufficient as regards the effect at distances from the conductors that are great compared with the dimen-

electrons thus must be present in the circuit.

all along its circuit by radiational action across the medium. The dynamical experiments, gyrostatic and electromotive, there foreshadowed, some of them since carried through, would (§ 574) transfer steady electrical action from "a phenomenon due to an unknown cause" to "the result of known motions of known portions of matter, in which not only the total effects and final results but the whole intermediate mechanism and details of the motion are taken as the objects of study."

sions of their cross-sections. But if the details of the distribution round the section are required, the term in the energy must be more minutely specified as a surface integral due to the interaction of the different elementary filaments of the flow which are situated round the periphery of the section, much as the energy of a vortex sheet is introduced in the theory of discontinuous fluid motion; and its variation will now lead to electrodynamic equations of continuous electric flow in the ordinary manner. There is no difficulty in extending this view to cases in which the breach of circuital character of the displacement current $d/dt\ (f, g, h)$ may have to be made up by an ideal distribution of flow throughout the volume, that is, by a volume instead of a surface distribution of electric currents, as in an actual conductor of finite resistance.

Current sheets as the terminal complement of changing displacement.

(51. June 14. The velocity of a fluid is derivable in hydrodynamics, by kinematic formulae, from the vorticity of its flow, provided we suppose the vorticity to include the proper vortex sheets spread over the surfaces of discontinuity of flow, if such exist; in the same way the magnetic force is derivable as above from the displacement current, provided this current includes the proper current sheets over the surfaces of the conductors or other surfaces of discontinuity of the magnetic field.

Charging must be a discontinuous process:

Let us consider an isolated uncharged conductor, and imagine an electric charge imparted to it. This charge is measured by the integral of the electric displacement (f, g, h) taken over any closed surface surrounding the conductor. Now if this rotational displacement were produced by continuous motion in the surrounding medium, its surface integral over any open sheet would be equal to the line integral of the linear displacement of the medium taken round the edge of the sheet. In a closed sheet the surface integral would therefore be null; thus a charge cannot be imparted to a conductor without some discontinuous motion, or slip, or breach of rotational elasticity, in the medium surrounding it. If we imagine the charge to be imparted by means of a wire, the integral of electric displacement over any open surface surrounding the conductor and terminated by the wire is equal to the line integral of the linear displacement of the medium round the edge of this surface where it abuts on the wire. If the wire is thin, this line integral is therefore the same at all sections of it, and thus involves a constant circulatory displacement of the medium around it. If the wire is a perfect conductor, there is no elasticity and therefore no rotational displacement of the aether inside its surface; thus there is slip in the medium at the surface of the wire; and if we desire to retain the formulae of continuous analysis, we must contemplate a very rapid transition by means of a vortex sheet at

as by flux of electron sheet over a perfect conductor.

the surface, in place of this discontinuity. This vortex sheet is in the present example continuous with *rotational* motion in the outside medium; the tubes of changing vorticity, *i.e.* of electric current, are completed and rendered circuital by displacement currents in the surrounding dielectric. But in the case of the condenser circuit above considered, the alteration of the density of the vortical lines between a pair of plates, which is produced by separating them, involves a translational circulatory movement around the edge of the condenser and throughout the medium outside, which is almost entirely of *irrotational* type, except at the surface of the conducting wire where a vortex sheet has to be located in order to avoid discontinuity. The irrotational motion in the surrounding medium, which is thus continuous with the vortex sheet, and therefore determined by it, represents the magnetic field of the current flowing in the wire. On the other hand, in the illustration of this section, the motion in the medium is not irrotational, for it represents the field determined by the displacement currents in the medium and the conduction current in the wire, taken together.)

The magnetic field located in surrounding region:

52. To return to our condenser illustration; it does not follow from the superficial character of the current $d/dt\,(f, g, h)$ that the velocity vector $d/dt\,(\xi, \eta, \zeta)$ is also very small throughout the field except at the very surface of the wire. We have in fact $(f, g, h) = \text{curl}\,(\xi, \eta, \zeta)$, therefore

$$\nabla^2\,(\xi, \eta, \zeta) - \left(\frac{d}{dx}, \frac{d}{dy}, \frac{d}{dz}\right)\left(\frac{d\xi}{dx} + \frac{d\eta}{dy} + \frac{d\zeta}{dz}\right) = -\,\text{curl}\,(f, g, h);$$

so that, the compression $\left(\dfrac{d\xi}{dx} + \dfrac{d\eta}{dy} + \dfrac{d\zeta}{dz}\right)$ being null, $d/dt\,(\xi, \eta, \zeta)$ are the potentials of certain ideal mass-distributions close to the surface of the wire; therefore they are of sensible magnitude throughout the surrounding field.

It appears from the surface character of the disturbance of the electric displacement (f, g, h) which is thus introduced for current systems flowing in complete circuits, that if we transform the kinetic-energy function

the distribution of magnetic energy:

$$T = \tfrac{1}{2}\int\left(\frac{d\xi^2}{dt^2} + \frac{d\eta^2}{dt^2} + \frac{d\zeta^2}{dt^2}\right) d\tau,$$

in which it is convenient to take the density to be unity, so that it shall be expressed in terms of the current $d/dt\,(f, g, h)$, at the same time treating the rotational displacement of the medium as continuous, we shall have practically reduced it to a surface integral along the wire. To effect this, let (F, G, H) be the potentials, throughout the region, of ideal mass-distributions of densities $d/dt\,(f, g, h)$: so that

can also be expressed in terms of the currents, thus with different location.

$$(F, G, H) = \int \frac{d\tau'}{r'}\frac{d}{dt}(f', g', h'),$$

where r' is the distance from the element of volume $d\tau$ to the point considered; then

$$\frac{dG}{dx}-\frac{dF}{dy}=-\frac{d}{dt}\iint\left\{\nabla^2\zeta-\frac{d}{dz}\left(\frac{d\xi}{dx}+\frac{d\eta}{dy}+\frac{d\zeta}{dz}\right)\right\}\frac{d\tau}{r}$$

$$=4\pi\frac{d\zeta}{dt},\ \text{as}\ \frac{d\xi}{dx}+\frac{d\eta}{dy}+\frac{d\zeta}{dz}\ \text{is null.}$$

Thus*

$$T=\frac{1}{8\pi}\int\left\{\frac{d\xi}{dt}\left(\frac{dH}{dy}-\frac{dG}{dz}\right)+\frac{d\eta}{dt}\left(\frac{dF}{dz}-\frac{dH}{dx}\right)+\frac{d\zeta}{dt}\left(\frac{dG}{dx}-\frac{dF}{dy}\right)\right\}d\tau$$

$$=\frac{1}{8\pi}\int\left\{F\frac{d}{dt}\left(\frac{d\zeta}{dy}-\frac{d\eta}{dz}\right)+G\frac{d}{dt}\left(\frac{d\xi}{dz}-\frac{d\zeta}{dx}\right)+H\frac{d}{dt}\left(\frac{d\eta}{dx}-\frac{d\xi}{dy}\right)\right\}d\tau$$

on integrating by parts. The medium is supposed here to be mathematically continuous as above, thus avoiding separate consideration of the conducting channels,—though its structure may change with very great rapidity in crossing certain interfaces; and it is taken to extend through all space, so that the surface integral terms may be omitted, no active parts of the system being supposed to be at an infinite distance. Thus

$$T=\frac{1}{8\pi}\int\left(F\frac{df}{dt}+G\frac{dg}{dt}+H\frac{dh}{dt}\right)d\tau$$

$$=\frac{1}{8\pi}\iint\frac{1}{r}\left(\frac{df}{dt}\frac{df'}{dt}+\frac{dg}{dt}\frac{dg'}{dt}+\frac{dh}{dt}\frac{dh'}{dt}\right)d\tau\,d\tau',$$

which is the form required, expressed as a double integral throughout space.

For a network of complete circuits carrying currents $\iota_1, \iota_2, \ldots$ we may express this formula more simply as

$$4\pi T=\tfrac{1}{2}\iota_1{}^2\iint\frac{\cos\epsilon_1}{r_1}ds_1ds_1+\ldots+\iota_1\iota_2\iint\frac{\cos\epsilon_{12}}{r_{12}}ds_1ds_2+\ldots,$$

where ϵ is the angle between the directions of the two elements of arc; which is Neumann's well-known form of the mechanical energy of a system of linear currents. The currents are here simply mathematical terms for such flows of electric displacement along each wire as would be required to make the displacement throughout the field perfectly circuital, if the effective elasticity were continuous in accordance with the explanation above. **Its form for a system of linear currents, as here visualized.**

53. Now if two wire circuits carry steady currents, generated from condensers in this manner, and are displaced relatively to each other

* A factor 4π affects T throughout this section which arises from defining (f, g, h) as curl (ξ, η, ζ) and taking the aether density to be unity, without the factors which adjust to electromagnetic units.

When radiation and friction are excluded the system is self-contained and dynamical.

with velocities not considerable compared with the velocity of propagation of electromotive disturbances, the electric energy of the medium is thereby altered. There is supposed to be no viscous resistance in the system, and no sensible amount of radiation; therefore the energy that is lost by the medium must be transferred to the matter. This transfer is accomplished by the mechanical work that is required to be done to alter the configuration of the wires against the action of electrodynamic forces operating between them; for these mechanical changes have usually a purely statical aspect compared with the extremely rapid electric disturbances. The expression T, with its sign changed, is thus the potential energy of mechanical electrodynamic forces acting between the material conductors which carry the currents.

Furthermore, as above observed, the electrokinetic energy and the electrodynamic forces at which we have arrived are expressed in terms of the total current flowing across any section of the wire supposed thin, and do not involve the distribution of the current round the contour of the section to the neighbourhood of which it is confined, nor the area or form of the section itself. It therefore does not concern us whether the wire is a perfect conductor or not; the previous argument from the circuital character of the rotation (f, g, h) shows that the total current is still the same across all sections of the wire, and that the energy relations are expressed in the same manner as before in terms of the total current.

Electrons groped after to complete the scheme.

The electrodynamic forces between linear current systems are thus fully involved in the kinetic-energy function of the aethereal medium. The only point into which we cannot at present penetrate is the precise nature of the surface action by which the energy is transferred (just as in § 45) from the electric medium to the matter of the perfect conductor; all the forces of the field are in fact derived from their appropriate energy-functions, so that it is not necessary, though it is desirable, to know the details of the interaction between aether and matter, at the surface of a conductor.

Mathematical Analysis of Electrokinetic Forces and their reaction on the Material Medium.

54. We have shown that the electrokinetic energy of a system of linear electric currents may be expressed in the form

$$T = \tfrac{1}{2}\Sigma\iota_1^2 \iint \frac{\cos\epsilon_1}{r_1}\, ds_1 ds_1 + \Sigma\iota_1\iota_2 \iint \frac{\cos\epsilon_{12}}{r_{12}}\, ds_1 ds_2,$$

the velocity system which they involve being sufficiently described by the set of velocity coordinates $\iota_1, \iota_2, \ldots$ combined with the kinetic constraints derived from the constitution of the aether. To mark that these quantities are dynamically velocities, let us denote $\iota_1, \iota_2, \ldots$ by $de_1/dt, de_2/dt, \ldots$ so that $e_1, e_2, \ldots$ will be taken as electric coordinates of position. The general variational equation of motion may be expressed in the form

$$\delta \int T dt = \int \delta W_1 dt + \int (E_1 \delta e_1 + E_2 \delta e_2 + \ldots)\, dt,$$

where E_1 is by definition such that $E_1 \delta e_1$ is the work done in the system during a displacement δe_1, so that in electric phraseology E_1 with sign changed is the electric force integrated round the circuit 1, or the electromotive force in that circuit. Also W_1 is any other potential energy the system may possess; the energy of electric strain throughout the medium being now very small, as there are no static electrifications, and the motions are supposed slow compared with the velocity of radiation. Thus, adopting the notation of coefficients of electrodynamic induction, so that

Cyclic line integral is of electric force.

$$T = \tfrac{1}{2} L_1 \frac{de_1{}^2}{dt^2} + \tfrac{1}{2} L_2 \frac{de_2{}^2}{dt^2} + \ldots + M_{12} \frac{de_1}{dt}\frac{de_2}{dt} + \ldots,$$

$L_1, L_2, \ldots, M_{12}, \ldots$ depending on the configurations of the circuits, we have

$$\delta \int T dt = \Sigma \int \left(L_1 \frac{de_1}{dt} + M_{12} \frac{de_2}{dt} + \ldots \right) \frac{d\delta e_1}{dt} dt + \int \delta_1 T dt,$$

where in the last term $\delta_1 T$ refers to the change of T due to change of material configuration only. Hence

$$\delta \int T dt = | \Sigma (L_1 e_1 + M_{12} e_2 + \ldots)\, \delta e_1 |$$

$$- \Sigma \int \frac{d}{dt} \left(L_1 \frac{de_1}{dt} + M_{12} \frac{de_2}{dt} + \ldots \right) \delta e_1 dt + \int \delta_1 T dt,$$

the terms in | ... | referring to the beginning and end of the time.

Thus as the variations are all independent we derive, and that in Maxwell's manner but rather more rigorously, Faraday's law of the induced electromotive force E [reckoned as among forces acting on the system] under the form

Induced electromotive force in a circuit.

$$E_1 = - \frac{d}{dt} (L_1 \iota_1 + M_{12} \iota_2 + \ldots) = - \frac{d}{dt} \frac{dT}{d\iota_1}.$$

55. As already mentioned, for currents flowing round complete conducting circuits devoid of viscosity, the values of $\iota_1, \iota_2, \ldots$ are constant, by a sort of constraint or rather by the constitution of the medium, throughout all time; and the electromotive forces $E_1, E_2, \ldots$ here determined have no activity. But if, as in actual electric currents, the strengths are capable of change owing to the circuits being completed by displacement currents in the dielectric or across a voltaic

Even permanent currents must be discrete if due to induction:

battery thus constituting gaps through which additional displacement can so to speak flow into the conductors, or owing to viscous effects in the conductors carrying them which must also involve such discontinuity, then the forces E_1, E_2, ... here deduced from the energy-function will have an active existence, and the phenomena of electrodynamic induction will occur. Alteration of the strength of a current implies essentially incompleteness of the inelastic circuit round which it travels, and may be produced either by change of displacement across a dielectric portion of the circuit, or through the successive breaches of the effective elasticity of the aether which are involved in electric transmission across an electrolyte, and also probably in transmission through ordinary media which are not ideal perfect conductors. In short, the existence of electrodynamic induction leads to the conclusion that currents of conduction always flow in open circuits; if the circuit were complete, there would be no means available for the medium to get a hold on the current circulating in it. On this view the Amperean current circulating in a vortex atom is constant throughout all time, and unaffected by electrodynamic induction, so that there is apparently no room for Weber's explanation of diamagnetism.

onic currents:

discreteness must apply also to Amperean intra-atomic currents.

Nature of electric transfer.

56. The vorticity in a circuit, that is, the current flowing round it, can thus be changed only by an alteration of the displacement across a break in the conducting quality of the circuit, or by the transfer of electric charge across an electrolyte, in which case it is elastic rupture of the medium that is operative. Such an alteration of current will be evidenced by, and its amount will be derivable from, the change in the energy-function of the dielectric medium, in the manner above described. When there is no break in the conducting circuit, the current in it is restricted by the constitution of the medium to remain constant; and therefore an electromotive force E round the circuit, of the kind here determined, can do no work; it is not operative in the phenomena. The induction of a current on itself, due to change of form of its circuit, is bound up with the continued maintenance of the current by feed from batteries or other sources included in the circuit, in opposition to dissipation in the conductors which is connected with a sort of transfer by discharge from molecule to molecule within their substance: in an ideal perfectly conducting circuit there would be no such induction. A case which strikingly illustrates these remarks is the maintenance of a continuous current by a dynamo without any source other than mechanical work. The very essence of this action consists in the rhythmical make and break of the two circuits of the dynamo in synchronism with their changes

Strength of a vortex unalterable.

of form, so that they are interlocked during one portion of the cycle and unlocked during the remainder. Such lockings and unlockings of the circuits may of course be produced by sliding contacts, but these are equivalent for the present purpose to breaches in the continuity of the conductors. The original apparatus of Faraday's rotations (Maxwell, *Treatise*, Vol. II. § 486), which was the first electromotor ever constructed, and which driven backwards would act also as a dynamo, illustrates this point in its simplest form. Without some arrangement which allows the two circuits to cut across each other in this manner, there could be no induction of a continuous current, but only electric oscillations in the dielectric field, which could however be guided along conducting wires, as in alternate-current dynamos*. The phenomena of electric currents in ordinary conducting circuits are thus more general than the phenomena of vortex rings in hydrodynamics, or [not] of atomic electric currents, in that the strengths of the currents in them are not constrained to remain constant; an additional displacement current can, so to speak, flow into a conductor at any of its breaches of continuity. The variables of the problem are thus more numerous, and the energy-function leads to more equations connecting them.

The broken circuits of continuous-current dynamos:

Faraday's electromotor:

alternating motors and dynamos excepted.

Hydrodynamic analogy to currents therefore excluded,

57. We might now attempt to proceed by including the mechanical energy of the material conductors in the same function as the electrokinetic energy, thus deducing that the energy gained by altering the coordinate ϕ_1 is $(dT/d\phi_1)\,\delta\phi_1$, in other words, that the displacement $\delta\phi_1$ is *opposed* by a force equal to $dT/d\phi_1$. This would make currents flowing in the same direction along parallel wires repel each other, and in fact generally the force thus indicated is just the opposite to the reality.

the mechanical forces coming out of wrong sign:

The expression T represents completely the energy of the system so far as electromotive disturbances are concerned, as has been proved above. But we have no right to assume that the energy of

the Action specification by circuits therefore incomplete:

* The original law of Faraday, that the electric force integrated round any circuit moving with the matter is $-dN/dt$, where N is the number of magnetic tubes that it encloses, is fundamental to all theories, including radiation: cf. p. 232 *supra*. As N cannot increase indefinitely, it involves that all natural currents are alternating, so that natural electrodynamics is a theory of electrical oscillations. Even in a field of X-rays from an arrested cathode stream, no continuous currents could be excited in any adjacent circuits. But by the artificial device of a commutator, first employed by Faraday himself, the fall of potential generated in a part of a circuit can be shunted into another circuit so as always to keep the same direction along it, thus leading to the original continuous dynamos and motors: cf. *supra*, p. 110, footnote. The discovery that the dynamo circuit may be electrodynamically unstable, giving rise to self-starting dynamos, came to Wilde and to Siemens.

the system, so far as to include movements of the conductors and mechanical forces, can be completely expressed by this formula with only the electric coordinates and the sensible coordinates of the matter involved in it; for the mechanism that links them together is too complicated to be treated otherwise than statistically. We may, however, proceed as in the electrostatic problem; a displacement increases T by δT; this increase must come from some source; as there is supposed to be no dissipation it must come ultimately from the energy of the material system. During the displacement the electromotive system is at each moment sensibly in an equilibrium condition, so that there is practically no interaction between the kinetic energies of the electromotive and the material systems such as would arise from mixed terms in the energy-function involving both their velocities,—a fact verified experimentally by Maxwell[1]. Thus somehow by means of unknown connecting actions, the displacement alters the mechanical energy of the system by an amount $-\delta T$, and of this, considered as potential energy, the mechanical forces are the result. The mechanical force acting to *increase* the coordinate ϕ_1 is therefore $dT/d\phi_1$. In fact, instead of considering the material system to be represented by the coordinates ϕ_1, ϕ_2, ... which enter into the electrokinetic energy, we must consider it to be an independent system linked on to the electrokinetic system by an unknown mechanism, which however is of a statical character, so that energy passes over from the electrokinetic system to the other one as mere statical work, without any complication arising from the effects of mixed kinetic reactions. In the discussion in Maxwell's *Treatise*, § 570, this idea of action and reaction between *two* interlocked systems, the electromotive one and the mechanical one, has in the end to be introduced to obtain the proper sign for the mechanical force. The energy T is electrokinetic solely; no energy of the material system is included in it.

the linked material system (electronic) foreshadowed.

58. This deduction of the electrostatic and the electrodynamic mechanical forcive may now be re-stated in a compact form, which is also noteworthy from the circumstance that it embodies perhaps the simplest method of treatment of the energy-function in all such cases. Let us consider the dynamical system under discussion to be the purely electric one, that is, to consist of the dielectric medium only, so that it has boundaries just inside the surfaces of the conductors, which are supposed to be perfectly inelastic. The energy function $T + W$ remains as above stated, for all the energy is located

A formulation with charged conductors in place of electrons.

[1] Maxwell, *Treatise*, Part IV, "Electromagnetism," Chap. VI. The apparatus was constructed as early as 1861.

Maxwell's inertia test for currents.

in the dielectric; the electrokinetic part T arises from motion of the medium, and the electrostatic part W from its rotational strain. But in the equation of Least Action we must also take account of tractions which may be exerted by the matter of the conductors on the boundary of this dielectric system. If $\delta w dS$ denote the work done on the dielectric by these tractions extended over the element dS of the surface, the equation of Action will be

$$\delta \int (T - W)\, dt - \int dt \int \delta w\, dS = 0,$$

the time of passage from initial to final position being unvaried. When the disturbances considered are, as usually taken, too slow to generate sensible waves in the dielectric, and even when this restriction is not imposed, it equally follows that the tractions of the conductors on the dielectric system are derived from a potential energy function $T - W$, only in the latter case the value of this function is more difficult to determine; hence the tractions of the dielectric on the conductors are derived from a potential energy-function $-(T - W)$. Of this potential function the first part gives the electrodynamic forces acting on the conductors, the second part the electrostatic forces. This mode of treatment is clearly perfectly general, and applies, for instance, with the appropriate modification of statement, to the determination of the electrodynamic forces of an element of a continuous non-linear current flowing through a conducting medium; it will be shown presently that the electric dissipation-function can contribute nothing to the ponderomotive forcive.

Location of electrostatic energy considered.

That the part of the forcive which is due to the variation of this potential energy W is correctly expressible by means of the electrostatic traction $KF^2/8\pi$ on the surfaces of the conductors, may be verified as follows. Suppose an element of surface dS of the conductor to encroach on the dielectric by a normal distance dn; the energy that was in the element of volume $dS\,dn$ of the dielectric has been absorbed; and in addition the energy of the mass of the remaining dielectric has been altered by the slight change of form of the surface of the conductor in the neighbourhood of the element dS. Now the dielectric is in internal equilibrium, therefore its internal energy in any given volume is a minimum; therefore the change produced in that energy by any small alteration of constraint, such as the one just described, is of the second order of small quantities. Hence the encroachment of the element dS of the conductor diminishes the total energy W simply by the amount contained in the volume $dS\,dn$; and therefore that encroachment is assisted somehow by a mechanical traction equal to the energy per unit volume of the dielectric at the place, that is, of intensity $KF^2/8\pi$.

Electrodynamic Effect of Motion of a Charged Body.

Current of convection.

59. When a charged body moves relatively to the surrounding aether, with a velocity small compared with the velocity of electric propagation, it practically carries its electric displacement system (f, g, h) along with it in an equilibrium configuration. Thus the displacement at any point fixed in the aether will change, and we shall virtually have the field filled with electric currents which are completed in the lines of motion of the charged elements of the body, so long as that motion continues. On this view, Maxwell's convection current is not differentiated from conduction current in any manner whatever, if we except the fact that viscous decay usually accompanies the latter.

Rowland's spinning disc: the charge must not slip back over it.

A metallically coated glass disc, rotating in its own plane without altering its position in space, would on this theory produce no convection current at all; but if the coating of the disc is divided into isolated parts by scratches, as in Rowland and Hutchinson's experiments[1], or even if there is a single line of division, each portion will carry its field of electric displacement along with it, the field preserving its statical configuration under all realizable speeds of rotation. If the scratches did not run up to the centre of the disc, the field of displacement due to the central parts would be quiescent, and the displacement currents would be altered in character[2]. The dielectric displacement in the experiments above mentioned, with two parallel rotating gilt glass condenser discs having radial scratches, is across the field from one disc to the other, and is steady throughout the motion; so that the convection currents are completely represented by the simple convection of the electric charges on the discs, and are not spread over the dielectric field.

Current of dielectric convection:

60. The motion of a dielectric body through a field of electric force ought also to carry its system of electric displacement along with it.

[1] H. A. Rowland and C. T. Hutchinson, "On the electromagnetic effect of Convection currents," *Phil. Mag.* June 1889, p. 445.

Speculation as to cause of failures to repeat the Rowland experiment (now cleared up).

[2] (The statement in the text is certainly true if we can regard the disc as a perfect conductor; on the other hand if it is an insulator, the charge will be carried along with it. It has been suggested that it is open to question whether the conductivity of a coating of gold-leaf is great enough to come practically under the first of these types. But if we are to adhere to the ordinary idea that the free oscillations of an electric charge on such a conductor are absolutely unresisted by any superficial viscosity, as they are certainly independent of ohmic resistance, we must, it would seem, regard a metallic disc as practically equivalent for the present purpose to a perfect conductor. This view would also suggest an explanation of the circumstance that some experimenters have not been able to verify the existence of the Rowland effect.)

It appears that Röntgen[1] has detected an effect of convection currents when a circular dielectric disc is spun between the two plates of a charged horizontal condenser. In this case, however, the displacement system in the field maintains its configuration in space absolutely unchanged; and according to the present view no effect of the kind should exist unless it be really caused by convection of an actual charge on the rotating dielectric plate (unless we find in it a proof of the convection of actual paired ions, of which the material dielectric is constituted. See § 125). **demands paired ions.**

On Vortex Atoms and their Magnetism.

61. Suppose, in the condenser system described above (§ 46), that a current is started round the circuit by a change of capacity of one of the condensers, and that then the two condensers are instantaneously taken out and the wire made continuous; the current, in the absence of resistance in the wire, will now be permanent. A permanent magnetic element will thus be represented by a circuital cavity or channel in the elastic aether, along the surface of which there is a distribution of vorticity; it will in short be a vortex ring with a vacuum (or else a portion of the fluid devoid of rotational elasticity) for its core. An arrangement like this must be supposed, in accordance with Ampère's theory[2], to be a part of the constitution of a molecule in iron and other magnetic metals. As a fundamental structure like the present can hardly be supposed to be broken up at the temperature at which iron becomes non-magnetic, to appear again on lowering the temperature, we must postulate that a permanent electric current of this kind is involved in the constitution of the atom; that in iron the atoms group themselves into aggregates with their atomic currents directed in such a way as not absolutely to oppose each other's action; while at the temperature of recalescence these groups are broken up and replaced by other atomic groups, for each of which the actions at a distance of the different atomic currents are mutually destructive. In a material devoid of striking magnetic properties, we may imagine the atoms as combined into molecules in this latter way. **Tentative Amperean magnetic vortex.** **Atomic re-groupings in ferromagnetics.**

62. If we imagine a vortex-ring theory of atoms, in which the velocity of the primeval fluid represents magnetic force, and the atoms are ordinary coreless vortices, we shall have made a step towards a consistent representation of physical phenomena. In such a fluid the vortices will join themselves together into molecules and **Tentative atomic analogy by coreless vortices:**

[1] Rowland, *loc. cit.* p. 446; Röntgen, *Wied. Ann.* Vol. xxxv. 1888.

[2] Maxwell, *Treatise*, Vol. II. Chap. XXII.

molecular groups; the vortices of each group will, however, tend to aggregate in the same way as elementary magnets, so that instead of neutralizing each other's magnetic effects, they will reinforce one another; on this view substances ought to be about equally magnetic at all temperatures, instead of showing as iron does a sudden loss of the quality. We must therefore find some other bond for the atoms of a molecule, in addition to the hydrodynamic one and at least of the same order of magnitude. This is afforded by the attractions of the electric charges of the atoms, which are required by the theory of electrolysis. But even now about half of the molecules would be made up so that the atoms in them assist each other's magnetic effects, unless we suppose each molecule to contain more than two atoms, arranged in some sort of symmetry. There is however no course open but to take all matter to be magnetic in the same way, the only difference being in some very special circumstance in the aggregation of the molecules of iron compared with other molecules. The small magnetic moment of molecules of most substances may in fact be explained more fully on the same lines as their small electric moment (§ 64). The vortices will be quite permanent as regards both atomic charge and electric intensity, so that the explanation of diamagnetic polarity given by Weber, on the basis of currents induced in the atomic conducting circuits, cannot now stand[1].

cohering by electric charges on them.

We have hitherto chosen to take the vortex atoms with vacuous cores, so that the currents must be represented by the vortex sheets on their surfaces; and this was in order to have an exact representation of the circumstances of perfect conductors. If we assigned a rotating fluid core, devoid of elasticity, to the vortex atom, not many essential differences would be introduced. The circumstances of an ordinary electric current flowing steadily round a channel which is not an ideal perfect conductor are somewhat more closely represented by supposing the channel to be the core of the ring, filled with fluid

Continuous currents ultimately ruled out.

[1] June 14. It has been suggested that the atomic electric charge might circulate round the ring under the influence of induction. It would appear however that such a circulation could have no physical meaning, for it would not at all alter the configuration of strain in the surrounding medium, which is the really essential thing.

It is otherwise with the motion of translation of a small charged body: the intrinsic twist of the surrounding medium is carried on with it, and the effect of the movement is thus to impose an additional twist or rotation round the line of motion (§ 59). Thus if we imagine an endless chain of discrete electrified particles, which circulate round and round, each particle of it will carry on independently its state of strain and so be subject separately to forcive; and we shall have the dynamical phenomena illustrated by a current of purely convective character, involving no electric displacement in the dielectric, and no generator.

A steady current in a wire must be of ions (electrons).

whose rotation is uniform across each section; this uniform distribution of the current across the channel is however primarily an effect of viscous retardation, due to the succession of discharges across intermolecular aether by which the propagation is effected.

Electrostatic Induction between Aggregates of Vortex Atoms.

63. When a piece of matter is electrified, say by means of a current conducted to it by a wire, what actually happens according to dynamical analysis on the basis of our energy-function, is that an elastic rotational displacement is set up in the aether surrounding it, the absolute rotation at each point representing the electric displacement of Maxwell. If there is no viscosity, *i.e.* if the matter and the wire are supposed to be perfect conductors, this result is a logical consequence of the assumed constitution of the aethereal medium; and of course the circumstances of the final equilibrium condition are independent of any frictional resistance which may have opposed its development, so that the conclusion is quite general.

The aether active in charge by conduction.

We may now construct a representation of the phenomena of electrostatic induction. A charged body exists in the field, causing a rotational strain in the aether all round it; consider the portion of the aether inside another surface, which we may suppose traced in the field, to lose its rotational elasticity as the result of instability due to the presence of molecules of matter; the strain of the aether all round that surface must readjust itself to a new condition of equilibrium; the vortical lines of the strain will be altered so as to strike the new conductor at right angles,—and everything will go as in the electrostatic phenomenon. But there will be no aggregate electric charge on the new conductor; for the electric displacement (f, g, h) is a circuital vector, and therefore its flux into any surface drawn, wholly in the aether, to surround the new conductor, cannot alter its value from null which it was before. Now suppose a thin filament of aether, connecting the two conductors, to lose its rotational elasticity; the conditions of equilibrium will again be broken, and the effect throughout the medium of this sudden loss of elasticity will be the same as if a wave of alternating vorticity were rolling along the surface of this filament from the one conductor to the other, with an oscillation backwards and forwards along it which will persist unless it is damped by radiation or viscous action. The final result, after the decay of the oscillations, will be a new state of equilibrium, with charges on both the conductors, precisely as under electrostatic circumstances.

Rotational aether analogue for oscillatory discharge.

Dielectrics: involve bipolar elements, or else small conducting regions.

64. The phenomenon of specific inductive capacity has been explained or illustrated at different times by Faraday, Mossotti, Lord Kelvin, and Maxwell, by the behaviour of a medium composed of small polar elements which partially orientate themselves under the action of the electric force; and these *quasi*-magnetic elements have been identified with the molecules, each composed of a positive and a negative ion. Another illustration[1] which leads to the same mathematical consequences supposes the dielectric field to be filled with small conducting bodies, in each of which electric induction occurs, thus making it a polar element so long as it is under the influence of the electric force. The *quasi*-magnetic theory is adopted by von Helmholtz in his generalization, on the notions of action at a distance, of Maxwell's theory of electrodynamics; and it is shown by him that such a hypothesis destroys the circuital character of the electric current, a conclusion which may also be arrived at by elementary reasoning[2]. The molecules must therefore on such a theory be arranged with their positive and negative elements in some form of symmetry so that they shall have no appreciable resultant electric moments[3]; and the specific inductive capacity must be wholly due to diminution of the effective elasticity of the medium. The hexagonal structure imagined for quartz molecules by J. and P. Curie, and independently by Lord Kelvin[4], in order to explain piezo-electricity, or any other symmetrical grouping, exactly satisfies this condition; the molecule in the state of equilibrium has no resultant electric moment; but under the influence of pressure or of change of temperature a deformation of the molecule occurs, which just introduces the observed piezo-electric or pyro-electric polarity.

Ionic dielectric current not circuital by itself. Can it be excluded?

Quasi-current of aethereal displacement the true solution.

(Added June 14. On the present view however there is absolutely no room for von Helmholtz's more general theory of non-circuital currents. The displacement of an electric charge constitutes a rotation in the medium round the line of the displacement, but the electric field which causes the displacement is here also itself a rotation round an axis in the same direction; whereas in von Helmholtz's theory the inducing electric force is not considered to have any intrinsic electric displacement of its own. When both parts are taken into account, the electric displacement becomes circuital throughout the field. There is thus nothing in the postulate of circuital currents that would require us to make the electric moment of a molecule indefinitely

[1] Employed by Maxwell, "Dynamical Theory," § 11, *Phil. Trans.* 1864.

[2] "On the theory of Electrodynamics," *Roy. Soc. Proc.* 1890.

[3] The term electric moment is employed, after Lord Kelvin, as the precise analogue of magnetic moment.

[4] Lord Kelvin, "On the piezo-electric quality of Quartz," *Phil. Mag.* Oct. 1893, Nov. 1893.

small; so that specific inductive capacity might still, if necessary, be explained or illustrated in the manner of Faraday and Mossotti.)

Cohesive, Chemical, and Radiant Forces.

65. If we consider a system of these vortex atoms, each of them will be subject to pulsations or vibrations, some comparatively slow, under the hydrodynamic influences of its neighbours in its own molecule; and each molecule will be subject to still slower vibrations under the influence of disturbances from the neighbouring molecules. In the former class we may possibly see the type of chemical forces, while the latter will have to represent phenomena of material cohesion and elasticity. But in addition to these purely hydrodynamical vibrations due to the inertia simply of the aether, there will be the types which will involve rotational distortion of the medium; that is, there will be the electrical vibrations of the atoms owing to the permanently strained state of the aether surrounding them which is the manifestation of their electric charges; the vibrations of this type will send out radiations through the aether and will represent the mechanism of light and other radiant energy. The excitation of these electric vibrations will naturally be very difficult; it will usually be the accompaniment of intense chemical action, involving the tearing asunder and re-arrangement of the atoms in the molecules. It is well known that the vibrations of an electrostatic charge on a single rigid atom, if unsustained by some source of vibratory energy, would be radiated so rapidly as to be almost dead-beat, and so would be incompetent to produce the persistent and sharply-marked periods which are characteristic of the lines of the spectrum. But this objection may be to some extent obviated by considering that all the vibrational energy due to any very rapid type of molecular disturbance must finally be transformed into energy of electric strain and in this form radiated away[1].

Possible Kelvin vortex-atom scheme (in addition to electrons):

but now with the fluid medium rotationally elastic,

thus introducing radiation.

Voltaic Phenomena.

66. According to this theory a transfer of electricity can take place across a dielectric by rupture of the elastic structure of the medium, and only in that way; and this is quite in keeping with ordinary notions. Further, an electrolyte is generally transparent to light, or if not, to some kind of non-luminous radiation, so that such a substance has the power of sustaining electric stress; it follows therefore that transfer of electricity across the electrolyte in a volta-

[1] I understand that a suggestion of this nature has already been made by G. F. FitzGerald.

All discharge is by ionic convection. meter, between a plate and the polarized atoms in front of it, can only occur along lines of effective rupture (such as may be produced by convection of an ion) of its aethereal elastic structure.

When two solid dielectrics are in contact along a surface, the superficial molecular aggregates will be within range of each other's influence, and will exert a stress which is transmitted by the medium between them. The transmission will be partly by an intrinsic hydrostatic pressure, as in Laplace's theory of capillarity, and partly by tangential elastic tractions produced by rotation of the elements of the medium. This rotation is the representative of electric force, or rather its effect electric displacement, in the medium; and, in so far as it is not along the interface, its line integral from one body to the other will account for a difference of electric potential between them. The electric force must be very intense, as in fact are all molecular forces, in order to give rise to a finite difference of potential in so short a range. If the bodies in contact are conductors, instead of dielectrics, similar considerations apply, but now the internal equilibrium of each conductor requires that the potential shall be uniform throughout it; therefore the surface stress must so adjust itself that the difference of potentials between the conductors is the same at each point of the interface.

Interfacial fields very intense: very steep fall of potential.

The contact phenomena between a solid and a liquid are different from those between two solids; for the mobility of the liquid allows, after a sufficient lapse of time, an adjustment of charged dissociated ions along its surface so as to ease off the internal stress; and thus the boundary of the liquid becomes completely and somewhat permanently polarized. If we consider for example blocks of two metals, copper and zinc, separated by a layer of water, the electric stress in the interior of the water becomes null, and the difference of potential between the two metals is the difference of the potential differences between them and water. That will not be the same as their difference of potential when in direct contact; but according to Lord Kelvin's experiment it is sensibly the same as the difference between them and air,—owing in Maxwell's opinion to similarity in the chemical actions of air and water. In this experiment the electric stress is not transmitted through either of the metals; its seat is the surrounding aether, and the function of the metals is so to direct it, owing to the absence of aethereal elasticity inside them, that the axis of the rotation of the aether shall be, at all points of their surfaces, along the normal.

Model for a voltaic cell.

67. Let us imagine a Volta's chain of different metals, forming a complete circuit, to be in electric equilibrium, as it must be, in the absence of chemical action and differences of temperature, by the

principles of Thermodynamics. There is no electric stress transmitted through any metallic link of the chain; the stress is transmitted through the portion of the aether surrounding each metal, consisting in part of the interfacial layers separating it from the neighbouring metal, and in part of the atmosphere which surrounds its sides. In the equilibrium condition the potential in the aether all round the surface of the same metal is uniform; and this uniformity applies to each link in the chain. Therefore the sum of the very rapid changes of potential which occur in crossing the different interfaces, is, when taken all round the chain, strictly null: and we are thus led to Volta's law of potential differences for metallic conductors. Now suppose some cause disturbs this equilibrium, say the introduction of a layer of an electrolyte at an interface; this will introduce a store of chemical potential energy which can be used up electrically, and so equilibrium need no longer subsist at all. The uniformity of potential in the dielectric all round the surface of each metal will be disturbed, and a change of the electric displacement, *i.e.* of the absolute rotation in the aether, will be set in action in the surrounding medium. If the metals are perfect conductors the effective flow of displacement will be confined to the surface, and will involve simply a vortex sheet along the surface of each metal; but if the conducting power is imperfect the disturbance will diffuse itself into the metals, and the final steady condition will be one in which it is uniformly distributed throughout them, forming an ordinary electric current obeying Ohm's law*.

Volta's law of compensating chains for metals:

disturbed by an ionizable member in the circuit.

68. On the present theory, high specific inductive power in a substance is equivalent to low electric elasticity of the aether; it in fact stands to reason that an elastic medium whose continuity is broken by the inelastic and mobile portions which represent the cores of vortex atoms may from this cause alone have its effective elasticity very considerably diminished.

Moreover it has been ascertained that, in electrolytic liquids, the specific inductive capacity attains very great values; the aether in these media interposes a proportionally small resistance to rotation, and the mobility or some other property of the vortex molecules in it has brought it so much the nearer to instability; it is thus the easier to see why such media break down under comparatively slight electric stress†. Such a medium also frees itself, as described below, from electric stress, without elastic rupture, in a time short compared with

High dielectric capacity involves loose aggregation in the molecule.

* For reasons completely involved in electron theory: cf. Vol. II. *infra, Math. Congress Lecture,* 1912.

† Also why they are powerful solvents: cf. *infra.*

ordinary standards, but in most instances long compared with the periods of light-vibrations; while in metallic media the period of decay of stress is at least of the same order of smallness as the periods of light-waves.

Electric model of an ion (or electron).

69. An atom, as above specified, would be mathematically a singular point in the fluid medium of rotational elastic quality. Such a point may be a centre of fluid circulation, and may have elastic twist converging on it, but it cannot have any other special property besides these; in other words this conception of an atom is not an additional assumption, but is the unique conception that is necessarily involved in the hypothesis of a simple rotationally elastic aether.

Work of mechanical forces between ions exhausts aethereal energy:

The attraction of a positively-charged atom for a negatively-charged one, according to the law of inverse squares, has already been elucidated. If the two atoms are moved towards each other so slowly that no kinetic energy of the medium is thereby generated, the potential energy of the rotational strain between them is diminished; and this diminution can be accounted for, in the absence of dissipation, only by mechanical work performed by the atoms or stored up in them in their approach. It has been observed by von Helmholtz that the phenomena of reversible polarization in voltameters involve no sensible consumption of energy, but that it is the actions which effect the transformation of the electrically charged ions into the electrically neutral molecules that demand the expenditure of motive power; and he draws the conclusion that energy of chemical decomposition is chiefly of electrical origin. In the explanation here outlined, the chemical (hydrodynamic) forces between the component atoms of the molecule are required to be, in the equilibrium position, of the same order of intensity as the electrical forces (elastic stress); but then they are of much smaller range of action as their intensity depends on the inverse fourth power of the distance, so that the work done by them during the formation of the molecule will probably be very small compared with the work done by the electric forces.

while chemical action involves ionic rupture (or transfer),

which accounts for the large energy-change.

The molecule is self-sufficing:

(70. Added June 14. The charged atoms will tend to aggregate into molecules, and when this combination is thoroughly complete, the rotational strain of each molecule will be self-contained, in the sense that the lines of twist proceeding from one atom will end on some other atom of the same molecule. If this is not the case, the chemical combination will be incomplete, and there will still be unsatisfied bonds of electrical attraction between the different molecules. A molecule of the complete and stable type will thus be electrically neutral; and if any cause pulls it asunder into two ions, these ions will possess equal and opposite electric charges.

while an ion is unsaturated.

In the theory as hitherto considered, electric discharge has been represented as produced by disruption of the elastic quality of aether along the path of the discharge; and this is perhaps the most unnatural feature of the present scheme. If, however, we examine the point, it will be seen that the phenomena of electric flow need involve only convection of the atomic charges without any discharge across the aether, with the single exception of electrolysis. An attempt may be made (cf. abstract, *supra*, p. 406) to account for the uniformity of the atomic charges thus gained or lost, from the point of view of the establishment of a path of disruptive discharge from one atom to another. But it seems preferable to adopt a more fundamental view.

Electric transfer is not by rupture.

The most remarkable fact about the distribution of matter throughout the universe is that, though it is aggregated in sensible amounts only in excessively widely separated spots, yet wherever it occurs, it is most probably always made up of the same limited number of elements. It would seem that we are almost driven to explain this by supposing the atoms of all the chemical elements to be built up of combinations of a single type of primordial atom, which itself may represent or be evolved from some homogeneous structural property of the aether[1]. It is, again, difficult to imagine how the chemical elements should be invariably connected, through all their combinations, with the same constant of gravitation, unless they have somehow a common underlying origin, and are not merely independent self-subsisting systems. We may assume that it is these ultimate atoms, or let us say monads, that form the simple singular points in the aether; and the chemical atoms will be points of higher singularity formed by combinations of them. These monads must be taken to be all quantitatively alike, except that some have positive and others negative electrifications, the one set being, in their dynamical features, simply perversions or optical images of the other set. On such a view, electric transfer from ion to ion would arise from interchange of monads by convection without any breaking down of the continuity of the aether.

A single type of primordial structure for matter.

Gravitation fundamentally uninfluenced by aggregation.

The ultimate atoms (electrons): all identical, positive the structural images of negative.

But a difficulty now presents itself as to why the molecule say of hydrochloric acid is always $H + Cl -$, and not sometimes $H - Cl +$. This difficulty would however seem to equally beset any dynamical theory whatever of chemical combination which makes the difference between a positive and a negative atomic change* representable wholly by a difference of algebraic sign.)

Yet structural images of chemical molecules do not exist.

[1] Cf. Thomas Graham's *Chemical and Physical Researches*, Introduction, and p. 299.

* The solvent influence of the environment has been invoked as the necessary additional element.

The Connection between Aether and Moving Matter:
[Convection of Rays].

71. A mode of representation of the kind developed in this paper must be expected to be in accord with what is known on the subject of the connection between aether and matter, both from the phenomena of the astronomical aberration of light, and from recent experimental researches[1] on the motion of the aether relative to the Earth, and relative to transparent moving bodies.

The data of the problem of material convection through the field.

Let us consider a wave of light propagated through the free aether with its own specific velocity, and let it be simultaneously carried onward by a motion in bulk of the aether which is its seat. That motion will produce two effects on a wave; the component along the wave-normal of the velocity of the aether will be added on to the specific velocity of the wave; while the wave-front will be turned round owing to the rotational motion of the medium. The second of these effects will result in the ray being turned out of its natural path; in order that the motion of the medium may not affect the natural path of the ray, it must therefore be of irrotational character. This will be the case as regards all motions of the free aether so long as we consider it to be hydrodynamically a frictionless fluid; and the phenomenon of astronomical aberration is, after Sir George Stokes, explained, so far as it may depend on motion of the external aether.

Rays not affected by irrotational motion of their medium.

Stokes on aberration of light.

72. The motion of the Earth through space may however be imagined as the transference of a vortex aggregate through the quiescent aether surrounding it and permeating it; the velocity of translation of the aether will then be null, and consequently in the comparatively free aether of the atmosphere the velocity of the light will be unaffected, to the first order of approximation. But what should happen in transparent material media it is apparently not easy to infer. On the present view of Optics, the density of the aether is constant throughout space, the mere presence of mobile electrified vortices in it not affecting the density though the effective elasticity is thereby altered. The nature of the further slight alteration of this elasticity produced by a motion of the matter as a whole, there appears to be no easy means of directly determining (see § 124); but the experiments may be taken as verifying Fresnel's hypothesis that its effect is to add on to the velocity of propagation of the light the fraction $1 - \mu^{-2}$ of the velocity of the matter through which it is moving, where μ represents the index of refraction.

Aether need not be disturbed except locally, by a moving vortex aggregate.

Difficulties of convection of light (until electrons are introduced).

[1] A. A. Michelson and E. W. Morley, *American Journal of Science*, 1881 and 1886, also *Phil. Mag.* Dec. 1887; O. J. Lodge, *Phil. Trans.* A, 1893.

This formula of Fresnel[1], for the change of the velocity of propagation in a moving ponderable medium, was specially constructed so as to insure that the laws of reflexion and refraction of the *rays* shall be the same as if the media were at rest, a circumstance which must be intimately connected with the dynamical reason for its validity. The laws of reflexion and refraction of rays can be deduced from the theory of exchanges of radiation, on the single hypothesis that a condition of equilibrium of exchanges is possible in an enclosure containing transparent non-radiating bodies. One interpretation of Fresnel's principle is therefore that the exchange of radiation between the walls of an enclosure containing transparent bodies is not affected by any motion imparted to these bodies, a conclusion which may be connected with the law of entropy. [See § 77.]

Laws of reflexion unaffected by convection: as is necessary for a theory of exchanges.

73. On the present theory, magnetic force or rather magnetic induction consists in a permeation or flow of the primordial medium through the vortex aggregate which constitutes the matter; apparently it has not been tried (see however § 81) whether light-waves are carried on by this motion of the medium and their effective velocity is thereby altered, as we would be led to expect. It has been shown, however, by Wilberforce[2] that the velocity of light is not sensibly altered by motion along a field of electric displacement, so far negativing any theory that would connect electric displacement with considerable velocity of the aether; and it has also been verified, by Lord Rayleigh, that the transfer of an electric current across an electrolyte does not affect the velocity of light in it.

Light not convected by electric displacement or current.

As motion of the aether represents magnetic force, the fact that the magnetic permeability is almost the same in all sensibly non-magnetic bodies as in a vacuum must be taken to indicate that the aether flows with practically its full velocity in all such media, so that there is very little obstruction interposed by the matter; it follows that, in the motion of a body through the aether, the outside aether remains at rest instead of flowing round its sides. The aether we thus assume to be at rest in any region, except it be a field of magnetic force, even though masses are moving through the region; so that the coefficient of Fresnel, which is null for free aether and very small for but slightly ponderable media, would represent simply a change of velocity due to slight unilateral change of effective elasticity somehow produced by the motion through the quiescent medium of the vortices constituting the matter.

Matter is a very open structure in aether.

74. The notion of illustrating magnetic induction by the permeation

[1] A. Fresnel, letter to Arago, *Annales de Chimie*, IX. 1818.

[2] L. R. Wilberforce, *Trans. Camb. Phil. Soc.* Vol. XIV. 1887, p. 170.

Kelvin's permeation idea for magnetic flux.

of a fluid through a porous medium containing obstacles to its motion has been shown by Lord Kelvin[1] to lead to a complete formal representation of the facts of diamagnetism; and such an idea of very slightly obstructed flow might possibly be made to serve as a substitute for Weber's theory, if we are unable to retain it. [See § 114.]

75. The motion of a material body through the aether must, in any case, either carry the aether with it, or else set up a backward drift of the aether through its substance, so that the vortex cores (which might be vacuous and therefore merely forms of motion) would be carried on, while the body of the aether remained at rest.

Coreless vortices: but see § 114.

Convection and magnetism: alternatives.

On the first view, the motion of the body must produce a field of irrotational flow in the surrounding aether, in other words a magnetic field. Whether this would be powerful enough to be directly detected depends on the order of magnitude of the aethereal velocities which represent ordinary magnetic forces, and thus ultimately on the value of the density of the aethereal medium. But if the density were small, the square of the velocity would be large in proportion, and the influence of magnetization on the velocity of light should be the greater; so that on this account also the first of the above views must, on the present theory, be rejected. We should, however, expect an actual magnetic field like the Earth's to affect very slightly both the velocity of propagation and the law of reflexion.

76. The second view is, as we have stated, the one formulated by Fresnel, and it would be strongly confirmed if the velocity of light-waves were quite unaffected by passing near a moving body, so shaped that it would on the other hypothesis cause a current in the perfectly fluid aether; but it is sometimes held (see however § 80) to be against the evidence of the null result of Michelson's experiments on the effect of the Earth's motion on the velocity of transmission of light through air.

Surface slip to be avoided:

There is also the fact noticed by Lorentz that an irrotational disturbance of the surrounding aether, caused by the motion of an impermeable body through it, would necessarily involve slip along the surface, which could not exist in our fluid medium; this would at first sight compel us to recognize that the surrounding aether, instead of flowing round a moving body, must be taken to flow through it, or rather into it, at any rate to such an extent as will be necessary

[1] Lord Kelvin (Sir W. Thomson), "Hydrokinetic analogy for the magnetic influence of an ideal extreme diamagnetic," *Proc. R. S. Edin.* 1870, *Papers on Electrostatics and Magnetism*, pp. 572–583; "General hydrokinetic analogy for Induced Magnetism," *Papers on Electrostatics and Magnetism*, 1872, pp. 584–592.

in order to make the remaining motion outside it irrotational, without discontinuity at the surface.

It has been shown however by W. M. Hicks that a solitary *hollow* vortex in an ordinary liquid carries along with it a disc-shaped mass of fluid and not a ring-shaped mass, unless its section is very minute; thus it is possible that the vortex aggregate constituting a moving solid may completely shed off the surrounding fluid without allowing any permeation through its substance, and without any such discontinuity at the surface as would be produced by the motion of an ordinary solid through liquid. How far the electric charges on the vortex atoms, or their combination into molecules, would negative such a hypothesis seems a difficult inquiry. But however that may be, a *consensus* of various grounds seems to require the aether to be stationary on the present theory. Thus if the motion of solids moved the surrounding aether, two moving solids would act on each other with a hydrodynamic forcive, which would be of large amount if we are compelled to assume a considerable density for the aether. Again, such a view would disturb the explanation, as above, of the fact that the forcive on a charged conductor in an electric field is a surface traction equal at each point of the surface to the energy in the medium per unit volume. There is in any case nothing contradictory in the hypothesis of a stationary aether; if the fluid is not allowed to stream through the circuits of the atoms, we have only to make the ordinary supposition that the molecules are at distances from each other considerable compared with their linear dimensions, and it can stream past between them.

as is possible for a convected vortex aggregate.

Other reasons for a stationary aether.

77. Let us test a simple case of motion of a body through the aether, with respect to the theory of radiation. Consider a horizontal slab of transparent non-radiating material, down through which light passes in a vertical direction; the equilibrium of exchanges of radiation would be vitiated if the amount of light transmitted by the slab when in motion downwards with velocity v were different from the amount transmitted when it is at rest. Let V be the velocity of the light outside the slab, and $V/\mu + v - v'$ the velocity in the moving slab. For an incident beam, of amplitude of vibration which we may take as unity, let r be the amplitude of the reflected beam, and R of the transmitted beam. The conditions governing the reflexion are continuity of displacement at the surface, and continuity of energy, estimated in MacCullagh's manner as proportional to the square of the amplitude; thus the conditions at the first incidence are

Transmission of light across a slab must be unaltered by convection:

$$1 + r = R,$$

$$V - v - (V + v)\, r^2 = (V/\mu - v')\, R^2.$$

On neglecting squares of v/V and v'/V, these equations lead to

$$R = \frac{2\mu}{\mu + 1}\left\{1 - \frac{v}{V}\left(\frac{\mu}{\mu + 1} + \frac{\mu - 1}{2\mu}\right) + \frac{v'}{V}\frac{\mu}{\mu + 1}\right\}.$$

The ratio R', in which the amplitude is changed by transmission at the lower surface of the slab, is derived from the above by replacing V by V/μ, and μ by $1/\mu$, and interchanging v and v'; thus

$$R' = \frac{2}{\mu + 1}\left\{1 - \frac{\mu v'}{V}\left(\frac{1}{\mu + 1} - \frac{\mu - 1}{2}\right) + \frac{\mu v}{V}\frac{1}{\mu + 1}\right\}.$$

Hence

$$RR' = \frac{4\mu}{(\mu + 1)^2}\left\{1 - \frac{v}{V}\frac{\mu - 1}{2\mu} + \frac{\mu v'}{V}\frac{\mu - 1}{2}\right\}.$$

leading to Fresnel's coefficient.

That the amount of the light transmitted should not be altered by the motion of the slab requires that $v' = v/\mu^2$, which is Fresnel's law; it has been assumed in the analysis that the light is propagated down to the slab as if the aether were at rest, in accordance with Fresnel's hypothesis. It will be observed that the amplitudes of the refracted and reflected light, at either surface separately, are disturbed by the movement of the slab, though there is no loss of energy; thus on direct refraction into a slab moving away from the light with velocity v,

Influence of convection on reflexion.

$$R = \frac{2\mu}{\mu + 1}\left\{1 - \frac{3}{2}\frac{v}{V}\frac{\mu - 1}{\mu}\right\}, \quad r = \frac{\mu - 1}{\mu + 1}\left(1 + \frac{3}{2}\frac{v}{V}\right).$$

If Fresnel's law is not fulfilled, it would apparently be possible to concentrate the radiation from the walls of an enclosure of uniform temperature by a self-acting arrangement of moving screens and transparent bodies inside the enclosure; and this would be in contradiction to the Second Law of Thermodynamics[1].

Rays: characteristic function of a coherent beam.

78. The whole theory of rays is derived from the existence of the Hamiltonian characteristic function U, the path of a ray from one point to another in an isotropic medium being the course which makes δU or $\delta\int \mu ds$ null, where μ is a function of position which is equal to the reciprocal of the effective velocity of the light. The general law of illumination may be shown to follow from this, that if two elements of surface A and B are radiating to each other across any transparent media, the amount of the radiation from A that is received by B is equal to the amount of radiation from B that is received by A; with the proviso, when different media are just in front of A and B, that the radiation of a body is *caeteris paribus* to be taken as proportional to the square of the refractive index of the

Interchange of energy of distant bodies by rays.

[1] Cf. Clausius, "On the Concentration of Rays of Light and Heat, and on the Limits of its Action," *Papers on the Mechanical Theory of Heat*, translated by W. R. Browne, pp. 295–331.

medium into which it radiates. Now if the part $\mathbf{v}$ of the velocity of the light, which is produced by motion through the medium of the bodies contained in it, make an angle θ with the element of path ds, this equation will assume, after H. A. Lorentz and O. J. Lodge[1], the form

Law of stationary time of transit determines path of ray:

$$\delta \int (V + v \cos \theta)^{-1} ds = 0,$$

which is to a first approximation

$$\delta \int V^{-1} ds + \delta \int V^{-2} (u dx + v dy + w dz) = 0,$$

where V is the ordinary velocity of the light, and (u, v, w) are the components of $\mathbf{v}$. In order that the paths of the rays in a homogeneous isotropic moving medium may remain the same as when the medium is at rest, the additional terms in the characteristic function must depend only on the limits of the integral, and therefore $u dx + v dy + w dz$ must be an exact differential; that is, the part thus added to the velocity of the light must be of irrotational character. If this part of the velocity were rotational, the law of illumination would not hold, as the type of the characteristic equation of the rays would thereby be changed. Thus the equilibrium of exchanges of radiation which would subsist in an enclosure with the free aether in it at rest, would be violated were the aether put into a state of rotational motion. Now any modification of the laws of emission and absorption would be conditioned only by the motion of the aether close to the radiating surface; and the motion at the surface by no means determines the motion throughout the enclosure, unless it is confined to be irrotational. Hence the theory of exchanges seems to require that any bodily motion that can be set up in the free aether must be of the irrotational kind.

its form thus unaffected by irrotational convection:

as a Law of steady Exchanges requires:

79. This modified characteristic equation of the rays also shows that in a heterogeneous isotropic medium containing moving bodies, the paths of the rays will be unaltered to a first approximation provided $\mu^2 (u dx + v dy + w dz)$ is everywhere continuous and an exact differential; and this condition virtually implies (Lodge, *loc. cit.*) Fresnel's hypothesis. The interchange of radiation now depends partly on the reflexion and refraction at the different interfaces in the medium, as in the simple case calculated above; but we may take advantage of a device which has been employed in other connections by Lord Rayleigh, and suppose the transitions to be gradual, that is to be each spread over a few wave-lengths; the reflexions will

[1] O. J. Lodge, "Aberration Problems," *Phil. Trans.* A, 1893, pp. 748–753.

then be insensible, and the rays will thus be propagated with undiminished energy. We thus attain a general demonstration that the theory of exchanges of radiation demands Fresnel's law of connection between the velocity of the matter through the field of stationary aether and the alteration in the velocity of the light that is produced by it; while it also requires that any motion of the aether itself, such as occurs in a field of magnetic force, must be of irrotational type.

leading to Fresnel's coefficient.

80. This theory has been developed up to and including the first order of small quantities; it seems plain therefore that the experiments of Michelson on the effect produced by the motion of the earth on transmission through air are not in contradiction with it, for these experiments relate to terms of the second order of small quantities. To explain the remarkable, because precisely negative, result arrived at by Michelson would require the elaboration of a theory including the second order of small quantities. For example, when light is reflected, as in those experiments, at the surface of a body which is moving towards it through the stationary aether, the wave-length of the reflected light is diminished so as just to make up, to the first order of approximation, for the acceleration of phase caused by the reflector moving up to meet it. The mechanism involved in this alteration of wave-length is not known, nor what is going on at the surface of the advancing reflector; and it seems to be a very uncertain step to assume that when terms of the second order are included, this effect on the wave-length is not subject to correction. As the circumstances of the reflexion are thus not known with sufficient exactness, it is necessary to fall back on general principles. Now Professor Lodge has emphasized the fact that, when a beam of light traverses a complete circuit in a medium containing moving bodies but devoid of magnetic intensity, the change of phase produced by their motion is null to the first order of small quantities. If it were exactly null, or null to the second order, the result of Michelson would follow; and it would seem also that Michelson's result favours somewhat the exact validity of this principle. The exactness of this circuital principle seems to be required also by the argument (§ 79) from the equilibrium of exchange in an enclosure. For if when a system of rays pass from a point to its image point their relative differences of phases were not the same to a small fraction of a wave-length whether the bodies are at rest or in motion, it would follow that the distribution of the energy in the diffraction pattern which forms the physical image would depend on the movement of the bodies. Thus concentration of the radiation might be produced by movements of the transparent bodies, which are subject to control.

Michelson's null experiment of convection held over:

second-order intricacies.

Relativity considerations.

The present discussion supposes the motion of the transparent bodies to be practically uniform; the condition $\mu^2 (udx + vdy + wdz)$ an exact differential would be violated inside a transparent body in rapid rotation, but then (§ 98) the formula of Fresnel would require correction owing to the space-rate of variation of the velocity of the material medium*.

Experiments by Professor Oliver Lodge.

81. Since this account of the theory was written, Professor Lodge has kindly made some experiments on the effect produced by a magnetic field on the velocity of light, which considerably affect its aspect. By surrounding the path of the beam of light in his interference apparatus[1] by coils carrying currents, he realized what was equivalent to a circuit of 50 feet of air magnetized to ± 1400 C.G.S.; and he would have been able to detect a shift in the fringes, between beams of light traversing this circuit in opposite directions, of $\frac{1}{80}$ of a band, or say with absolute certainty $\frac{1}{20}$ of a band, either way. Four coils were employed, each 18 inches long and with 7000 turns of wire; and they were excited by a current of 28 amperes at 230 volts, involving nearly 9 horse-power. The result was wholly negative; and in consequence the velocity of light cannot be altered by as much as 2 millimetres per second for each C.G.S. unit of magnetic intensity. The cyclic aethereal flow in a magnetic field must therefore be very slow; but the radiation traversing it is of course very fast.

Does a magnetic field convect radiation?

Lodge's result negative up to less than 2 mm. per sec. per gauss:

To bring this result into line with the present theory we are compelled to assume that the density of the aether is at least of the same order of magnitude as the densities of solid and liquid matter, at any rate if we must adhere to the view that the motion of the aether carries the light with it. This hypothesis is of a somewhat startling character; the density under consideration belongs, however, to an intangible medium and is not apparently amenable in any way to direct perception; it is on a different plane altogether from the density of ordinary matter, and is in fact most properly considered simply as a coefficient of inertia in the analytical expression for the energy.

demands that inertia of a rotational aether be very great,

82. The maximum electric force which air can sustain at ordinary temperatures and pressures is about 130 C.G.S.; and on Pouillet's data the maximum electric force involved in the solar radiation, near the

* *E.g.* in a transparent whirling body, like the sun, rays from the centre would bend slightly in the direction of the whirl to a form determinable from stationary time of transit as *supra*. The obliquity of their back-pressure is unimportant as regards solar rotation period.

[1] O. J. Lodge, "Aberration Problems," *Phil. Trans.* A, 1893. (There are also some earlier experiments by Cornu.)

and so amplitude of waves extremely small, in keeping with the perfect transmission across space.

Sun's surface, is about 30 C.G.S., a value which would be much increased on more recent estimates. One result of taking a high value for the aethereal density would be that in the most intense existing field of radiation we are certain of being still far from the limits of perfect elasticity of the comparatively free aether.

Lower limit for inertia of the aether.

The kinetic energy in the free aether is the square of the magnetic intensity divided by 8π; and this must be $\frac{1}{2}\rho v^2$, where ρ is its density and v its velocity. Now from Professor Lodge's result the velocity corresponding to the C.G.S. unit of magnetic force is less than ·2 centimetre per second; hence the inertia of the aether must exceed twice that of water. The elasticity must of course be taken large in proportion to the density, in order to preserve the proper velocity of radiation. In view of the very great intensity of the chemical and electrical forces acting between the atoms in the molecule, values even much greater than these would not appear excessive. But on the other hand such a value of the density requires us to make the aether absolutely stationary except in a magnetic field, in order to avoid hydrodynamical forcives between moving bodies. The residual forcive between bodies at rest in a field of aethereal motion, due to very slight defect of permeability, has already been shown, after Lord Kelvin's illustration, to simulate diamagnetism; and the fact that there exist no powerfully diamagnetic substances is so far a confirmation of the present hypothesis. The view that the magnetic field of a current involves only slight circulation of the fluid aether is also in keeping with the account which has been given (§ 46) of the genesis of such a field.

Very strong resilience natural in view of the extremely intense intra-molecular fields.

Diamagnetism always weak.

The magnetic vortex of a current extremely slow.

On Magneto-Optic Rotation.

Interaction of light with possible material vortices:

83. The rotation of the plane of polarization of light in a uniform magnetic field depends on the interaction of the uniform velocity of the aether, which constitutes that field, with the vibrational velocity which belongs to the light-disturbance. The uniform flow in the medium we may consider to be connected with a partial orientation of the vortex molecules; the chemical or hydrodynamic vibrations, in other words vibrations of the magnetism, can now be propagated in waves, and it is natural to expect that the propagation of the light will be somewhat affected by this regularity. Now for the light-waves the motion that is elastically effective is the rotation $d/dt\,(f, g, h)$; and the varying part of the velocity of an element of volume containing the rotational motion of the magnetic vortices which is to some extent interlinked with the motion of the light-waves, is proportional to

$$\frac{d}{d\theta}(\xi, \eta, \zeta), \text{ where } \frac{d}{d\theta} = \alpha_0 \frac{d}{dx} + \beta_0 \frac{d}{dy} + \gamma_0 \frac{d}{dz},$$

$(\alpha_0, \beta_0, \gamma_0)$ being the imposed magnetic field. This variation is caused by alteration of the vibrational velocity of a particle owing to its change of position as it is carried along in the magnetic field, analogously to the origin of the corresponding term in the acceleration of an element of the medium, in the equations of hydrodynamics. There may exist a term in the energy, resulting from this interaction, of the form

analysis of type for Maxwell's tentative scheme of magneto-optic rotation.

$$C'\left(\frac{d\xi}{d\theta}\frac{df}{dt}+\frac{d\eta}{d\theta}\frac{dg}{dt}+\frac{d\zeta}{d\theta}\frac{dh}{dt}\right);$$

and I have elsewhere[1] tried to show that, on a *concensus* of various reasons, this term, originally given by Maxwell, must be taken as the correct representation of the actual magneto-optic effect. The term is extremely small, and is distinct from the direct effect of the motion of the aether (§ 79), which is irrotational; it leads to an acceleration of one kind of circularly polarized light, and a retardation of the other kind, which are of equal amounts.

The magnetic whirl is of the ions not of aether.

It was this phenomenon of magneto-optic rotation that gave the clue to Maxwell's theory of the electric field. As has recently been remarked by various authors[2], the deduction from it, that magnetic force must be a rotation of the luminiferous medium, is too narrow an interpretation of the facts; the identification of magnetic force with rotation has however hitherto been retained as an essential part of most theories of the aether.

Difficulty in problems of reflexion (until electrons are introduced).

84. It is to be observed that the magneto-optic terms in the energy of the medium do not depend essentially on any averaging of the effect of molecular discreteness, in the same way as dispersive terms or structural rotatory terms. The problem of reflexion is, in the magnetic field, perfectly definite; and the boundary conditions at the interface can all be satisfied, provided we recognize a play of electromotive pressure at the interface, which assists in making the stress

[1] "On Theories of Magnetic Action on Light...," *Report of the British Association*, 1893, *supra*, p. 324. Any other energy-term containing the same differential operators would however equally satisfy these conditions; for example $d\xi/d\theta \,.\, df/dt$ might be replaced by $d\xi/dt \,.\, df/d\theta$ or even by $f \,.\, d^2\xi/d\theta dt$, so far as the equations of bodily propagations are concerned. Such forms would be discriminated by the theory of reflexion. As the term in the energy is related to the motion of the medium, it must involve $d/d\theta$; and this circumstance, combined either with the character of the optical rotation produced, or with the present hypothesis which requires that the term involves (f, g, h), suffices to limit it to one of these types; cf. *loc. cit.* § 3.

[2] *E.g.* H. Lamb, "On Reciprocal Theorems in Dynamics," *Proc. Lond. Math. Soc.* Vol. XIX. 1888, where the remark is actually made that a distribution of vortices with their axes along the direction of the field might account for the magnetic rotation of the light.

continuous[1], and which is required on account of this interaction of the linear motion of the medium with the rotational motion of the waves. The chief obstacle in the way of a complete account of the magnetic phenomena of reflexion appears to be the uncertainty with respect to the proper mathematical representation of ordinary metallic reflexion*.

On Radiation.

85. In accordance with this theory, radiation would consist of rotational waves sent out into the aether from the vibrations somehow set up in the atomic charges. It has been observed (§ 65) that the characters and periods of these electric vibrations, and of the radiations they emit, depend only on the relative positions and motions of the vortex atoms in the molecule, and are quite unaffected, except indirectly, by irrotational motion (magnetic intensity) in the aether which they traverse. The mode of propagation of electric vibrations in free aether cannot be interfered with by the bodily motion of the medium, however intense, except in so far as the motion of the medium carries the electrical waves along with it; a result justifying the Doppler principle which is applied to the spectroscopic determination of stellar motions. It also follows that radiation will not be set up by motions of the surrounding free aether, except in so far as the molecules are dissociated or their component atoms violently displaced with respect to each other. To allow the radiation to go on, such displacement must result on the whole in the performance of work against electric attractions, at the expense of the heat energy and chemical energy of the system, which must thus be transformed into electrical energy before it is radiated away. The radiation of an incandescent solid or liquid body is maintained by the transfer of its motion of agitation into electrical energy in the molecules, and thence into radiation. This action goes on until a balance is attained, so that as much incident radiation is absorbed by an element of volume as it gives out in turn; when this state is established throughout the field of radiation the bodies must be† at the same temperature.

No magnetic effect possible on spectra without discrete electrons.

Translational atomic energy can pass into radiation.

Equilibrium of radiation must agree with equality of material temperatures.

[1] J. Larmor, *Report of the British Association*, 1893; G. F. FitzGerald, *Phil. Trans.* 1880. Professor FitzGerald informs me that he has for some time doubted the view that the magnetic force can be solely a rotation in the medium, on the ground that the magnetic tubes of a current system are circuital and have no open ends, making it difficult to imagine how alteration of the rotation inside them could be produced; also that a flow along these tubes need not produce any disturbance in the other properties of the electric field; also that the magnetic rotation being a purely material phenomenon, whose direction is not subject to any definite law, it must be of a secondary character.

Nature of magnetic tubes.

* The subject has been thoroughly analyzed by J. G. Leathem, *Phil. Trans.* 1897; cf. *supra*, pp. 331, 339.

† A main inducement to recent theories, which make radiation atomic as

Conversely, the absorption of incident radiation by a body results finally in a diffusion of its energy into irregular material motions or heat, directed motion always implying magnetic force.

86. There appear to be experimental grounds for the view that a gas cannot be made to radiate (at any rate with the definite periods peculiar to it) by merely heating it to a high temperature, so that radiation in a gas must involve chemical action or, what is the same thing, electric discharge. This would be in agreement with the conclusion that motion of a molecule through the aether, however the latter is disturbed, will not appreciably set up electric vibrations, unless it comes well within range of the chemical forces of another molecule; and it implies that the encounters of the molecules that are contemplated in the kinetic theory of gases are not of so intimate a character[1] as the encounters in a solid or liquid mass; in the latter case there is perhaps not sufficient space for free repulsion, and the molecules become so to speak jammed together. In the theory of exchanges of radiation, a gas would thus act simply as a medium for the transfer of radiations from one surface to another without itself adding to or subtracting from them.

Radiation associated with atomic disruption.

Presumed soft encounters in gases.

It follows from the second law of Thermodynamics that the heat-equivalent of the radiation of a given substance rises with the temperature, and this may be extended to each separate period in the radiation; this is, however, a theorem of averages not directly applicable to single molecules.

The thermodynamic value of radiation.

It seems a noteworthy consequence of the foregoing that the kinetic theory of gases is valid without taking any account of radiation. Without some tangible mode of presentation such as the mechanism of radiation here put forward, there would be a strong temptation to assume that the interchange of energy in that theory must take place not only between the different free types of vibration of the molecule (*i.e.* hydrodynamical vibrations of the vortices), but that also there is even in the steady state continual interchange with the aether. According to the present views such interchange would involve dissociation in the molecules; and there exist in fact observations relative to the action of ultra-violet radiations in producing discharge of electricity across a gas and consequent luminosity in it, a phenomenon which very probably depends on dissociation. Whether the ideas here indicated turn out to be tenable or not, they at all events may serve to somewhat widen our range of conceptions.

No frittering of intrinsic atomic energy into radiation without dissociation.

well as matter, is that the validity of this double approach to the steady thermal state then admits of ready understanding.

[1] The difficulty of chemical combination of dry gases confirms this conclusion; as also for example the fact that molecular impacts do not explode a mixture like hydrogen and chlorine.

Spectral periods unaffected by temperature:

87. The result that the electric vibrations of a molecule depend on its configuration and the relative motion of its parts, not directly on its motion of translation through the aether, seems also to be of importance in connection with the fundamental fact that the periods of the radiations corresponding to the spectral lines of any substance are precisely the same whatever be its temperature. The lines may broaden out owing to frequency of collisions due to increase of density or rise of temperature of the substance, but their mean period does not change. If we consider a system of ordinary hydrodynamical isolated vortex atoms, a rise of temperature is represented by increase of the energy, and that involves an expansion of each ring and a diminution of its velocity of translation; such an expansion of the ring would in turn alter the periods of its electric vibrations. The question arises, how far the action of the atomic charge will modify or get rid of these two fundamental objections to a vortex-atom theory of gases. Independently of this, it seems quite reasonable to hold that in the case of atoms paired together into molecules by their electrical and chemical forces, the size and configuration of the rings will be determined solely by these forces, which are far more intense than any forces due to mere translation through the medium; and then, when radiation occurs as the result of some violent disturbance, or of dissociation of the molecule, it will have subsided before any sensible change of size due to slowly-acting hydrodynamical causes could have occurred. As was pointed out by Maxwell, the definiteness of the spectral lines requires that at least some hundreds of vibrations of a molecule must be thrown off before they are sensibly damped; and on this view there is ample margin for such a number.

is a difficulty for vortex atoms,

unless radiation may occur in pulses.

On these ideas the velocity of translation of a molecule in a gas would not be connected with the natural hydrodynamical velocity of a simple vortex atom, but would rather be determined by the circumstances of collisions, as in the ordinary kinetic theory of gases. The configuration of a molecule, which determines its electric periods, would also be independent of the movements of translation and rotation, which constitute heat and are the concern of the kinetic theory of gases.

Introduction of the Dissipation Function.

88. The original structure of Analytical Dynamics, as completed by the work of Lagrange, Poisson, Hamilton, and Jacobi, was unable to take a general view of frictional forces; one of the most important extensions which it has since received, from a general physical standpoint, has been the introduction of the Dissipation Function by

Lord Rayleigh. He has shown[1] that in all cases in which the frictional stress between any two particles of the medium is proportional to their relative velocity, when the motion is restricted to be such as maintains geometrical similarity in the system—*i.e.* in all cases in which, $(x_1 y_1 z_1)$ and $(x_2 y_2 z_2)$ being the two particles, the components of the frictional stress between them are

$$\mu_x (\dot{x}_1 - \dot{x}_2), \ \mu_y (\dot{y}_1 - \dot{y}_2), \ \mu_z (\dot{z}_1 - \dot{z}_2),$$

where μ_x, μ_y, μ_z are any functions of the coordinates—the virtual work of the frictional forces in any geometrically possible displacement may be derived from the variation of a single function $\mathfrak{F}$. The virtual work for the two particles just specified is in fact

The Rayleigh dissipation function:

$$\mu_x (\dot{x}_1 - \dot{x}_2)\, \delta (x_1 - x_2) + \mu_y (\dot{y}_1 - \dot{y}_2)\, \delta (y_1 - y_2) + \mu_z (\dot{z}_1 - \dot{z}_2)\, \delta (z_1 - z_2);$$

and for the whole system it will be found by addition of such expressions as this. Now if we form the variation, with respect to the velocities alone, of the expression

$$\mathfrak{F} = \tfrac{1}{2}\Sigma \{\mu_x (\dot{x}_1 - \dot{x}_2)^2 + \mu_y (\dot{y}_1 - \dot{y}_2)^2 + \mu_z (\dot{z}_1 - \dot{z}_2)^2\},$$

and in it replace the variations of the velocities by the variations of the corresponding coordinates, we shall have just obtained this virtual work. This function $\mathfrak{F}$ may now be expressed in terms of any generalized coordinates that may be most convenient to represent the configuration of the system for the purpose in hand, and the virtual work of the viscous forces for any virtual displacement specified by variations of these coordinates will still be derived by this rule. "But although in an important class of cases the effects of viscosity are represented by the function $\mathfrak{F}$, the question remains open whether such a method of representation is applicable in all cases. I think it probable that it is so; but it is evident that we cannot expect to prove any general property of viscous forces in the absence of a strict definition which will enable us to determine with certainty what forces are viscous and what are not."[2]

its limitations:

89. The general variational equation of motion of the viscous system will in fact be

introduced symbolically into the Least Action invariant procedure.

$$\int (\delta T - \delta W - \delta' \mathfrak{F})\, dt = 0,$$

[1] *Proc. Lond. Math. Soc.* 1873; *Theory of Sound*, Vol. I. 1877, § 81. (An analytical function of this kind occurs however incidentally in the *Mécanique Analytique*, Section VIII. § 2.)

[2] Lord Rayleigh, *Theory of Sound*, § 81. An extension of the range of the function is easy after the method of Lagrange, *loc. cit.* It is worthy of notice that we can also formulate a function of mutual dissipation between two interacting media.

wherein δ represents variation with respect to the coordinates and velocities of the system, while δ' represents variations with respect to the velocities only, the differentials of the velocities being in the result of the latter variation replaced by differentials of the corresponding coordinates[1].

90. The importance of this analysis in respect to problems in the theory of radiation is fundamental. If a radiation maintains its period of vibration unaltered in passing through a viscous medium, it follows necessarily that the viscous forces of the medium are of the type above specified. If the elastic forces were not linear functions of the displacements and the viscous forces linear functions of the velocities, the period of a vibration would be a function of its amplitude; and thus a strong beam of homogeneous light, after passing through a film of metal or other absorbing medium, would come out as a mixture of lights of different colours. So long as we leave on one side the phenomena of fluorescence, we can therefore assert that the laws of absorption must be such as are derivable from a single dissipation function, of the second degree in the velocities, which is appropriate to the medium.

Wide scope of the dissipation function in optics and electrics.

Recapitulation of the Vibrational Qualities of the Aether.

91. On the present extension of MacCullagh's scheme, the properties of the aether in a ponderable medium, as regards those averaged undulations which constitute radiation, are to be derived from the following functions; its kinetic energy

Averaged energy-densities in aether: kinetic and potential:

$$T = \tfrac{1}{2}\rho \int \left(\frac{d\xi^2}{dt^2} + \frac{d\eta^2}{dt^2} + \frac{d\zeta^2}{dt^2}\right) d\tau,$$

[and for rectangular symmetry] its potential energy

$$W = \tfrac{1}{2}\int (a^2f^2 + b^2g^2 + c^2h^2)\, d\tau, \text{ where } (f, g, h) = \operatorname{curl}(\xi, \eta, \zeta),$$

its dissipation function, representing decay of the regularity of the motion,

rate of dissipation.

$$\mathfrak{F} = \tfrac{1}{2}\int \left(a'^2\frac{df^2}{dt^2} + b'^2\frac{dg^2}{dt^2} + c'^2\frac{dh^2}{dt^2}\right) d\tau.$$

Power and facility of the variational method.

[1] It may be observed that the use of this variational equation would form the most elegant method of deriving the ordinary equations of motion of material dissipative systems in which the value of $\mathfrak{F}$ is known. For example the equations of motion of a viscous fluid in cylindrical, polar, or any other type of general coordinates, may be derived at once from the expressions for the fundamental functions in these coordinates, without the necessity of recourse to the complicated transformations sometimes employed. Cf. "Applications of Generalized Space-Coordinates to Potentials and Isotropic Elasticity," *Trans. Camb. Phil. Soc.* Vol. XIV. 1885: *supra*, p. 113.

We may add as subsidiary terms the magneto-optic energy

$$T' = \int\left(\alpha^2 \frac{d\xi}{d\theta}\frac{df}{dt} + \beta^2 \frac{d\eta}{d\theta}\frac{dg}{dt} + \gamma^2 \frac{d\zeta}{d\theta}\frac{dh}{dt}\right) d\tau,$$

Magneto-optic energy:

where
$$\frac{d}{d\theta} = \alpha_0 \frac{d}{dx} + \beta_0 \frac{d}{dy} + \gamma_0 \frac{d}{dz},$$

$(\alpha_0, \beta_0, \gamma_0)$ being the intensity of the imposed magnetic field; and the optical rotational energy

$$W' = \int(\alpha'^2 f\nabla^2\xi + \beta'^2 g\nabla^2\eta + \gamma'^2 h\nabla^2\zeta)\, d\tau.$$

structural chiral energy.

And there are also to be included the terms in W of higher orders [MacCullagh], that produce regular (*i.e.* sensibly non-selective) dispersion of various kinds, of which the chief is

$$W_1 = \int\Phi\,\{(f, g, h),\ \nabla^2(f, g, h)\}\, d\tau,$$

Conceivable static dispersional types of energy:

where the symbol Φ in the integral denotes a lineo-linear function.

Throughout these equations, the elastic properties of the aether retain their purely rotational character; its internal elastic energy, its dissipation, and its connections with other interlinked motions, depend on the rotation of its elements and not on their distortion or compression. A partial exception occurs in the magneto-optic terms, which represent interaction with a motion of partly irrotational character; and this exception is evidenced by the necessity which then arises of taking explicit account of incompressibility in order to avoid change from rotational to longitudinal undulation in a heterogeneous medium.

magneto-optic reflexion.

92. The question occurs, how far the form of these functions may be susceptible of alteration, so as thereby to amend those points in which the account given by the electric theory of light is [conceived to be] at variance with observation, for example, in the problem of metallic reflexion. The form of the function $\mathfrak{F}$ is derived from the phenomena of electrical dissipation when the currents are steady or changing with comparative slowness; as in other cognate cases, it may be subject to modification when the rate of alternation is extremely rapid. But as the elastic quality of the medium is assumed to be determined by the components of its rotation, and not at all by distortion or compression, it seems natural to infer that the viscous resistance to change of the strain is determined in terms of the same quantities and therefore by a quadratic function of $d/dt\,(f, g, h)$. This argument, if granted, will carry with it the assertion of Ohm's law of linear conduction in its general form, though probably with

Forms of the functions: analogy of rotational medium.

coefficients depending on the period, for disturbances of all periods however small.

In the expressions for $\mathfrak{F}$ and W, as given above, the principal axes of the aeolotropic conductivity are taken to coincide with the principal axes of the aeolotropic electric displacement, a simplification which need not generally exist.

The fact that the electric dissipation function does not involve the velocities of the material system shows that the forces derived from it are solely electromotive.

Supposed trouble about metallic optical reflexion.

93. It seems clear that viscous terms alone could not possibly in any actual medium be so potent as to reduce the real part of the complex index of refraction suitable to metallic media to be a negative quantity*. Such a state of matters arising from purely internal action involves instability; while on the contrary the general influence of viscosity is to improve rather than to diminish the dynamical stability of a system. This phenomenon, if indeed it is here properly described, must therefore be due to the support and control of some other vibrating system; an explanation which has been proposed is to adopt the views of Young and Sellmeier, and ascribe its origin to a near approach between the periods of hydrodynamical vibrations of the atoms in the molecule and the simultaneous rotational vibrations of the aether produced by the light-waves. A theory like this is, however, usually held as part of the larger view which represents ordinary refraction as the result of synchronism of periods and consequent absorption in the invisible part of the spectrum; while, in the above, the main part of the refraction is ascribed to defect of elasticity due to mobile atomic charges. It seems natural therefore to look for some other explanation of the discrepancies between theory and observation in ordinary metallic reflexion; and the idea suggests itself that if the opacity near the surface were so great as to cause sensible absorption in a very small fraction of a wave-length, the analytical formulae might be entirely altered.

Possible influence of transitional layer (not required).

Sir George Stokes[1] has, however, supported the view that besides the effects due to simple absorption, metals probably also show reflexion phenomena involving change of phase, such as were originally discovered by Airy for the diamond, and were afterwards found in other highly refractive substances. These effects, which were extended by Jamin to ordinary media, have been eliminated by Lord Rayleigh for the case of water by cleansing of the surface, by which means the

* There is no difficulty when the medium is compound; it is thus removed by the electron theory as in Part II, *infra*, § 11.

[1] In a note appended to a paper by Sir J. Conroy, "Some experiments on Metallic Reflexion," *Roy. Soc. Proc.* Feb. 1893.

sharpness of the optical transition would be improved. The phenomena for the case of diamond were long ago classed by Green[1] as a result of gradual transition; and this might be expected to be more marked between hard substances whose optical properties are very different. On this view we may not be driven to try the hypothesis of extreme absorption in the interfacial layer, which is unsatisfactory for the same reasons as apply to Kirchhoff's doctrine of extraneous forces; the quality above mentioned, for which Sir George Stokes proposes the name of the adamantine property, being sufficient.

Reflexion by Partially Opaque [Metallic] Media.

94. The ordinary formulae for reflexion at the surface of an absorbing medium may now be derived from the analytical functions which express the averaged dynamical constitution of the aether for the case of its vibrations in ponderable bodies. If the general argument is correct, it is to be expected that these formulae would be verified for reflexion at the surfaces of such media as are not too highly absorbent in comparison with the length of the wave. There are in fact two extreme cases; first the reflexion of electromagnetic waves of sensible length from metallic surfaces, where the reflexion is complete and there is no absorption at all; and second the reflexion of waves from perfectly transparent media, where the reflexion is incomplete because part of the energy goes on in the transmitted wave. The reflexion of light from metals may conceivably be more nearly akin to the first of these limiting cases than to the second; but for media more transparent than metals we should expect closer agreement with the ordinary theory, now to be developed.

Absorbing media with linear relations.

95. The general variational equation of the motion is

Their variational equation:

$$\int (\delta T - \delta W - \delta' \mathfrak{F})\, d\tau = 0,$$

leading (§ 91) to

$$\int dt \left[\rho \int \left\{ \frac{d\xi}{dt}\frac{d\delta\xi}{dt} + \frac{d\eta}{dt}\frac{d\delta\eta}{dt} + \frac{d\zeta}{dt}\frac{d\delta\zeta}{dt} \right\} d\tau \right.$$
$$- \int \left\{ a^2 f \left(\frac{d\delta\zeta}{dy} - \frac{d\delta\eta}{dz} \right) + b^2 g \left(\frac{d\delta\xi}{dz} - \frac{d\delta\zeta}{dx} \right) + c^2 h \left(\frac{d\delta\eta}{dx} - \frac{d\delta\xi}{dy} \right) \right\} d\tau$$
$$- \int \left\{ a'^2 \frac{df}{dt} \left(\frac{d\delta\zeta}{dy} - \frac{d\delta\eta}{dz} \right) + b'^2 \frac{dg}{dt} \left(\frac{d\delta\xi}{dz} - \frac{d\delta\zeta}{dx} \right) \right.$$
$$\left. \left. + c'^2 \frac{dh}{dt} \left(\frac{d\delta\eta}{dx} - \frac{d\delta\xi}{dy} \right) \right\} d\tau \right] = 0.$$

[1] G. Green, "Supplement to a Memoir on the Reflexion and Refraction of Light," *Trans. Camb. Phil. Soc.* May 1839.

On integrating by parts so as to eliminate the differential coefficients of the variation $\delta\,(\xi, \eta, \zeta)$, and neglecting the terms relating to the limits of the time, this gives the integral with respect to time of the expression

$$-\int\rho\left\{\frac{d^2\xi}{dt^2}\,\delta\xi+\frac{d^2\eta}{dt^2}\,\delta\eta+\frac{d^2\zeta}{dt^2}\,\delta\zeta\right\}d\tau$$

$$-\int\left\{\left(\frac{dc^2h}{dy}-\frac{db^2g}{dz}\right)\delta\xi+\left(\frac{da^2f}{dz}-\frac{dc^2h}{dx}\right)\delta\eta+\left(\frac{db^2g}{dx}-\frac{da^2f}{dy}\right)\delta\zeta\right\}d\tau$$

$$+\int\{(mc^2h-nb^2g)\,\delta\xi+(na^2f-lc^2h)\,\delta\eta+(lb^2g-ma^2f)\,\delta\zeta\}\,dS$$

$$-\int\left\{\frac{d}{dt}\left(\frac{dc'^2h}{dy}-\frac{db'^2g}{dz}\right)\delta\xi+\frac{d}{dt}\left(\frac{da'^2f}{dz}-\frac{dc'^2h}{dx}\right)\delta\eta\right.$$
$$\left.+\frac{d}{dt}\left(\frac{db'^2g}{dx}-\frac{da'^2f}{dy}\right)\delta\zeta\right\}d\tau$$

$$+\int\left\{\frac{d}{dt}\,(mc'^2h-nb'^2g)\,\delta\xi+\frac{d}{dt}\,(na'^2f-lc'^2h)\,\delta\eta\right.$$
$$\left.+\frac{d}{dt}\,(lb'^2g-ma'^2f)\,\delta\zeta\right\}dS.$$

leading to the equations of propagation, Hence the equations of propagation of vibrations are of the type

$$\rho\frac{d^2\xi}{dt^2}+\frac{dc^2\zeta}{dy}-\frac{db^2\eta}{dz}+\frac{d}{dt}\left(\frac{dc'^2\zeta}{dy}-\frac{db'^2\eta}{dz}\right)=0,$$

that is
$$\rho\frac{d^2\xi}{dt^2}+\frac{dc_1{}^2\zeta}{dy}-\frac{db_1{}^2\eta}{dz}=0,$$

where
$$(a_1{}^2, b_1{}^2, c_1{}^2)=\left(a^2+a'^2\frac{d}{dt},\ b^2+b'^2\frac{d}{dt},\ c^2+c'^2\frac{d}{dt}\right).$$

covering the simpler crystalline types. Thus on the assumption that the principal axes of the dissipation function are the same as those of the optical elasticity, the equations of propagation in absorptive crystalline media differ from those of transparent media only by the principal indices assuming complex values.

96. To determine how the absorption affects the interfacial conditions on which the solution of the problem of reflexion depends, let us transform the axes of coordinates so that the interface becomes the plane of yz, and $(l, m, n) = (1, 0, 0)$. The potential energy-function and the dissipation function will now be quadratic functions of the rotation and its velocity respectively, U and U' say, as in § 14; and we can now incidentally extend our view to the case in which these functions have not the same principal axes. The variational equation

of motion is represented by the vanishing of the time-integral of the expression

General crystalline medium.

$$-\int\rho\left\{\frac{d^2\xi}{dt^2}\delta\xi+\frac{d^2\eta}{dt^2}\delta\eta+\frac{d^2\zeta}{dt^2}\delta\zeta\right\}d\tau$$

$$-\int\left\{\left(\frac{d}{dy}\frac{dU}{dh}-\frac{d}{dz}\frac{dU}{dg}\right)\delta\xi+\left(\frac{d}{dz}\frac{dU}{df}-\frac{d}{dx}\frac{dU}{dh}\right)\delta\eta+\left(\frac{d}{dx}\frac{dU}{dg}-\frac{d}{dy}\frac{dU}{df}\right)\delta\zeta\right\}d\tau$$

$$+\int\left\{\left(m\frac{dU}{dh}-n\frac{dU}{dg}\right)\delta\xi+\left(n\frac{dU}{df}-l\frac{dU}{dh}\right)\delta\eta+\left(l\frac{dU}{dg}-m\frac{dU}{df}\right)\delta\zeta\right\}dS$$

$$-\int\left\{\frac{d}{dt}\left(\frac{d}{dy}\frac{dU'}{dh}-\frac{d}{dz}\frac{dU'}{dg}\right)\delta\xi+\frac{d}{dt}\left(\frac{d}{dz}\frac{dU'}{df}-\frac{d}{dx}\frac{dU'}{dh}\right)\delta\eta+\frac{d}{dt}\left(\frac{d}{dx}\frac{dU'}{dg}-\frac{d}{dy}\frac{dU'}{df}\right)\delta\zeta\right\}d\tau$$

$$+\int\left\{\frac{d}{dt}\left(m\frac{dU'}{dh}-n\frac{dU'}{dg}\right)\delta\xi+\frac{d}{dt}\left(n\frac{dU'}{df}-l\frac{dU'}{dh}\right)\delta\eta+\frac{d}{dt}\left(l\frac{dU'}{dg}-m\frac{dU'}{df}\right)\delta\zeta\right\}dS.$$

The equations of propagation are therefore of type

$$\rho\frac{d^2\xi}{dt^2}+\frac{d}{dy}\frac{dU_1}{dh}-\frac{d}{dz}\frac{dU_1}{dg}=0$$

where
$$U_1=U+\frac{d}{dt}U'.$$

The boundary condition demands in general the continuity of the expression

The inter-facial relations.

$$\int\left\{\left(m\frac{dU_1}{dh}-n\frac{dU_1}{dg}\right)\delta\xi+\left(n\frac{dU_1}{df}-l\frac{dU_1}{dh}\right)\delta\eta+\left(l\frac{dU_1}{dg}-m\frac{dU_1}{df}\right)\delta\zeta\right\}dS$$

in crossing the interface; for the special case of $(l, m, n) = (1, 0, 0)$, this involves continuity in η, ζ, dU_1/dg and dU_1/dh.

Widest formal scheme.

Thus, under the most general circumstances, the inclusion of opacity is made analytically by changing the potential energy-function from U to U_1, where U_1 is still a quadratic function, but with complex coefficients. If U and U' have their principal axes in the same directions, a change of the principal indices of refraction of the medium from real to complex values suffices to deduce the circumstances both of propagation and of reflexion of light in partially opaque substances from the ones that obtain for perfectly transparent media. In all cases, however, the function U_1 has three principal axes of its own, whose position depends on the period of the light.

Dynamical Equations of the Primordial Medium:
[*in further reference to the rotational fluid aether model*].

97. The medium by means of which we have been attempting to coordinate inanimate phenomena is of uniform density, if there be excepted the small volumes occupied by possibly vacuous cores of the vortex atoms. Its motion is partly hydrodynamical and irrotational, and is partly of rotational elastic quality. Its equations of motion are, for the averaged displacements which represent the general circumstances of crystalline quality,

Equations of a rotationally elastic fluid medium.

$$\rho \frac{D^2\xi}{dt^2} + \frac{dc^2h}{dy} - \frac{db^2g}{dz} + \frac{dp}{dx} = 0,$$

$$\rho \frac{D^2\eta}{dt^2} + \frac{da^2f}{dz} - \frac{dc^2h}{dx} + \frac{dp}{dy} = 0,$$

$$\rho \frac{D^2\zeta}{dt^2} + \frac{db^2g}{dx} - \frac{da^2f}{dy} + \frac{dp}{dz} = 0,$$

where (ξ, η, ζ) is the linear displacement, (f, g, h) is its vorticity or curl, and p is a hydrostatic pressure in the medium, the symbol D^2/dt^2 denoting the acceleration of a moving particle as contrasted with the rate of change of velocity at a fixed point.

98. These equations represent the general circumstances of the propagation of radiation through the medium; and in them the velocity of translation of the medium due to vortices in it has been averaged. But if we desire to investigate in detail the motion and vibrations of a single vortex ring or a vortex system in a rotationally elastic fluid medium, it is of course not legitimate to average the motion of translation near the ring. The determination of the circumstances of the influence of a moving medium on the radiation also requires a closer approximation. Considering therefore the free aether, which is devoid of crystalline quality, and substituting

Convection separated from vibration.

$$\frac{d}{dt}(\xi, \eta, \zeta) = (u + u_1,\ v + v_1,\ w + w_1),$$

so as to divide the velocity into two parts, one of which represents the translation of the medium and the other its vibration, we have

$$\frac{D}{dt} = \frac{d}{dt} + (u + u_1)\frac{d}{dx} + (v + v_1)\frac{d}{dy} + (w + w_1)\frac{d}{dz},$$

so that $$\frac{D}{dt}(u + u_1) = \frac{\delta u}{dt} + \frac{\delta u_1}{dt} + u_1\frac{du}{dx} + v_1\frac{du}{dy} + w_1\frac{du}{dz}$$

very approximately, where

$$\frac{\delta}{dt} \text{ represents } \frac{d}{dt} + u\frac{d}{dx} + v\frac{d}{dy} + w\frac{d}{dz}.$$

Hence separating the hydrodynamical part in the form

$$\rho \frac{\delta}{dt}(u, v, w) = -\left(\frac{d}{dx}, \frac{d}{dy}, \frac{d}{dz}\right) p_0,$$

which represents irrotational motion except in the vortices, there remain vibrational equations of the type

$$\rho\left(\frac{\delta u_1}{dt} + u_1\frac{du}{dx} + v_1\frac{du}{dy} + w_1\frac{du}{dz}\right) + a^2\left(\frac{dh_1}{dy} - \frac{dg_1}{dz}\right) + \frac{dp_1}{dx} = 0.$$

In a region in which the velocity of translation (u, v, w) is uniform, the radiation is thus simply carried on by the motion of the medium.

99. The vibrational motion which is propagated from an atom is interlinked with the motion of translation of the medium, only through the hydrostatic pressures which must be made continuous across an interface; the form of the free surface has in fact to be determined so as to adjust these pressures at each instant. To fix our ideas, let us consider for a moment the problem of the vibrations of a single ring with vacuous core, moving by itself through the medium, in the direction of its axis, with a given atomic electric charge on it. To obtain a solution we assume that the radius vector of the cross-section of the core varies with the time according to the harmonic function suitable to its types of simple vibration; and we determine the irrotational motion in the medium that is produced by this motion of the surface of the core, and calculate the pressure p_0 at the free surface. Next we determine the vibrational rotation (f, g, h) that is conditioned by the same vibratory movement of the surface of the core, while it is independent of the inertia of the hydrodynamical motion in the medium; this has also to satisfy the condition that the tangential components of the rotation are null all over the surface, so that there may be no electromotive tangential traction on it. In order to satisfy all these surface conditions it will usually be necessary to introduce an electromotive pressure p_1 into the equations of vibration, although this was not required in the problem of reflexion at a fixed interface; in other words the pressure in that problem was quite unaffected and therefore left out of account. The magnitude of this pressure is then to be calculated from the solution; and the condition that it is equal and opposite at the free surface, to the pressure p_0 of hydrodynamical origin, gives an equation for the period of the vibrations of the type assumed. If, on the other hand, the core is taken to consist of spinning fluid devoid of rotational elasticity, instead of vacuum, the conditions at its surface will be modified.

An electromotive pressure arises.

Free periods of radiation.

100. If the form of the ring is such that the period of its hydrodynamic vibration is large compared with that of the corresponding electric vibration, an approximate solution is much easier; it is now only necessary to suppose that on each successive configuration of the core there is a distribution of static electricity in equilibrium, and to allow for the effect of this distribution on the total pressure which must vanish at a free surface.

Sustained radiation from a charged vortex atom:

In this case the electric vibrations will continue for a comparatively long time, until all the energy of the disturbance in the molecule is radiated away, but they will be of very small intensity. The vibrations of an electric charge over a conducting atom which is not a vortex ring are practically dead-beat, and could not give rise to continued radiation of definite periods: but the case is different here, and the vibrations will go on until the energy of the disturbance of the steady motion of the vortex-ring atom has all been changed into electrical waves.

estimates of periods and intensities.

Now the periods of the principal hydrodynamical vibrations of a single ring may be regarded as the times that would be required for disturbances of the different permanent types to move round its core with velocities of the same order of magnitude as the actual velocity of translation of the ring through the medium; while the periods of the electrical vibrations are the times that would be required for electric disturbances to move round the core with velocities of the same order as the velocity of radiation. The first of these periods is for an isolated ring very much the greater, so much so that electric vibrations could hardly be excited at all by vibrations of the atom comparatively so slow. But in the case of a molecule there would also be much smaller hydrodynamical periods, due to the interaction between neighbouring parts of the paired rings, which may be expected to maintain electrical vibrations in the manner above described; and in the case of an isolated ring the periods which involve crimping of the cross-section may produce a similar effect, though they cannot involve a sensible amount of energy.

Types.

When the core is of the same density as the surrounding fluid, and there is no slip at its surface, the hydrodynamical pressure across the interface will be continuous in the steady motion of the ring; therefore the above electric pressure must be uniform all over the interface; that is, the electric force must be constant over it, as well as the electric potential. These conditions determine the form of the interface in the steady motion; and the rotational motion of the core is then determined, through its stream function, so as to have given total amount and to be continuous with the circulatory irrotational motion just outside it.

On Gravitation and Mass.

101. The hypothesis of finite, though very small, compressibility of the aether has occasionally been kept in view in the foregoing analysis, in the hope that it may lead to results having some affinity to gravitation. There does not appear, however, to be any correspondence of this kind. A tentative theory has already been proposed and examined by W. M. Hicks, which makes gravitation a secondary effect of those vibrations of vortices in an incompressible fluid which consist in pulsations of volume of their vacuous cores. But the periods of such vibrations are not very different from the periods of their other types; and the theory cannot be said to be successful, the objections to it being in fact fully stated by its author[1].

Pulsation theory of gravitation tried:

Let us now consider the effect of a compressional term in the potential energy of the medium, of the form

$$\tfrac{1}{2}A\int\left(\frac{d\xi}{dx}+\frac{d\eta}{dy}+\frac{d\zeta}{dz}\right)^2 d\tau, \text{ say } \tfrac{1}{2}A\int \varpi^2 d\tau,$$

compressile theory tried.

where $-\varpi$ is the compression in the medium. The variation of this term will be

$$A\int \varpi\,(l\delta\xi+m\delta\eta+n\delta\zeta)\,dS - A\int\left(\frac{d\varpi}{dx}\,\delta\xi+\frac{d\varpi}{dy}\,\delta\eta+\frac{d\varpi}{dz}\,\delta\zeta\right)d\tau.$$

Thus there will be added to the right-hand side of the equations of vibration new terms, giving in all

$$\rho\frac{d^2\xi}{dt^2}+\frac{dc^2h}{dy}-\frac{db^2g}{dz}-A\frac{d\varpi}{dx}=0,$$

$$\rho\frac{d^2\eta}{dt^2}+\frac{da^2f}{dz}-\frac{dc^2h}{dx}-A\frac{d\varpi}{dy}=0,$$

$$\rho\frac{d^2\zeta}{dt^2}+\frac{db^2g}{dx}-\frac{da^2f}{dy}-A\frac{d\varpi}{dz}=0.$$

It follows that ϖ satisfies the equation

$$\rho\frac{d^2\varpi}{dt^2}=A\nabla^2\varpi;$$

so that the compressional wave is propagated independently of the rotational one, of which the circumstances are given by equations of the type

Waves of compression would be independent of radiation:

$$\rho\frac{d^2f}{dt^2}=-\frac{d}{dx}\left(\frac{da^2f}{dx}+\frac{db^2g}{dy}+\frac{dc^2h}{dz}\right)+\nabla^2a^2f.$$

[1] W. M. Hicks, *Proc. Camb. Phil. Soc.* 1879; *Roy. Soc. Proc.* 1883; also *Phil. Trans.* 1883, p. 162.

In the discussion of the reflexion of light it has been shown that the same absolute separation of compression and rotation is manifested in the passage across an interface into a new medium; so that however heterogeneous the medium be rendered by the presence of vortex atoms, these two types of disturbance are still quite independent of each other.

they may be instantaneous, establishing a static pressure.

The alteration in the electrostatic equations which would be produced by this compressional quality has already been given; if the value of the modulus A is extremely great, this alteration will be quite unnoticeable. In that case, waves of compression will be propagated with extremely great velocity, so that as regards compression the medium will assume almost instantly an equilibrium condition, for which therefore $\nabla^2\varpi = 0$.

It follows that the value of the integral $\int d\varpi/dn \,.\, dS$ is the same for all boundaries which contain inside them the same atoms. If we want to make this integral constant throughout time, we may imagine that the medium was originally in equilibrium without compression, and was then strained by altering the volume of each electrically charged atom by a definite amount. The state of strain thus represented *in the aether* has a pressure at each point equal to A multiplied into the gravitation potential of a mass equal to this constant, supposed placed at the atom. Its energy is however

Gravitation potential would then be pressure:

$$\tfrac{1}{2}A\int\varpi^2 d\tau, \text{ instead of } -\tfrac{1}{2}A\int\left(\frac{d\varpi^2}{dx^2}+\frac{d\varpi^2}{dy^2}+\frac{d\varpi^2}{dz^2}\right)d\tau,$$

but the theory fails.

which it ought to be[1] if it were gravitational energy; so that there is no means of explaining gravitation here.

Unequal positive and negative electrons tried.

102. If we could imagine for a moment that the electric charges of the two ions in a molecule do not exactly compensate each other, but that there is a slight excess always of the same sign, we should have a *repulsive* force of gravitational type, transmitted by a stress in a rotational aether. A term of this form in the energy, if it were kinetic instead of potential, would account for gravity. The question thus suggested is, whether the kinetic energy of the primordial medium has been sufficiently expressed, in view of the inherent rotational quality in its elements. It was proved by Laplace that the velocity of gravitation must be enormously great compared with

[1] Cf. Maxwell, "A dynamical theory of the Electromagnetic Field," § 82, *Phil. Trans.* 1864. [Maxwell makes a general objection on the ground that strain energy must be positive: if, however, the energy were kinetic, it would enter into $L\ (= T - W)$ with changed sign; and in the general Action scheme there is nothing to restrict the sign of A.]

that of light*; so that the gravitational energy, whatever its origin, must preserve a purely statical aspect with respect to all the other phenomena that have been here under discussion.

The objection has been raised, by Clerk Maxwell and others, to the vortex-atom theory of matter, that it can give no account of mass for the case of sensible bodies. But it may be urged that mass is a dynamical conception, which in complicated cases it would be hard to define exactly or give an account of. The clearest view of dynamics would appear to be the one maintained by various writers, notably by L. N. M. Carnot and by Kirchhoff, that the function of that science is to correlate, or give a general formula for, the sequence of physical phenomena. The ultimate formula which is, it is hoped, to embrace the physical universe is the law of Least Action; and the ultimate definition of mass is to make it a coefficient in the kinetic part of the energy-function of the matter in that formula. As the theories here discussed are referred to the single basis of this law of Least Action, the objection that they do not take account of mass can hardly be prohibitive; though they may not be able to explain how the idea of mass is originated by aggregation of terms in that equation.

A definite mass not a property of vortices (but is of electrons).

Ultimate mass.

103. It is conceivable that the rotational elasticity of the fundamental medium is really due to a rotatory motional distribution in it, which resists disturbance from its steady equilibrium state with excessively great effective elasticity, while the tractions necessary to equilibrate a free boundary are non-existent. Such a hypothesis looks like explaining one aether by means of a new one, but it is perhaps not really more complicated than the facts; on our present principle of interpretation, the change of gravitation in the field due to a disturbance at any point must have been propagated somehow, while in the machinery that transmits electric and luminiferous disturbances no elasticity has yet been recognized anywhere near intense enough to take part in such a propagation.

Vortex sponge rotational aether.

Is change of gravitational field propagated?

We may not surmount the difficulty by the assumption that, in addition to the finite resistance to rotation which is the cause of the propagation of the radiation, the medium also possesses an enormously greater static resistance to rotations of some more fine-grained structure, and that the surface integral of the rotation over any surface enclosing a vortex atom is a positive constant, of course definite and unchangeable in value for each atom; for this would

* On the supposition, however, that the rays of the solar gravitation when they arrive at the planet are taken to affect it as subject to an uncompensated Bradley aberration.

Being attractive a kinetic origin is suggested.

lead to gravitational repulsion instead of attraction. The term must be in the kinetic energy, not in the potential energy of the medium.

104. In a representation of a magnetic or other medium[1], imagined to be composed of gyrostatic elements spinning indifferently in all directions, and linked into a system by an arrangement like idle-wheels between them, in fact by an ideal system of universal ball-bearings, the kinetic energy function would have a rotatory part

Maxwell's aether model with displacement as motion of ball-bearings (improved by FitzGerald).

$$T = \tfrac{1}{2}C \int \left(\frac{df^2}{dt^2} + \frac{dg^2}{dt^2} + \frac{dh^2}{dt^2}\right) d\tau,$$

where (f, g, h) is the absolute rotation of an element, which is supposed from the connecting mechanism to be a continuous function of position in the system.

Rotational kinetic energy:

We would have therefore

$$\delta T = C \int \left\{\frac{df}{dt}\frac{d}{dt}\left(\frac{d\delta\zeta}{dy} - \frac{d\delta\eta}{dz}\right) + \frac{dg}{dt}\frac{d}{dt}\left(\frac{d\delta\xi}{dz} - \frac{d\delta\zeta}{dx}\right) + \frac{dh}{dt}\frac{d}{dt}\left(\frac{d\delta\eta}{dx} - \frac{d\delta\xi}{dy}\right)\right\} d\tau$$

$$= \int \{\ldots\} dS - C \int \left\{\frac{d^2}{dt^2}\left(\frac{dh}{dy} - \frac{dg}{dz}\right)\delta\xi + \frac{d^2}{dt^2}\left(\frac{df}{dz} - \frac{dh}{dx}\right)\delta\eta + \frac{d^2}{dt^2}\left(\frac{dg}{dx} - \frac{df}{dy}\right)\delta\zeta\right\} d\tau.$$

its local reaction,

Thus the kinetic forcive which is the equivalent of the actual applied forcive in the medium per unit volume, arising from its potential energy and such extraneous forces as act on it, is

$$C\frac{d^2}{dt^2}\,\mathrm{curl}\,(f, g, h), \quad \text{or} \quad -C\frac{d^2}{dt^2}\nabla^2(\xi, \eta, \zeta).$$

confined to the spinning parts.

If we suppose the displacement (ξ, η, ζ) to be originally derived from a potential function ϕ, this kinetic forcive exists only where there is some portion of the ideal mass-system of which ϕ is the potential; the spin in the medium thus produces no forcive anywhere except in the spinning parts.

We may imagine this medium to be a hydrodynamical one such as could sustain vortex motion; then this kinetic forcive is confined to the vortices. Throughout a small volume containing a vortex, the aggregate of this forcive is

$$-C\int \frac{d^2}{dt^2}\nabla^2(\xi, \eta, \zeta)\, d\tau;$$

of which the part outside the core of the vortex is

$$-C\int \frac{d^2}{dt^2}\left(\frac{d}{dx}, \frac{d}{dy}, \frac{d}{dz}\right)\nabla^2\phi\, d\tau,$$

and is therefore null, so that this quantity $\int\nabla^2(\xi, \eta, \zeta)\, d\tau$ may be

[1] Cf. Maxwell's "Hypothesis of Molecular Vortices," *Treatise*, §§ 822–827.

taken as an intrinsic constant for any particular isolated vortex throughout all time. Again, its value is the same for the regions bounded by all surfaces which include the same vortices; thus there is a kinetic reaction proportional to the second differential coefficient with respect to time of the amount of this particular constant thing that is carried by the vortices contained in the element of volume. If we attach in thought this forcive to a moving element of volume containing the vortices, instead of to the fixed element of volume, it will vary jointly as the amount of this thing that belongs to the vortex group, and the acceleration of the element of volume in space; and its aggregate amount will not be affected by interaction between the vortices of the group. This appears to introduce the dynamical notion of mass and acceleration of matter; and this illustration has been furnished by a function representing energy of spin in the medium, which exists only where that spin is going on, *i.e.* in the vortices. The remaining part of the kinetic energy of the medium, which is the whole of the kinetic energy of that part of the medium not occupied by vortices, is translational as above and equal to

Mass as conceivably resulting from spin-energy in the vortices.

$$\tfrac{1}{2}\rho\int\left(\frac{d\xi^2}{dt^2}+\frac{d\eta^2}{dt^2}+\frac{d\zeta^2}{dt^2}\right)d\tau.$$

105. To make a working scheme we must suppose a layer of the medium, possessing actual spin, to cover the surface of each coreless vortex atom; we might imagine a rotationless internal core which allowed no slipping at the surface, and this spin would be like that of a layer of idle-wheels which maintained continuity between this core and the irrotational circulatory motion of the fluid outside. A gyrostatic term in the kinetic energy thus appears to introduce and be represented by the kinetic idea of mass of the matter; it enters as an aeolotropic coefficient of inertia for each vortex, but when averaged over an isotropic aggregate of vortices, it leads to a scalar coefficient for a finite element of volume.

Not isotropic.

If the core of the vortex atom is not vacuous but consists as in ordinary vortices of spinning fluid, here devoid of rotational elasticity, the rotational kinetic energy of the vortex as distinguished from translational energy will be a possible source of the phenomena of mass; but to possess such energy the medium must have some ultimate structure, for in an infinitely small homogeneous element of volume the ratio of the rotational to translational part of the kinetic energy would be infinitely small. Such a structure, confined to the cores of the vortices, need not be in contradiction with Maxwell's principle that the constitution of a perfect fluid cannot be molecular.

Structure inside vortex cores.

(*Added June* 14, 1894.) *On Natural Magnets.*

106. Lord Kelvin[1] has pointed out that the forcive between a pair of rigid cores in a fluid, with circulatory irrotational motion through their apertures, is equal but opposite to the forcive between the corresponding steady electric currents as expressed by the electrodynamic formulae. The reason of this difference lies in the circumstance that the connections and continuity of the fluid system prevent the circulation round any core from varying, so long as that core is unbroken; while the constraints must be less complete in the electrodynamic problem, because the currents change their values by induction. These constant circulations are of the nature of the constant momenta belonging to cyclic motions of dynamical systems; and it is known that when such constant momenta are introduced into the expression for the energy in place of the corresponding velocities, the type of the general dynamical equation is thereby altered[2]. The modification which the equation of Least Action must undergo under these circumstances has been investigated on a previous occasion[3]. In the case of fluid circulation, when the cores are so thin as to interpose no sensible obstacle to the flow, the sign of that part of the kinetic energy which involves the cyclic constant of the motion has merely to be changed; in other words, this energy is for the purpose of the modified dynamical equations to be treated as potential instead of kinetic. In all cases in which coordinates of a dynamical system can be ignored by elimination in this manner the energy-function consists of two parts, one a quadratic function of the velocities of the bodies, the other a quadratic function of the constant momenta: in the case just mentioned the former part is negligible, so that the part whose sign is to be changed is practically the total energy.

Electric currents cannot be continuous,

as are vortex rings.

The Modified Action function of vortices:

for solid rings it involves the momenta as constants in a force-function, of the wrong sign for an analogy with currents.

The validity of the application of the Lagrangian equations in the unmodified form to electric currents, as in the discussion in this paper, thus requires that there is no intrinsic cyclosis in the motions which exist in the electrodynamic field. The conductors must therefore all form practically incomplete circuits, in which the flow may be maintained and altered by means of what are effectively breaches in

broken circuits:

[1] Lord Kelvin (Sir W. Thomson), "Hydrokinetic Analogy," *Proc. Roy. Soc. Edin.* 1870; *Papers on Electrostatics and Magnetism*, p. 572. Also Kirchhoff, *Crelle*, 1869.

[2] Routh, *Stability of Motion*, 1877, Chap. IV, §§ 20 *seq.*; Thomson and Tait, *Natural Philosophy*, 2nd ed., 1879, §§ 319, 320; von Helmholtz, "On Polycyclic Systems," *Crelle*, 1884–1887.

[3] "Least Action," *Proc. Lond. Math. Soc.* Vol. XV. March 1884. On p. 182 the electrodynamic energy is quoted with the wrong sign [corrected *supra*].

the continuity of the medium; and as a further consequence, arising from such breaches of continuity, the mechanical forcives between the conductors will not now be wholly due to ordinary fluid pressure.

In an ordinary electric circuit, the circulation of the medium is thus maintained around the conducting part of the circuit by electric convection or displacement across the open or electrolytic part, by means of a process in which the rotational elasticity of the medium is operative. We may imagine this electric convection to be performed mechanically, and to be the source of the energy of the current: the force-component corresponding to the dynamical velocity which represents the current will then be the electric force which does work in the convection of charged ions. If this convection ceased, the circulatory motion which constitutes the magnetic field of the current (*i.e.* its momentum) would be stopped by the elasticity of the medium; and by altering the velocity of this convection, we have the means of adding to or subtracting from the circulatory motion, the change of kinetic energy so produced being derived from the electric force which resists convective displacement. This mode of mechanical representation suffices to include all the phenomena of ordinary electric currents. On the other hand, in a molecular circuit there is no electric convection, but only a permanent fluid circulation through it, such as would be self-subsisting, by aid of fluid pressure only, when the core is fixed, and could not in any case be permanently altered, on account of the rotational elasticity.

mechanical work on the ions transformed into energy of cyclic magnetic field.

An Amperean continuous current contrasted with orbital electrons.

In the establishment of an ordinary current in an open circuit, the rotational elasticity of the medium acts very nearly as a constraint, on account of the great velocity of electric propagation; and there is therefore at each instant only an insignificant amount of energy involved in it. But notwithstanding, if there are other open conducting circuits in the neighbourhood the action of this elasticity in establishing the current will be partly directed by them and relieved by circulation round them. The final result for maintained currents is however irrotational motion through the circuits; the kinetic energy is sufficiently represented, for slow changes, by the ordinary electrodynamic formula for linear currents; and it is directly amenable to the Lagrangian analysis. If the currents move in each others' fields, with external agencies to prevent their strengths from altering, these agencies must supply twice as much energy as is changed into mechanical work in the movement, in accordance with a theorem of Lord Kelvin's.

Work done against the magnetic forces increases intrinsic energy of field (except for dynamo).

Conversely, assuming that the electromagnetic energy is kinetic, it would seem that we are required by Lenz's law to take the currents in ordinary electric circuits to be of the nature of velocities, in the dynamical theory; though in the essentially different configuration of an Amperean magnetic molecule, the circulation which corresponds most closely to the current is more allied to a generalized momentum.

The energies of magnetic vortex atoms would have to be introduced with changed sign into the modified equation of Least Action, and this will involve the presence in the modified function of terms containing the electric generalized velocities in the first degree. Unless the cross-sections of the rings are very small compared with their diameters, there will also occur terms involving products of the strengths of the vortices and the velocities of the movements of the rings. For two stationary thin rigid cores of very narrow section, the mutual forcive due to fluid pressure will thus be equal but opposite to the forcive between the corresponding electric currents; the general features of this result are in fact easily verified by consideration of the distribution of velocity, and therefore of pressure, in the steady fluid motion of the medium.

Magnets cannot be continuous vortexes.

107. The serious difficulty presents itself that the mutual attractions of natural magnets are actually in the same direction as those of the equivalent electric currents, and not, as would appear from this theorem, in the opposite direction. In the first place, however, the theorem is proved only for rigid cores, held in the circulating fluid medium, and the forcive in question is simply the resultant of fluid pressures over the surfaces of the cores. In the case of vortex atoms with vacuous cores, such a pressure would not exist at all. And when we consider individual molecules, the question is also mixed up with the unsolved problem of the nature of the inertia of a vortex molecule.

Can they be analogous to open tubes with the medium circulating through them (Kelvin)?

It may be of use to examine separately the distribution of kinetic energy which the presence of two vortex aggregates implies in the medium surrounding them and between them, as distinguished from the kinetic energy inside them which is in direct relation with intermolecular forces. Let us take Lord Kelvin's illustration, a set of open rigid tubes in a frictionless fluid, through each of which there is circulatory motion. "When any change is allowed in the relative positions of two tubes by which work is done, a *diminution* of kinetic energy of the fluid is produced within the tubes, and at the same time an *augmentation* of its kinetic energy in the external space. The former is equal to double the work done; the latter is equal to the work done; and so the loss of kinetic energy from the whole liquid is equal to the work done[1]." The distribution of energy in the medium, out-

[1] Lord Kelvin, *Electrostatics and Magnetism*, 1872, § 737.

side two vortex aggregates, thus varies in the same way and with the same sign as the energy of the field of the corresponding magnets, as of course it ought to do. And the question is suggested, are we allowed to turn the difficulty as to the nature of the inertia of the vortex atoms by considering the magnetic forcive between two permanent aggregates as derived from the transformation of the kinetic energy in the medium between them?

The motion of the medium between them may be set up by the proper impulsive pressures over the surfaces of the aggregates, just as the magnetic field is determined by the distribution of magnetic intensity over the outer boundary of the magnets. And the principle of energy by itself shows that if we bound the two aggregates by moving surfaces which always pass through the same particles of the medium, the increment of the kinetic energy outside is equal to the work done *in the actual motion* by the pressures transmitted across the surfaces of the two aggregates; though we are unable to extend this result to arbitrary virtual displacements of the surfaces. Nor is the method of § 58 now applicable to complete the proof, because it is impossible to have an equipotential surface surrounding a magnetic system.

Vortex analogy fails for a magnet.

108. In all theories which ascribe the induction of electric currents to elastic action across the intervening medium, a discrepancy arises when the induction is produced by movement through a steady magnetic field: for in such cases there is no apparent play of electric force across the field. This difficulty may perhaps disappear, on the present view, when we regard such a field, not as an absolutely steady motion like fluid circulation round fixed cores, but as the statistically steady residue of elementary elastic disturbances sent out through the medium by the molecular discharges which maintain the inducting currents, or by changes of orientation and other disturbances of the molecules of the permanent magnets, such as are involved in any kinetic theory of matter. These elastic disturbances do not spread out indefinitely as waves, but come to an end when the medium has attained a new steady state which they have been instrumental in forming. The progress and decay of each small disturbance generates a current on the secondary system, whose integrated amount would be null if that system were at rest: but in the actual circumstances of movement during the progress of the induction there will be a residual value. The aggregate of such differences between elementary direct and reverse induced currents would constitute the observed total current. Thus as regards induction, change of the magnetic field of a permanent magnet would act in the same way as that of an

Electric induction in uniform magnetic field: apparent paradox.

ordinary current, notwithstanding that if each molecule of the magnet were held fixed there might (§ 106) be no induction.

On these grounds, the field of a permanent magnet would be regarded, not as a steady circulation of the aether, absolutely devoid of elastic reaction, but as the statistically steady resultant of the changing fields of the incessantly moving molecules which make up the magnet. The steady field of motion associated with a fixed magnetic molecule would be maintained by fluid pressure alone: but when the molecule is rotated, some agency is required to prevent slip during the establishment of the new steady motion; and in this way the elasticity may come into play. In ordinary hydrodynamics, the process of the establishment of a fluid motion is kept out of sight: it is simply assumed that the motion can be set up without slip, and that it is set up practically instantaneously throughout the field. In the present problem, on the other hand, something formally equivalent to slip does occur across the dielectric gaps in each electric circuit; and this circumstance modifies the process of establishment of the motion.

Origins.

This explanation, if valid, would carry with it, by virtue of the principle of energy, the observed law of attraction of a permanent magnet on an ordinary electric current; and also, provided we could assume the law of action and reaction to be applicable, that of a magnetic field on the aggregate constituting a permanent magnet. And as in the case of currents maintained steady, when two permanent magnets move each other the energy in the medium surrounding them is increased by the mechanical work done, but the energy in their interiors is diminished by twice that amount.

Permanent magnets anomalous,

Whatever be the value of these remarks, it would seem that the difficulty with respect to permanent magnets can hardly be insuperable, as it must attach in some form to any theory which makes magnetic energy kinetic. For, on that hypothesis, this energy must be wholly cyclic when there are only permanent magnets on the field; and its sign would therefore have to be changed, just as above, in forming dynamical equations which take separate account of each magnetic molecule. If, on the other hand, the statistical view above adopted is allowed, the complication introduced by intermolecular actions will be avoided, and only the averaged action between the two systems will remain.

if purely cyclic.

On the Electrodynamic Equations.

109. The kinetic energy of the electric medium is

$$T = \tfrac{1}{2} \int \left(\frac{d\xi^2}{dt^2} + \frac{d\eta^2}{dt^2} + \frac{d\zeta^2}{dt^2} \right) d\tau.$$

Let us transform this expression to new variables (f, g, h) which represent the components of the absolute rotation at each point; and let us suppose that there is nowhere any discontinuity or defect of circuital character in these quantities. We must therefore assign to them very large but not infinite values in an indefinitely thin superficial layer of the conductors, which shall be continuous with their actual values outside and their null value inside that surface[1]. The object of doing this is to abolish all surface-integral terms which would otherwise enter, on integration by parts, at each interface of discontinuity; the surface-integral terms that belong to the infinitely distant boundary need not concern us, except in cases where radiation plays a sensible part.

We may show as in § 52 that under these circumstances

Magnetic energy expressed in electric variables.

$$T = \frac{1}{8\pi}\int\left(\frac{df}{dt}F + \frac{dg}{dt}G + \frac{dh}{dt}H\right)d\tau,$$

where

$$(F, G, H) = \int\frac{d\tau'}{r'}\frac{d}{dt}(f', g', h'),$$

r' being the distance of the element $d\tau'$ from the element $d\tau$.

It is of necessity postulated throughout that (f, g, h) is circuital, for it is the curl of (ξ, η, ζ); that is, the proper current sheet must always be taken to exist at the surface of the conductor in order to complete the electric displacement in the medium. It follows as in § 57, but only under this proviso, that the magnetic force is the curl of Maxwell's vector potential (F, G, H) of the current system.

The transformation of the kinetic energy T to the directly elastic coordinates (f, g, h) is thus established; and the dynamical equation of the medium is

$$\delta\int(T - W)\,dt = 0,$$

Variation of Action subject to limitation on new electric variables:

in which the time is to remain unvaried. In order, however, to obtain equations wide enough to allow of the restriction of (f, g, h) to circuital character, which is now no longer explicitly involved, we must incorporate this restriction in the variational equation after the manner of Euler and Lagrange, and so make

$$4\pi\delta\int dt\,(T - W) + \delta\int dt\int d\tau\psi\left(\frac{df}{dx} + \frac{dg}{dy} + \frac{dh}{dz}\right) = 0,$$

[1] The procedure of this section leaves out dissipation, and so confines the currents to the surfaces of the conductors. [A complete Action theory, including the electrons interacting with the aether, is developed in *Aether and Matter* (1900), Chap. VI.]

and restrict the function of position ψ subsequently so as to satisfy the circuital relation. Thus

$$\delta \int dt \int \left\{ \tfrac{1}{2} \left(F \frac{df}{dt} + G \frac{dg}{dt} + H \frac{dh}{dt} \right) - W + \psi \left(\frac{df}{dx} + \frac{dg}{dy} + \frac{dh}{dz} \right) \right\} d\tau = 0.$$

Now in all cases in which the kinetic energy of a dynamical system involves the velocities but not the coordinates, the result of its variation is the same as if the momenta, such as F, G, H, in the expression in terms of momenta and velocities, were unvaried, and the result so obtained were doubled. Thus we have here

$$\int dt \int \left\{ \left(F \frac{d\delta f}{dt} + G \frac{d\delta g}{dt} + H \frac{d\delta h}{dt} \right) + \psi \left(\frac{d\delta f}{dx} + \frac{d\delta g}{dy} + \frac{d\delta h}{dz} \right) \right\} d\tau = \int dt \delta W;$$

or, integrating by parts and omitting the boundary terms for the reasons above given,

$$\int dt \int \left\{ \left(-\frac{dF}{dt} - \frac{d\psi}{dx} \right) \delta f + \left(-\frac{dG}{dt} - \frac{d\psi}{dy} \right) \delta g + \left(-\frac{dH}{dt} - \frac{d\psi}{dz} \right) \delta h \right\} d\tau = \int dt \delta W.$$

leads to formula for the aethereal electric force.

Therefore throughout the system the forcive corresponding to the displacement (f, g, h) is

$$(P, Q, R) = -\left(\frac{dF}{dt} + \frac{d\psi}{dx}, \frac{dG}{dt} + \frac{d\psi}{dy}, \frac{dH}{dt} + \frac{d\psi}{dz} \right).$$

109*. When, however, we consider the case of conductors in motion, so that their current sheets, instead of being referred to fixed axes, are carried on along with them, we shall have to refer the medium and therefore also the above variational operation to a moving scheme of axes or more generally to a moving space; and this will be accomplished if we include in $d/dt\,(F, G, H)$ not only ordinary partial differential coefficients with respect to the time, but also the rate of change due to alteration of position of the point considered owing to the movement of the space to which it is referred.

Equations *relative* to a moving material system:

The result of this reference to moving space, for the case in which it moves like a body of invariable form, is worked out as in Maxwell, *Treatise*, § 600, and leads to his well-known equations of electric force. These equations are, however, expressed with equal generality by eliminating the adjustable quantity ψ, thus obtaining for any complete circuit, with this extended meaning of d/dt,

$$\int (P dx + Q dy + R dz) = -\frac{d}{dt} \int (F dx + G dy + H dz)$$
$$= -\frac{d}{dt} \iint (la + mb + nc)\, dS.$$

As this relation retains the same form whether referred to fixed or to moving space, it expresses the Faraday-Maxwell law that under all circumstances the electromotive force referred to a circuit, fixed or moving, is equal to the rate of diminution of the magnetic induction through its aperture.

Faraday law of induction for circuits applies universally.

The expressions for the electric force thus determined are merely *formulae* for the kinetic reaction of the disturbed medium, which must be at each instant balanced by the forces of the elastic strain which is the other aspect of the efficient cause of the phenomena. Thus they do not imply any conclusion that in all material dielectrics, whether gaseous or liquid or solid, the motion of the matter produces an electric effect which is objectively the same for all; the equations referred to moving space apply in fact quite as readily to the free aether itself as to a moving material medium, provided the currents as well as the electric force are referred to the moving space.

In any actual problem, the quantity ψ, which enters into the electric force, is made determinate by means of the circuital condition to be satisfied by the currents throughout the dielectric: as a matter of convention we may if we please take ψ to include the electric potential of charges on the conductors which are the terminal aspects of the elastic strain in the dielectric, but nothing essential is perhaps gained by such a course, unless in the case of slow movements.

Electrostatic potential.

110. If, however, we were to adopt, on the lines of Helmholtz's theory of 1870, a different procedure and assume that the vector (F, G, H) is a physical entity as distinct from a mathematical expression, and so assign a definite physical formula for it, which must from our actual knowledge be of the type

$$F = \int \frac{u}{r} d\tau + \int \left(B \frac{d}{dz} - C \frac{d}{dy} \right) \frac{1}{r} d\tau,$$

it would follow that the circumstances of the induced electric force are not determined merely by the distribution of magnetic induction in the field, but involve the actual distribution of electric current and of magnetism throughout all space. For there are very various distributions of electric current and magnetism in the more distant parts of space which lead to the same distribution of magnetic induction in the neighbourhood of the system in which the currents are induced: these would be equivalent as regards the magnitudes of induced currents, but not as regards the distribution of induced electric force.

Apparent paradox of unretarded potentials.

This state of things would not be inconsistent with general principles*. The electric influence arising from a disturbance of one system is propagated elastically to other systems across the intervening medium, the propagation being nearly instantaneous without showing any sensible trace of the disturbance during its transit through the medium, and this on account of the high elasticity and consequent great velocity of propagation. The magnetic field is a residual effect of this propagation; that field is sufficient to represent the aggregate features of the result in cases in which the current is mostly conducted, but it need not represent the features of the propagation in detail. There are in fact cases in which induction takes place across a space in which there is at no time any sensible electric or magnetic force at all: for example, the starting of a current in a ring electromagnet induces in this way a current in any outside circuit which is linked with the ring: the elastic propagation here leaves no trace in the form of motion of the aether or magnetic force.

Magnetic field as a residue.

In ordinary electro-dynamics too slow for radiation the aether may be latent.

111. When the velocity of electric propagation may be taken as indefinitely great compared with the velocities of the conductors in the field, the phenomena of induced currents will depend only on the *relative* motion of the inducing and induced systems; thus we may simplify the conditions by taking the induced system at rest subject to the electric influences sent out from an inducing system in motion and otherwise changing. Now in this simpler case the electric intensity consists of two parts, one of them required to keep the current going against the viscous resistance of the conductor and the elastic resistance of the dielectric, and the other a free disturbance which will be continually cancelled with the velocity of radiation as fast as it is produced. The latter part therefore practically does not exist in ordinary problems of induction, in which the movements are slow compared with the velocity of light. Thus the elastic displacement of the electric medium may be taken as in internal equilibrium by itself in all such cases; there can be no free electric force inside a conductor, and the electric charge, if any, will reside on its surface. The amount of this superficial charge will be the time-integral of the displacement current which is involved in the total current, and which is wholly in the outside dielectric. Now the determination of the complete current is a perfectly definite problem, on the principles of Ampère and Faraday: thus the electric force at any point and the static electrification on the conductors are also on the same principles

* The true ultimate explanation is that static potential formulae are appropriate in terms of Maxwell's total current, propagated potentials in terms of true currents of moving electrons. Perhaps this has been one of the main sources of the obscurity said to inhere in Maxwell's formulations.

definite and determinate, subject to this proviso of slow movement of the bodies concerned.

Conclusion.

112. The foundation of the present view is the conception of a medium which has the properties of a perfect incompressible fluid as regards irrotational motion, but is at the same time endowed with an elasticity which allows it to be the seat of energy of strain and to propagate undulations of transverse type; and the question discussed is how far such a simple type of medium affords the means of co-ordination of physical phenomena. This idea of a medium with fluid properties at once disposes of the well-known difficulties which pressed on all theories that imposed on the aether the quality of solidity. If the objection is taken, which has been made against the ordinary vortex-atom theory of matter—that a perfect fluid is a mathematical abstraction which does not exist in nature, and the objective existence of which has not been shown to be possible,—the conclusive reply is at hand that the rotational elasticity with which the medium is here endowed effectually prevents any slip or breach such as would be the point of failure of a simple fluid medium without some special quality to ensure continuity of motion. On this head it will be sufficient to refer to some remarks of Sir G. G. Stokes[1] on a cognate topic. If therefore it is objected that we have no experience of a medium whose elasticity depends on rotation and not on distortion, the reply is that we can form no notion of the structure of a continuous frictionless fluid medium, unless we endow it with just some such elastic property in order to maintain its continuity.

Scheme of the illustrative medium: fluid, but ensured against slip, as in Stokes' aether.

The idea of representing magnetic force in the equations of electrodynamics by the velocity of the electric medium has been tried already, for example by Heaviside and by Sommerfeld, not to mention Euler. The objection, however, has been taken by Boltzmann and also by von Helmholtz that it would be impossible on such a theory for a body to acquire a charge of electricity. A cardinal feature in the electrical development of the present theory is on the other hand the conception of intrinsic rotational strain constituting electric charge, which can be associated with an atom or with an electric conductor, and which cannot be discharged without rupture of the continuity of the medium. The conception of an unchanging configuration which can exist in the present rotational aether is limited to a vortex ring with such associated intrinsic strain: this is accordingly our specification of an atom. The elastic effect of convection through the medium

Historical.

Nature of charge (*infra*, § 113): indestructible.

[1] Sir G. G. Stokes, "On the Constitution of the Luminiferous Aether," *Phil. Mag.* 1848, *Collected Papers*, Vol. II. p. 11.

Vortex atom (excluded in favour of electronic orbital atom, § 117).

of an atom thus charged is equivalent to that of a twist round its line of movement: such a twist is thus a physical element of an electric current.

Aether links light with electricity: Maxwell.

113. The chief result of the discussion is that a rotationally elastic fluid aether gives a complete account of the phenomena of optical transmission, reflexion, and refraction, in isotropic and crystalline media, coinciding in fact formally in its wider features with the electric theory of light; and that it gives a complete account of electromotive phenomena in electrostatics and electrodynamics. It assigns correctly the magnetic rotatory action on light to a subsidiary term of definite type in the energy-function of a material medium; while to avoid a magnetic translatory action of such amount as would be detectable, it is compelled to assign a high value to the coefficient of inertia of the free aether. In unravelling the detailed relations of aether to matter it is not very successful, any more than other theories; but it suggests a simple and precise basis of connection, in that form of the vortex-atom theory of matter to which it leads; and even should the present mode of representation of the phenomena become on further development in this direction definitely untenable, it may still be of use within its limited range as illustrating wider views of possibilities in that field. The theory also leads to the correct expressions for the ponderomotive class of electrostatic and electrodynamic phenomena, or rather it is not in disagreement with them; for here again knowledge of the details of the relation between the aether and the matter is defective, and thus, for example, the law of the attraction between permanent magnets is left unexplained. It supplies also a more definite view of the essentially elastic origin of all electrodynamic action than has perhaps hitherto been obtained, especially in cases of induction by motion across a steady magnetic field.

Present model implies very great inertia of aether: is provisional.

Utility of models.

(*Added August* 13, 1894.) *Introduction of Free Electrons**.

Currents could not be continuous vortex rings in aether:

114. The conclusion to which we are led in § 107 is that a simple vortex-atom theory is not in a position to attempt to explain the law of the forcive between permanent magnets, if only for the reason that

* The electron having thus been necessitated, and fortified later by a working model of its aethereal intrinsic constitution, the tendency in what follows is to get rid altogether of the free vortex rings that yet could also subsist in this aether. The atom is an orbital system of electrons, made much simpler afterwards by Rutherford's introduction, on due experimental evidence, of his massive proton as an immobile centre. But all such free electronic orbits are necessarily dissipative by radiation, slowly, however, if they are densely distributed; while a vortex ring in aether is not. Various types of structures of pure constraint in the aether with associated electrons are thus available

on such a theory no explanation of the inertia of matter has yet been developed. This difficulty is, however, not peculiar to the present special view of the electric field; any representation of a magnetic molecule, which assigns to it a purely cyclic motional constitution, is subject to an equal or greater difficulty in explaining why it is that the law of the forcive between magnets is the same as between currents, and not just the reverse.

What is required in order to obtain a decisive positive result is, that the assumption of a purely cyclic character for the motions associated with permanent magnets shall be avoided by giving the elasticity of the medium some kind of grip on them. The movements of rotation and vibration of the simple vortices which constitute a vortex aggregate are not competent to secure this, however sudden they may be, for in the irrotational fluid motion the constraint of the rotational elasticity has only to reduce a labile condition of the medium into a stable one; thus there is no sensible play of elastic energy introduced, such as would be required to explain induction in a steady magnetic field.

therefore must be due to convection of discrete charges:

One way of bringing about this desired interaction of magnetic with elastic energy, at the same time safeguarding the permanence of the atomic current, would be to make it a current of convection, *i.e.* to suppose the core of the vortex ring to be made up of discrete electric nuclei or centres of radial twist in the medium. The circulation of these nuclei along the circuit of the core would constitute a vortex which can move about in the medium, without suffering any pressural reaction on the circulating nuclei such as might tend to break it up; the hydrodynamic stability of the vortex, in fact, suffices to hold it together. But its strength is now subject to variation owing to elastic action, so that the motion is no longer purely cyclic. A magnetic atom, constructed after this type, would behave like an ordinary electric current in a non-dissipative circuit. It would, for instance, be subject to alteration of strength by induction when under the influence of other changing currents, and to recovery when that influence is removed; in other words the Weberian explanation of diamagnetism would now hold good.

thus effectively vortex rings variable by induction.

The essential diamagnetism of the atom.

The monad elements (§ 70) out of which a magnetic molecule of

for the further imagining of types of permanent atoms. Vortex rings in its structure would probably now be not unwelcome to spectroscopic theorists, as providing the magnetic moment which they aspire to, as well as the cyclosis that the dynamical interpretation of quanta is found to involve.

For the construction of electric theory on the basis simply of ions, as electric quanta experimentally recognized, by Professor Lorentz, see *Versuch einer Theorie...*, Leiden, 1895, especially p. 8, referring back to earlier memoirs of 1892 and, on Weberian ideas, of 1878.

The electron definitely introduced into the theory.

The atom as a secular electronic structure:

its Gaussian scheme.

this kind is built up are electric centres or nuclei of radial rotational strain. From what is known of molecular magnitudes, in connection with electrochemical data, it would appear that to produce an intensity of magnetization of 1700 C.G.S., which is about the limit attainable for iron, these monad charges—or *electrons*, as we may call them, after Dr Johnstone Stoney—must circulate very rapidly, in fact with velocities not many hundred times smaller than the velocity of radiation[1]. Even a single pair of electrons revolving round each other at such a rate as this would produce a practically perfect secular vortical circulation in the medium; so that a magnetic molecule may quite well be composed of a single positive or right-handed electron and a single negative or left-handed one revolving round each other in this manner*. We may in fact rigorously apply to the present problem the principle used by Gauss for the discussion of secular effects in Physical Astronomy. Instead of proceeding by addition of the elementary effects produced by a planet as it moves from point to point of its orbit, Gauss pointed out that the secular results as distinguished from mere periodic alternations are the same as if the mass of the planet were supposed permanently distributed round its orbit so that the density at any point is inversely proportional to the velocity the planet would have when at that point. Just in the same way here, the steady flow of the medium, as distinguished from vibrational effects, is the same as if each electron were distributed round its circular orbit, thus forming effectively a vortex ring, of which, however, the intensity is subject to variation owing to the action of other systems[2].

High speed of orbital electrons in the atom.

[1] Let q be the ionic charge, v its velocity, A the area of the orbit and l its length, n the number of atoms in 1 cub. centim.; then $n \,.\, q/l \quad v \,.\, A = 1700$. From electrochemical data we may take $nq = 10^3$, and from molecular dimensions $A/l = \frac{1}{2} \,.\, 10^{-8}$; whence $v = 3 \,.\, 10^8$, which is of the order of about one-hundredth of the velocity of radiation. This would make the periodic time come out about 10 times the period of luminous radiations.

* The positive electron, which ought to exist as the optical image of the negative, unless some fundamental feature has not yet come to light, has not been discovered: in the current schemes of physical atomic constitution, the positive element or proton is identified with the hydrogen ion, which, as found by J. J. Thomson, is 1800 times more massive than the negative element or electron. A single electron circulating round a far more massive proton, as in § 118, is taken to represent a hydrogen atom—with, however, the recent Bohr developments and restrictions.

[2] It may be observed that for the case of a simple diad molecule, composed of two equal and opposite electrons rotating round each other in equal orbits, their secular effects just cancel each other, so that the molecule as a whole is non-magnetic. This exact cancelling will not however usually occur when there are more than two electrons in the molecule, or when a number of molecules are bound together in a group as in the case of an iron magnet. Similar considerations also apply as regards the average electric moment of a molecule, which is in fact the electric moment of the Gaussian secular equivalent above described.

This mode of representation would leave us with these electrons as the sole ultimate and unchanging singularities in the uniform all-pervading medium, and would build up the fluid circulations or vortices—now subject to temporary alterations of strength owing to induction—by means of them.

115. It may be objected that a rapidly revolving system of electrons is effectively a vibrator, and would be subject to intense radiation of its energy. That, however, does not seem to be the case. We may, on the contrary, propound the general principle that whenever the motion of any dynamical system is determined by imposed conditions at its boundaries or elsewhere, which are of a steady character, a steady motion of the system will usually correspond, after the preliminary oscillations, if any, have disappeared by radiation or viscosity. A system of electrons moving steadily across the medium, or rotating steadily round a centre, would thus carry a steady configuration of strain along with it; and no radiation* will be propagated away except when this steady state of motion is disturbed.

The atomic radiation difficulty:

implies steady states:

It is in fact easy to investigate the characteristics of this strain configuration when the electric system is moving with constant velocity, say in the direction of the axis of x with velocity c. By § 97, the dynamical equations of the surrounding medium are

$$\left(\frac{d^2}{dt^2} - a^2\nabla^2\right)(f, g, h) = 0,$$

referred to coordinates fixed in space. The equations determining the disturbance relative to the electric system are derived by changing the coordinate x to a new relative coordinate x', equal to $x - ct$; this leaves spacial differentiations unaltered, but changes d/dt into $d/dt - c\,d/dx'$, thus giving [, a being here velocity of radiation and c that of convection,]

$$\left\{(a^2 - c^2)\frac{d^2}{dx'^2} + a^2\frac{d^2}{dy^2} + a^2\frac{d^2}{dz^2}\right\}(f, g, h) = \left(\frac{d^2}{dt^2} - 2c\frac{d^2}{dx'dt}\right)(f, g, h).$$

In a steady motion the right-hand side of this equation would vanish†; and the conditions of steady motion are thus determined by the solution of the ordinary potential equation for a uniaxial medium. The constants involved in the values of f, g, h so determined are connected by the fact that at a boundary of the elastic medium the rotation

verified for uniform convection.

* There will be radiation determined as that due to the secular electric moment, if any, of the atom. In a self-balanced atom, especially if complex, that will always be very small: see *Aether and Matter* (1900), § 151: and constraints too slight to otherwise disturb the theory may perhaps be imagined as in the text that would suppress it.

† But the motion of a ring of electrons round a nucleus could not be strictly steady, on account of radiation, though it is the more nearly so, with rapid approximation, the greater their number.

(f, g, h) must be directed along the normal. It follows at once, for example, that for a spherical nucleus[1] the rotation is everywhere radial. As the velocity of the electric system is taken greater and greater the permeability, in the direction of its motion, of the uniaxial medium of the analogy becomes less and less, and the field therefore becomes more and more concentrated in the equatoreal plane. When the velocity is nearly equal to that of radiation, the electric displacement forms a mere sheet on this plane, and the charge of the nucleus is concentrated on the inner edge of this sheet. The electrokinetic energy of a current system of this limiting type is infinite (§ 52), and so is the electrostatic energy; thus electric inertia increases indefinitely as this state is approached, so that the velocity of radiation is a superior limit which cannot be attained by the motion through the aether of any material system.

Limit to possible velocities of matter.

Again, the steady electric field carried along with it by a system rotating about a fixed axis with angular velocity ω is to be obtained by changing d/dt in the elastic equations into $d/dt - \omega d/d\theta'$, where θ' denotes relative azimuth around the axis; they therefore assume the form

$$\left(\nabla^2 - \frac{\omega^2}{a^2}\frac{d^2}{d\theta'^2}\right)(f, g, h) = 0,$$

of which the solution would be difficult. And the equations of the relative steady field for the most general case of uniform combined translation and rotation of an electric system, supposed still of invariable shape, are expressed in like manner, by taking the central axis of the movement as the axis of x, in the form

Hypothetical general non-radiating system: a cyclic feature.

$$\left(\nabla^2 - \frac{c^2}{a^2}\frac{d^2}{dx'^2} - \frac{\omega^2}{a^2}\frac{d^2}{d\theta'^2}\right)(f, g, h) = 0.$$

The circuital character of (f, g, h) will allow us to reduce these three variables in cases of symmetry to a single stream function, of which the slope along the normal at the surface of the nucleus must be null.

Any deviation from this steady motion of a molecule, produced by disturbance, will result in radiation which will continue until the motion has again become steady*. If we roughly illustrate by the phenomena of the Solar system, the mean circular orbits of the planets

Planetary analogy (imperfect).

[1] J. J. Thomson, *Recent Researches*..., 1893, pp. 16–22, where the existence of a superior limit (*infra*) to possible velocities was first pointed out: also Heaviside, *Phil. Mag.* 1889, cf. *Electrical Papers*, Vol. II. pp. 501 *seqq.* The problem of the dynamics of moving charges appears to have been first attacked on Maxwell's theory by J. J. Thomson, *Phil. Mag.* 1881.

* Cf. the transition by fall from one steady configuration to another, each presumably one of minimal energy, in the modern symbolic orbital scheme, developed by Niels Bohr, of atomic radiation in relation to its spectrum.

Expressed in the ideas of recent *quantum* theory (Schrödinger) the criterion for steady states would be whether they are "eigenwerthe" for ω which determine solutions finite everywhere except at the pole.

will represent the steady motion, while disturbances introduce planetary inequalities which would give rise to radiation of corresponding periods. An apparent obstacle to the application of this hypothesis to the theory of the spectrum is that such a steady motion is not unique, its periods depend on the energy of the system; but, from whatever cause, the chemical energy of a molecule (which is *electric*, therefore aethereal) has a definite value quite independent of the amount of *material* kinetic energy that may be involved in its temperature and capacity for heat. The periods of the vibrations would thus be fixed by the electric energy; while the prevailing character of the disturbances, which determines the relative intensities of the radiations, would depend on temperature. If there are lines in any spectrum which have this kind of origin, we should expect to find simple linear relations between the *reciprocals* of their periods or wave-lengths, as in the Planetary Theory. On the other hand, the sharpness of the spectral lines shows that the waves in the aether are absolutely simple harmonic, and this would point to atomic rather than molecular vibrations, were it not that the molecule is so small compared with a wave-length and also the periods far too great for such an origin[1].

The intrinsic atomic configurations: vibrations of the atom.

116. A difficulty has been felt as to how the centre of rotational strain which represents an electron is possible without a discrete structure of the medium; the following explanations may therefore be pertinent*. In the first place, it is essential to any simple elastic theory of the aether that the charge of an ion shall be represented by some permanent state of strain of the aether, which is associated with the ion and carried along by it. Such a strain configuration (in

[1] See G. Johnstone Stoney, "On the Cause of Double Lines and of Equidistant Satellites in the Spectra of Gases," *Trans. Roy. Dublin Soc.* 1891. [The sentence in the text is on the implication that, so far as was then known, the electron would be chemically an atom of very simple type such as, *e.g.*, that of hydrogen. Cf. p. 525 *infra*.]

* The model of an electron, limited in scope though it may be, which here comes in for the first time (cf. also Part III, 1897, §§ 3–5) is an essential feature of an electric theory of atomic matter; and indeed the point of view contains an explanation of the existence of matter as discrete atoms, as did previously the vortex-atom theory. The alternative to a direct theory of electrons, as developed here and later, and of atoms as possibly orbital systems of electrons, is a theory of a continuous density of electricity which has no structure and so is incapable of development.

In the development of the Lorentz transformation later, as the explanation of optical and electrical relativity (*Aether and Matter*, 1900, Chap. XI), it is this orbital structure of atoms, thus built out of electronic singularities purely aethereal, which restricts the demonstration to the second power in v/c. Beyond that order it is only a tentative exploring postulate, as in the modern relativity theory.

the light of what follows) can hardly be otherwise than symmetrical all round the ion; even if the nucleus be not itself symmetrical, this symmetry will be attained at a sufficient distance away from it. Now in an isotropic medium a steady configuration of strain of this kind must consist of a radial displacement such as we could imagine to be produced by an intrinsic pressure in the nucleus, or of a radial twist as above described, or it may combine the two. But for a great variety of reasons, electric and optical phenomena have no relation to any compression of the aether; therefore the notion of an intrinsic radial twist is the only representation that is available. An ideal process for the creation of such a twist centre has already been described in § 51 for the case of the rotational aether. A filament of the aether ending at the nucleus is supposed to be removed, and the proper amount of circulatory motion is to be imparted to the walls of the channel so formed, at each point of its length, so as to produce throughout the medium the radial rotational strain that is to be associated with the electron; when this has been accomplished the channel is to be filled up again with aether which is to be made continuous with its walls. On now removing the constraint from the walls of the channel, the circulation imposed on them will tend to undo itself, until the reaction against rotation of the aether with which the channel has been filled up balances that tendency, and an equilibrium state thus supervenes with intrinsic rotational strain symmetrically surrounding the nucleus. If, on the other hand, the aether had the properties of an elastic solid, and resisted shear but not rotation, the equations of *bodily* elasticity would remain just the same (§ 19); but the surfaces of shear of such a nucleus would be conical, with the channel by which the shear is introduced as their common axis, and when the constraint is removed the rotation imposed on the surface of this channel will undo itself and the shear thus all come out again, because the medium with which the channel is now filled up opposes no resistance to being rotated. Thus an elastic solid aether does not admit of any configuration of intrinsic strain such as would be required to represent an electric charge; and this forms an additional ground for limitation of that medium to a rotationally elastic structure. For an isotropic medium must be either elastic like a solid or fluid, or rotationally elastic, or it may combine these two properties; there is no[1] other alternative.

Model of an electron and its field, as a mobile strain centre supernaturally created:

requires a rotationally elastic aether,

such as could not be confounded with matter.

Gyrostatic rotational elasticity.

As to the intrinsic nature of the rotational elasticity of the free

[1] Professor FitzGerald remarks that it might, conceivably, resist absolute linear displacement. An hypothesis of this sort, which is on a lower plane than those mentioned above, is in fact involved in the usual expositions of Fresnel's dynamics of double refraction.

aether, although it is an important corroboration of our faith in the possibility of such a medium to have Lord Kelvin's gyrostatic scheme by which it might be theoretically built up out of ordinary matter, yet we ought not to infer that a rotational free aether is necessarily discrete or structural in its ultimate parts, instead of being a *continuum*. As a matter of history, the precisely similar argument has been applied to ordinary solids; the fact that deformation induces stress has been taken, apparently with equal force, as evidence of molecular structure in any medium which exhibits ordinary elasticity. It is necessary to put some limit to these successive refinements; there must be a final type of medium which we accept as fundamental without further analysis of its properties of elasticity or inertia: and there seems to be no adequate reason why we should prefer for this medium the constitution of an elastic solid rather than a constitution which distortion does not affect—perhaps there is just the reverse.

Matter is necessarily discrete.

117. The fluidity of the medium allows us to apply the methods of the dynamics of particles to the discussion of the motions through it of these electrons or strain configurations, and their mutual influences. The potential energy of a system of moving electrons will be the energy of the strain in the medium; unless their velocities are appreciable compared with the velocity of radiation, this will be a function of their relative positions alone. The kinetic energy is that of the fluid circulation of the medium, which will, under the same circumstances, be a quadratic function of the velocity components of the electrons, with coefficients which are functions of their relative positions. When, however, their velocities approach that of radiation the problem must be treated by the methods appropriate to a *continuum*, and cannot be formulated merely in terms of the positions of the electrons at the instant. It will suffice for the present to avoid the difficulties of the general case by supposing the velocities to be small, and the strain configuration of each electron therefore carried on unaltered by it; as the correction required depends on $(c/a)^2$ it will possibly be negligible for any actual problem.

Methods of planetary theory applicable to dynamics of an atom.

Let us then consider a single electron represented by a charge e moving along the direction of the axis of x with velocity v. The components of rotation in the medium due to its presence are at any instant $-e\,(d/dx,\ d/dy,\ d/dz)\,r^{-1}$, and those of the displacement current are derived from them by operating with the factor $-vd/dx$. This displacement current is the curl of the velocity of the medium, whence it may be easily verified that this velocity is

Model of the electron:

$$ev\left(0, -\frac{d}{dz}, \frac{d}{dy}\right)r^{-1},$$

being a circulation round the line of motion of the electron[1]. The kinetic energy* is thus

its energies.

$$\tfrac{1}{2}(ev)^2 \int (y^2 + z^2)\, r^{-6}\, d\tau;$$

which is equal to $4\pi/3a \,.\, (ev)^2$, if the nucleus which bounds internally the strained medium is spherical and of radius a. The potential energy of elastic strain in the medium is, on the same supposition, by the ordinary electrostatic formula, $\tfrac{1}{2}(eV)^2/a$, where V is the velocity of electric propagation. We assume that the nucleus of the electron has no other intrinsic inertia of its own, and no other potential energy of its own; under these circumstances its potential and kinetic energies will be of the same order of magnitude only when its velocity is comparable with that of radiation. In that case the present formulae are not applicable, except merely to indicate the orders of magnitude; but we can conclude that, in a steady molecular configuration of electrons, where there must be an increase of kinetic energy equal to the potential energy which has run down in their approach, the velocities of the constituent electrons must be comparable with that of radiation, just as the above estimate from magnetic data suggested.

Electronic inertia postulated to be purely aethereal.

The high orbital speeds of the electrons in the atom.

Suppose there are two electric systems in the field producing velocities (u, v, w) and (u', v', w') respectively. The kinetic energy is now

$$\tfrac{1}{2}\int \{(u + u')^2 + (v + v')^2 + (w + w')^2\}\, d\tau,$$

of which the part that involves their mutual action is

$$\int (uu' + vv' + ww')\, d\tau.$$

Structure of the electron is latent.

[1] It is to be observed that we cannot expect to obtain an expression for the displacement in the medium which is due to an electron; for the electron is part of the original constitution of the medium, and we cannot imagine it to be removed altogether. It may, however, be moved on into a new position, and we can then determine, as above, the displacement in the medium produced by this change of its locality.

* The density ρ of the aether has accidentally dropped out: with e in the Heaviside rational electromagnetic units which fit here, its numerical value comes out to be 4π. This charge e is the charge e' in standard electromagnetic units divided by 4π. Thus the kinetic energy integrates to $\tfrac{1}{3}a \,.\, (e'v)^2$ corresponding to a mass $2e'^2/3a$ as is now familiar. The potential energy, *infra*, would be $\tfrac{1}{2}(e'V)^2/a$, where e' is $4\pi e$ and V stands for the more usual c: with the modern relation between electric energy and inertia this would give a mass $e'^2/2a$. The discrepancy of this latter, the true value, from the previous (and still usual) value, as here first introduced, is due to this calculation of the field having ignored the correction for relativity shrinkage of the spherical model, which depends, as does the mass, on v^2/c^2.

If the velocity (u, v, w) belongs to an electron (e, v) as above, the mutual part of the kinetic energy is

$$ev\int\left(-v'\frac{d}{dz}+w'\frac{d}{dy}\right)r^{-1}\,d\tau,$$

or on integration by parts

$$-ev\int(v'n-w'm)\,r^{-1}dS-ev\int\left(\frac{dw'}{dy}-\frac{dv'}{dz}\right)r^{-1}d\tau,$$

Mutual energy of two moving electrons

of which the former part is null when the external boundary is very distant. Thus the mutual electrokinetic energy is

$$-ev\int r^{-1}\frac{df'}{dt}\,d\tau,$$

where f' is the component parallel to v of the electric displacement belonging to the other system.

If the other system is also an electron (e', v') the total electrokinetic energy is

$$T=\tfrac{1}{2}L\,(ev)^2+\tfrac{1}{2}L'\,(e'v)^2+M\,.\,ev\,.\,e'v',$$

where L, L' are as determined above, having the values $8\pi/3a$, $8\pi/3a'$* when the nuclei are spherical, while

$$M=r^{-1}\cos\,(ds\,.\,ds')+\tfrac{1}{2}\,d^2r/ds\,ds',$$

in which ds, ds' are in the directions of v, v', and r is the distance between the monads[1]. The potential energy is

$$W=\tfrac{1}{2}A\,(eV)^2+\tfrac{1}{2}A'\,(e'V)^2+B\,.\,eV\,.\,e'V,$$

where A and A' are as determined above, being the reciprocals of the radii† when the nuclei are spherical, and $B=r^{-1}$. The equations of motion of the two electrons may now be formed in the Lagrangian manner, and will hold good so long as the motions are fairly slow compared with radiation.

A purely electric theory of matter introduced.

The question, however, arises whether we should not associate with the electric inertia of an ion of this kind a much greater inertia of matter to which the ion belongs. When we trace as above the consequences of refraining from doing so, we arrive at the result that these free electrons can be projected by their mutual actions, with velocities which are a considerable fraction of that of radiation. Bearing in mind the phenomena of the Solar corona and of comets' tails, and certain electric phenomena in vacuum tubes[2], where some

* Corrected *supra* to $2/3a$, $2/3a'$.

1 The calculation of M is given concisely by H. Lamb, *Proc. Lond. Math. Soc.* June 1883, p. 407; the result is given also by Heaviside, *Electrical Papers*, Vol. II. p. 501. [The velocities v, v' are supposed small compared with the velocity of radiation.]

† Divided by $16\pi^2$.

2 Professor FitzGerald suggests the addition to this list of auroras and magnetic storms.

modification of the aether which affects light by reflexion and otherwise is projected with velocities of that order, there seems to be no reason for the summary exclusion of such an hypothesis as the present[1], especially as an electrically neutral molecule could attain no such velocities, and would comport itself more like ordinary matter.

Dynamics of an orbital system of two ions.

118. The circumstances of steady motion may be illustrated by a calculation for the case of two electrons; the same method would clearly also apply to a greater number. The kinetic energy of two electrons e_1 and e_2, whose coordinates are $(x_1 y_1 z_1)$ and $(x_2 y_2 z_2)$, moving under their mutual influence, is, by § 117,

$$T = \tfrac{1}{2}L_1 e_1^2 (\dot{x}_1^2 + \dot{y}_1^2 + \dot{z}_1^2) + \tfrac{1}{2}L_2 e_2^2 (\dot{x}_2^2 + \dot{y}_2^2 + \dot{z}_2^2) + \frac{e_1 e_2}{2r}(2\dot{x}_1\dot{x}_2 + \dot{y}_1\dot{y}_2 + \dot{z}_1\dot{z}_2),$$

the axis of x being parallel to their mutual distance r.

Let us take the case when they revolve steadily in the plane of xy with angular velocity ω round a common centre, at distances r_1, r_2 from it, where $r_1 + r_2 = r$. The kinetic reaction on e_1 resolved parallel to x is

$$\frac{d}{dt}\frac{dT}{d\dot{x}_1} - \frac{dT}{dx_1} = L_1 e_1^2 \ddot{x}_1 + e_1 e_2 \frac{d}{dt}\left(\frac{\dot{x}_2}{r}\right) - \frac{e_1 e_2}{2r^2}\cos\theta\,(2\dot{x}_1\dot{x}_2 + \dot{y}_1\dot{y}_2 + \dot{z}_1\dot{z}_2)$$

in which θ is null when r is taken along x, while

$$\ddot{x}_1 = \omega^2 r_1,\ \ddot{x}_2 = -\omega^2 r_2,\ \dot{y}_1 = \omega r_1,\ \dot{y}_2 = -\omega r_2.$$

On equating this reaction to the electrostatic attraction, we have

$$\left(-L_1 e_1^2 r_1 + e_1 e_2 \frac{r_2}{r} + e_1 e_2 \frac{r_1 r_2}{2r^2}\right)\omega^2 = e_1 e_2 \frac{V^2}{r^2}.$$

Similarly

$$\left(-L_2 e_2^2 r_2 + e_1 e_2 \frac{r_1}{r} + e_1 e_2 \frac{r_1 r_2}{2r^2}\right)\omega^2 = -e_1 e_2 \frac{V^2}{r^2}.$$

Hence $$\left(L_1 e_1 + \frac{e_2}{r}\right) e_1 r_1 = \left(L_2 e_2 + \frac{e_1}{r}\right) e_2 r_2,$$

which determines the ratio of r_1 to r_2 in the steady motion; and then the value of ω gives the period of the rotation*.

[1] Professor J. J. Thomson informs me that he finds the velocity of the negative rays in vacuum tubes to be about 2×10^7 C.G.S.

* Writing $m_1 = L_1 e^2 = e^2/2a$, $m_2 = km_1$, a being the effective radius of an electron and k about 1700 for a proton, the last equation reduces to $r_1 - kr_2 = \tfrac{1}{2}a$, where $r_1 + r_2 = r$.

The orbit would be made definite by imposing the quantum of cyclic momentum of Bohr, namely $2\pi(m_1 r_1^2 + m_2 r_2^2)\,\omega = nh$, where n is an integer: or the alternative form that the *kinetic* energy, not the change of the whole energy, is $n \,.\, \tfrac{1}{2}h\nu$.

The field-inductance term here introduced is a sensible fraction of the Bohr

For example, when the electrons are equal and opposite $e_1 = -e_2$, $L_1 = L_2$, and $r_1 = r_2$: thus the square of the velocity of either, $(\frac{1}{2}\omega r)^2$, is equal to $V^2/(2Lr - \frac{5}{4})$. For the case of a spherical nucleus of radius a, $L = 8\pi/3a$ [now corrected to $2/3a$]; thus [when $r > 5a$] the velocity of either must be considerably less than $\frac{1}{8}V$, which is small enough to allow this method to approximately represent the facts for that case.

System of two conjugate pure electrons.

It may be observed that in the general problem of the dynamics of a system of n electrons, the equations of conservation of momentum assume the forms

Conserved momentum:

$$\frac{dT}{d\dot{x}_1} + \frac{dT}{d\dot{x}_2} + \ldots + \frac{dT}{d\dot{x}_n} = \text{const.},$$

with similar equations in y and z. For the case of two electrons moving in the same line, the equations of energy and momentum determine the motion completely; their forms illustrate the complexity of the electric inertia which is involved.

simplest case.

119. In the general theory of electric phenomena it has not yet been necessary to pay prominent attention to the molecular actions which occur in the interiors of conductors carrying currents: it suffices to trace the energy in the surrounding medium, and deduce the forces acting on the conductors, considered as continuous bodies, from the manner in which this energy is transformed. The calculations just given suggest a more complete view, and ought to be consistent with it; instead of treating a conductor as a region effectively devoid of elasticity, we may conceive the ions of which it is composed as free to move independently, and thus able to ease off electric stress; the current will thus be produced by the convection of ionic charges. Now if all the atoms took part equally in this convection, their velocity would be exceedingly small; a current of i amperes per square centimetre would imply a velocity of about $10^{-4}\,i$ centimetres per second. The kinetic energy of an ion due to intrinsic electric inertia is, according to the formula above, $\frac{1}{2}2/3a \, . \, (ev)^2$, where a is of order $< 10^{-8}$, e of order 10^{-21}; this would imply as above a centrifugal electric force of intensity $2/3a \, . \, e \, . \, v^2/R$, which may be of order $10^{-19}\,i^2$, acting on this particular ion when it is going round a curve of radius R. Now even if the conductor were of copper, the slope of potential along it would be, with this current intensity, as much as $164i$. The effects of the intrinsic electric inertia are therefore so far quite beyond

Conduction is by convection of electrons,

which may be a very slow drift.

correction due to finiteness of the nuclear mass only, when r is comparable with a, that is only for the x-ray part of the spectrum.

As between a proton nucleus and an electron it is only about 10^{-3} of the spectroscopic term of Sommerfeld arising from varying mass: but as between two electrons it is of the same order.

Intrinsic inertia of currents usually negligible.

the limit of observation. We have, however, been taking the electric drift v to be the only velocity of the ions or electrons. If they possess a velocity of their own in fortuitous directions of order V, the average centrifugal electric force on an electron due to the current will possibly be as high as $2/3a \cdot e \cdot vV/R$, because change of sign of V does not change the sign of the force. This would still hardly be detectable even if V were comparable with the velocity of radiation*.

But an *electric* force of a cognate kind has in fact already been looked for and detected by E. H. Hall. When the current is moving in a field of magnetic force H at right angles to itself, there must be an electric force at right angles to both, acting on each particular ion, of which the intensity is vH[1]. For example, if H were 10^3 C.G.S., this electric force would be $10^3 v$ C.G.S. or $10^{-5} v$ volts; in the rough estimates of the last paragraph it would be of order $10^{-1} i$, as compared with a slope of potential along the conductor of $164 i$; therefore it is quite amenable to observation, so that we must consider it more closely. As there are also an equal number of negative ions moving in the opposite direction, they must give rise to an opposite electric force acting on them; thus the total transverse electric force, as observed, will be reduced from the above value in the ratio

Preliminary aspect of the Hall effect: in electrolytes with two kinds of ions:

$$(v_2 - v_1)/(v_2 + v_1),$$

where v_2 and v_1 are the velocities of drift of the positive and negative ions, which may be different just as Kohlrausch found them to be in ordinary electrolysis. The absolute velocity V of an ion does not affect the result in this case. This view would therefore make the sign of the Hall effect depend on whether positive or negative ions conveyed most of the current.

its sign.

* The mechanical impulse arising from abrupt arrest of the electrons of a current has more recently been detected in America, also the converse initiation of a current by shock.

[1] It is assumed here that all forces of electric origin acting on the moving atomic charges are primarily electric forces; in accordance with the previous theory (§ 57) it is only the part of the energy change which cannot be compensated by electromotive work, that reveals itself ultimately as a forcive working mechanically on the aggregates which constitute conducting bodies, or as heat in case it is too fortuitously constituted to admit of transformation into a regular mechanical working forcive. This ultimate destiny is independent of any question as to the origin of the inertia of the atoms. Thus the steady and unlimited fall of the electric resistance of metals with lowering of the temperature, found by Dewar and Fleming, shows that the frittering away of electric energy into heat in a metallic conductor depends upon the velocity of fortuitous agitation of the molecules, and would disappear when it ceased. The regular transfer of the electrons would thus involve no degradation of electric energy (§ 115), except so far as it is disturbed and mixed up by the thermal agitations of the molecules of the conductors. In electrolytes the dependence of the degree of ionization on the temperature may mask the direct effect of the thermal agitations.

Super-conductance in metals.

120. The electromagnetic or mechanical forces acting on the conductors conveying the currents are on the other hand to be derived from the energy-function, considered as potential after change of sign as in § 57, by the method of variations. For the reasons given above, the effect of the term $\Sigma\frac{1}{2}Le^2$, involving intrinsic electric inertia, is in the present problem inappreciable, except as giving a kind of internal gaseous pressure if the velocities of free electrons were comparable with that of radiation. The total electrokinetic energy is thus practically

The type of mechanical energy-function for currents regarded as continuous:

$$\iint M\,ids\,i'ds', \text{ where } M = r^{-1}\cos(ds, ds') + \tfrac{1}{2}d^2r/ds\,ds';$$

and on the present hypothesis the energy may be considered to be correctly localized in this formula.

location of the energy:

If the currents are uniform all along the linear conductors, the second term in M integrates to nothing when the circuits are complete, and we are thus left with the Ampère-Neumann expression for the total energy of the complete currents, from which the Amperean law of force may be derived in the known manner by the method of variations. But it must be observed that, as the localization of the energy is in that process neglected, the legitimate result is that the forcive of Ampère, together with internal stress as yet undetermined between contiguous parts of the conductors, constitute the total electromagnetic forcive: it would not be justifiable to calculate the circumstances of internal mechanical equilibrium from the Amperean forcive alone, unless the circuits are rigid. For example, if we suppose that the circuits are perfectly flexible, we may calculated the tension in each, in the manner of Lagrange, by introducing into the equation of variation the condition of inextensibility. We arrive* at a tension $i\int Mi'ds'$, where i is the current at the place considered; whereas the tension as calculated from Ampère's formula for the forcive would in fact be constant, the forcive on each element of the conductor being wholly at right angles to it.

the derived forces:

leads to unverified internal stresses: so currents must be due to free discrete electrons.

The general case when the currents are not linear is also amenable to simple analysis. The energy associated with any linear element ids is $ids\int Mi'ds'$; which is equal to ids multiplied by the component of the vector potential of the currents in the direction of ds, when the conduction and convection currents move round complete circuits. Thus, changing our notation, the energy associated with a current (u, v, w) in an element of volume $d\tau$ is $(Fu + Gv + Hw)\,d\tau$.

Energy-density in electronic system:

* This argument implies a continuous braced-up medium, to which the energy belongs. The experiments instituted by FitzGerald and by Lodge disproved the prediction (Part II, *infra*, p. 543, § 3), and so involve the inference, already forced upon us, that the current is made up of electrons independently mobile.

in terms of vector potential:

In this expression (F, G, H) is the vector potential of the currents; if there is also magnetism in the field, there will be a part of this vector potential due to it, which may be calculated from the equivalent Amperean currents. Thus for a single Amperean circuit*,

$$F = i\int r^{-1}\,dx,$$

thus by Stokes' theorem

$$F = i\iint\left(\mu\frac{d}{dz} - \nu\frac{d}{dy}\right) r^{-1}dS,$$

where (λ, μ, ν) is the direction vector of the element of area dS; hence the magnetic part of the vector potential is

the magnetic part:

$$\left(B\frac{d}{dz} - C\frac{d}{dy},\ C\frac{d}{dx} - A\frac{d}{dz},\ A\frac{d}{dy} - B\frac{d}{dx}\right) r^{-1},$$

which agrees with the assumption in § 110. It will be observed that in the vector potential of the field, as thus introduced, there is no indeterminateness; it is defined by the expression for the energy, as above.

We may complete this mode of expression of the energy by including the energy of the magnetism in the system, due to the field in which it is situated. For a single Amperean atomic circuit it is

$$i\int(F dx + G dy + H dz),$$

which is by Stokes' theorem

part due to external magnetic field.

$$i\iint\left\{\lambda\left(\frac{dH}{dy} - \frac{dG}{dz}\right) + \ldots + \ldots\right\} dS;$$

thus the energy of the magnets is

$$\int(A\alpha + B\beta + C\gamma)\,d\tau,$$

where (α, β, γ) is the magnetic force due to the external field as usually defined; this follows from the formulae for (F, G, H) already obtained. There is also the intrinsic energy of the magnets due to their own field; by the well-known argument derived from the work done in their gradual aggregation, the coordinated part of this is

Intrinsic magnetic energy.

$$\tfrac{1}{2}\int(A\alpha_0 + B\beta_0 + C\gamma_0)\,d\tau,$$

where $(\alpha_0, \beta_0, \gamma_0)$ is the force of their own field. These terms will add on without modification to the other part of the electrokinetic energy for the purpose of forming dynamical equations, provided we assume as above that the magnetic motions are not of a purely cyclic character. This sketch will give an idea of how magnetism enters in a

* In a field involving sensible radiation part of (F, G, H) would arise from the Maxwellian displacement current.

dynamical theory which starts from the single concept of electrons in movement.

The energy being thus definitely localized, and all the functions precisely defined, we derive in the Lagrangian manner the electric force

$$(P, Q, R) = -\left(\frac{dF}{dt}+\frac{d\Psi}{dx}, \frac{dG}{dt}+\frac{d\Psi}{dy}, \frac{dH}{dt}+\frac{d\Psi}{dz}\right)$$

The field of induced electric force.

when Ψ is some function of position as yet undetermined, whose value is to be adjusted to satisfy the restriction to circuital flow which the present analysis for conduction and convection currents involves. The electrodynamic forcive acting on the conductors carrying the currents is

$$(X, Y, Z) = -\left(u\frac{dF}{dx}+v\frac{dG}{dx}+w\frac{dH}{dx}, u\frac{dF}{dy}+v\frac{dG}{dy}+w\frac{dH}{dy}, u\frac{dF}{dz}+v\frac{dG}{dz}+w\frac{dH}{dz}\right);$$

Tentative application to force on the material structure, based on continuous specification of currents,

but this involves, in addition to the usually recognized forcives of Ampère's law and Faraday's rule, a forcive in the direction ds of the resultant current Γ and equal to $-\Gamma dN/ds$, where N is the component of the vector potential in the direction of ds. This additional forcive may be represented as balanced by a tension iN, in each filament or tube of flow carrying a current i, just as above. The existence of this tension seems to admit of easy test by a suitable modification of Ampère's third crucial experiment.

leads to mechanical tension along currents: (which does not exist, Part II).

It is now a simple matter to complete this theory, which at present applies to circuital convection and conduction currents, so as to include the effect of convection without this restriction. It will suffice to consider a uniform current i' flowing in an open path, thus accumulating electrification at one end and removing it from the other end. The second term in M when integrated with respect to ds' yields $ids\,.\,\tfrac{1}{2}i'\left|\frac{dr}{ds}\right|_1^2$; thus in the energy of the element of ids there is a term $ids\,.\,\tfrac{1}{2}\int\frac{d\rho}{dt}\cos\theta\,d\tau$, where θ is the angle between ds and the distance r of $d\tau$ from it, and $d\rho/dt$ is the rate of increase of the density of electrification at the element $d\tau$. Thus [replacing $ids\cos\theta$ by the dynamical $e\,(l\dot{x}+m\dot{y}+n\dot{z})$, where l is x/r,] there is an additional electric force

A continuous flow would involve forces between terminals.

$$-\tfrac{1}{2}\frac{d}{dt}\int\left(\frac{x}{r}, \frac{y}{r}, \frac{z}{r}\right)\frac{d\rho}{dt}\,d\tau,$$

and an additional electromagnetic force

$$\tfrac{1}{2}\int\left(\frac{y^2+z^2}{r^3}, \frac{z^2+x^2}{r^3}, \frac{x^2+y^2}{r^3}\right)\frac{d\rho}{dt}\,d\tau,$$

where (x, y, z) have reference to the element $d\tau$ as origin. These expressions are appropriate where, in place of following the convection of single electrons, we contemplate the change of electric density at a point in space; they suffer from an apparent want of convergency, which would be real were it not that $\int \rho d\tau$ is null.

Electric theory of matter: replaces vortex atoms.

121. It may be observed finally, that the question as to how far it is permissible to entertain the view that the non-electric properties of matter may also be deducible from a simple theory of free electrons in a rotationally fluid aether, has hardly here been touched upon. The original vortex-atom theory of matter has scarcely had a beginning made of its development, except in von Helmholtz's fundamental discovery of the permanence of vortices, and the subsequent mathematical discussions respecting their stability. How far a theory like the present can take the place of or supplement the vortex theory, is therefore a very indefinite question. In the absence of any such clue, a guiding principle in this discussion has been to clearly separate off the material energy involving motions of matter and heat, from the electric energy involving radiation and chemical combination, which alone is in direct relation to the aether. The precise relation of tangible matter, with its inertia and its gravitation, to the aether is unknown, being a question of the structure of molecules; but that does not prevent us from precisely explaining or correlating the effects which the overflow of aethereal energy will produce on matter in bulk, where alone they are amenable to observation.

Activity of aether is initiated by matter: interaction of the two media.

Optical Dispersion; and Moving Media.

122. The [formal hypothetical] view of optical dispersion developed in the first part of this paper, on the basis of MacCullagh's analysis, has its foundation in the discreteness of the medium, the dispersion being assigned to residual terms superposed on the average refraction. The cause of the refraction itself is found in the influence of the contained molecules, which are constituted in part at least of mobile electrons and so diminish the effective elasticity of the medium. Now if these molecules formed a web permeating the medium, with connections of its own, this web would act as an additional support, and the optical elasticity would, if affected at all, be increased. But it is different if the molecules are so to say parasitic, that is, if they are configurations of strain in the aether itself, and their energy is thus derived directly from the aether and not from an independent source. To more clearly define the effective elasticity in that case, let us suppose a uniform strain of the type in question to be imparted to the medium by the aid of constraints; it follows from the linearity of the elastic relations that the stress involved in this superposed

The electrons interact only through the aether:

strain will be that corresponding to the elastic coefficient of the free aether, for there is by hypothesis no web involved with extraneous elasticity. Now suppose the constraints required to maintain this pure strain to be loosened; the molecules will readjust themselves into a new equilibrium position which involves less energy, and this diminution of the total energy of the strain implies a diminution of the corresponding effective elastic coefficient. This analysis has to do with the statical elasticity; in electrical terms it corresponds to the explanation of Faraday and Mossotti as to how it is that the ratio of electric force to electric induction is diminished by the presence of polarized molecules. If, however, in a problem of vibration, the displacement of the medium involved in the molecules thus settling down into a new conformation of equilibrium, after the constraints are removed, is comparable with that involved in the original strain, the kinetic energy of the medium will be affected by the molecules as well as the strain energy, and the circumstances of propagation will depend on the period of the waves. As the present theory involves altered effective elasticity but unaltered effective inertia, this dependence can be but slight; in other words, the orientation of the molecules does not involve any considerable additional kinetic energy of displacement of the medium in comparison with the work done by electric forces; just as was to be anticipated from § 117, where it has been shown that to produce a comparable motional effect very great velocity of translation or rotation of the molecule is requisite, not the comparatively small velocity of movement of the elements of the medium caused by a wave passing over it.

and produce dispersion by electric not inertial interaction.

This amounts in fact to asserting that it is only the electric inertia of the molecules that affects the electric waves. Their material inertia is quite a different and secondary thing from the inertia of the aether[1]; on an electric theory it can have no direct influence on the radiation.

It seems clear also that if the molecules, in their relations to the aether, behave as systems of grouped electrons, their presence cannot disturb the fluidity of that medium, so that the foundation given above (§ 28) for MacCullagh's dispersion theory remains valid.

123. Let us contrast the merits of this view of dispersion with those of the type of theory in which it is ascribed to imbedded ponderable molecules. It has been shown[2], that for an elastic-solid theory (or any theory treated by the method of rays, § 22) to give an account of the observed laws of reflexion at the surfaces of transparent media, the inertia may be supposed to vary from one medium to another,

Contrast with an inertial optical theory:

[1] Cf. Lord Kelvin, *Baltimore Lectures on Molecular Dynamics*, 1884, Lecture xx.

[2] Lord Rayleigh, *Phil. Mag.* Aug. 1871.

34-2

or else the rigidity, but not both. Thus, setting aside the latter alternative for other reasons, the molecules must act simply as a load upon the vibrating aether; this requires that their free periods must be very long compared with the period of the waves, which is a very reasonable hypothesis. But if the optical rigidity is absolutely the same for all media, we are bound to explain not only the dispersion, but the whole refraction, by the influence of the inertia of the load of molecules; thus to explain dispersion we have to take refuge in Cauchy's doctrine of simple discreteness of the medium.

whose insufficiency is explored by illustrations.

Now let us formulate the problem of wave-propagation in a discrete medium of this kind. It will be a great simplification to consider stationary vibrations instead of progressive undulations; let us therefore combine two equal wave-trains travelling in opposite directions, and so obtain nodes and antinodes. We may imagine the continuity of the medium severed at two consecutive antinodes; thus the problem before us is to find the gravest free period of a block of the medium, forming half a wave-length, with its imbedded molecules. To represent in a simple manner the general features of this question, let us take Lagrange's problem of the vibrations of a stretched cord with n equidistant beads fixed on it. This will be a sufficient model of the case now in point, where the molecules act simply as a load; but if we are to consider possible influence of their free periods, so as to include anomalous dispersion as well as ordinary dispersion, we must also endow the beads of the model with free periods, which may be done by imposing an elastic restoring force on each[1]. In this latter case, however, the difficulty of representing the nature and origin of the restoring force detracts very seriously from the efficiency and validity of this mode of representation.

Loaded discrete atoms cannot account for actual dispersion of light:

Fortunately the simpler and more definite case is all we now require; when the mass is all concentrated in the beads, Lagrange finds that the velocity of propagation of a wave whose length contains n beads is $(V_0 \sin \pi/2n) \div \pi/2n$.[2] For the case of an ordinary light-wave there are about 10^3 molecules in a wave-length, so that the dispersion for an octave should by this formula be about $\frac{1}{6}(\pi/2000)^2$ of the velocity, which is enormously smaller than the corresponding dispersion, usually about one per cent., of actual optical media.

elasticity of free periods essential.

Thus we must conclude that, while the present form of MacCullagh's theory ascribes refraction to the defect of elastic reaction of the molecules, and dispersion to the influence of their free periods, so also the elastic-solid theory must ascribe refraction to loading by the

[1] Cf. Lord Kelvin, *Baltimore Lectures*, 1884.

[2] Lagrange, *Méc. Anal.* II. 6, § 30; Rayleigh, *Sound*, § 120; Routh, *Dynamics*, Vol. II. § 402.

mass of the molecules, and dispersion to the influence of their free periods. In these respects the two theories run parallel, and there is not much to choose between them; a model constructed on either basis would fairly represent the phenomena of dispersion. The latter ascribes the influence of the matter to nodules of mass, in the aethereal, not by any means the material or gravitative sense, supposed distributed through the medium; the former finds its cause in the properties of the nuclei of intrinsic strain, or electrons. On either view, Fresnel's laws of reflexion are a first approximation obtained by neglecting dispersion, and are as we know departed from by a medium which produces anomalous dispersion of the light, even for wave-lengths which suffer no sensible absorption[1].

[1] The most definite form which the Young-Sellmeier type of theory has yet assumed is that of Lord Kelvin (*Baltimore Lectures*, 1884). The author begins with an illustrative molecule, consisting of a core of very high inertia joined by elastic connections to a chain of outlying satellites of which the last only is in connection with the aether. The core being thus practically unmoved, the whole system is so to speak anchored to it, and the mass of the core does not come into account. Such an illustration gives very vivid representations of absorption and fluorescence. After working out the formula for the index of refraction in the manner of Lagrange's dynamics of linear systems, a transformation is suggested by consideration of the zeros and infinities of the function representing the index, which gives *à priori* a result whose validity is far wider than any special illustration, in the form

A loaded aether as a tentative dispersive model:

$$\mu^2 = 1 + \frac{c_1\tau^2}{\rho}\left\{-1 + \frac{q_1\tau^2}{\tau^2 - \kappa_1^2} + \frac{q_2\tau^2}{\tau^2 - \kappa_2^2} + \dots\right\},$$

a Kelvin theory.

where τ is the period of the waves, $\kappa_1, \kappa_2 \dots$ are the free periods of the molecule, and the coefficients $q_1, q_2 \dots$ depend on the distribution of the energy of the steadily vibrating molecule amongst these periods. On this theory the aether is *not* simply loaded by the molecule, but the coefficient c_1 depends on the manner in which the molecule is anchored in space; the theory is accordingly in difficulties with regard to double refraction and reflexion (*loc. cit.* Lecture XVI), of which the former is not a dispersional phenomenon.

The analogous electric theory explained above appears to be free from these difficulties. The relation of the average disturbance of the molecule to the disturbance of the aether is there introduced simply by means of an experimental number, the specific inductive capacity of the medium. The correlative mechanical hypothesis would require us, not to anchor a massive core of the molecule in space, but to introduce a coefficient to express the ratio of the displacement of the molecule to the displacement of the medium on some appropriate kind of equilibrium theory,—thus in fact to directly load the aether, and refer only the variable part of dispersion to the free periods of the molecule; but such an idea would introduce all kinds of difficulties with respect to the kinetic theory of gases and material motions in general. In the electric theory these difficulties are evaded by the principle that the inertia of matter is different in kind from the inertia of aether; the one is subject to electromagnetic forcive, the other to electromotive forcive.

Contrast with electric theory.

The recent discovery of an upper limit beyond which radiations that can travel in a vacuum do not travel across air, has an important bearing on the present subject. [It is due to ordinary absorption.]

124. The analogy just mentioned suggests fresh search for a purely dynamical explanation of Fresnel's formula for the influence of motion of the medium on the velocity of light, of which we had previously to be content with an indirect demonstration on the basis of the law of entropy. In the first place, we shall consider the usually received proof[1], on the theory of a loaded mechanical aether. Let ρ be the density of the aether and ρ' that of the load, and let ϑ be the displacement of the medium; the equation of propagation for the medium at rest is $(\rho + \rho')\, d^2\vartheta/dt^2 = \kappa d^2\vartheta/dx^2$; the equation for a medium in which the load ρ' is moving on with velocity v in the direction of propagation is

Optical convection for a loaded mechanical aether:

$$\rho \frac{d^2\vartheta}{dt^2} + \rho' \left(\frac{d}{dt} + v\frac{d}{dx}\right)^2 \vartheta = \kappa \frac{d^2\vartheta}{dx^2}.$$

We have clearly $\kappa/\rho = V^2$, $\kappa/(\rho + \rho') = V^2/\mu^2$, where V is the velocity of propagation in free aether; and on substituting $\vartheta = A \exp 2\pi/\lambda \,.\, \iota\, (x - V_1 t)$, we find for V_1 the velocity of propagation in the moving medium the value $V\mu^{-1} + v\,(1 - \mu^{-2})$, which is Fresnel's expression. This explanation precisely fits in with our previous conclusion, that on a mechanical theory the matter must affect the inertia but not at all the elasticity of the medium, except as regards the dispersion; and conversely, it may be used as independent evidence for that assumption.

contrasted with the rotational or electric aether.

The treatment of the same problem on the theory of a rotational aether follows a rather different course. By the hypothesis, the electric displacement or strain ϑ_2 due to orientation of the molecules may be treated as derived, by an equilibrium theory, from the inducing displacement ϑ_1 which belongs to the waves and provides the stress by which they are propagated. That part ϑ_2 of the electric displacement is in internal equilibrium at each instant with the displaced position of the molecules, and so furnishes no stress for the wave-propagation. The relation between ϑ_1 and the total displacement $\vartheta_1 + \vartheta_2$ is that of electrostatics, $\vartheta_1 + \vartheta_2 = K\vartheta_1$, where K is the effective specific inductive capacity of the medium. The equation of propagation when the medium is at rest is $\rho d^2(\vartheta_1 + \vartheta_2)/dt^2 = \kappa d^2\vartheta_1/dx^2$, showing that the velocity of the waves is $(\kappa/K\rho)^{\frac{1}{2}}$, so that $K = \mu^2$. The equation of propagation when the molecules are moving through the stationary aether with velocity v in the direction of the wave-motion, is

$$\rho \frac{d^2\vartheta_1}{dt^2} + \rho \left(\frac{d}{dt} + v\frac{d}{dx}\right)^2 \vartheta_2 = \kappa \frac{d^2\vartheta_1}{dx^2},$$

where $\vartheta_2 = (\mu^2 - 1)\,\vartheta_1$ as above. Thus, V_1 being the velocity of the

[1] Cf. Glazebrook, "On Optical Theories," *Brit. Assoc. Report*, 1882.

wave, and V the velocity of propagation in free aether, we have just as before

$$V_1^2 + (V_1 - v)^2 (\mu^2 - 1) = V^2,$$

giving very approximately $V_1 = V\mu^{-1} + v(1 - \mu^{-2})$, which is Fresnel's law.

The exact expression for V_1 merely modifies the first term of Fresnel's approximation by a correction involving $v^2(1 - \mu^{-2})$, which does not change sign with v; thus in the application to Michelson's second-order experiment there is no essential modification, and his negative result remains outside the scope of this analysis*. [See Part II, § 13.]

But the Fresnel formula is not thus obtained in exact form.

125. An important corollary to the present theory is suggested and confirmed by the experiments of Röntgen on the convection of excited dielectrics, mentioned above (§ 60). When a material dielectric is moved across an electric field, each ion of the group which constitutes one of its molecules produces its own convection current, composed partly of change of electric displacement in the surrounding free aether, but completed and made circuital by the actual convection of the ionic charge itself. When, as in Röntgen's experiment of a spinning dielectric, the configuration in space does not change by the motion, so that there is no displacement current in the surrounding aether, it is easy to see that the total electromagnetic effect is the same as if the dielectric were magnetized to an intensity which is at each point the vector product of its velocity of movement and its electric moment per unit volume, the latter being $(K - 1)/4\pi$ times the electric force at the place. We have just seen (§ 124) that this is in accord with the optical aspect of convection of transparent matter.

Convective effect of Röntgen's spinning dielectric plate.

I have much pleasure in expressing my deep obligation to Professor G. F. FitzGerald for a very detailed and instructive criticism of this paper with which he has favoured me. I have been much guided by his comments in revising the paper, and would have made still more use of them but for the length to which it had already run. I need hardly state, however, that he is not to be held responsible for any of the views herein expressed.

My best thanks are also due to Mr A. E. H. Love for a criticism at an earlier stage, from which I derived much advantage.

* This refinement is however nugatory. The Fresnel coefficient had been corrected by Lorentz to $1 - \frac{1}{\mu^2} - \frac{d \log \mu}{d \log \tau}$, on account of the change produced by convection in the period (τ) of the light relative to the dispersive material substance: and the correction has recently been verified very closely by the experiments of Zeeman.

39

A DYNAMICAL THEORY OF THE ELECTRIC AND LUMINIFEROUS MEDIUM.—PART II: THEORY OF ELECTRONS.

(*Abstract.*)

[*Proceedings of the Royal Society*, Vol. LVIII. (May 16, 1895) pp. 222–228.]

Rotational fluid aether model.

IN a previous paper on this subject[1], it has been shown that by means of a rotationally elastic aether, which otherwise behaves as a perfect fluid, a concrete realization of MacCullagh's optical theory can be obtained, and that the same medium affords a complete representation of electromotive phenomena in the theory of electricity. The ponderomotive electric forcives were, on the other hand, deduced from the principle of energy, as the work of the surplus energy in the field, the motions of the bodies in the field being thus supposed slow compared with radiation. It was seen that in order to obtain the correct sign for the electrodynamic forcives between current systems, we are precluded from taking a current to be simply a vortex ring in the fluid aether; but that this difficulty is removed by taking a current to be produced by the convection of electrons* or elementary electric charges through the free aether, thus making the current effectively a vortex of a type whose strength can be altered by induction from neighbouring currents. An electron occurs naturally in the theory as a centre or nucleus of rotational strain, which can have a permanent existence in the rotationally elastic aether, in the same sense as a vortex ring can have a permanent existence in the ordinary perfect fluid of theoretical hydrodynamics.

Difficulties of vortical currents: led to theory of free electrons; with natural models, that can exist in a rotational aether.

In the present paper a further development of the theory of electrons is made. As a preliminary, the consequences, as regards ponderomotive forces, of treating an element of current $\iota\delta s$ as a separate [unresolved] dynamical entity, which were indicated in the previous paper, are here more fully considered. It is maintained that a hypo-

[1] *Roy. Soc. Proc.* November 1893; *Phil. Trans.* 1894, A, pp. 719–822.

* The idea of electrons, so named at the suggestion of G. F. FitzGerald after G. J. Stoney's term, was introduced later in the complete previous memoir (*supra*, pp. 514 *seq.*) after the abstract (*supra*, p. 389) was published, and their inertia was discussed leading to a purely electric scheme of matter.

thesis of this kind would lead to an internal stress in a conductor carrying a current, in addition to the forcive of Ampère which acts on each element of the conductor at right angles to its length. Though this stress is self-equilibrating as regards the conductor as a whole, yet when the conductor is a liquid, such as mercury, it will involve a change of fluid pressure which ought to be of the same order of magnitude as the Amperean forcive, and therefore capable of detection whenever the latter is easily observed. Experiments made by Profs. FitzGerald and Lodge on this subject have yielded purely negative results, so that there is ground for the conclusion that the ordinary current element $\iota\delta s$ cannot be legitimately employed in framing a dynamical theory.

Amperean current element must give way to electronic:

FitzGerald: Lodge.

This result is entirely confirmed when we work out the properties of the field of currents, considered as produced by the convection of electrons. There can be no doubt that a single electron may be correctly taken as an independent element of the medium for dynamical purposes; so that electrodynamical relations deduced from a statistical theory of moving electrons will rest on a much surer basis than those derived from the use of a hypothetical current element of the ordinary kind, in cases where they are in discrepancy.

Now it is shown that an intrinsic singularity in the aether, of the form of an electron e, moving with velocity $(\dot{x}, \dot{y}, \dot{z})$ relative to the quiescent mass of aether, is subject to a force $e\,(P, Q, R)$, given by equations of the form

Electro-dynamic force deduced

$$P = c\dot{y} - b\dot{z} - dF/dt - d\Psi/dx;$$

in which (a, b, c) is the velocity of flow of the aether where the electron is situated, and is equal to the curl of (F, G, H) in such way that the latter is Maxwell's vector potential given by the formulae of the type [an unretarded potential because (u, v, w) is total current, not true flux (u', v', w') of electrons]

$$F = \int \frac{u}{r}\,d\tau + \int \left(B\frac{d}{dz} - C\frac{d}{dy}\right)\frac{1}{r}\,d\tau;$$

in terms of unretarded vector potentials:

and where Ψ is the electrostatic potential due to the electrons in the field, so that $\Psi = c^2\Sigma e/r$, where c is the velocity of radiation. These equations are proved to hold good, not merely if the motions of the electrons are slow compared with radiation, as in the previous paper, but quite irrespective of how nearly they approach that limiting value; thus the phenomena of radiation itself are included in the analysis.

An element of volume of an unelectrified material medium contains as many positive electrons as negative. This force (P, Q, R) tends to produce electric separation in the element by moving them in opposite directions, leading to an electric current in the case of a conductor whose electrons are in part free, and to electric polarization in the case of a dielectric whose electrons are paired into polar molecules. In the former case, the rate at which this force works on a current of electrons (u', v', w') is $Pu' + Qv' + Rw'$; it therefore is identical with the electric force as ordinarily defined in the elementary theory of steady currents. In the case of a dielectric it represents the ordinary electric force producing polarization. So long as a current is prevented from flowing, the ponderomotive force acting on the element of volume of the medium is the one of electrostatic origin due to such polarization as the element may possess, for as the element is unelectrified it contains as many positive electrons as negative. But if a current is flowing, the first two terms of (P, Q, R), instead of cancelling for the positive and negative electrons, become additive, as change of sign of the electron is accompanied by change of sign of its velocity; so that there is an electrodynamic force on the element of volume,

but it acts only on true current.

$$(X, Y, Z) = (v'c - w'b,\ w'a - u'c,\ u'b - v'a),$$

Law of mechanical force obtained.

where, however, (u', v', w') is the *true* current composed of moving electrons, not the total circuital current (u, v, w) of Maxwell, which includes the rotational displacement of the free aether in addition to the drift of the electrons.

Agreement with Maxwell's original forms derived from his model:

The electric force (P, Q, R) as thus deduced agrees with the form obtained originally by Maxwell[1] from the direct consideration of his concrete model of the electric field, with idle wheels to represent electrification. It has been pointed out by von Helmholtz and others, that the abstract dynamical analysis given in his *Treatise* does not really lead to these equations when all the terms are retained; this later analysis proceeds, in fact, by the use of current elements, which form an imperfect representation, in that they give no account of the genesis of the current by electric separation in the element of volume of the conductor.

dynamics by current elements defective.

The ponderomotive force (X, Y, Z) is at right angles to the direction of the true current, and is precisely that of Ampère in the ordinary cases where the difference between the true current and the total current is inappreciable. It differs from Maxwell's [formula] in involving true current instead of total current; that is, the forcive ends to move an element of a material body, but there is no such forcive tending to move an element of the free aether itself. In this respect it differs also from the hypothesis underlying von Helmholtz's recent treatment of the relations of moving matter to aether.

[1] Maxwell, "On Physical Lines of Force," *Phil. Mag.* 1861–62; *Collected Papers*, Vol. I. pp. 450–512.

When we treat of a single electron, (a, b, c) is the flow of the aether where it is situated. When we treat of an element of volume with its contained electrons, (a, b, c) becomes the smoothed out, or averaged, flow of the aether in the element of volume; it is circuital because the aether is incompressible, and thus it represents the magnetic induction of Maxwell.

Analogues of the magnetic vectors.

When magnetic polarization of the medium contributes to the forcive, it is necessary to divide (a, b, c) into two parts, one part (α, β, γ) contributed by the medium as a whole, and independent of the surroundings of the element, and the other representing the effect of the polarization in the immediate neighbourhood; the former part is, of course, the magnetic force of Maxwell. Similar considerations apply as regards the electric force in a polarized dielectric; it is clearly proper to define it so as to correspond to magnetic induction, not to magnetic force. It is then shown from the direct consideration of the orbital motions of electrons, that there is, in addition to the electrodynamic force on the element of volume of the material medium, a magnetic force derived from a potential function $\frac{1}{2}\kappa(\alpha^2+\beta^2+\gamma^2)$, and a force of electric origin derived from a potential $(K-1)/8\pi c^2 \,.\, (P^2+Q^2+R^2)$. If the element carries an electric charge of density ρ, there is also the force $\rho\,(P, Q, R)$. In addition to these latter forces on the polarized element, there are also stresses due to interaction between neighbouring parts, in which are to be found the main explanation of the phenomena of electrostriction and magnetostriction.

The magnetic force defined:

also the electric force within a dielectric.

Stresses:

As an example of these ponderomotive forces, the mechanical pressure produced by radiation is examined later on, with a result half that of Maxwell when the light is incident on an opaque body, and which gives pressures on the two sides of the interface each equal to Maxwell's expression multiplied by $\frac{1}{2}(1-\mu^{-2})$, when the interface separates two transparent media*.

relevant to optical theory.

The distinction between true current and total current is practically immaterial, except in questions relating to electrical vibrations and to optics. The remaining part of the theory is therefore developed more particularly with a view to optical applications. At the end of the previous paper a brief outline of the method of treating optical dispersion was given; and it was shown that the same principles led directly to Fresnel's formula for the effect on the velocity of light produced by motion, through the aether, of the material medium

Partial convection of light-waves.

* The Maxwell pressure, as the result of momentum convected by the radiation, is established for rays in free space: when they travel in a medium as here, part of the mechanical force is exerted on the electrons of the medium, and ambiguities arise, as here, which are cleared up by introduction of the stress-momentum tensor.

which transmits it. In the latter respect the theory is in agreement with a more recent discussion by H. A. Lorentz, of the propagation of electrical and optical effects through moving media.

Continuous theories of dispersion inadequate.

A detailed theory of optical propagation in transparent and opaque ponderable media is given, on the basis that it is the contained electrons that are efficient in modifying the mode of propagation from that which obtains in free aether. The dispersive theory of MacCullagh had been physically interpreted in the earlier part of the previous paper; but it appears from the same train of reasoning as was there applied to Cauchy's theory, that molecular magnitudes are too small compared with the wave-length to allow any considerable part of the actual dispersion to be accounted for statically in that way. The rotatory dispersions, both natural and magnetic, are, however, structural phenomena; and this accounts for their smallness compared with ordinary dispersion.

Molecular theory:

As regards ordinary dispersion, a formula is obtained for the case of perfectly non-conducting media, namely, $\mu^2 = 1 + A/(\beta^2 - p^2)$, where $2\pi/p$ is the period, of the same type as one recently deduced by von Helmholtz by an abstract process based on the principle of Least Action combined with a theory of electrons, which, however, does not correspond with the views here developed. That this formula is a good representation of the experimental facts for ordinary transparent media is generally recognized; especially as it may, in case of necessity, be modified by the inclusion of slight non-selective opacity, due to drift of free electrons, after the manner of ordinary conduction.

includes the special features in metals.

When this kind of general opacity is predominant, the result obtained in the paper conforms to the main features of metallic propagation; thus, with sufficient conductivity the real part of the square of the refractive index becomes negative, and the real part of the index itself may become less than unity, while the dispersion is usually abnormal.

The Michelson experiment.

When the phenomena of moving media are treated, dispersion may, for simplicity, be left out of account. It is shown that, if the view described in the previous paper, that all the dynamical properties of matter are to be derived from the relations of electrons, with or without intrinsic inertia*, in a rotationally elastic fluid aether, is entertained, the null result of the Michelson-Morley second-order experiment on the effect of the Earth's motion on the velocity of light becomes included in the theory; in fact, according to a suggestion thrown out by FitzGerald and Lorentz, and developed somewhat in

* The former alternative (of non-electric inertia) might, *e.g.*, include the positive ions or protons of recent theory; but it would not involve the Michelson effect without auxiliary hypotheses that other forces behave like electric forces.

this manner by the latter, the second-order optical effect is just compensated by a second-order effect on the lengths of the moving arms of Michelson's apparatus, which is produced by its motion along with the Earth through the aether.

As mixed dynamical and statistical theories of electrons or other objects require delicate treatment, especially when pushed, as here, to the second order of small quantities, the formulae of this part of the paper are deduced independently by two very different analytical methods. In the first place, there is the usual process of extending the fundamental circuital relations of the free aether which express its dynamical relations as differential equations of the first order, by suitable modification of the significance of the vectors involved in them, so that the same equations shall apply to ponderable media as well, the vectors then representing averages taken over the element of volume. The other method consists in working out the dynamics of a single electron, and applying the results statistically to the inclusion of the various ways in which the electric current arises from the movements of the electrons in ponderable media.

Analysis by averaging:

by dynamics of the electrons.

The theory as thus developed from the electron as the fundamental element, may be stated in a form which is *independent of* the dynamical hypothesis of a rotational aether. Maxwell's formal equations of the electric field may take the place of that hypothesis, though it may, I think, be contended that an abstract procedure of that kind will neither be so simple nor so graphic, nor lend itself so easily to the intuitive grasp of relations, as a more concrete one of the type here employed.

Rotational model not necessary: but useful.

The exact permanence of the wave-lengths in spectra, under various physical conditions, may be ascribed to the influence of radiation on the molecule, which keeps it in, or very close to, a constant [non-radiating] condition of steady motion, of minimum total energy corresponding to its pre-determined constant momenta. It is also pointed out, from the analogy of physical astronomy, that the harmonic oscillations into which the spectroscope divides the radiation from a molecule, may be far more numerous than the coordinates which specify its relative motions; that, therefore, relations of a semi-dynamical character may be discovered among the spectral lines, without its being rendered likely that we can ever penetrate from them back to the actual configuration of the molecular system [except in symbolic manner].

Atom must have non-vibrating configurations of minimal energy.

Reverting finally to matters relating purely to a rotational aether theory, with electrons as the sole foundation for matter, it is possible to identify the inertia of matter with the electric inertia of the electrons, if only we may assume their nuclei to be small enough, or

Inertia can be wholly electric.

sufficiently numerous. And the fact that these nuclei have free periods of elastic radial vibration in the fluid aether, not subject to damping by radiation, reminds us that a pulsatory theory of gravitation has been developed by Hicks and Bjerknes. There is no recognized fundamental interaction* of electric and radiative phenomena with gravitation, so for present purposes we are not bound to produce a precise explanation of gravitation at all. The scope of this remark is restricted to merely showing that a rotational aether is not incompetent to include such an action among its properties.

Gravitation might be pulsatory.

* Such interaction, now probably verified astronomically, is the most fundamental result of Einstein's recent brilliant syntheses.

40

A DYNAMICAL THEORY OF THE ELECTRIC AND LUMINIFEROUS MEDIUM.—PART II: THEORY OF ELECTRONS.

[*Philosophical Transactions of the Royal Society*, Vol. CLXXXVI, pp. 695–743. Received *May* 16, read *June* 20, 1895.]

The theory of electrons.

1. In a previous paper the concrete representation of electrical and optical phenomena by means of a rotationally elastic fluid aether has been discussed[1]. In an Appendix it has been shown that whenever there is direct question of interaction between the molecules of matter and the aether, whether it be in the phenomena of magnetism or in the optical phenomena of dispersion and moving material media, the consideration of groups of electrons or permanent strain centres in the aether, which form a part of, or possibly the whole of, the constitution of the atoms of matter, suffices to lead to a correlation of the various modes of activity; while this scheme seems to be free from the chief difficulties which have pressed on other methods of representation.

Electric currents must be essentially convective: their field.

The present paper is chiefly concerned with the further development of the molecular aspect of this theory*. As a preliminary, it is maintained that a dynamical theory of electric currents, based on the ordinary conception of a current element, must lead to expressions for the electrodynamic forces which are at variance with the facts. On the other hand, a theory which considers moving electrons to be the essential elements of the true currents in material media, gives a definite account of the genesis and the mutual relations of both types of forcive, the electromotive and the ponderomotive, and gives formulae for them, which correspond in the main with those originally deduced by Maxwell from consideration of the properties of his concrete model of the electric field, though they are not substantiated by his later abstract theory based [directly, cf. § 2] on current elements. That theory is held to be defective, in the first place

[1] *Phil. Trans.* 1894, A, pp. 719–822.

* At this date the theory of electrons had still to make its way on indirect evidence. Rapid free negative electrons were identified in the cathode stream and their mass determined as very small and uniform by J. J. Thomson in 1897; and their inertia soon became recognized as purely of electric origin, as the present models imply.

on account of the discrepancy with experiment above mentioned, and in the second place, because it is not competent to describe the mode of genesis of a conduction current by electrical separation produced in the element of volume of the conductor under the influence of the field of force; it is thus an incomplete formulation of the phenomena.

Optical development:

The application of the method of electrons to vibrational phenomena leads to formulae for optical dispersion, and optical propagation in metals, which are in general agreement with experimental knowledge. These subjects have already been treated from a similar point of view by von Helmholtz in 1892, but his equations, which are arrived at in a more abstract manner, are fundamentally at variance with those of the present theory. They lead, however, to the same type of dispersion formula when the medium is transparent, a type which has been generally accepted as in good agreement with exact measurements of dispersion.

in convected media.

The application to the optical properties of moving media leads to Fresnel's well-known formula, as had already been shown in the previous paper. If the theory of the constitution of matter which is suggested in that paper is allowed, it also leads to an explanation of the null result of the well-known second-order experiment of Michelson and Morley, of which previous theoretical discussions have quite failed to take cognizance.

As the statistical processes connected with molecular theory are of a very delicate character, and subject to suspicion especially when pushed to the second order of small quantities, the formulae are derived by two independent methods, which supplement and illuminate each other but are analytically of very different types. One of these methods, which lies nearer to recent procedure in electrodynamics, and has been used by Lorentz in a comprehensive discussion of the subject of moving media, published at the beginning of the present year, is to adapt the fluxes and forces occurring in the fundamental circuital equations of the free aether, so as to make similar equations apply to the aether in ponderable media. The other method is to investigate the forcive acting on a single electron, and then to build up the electric and optical phenomena out of the various types of movements of electrons that can occur in dielectric and conducting media. It is only in this latter way that a rational detailed account of the ponderomotive forces in media which are transmitting currents, or are electrically or magnetically polarized, can be derived; as a special example, a formula for the ponderomotive pressure due to radiation is obtained.

The mechanical forces: constructed on an electron basis.

The discussion of the topics above mentioned, with the exception of the Michelson-Morley experiment, is quite independent of any speculation as to the nature of electrons or the relation of aether to matter. It may be founded directly on Maxwell's equations of the electric field in free aether, as now experimentally confirmed.

The theory not dependent on any special aether model:

But when we go further, it seems to be a strong argument in favour of a rotational aether, or at any rate of an aether whose actual properties are represented with close fidelity by the scheme of a rotationally elastic fluid medium, that in this way we derive an actual physical structure for an electron, and that by the orbital motions of electrons in the atom we derive a representation of an atom as a fluid vortex, an idea which has always been present to physical speculation from Leucippus, through Descartes, down to the recent definite dynamical conceptions of vortices. To justify tentative adhesion to such a view, it is not necessary to be able to produce an explanation of the fundamental properties of mass and gravitation in matter: in particular these are not in necessary connection* with the electrical and optical phenomena. But it is easy to see that a rotational aether is not inadequate to including such properties among its relations: if the nuclei of the electrons are supposed small enough, the inertia of matter would be definitely represented by the electric inertia of the electrons; and as the electrons may then have vacuous nuclei and have each a free period of radial vibration in the fluid aether, which is not subject to damping by radiation (but subject, however, to a certain instability unless the free spherical form of surface is fortified by some kind of constraint), the gravitation between them may be represented or illustrated by the hydrodynamical pulsatory theory of Bjerknes and Hicks.

but the fluid rotationally elastic model can be instructive,

as regards nature of electrons and atoms.

Suggestion that atomic mass is wholly of the electric field. Pulsatory hydrodynamic attractions not excluded in the model.

Examination of Theories involving the Electrodynamic Potential Function.

2. Every scheme for reducing the phenomena of electric currents to a purely dynamical basis must start ultimately from the formula for the electrodynamic potential of a system of ordinary currents, discovered and first extensively applied by F. E. Neumann.

Formula for energy of a system of currents.

In a preliminary tentative discussion of this formula, the only course open appears to be to take the current flowing in an element of volume, and the coordinates of the element, as the independent variables of a continuous analysis extended over the field of currents.

If the function is treated merely as a potential of the forces of the field, we may employ the principle of Neumann, afterwards elaborated and extended by von Helmholtz, but still hypothetical in its dy-

Its variation, when regarded as a potential of the electromotive and the mechanical forces:

* This still holds: even on the Einstein hypothesis gravitation might be of arbitrary strength or non-existent.

namical aspect, that a variation of the potential, due to alteration of the positions of the conductors, but without any variation of the strength of the current flowing along any linear element of a tube of flow, leads to the ponderomotive forces tending to alter their positions: while a variation due to alteration of the strengths of the currents, the conductors being fixed or moving, leads to electromotive forces tending to change the strengths of the said currents, but not to alter the motion of the bodies.

but it is really kinetic energy of the field,

The dynamical theory of Maxwell aims at going further: the potential function with sign changed is assumed to be the kinetic energy residing in the latent or impalpable active medium which is associated with the current system. The induced electric forces are now taken to be reversed kinetic reactions corresponding to the currents considered as generalized electric velocity components, in accordance with the Lagrangian formula

$$-\left(\frac{\delta}{dt}\frac{dT}{d\dot{\theta}}-\frac{dT}{d\theta}\right);$$

and these together with the applied electromotive forces make up the total ones that drive the current in conformity with Ohm's law. In the same way the ponderomotive forces acting on the conductors are the reversed kinetic reactions corresponding to change of position of the material system, which require to be compensated by equal and opposite applied forces if mechanical equilibrium of the conductors is to subsist. It was one of the discoveries of Maxwell that, for a system of *circuital* conduction currents, the potential function can actually be formally represented as the kinetic energy of a latent moving system, coupled with the palpable conductors and thereby influencing them—of which system the currents are to be treated as generalized velocity components corresponding to electric coordinates which do not appear themselves in the function—without thereby introducing any discrepancy into the general scheme, as already experimentally determined, of the equations of the electric field.

as verified by Maxwell for complete current circuits.

Spatial specification of the energy of electric flow regarded as continuous:

Whichever of these methods of development of the potential function is essayed, it is a necessary preliminary with a view to a complete analysis to take the strength of the current flowing in each element of volume, or it may be in a linear element of a tube of flow, as dynamically a separate electric entity. The object of the following discussion is to consider how far this is consistent with a more detailed examination of the forcives thus derivable, particularly with the nature of that part of the internal stress in a conductor carrying a current which tends to alter its shape but not to produce motion of translation of the conductor as a whole.

3. According then to the type of theory which considers a current system to be built up of physical current elements of the form $(u, v, w)\,\delta\tau$, the energy associated with an element of volume $\delta\tau$, as existing in the surrounding field and controlled by the element, is

$$T = (Fu + Gv + Hw)\,\delta\tau.$$

The ponderomotive force acting on the element will be derived from a potential energy-function $-T$, by varying the coordinates of the material framework: it must in fact consist, per unit volume, of a force leads to a forcive differing from Ampère's,

$$\left(u\frac{dF}{dx} + v\frac{dG}{dx} + w\frac{dH}{dx},\ u\frac{dF}{dy} + v\frac{dG}{dy} + w\frac{dH}{dy},\ u\frac{dF}{dz} + v\frac{dG}{dz} + w\frac{dH}{dz}\right),$$

and a couple

$$(vH - wG,\ wF - uH,\ uG - vF),$$

the former being derived from a translational, the latter from a rotational virtual displacement of the element[1]. We may simplify these expressions by taking the axis of z parallel to the current in the element $\delta\tau$, so that u and v become null; then we have

$$\text{a force } \left(w\frac{dH}{dx},\ w\frac{dH}{dy},\ w\frac{dH}{dz}\right) \text{ and a couple } (-wG,\ wF,\ 0).$$

According to the Ampère-Maxwell formula, there should be simply a force at right angles to the current, specified by the general formula

$$(vc - wb,\ wa - uc,\ ub - va),$$

which becomes for the present special axes of coordinates

$$\left\{-w\left(\frac{dF}{dz} - \frac{dH}{dx}\right),\ w\left(\frac{dH}{dy} - \frac{dG}{dz}\right),\ 0\right\}.$$

The forcive at which we have here arrived thus differs from the Ampère-Maxwell one by

$$\text{a force } \left(u\frac{dF}{dz},\ v\frac{dG}{dz},\ w\frac{dH}{dz}\right) \text{ and a couple } (-wG,\ wF,\ 0);$$

these are equivalent to forces acting on the ends of each linear current element, equal at each end numerically to (wF, wG, wH) per unit of cross-section, positive at the front end and negative at the rear end. They are thus of the nature of an internal stress in the medium, and are self-equilibrating for each circuital current and so do not disturb the resultant forcive on the conductor as a whole due to the field in which it is situated. From Maxwell's stress standpoint they would form an equilibrating addition to the stress specification in the conductor which is the formal equivalent of the electrodynamic forcive. by an internal mechanical stress,

[1] In the previous paper, § 120 [*supra*, p. 529], the couple was omitted.

According to the Ampère-Maxwell formula, the forcive on an element of a linear conductor carrying a current is at right angles to it, so that the tension along the conductor is constant so far as that forcive is concerned. The traction in the direction of the current, arising from the above additional stress, would introduce an additional tension, equal to the current multiplied by the component of the vector potential in its direction, which is not usually constant along the circuit, and so may be made the subject of experimental test with liquid conductors, as it would introduce differences of fluid pressure. There will also be an additional transverse shearing stress which should reveal itself in experiments on solid conductors with sliding contacts.

now shown to be non-existent experimentally,

In particular these additional forces should reveal themselves in the space surrounding a closed magnetic circuit, where the ordinary Amperean force vanishes because the magnetic field is null; in that case (F, G, H) may be interpreted as the total impulsive electric force induced at any point by the making of the circuit. Professor G. F. FitzGerald has devised an experiment in which the behaviour of a thread of mercury carrying a strong current and linked with a complete magnetic circuit was closely observed when the circuit was made and broken. No movement was detected, whereas, when the magnetic circuit was incomplete, the ordinary Amperean forces were very prominent. According to the above analysis, the two types of forcive should be of the same order of magnitude in such a case: the result of the experiment is therefore against this theory. A like negative result has also attended an experiment by Professor O. J. Lodge, in which he proposed to detect minute changes of level along the upper surface of a uniform mercury thread by an interference arrangement on the principle of Newton's rings: when the current was turned on, the section of the thread became more nearly circular owing to the mutual attractions of the different filaments of the current, but there was no alteration in the direction of its length.

by FitzGerald,

by Lodge.

A theory of mobile electrons avoids such stress.

This experimental evidence, combined with the fact (*infra*) that a theory of moving electrons, which are certainly independent physical entities, leads simply to the Ampère-Maxwell forcive, seems to justify the conclusion that either the above analysis is wrong, or else the ordinary treatment of electrodynamics in terms of a specification by current elements is physically untenable. As tending to the exclusion of the first alternative, and also as of independent critical interest, the following discussion by aid of the more usual analytical method employed by Ampère, Neumann, and von Helmholtz is given.

4. Assuming that current elements serve as a sufficient physical specification, it is known that the mutual energy (kinetic) of two such elements, $\iota\delta s$ and $\iota'\delta s'$, must be

An alternative analysis by Amperean methods.

$$\iota\delta s \,.\, \iota'\delta s' \left(\frac{1}{r}\frac{dr}{ds}\frac{dr}{ds'} + \frac{d^2\phi(r)}{ds\,ds'}\right),$$

where $\phi(r)$ is some function of their distance apart.

The variation of this energy, due to mutual displacement of the elements, is

$$\iota\delta s \,.\, \iota'\delta s' \left(-\frac{\delta r}{r^2}\frac{dr}{ds}\frac{dr}{ds'} + \frac{1}{r}\frac{d\delta r}{ds}\frac{dr}{ds'} + \frac{1}{r}\frac{dr}{ds}\frac{d\delta r}{ds'} + \frac{d^2\delta\phi(r)}{ds\,ds'}\right).$$

On substituting

$$\frac{d}{ds}\left(\frac{1}{r}\frac{dr}{ds'}\,\delta r\right) - \frac{d}{ds}\left(\frac{1}{r}\frac{dr}{ds'}\right)\delta r \text{ for } \frac{1}{r}\frac{d\delta r}{ds}\frac{dr}{ds'},$$

and similarly for the following term, the variation becomes

$$\iota'\delta s' \,.\, \iota\delta s\,\frac{d}{ds}\left(\frac{1}{r}\frac{dr}{ds'}\,\delta r\right) + \iota\delta s \,.\, \iota'\delta s'\,\frac{d}{ds'}\left(\frac{1}{r}\frac{dr}{ds}\,\delta r\right) + \iota\delta s \,.\, \iota'\delta s'\,\frac{d^2\delta\phi(r)}{ds\,ds'}$$
$$+ \iota\delta s \,.\, \iota'\delta s'\left(\frac{1}{r^2}\frac{dr}{ds}\frac{dr}{ds'} - \frac{2}{r}\frac{d^2 r}{ds\,ds'}\right)\delta r.$$

If we take the aggregate for a series of elements $\iota'\delta s'$ which form a circuital current, the second and third terms vanish; and the forcive on the element $\iota\delta s$, due to a circuital current system, is thus compounded of

$$\iota\delta s \,.\, \iota'\delta s'\left(\frac{1}{r^2}\frac{dr}{ds}\frac{dr}{ds'} - \frac{2}{r}\frac{d^2 r}{ds\,ds'}\right)$$

towards each element $\iota'\delta s'$ of the influencing current system, which is the ordinary forcive of Ampère, together with a force acting on each end of $\iota\delta s$, positive on the forward end and negative on the rearward end, and for each end compounded of $\iota'\delta s' \,.\, \iota r^{-1} dr/ds'$ acting towards each element $\iota'\delta s'$ of the influencing system. The components of the aggregate of this latter [terminal] force, by which the present result differs from the Amperean one, are

$$\iota\int \iota'\frac{x}{r^2}\frac{dr}{ds'}\,ds', \quad \iota\int \iota'\frac{y}{r^2}\frac{dr}{ds'}\,ds', \quad \iota\int \iota'\frac{z}{r^2}\frac{dr}{ds'}\,ds',$$

where (x, y, z) are the coordinates of the element $\iota'\delta s'$ with reference to an origin situated at the element under consideration $\iota\delta s$. On integration by parts round the complete current circuit these become

$$\iota\int\frac{\iota'}{r}\frac{d}{dx}\left(r\frac{dr}{ds'}\right)ds', \quad \iota\int\frac{\iota'}{r}\frac{d}{dy}\left(r\frac{dr}{ds'}\right)ds', \quad \iota\int\frac{\iota'}{r}\frac{d}{dz}\left(r\frac{dr}{ds'}\right)ds',$$

that is, by definition, $\iota F, \iota G, \iota H,$

which is the same expression as we had previously arrived at.

Apart from detailed explanation, the previous mode of variation may be cast into the briefer analytical form

Procedure by formal variation.

$$\delta\iota\,(F\,dx + G\,dy + H\,dz) = \iota\left(\frac{dF}{dx}\,dx + \frac{dG}{dx}\,dy + \frac{dH}{dz}\,dz\right)\delta x + \ldots + \ldots$$
$$+\ \iota\,(Fd\delta x + Gd\delta y + Hd\delta z)$$

of which the last term

$$= \iota \mid F\delta x + G\delta y + H\delta z \mid - \iota\delta x\left(\frac{dF}{dx}\,dx + \frac{dF}{dy}\,dy + \frac{dF}{dz}\,dz\right) - \ldots - \ldots,$$

so that the whole

$$= \left\{\left(\frac{dG}{dx} - \frac{dF}{dy}\right)\iota dy - \left(\frac{dF}{dz} - \frac{dH}{dx}\right)\iota dx\right\}\delta x + \ldots + \ldots$$
$$+ \mid \iota F\delta x + \iota G\delta y + \iota H\delta z \mid,$$

which represents the work of the Amperean forcive on the linear element together with a traction $\iota\,(F, G, H)$ on its ends.

Discrimination between Velocities and Momenta in Generalized Dynamics.

Currents contrasted with continuous vortices in fluid.

5. In the dynamics of systems whose internal connections are only partially known, it is essential to have a clear view of the circumstances which determine whether the various quantities which enter into the specification of the energy are to be classed as coordinates, velocities, or momenta. For example, in determining the forcives between cores in problems of cyclic motion, the circulations must be treated as generalized momenta, while in the Maxwellian electrodynamics of complete circuits the currents flowing in them are rather to be classed as velocity components. The basis of the distinction between these two classes of quantities is of course fundamental; and it is to be found in the way in which they occur in the Hamiltonian analysis. The essential property of a velocity is that it is a perfect differential coefficient, with respect to the time; any function involving rate of change of configuration, which enjoys this property so that its time integral is a function of position only, may be taken to be a velocity, provided we, if need be, contemplate also a corresponding forcive. On the other hand, any such function of the rate of change of configuration, even though it be a perfect differential with respect to the time must be treated as a momentum if it is known to remain constant with the time while no applied forcive controls it; for if it were a velocity, linked up with other velocities, its constancy in the free motion could not usually fit in with the analytical theory.

Dynamical specification of the field of currents: identifying the generalized velocities: and momenta.

Latent cyclic momenta may be involved.

In the theory of cyclic fluid motion, the circulations, being constant, must thus be taken as momenta; and when the energy is expressed in terms of them, it must be modified before the forcives can be derived from it in the manner of Lagrange and Hamilton. In the theory of electrodynamics the currents are not unalterable with the time; and as they are differential coefficients with respect to the time of definite physical quantities, the charges of electricity, they may be taken as velocities, provided we recognize the play of corresponding forcives. In the electrodynamics of complete circuits, after Maxwell, there is no reason, in that theory taken by itself, why the functions designated as the electrokinetic momenta should not be taken as velocities instead*, if so desired; for they satisfy all the above conditions, though of course the corresponding forcives would be of quite different types from the usual ones. This remark is in illustration of the fact that the distinction between momenta and velocities is to a certain extent one of convenience.

The problem of permanent magnetism.

A theory which would regard the current flowing in a complete circuit as a generalized momentum of the latent aethereal motion labours under many difficulties. Such a hypothesis seems tempting at first sight, and, for example, lends itself easily to the inclusion of permanent magnetism in a dynamical system: it has accordingly been developed, amongst others, by von Helmholtz, who formulated the equations on this basis in his Memoir on Least Action. The electrodynamic potential function would then represent the kinetic energy modified in Routh's manner by the substitution of electric momenta for velocities, which in the present case simply changes its sign: and this would seem to require that the kinetic energy of the electric field should in ordinary circumstances be negative, which could hardly be correct.

Forces between vortices the opposite of those between currents.

In the analogous problem of cyclic irrotational motion in fluids, with the circulation guided by rigid ring-cores, the circulations which correspond formally to the electric currents remain constant in the free motion, and so must be analytically momenta, and not velocities. The energy in this vortex problem is given by the same formula as in the electric problem, but it must be modified by change of its sign before the forcives are derived from it; and thus the circumstance, first emphasized by Lord Kelvin, that the forcives between such rigid ring-cores are equal but opposite to the forcives between the analogous rigid conductors carrying electric currents, must find its explanation in the fact that the electric currents are of the type of velocities, while the fluid circulations are of the type of mementa.

6. In the dynamics of the circulation of a fluid through ring-shaped solids, the expression for the energy in terms of vortex filaments,

$$T = \tfrac{1}{2}\Sigma \iint \sigma_1 \sigma_2 r^{-1} \cos \epsilon \, ds_1 \, ds_2,$$

* Provided radiation can be neglected as in ordinary electrodynamics; or provided the velocity of its waves is taken as infinite, when it will not be present at all.

represents its proper distribution with respect to each solid as a whole, but does not represent its distribution as regards each element of a vortex filament. Thus, when the solids are treated as rigid bodies, and are referred to the appropriate number of coordinates (six for each), this expression for the energy is competent to determine the aggregate forcive on each solid. But it is not competent to express the actual forcive on an element of the solid, because, in its formation by integration from the actual distribution of kinetic energy throughout the elements of volume of the fluid, a process of integration round the apertures of the solids has been employed, which supposes that each element of a vortex filament is connected with other elements so as to be part of a complete circuit.

Elimination of internal stresses in general dynamical method:

In any such case the distribution of the resultant forcive throughout the solid has an undetermined part which is of the nature of an internal stress. In the example of cyclic fluid motion there is however only needed, in addition to this expression for the energy, the property that the forcive on the solids is distributed as a fluid pressure over their surfaces, in order to obtain from this form of the energy a quite definite result. The distribution of forcive on a flexible core is thus at each point at right angles to its axis, and the tension in such a core is therefore constant all along it.

applied to a system of currents regarded as continuous.

If we attempt to deduce electrodynamic forcives, in cases of circuital electric flow, from the cognate expression for the energy, with current in place of vorticity but with different sign, the applicability of the method is limited in like manner. An additional hypothesis, or experimental principle, that the electrodynamic forcive on an element of a conductor is at right angles to the current which it carries, will, however, make the problem definite*.

But if we were to remove the restriction that the flow of true electricity (*i.e.* of electrons) must be circuital, and to treat for each element of the conductor the strength of the current as dynamically an electric velocity corresponding to a single definite electric coordinate which does not explicitly appear, the energy-function would now be adequate by itself to determine the distribution of the forcive, and the result would be as given above, § 3. The discrepancy between this result and experiment thus seems to show that it is not legitimate to consider the current as a generalized velocity distributed over the elements of the conductor and independent for each of them. Nor is this in any way surprising, when we bear in mind that the velocity of a *single* electron is certainly of the type required for a generalized

* Physical current elements are constituted of drifting electrons: for the mutual energy of two such elements, from which their forces are derivable, see p. 523 *supra*.

dynamical velocity component, and when we infer, as below (§ 23), that the forcive on a current considered as made up statistically from the moving electrons is different from what would be obtained by taking the current element itself as a generalized velocity component. Our conclusion is, then, that the only proper basis for the dynamical analysis of the phenomena of currents flowing in conductors, in fact, of all cases of the flow of true electricity, is to treat the currents as the statistical aggregates of the movements of the electrons.

Dynamically a flow of electrons is demanded.

Two different Methods of Analysis.

7. With a view to the analytical formulation of the properties of an aether pervaded by electrons, it is necessary in the first place to take a survey of the various ways in which these electrons are distributed. In statical circumstances the great bulk of them are grouped together into polar molecules, which may be either in totally irregular orientation as in an unpolarized dielectric, or may possess features of regularity which can be represented in a statistical manner by the type of theory first developed by Poisson with respect to induced magnetism. If the element of volume possesses an electric charge, it must also contain free electrons not so grouped; but their number is excessively small compared with the former class.

Static polar molecular distributions: Poisson, Kelvin.

Proceeding now to kinetics, the motions of these various classes of electrons constitute true currents of various types. A drift of free electrons constitutes an ordinary conduction current, and Faraday's fundamental law of electrolysis shows that when there is an electrolyte (or a dielectric) in the circuit of a steady current, the current must be made up half by a drift of positive electrons*, and half by a drift of negative ones. The simplicity of the relations of a steady circuital current is due to the fact that it involves circulation, but not strain, in the surrounding aether. As a sub-class there is the so-called convection current, due to the transfer of a charge of electrons along with a material body, in its motion through the aether. There is also the current of displacement when the polar dielectric is excited, consisting of a drift within small range of the opposed positive and negative electrons in each molecule, owing to its orientation by the electric field. There is also a current due to convection of an excited dielectric, which in the case of uniform velocity may be represented as formally equivalent to a magnetization of the medium. Then again there is the molecular current, due to the orbital motions of the electrons in the molecule; so long as we avoid the dynamics of molecular structure we can only consider the time average of this current, and that is most conveniently represented as the magnetic

Currents: of conduction as essentially electronic convection: of convection of a charged body: of changing polarization: of convection of a polarized dielectric: Amperean intra-atomic currents:

* Rather atoms or ions: the electron was isolated free by J. J. Thomson in 1897 and shown to have a mass much smaller than any atom.

polarity of the molecule; the displacement current in a dielectric above mentioned involves a similar process of averaging with respect to time. There is also the current of convection of electrons which re-arranges the electric charge of a conductor. Of this last practically nothing is known*; the subject has wholly eluded experimental inquiry, except in the case of electrolytes, where the modern theories of solution and electrolysis, in the hands of Hittorf, Kohlrausch, Arrhenius, and Nernst, have yielded at any rate the first approximation to an analytical theory. As a matter of fact, the empirical and variable character of conductivity as depending on physical state, and its overwhelming importance as compared with the other phenomena in conductors of metallic type, restrict within narrow limits the problems in which a knowledge of their relations of convection and polarization would be of practical import.

current of re-arrangement of charge.

Finally, there is one case in which an experimentally recognized result in the medium involves the dynamics of the molecule; when the magnetic field is increased the orbital motions in the molecule are altered so as to oppose the change, and this induced alteration in the molecular current constitutes diamagnetism.

Diamagnetism a universal quality.

8. In the face of all these various types of motion of electrons that have to be taken into account, there are two distinct modes of procedure open to us; and they have both been used by Maxwell, without perhaps that sharp demarcation between essentially different methods which is necessary to the perfect grasp of either of them.

We may start from the dynamical equations of the free aether, which are known, and which apply exactly to a sub-element of volume which does not contain any electrons; we can make use of the various forces and fluxes which have proved useful in forming a representation of electrical phenomena, and by their help we can replace these dynamical equations which are of the second order in differentiations, by the two circuital relations which are each of the first order. The advantages, both for analysis and for explanation, of a transformation of this kind have long been recognized in pure dynamics. Now we can pass, by integration of these relations, from the sub-element of volume which contains no electrons, to the effective element of volume which contains a number of electrons large enough to enable us to smooth out their individual peculiarities and so retain a continuous differential analysis. The transformation is most conveniently effected so that the circuital relations shall remain unaltered in form, while the meanings of the various quantities that enter into them

The smoothed-out circuital equations of the field.

* In metal it may be presumed to be a current of ordinary conduction, due to an initial distribution of electric force which becomes uncompensated when the electrons on which it is anchored are released.

are modified by taking account of the presence of the distribution of electrons in the enlarged element of volume. This method of smoothing out the molecular discreteness in the aether is the one that has been most in favour in recent years[1]; but when we come to the consideration of moving media and convection currents, and particularly of ponderomotive forces, it will be necessary to supplement it by the more direct procedure now to be described.

The other method of analysis involves the utilization of the ideas of the electrodynamic potential in an amended form. It is concerned primarily with the dynamics of a single electron in an electric field given as regards motion and strain, or kinetic and potential energy; it forms the kinetic energy T, and the potential energy W, of the electron, and thence its Lagrangian function or reversed electrodynamic potential $T - W$. The relations between the motion and the forcives of the electron are now to be deduced by dynamical methods; and there only remains a process of summation in order to determine the equations that are appropriate to the various modes of co-ordinated groupings and movements of electrons that have been enumerated above. This process has a fundamental theoretical superiority over the previous one, in that it includes in its scope the ponderomotive forces which act on the bodies in the field. The various discussions in Maxwell's *Treatise*, which proceed by use of the vector potential of electrodynamic induction, are to be classed as essentially related to this point of view.

The underlying dynamics of the electron:

proceeds by aid of Maxwell's vector potential.

Method of averaged Forces and Fluxes: the Circuital Relations: Optical Dispersion.

9. The aether is to be regarded as containing a distribution of electrons, that is of intrinsic centres or nuclei from each of which a configuration of rotational strain spreads out into the surrounding space. An additional strain may be considered as imparted to this medium in two distinct ways, (i) as the result of tractions applied

Electrons as the strain-nuclei of their fields:

[1] It appears to have escaped notice that the expression of the equations of the electric field in the form of the two circuital relations, together with linear equations (involving the constitution of the medium) which connect the corresponding forces and fluxes in these relations, had been given by Maxwell himself. At the end of the Memoir "On a Method of Making a Direct Comparison of Electrostatic with Electromagnetic Force...," *Phil. Trans.* 1868, he alludes to the difficulty that had been felt in grasping the basis on which the electric theory of light is founded; and he proceeds to attempt to diminish this difficulty by setting down the formally simplest foundation of hypothesis which will lead to the theory at which he had previously arrived in a more dynamical and inductive manner. The form which he thus gives to the theory is precisely the one into which its equations have more recently been thrown by Heaviside and by Hertz. Aug. 25.

Formulated originally by Maxwell himself.

over an outer boundary surface, the electrons being supposed held fixed, and (ii) as the result of movement of the electrons, each of which will carry its atmosphere of strain along with it, practically without alteration unless the velocity of the electron is so great as to approximate to the velocity of radiation. A rotational strain imparted in the first manner is circuital from its nature, as depending on absolute rotation of the element of aether. But the circuital character of a strain of the second kind is vitiated by the existence of the nuclei or intrinsic singularities that belong to it; the surface integral of the rotation, taken over any interface, remains constant and therefore keeps its initial zero value only so long as no electrons cross that interface. But we can retain the circuital property for this kind of strain also, if we associate an ideal rotational displacement with the movement of each electron, in such manner that the integral of this rotation taken (vectorially) throughout a small volume including the initial and final positions of the electron is equal to the strength of the electron multiplied by its linear displacement. This is expressed electrically by saying that a moving electron constitutes a convection current, which must be added to the flux of elastic rotation in the aether in order to obtain a total current which shall possess the circuital property. But this convection current is, so far, only a kinematic fiction; it allows us to retain the circuital relation between the total electric current and the velocity of the aether, but it is not to be counted in estimating the stress in the aether, from which the static dynamical effects result.

which can be regarded as static for ordinary travelling electrons.

Current of aethereal displacement not circuital,

unless there is added to it the true electric current of electronic convection: thus giving the Maxwellian *total* current, which is thus a formal conglomerate.

In ordinary electrodynamics, what we may call the *true current*, whether of convection or conduction, is measured in this way per unit volume, by the aggregate directed drift of the electrons in the element of volume. It is completed or rendered circuital by the *displacement current*, or rate of change of rotational strain, in the aether; but in most cases of ordinary occurrence the magnitude of this displacement current is so very small compared with that of the true current, that it is allowable to neglect the former, and therefore make the true current flow in complete circuits. The *total current* of Maxwell, which consists of both parts taken together, is exactly circuital in all cases.

The *true* current, as in electrotechnics.

The aethereal *quasi*-current, sensible only in electric radiation.

In circumstances of conduction, though the electric displacement (*i.e.* rotational strain) in the medium is absolutely negligible, yet the drift of the electrons which constitutes the true current causes an irrotational flow of the medium (the magnetic field) which is related to the current in precisely the same way as the flow in a perfect fluid is related to the vortex filaments which suffice to specify it; the energy of the current system is thus the kinetic energy of this irrotational

The magnetic field imaged as hydrodynamic flow, therefore irrotational:

flow, the rotational flow arising from strain being in comparison inappreciable. As here primarily introduced, the true current was a fiction, so far as elastic stress in the medium is concerned; but it has now acquired an objective meaning as the mathematical quantity that serves to completely specify the energy of the flow of the medium which is associated with movement of electrons in bulk, that is, the energy of the magnetic field. The ordinary electrodynamics of conduction currents is a dynamical problem of the aether in which the kinetic energy is a function of the true current, and the potential energy, when there is such in the field, is a function of the rotational strain of the medium, that is, of the aethereal displacement current only.

is determined by the current system, when radiation is absent:

thus cyclic, being its kinetic manifestation.

10. We proceed to apply these ideas to the comparatively simple circumstances of the mode of transmission of regular vibrations by a medium thus constituted. Using the ordinary vector notation for brevity, let $\mathfrak{D}$ denote the actual rotational strain in the aether, so that $d\mathfrak{D}/dt$ is the displacement current; let $d\mathfrak{D}'/dt$ denote the true current due to movement of electrons, and let $\mathfrak{B}$ denote the velocity of the flow of the aether, that is, the magnetic induction. Each of these quantities is supposed to have its averaged value per unit volume, the irregularity of distribution due to the presence of electrons in the element of volume under consideration having been smoothed out in the analysis. The kinematic relation, which the introduction of the true current was here primarily intended to conserve, is

The vectors concerned.

$$\operatorname{curl} \mathfrak{B} = 4\pi \frac{d}{dt}(\mathfrak{D} + \mathfrak{D}').$$

Again, as the electric force at any point in the aether is $4\pi c^2\mathfrak{D}$, c being the velocity of radiation, [$4\pi c^2$ thus being the reciprocal of the elasticity that is involved] the ordinary dynamical equation of the free aether,

$$-\frac{d\mathfrak{B}}{dt} = 4\pi c^2 \operatorname{curl} \mathfrak{D},$$

applies exactly throughout any sub-element of volume which does not contain electrons. On summing up for all the sub-elements which go to make up the ordinary element with its contained electrons, we derive the same equation in which $\mathfrak{B}$ and $\mathfrak{D}$ are now defined as averages taken over the volume of the element, as in the kinematic equation above.

It remains to specify the relation between $\mathfrak{D}$ and $\mathfrak{D}'$. For the case of a dielectric medium in which the positive and negative electrons are combined into polar molecules so that in each molecule they exactly compensate each other, the statical effect of an applied electric force, $4\pi c^2 \mathfrak{D}$, is to induce a polarity, $\mathfrak{D}'$, so that the total circuital electric displacement is $\mathfrak{D} + \mathfrak{D}'$; thus

$$\mathfrak{D} + \mathfrak{D}' = K/4\pi c^2 \,.\, 4\pi c^2 \mathfrak{D} = K\mathfrak{D},$$

where K is a quantity which enters in electrostatics as the inductivity of the complex medium; we take it to be a constant independent of $\mathfrak{D}$, for sufficient reason *à priori*, at any rate for small values of $\mathfrak{D}$, such as occur in ordinary electric vibrations. If we suppose that the periods of the free oscillations of the electric polarity of a molecule are very high compared with the periods of the vibrations of the medium that are under consideration, this equation will still hold for the problem of a vibrating medium, as an equilibrium theory will apply to the molecules. But when the medium shows dispersive quality, its vibrations must excite sensibly the independent vibrations of the molecules, and the equation of equilibrium for the induced polarity of the molecules must involve the kinetic reaction of their vibrations. Thus

Theory of slow electric change.

$$\mathfrak{D} - \frac{\mathfrak{D} + \mathfrak{D}'}{K} = \frac{i}{K}\frac{d^2\mathfrak{D}'}{dt^2},$$

where i/K is a coefficient of inertia, of a kind which we do not need at present to further particularize. As the element of volume resists rotation, the uncompensated statical forcive tends to produce absolute rotation—not relative strain as in the case of an elastic solid—and is used up in accelerating it.

Spatial equations valid except for X-rays,

This is the general form of the relation between $\mathfrak{D}$ and $\mathfrak{D}'$ when vibrations are transmitted across a dielectric medium which contains a large number of molecules to the wave-length; the actual number for light-waves, about 10^3 for solids and liquids and 10^2 for gases, is amply sufficient to justify the process of averaging which has been employed. To retain the present simple form, the analysis must, however, be restricted to molecules with a single efficient free period. And the hypothesis implicitly involved, that no portions of free aether (supposed perfectly continuous, *i.e.* not itself molecular) are so effectively enclosed by surrounding molecules as to have efficient free periods of their own, involves an assumption that the volume occupied by the nuclei of electrons is small compared with the whole space. If this were not the case, the relation connecting $\mathfrak{D}$ and $\mathfrak{D}'$ would contain second differential coefficients of $\mathfrak{D}$ with respect to time, as well as of $\mathfrak{D}'$, so that in the formula for K' in § 11 (*infra*), the first term would no longer be unity. First differential coefficients of $\mathfrak{D}$ are always excluded because they would imply viscosity in the free aether.

which may be in part due to free vibrations in aether regions within the atom, like sound in resonators.

As yet the true current is a current of dielectric molecular displacement which is perfectly reversible, so there is no dissipation of energy such as would cause absorption of the vibrations in the medium and consequent opacity. This will, however, occur should the paired electrons of the polar molecules occasionally get separated from each other when under the action of the electric force; such electrons would then travel different ways across the field and thus give rise to a conduction current. It is definitely known that this is what happens in the case of electrolytes under experimental conditions; and it is very probable (§ 23 *infra*) that the transfer of electricity by metals is of a similar nature. It is therefore necessary that we should include a similar agency in the case of light propagation; only it must be borne in mind that under a force which reverses with such tremendous rapidity, there are not the same opportunities for diffusion of momentum by action between contiguous molecules as there are in cases of steady force, so that the coefficient of opacity must be expected to be very much smaller than the conductivity of the substance as ordinarily measured by aid of Ohm's law.

Opacity implies ionization:

but with reduced effective conductances for rapid periods.

Propagation in Metals.

11. In circumstances of statical strain, the relation between the total strain $\mathfrak{D}$ communicated to a given portion of the aether from its surroundings, and the polarization $\mathfrak{D}'$ of the molecules that are excited by it, is $\mathfrak{D} + \mathfrak{D}' = K\mathfrak{D}$.

Under vibrational circumstances there will be a kinetic reaction, $id^2\mathfrak{D}'/dt^2$, which will take part in this equilibrium, and the corresponding equation is

Kinetic reaction of aether:

$$i\frac{d^2\mathfrak{D}'}{dt^2} + \mathfrak{D}' = (K - 1)\,\mathfrak{D}.$$

When there are free electrons in the medium, as well as polar molecules, there will also be a conduction current; if m be the translational inertia of a free electron e, the electric force being $4\pi c^2\mathfrak{D}$, its equation of motion will be $m\ddot{x} + \sigma'\dot{x} = e\,.\,4\pi c^2\mathfrak{D}$, not indeed in the literal sense that there is a frictional force $\sigma'\dot{x}$ acting on each electron, but in the sense that on the average of a large number of electrons the part of the electric force that is not used up in accelerating their motion is spent in maintaining their steady drift, σ' being thus a coefficient of resistance to migration through the medium, whose value may, however, be dependent on the period; the conduction current $\mathfrak{C}$ will therefore be of the type

of its electrons.

$$\mathfrak{C} = k\,(md/dt + \sigma')^{-1}\,\mathfrak{D},$$

Equations of metallic conduction:

where $k = eN \cdot 4\pi C^2$, N being the number of ions of mass m per unit volume.

of propagation.

The circuital relations which express the properties of the aether will be

$$\frac{1}{4\pi}\operatorname{curl}\mathfrak{B} = \frac{d}{dt}(\mathfrak{D} + \mathfrak{D}') + \mathfrak{C}$$

$$-\frac{d\mathfrak{B}}{dt} = 4\pi C^2 \operatorname{curl}\mathfrak{D}.$$

It is thus only the first of these relations that is modified; but we may reduce it back again to the standard form which obtains for free aether. Eliminating $\mathfrak{D}'$ from it, and writing $-p^2$ for d^2/dt^2, so that $2\pi/p$ represents the period of the vibrations, we have

$$\frac{1}{4\pi}\operatorname{curl}\mathfrak{B} = \frac{d}{dt}\left(1 + \frac{K-1}{-ip^2+1}\right)\mathfrak{D} + k\,\frac{-md/dt + \sigma'}{m^2p^2 + \sigma'^2}\,\mathfrak{D},$$

which is now of the standard form

$$\frac{1}{4\pi}\operatorname{curl}\mathfrak{B} = K'\frac{d\mathfrak{D}}{dt},$$

provided

$$K' = 1 + \frac{K-1}{1-ip^2} - \frac{km}{m^2p^2+\sigma'^2} - \frac{kp^{-1}\sigma'\iota}{m^2p^2+\sigma'^2}, \quad k = eN \cdot 4\pi C^2,$$

Effective dielectric constant.

one constant K' characteristic of the metal being thus a complex quantity.

The interfacial conditions are, as usual, that the tangential components of $\mathfrak{D}$ and $\mathfrak{B}$ are continuous (the medium being non-magnetic), so that this theory [also necessarily linear] differs from the one applicable to transparent media simply in the complex value of K'.

Law of dispersion:

When the medium is perfectly transparent the square of the refractive index assumes the form $1 + A/(\beta^2 - p^2)$, which agrees with a formula [Sellmeier's] given by von Helmholtz and generally recognized to be a good representation of the dispersive properties of ordinary media; as that author has remarked, it may be helped out of any outstanding discrepancy by assuming slight non-selective opacity due to ordinary conduction, such as would not sensibly affect Fresnel's laws of reflexion. In this theory there is only one absorption band in the ultra-violet, corresponding to $p = \beta$; if, however, the molecule has several free periods, the square of the index must still, irrespective of special theory, be a rational function of p^2, and therefore when expressed in partial fractions must assume the form $1 + \Sigma A_r/(\beta_r^2 - p^2)$; while in the case of opaque media the constants typified by A will be complex, and other partial fractions not corresponding to absorption bands will also enter.

is determined by the absorption bands, except for general opacity.

In the case of opaque media, a peculiarity of this analysis, as compared with that of von Helmholtz[1] and others, is that the conduction current is derived from the electric displacement by the inverse operator $k\,(md/dt + \sigma')^{-1}$ instead of a direct operator involving a coefficient of conductivity in place of the coefficient of [direct frictional] resistance σ'. Notwithstanding that there are as many as four adjustable constants, it is possible, to some extent, to crucially compare the resulting formula for K' with the experimental facts for metallic media. It is known that the broad type of formal theory [Cauchy, MacCullagh] which simply assigns to the metals a complex index of refraction $K'^{\frac{1}{2}}$ is confirmed by the general agreement between results deduced from calculations relating to reflexion experiments by Beer, Voigt, Drude and others, and those obtained directly from deviation experiments with thin metallic prisms by Kundt[2]. It turns out, however, that the real part of K' is invariably negative for metallic media; and this is a fundamental difficulty in ordinary elastic theories, as it implies instability of the optical medium. On the present theory it implies that k is sufficiently large to allow the third term of K' to outweigh the first two terms[*]. It is also found (by both

Features of the metallic refractive indexes:

[1] Von Helmholtz, "Electromagnetische Theorie der Farbenzerstreuung," Wied. *Ann.* XLVIII. 1893. The analysis of von Helmholtz consists, as usual with him, in a tentative process of fitting known electric laws into a minimum theorem which is an extended form of the Principle of Least Action. For this purpose he uses two sets of variables to represent what we have here called the true current and the displacement current of the free aether; and he varies them independently of each other. There is no distinction drawn between the polarization and the conduction parts of the true current; so that the current of conduction appears in the potential energy-function, thus being assumed to imply elastic strain of the medium, in opposition to the views that have been here set forth. It has been shown by Reiff (Wied. *Ann.* I. 1893, p. 361) that the theory of von Helmholtz does not lead to Fresnel's formula for the influence of moving media on the velocity of propagation, unless the aether is supposed partially to partake of the motion of the material medium; but that when certain terms are omitted from his potential energy-function, it is not necessary to assume that the aether is moved with the matter. But it does not appear that any reason is assigned for such modification of the theory, which, as already remarked, seems to be intrinsically at variance with the view here taken.

Critique of Helmholtz's formulation.

[2] Kundt, *Phil. Mag.* 1888 (2). It had been already shown by Voigt (Wied. *Ann.* XXIV. 1885), that the ordinary optical formulae for prismatic deviation apply when the metallic prism is of very minute angle: also proved in Part III, *infra*, § 34.

[*] The first two terms of K' amount to a small multiple of unity: the third is less than k/mp^2, where $k = eN \,.\, 4\pi c^2$, $p = 2\pi c/\lambda$, which gives for it $Ne/m \,.\, \lambda^2\pi$, while for electrons e/m is now known to be $\frac{7}{4}10^7$, for yellow light λ is $\frac{1}{2}10^{-5}$, giving $\frac{1}{4}N10^{-3}$. Thus if the density of ions N were even as small as 10^5 the third term would still much predominate, and determine a negative sign for the real part of K'. As the effective value of N must be vastly greater, we must, to avoid excessive values of K', take the time of free path of the electron to be much smaller than the period of the radiation, so that its motion cannot be free as here assumed in its equation but impeded by frequent collisions.

methods) that if $K'^{\frac{1}{2}} = n(1 - \iota\kappa)$, then for the better conducting metals, silver, gold, copper, n is less than unity, involving velocity of propagation greater than in a vacuum, while κ is a considerable number[1]; this implies that the fourth term in the formula for K' is in these cases small compared with the third, which is just what is to be expected from the smaller value of the resistance coefficient σ'. This point is confirmatory of the present scheme; for if the conduction current were derived from the displacement $\mathfrak{D}$ by a direct operator involving conductivity instead of an inverse one involving resistance as here, the opposite effect would be indicated.

ready confirmation for the noble metals.

Further confirmations.

When the third and fourth terms in K' are predominant, the dispersion will usually be in the abnormal direction, as it is known to be for the great majority of metals, a few of the more conducting ones being, however, exceptions; and the value of κ will usually increase with the frequency, which is also in accordance with fact.

Electric instability avoided only by introducing free inertias.

(Aug. 25. The introduction of effective inertia into the equation of conduction, as above, thus appears to be essential to the theory. For the actual negative value of the real part of K' in metals cannot arise from dielectric polarity of the material medium; that, as has often been remarked, would imply instability and consequent destruction of such polar structure. To make the other agency which is at work, namely conduction, effective for that purpose, its equation must involve inertia: and the influence of this inertia appears in fact also in other ways; thus the work done by the electric force on the ions in an electrolyte must be used up proximately in accelerating their velocities, while the increased average velocity reveals itself as Joulean heat. If there were a large number of dissociated ions along a wave-length, which is indeed the condition that the above continuous analysis be literally applicable, it is easy to see by consideration of molecular magnitudes that the effective inertia of an ion would have to be very much greater than its actual mass, or else the effect would be excessive*. But this difficulty is only apparent. When the ions are more thinly scattered through the medium, there will be two sets of waves propagated; the waves of free aether modified somewhat by the presence of the ions, but not extremely different from what they would be if the ions were held fixed in the medium; and much slower waves propagated from ion to ion with the intervening aether nearly in an equilibrium condition at each instant.

On the number of free ions:

[1] Cf. the numbers quoted from Drude in Professor J. J. Thomson's *Recent Researches*..., § 355.

* See previous footnote. The influence of free ions, few in number, is in fact the cause of the bending of the *long* electric waves of wireless signalling round the Earth's surface. See *Phil. Mag.* Dec. 1924, or *infra*. The alternative to high inertia is obstructed motion.

The former class alone would be sensibly excited by optical means: it may be formally represented by the above scheme of equations with the inertia coefficient large, and it is wide enough to include the phenomena both of transparent dielectrics and of opaque metallic media. It is found that, for the wave-lengths of luminous radiation, the real and imaginary parts of the square of the index of refraction are of about the same order of magnitude, for all metals. This possibly indicates that the depth to which the light can penetrate in the metal, and therefore also the coefficient of absorption, depends essentially on the ratio of molecular magnitudes to wave-length, which would be about the same for all.)

influence on short optical waves.

Refraction distinct from Dispersion.

12. In the former paper, a physical foundation [rotationally elastic] was assigned for MacCullagh's theory of dispersion. That theory being a statical one, must rest on the discreteness of the medium being comparable with the wave-length of the radiation; but considerations similar to those given in that paper (§ 123) in connection with Cauchy's theory show that the number of molecules in the wave-length is too great to allow this cause to account for the magnitude of the actual dispersion. Thus by far the greater part of ordinary dispersion, as distinct from refraction, is to be assigned to the sympathetic vibrations of the molecules as here discussed.

Dispersion dependent on free periods:

(Aug. 25. For waves of long period and therefore great length, p approaches the value zero; thus K' approaches the limiting value K and a statical theory represents the phenomena, the molecules being at each instant in the equilibrium position corresponding to the strained state of the surrounding aether. In any case, we may conveniently designate this constant part of K', the square of the index, by the name of the *refraction*, and the variable part, which depends on the period and ultimately vanishes when the period is long, by the name of the *dispersion**. In the very wide class of media for which the specific inductive capacity, that is the square of the index of refraction for very long waves, is nearly equal to the square of the index for ordinary light-waves, the dispersion is thus small

in general is small.

* The general theory of dispersion is developed *infra* in Part III of this memoir on the general analogy of sources of electric vibration with sources in the theory of sound (Stokes), sympathetic vibration and emission alone being regarded as involved. This holds for antennae and other vibrators constituted of matter in bulk. But a wider theory, involving definite capacities for storage of energy, is now found to be necessary for molecular sources: and much progress has been made towards its formal expression, on the initiative of N. Bohr and involving Planck's quanta, on the basis of the experimental laws of spectra.

MacCullagh's static formulation of dispersion in crystals:

compared with the refraction. In all such cases a statical theory of refraction, which, according to the argument of the previous paper, must [formally] be MacCullagh's theory, will certainly be correct; and there is ground for making this conclusion general. Thus, in particular, MacCullagh's theory of the double refraction in crystalline media will hold good as the first approximation. But just like ordinary refraction, this crystalline refraction is subject to dispersive variation when the wave-length of the light is altered. This dispersion of the optic axes in crystals is usually small compared with the double refraction itself, which justifies the present mode of treatment of it as a subsidiary effect to be joined to the main part of the double refraction. In order to obtain equations of propagation in which it shall be included, we have only to add to the effective coefficient of inertia of the aether in MacCullagh's equations a subsidiary aeolotropic part. This part may be complex instead of real when the medium is not perfectly transparent, thus including the effect of conduction arising from the presence of free ions; the general equations of absorbing (pleochroic) doubly refracting media may in fact be formulated without difficulty on the lines of § 11.

preserves Fresnel's wave-surface.

For transparent media, this generalization of MacCullagh's theory preserves the wave-surface, corresponding to any given period of the light, exactly Fresnel's. The character of the laws of crystalline reflexion also remains unaltered, but the constants that are involved in them are no longer exactly the same constants that occur in the equation of the wave-surface.

Dispersion an essential feature of rotatory polarization.

The asymmetric refraction of higher order, which evidences itself by rotation of the plane of polarization, is obviously of a structural kind, and so is correctly represented by MacCullagh's terms. It is itself a highly dispersive phenomenon, on account of the higher differential coefficients on which it depends; thus the dispersion due to variation of its constants with the wave-length may usually be neglected in comparison.)

Influence of Motion of the Medium on Light Propagation.

Propagation of light in convected bodies.

13. To deduce Fresnel's law for moving media directly from these principles we have to remember that a movement of the material dielectric through the aether with velocity v parallel to the axis of x produces an additional displacement current at any point fixed in the aether, which (§ 31, *infra*) is equal to $vd\mathfrak{D}'/dx$. Thus in Ampère's circuital relation $d\mathfrak{D}'/dt$ must be replaced by $(d/dt + vd/dx)\,\mathfrak{D}'$. Again, from the mode in which the other circuital relation appears in the dynamical theory of the medium, $d\mathfrak{B}/dt$ must mean the total acceleration of velocity of the aether, due in part to change of time

and in part to movement of the material dielectric; thus this d/dt also, when it operates on $\mathfrak{D}'$, must be replaced by $d/dt + v\,d/dx$. In the present connection we may neglect dispersional phenomena and so take $\mathfrak{D} + \mathfrak{D}' = K\mathfrak{D}$. Thus finally

$$c^2\nabla^2\mathfrak{B} = \frac{d^2\mathfrak{B}}{dt^2} + (K - 1)\left(\frac{d}{dt} + v\frac{d}{dx}\right)^2 \mathfrak{B},$$

which leads directly to Fresnel's formula as in § 124 of the previous paper.

14. A clear appreciation of what is involved in the unexplained null result of Michelson and Morley may be obtained by transferring this equation of propagation to coordinates fixed in the moving material system. Let us assume new coordinates x', t' given by

The type of an equation of propagation.

$$x' = x - vt, \quad t' = t + \alpha x,$$

while y and z are unaltered. Then

$$\frac{d}{dx} = \frac{d}{dx'} + \alpha \frac{d}{dt'}, \quad \frac{d}{dt} = \frac{d}{dt'} - v\frac{d}{dx'}.$$

Thus the equation of propagation becomes, on dropping the accents,

$$\left(\frac{d}{dt} - v\frac{d}{dx}\right)^2 \mathfrak{B} + (K - 1)(1 + \alpha v)^2 \frac{d^2}{dt^2}\mathfrak{B}$$
$$= c^2\left\{\frac{d^2}{dy^2} + \frac{d^2}{dz^2} + \left(\frac{d}{dx} + \alpha\frac{d}{dt}\right)^2\right\}\mathfrak{B}.$$

This will be again of the type of an equation of simple wave-propagation if the coefficient of $d^2\mathfrak{B}/dx\,dt$ vanishes, that is if $v = -c^2\alpha$; it then becomes

Convection inoperative up to the first order.

$$\left(1 - \frac{v^2}{c^2}\right)\left\{1 + (1 - K^{-1})\frac{v^2}{c^2}\right\}K\frac{d^2\mathfrak{B}}{dt^2} = c^2\left\{\frac{d^2}{dy^2} + \frac{d^2}{dz^2} + \left(1 - \frac{v^2}{c^2}\right)\frac{d^2}{dx^2}\right\}\mathfrak{B}.$$

This means that, referred to a standard of time that varies from point to point, corresponding times at different points being those for which $t - vx/c^2$ is the same, the propagation relative to the moving material homogeneous isotropic medium is precisely the same as if that medium were at rest with respect to the aether[1], provided it be taken slightly aeolotropic, in the sense of being less elastic in the direction of the velocity v than in other directions, the difference depending on v^2/c^2, and so being excessively minute. This different reckoning of time at different points does not affect the paths of the

Outstanding second-order effect.

[1] Fresnel's law for the effect of motion of the medium is involved in this statement, as has been shown by Lorentz, *loc. cit.* § 39 *infra*.

rays, which are determined by Fermat's principle of minimum time of transit; thus the paths of the rays relative to the moving material medium are precisely the same as if that medium were fixed, and at the same time made very slightly aeolotropic. But so long as this aeolotropy is not compensated by some other second-order correction, the time of passage of a ray from one point to another and back again will depend on the direction as well as the length of the line joining these points, which is in opposition to the experimental result.

FitzGerald—Lorentz: annulled completely by a shrinkage in space if steady in time:

Following up a conjecture already thrown out by G. F. FitzGerald and by H. A. Lorentz[1], we can find this compensating second-order correction in a change of the dimensions of a material body produced by motion through the aether. For this purpose we must assume, in conformity with the considerations already given in the previous paper (§ 115), that material systems are built up solely out of singular points in the aether which we have called electrons and that atoms are simply very stable collocations of revolving electrons. Now the equation already given is the one which governs phenomena, when the axes of coordinates as well as the material system are moving through the aether with velocity v parallel to the axis of x. If we write $x' = x\,(1 - v^2/c^2)^{-\frac{1}{2}}$, it assumes the form

$$\left(1 - \frac{v^2}{c^2}\right)\left\{1 + (1 - K^{-1})\frac{v^2}{c^2}\right\} K \frac{d^2\mathfrak{B}}{dt^2} = c^2 \left(\frac{d^2}{dx'^2} + \frac{d^2}{dy^2} + \frac{d^2}{dz^2}\right) \mathfrak{B};$$

Formulation of the exact correlation:

and this is now of the same type as if the axes of coordinates were not moving. That is, if we know an actual configuration, steady or varying, of matter at rest in the aether, we derive from it a configuration of matter moving through the aether with velocity v by compressing it in the direction of this velocity in the ratio of $(1 - v^2/c^2)^{\frac{1}{2}}$ to unity, and at the same time adopting a standard of time which varies from point to point so that each point has its own origin from which time is reckoned*. Now the movable framework of Michelson's experiments forms a steady configuration of matter when it is at rest, the element of time only entering in connection with molecular motions; hence when it is moving, the condition of continued steadiness requires that it should change its steady dimensions in such way that dx becomes dx'. Thus when linear measurements are estimated with respect to this moving framework the equation which governs the relative phenomena is of precisely iso-

giving the Michelson result.

[1] Cf. especially Lorentz, *loc. cit.* § 92, who suggests a possible explanation which comes to the same as the one here advanced. It would necessitate an annual inequality of excessively minute amount, in the length of the sidereal day, owing to the Earth's orbital motion.

* See Appendix II at end of this volume.

tropic type; and the negative result of Michelson is explained. This transformation is independent of K altogether; it will therefore hold good equally if the atoms are supposed to have intrinsic inertia in addition to the electric inertia which arises directly from the relations of the electrons to the surrounding aether.

Cognate transformation possible for an atomic medium.

If this argument be allowed, it follows (§ 80, previous paper) that movements of perfectly transparent matter, if without rotation, cannot affect the course of a train of light-waves which passes across that matter, in conformity with the result previously deduced from the law of entropy. It is to be noted, however, that if the radiation exerted a resultant mechanical forcive on the moving matter so as to do work on it in its motion, the argument from entropy would break down. This illustrates the limitations of that kind of argument; but in the present case we shall find (§ 28) that the argument is justified, the resultant forcive in question being null when the matter is perfectly transparent.

Convection not detectable optically, up to second order.

Radiation pressures not here relevant.

Method of Separate Electrons.

15. In the Appendix to the previous paper [*supra*, p. 514], the dynamics of the mutual actions of electrons was briefly sketched, under the restriction that the velocities of the individual electrons* did not approach the velocity of radiation. If these velocities do not exceed say one per cent. of the velocity of light, the strain in the aether will at each moment be practically in the steady equilibrium state which it would have exactly were the velocity of radiation infinite, and the whole circumstances of the field will depend solely on the coordinates which specify the positions of the electrons. In an approximate theory of this kind the phenomena of radiation cannot, of course, themselves be included.

Introduction of electrons:

theory simple for slow motions, neglecting radiation.

To obtain an exact theory, free from this restriction, it will be necessary to express the distribution of energy in the aether in terms of coordinates of a kind which will allow us to include among them the position coordinates of the various electrons; we must, therefore, transform the kinetic energy into an expression involving change of the rotational strain (f, g, h) in the aether instead of its translational velocity $(\dot{\xi}, \dot{\eta}, \dot{\zeta})$. This procedure will be of fundamental importance, because the relation of matter to aether must resolve itself into the relation of electrons to aether, and the various forcives between aether and matter must be derivable in that way.

Now extended by suitable choice of variables,

to express their fields.

* Positive as well as negative electrons are implied throughout: but the actual absence of positive except in association with high inertia involves no essential change.

If, for facility of interpretation, we adhere to the ordinary electromagnetic system of units, we have

$$T = \frac{1}{8\pi}\int(\dot\xi^2 + \dot\eta^2 + \dot\zeta^2)\,d\tau$$

$$= \frac{1}{8\pi}\int\left\{\left(\frac{dH}{dy} - \frac{dG}{dz}\right)\dot\xi + \left(\frac{dF}{dz} - \frac{dH}{dx}\right)\dot\eta + \left(\frac{dG}{dx} - \frac{dF}{dy}\right)\dot\zeta\right\}d\tau$$

$$= \frac{1}{8\pi}\int\left\{F\left(\frac{d\dot\zeta}{dy} - \frac{d\dot\eta}{dz}\right) + G\left(\frac{d\dot\xi}{dz} - \frac{d\dot\zeta}{dx}\right) + H\left(\frac{d\dot\eta}{dx} - \frac{d\dot\xi}{dy}\right)\right\}d\tau,$$

on integration by parts, a surface integral over the infinite sphere being neglected, as usual, because the energy is all considered to be in a finite region. Now, in free aether, the curl of the velocity of the aether $(\dot\xi, \dot\eta, \dot\zeta)$, divided by 4π, is equal to the displacement current; in ponderable media it is, on the other hand, equal to the total current composed of this displacement current and the drift of electrons, that total current being circuital in accordance with this relation. If the total current be (u, v, w), we have, therefore,

$$T = \tfrac{1}{2}\int(Fu + Gv + Hw)\,d\tau.$$

In these formulae $(\dot\xi, \dot\eta, \dot\zeta)$ is the curl of the vector (F, G, H), so that

$$\nabla^2 F - \frac{d}{dx}\left(\frac{dF}{dx} + \frac{dG}{dy} + \frac{dH}{dz}\right) = -4\pi u,$$

with two similar equations. The unique solution of these equations satisfying the condition $dF/dx + dG/dy + dH/dz$ null* is

$$(F, G, H) = \int(u, v, w)\,r^{-1}\,d\tau.$$

Thus (F, G, H) is the vector potential of the total current, including, when necessary, Amperean molecular currents; and

$$T = \Sigma\Sigma\,(uu' + vv' + ww')\,r^{-1}\delta\tau\delta\tau'.$$

If an electron e, moving with velocity v, say $(\dot x, \dot y, \dot z)$, constitute a part of the total current, we have, setting out at length all the part of the energy depending on e,

Energy of interaction of an electron and the aether:

$$T = \tfrac{1}{2}Le^2(\dot x^2 + \dot y^2 + \dot z^2) + (e\dot xF + e\dot yG + e\dot zH) + \ldots.$$

Hence the motion provides a kinetic reaction which must be balanced by stress in the aether, just in the same way as the reversed mass-acceleration of a moving body balances the applied forcive by

* Here throughout (F, G, H) is the static vector potential of Maxwell, which includes that of the fictitious aethereal current. If, with more recent writers, it were the potential of the ions alone, it would have to be propagated, and instead of being null its divergence would be $c^{-2}/dVdt$. Cf. *supra*, p. 235. This contrast with Maxwell's usage has been a source of misunderstandings. In either case the electric field vector is of restricted type.

d'Alembert's principle; and the component of this kinetic reaction on the electron which has reference to translation along x, is in this case, by Lagrange's dynamical equation*,

$$P_1' = -\left(\frac{\delta}{dt}\frac{dT}{d\dot{x}} - \frac{dT}{dx}\right)$$

$$= - Le^2\ddot{x} - e\left(\frac{\delta F}{dt} + \dot{x}\frac{dF}{dx} + \dot{y}\frac{dG}{dx} + \dot{z}\frac{dH}{dx}\right),$$

in which δ/dt implies total rate of variation, due both to the change of time and to change of position of the electron in that time, so that

$$\frac{\delta}{dt} = \frac{d}{dt} + \dot{x}\frac{d}{dx} + \dot{y}\frac{d}{dy} + \dot{z}\frac{d}{dz}.$$

Hence finally,

$$P_1' = - Le^2\ddot{x} + e\left(c\dot{y} - b\dot{z} - \frac{dF}{dt}\right),$$

where $$(a, b, c) = \text{curl}\ (F, G, H).$$

It is simplest to split this up into two parts, and to say that the motional forcive (P_1, Q_1, R_1) typified by

leads to the motional forcive on the electron,

$$P_1 = e\left(c\dot{y} - b\dot{z} - \frac{dF}{dt}\right),$$

will go on increasing the acceleration of the electron, of effective mass Le^2, until the motion has brought the system of electrons into such a configuration that this forcive is compensated by traction due to the elastic strain of the surrounding aether, or in case of currents of conduction by Ohmic resistance to diffusion.

and its inertial reaction.

16. The potential energy of the system resides wholly in the aether and is

The strain energy:

$$W = 2\pi c^2 \int (f^2 + g^2 + h^2)\, d\tau.$$

The variation of W will give the statical forcive which would work in altering (f, g, h): but to this forcive must be added a constraining forcive required to maintain the condition of constraint between f, g, and h, which is represented by

$$\frac{df}{dx} + \frac{dg}{dy} + \frac{dh}{dz} = \rho,$$

where ρ is the density of free electrons. This constraint will be included if we conduct the variation of the equivalent expression

its variation in relation to the electrons,

$$W' = 2\pi c^2 \int (f^2 + g^2 + h^2)\, d\tau - \int \Psi \left(\frac{df}{dx} + \frac{dg}{dy} + \frac{dh}{dz} - \rho\right) d\tau,$$

* For the systematic formulation of this and the following argument from the foundation of Minimal Action see *Aether and Matter* (1900), Ch. VI.

and afterwards determine suitably this function of position Ψ. Now

$$\delta W' = \int\left\{\left(4\pi c^2 f + \frac{d\Psi}{dx}\right)\delta f + \ldots + \ldots\right\} d\tau + \int \Psi\,\delta\rho\, d\tau - \int \Psi\,(lf + mg + nh)\, dS,$$

after the usual integration by parts. Hence we have the forcive on free aether

giving a static forcive on the free aether:

$$(P, Q, R) = \left(4\pi c^2 f + \frac{d\Psi}{dx},\ 4\pi c^2 g + \frac{d\Psi}{dy},\ 4\pi c^2 h + \frac{d\Psi}{dz}\right).$$

involving a potential, to be suitably adjusted, which gives a force on each electron.

And we also observe, from the form of $\delta W'$, that Ψ represents the potential of the distribution of electrons in the sense that a change of density in the volume $\delta\tau$ increases the statical energy by $\Psi\delta\rho\delta\tau$; so that the movement of an electron e is assisted by a force

$$-e\left(\frac{d}{dx},\ \frac{d}{dy},\ \frac{d}{dz}\right)\Psi.$$

Its *rationale*:

To understand how this latter forcive works, let us consider an electron, in the electric field represented by a rotational strain (f, g, h) in free aether. The energy from which the traction is derived is

$$W = 2\pi c^2 \int (f^2 + g^2 + h^2)\, d\tau,$$

where in the immediate neighbourhood of the electron

the field near the electron,

$$(f, g, h) = \left(-\frac{e}{4\pi}\frac{dr^{-1}}{dx} + f',\ -\frac{e}{4\pi}\frac{dr^{-1}}{dy} + g',\ -\frac{e}{4\pi}\frac{dr^{-1}}{dz} + h'\right),$$

the first terms representing the part of the field due to the electron itself. Thus expressing only the terms which represent the interaction between the electron and the extraneous part of the field (f', g', h'),

local part of the energy.

$$W = -c^2 e \int\left(f'\frac{d}{dx} + g'\frac{d}{dy} + h'\frac{d}{dz}\right) r^{-1}\, d\tau + \ldots$$

$$= -c^2 e \int (lf' + mg' + nh')\, r^{-1} dS + c^2 e \int\left(\frac{df'}{dx} + \frac{dg'}{dy} + \frac{dh'}{dz}\right) r^{-1} d\tau + \ldots$$

$$= c^2 e \Sigma \frac{e'}{r} + \ldots$$

[assuming a state practically steady as regards aethereal displacement], provided the nucleus of the electron sensibly maintains its spherical form*. The resultant traction of the medium over the surface of the electron is obtained by varying W, and is therefore a forcive

Static resultant forcive on the electron:

$$-e\left(\frac{d}{dx},\ \frac{d}{dy},\ \frac{d}{dz}\right)\Psi, \text{ where } \Psi = c^2\Sigma\frac{e'}{r},$$

which helps towards compensating the kinetic reaction aforesaid.

* This restriction is violated very slightly by the shrinkage arising from convection: it has been explained, *supra*, p. 522, how that accounts for the discrepancy between the historical magnetic-field mass and the energy mass as derived directly from the energy.

This result has been obtained by separating away the part of the stress which increases indefinitely as the electron under consideration is approached, and which therefore represents the field of that electron itself in the neighbouring aether. It is noteworthy that the resultants of the stresses between the electrons at finite distances apart are represented simply by their electrostatic attractions, whether the field of aether is in equilibrium or is disturbed in any manner whatever [provided radiation is not involved to a sensible degree]; a result which depends [not] on the smallness of the nuclei, and the consequent intensity of the permanent strain near their surfaces.

is due to the intense adjacent aether stress thus separated off.

The forcive thus derived, partly kinetic and partly static, is named the electric force, because in the collocation of polarized molecules and free electrons forming an unelectrified body, it tends to produce electric separation by driving the positive electrons one way and the negative ones the opposite way, while it has no tendency to move the element of volume as a whole. But if there is an excess of one kind of electrons over the other in an element of volume of the body, so that the element has a charge q, it also provides a force, equal to q multiplied by its intensity, acting on the element of volume of the charged body. This mechanical forcive may in certain cases be represented as the unbalanced part of a stress in the manner of Maxwell's stress in the medium; but in our present order of ideas it is the force itself that is the reality, and the stress is only a mathematical mode of representing it, which has no physical significance as it does not represent the actual stress either in the aether or in the material medium*.

The electric force as distinct from aethereal: involves velocity of the set of electrons on which it acts.

The Maxwell quadratic stress, as a very convenient synthesis.

In one case this analysis must go deeper, namely in deducing the traction on an element of surface of the vacuous core of an electron. One mode of procedure would be to imagine the aether to be continuous throughout the core, but unstrained inside it, thus avoiding internal boundaries—a method which has been already applied (previous paper, § 51) in more intricate cases. We must then imagine the electric charge as freely distributed over the surface of the core in order to maintain this state of equilibrium, and the traction will simply be the usual forcive on this electric charge, namely, a normal pull equal to $2\pi\sigma^2$ per unit area, where σ is its surface density, as above [in electrostatic units]. The same result would be more directly

The shell type of model of an electron.

* Its elegant and very fundamental origin from variation of the Action, as the scheme naturally expressive of the mechanical forces across a medium whose internal equilibration has already been secured, is sketched after Part III. Cf. also *Proc. R. S. Edin.* 1927. A distribution of quadratic aethereal momentum is also involved, as well as stress, the fourfold symmetrical invariant stress-tensor of the relativity formulation being the final outcome.

derived by varying the energy in the aether with respect to a displacement of the surface of the core.

Kinetic part of aethereal force:

17. It remains to find the electrokinetic part of the forcive tending to increase the electric displacement (f, g, h) in an element $\delta\tau$ of free aether. The kinetic energy associated with the element is

$$T_2 = (\dot{f}F + \dot{g}G + \dot{h}H)\,\delta\tau;$$

therefore $$P_2 = -\frac{\delta}{dt}\frac{dT_2}{d\dot{f}} = -\frac{dF}{dt}.$$

No difference is here introduced by motion of the aether such as represents magnetic induction: in fact we have seen that the kinetic reaction on an electron e moving along with the aether is $-edF/dt$ simply, and to a doublet moving in this way the electric displacement in the element of aether itself may be assimilated. This forcive (P_2, Q_2, R_2) strains the aether, and is compensated jointly by the stress which it thus calls forth and the kinetic reaction of the motion involved in change of strain.

its reaction.

Discrimination of the contributions to total current:

18. We have now to formulate the intensity (u, v, w) of the total circuital current in an extended body, as made up of conduction current, polarization current, aethereal displacement current, convection currents of free electrons and of polarized molecules, and the current which produces redistributions of electric charges in a conductor. If the vector $\mathfrak{D}'$ denote the aggregate result of the polarization in the body, made up of orientation of the electrons with the aethereal displacement bound to them, and $\mathfrak{D}$ the free displacement in the aether which excites it, the Poisson-Mossotti theory of polarization will give a relation $\mathfrak{D} + \mathfrak{D}' = K\mathfrak{D}$, where the constant K denotes specific inductive capacity; thus of the total displacement $\mathfrak{D} + \mathfrak{D}'$, the fraction $1 - K^{-1}$ is bound to the polarization of the molecules and only the fraction K^{-1} is free elastic displacement.

the elastic aethereal part.

When an element of volume of a conductor moves across a magnetic field with velocity (p, q, r), the force (P, Q, R) acting on its contained electrons, supposed at rest in it, is of the type

The electric force, on the convected electrons.

$$P = cq - br - \frac{dF}{dt} - \frac{d\Psi}{dx},$$

acting on the positive electrons one way, on the negative ones the reverse way. When the electrons are in motion forming a current, it might at first sight appear that the ones that drift in the same direction as the conductor moves would experience the greater force, and so carry most of the current. But no difficulty of this kind occurs; for it is only the component of the velocity of the conductor at right angles to the direction of the current that is effective, and

The conduction current not affected by convection.

this acts equally on both sets of electrons. When a material body is in motion, the force tending to produce electric separation of its electrons, that is, the electric force exerted in it, is given by this formula whether it carries a current or not. Strictly, the velocity (p, q, r) which occurs in this formula is relative to the aether, so that it will involve the velocity of the aether itself which [on the present aether model] constitutes the magnetic field: but it follows from optical experiments that the latter is extremely minute in comparison with ordinary velocities of material bodies.

Velocities primarily relative to the aether (except for relativity).

On the other hand, the force tending to produce free rotational displacement in the aether is (P', Q', R'), where

$$P' = -\frac{dF}{dt} - \frac{d\Psi}{dx}.$$

The former force acts on $\mathfrak{D}'$, this one on $\mathfrak{D}$. Under steady circumstances the corresponding parts of the current must have for x component

Formula for total current in a dielectric:

$$\frac{K-1}{4\pi c^2}\frac{\delta P}{dt} + \frac{1}{4\pi c^2}\frac{dP'}{dt};$$

in which $\frac{\delta}{dt}$ represents $\frac{d}{dt} + p\frac{d}{dx} + q\frac{d}{dy} + r\frac{d}{dz}$, as appears from the fact that mere convection of a steady polarization by motion of the matter through the aether itself constitutes a current*.

The total current is thus (u, v, w) given by

for the total intensity of current in general.

$$u = \sigma P + \frac{K-1}{4\pi c^2}\frac{\delta P}{dt} + \frac{1}{4\pi c^2}\frac{dP'}{dt} + p\rho + u_0,$$

where $$4\pi\rho = \frac{dP'}{dx} + \frac{dQ'}{dy} + \frac{dR'}{dz} = -\nabla^2\Psi;$$

this is on the hypothesis that the material polarization is all induced and therefore circuital, and so adds nothing to the convergence of (P', Q', R'). If the velocity of the material medium is supposed uniform, we have

Permanent polarization absent.

$$\frac{dP}{dx} + \frac{dQ}{dy} + \frac{dR}{dz} = -4\pi(pu + qv + rw) - \nabla^2\Psi;$$

hence, when the material medium is at rest, the condition of circuitality of the current (u, v, w) is

No body-charge in a conductor at rest.

$$0 = \left(4\pi\sigma + Kc^{-2}\frac{d}{dt}\right)\rho - \frac{d\rho}{dt};$$

so that, if initally there is any volume density of electrification, it at once diffuses on to the interfaces in the case of conducting media.

* See *Aether and Matter* (1900), § 63; where, however, amendment is necessary as *infra*, Part III, § 13, footnote.

The only part of this scheme of equations that is incomplete is the specification of (u_0, v_0, w_0), that portion of the current which redistributes free electrifications. In an ordinary conductor this current is of the same order of magnitude as the polarization current, which is itself wholly masked by the current of conduction, except in the case of optical phenomena. In a dielectric this current does not exist at all. In an electrolyte we may form a provisional scheme of it on the lines of Nernst's theory of migration of the ions, by putting $(u_0, v_0, w_0) = k\rho\,(P, Q, R)$, on the supposition that the medium is uniform, where k is analogous to a coefficient of ionic mobility: but we shall thereby destroy the linearity of the system of equations. The fact that free electrons act on each other simply by their electrostatic attractions, however the aether between them be disturbed, shows that they will tend to drive each other to the surfaces of the conductors, so that the current (u_0, v_0, w_0) will usually be a transient phase at the beginning of the settling of the disturbance, in agreement with the above, if it ever exist at all; and once the free charges are on the surfaces of the conductors they will remain there and be redistributed by ordinary conduction currents. This part of the current may therefore possibly [usually] be left out of account; that is, the electric density in a good conductor at rest may be taken to be always null.

Current of redistribution negligible.

Electrolytic current.

19. It is the migration of the positive and negative ions through the unelectrified material medium in opposite directions that constitutes the conduction current: a movement of the medium itself carries as many positive as negative electrons along with it, and so adds nothing to the current: the medium in a sense moves through the current, without carrying it along—in opposition to the assumption usually implied in the notion of a current element. This is true irrespective of the manner in which the current is distributed between flux of positive electrons one way, and flux of negative electrons the opposite way. In all cases, however, in which there is an electrolyte in the circuit, the law of Faraday shows, not that the current is equally divided between positive and negative ions, but that the numbers of these ions crossing in opposite directions any section of the steady current are equal—although this necessitates extensive changes of concentration in the electrolytic solution after the current has become steady, when the ions have different rates of migration, as has been demonstrated by Hittorf. If we were to assume that such changes of concentration of ionic electrons are not important in the metallic portion of the circuit, the interface between two metallic media not being sensibly polarizable, it would follow that the velocities

Electric current is relative to the conductor.

Different velocities of the ionic drifts, involving concentration gradient, in electrolytes.

Nature of current in metals.

of migration of the positive and negative electrons in such conductors would be sensibly equal*; and this equality would not be altered by the presence of a magnetic field. (Cf., however, § 23.)

The conduction current does not involve elastic displacement; if it flows in a complete circuit so that electrons are not allowed to accumulate and exert a back electric force, it will go on permanently, a limit being set to it only by the *quasi*-frictional resistance to the motions of the ions through the medium in the sense of the kinetic theory of gases, which is expressed by the law of Ohm†. **Ideal permanent currents.**

The relation to the principle of energy of the force (P, Q, R) which tends to produce electric separation in a conductor is expressed by the formula that, if there is a drift of positive electrons one way and negative the opposite way, so as to form a *true* current (u', v', w'), the time-rate at which the potential energy of the field is thereby exhausted is $u'P + v'Q + w'R$ per unit volume. The electric force as here introduced therefore agrees with its formulation in the ordinary elementary theory of true currents, in connection with Ohm's law. **Exhaustion of actual current.**

20. The system of electromotive equations may now be collected together. The dynamical equations in the free aether are of type **Scheme of the electromotive equations:**

$$4\pi C^2 f + \frac{d\Psi}{dx} = -\frac{dF}{dt}, \quad \text{where} \quad F = \int \frac{u}{r}\,d\tau + \int \left(B\frac{d}{dz} - C\frac{d}{dy}\right)\frac{1}{r}\,d\tau.$$

The force acting on an electron e, moving with velocity $(\dot{x}, \dot{y}, \dot{z})$, is of type

$$eP = e\dot{y}c - e\dot{z}b - e\frac{dF}{dt} - e\frac{d\Psi}{dx}, \quad \text{where} \quad (a, b, c) = \text{curl}\,(F, G, H);$$

and its coefficient of electric inertia is Le^2, where L depends on the radius of its nucleus.

The total current in an element of matter moving with velocity (p, q, r) through the aether is of type

$$u = \sigma P + \frac{K - 1}{4\pi C^2}\frac{\delta P}{dt} + \frac{df}{dt} + p\rho + u_0,$$

where
$$P = qc - rb - \frac{dF}{dt} - \frac{d\Psi}{dx},$$

* The alternative, that must be accepted, is that the negative electron-carriers do not merely drift, but have independent statistical velocities as in gas-theory.

† In the state of perfect conduction which does actually arise in some metals near the absolute zero of temperature, this free *quasi*-gaseous diffusion of the ions must be somehow inhibited.

and $\frac{\delta}{dt}$ represents $\frac{d}{dt}+p\frac{d}{dx}+q\frac{d}{dy}+r\frac{d}{dz}$; while

$$\rho=\frac{df}{dx}+\frac{dg}{dy}+\frac{dh}{dz},$$

where $$f=\frac{1}{4\pi c^2}P'=-\frac{1}{4\pi c^2}\left(\frac{dF}{dt}+\frac{d\Psi}{dx}\right).$$

extended to a convected material system:

21. When the material system is moving through the aether with uniform velocity p parallel to the axis of x, the equations become

$$(P,Q,R)=\left(\frac{dF}{dt}-\frac{d\Psi}{dx},\quad -\frac{dG}{dt}-\frac{d\Psi}{dy}-pc,\quad -\frac{dH}{dt}-\frac{d\Psi}{dz}+pb\right).$$

Thus $$\frac{dP}{dx}+\frac{dQ}{dy}+\frac{dR}{dz}=p\nabla^2F-\nabla^2\Psi;$$

also, in all cases $$\frac{dF}{dx}+\frac{dG}{dy}+\frac{dH}{dz}=0^*.$$

As the total current (u, v, w) is circuital, we must have

$$0=\left(\sigma+\frac{K-1}{4\pi c^2}\frac{\delta}{dt}\right)(p\nabla^2F-\nabla^2\Psi)-\frac{1}{4\pi c^2}\frac{d}{dt}\nabla^2\Psi+\left(\frac{\delta\rho}{dt}-\frac{d\rho}{dt}\right)-\frac{\delta\rho}{dt}.$$

It will suffice at present to confine ourselves to a non-conducting medium, so that σ, ρ, and (u_0, v_0, w_0) are null; then this condition of circuitation gives

$$\left\{\frac{d}{dt}+(1-K^{-1})\,p\frac{d}{dx}\right\}\Psi=(1-K^{-1})\,p\left(\frac{d}{dt}+p\frac{d}{dx}\right)F,$$

or neglecting cubes of p/c,

$$\Psi=(1-K^{-1})\,p\left(1+K^{-1}p\frac{d/dx}{d/dt}\right)F.$$

We have also, when all the magnetism is induced,

$$\nabla^2(F,G,H)=-4\pi\mu\,(u,v,w);$$

thus

$$\frac{c^2}{4\pi\mu}\nabla^2F=-\frac{K-1}{4\pi}\left(\frac{d}{dt}+p\frac{d}{dx}\right)\left(-\frac{dF}{dt}-\frac{d\Psi}{dx}\right)-\frac{1}{4\pi}\frac{d}{dt}\left(-\frac{dF}{dt}-\frac{d\Psi}{dx}\right)$$
$$=\frac{K}{4\pi}\frac{d^2F}{dt^2}+\frac{K-1}{4\pi}\left\{\frac{d^2\Psi}{dx\,dt}+p\frac{d^2F}{dx\,dt}+p\frac{d^2\Psi}{dx^2}\right\}+\frac{1}{4\pi}\frac{d^2\Psi}{dx\,dt},$$

so that

$$(K\mu)^{-1}c^2\nabla^2F=\frac{d^2F}{dt^2}+2\,(1-K^{-1})\,p\frac{d^2F}{dx\,dt}+(1-K^{-1})\,p^2\frac{d^2F}{dx^2}.$$

Again

$$\frac{c^2}{4\pi\mu}\nabla^2G=-\frac{K-1}{4\pi}\left(\frac{d}{dt}+p\frac{d}{dx}\right)\left\{-\frac{dG}{dt}-\frac{d\Psi}{dy}-p\left(\frac{dG}{dx}-\frac{dF}{dy}\right)\right\}$$
$$-\frac{1}{4\pi}\frac{d}{dt}\left(-\frac{dG}{dt}-\frac{d\Psi}{dy}\right);$$

* See footnote to § 15, *supra*, p. 568.

so that

$$(K\mu)^{-1}c^2\nabla^2 G = \frac{d^2G}{dt^2} + 2(1 - K^{-1})\,p\,\frac{d^2G}{dx\,dt} + (1 - K^{-1})\,p^2\,\frac{d^2G}{dx^2},$$ equation of vector potential:

and a similar equation holds for H.

Thus F, G, H all satisfy equations of precisely the same type up to and including terms of the second order in p/c; so that if the vector $\mathfrak{A}$ represent the vector potential, we have verified.

$$(K\mu)^{-1}c^2\nabla^2\mathfrak{A} = \frac{d^2\mathfrak{A}}{dt^2} + 2(1 - K^{-1})\,p\,\frac{d^2\mathfrak{A}}{dx\,dt} + (1 - K^{-1})\,p^2\,\frac{d^2\mathfrak{A}}{dx^2},$$

which is the same equation (with p substituted for v) that was found (§ 13) for the [resulting] magnetic induction $\mathfrak{B}$ by the method of averaged fluxes.

22. Hitherto we have omitted the complication which arises in optical applications owing to dispersion. In the case when the material medium is at rest, it is, however, easy to include the effects of dispersion and opacity. The total current is made up, as before, according to the specification Extension to conducting and dispersing media.

$$\mathfrak{C} + \mathfrak{D} + \mathfrak{D}' + (u_0, v_0, w_0),$$

where, as in § 11,

$$\mathfrak{C} = k\left(m\frac{d}{dt} + \sigma'\right)(P, Q, R), \quad P = -\frac{dF}{dt} - \frac{d\Psi}{dx},$$

$$\mathfrak{D} + \mathfrak{D}' = K\mathfrak{D} - Ki\frac{d^2\mathfrak{D}'}{dt^2},$$

$$\nabla^2\Psi = -4\pi\rho, \quad \frac{du_0}{dx} + \frac{dv_0}{dy} + \frac{dw_0}{dz} = -\frac{d\rho}{dt}.$$

The condition of circuitation shows, as in § 18, that a volume density ρ subsides in a conducting medium in a non-vibrational manner, so that Ψ and ρ may be left out of account. Then Instant subsidence of free density.

$$(P, Q, R) = -\frac{d}{dt}(F, G, H),$$

and we arrive at the same equations as previously in § 11.

Ponderomotive Forces.

23. When a conductor carrying a current is in a magnetic field, the positive and negative electrons[1] are urged equally by the magnetic force in a common direction perpendicular to the current and to the field, when they are themselves drifting with equal velocities in opposite directions. A difference in their velocities of migration,

[1] The argument in these sections is equally applicable, whether the current in metallic conductors is supposed to be carried by material ions, or by electrons considered as immaterial.

Effects of magnetic field on current: a differential Hall effect: involves a resistance effect: such as has actually been found to exist in electrolytes, would lead to a greater force on one kind than the other, and so produce a small transverse electric force representing the Hall effect. When the flow in the magnetic field has become steady, the distribution of drifting free positive and negative electrons across the section will not be quite uniform, thus involving also change of resistance.

the Amperean mechanical force is on the true current only. In any case the aggregate of the transverse forcives acting on the free electrons will constitute a mechanical electrodynamic forcive per unit volume,

$$(X, Y, Z) = (v'c - w'b,\ w'a - u'c,\ u'b - v'a),$$

acting on the conductor, where however (u', v', w') is the ***true*** current, namely $(u - \dot{f},\ v - \dot{g},\ w - \dot{h})$. There will in addition be a mechanical forcive $\rho\ (P, Q, R)$, if the element of volume contain an excess of one kind of electrons, including as a limiting case, a normal traction $\frac{1}{2}\sigma N$ over the surface of each charged conductor; and there will be the forcives acting on the magnetic and electric polarization of the element derived respectively from potential functions $\frac{1}{2}\kappa\ (\alpha^2 + \beta^2 + \gamma^2)$ and $(K - 1)/8\pi\ .\ (P^2 + Q^2 + R^2)$, leaving out of account the part which merely produces [local or constitutive] molecular stress; while, if part of the magnetic or electric polarity is permanent, there will also be couples as in §§ 33, 35.

Forces of electric type.

Contrast with other Electronic Theories.

Summary of theory of electrons. 24. We have thus attained to a complete scheme of equations of the electric field, simply on the assumption that a material medium contains electrons, as many positive as negative in the element of volume unless it is electrified; that these elctrons are in part combined into systems which are, or belong to, the molecules, some of these molecules being neutral, and some having an excess of positive or negative electrons, being therefore ions. The only other assumption is that the nuclei of the electrons occupy a negligibly small part of the whole space. As regards the manner in which the electromotive and the ponderomotive forcives in the electric field are accounted for, a similarity with Weber's molecular theory may be remarked[1]. In that theory electric molecules act on one another directly at a distance according to a law of force which involves their relative velocity; on the present theory actions are transmitted from one moving electron to another solely by the intervention of the aether. The Weberian theory has been subjected to destructive criticism by von Helmholtz, on the ground that it implies the possibility of a perpetual motion;

Action at a distance.

[1] W. Weber, *Electrodynamische Maasbestimmungen*; Maxwell, *Treatise*, §§ 846–860.

the mode of genesis of the present theory obviates such a criticism. The features of the Weberian theory above mentioned were characterized by Maxwell as "eminently successful" (*Treatise*, § 856), although this commendation is afterwards limited by the assertion that these features are necessarily connected by the principle of energy. It is now, however, recognized that this principle cannot by itself furnish more than one relation between the various quantities that enter into the problem of electrodynamic induction, so that the fact that any theory, not otherwise discredited, accounts for the two types of forcive in the electric field, ought to weigh strongly in its favour. The range of a theory of moving electrons of the present type, with its underlying aether, is of course much wider than that covered by Weber.

Weber's electric particles: successes of that theory: contrasted:

When the mechanical forcive is exerted only on the discrete electrons contained in the element of volume, it clearly cannot involve directly in its constitution an equilibrating internal stress, such as we found must be included if we take the current element as a connected physical entity; it is for this reason that Weber's theory, though in other respects different from the present one, agrees with it in giving the Ampère-Maxwell ponderomotive force, involving the *true* current however, not the total current as in Maxwell's formula.

correspondences.

Unipolar Induction.

25. The phenomenon of unipolar induction, in which a current is induced when a magnet revolves round its axis of symmetry through its own field of force, is deprived of all difficulty or ambiguity when it is considered under the present point of view. The electrons in the magnet, as they are moved across the magnetic field of the aether, are each subject to a forcive which is proportional to the component magnetic intensity in the meridian plane, and which produces electric separation by drifting the positive ions towards the axis and in the direction of the length of the magnet one way, and negative ions the opposite way. This constitutes an electromotive force along the revolving magnet.

Discrete electrons demanded by unipolar currents.

It follows for instance that a magnet symmetrical around its principal axis will, on rotation round that axis in its own field, acquire an electrification of excessively minute amount when the circuit is incomplete, but still sufficient to compensate the electric force induced by the motion. We can utilize Maxwell's equations of electric force, modified so as to refer to a system of axes moving through the aether*, to infer at once that for a *solid* magnet of any form, in

Spinning of a magnet entirely compensated by an induced charge on it:

* *Treatise*, § 600, or as in *Phil. Mag.* Jan. 1884, p. 12; *supra*, p. 18.

motion of any type, the induced electric force is derived from a potential $-(Fp + Gq + Hr)$, where (p, q, r) is the velocity through the aether of the element of the magnet at the point considered; so that it can at each instant be compensated by the static force due to a minute induced electrification. The maximum difference of potential to be thus compensated, between the axial and circumferential parts of the rotating magnet, is of the order 10^{-4} volt in the case of the Earth. If the force were not thus derivable from a potential, a magnet in motion would induce currents in itself even when there is no extraneous field, and the energy of the absolute motion through the aether, of the Earth and all other magnets, would gradually be converted into heat owing to this cause; as things are, only the energy of relative motion of magnets is subject to dissipation in this way.

e.g. the Earth as a spinning magnet.

Relativity essential.

(Aug. 3, 1895. 26. The case of a solid conductor spinning steadily round an axis of symmetry, in a magnetic field which is also symmetrical round that axis, has an important bearing on theory on account of the simplicity of the conditions. If it were legitimate to specify the electrodynamic energy of the system in terms of current elements and their mutual configurations, then in this case the energy belonging to a current element associated with an element of volume of the conductor would remain invariable, on account of the steady configuration of the motion; therefore there would be no electric force induced in such an element. Thus, for a conductor spinning in this manner there could then be no differences of electric potential caused by the motion. And, moreover, if a so-called unipolar circuit is completed by means of a fixed conducting wire, attached to the spinning conductor by sliding contacts at its equator and one of its poles, there could be no electric force induced in this wire either; and the electromotive force of the current which actually flows round this circuit would have to be sought for wholly in the sliding contacts, which, considering the definiteness of this electromotive force and the great variety of types of contact that are possible, seems to be an untenable alternative.

Electrons essential:

also sliding contacts, for a continuous-current dynamo or motor.

On the other hand, the formulae of the present paper, which considers an electrification to be made up of discrete elements each surrounded by free aether, make the induced electromotive force along any open line of material particles consist of two parts, (i) a part due to motion of this line of particles with respect to the quiescent aether, and equal to the time-rate at which it cuts across the tubes of magnetic induction, themselves supposed to be stationary in this computation, and (ii) a part equal to the line integral of $-d/dt\,(F, G, H)$, to be computed by the integral formula for (F, G, H) given above, which represents the effect of change in the inducing system. When the electromotive force is taken round a complete circuit, all theories are of course in agreement. In a case of steady motion, such as is now under consideration, the second of these parts is null, and the electromotive force induced even in an open circuit is given by Faraday's original rule, in terms of the number of tubes of magnetic induction which cut across the current.

Faraday's rule of induced force valid even for open circuits, if steady.

In the present order of ideas, a distinction has to be observed between (i) the electric stress in the aether, which is the tangential shear derived from its potential energy and represents the whole forcive acting on free aether, and (ii) the electric force which acts on electrons and moves them through the aether, thus polarizing dielectric media and producing true electric currents in conductors. In the hands of Faraday and Maxwell, the current of conduction was completed or rendered circuital by an effect across dielectrics, which was equivalent to a current, and there was no difference contemplated in this effect depending on whether the dielectric was a material substance or free aether, both being considered to be merely polarizable; on the present more complete view this dielectric action has to be divided into the true polarization current in the material dielectric, which is excited by the electric force orientating its polar molecules, and the rotational strain in the aether which is regulated by the laws of elasticity of that medium. Thus, the question whether there is electric force induced in the free aether itself is, on our present view, nugatory, there being no electrons on which it could operate.

Maxwell's circuital total current dissected.

Aether strain and electric force.

The calculations given in *Phil. Mag.*, Jan. 1884 [*supra*, p. 10], of the differences of potential, and the consequent electrification, induced in a sphere or other conductor of revolution, rotating in a symmetrical magnetic field, will thus, on the present view, be absolutely correct when the conductor is moving in a vacuum. When it is rotating in a gaseous dielectric, like air, there will be aerial flow produced owing to viscosity, after the manner of the action of a fan, and there will therefore be electric force induced in the moving portions of the air; but the capacity of air for dielectric polarization is so small that the polarity thereby induced will not sensibly affect the state of electrification of the system. If, on the other hand, the conductor were rotating in a liquid dielectric, the polarity induced in the moving parts of the liquid would depend in part on its motion, which would thus very materially influence the distribution of electrification in a manner which it would not be difficult to calculate if it were necessary to do so. The unipolar current obtained on completing the conducting circuit will, of course, in any case, unless the

Globe spinning in a magnetic field:

influence of disturbed dielectric.

Unipolar dynamo:

conductivity is almost evanescent, be practically independent of the nature of the surrounding dielectric.

impossible on a continuous-current theory. The analytical theory [formulae] of Maxwell's *Treatise* being based on current elements, that theory should, when correctly developed, give in the rotating conductor the null electrification of the beginning of this section, instead of, as here, an electrification based on Faraday's rule. And this is easily verified if to Maxwell's original equations of electric force are added, in accordance with von Helmholtz's correction, terms derived from a potential function which is the scalar product of the vector potential and the velocity of the material medium; in computing the radial force, which is in cases of symmetrical rotation the total force, these terms exactly cancel the Faraday part.

Helmholtz's crucial experiment: The discrepancy between these two theories is put to a test in a classical experiment made by von Helmholtz[1]. He found that when a conductor was spun in a magnetic field symmetrical around the axis of rotation, there was a difference of potential induced between the axial and circumferential parts as evidenced by resulting electrification, which agreed with Faraday's rule within three per cent., a quantity well inside the limits of uncertainty of measurement. He drew the conclusion that an electrodynamic potential theory of the Neumann type is thus proved inadmissible unless it recognizes polarization currents in the dielectric. The considerations stated at the beginning of this section seem to show that it could not even thereby be helped out; and that the true inference from the experiment must be a wider one, namely the abandonment of a theory of currents ultimately continuous in favour of one which regards them as made up of discrete electrons separated from each other by free aether. **interpretation.**

The conjugate Rowland effect: 27. The electrification induced in a conductor by rotation in a symmetrical magnetic field has just been examined; we pass on naturally to the conjugate problem of the magnetic field induced by a rotating electrified conductor, where similar considerations must crop up. The difficulties involved in the interpretation of the results of Rowland's experiment, on the hypothesis that an electric charge

Maxwell's crucial test for inertia of currents. [1] *Berlin Monatsber.* 1875; *Ges. Abhandl.* Vol. I. p. 783. Maxwell had very early considered the question whether in his total kinetic energy as specified in terms of the current elements and the velocities of the conductors, there are any terms which involve products of these quantities, and he had drawn a negative conclusion from experiment; *Treatise*, Vol. II. Chap. VI. secs. 568–577. The electrification in von Helmholtz's crucial experiment might, however, be formally expressed on that system as due to energy terms of this mixed type, but, of course, extremely minute.

in a conductor is a continuous distribution of electricity, and an electric current a continuous flux, have been already considered in the previous paper[1], and proved surmountable only on the extremely precarious assumption that the rotating gilded glass discs of the experiments were divided into mutually insulated segments by the scratches which were intended to prevent Foucault currents. The translation of an isolated electric charge carries on its own surrounding electric field and so alters the electric intensity at each point in the aether; and can therefore certainly induce a magnetic field with consequent reaction on the moving charge. But the steady rotation round its axis of a charged conductor of revolution in no respect affects the field of electric strain in the surrounding aether; that remains steady, and therefore no magnetic, that is kinetic, energy can be locally generated anywhere in it. And there is also the related difficulty previously enforced, that if a charged conductor connotes merely a field of self-locked electric strain in the surrounding dielectric, the elasticity breaking down when the surface of the conductor is reached, the rotation of the conductor round its axis of symmetry could exert no grip on this field of strain, which would therefore not be affected at all. These difficulties will not vanish unless the electric charge on the conductor is made up of discrete portions, separated by dielectric spaces however narrow. The circumstances will then be no longer symmetrical with respect to the axis of rotation so far as these spaces are concerned; the magnetic field induced, at a place at finite distance from the conductor, will depend not only on the change of electric intensity at that place, which is null as before, but also on the surface conditions which obtain along the boundaries of the dielectric region.

demands electrons.

The phenomena are thus accounted for on the hypothesis that the electric charge on a conductor consists of a distribution of discrete electrons over its surface. It is true that the facts of electric vibrations on conductors show that these electrons must be extremely mobile and sensitive to electric force: this is because of the very strong charges involved in them, but it does not imply that when the conductor is rotated, the electrons will slip backward over its surface and remain where they were, owing to the electric inertia. They will not take up the motion of the conductor just at once; but there is no electric force which would tend to prevent that result, and the same steady viscous agencies that produce electric resistance when a steady current of electrons is flowing in a fixed conductor, will ultimately make them move with the rotating conductor, after a time relatively long but probably absolutely very short.

Nature of surface charge:

its convection complete, with very slight time-lag.

[1] *Phil. Trans.* A, 1894, p. 764; *supra*, p. 466.

The case for electrons enforced: It does not, in fact, appear that there are any of the hitherto outstanding difficulties of pure electrodynamic theory that are not removed by the hypothesis of moving electrons, to which, from the consideration of several distinct classes of phenomena, and apart altogether from electrochemical theory, we have been compelled to resort. This hypothesis, in its wider aspect involving the nature of matter itself, seems also to have a philosophical necessity; for the location of causes of disturbance of the uniform all-pervading medium in permanent discrete singularities or nuclei of strain or motion, that belong to it and can move about through it, is the only way of avoiding the introduction into theory of either direct distance actions, or else those assumptions of independent media superposed in the same space and discharging different functions, which violate the maxims of modern physics. Without a precise conception of the causes which produce disturbances in it and form one side of the play of action and reaction, a theory of the aether can be merely descriptive; while any assumed causes that are not of the nature of singularities arising from the constitution of the medium itself, must introduce a foreign element and so deprive the theory of its interconnection and self-contained character.)

else multiple aethers or distance action.

Coherence of present scheme.

Mechanical Pressure of Radiation.

28. An application of the expression for the ponderomotive forcive will be to the examination of Maxwell's mechanical pressure of radiation (*Treatise*, §§ 792, 793). Let us consider a train of plane waves moving along the axis of x, with their magnetic induction c along z, and their current v along y. The circuital relations give

Plane-polarized wave-train:

$$4\pi\mu v = -\frac{dc}{dx}, \quad \frac{dQ}{dx} = -\frac{dc}{dt};$$

while the equation connecting electric force with current is

$$v = \sigma Q + \frac{K}{4\pi \mathrm{C}^2}\frac{dQ}{dt}.$$

Hence $$\frac{d^2c}{dx^2} = K\mu \mathrm{C}^{-2}\frac{d^2c}{dt^2} + 4\pi\sigma\mu\frac{dc}{dt};$$

leading to $$c = c_0 e^{-px}\cos(nt - qx),$$

where $$(p + \iota q)^2 = K\mu \mathrm{C}^{-2} n^2 - 4\pi\sigma\mu n\iota.$$

damped by absorption:

Of the total current v, the part $v' = v - \frac{1}{4\pi \mathrm{C}^2}\frac{dQ}{dt}$ is the true current, derived from motion of electrons and not including the aethereal rotation, C being the velocity of radiation in free aether. The mechanical forcive per unit volume of the material medium is thus $X = v'c$; hence

$$\int X\,dx = -\frac{c^2}{8\pi\mu} - \int \frac{c}{4\pi C^2}\frac{dQ}{dt}\,dx.$$ bodily forcive for a segment:

Let us first take the case of a transparent medium, for which

$$\frac{dQ}{dt} = \frac{4\pi C^2}{K}v = -\frac{C^2}{\mu K}\frac{dc}{dx};$$

so that $$\int X\,dx = -\frac{c^2}{8\pi\mu}\left(1 - \frac{1}{K}\right) = -\frac{c^2}{8\pi\mu}(1 - m^{-2}),$$

where m is the index of refraction.

The value of this indefinite integral, taken over an exact number of half wave-lengths in a homogeneous medium, gives a null result for the total forcive; but if an interface between two media is included in the volume of integration, there will also be terms in it which may be represented by averaged tractions $\frac{1}{2}\frac{c_0^2}{8\pi\mu}(1 - m^{-2})$ exerted by this interface on its two sides, or what is the same thing, pressures acting on the interface. In air this pressure practically vanishes. This is the result which replaces Maxwell's formula $c_0^2/8\pi\mu$ for the mechanical pressure produced by radiation falling on the surface separating two media*. if transparent reduces to tractions on the ends.

29. When the train of waves falls on an absorbing medium, the circumstances are different. From the value of c above given, it follows that [for the transmitted train]

$$\frac{dQ}{dt} = n^2 c_0 \frac{e^{-px}}{(p^2+q^2)^{\frac{1}{2}}}\cos(nt - qx - \epsilon'), \text{ where } \tan\epsilon' = \frac{q}{p};$$

$$c\frac{dQ}{dt} = \tfrac{1}{2}n^2 c_0^2 \frac{e^{-2px}}{(p^2+q^2)^{\frac{1}{2}}}\{\cos(2nt - 2qx - \epsilon') + \cos\epsilon'\},$$

$$\int \frac{c}{4\pi C^2}\frac{dQ}{dt}\,dx = \frac{n^2 c_0^2}{8\pi C^2}e^{-2px}\left\{\frac{\cos(2nt - 2qx - \epsilon')}{2(p^2+q^2)} - \frac{\cos\epsilon'}{2p(p^2+q^2)^{\frac{1}{2}}}\right\}.$$

Hence $$-\int X\,dx = \frac{c_0^2}{8\pi\mu}\cos^2 nt + \frac{n^2 c_0^2}{8\pi C^2}e^{-2px}\{\ldots\},$$

* These sections are left, in part, to illustrate the pitfalls of direct calculation in this subject. The last result is in error, because at a sudden transition a reflected wave-train must be set up, while across a gradual one the properties of the substance are functions of x.

The same applies to the next section. What is there determined is the mechanical reaction on the absorbing medium of the radiation that is transmitted into it, sustained by the medium but regarded as transmitted back to the interface. On the other hand, what the Maxwell stress gives for direct incidence is a resultant of traction on the interface from both sides equal to the difference of the energy-densities on the two sides, combined with a bodily traction on the absorbing medium equal to the gradient of the energy-density along it.

in which the lower limit is at the interface, and the upper one at a place where the radiation has been practically extinguished. Thus here there is an average pressure $\frac{1}{2}\frac{c_0^2}{8\pi\mu}$ acting on the interface towards the opaque material medium, which is half of Maxwell's result*.

Interfacial Conditions.

30. It is important to emphasize the dynamical distinction between electric force and electric displacement in a dielectric. The electric force is that vector which occurs in the rotational stress in the aether, and which, divided by the proper elastic coefficient $4\pi C^2$, gives the rotational strain or electric displacement in it. This is not circuital when there are moving electrons present, but by adding a fictitious electric displacement representing the drift of the electrons, a circuital total displacement is obtained. This fictitious displacement represents, however, the rotation in a hydrodynamical flow of the aether which the motion of the electrons sets up, which belongs to the kinetic energy and in ordinary electrodynamical applications constitutes practically the whole of it. At an interface separating media of different dielectric qualities, the electric stress in the aether must be continuous, and as this is tangential, equal to the tangential component of the electric force but at right angles to its direction, it follows that the tangential components of the electric force must be continuous.

Tangential forces and normal fluxes are continuous: electric:

Again, in a magnetic medium, the magnetic induction represents the smoothed-out velocity of flow due partly to motions communicated by distant disturbances, and partly to the circulatory motions of the magnetic vortices that are caused by the very rapid orbital rotations of the electrons in the molecule. The magnetic induction is therefore always circuital, so that its normal component is continuous at an

magnetic.

* Maxwell's law of radiation pressures on matter, as the result of deflection or obstruction of rays with their convected momentum, is established for free space. When the rays travel in a material medium, the momentum may be used up along the path to overcome resistances of the electrons of the matter, and complexities arise. This is obvious when there is absorption: then, also, the energy of the waves is no longer half potential and half kinetic.

For a *perfect* reflector consider the stress over the boundary of a thin layer including the interface. The magnetic field just in front is $2c_0 \cos nt$, the electric field null; and just behind there is no field. Thus the energy density is $\frac{4c_0^2}{8\pi}\cos^2 nt$: while the current in the interface is $\frac{2c_0}{4\pi}\cos nt$ subject to an average magnetic force $\frac{1}{2}2c_0 \cos nt$, giving a mechanical thrust $\frac{c_0^2}{2\pi}\cos^2 nt$. Thus Maxwell's law verifies for electric pressure on the interface. See the additions to lecture on "The Dynamics of Radiation" (1912), *infra*.

interface. But as regards the tangential component, we must divide the whole induction into two parts, one of them representing the effect of the vortices in the immediate neighbourhood of the point under consideration, and the other including all the rest of the flow. The latter part is the magnetic force, in the ordinary phraseology, and it is clearly only this part whose tangential component must be continuous, when we cross an interface into a region in which the distribution of the magnetic molecular vortices is different.

It is a check on the averaged form of the dynamical equations for a material medium that these six interfacial conditions regarding force and flux should be consistent with the incompressibility of the aether; and in fact the relations given above show that only four of the six are independent. It is on this account, for example, that the general requirements of the problem of physical optics are satisfied.

Compressional aether waves do not arise.

Molecular Current Systems replaced by Equivalent Continuous Currents.

31. To secure a perfectly homogeneous specification of the electrodynamic field in a dynamical theory, there is no alternative but to reduce magnetism to molecular electric currents. As, however, our analytical equations for an extended medium involve (u, v, w), the flow according to a volume specification, while the molecular currents are minute whirls not involving continuous flow in any direction, it is necessary to determine the volume specification of flow that shall be their equivalent when the element of volume is so great as to contain a large number of whirls of which the average effect only is required*. Such a specification is clearly possible; and when we bear in mind that the volumes occupied by the cores of the whirling electrons form only an excessively small part of the total space, it is also clear that the type of magnetic force which represents the velocity of the incompressible aether must be a circuital one, that is, the magnetic induction of Maxwell. There is no difficulty in verifying by direct analysis that the magnetic force at a point due to a system of molecular currents is the same as the magnetic induction due to the equivalent magnet; in fact it is shown (previous paper, § 120) that the said force is equal to the curl of a vector potential

Analogy for magnetism in aether flow.

Atoms are of very open structure.

$$\int\left(B\frac{d}{dz} - C\frac{d}{dy},\ C\frac{d}{dx} - A\frac{d}{dz},\ A\frac{d}{dy} - B\frac{d}{dx}\right) r^{-1}\, d\tau,$$

Vector potential of magnetism.

the same vector potential from which the magnetic induction of the equivalent magnet is derived.

* The previous statistical formulae for electrons in motion apply to free electrons. The orbital electrons in the atom require a different treatment as here, which is best evolved on a geometric basis for each atom.

Now let us consider the projections of the molecular circuits parallel to the plane of xy: let these projections swell out in area until they come into contact filling up the whole plane, the currents round them being reduced in the same ratio as their areas are increased. Along each edge common to two circuits there will be a differential current flowing, and by making the enlarged circuits rectangular of the form $\delta x \delta y$, it becomes clear that the aggregate of these differential currents make up a volume distribution of currents $(-dA/dy,\ dB/dx,\ 0)$ together with a current flowing round the boundary. Hence, adding up for the projections on all three coordinate planes, we obtain the distribution represented by $-$ curl (A, B, C) together with a current sheet on the external bounding surface equal to

Magnetism replaced by a current distribution,

along with boundary terms:

$$(Bn - Cm,\ Cl - An,\ Am - Bl)$$

per unit area. The validity of this substitution is verified by the analytical transformation

$$F = \int \frac{u}{r}\, d\tau + \int \left(B \frac{d}{dz} - C \frac{d}{dy} \right) \frac{1}{r}\, d\tau$$

$$= \int \frac{u}{r}\, d\tau + \int (Bn - Cm) \frac{1}{r}\, d\tau - \int \left(\frac{dB}{dz} - \frac{dC}{dy} \right) \frac{1}{r}\, d\tau.$$

the simple theory for linear induction.

Thus if the magnetism is wholly induced, so that

$$(A, B, C) = \kappa\, (\alpha, \beta, \gamma),$$

the equivalent volume distribution of currents is $4\pi\kappa\, (u, v, w)$, provided κ is constant throughout the field; hence the induced magnetism may be ignored if in estimating the induction we multiply the current system by $1 + 4\pi\kappa$ or μ.

More generally, if a part (A_0, B_0, C_0) of the magnetism is permanent, we have in a region of constant μ

$$\nabla^2 F = -4\pi\mu u + 4\pi \left(\frac{dC_0}{dy} - \frac{dB_0}{dz} \right);$$

of which the last term vanishes when there is no permanent magnetism, or when it is of lamellar type. At the interface between two different media (F, G, H) must be continuous. These relations are sufficient to determine (F, G, H) completely in terms of (u, v, w).

Again, a steady electric polarization (f', g', h') in a dielectric, moving across the aether with velocity (p, q, r), is equivalent in its kinetic effects (previous paper, § 125) to a molecular current system or magnetization

$$(rg' - qh',\ ph' - rf',\ qf' - pg')$$

together with a current sheet over the bounding surface, *only pro-*

vided the velocity is uniform. In any case, however, it is equivalent to the current system*

$$\left(p\frac{d}{dx}+q\frac{d}{dy}+r\frac{d}{dz}\right)(f', g', h').$$

The current equivalent of convected polarization.

Mechanical Forcives acting on Magnetically and Electrically Polarized Media.

32. We have still to calculate the steady forcive on a molecular current. The force on a revolving electron e has been shown to be $e\,(P, Q, R)$, where

$$P = c\dot{y} - b\dot{z} - \frac{dF}{dt} - \frac{d\Psi}{dx},$$

Averaged force on a magnetic atom,

$(\dot{x}, \dot{y}, \dot{z})$ being its velocity, and Ψ the potential of the static part of the electric field due to such free charges as may exist in the material media. The first two terms, involving the velocity of the electron, will give rise to magnetic forcive, the remaining part will make up into the electric forcive, in a polarized medium.

regarded as electronic.

The averaged value of the first two terms, over the orbit of a single electron, gives $\iota\int(c\,dy - b\,dz)$, where ι the equivalent current is equal to e multiplied by its velocity v and divided by the length of the orbit; this is the same as

$$\iota\int\left(l\frac{da}{dx}+m\frac{db}{dx}+n\frac{dc}{dx}\right)dS$$

over a surface bounded by the orbit, by Stokes' theorem in conjunction with the circuital character of (a, b, c). Summing up for all the orbits of electrons, this would give for the translational magnetic forcive per unit volume (X', Y', Z'), where

$$X' = A\frac{da}{dx}+B\frac{db}{dx}+C\frac{dc}{dx},$$

and (A, B, C) is the magnetization of the medium. Now the actual magnetic forcive must certainly, from the most cursory observation of its intensity for iron, involve the magnetic force (α, β, γ) and not the induction (a, b, c); whence then this discrepancy? The forcive on an electron involves correctly (a, b, c), for the velocity of the incompressible aether is circuital. But it must be borne in mind that the molecule contains as many negative revolving electrons as positive, and therefore that the forcive on the molecule is a differential one, positive and negative electrons pulling on the average opposite ways [though if the positives are massive protons they contribute very little]; so that in treating of single molecules we cannot take the actual velocity of the aether to be an averaged function of position such as (a, b, c).

Discrepancy:

* See Part III, *infra*, footnote, § 13.

analyzed: 33. To find the forcive on a molecule, we must in the first place divide this velocity into two parts, one independent of the immediate surroundings of the point considered and depending only on the general character of the field, the other representing the effect of local configuration. The former is the magnetic force as usually represented by (α, β, γ); it is made up of the force due to the distribution of ideal magnetic matter introduced by Poisson into the theory, together with that due to the distribution of electric flow. The former element in it is independent of local peculiarities, by the ordinary theory of the gravitation potential; the latter also has never any term involving the flow at the point considered, for similar reasons. The purely local portion of the velocity of the aether consists itself of two parts, a steady one, and one of varying type with very rapid alternation characteristic of the orbits of the electrons; the part last mentioned averages to nothing, while the former part adds on to the magnetic force to produce the total averaged velocity of flow of the aether, that is the magnetic induction (a, b, c).

Now even if instead of the actual velocity of the aether, rapidly varying from point to point, we substituted this average velocity (a, b, c), we should still obtain both a forcive which depends on the general character of the field, and one (though only the regular part of it) which depends on the interaction of the molecule with its immediate surroundings. The latter forcive, which cannot be completely expressed in terms of the quantities of the present theory, belongs to an intermolecular stress of cohesive type; it is self-equilibrating and contributes nothing to the forcive on the material medium as a whole, though, as we shall see later, it produces deformation of its parts. If we agree to consider it as a separate molecular forcive, we have for the magnetic forcive proper (X', Y', Z') the formula

local part rejected as constitutive, producing only structural change:

transmitted force:

$$X' = A\frac{d\alpha}{dx} + B\frac{d\beta}{dx} + C\frac{d\gamma}{dx}.$$

There will also be a magnetic couple (L', M', N') per unit volume, where

and torque.

$$L' = B\gamma - C\beta;$$

but there will clearly be no additional molecular couple acting in the element of volume.

A potential. 34. The steady magnetic force and couple thus obtained may be provisionally considered as derived from an energy-function (kinetic) $A\alpha + B\beta + C\gamma$ per unit volume, in which (A, B, C) is unvaried (cf. Maxwell, *Treatise*, § 639); the regular part of the molecular forcive may be considered as involving an internal energy-function

$$2\pi(A^2 + B^2 + C^2)$$

in which (A, B, C) is varied.

More fundamentally*, if we divide (α, β, γ) into a part $(\alpha' \beta', \gamma')$ due to the magnet itself and a part $(\alpha_0, \beta_0, \gamma_0)$ due to the inducing field, the actual energy associated with magnetic force is

The energy of the magnetic system:

$$\tfrac{1}{2}(A\alpha' + B\beta' + C\gamma') + A\alpha_0 + B\beta_0 + C\gamma_0,$$

in which (A, B, C) is now to be varied; and this, when there is no permanent magnetism, leads to the same mechanical forcive as an energy-function $\tfrac{1}{2}\kappa(\alpha^2 + \beta^2 + \gamma^2)$, where κ is the coefficient of magnetization, the part of it not thus compensated being connected with the internal work of orientation of the molecules. This energy-function is not the whole kinetic energy of the aether; that would be in the present units the space-integral of $(a^2 + b^2 + c^2)/8\pi$ together with a molecular part, that is, the space-integral of

$$(a\alpha + b\beta + c\gamma)/8\pi + \tfrac{1}{2}\kappa(\alpha^2 + \beta^2 + \gamma^2),$$

of which the latter term is exactly compensated by the above mechanical forces acting on the magnetized body, while the other term which remains over goes to produce electrodynamic effects, and in part to represent intrinsic energy of magnetization, and is in fact the electrodynamic energy formulated in Maxwell's scheme.

the electrodynamic part.

It is to be remarked that the magnetic vector potential, obtained in the previous paper as the vector potential of the Amperean currents in the form

Very fast velocities of electrons excluded.

$$F = \int\left(B\frac{d}{dz} - C\frac{d}{dy}\right) r^{-1} d\tau,$$

implies that the orbital velocities of the electrons do not approximate very closely to the velocity of radiation, a condition which in reality is sufficiently satisfied: if that were not so, this expression would require correction for an aethereal displacement part of the molecular current.

Magnetic relations affected by electric flow:

35. The magnetic forcive (X', Y', Z') acting on the actual magnetization I is different from the forcive that would act on the three components of the magnetization A, B, C, in case there are currents flowing in the magnet; so that in this connection it is not permissible to treat the magnetization as a vector and resolve it into components. The reason here is that we have actually to do with molecular currents; and if we replace a molecular circuit by its three components we thereby alter the character of its linking with the lines of flow of the current flowing across the medium. But there is a consideration of a less special kind which shows that this cannot be done: if we were to resolve the magnetization, we should obtain results for the forcive which would be physically different according to the system of coordinate axes that is adopted. If, however, we sum up forcives

* See also *supra*, p. 275.

criterion of invariance. on separate magnetic poles, instead of calculating forcives on resolved magnetic moments, we arrive again at the correct result, independent of the coordinate system: as will be illustrated immediately in connection with the forcive of electric origin.

36. The ponderomotive forcive of electric origin, acting on a dielectrically polarized medium, is made up of a bodily force (X, Y, Z) and a bodily couple (L, M, N), where

Forcive on a medium polarized electrically:

$$X = f'\frac{dP}{dx} + g'\frac{dQ}{dx} + h'\frac{dR}{dx}, \quad L = g'R - h'Q;$$

in these formulae (f', g', h') is the total polarization of the material medium, consisting of the ordinary induced polarization

$$(K - 1)/4\pi \,.\, (P, Q, R)$$

and, under circumstances of residual charge, also a permanent part. When the magnetic field is varying, so that (P, Q, R) is not derived from a potential, there will be delicate considerations concerning this force, similar to those discussed above in connection with magnetization. The formula here given for X is, however, correct, because it is what is directly obtained by grouping the actual electrons of the neutral molecule in pairs to form polar doublets; this is a legitimate procedure although the resolution of the averaged electric moment of an element of volume into three components proves not to be such. Thus the forcive on a doublet δD lying along the axis of x is a force $(\delta D \,.\, dP/dx, 0, 0)$ and a couple $(0, -\delta D \,.\, R, \delta D \,.\, Q)$; and when the axes of coordinates are changed so that δD becomes $\delta\,(f', g', h')$ these expressions change into the ones given above, because

its ultimate basis is electronic.

$$P\delta f' + Q\delta g' + R\delta h'$$

is an invariant function as regards change of axes[1].

37. The stress of molecular type, produced by magnetization, offers some points of interest. It has been remarked by von Helmholtz[2] long ago that in a polarized medium there exists a material tension

Contrast of electric and magnetic fields.

[1] (The circumstances of electrified and magnetized media are not parallel, in the sense that K corresponds to μ. In a magnetized medium the circuital vector, namely the induction (a, b, c), is the smoothed-out velocity of the aether; and the magnetic force (α, β, γ) is that part of it which is independent of the immediately adjacent vortices or Amperean currents. In a polarized dielectric (f, g, h), $= (P, Q, R)/4\pi$, is the smoothed-out *total* elastic rotation or electric displacement in the aether; (f', g', h') is the time integral of the polarization current of the material medium which is constituted of movements of orientation of the electrons; while it is the sum of these vectors that is now circuital, representing the total electric current arising from aethereal strain and movement of electrons combined.—August 25.)

[2] *Berlin Monatsber.* Feb. 1881; *Wissen. Abhandl.* Vol. I. p. 779.

along the lines of polarization and a pressure at right angles to them. Each of these is, however, proportional to the square of the susceptibility[1] of the medium, and they are not necessarily equal; for media of very high susceptibility they are thus far more intense than Maxwell's hypothetical stress which depends on its first power. Their origin is very conveniently exhibited in a chain of iron nails hanging end to end from a pole of a magnet: the nails hold together longitudinally but repel each other transversely. Now consider a longitudinally magnetized bar: this straining together of opposite polarities in neighbouring molecular groups will produce an internal stress in the bar proportional to I^2, which will usually tend to shorten it, but may conceivably in some cases do the reverse. Furthermore, the orientation of the molecular groups due to magnetization will directly alter the length to an extent depending on the first power of I. Hence the whole increase of length will be $AI + BI^2$, in which A and B may each be either positive or negative. When A and B have opposite signs there will be a value of I at which the total effect will reverse its sign by passing through a null value. It appears from experiment that for iron A is positive and B negative; for nickel, A is negative and B negative; for cobalt, A is negative and B positive[2].

Magneto-striction: in a hanging bunch of magnetized nails.

(Aug. 25. When A and B have opposite signs there will be an intensity of magnetization, $I = -A/2B$, which corresponds to maximum or minimum elongation. Near this intensity a small change in the magnetization will not affect the length of the bar; and therefore conversely a change of length produced by tension will have no influence on the magnetization, any changes in the two being independent of each other. It follows that the magnetization at which the influence of tension vanishes is only half the Villari critical magnetization at which the elongation of the bar is null. This is, of course, on the assumption that the phenomenon is a regular and reversible one, and that there is no sensible term in the elongation involving I^3.)

Interaction of magnetism and tension: two critical stages.

There must be similar strain effects connected with the polarization of a dielectric in an electric field, though of course they would be far more difficult to detect.

It seems probable that the greater part of the phenomena of electrostriction and magnetostriction is of this character, and that only a small part is due to strain of the material by the direct effect of electric and magnetic attractions between finite portions of the medium.

Electro-striction mainly local and constitutive.

[1] *Proc. Roy. Soc.* April 1892, p. 63: *supra*, p. 283.

[2] Cf. Ewing, *Magnetic Induction...*, Chap. IX.

General Considerations.

The present type of model demands high aethereal inertia:

38. It has been one aim of the present analysis to examine how far it is possible to identify, in the play of kinetic and potential energy between the electrons, the main phenomena of matter. In order to comply with the requirements of negative optical experiments by Lodge and others, it is necessary to assume the inertia of the aether to be at least comparable numerically with the inertia of derivative character which belongs to dense matter; and the velocities of its movements are thus extremely slow. On the other hand, in the descriptive electric theories that are now commonly held, it is usual to consider the aether to be like ordinary matter as regards its inertia, and also as regards capacity for electric and magnetic polarizations. Its density, for astronomical and other reasons, must then be excessively small; and therefore small forces, if unbalanced, will set it into very rapid motion. It has been recently shown by von Helmholtz[1] that it is not possible for an electrically polarizable medium of small inertia so to adjust itself by finite motions, that the electric and magnetic forcives on it shall be in internal equilibrium and thereby avoid producing very intense movements in it,—unless there exists finite slip at the surfaces of the moving bodies which set up the surrounding electric field; and it seems difficult to see how this slip could be allowed, or what circumstances would regulate its magnitude.

in contrast with previous ideas, found untenable by Helmholtz.

Magneto-optic rotation is due to the atomic whirl of electrons, not of aether.

On the present view, the inertia of matter is different in kind from that of the aether, and is possibly to be found in the electric inertia which is possessed by electrons. The well-known considerations advanced by Lord Kelvin, which find in the magnetic rotation of light evidence of a rotatory motion round the lines of magnetic force, still obtain a place here, but in a modified form; it is not the aether itself which is in rotation round the lines of force, but the electrons of the ponderable medium; rotation of the free aether would not here affect the elastic propagation except convectively. We should thus expect the magnetization of the *material* medium to rotate the plane of polarization without altering to any corresponding extent the mean velocity of radiation, just as in Lord Kelvin's theory: and the present theory is not open to objection on the score that the Faraday rotation is easy to observe, while convection of the radiation by a magnetic field has hitherto been sought for in vain.

Aether has to be absolutely frictionless,

It is to be observed that a primary condition for the permanence of material phenomena according to a scheme such as has here been sketched, is that the aether should be absolutely devoid of friction.

1 "Folgerungen aus Maxwell's Theorie über die Bewegungen des reinen Aethers," *Sitz. Berl. Acad.* July 1893, *Wiss. Abhandl.* III. p. 526.

If there were the slightest amount of dissipation, the motions by whose stability the system hangs together would gradually diminish, and finally the positive and negative electrons would fall into each other and thus suffer complete extinction; the whole material universe would in fact gradually vanish, and leave no trace behind.

All formulae with regard to the conservation, under certain circumstances, of linear or angular momentum, or of momenta of more general type as in the case of cyclic fluid motions, are bound up essentially with this absence of friction in the aethereal medium. If friction were present, the relative motions of parts of the system would always be transferring momentum as well as energy out of the system into the aether, and nothing could remain absolutely steady or permanent.

else no conservation of matter,

or its energy.

39. In a recent memoir by H. A. Lorentz[1], on electrical and optical phenomena in moving bodies, the author arrives, starting from the ordinary equations of the electric field, at fundamental equations of the same type as are given by the present theory. The method adopted by him is, as here, to find the cause of electrical phenomena in the motion of ions. He considers them merely as volume distributions of electricity* confined to limited spaces, and with this guiding idea determines the extension of Maxwell's fundamental equations that will most conveniently cover the extension of the problem to include their movements. By transforming, as in § 14, to axes of coordinates which partake of the uniform motion of the material system, various results relating to the independence between electrodynamic phenomena and such motion are deduced, and Fresnel's formula for the effect on the velocity of radiation is shown to be involved. The conception of a molecule, electrically polar by reason of its positive and negative electrons, of its magnetism due to the orbital motions of these electrons, and of its diamagnetism due to changes in the orbits, and the conception of conduction by a *quasi*-electrolytic separation, do not however occur, [the subject indeed hardly arising]. And in consequence of the absence of recognition of the fact that in an unelectrified element of volume there are present equal numbers of positive and negative ions, the relation between electromotive force and ponderomotive force is not attained. Professor Lorentz's conclusions from his system of equations as to the

The procedure of Lorentz's analysis.

[1] *Versuch einer Theorie...in bewegten Körpern*, Leiden, 1895. [See also footnote at beginning of Part III (1897) of the present memoir.]

* If the parts of the charge are supposed to be free to move, their mutual repulsion would prevent permanence. The electron must be of some coherent structure, as here, of the type of a vortex ring, not separable into parts, revealing itself electrically at a distance as a mere point singularity of electric charge.

untenability of Maxwell's scheme of stress [by itself, without momentum] are in agreement with the results of the present theory; but from the indications given, it is not clear that he would admit the form here deduced [now amended] for the mechanical forcive exerted by radiation. As this memoir of Lorentz only came to hand after the greater part of the present paper [Part II] had been developed, the agreement, so far as it goes, may be taken as additional evidence in favour of the validity of the method of investigation employed.

According to the present views, there is nothing paradoxical in the existence of unsaturated atoms or ions in which the two kinds of electrons do not compensate each other; and the properties of these ions will be very different from those of neutral molecules, as it is well known that they are. The fact that a neutral molecule, say HCl, dissociates so that the same element H is always positive, may possibly be ascribed to the catalytic action of the solvent without whose aid the splitting up would not occur. When the solvent is water, or it may be for a gaseous body merely traces of moisture, the decomposition is produced by the formation of an aggregate with the water molecules which splits easily in only one way; but when the solvent is changed, the characters of the charges of the ions might be inverted, as in fact sometimes occurs.

Ionization conditioned by the solvent: which may even be vaporous.

40. If we consider a molecule to be made up of, or to involve, a steady configuration of revolving electrons, it will follow that every disturbance of this steady motion will involve radiation and consequent loss of energy. The only steady configuration of which we can assert that it will remain permanent under all circumstances is, therefore, that one which possesses least total energy. Thus, by extending the conception of absolute stability, which already plays a part in the theories of material systems which lose sensible energy by viscosity, to systems which lose energy by radiation, we obtain a possible mode of accounting for the uniqueness of the atomic configuration and the invariability of its spectral lines. Such a view would also account for diffuse lines or bands being sharp at one edge and hazy at the other[1].

Inhibition of atomic radiation: thus securing structural definiteness for the atoms:

[1] In this connection a suggestive result has recently been obtained by Mr H. C. Pocklington, *Roy. Soc. Proc.* May 1895. As the energy associated with a simple vortex ring in perfect fluid increases, its velocity diminishes, so that a theory of gases in which the molecules are simple vortex rings fails to represent the facts. In the previous paper (§ 87), a theory of simple vortex atoms in a rotational aether, with the appropriate molecular electric charges, was noticed. It is now found that the effect of the electrification may be to reverse the above peculiarity: and, in particular, when the ring has nearly the configuration of minimum energy for its given vorticity and given charge, the total energy varies very nearly as the square of its velocity, that is as the temperature. But this configuration is precisely the one into which such a ring would settle down under the influence of radiation. Similar statements apply in a general

Kelvin vortex atoms become more amenable for gas-theory if electrified.

Minimal energy configurations. The electronic orbital atom.

It is easy to calculate the configuration of minimum energy for the simple but exceptional system of two electrons of opposite sign, considered in § 118 of the previous paper*; in that case, however, the distance between their centres comes out to be rather less than the diameter of one of them, so that they will gradually fall into each other owing to loss of their orbital energy. But this need not be the case if the system contained several electrons of the same sign; for example, if it were roughly of the type of a ring of positive electrons revolving round an inner ring of negative ones. In the special case of that investigation, also, an estimate of the diamagnetic coefficient may be obtained. It is easy to see that the result of imposing a magnetic field H at right angles to the circular orbit of an electron e is the same as if the effective inertia were increased from Le^2 to $Le^2 + eH/\omega$. This effect, though very minute, will be cumulative for all the electrons of the atom, whereas the change of orientation, which represents paramagnetism, will be a differential one as between the positive and negative orbits.

not attained for a pair of revolving electrons.

Model atom with rings of revolving electrons.

Diamagnetism:

paramagnetism.

By a representation like the present, the origin of the vast number of spectral lines of some molecules may be simply illustrated. Thus in the analogous astronomical system of the Sun, Earth, and Moon, which has only nine coordinates, three for each body, there exist the much larger number of periodic inequalities or oscillations that are discussed in the Lunar Theory. The fact is that the oscillations of the coordinates of the system are not themselves harmonic or even exactly periodic; it is only when they are analyzed by the mathematical processes suitable to vibrations, or by a physical instrument such as a spectroscope which yields the same results, that the lines of the spectrum come into existence. The radiation traverses the free aether as a homogeneous whole which is a function only of the coordinates of the vibrating molecule: it is only under the influence of dispersion or diffraction by material bodies that it is broken up into harmonic constituents, which are not independent elements of it in case they exceed in number the internal coordinates of the molecule.

Tentative astronomical view of spectra.

The aether transmits without change.

As in connection with the previous paper on the present subject, I am under great obligation to Professor G. F. FitzGerald, and also to Professor O. J. Lodge, for the readiness with which they have allowed me to consult and profit by their opinions on some of the questions here treated, and for their kindness in communicating the results of the experimental tests described in § 3 of this paper.

FitzGerald: Lodge.

way to the effective vortex made up of a system of revolving electrons, which is considered in the text, and which is not subject to the instability and the other difficulties that would beset an ordinary electrified vortex ring with hollow core.

* P. 524 *supra*. It is the mutual induction between the two electrons that prescribes a minimum. The effect of a magnetic field is simply to alter ω.

41

ON GRAPHICAL METHODS IN GEOMETRICAL OPTICS.

[*Proceedings of the Cambridge Philosophical Society*, Vol. VIII. (1895) pp. 307–313.]

1. The fundamental problem of geometrical optics relates to the modification produced in a filament of light on passing across a series of different media. From the geometrical standpoint the filament is made up of a narrow pencil of rays, which are straight when the medium is homogeneous: and it may be considered as defined by the focal lines of the pencil. The direct analytical method therefore hinges on the determination of these focal lines, after each refraction, from their already known positions before refraction: but the formulae, even for a single refraction, are complicated, and, when there is a question of combining a number of successive refractions, almost prohibitive. In such circumstances, however, as in other physical questions where we have to do with linear relations, the use of graphic methods will be found to lead to results of intrinsic simplicity, and thus give direct insight into the general relations of the subject.

Simpler than the formulae.

2. *Uniplanar System.* In the sketch here to be given of a geometrical method of treatment, it will be convenient to begin with the simple case of a narrow pencil of rays in one plane, and having therefore only one focus: that is, the optical system will at first be a columnar or cylindrical one.

Scheme of the conjugate foci:

Since a single focus now always corresponds to a single focus as image to object, and, the pencil being narrow, all its rays may be taken to a sufficient approximation as passing exactly through the focus, the elementary geometrical theory of Möbius and Maxwell may be applied to the filament. Thus we may determine on its axes at incidence and emergence a pair of principal points, and a pair of principal foci, and we may construct by aid of them the focus conjugate to any other assigned one[1].

determined by an auxiliary conic:

In this way, or by the more usual analytical processes, we see that conjugate foci are connected by a linear relation, as might have been expected *à priori*. Now let us draw the axis of the filament as incident on the optical system, and its axis as emergent from it. Conjugate foci on these two axes are homographically related; therefore the line connecting any pair of them envelopes a conic section which also has contact with the axes themselves. Suppose, by experiment or otherwise, that three pairs of conjugate foci have been determined:

[1] Cf. *Proc. Lond. Math. Soc.* Vol. XX. 1889, p. 182: *supra*, p. 182.

it will be possible to deduce all other pairs by linear construction without any knowledge of the internal constitution of the optical system. It has in fact been shown by Pascal how, given five points on a conic section, the other point in which any line through one of them meets it again may be determined by linear construction; and the conjugate or polar theorem, known as Brianchon's, solves the present problem. The lines connecting the three given pairs of conjugate foci, and the two axes of the pencil, form five tangents; and, simply by drawing three lines, the other tangent through any point on one of the axes is constructed, and this meets the other axis in the conjugate focus. **or by linear construction:**

3. There is an important case in which this general construction reduces to a very simple form. Suppose the problem relates to refraction of the filament of light at a single interface; or to passage across an optical system whose total thickness can be neglected in comparison with the focal distances of the pencil, for example, eccentric refraction across a thin lens, or across a thin plate of a medium with any kind of curvatures in its two faces. The incident and emergent axes meet on this plate, and if the incident focus is at their point of intersection, the conjugate one is obviously the same point. Thus the two axes of the pencil touch the conic at the same point; therefore the conic must be a finite line, and all tangents to it pass through one of its extremities. The line connecting conjugate foci in all such cases therefore passes through a fixed point. **which is very simple for any thin lens system.**

4. *Single Refraction in Primary Plane.* The position of this point is determined if the positions of two pairs of conjugate foci are known.

For the case of a single refraction at a curved surface (including of course reflexion) it is desirable to specify precisely its position, as this case is fundamental in the ordinary analytical theory. Now for a spherical interface there are always two exactly aplanatic points on each radius: the equation of the spherical surface can in fact be thrown into the form $r_1 - \mu r_2 = 0$, the origins from which r_1 and r_2 are measured being inverse points with respect to it. **A spherical interface:**

Let then $U_1 A U_2$ be the path of the central ray of an optical filament refracted at A, at a spherical interface whose centre of curvature is C.

We can find the aplanatic points L_1, L_2 on the path of the filament by inflecting the line CL_1 to make the angle at L_1 equal to the angle of refraction ϕ_2. These points will be exact conjugate foci, no matter how wide the optical beam may be. The easiest construction for these points is to bisect by CL_1 the angle between the radii CA_1 and CA_2 drawn to the points in which the axes of the filament again meet the spherical surface. **the exactly aplanatic points: constructed.**

As above, the point A is its own exact conjugate focus; for any pencil diverging from the point A continues to do so after refraction.

Further, we can find another pair of conjugate foci, not in this case exact, by constructing a beam such that the angles of incidence and refraction are the same for consecutive rays. For if the ray incident at B is to have the same angle of refraction as the ray incident at A, the circle drawn through C, A, B must meet the refracted ray AU_2 in the point where that ray intersects the ray refracted from B. Hence, passing to the limit when B ultimately coincides with A, the points in which the circle on AC as diameter meets an incident and refracted ray are conjugate foci; and the beam from one of these points diverges from the other, after refraction,—at the same angle, since the deviation of each ray is the same. This pair of conjugate foci

A conjugate pair for all indexes.

Method developed for a single refraction.

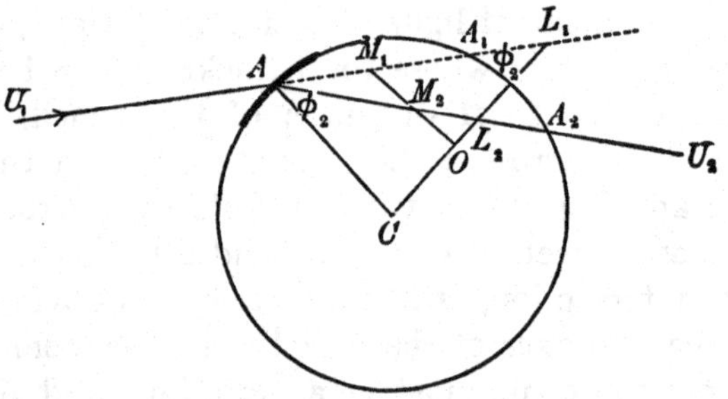

are M_1, M_2, the middle points of the intercepts made by the spherical surface on the paths of the central ray; and it is noticeable that they are conjugate foci whatever value the index of refraction may have.

Graphical solution completed:

As then A is its own conjugate, and the relation between conjugate foci is homographic, the line joining any pair of conjugates must pass through a fixed point. This is the point O in which the line connecting M_1 and M_2 meets CL_1, which is the perpendicular drawn to it from C. The march of conjugate foci in the primary plane is now open to inspection.

The primary focal lengths AF_1, AF_2 of the refracting surface for a pencil incident along the given direction U_1A are the intercepts made on each of the lines AU_1, AU_2 by vectors from O parallel to the other one.

due to Young.

The construction thus obtained for the primary focus after oblique refraction has, I find, been already stated without proof by Thomas Young[1]. "It approaches the nearest to Maclaurin's construction[2], but is far more convenient."

The following construction, ascribed to Newton, is given by Barrow,

[1] "On the Mechanism of the Eye," *Phil. Trans.* 1801.

[2] *Treatise on Fluxions*, § 413.

in his *Lectiones Opticae*[1], and is no doubt the earliest solution of the problem:

Draw AR at right angles to the incident ray meeting CP_1 in R, P_1 being the incident focus; and draw AQ perpendicular to the refracted ray on the side towards C, such that

The Newton-Barrow construction.

$$AQ : AR = AA_1 : AA_2;$$

then the primary focus conjugate to P_1 lies on CQ.

5. As a corollary we have a construction for the centre of curvature at any point P of a Cartesian oval of which S and H are foci. Draw any circle, centre O, touching the oval at P and meeting PS, PH in L, M; find the point R in which the line joining the middle points of these chords meets the bisector of the angle LOM; let SH meet PR in X; draw XF parallel to LM meeting PS in F, and draw FC perpendicular to PS meeting the normal in C, which is the centre of curvature required.

Rule for curvature of a Cartesian:

When the oval degenerates into a conic section, this becomes a well-known construction: from G the foot of the normal draw GF perpendicular to it, meeting PS in F, and draw FC perpendicular to PS meeting the normal in C, which is the centre of curvature required.

of a conic.

6. *Surface of Double Curvature: principal incidence.* The relation between the positions of conjugate foci in the secondary plane is of the same kind as above; they are in a line through C.

Extension to focal lines.

These results may be at once extended to the case of any beam incident in a plane of principal curvature of a non-spherical surface, provided the focal lines of the beam are in and at right angles to this plane. The relation between the conjugate primary foci is given by the above construction by aid of the circle of curvature in the plane of incidence; while the line joining conjugate secondary foci always passes through the centre of the circle of curvature in the perpendicular plane.

7. *General case of single interface.* The solution of the problem of the determination of the form of a general pencil after refraction at an interface of double curvature had been achieved long ago by Malus; but until Maxwell's adaptation[2] of the Hamiltonian method of the characteristic function of the pencil, the theory remained too lengthy and complicated to be of service in optical applications. It

General case:

[1] See also Newton's *Optical Lectures*, read in 1669, published posthumously in 1728.

[2] Maxwell, *Proc. Lond. Math. Soc.* Vol. IV. 1872; Vol. VI. 1874. [The method was entirely Hamilton's own: cf. Rayleigh's quotation, *Sci. Papers*, Vol. V. p. 456, from *Brit. Assoc. Report*, Cambridge Meeting, 1834: also Appendix, *infra*.]

is now well known, after Maxwell, that if the rays of the incident pencil are normals to the surface

$$z_1 - \frac{x_1^2}{2A_1} - \frac{y_1^2}{2B_1} - \frac{x_1 y_1}{C_1} + \ldots = 0,$$

and those of the refracted pencil normals to the surface

$$z_2 - \frac{x_2^2}{2A_2} - \frac{y_2^2}{2B_2} - \frac{x_2 y_2}{C_2} + \ldots = 0,$$

the origin being on the interface, the axis of z along the axis of the corresponding pencil and the axis of x in the plane of refraction, then A_1 is connected with A_2 and B_1 with B_2 by equations of the same form as in the special case of § 6, while C_1 is connected with C_2 by a similar equation. If lengths equal to A_1 and A_2 are laid off along the axes of the pencil, the line connecting their extremities passes through a fixed point; a similar statement applies to the other two pairs of points, and the coordinates of the three fixed points thus obtained have been given by Maxwell. The graphical solution may now be completed by assigning geometrical constructions for them. The fixed point connected with A_1 and A_2 is determined by Young's construction (§ 4) in which the circle is the circle of curvature of the section of the interface by the plane of refraction. The fixed point connected with B_1 and B_2 is the centre of curvature of the normal section perpendicular to the plane of refraction.

the graphical solution extended to it:

The connection between C_1 and C_2 is[1] through the equation

$$\frac{\cot \phi_1}{C_1} - \frac{\cot \phi_2}{C_2} = \frac{\cot \phi_1 - \cot \phi_2}{S},$$

where ϕ_1 and ϕ_2 are the angles of incidence and refraction, and S^{-1} is the coefficient of the product term in the equation of the interface. Thus, when C_1 is infinite, C_2 is equal to

$$S \sin \phi_1 \cos \phi_2 / \sin (\phi_1 - \phi_2);$$

and when C_2 is infinite, C_1 is equal to

$$- S \cos \phi_1 \sin \phi_2 / \sin (\phi_1 - \phi_2).$$

Hence, draw round the point of incidence a circle of radius S, in the plane of incidence: from the point where each axis meets it draw a perpendicular to the normal at the interface, and from its foot a parallel to the other axis of the pencil: these parallels meet in the fixed point required, which is in fact also on the line connecting the points in which the two axes meet the circle.

extended to any thin system.

8. *Any system of negligible thickness.* It follows (§ 3) from the elements of the theory of perspective, that when a general pencil

[1] Maxwell, *Proc. Lond. Math. Soc.* Vol. IV. 1872; Vol. VI. 1874.

crosses any optical system whose thickness is negligible, the refracted pencil may still be deduced from the incident one by the construction of § 7; except that the fixed points must now be determined either once for all experimentally, or else by a succession of linear constructions, one for each refraction.

A construction of this kind enables us with great facility to trace the course of the refraction as the incident focal lines gradually alter their positions. For example, we see that there is always one and only one position of the focus for an incident stigmatic pencil, along a given axis, such that the refracted pencil is also stigmatic; a proposition which is obviously of importance in a general theory of photographic combinations of lenses.

Only one stigmatic pair of planes for thin system.

9. *General Problem.* When the thickness of the optical system across which the filament of light passes is not negligible, the graphical treatment is not so simple. It has, however, been shown[1] that the optical effect of such a system may be precisely imitated by the effect, on a straight filament, of two definite thin astigmatic lenses mounted on the same axis at a definite distance apart; and the relations of this simple system can be represented in a graphical manner by aid of the constructions above. It will be more convenient, however, to combine deviation with the convergence produced by the two lenses, so that the axes of incident and emergent pencils may not be identical. The constants of the filament after refraction through the first thin lens may now be graphically constructed by aid of three fixed points; then they must be transformed to a new origin at the second lens; and finally by another linear construction the constants of the emergent filament may be obtained.

General case reduced to two thin systems.

The complexity inherent in the treatment of optical systems of sensible thickness arises solely from the trouble involved in transferring the analytical constants of the filament of light from one point to another of its axis as origin. This operation can be done graphically just as well as analytically, or better: but the process is necessarily clumsy. Probably the best available construction for this purpose is the one indicated by Maxwell[2].

Maxwell's construction.

(10. The development of the construction by Pascal's hexagram in § 2 places the theory in a very simple light. The problem under consideration is that of the conjugate focus in two-dimensional or cylindrical optical systems; or of the conjugate primary focal line in systems which have a plane of symmetry, or which possess the (for

The Pascal hexagram utilized for focal systems.

[1] *Proc. Lond. Math. Soc.* Vol. XXIII. p. 172 (1892): *supra*, p. 270.

[2] *Loc. cit.*

this purpose) equivalent property that the focal lines of all pencils which are stigmatic at incidence have common directions. We are supposed to know from observation the positions of three pairs of conjugate points, on the straight axes of the pencil in the uniform media at the two ends of the optical system. Let the three incident foci be marked off on any line at their proper distances apart, say A_1, B_1, C_1; and let A_2, B_2, C_2 represent their conjugates marked off at their proper mutual distances on any other line. The problem is to find pairs of points on the two lines which are in homography with these pairs. The graphical solution is to regard the three lines A_1A_2, B_1B_2, C_1C_2, the two given lines $A_1B_1C_1$ and $A_2B_2C_2$, and the unknown line through D_1, arranged in any order, as the sides of a hexagram, when the lines connecting opposite corners will meet in a point. We might then find the elongation corresponding to the conjugate points D_1 and D_2, by determining the conjugate E_2 of a point E_1 very near to D_1, and taking the limit of the ratio D_2E_2 to D_1E_1. And then the transverse magnification will be derived from the theorem, which might be conveniently named after Maxwell, that the elongation of any finite segment on the axis is equal to the ratio of the extreme indices multiplied by the product of the transverse magnifications corresponding to its two ends.

Simplified procedure. But there is a much simpler course open. Let the given foci be laid off so that a pair of them coincide at the point of intersection of the lines: then the lines connecting all other pairs will pass through a fixed point, which is at once determined. The principal foci will be obtained by drawing parallels to each of these lines through the fixed point; and the product of the principal focal lengths will be equal to the product of the distances of any pair of conjugate foci from the respective principal foci. As the ratio of the focal lengths is that of the extreme indices, their values, involving the positions of the Gaussian principal points, are known except as to sign. Whether the positive or negative sign is to be taken, requires the further knowledge of whether the image of some one object is erect or inverted; there being always two optical systems, with equal and opposite focal lengths, that give the same relation of conjugate foci along the axes. The cardinal points being thus determined, everything else follows as usual.)

42

ON THE ABSOLUTE MINIMUM OF OPTICAL DEVIATION BY A PRISM.

[*Proceedings of the Cambridge Philosophical Society*, Vol. IX. (1896) pp. 108–110.]

WHEN a ray of light crosses a prism in a principal plane, it is of course well known, and easy to verify graphically, that the deviation suffered by the ray is least when it crosses the prism symmetrically. It seems also to be recognized that no deviation smaller than this can be obtained when the ray does not pass in a principal plane; though I have not met with any valid demonstration[1] of this result. The following proof may therefore be worth recording.

Rays traversing a prism obliquely:

The inclinations η and η' of an incident and refracted ray to any plane normal to the refracting surface, for example their inclinations to the principal plane of the prism, obey the law of sines

$$\sin\eta = \mu \sin\eta'.$$

Hence, after passing across a prism, the emergent ray is inclined to the refracting edge at the same angle $\frac{1}{2}\pi - \eta$ as was the incident ray.

spherical construction for paths:

When directions are projected on to a spherical surface, let E represent the direction of the edge of the prism, and P, Q those of the incident and emergent rays. Then EP and EQ are each $\frac{1}{2}\pi - \eta$; and if the arcs EPp and EQq are each a quadrant, p and q will represent the projections of these rays on the principal plane of the prism.

These projected rays are refracted across the prism in accordance with the law of sines, the index being not μ but

$$\mu \cos\eta'/\cos\eta, \text{ where } \sin\eta = \mu\sin\eta'.$$

If D denote the actual deviation PQ, and d the projected deviation pq, then, from the isosceles spherical triangle,

$$\sin\tfrac{1}{2}D = \sin\tfrac{1}{2}d\cos\eta.$$

Thus, of all incident rays which have the same inclination $\frac{1}{2}\pi - \eta$ to the edge of the prism, that one has its projected deviation d, and therefore also its true deviation D, least, whose projection passes across the prism symmetrically. This least value is given by the equation

partial result.

$$\sin\frac{1}{2}(d + A) = \frac{\mu\cos\eta'}{\cos\eta}\sin\frac{1}{2}A,$$

where A is the angle of the prism.

[1] The proof quoted by Czapski, *Treatise*, p. 156, from R. S. Heath, *Treatise*, p. 31, does not seem to be valid.

As $\mu \cos \eta'/\cos \eta$ is greater than μ, it follows that d is greater than D_0, the minimum deviation for an actual ray passing in the principal plane: but as it is also greater than D, no inference can be drawn in this way as to the relative magnitudes of D and D_0.

The absolute minimum.

We may, however, find the absolute minimum of D by comparing with one another the rays, corresponding to different values of η, whose projections cross the prism symmetrically. In the annexed spherical diagram E represents the edge, N, N' the normals to the

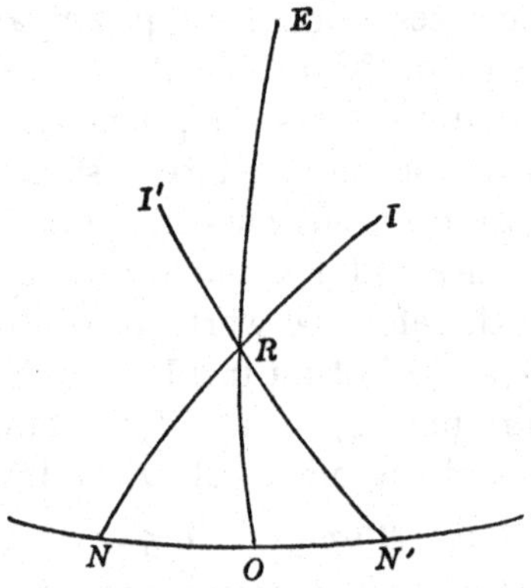

faces, I, I' the incident and emergent rays, and R the ray inside the prism. The pole of NN' is E, and EI, EI' are equal; the symmetry of the projection of the rays on the plane NN' requires that ON, ON' shall be also equal, so that ERO is the plane bisecting the external angle between the faces. Thus as $II' = D$, $NN' = A$, we have

$$\frac{\sin \frac{1}{2}D}{\sin \frac{1}{2}A} = \frac{\sin (i - r)}{\sin r},$$

where i, r are the angles of incidence and refraction IN, RN.

It is easy to see by geometrical construction that the right-hand side of this formula increases with r. It then follows that D is least when r is least, that is, when R is at O: thus the absolute minimum of deviation belongs to the ray passing in the principal plane of the prism.

43

ON THE GEOMETRICAL METHOD.

[Presidential Address to the Society for the Improvement of Geometrical Teaching, 1896.]

As I can pretend to only limited practical experience in the matter of teaching the foundations of Mathematics, and of Geometry in particular, I have thought that the time allowed for my remarks may be advantageously occupied in a rapid survey of what the geometrical method has accomplished, and the estimation in which it has been held at different periods of the history of the development of Mathematical Science. Thus, avoiding expert questions relating to Education, we may begin by considering from a general standpoint what is implied in the science of Geometry, for the right ordering of which, in its educational relations, this Association has been working for a long series of years. It would, I presume, be against the instincts of our members to interpret the subject in any narrow sense. Geometry is usually based on the partly mathematical, partly logical, framework which has descended to us from Greek times, and which is chiefly associated with its earliest great systematization in the work of Euclid. But modern development would perhaps render any definition of its scope incomplete that did not include all methods in formal and physical Mathematics which proceed by direct contemplation of the relations of the subjects considered, instead of through representations of them that are merely quantitative or numerical.

Enlarged geometry.

From this standpoint I think it will appear that the geometrical method has always been the one for which British predilection has strongly declared itself. After the early period, when Englishmen like Recorde assisted in forming the foundations of Algebra, it is noticeable how few of the improvers of the machinery of Analysis have been of British race. Even the Infinitesimal Calculus was directly based by its English founder on geometrical ideas, in marked contrast to the analytical notation which was a chief merit of Leibniz's rival scheme. During the whole of last century we had few cultivators of algebraical Analysis to place opposite the Bernoullis, Euler, Lagrange, and a host of Continental developers of that field. In the present century we have indeed been instructed in, and have taken full advantage of, the new weapons of analysis, but that at a time

Algebraic analysis mainly imported,

when they had nearly reached maturity of form, and opportunities of improving them were not so numerous. The fundamental analytical improvements due to Hamilton had intimate relation to the geometrical standpoint; and there perhaps remains in addition to these only the modern algebra of Cayley and Sylvester, which may be claimed as an analytical product chiefly British.

until Hamilton.

But alongside this national preference for graphical methods, which is associated no doubt with the practical and concrete character of the British genius, there is the remarkable fact that the narrower science of Pure Geometry proper has hardly formed, at any rate during the middle part of this century, any part of academic instruction in England. To take an illustration, during that time the geometrical portion of the Mathematical Tripos Examination at Cambridge was meagre in the extreme. It consisted of a few propositions from Euclid, to be demonstrated in the conventional form adopted by the English translations, and of a few original geometrical problems —riders as they were called—attached to the propositions, which were often more of the nature of puzzles than illustrations of geometrical ideas—a circumstance due in part to the absence of any organized treatment of geometrical method, and which were usually of excessive difficulty. It was, in fact, in the University of Dublin that the subject of Pure Geometry in the middle of this century showed organic growth, partly arising from the intimate connection maintained with the French school; and it was, no doubt, mainly through the influence of the writings of Dublin Mathematicians that Pure Geometry has regained an adequate place in the English educational system. Reference should, however, not be omitted to the continuous cultivation of Geometry in the Scotch Universities from the times of the Gregories and Maclaurin downwards. There is, perhaps, some ground for an induction that the Celtic mind is specially apt at the subtle kind of ingenuity which is necessary to the successful cultivation and explanation of the ramifications of this science; while the Teutonic mind is, with some exceptions, more attracted to searching after the ultimate philosophical basis of geometrical cognition than to exploring the immense landscape that is dominated by the axioms and postulates of Pure Geometry.

Systematic geometry hardly taught:

except in Ireland,

and Scotland.

I was about to express an opinion that Geometry is the queen of the Sciences; that dignity is, however, already bestowed. We may, at any rate, recall the recognition which has been granted in all ages to the claim that Geometry sets the example and points the method which every science involving long trains of deductive reasoning should follow. During the time of early development of a branch of knowledge, shorter and more empirical methods of explanation

Geometry the ideal method.

will prove economical; but the perfect form of the completed structure will usually approximate more and more to the geometrical ideal. According to Plato himself, the secrets of Philosophy are closed to those who do not bring with them a geometrical training. And in the modern age, when a Descartes or a Spinoza wishes to subject his speculation to the most searching scrutiny, or to claim for it irrefragable demonstration, he by instinct expresses it in severely geometrical form.

When it is an abstract survey of the chain of connection between thought and language that is aimed at, we are led to the study of the science of Logic and to a formulation of the principles of logical Analysis. But when it is the orderly development of a complicated system of reasoning that has to be undertaken, the desirable auxiliary is rather a perception of the actual scientific method which has served to express and to guide the growth of the most far-reaching and widely ramified train of pure deduction that the human mind has accomplished.

The revival of formal analysis in England:

It is, however, not difficult to understand the enthusiasm with which, three-quarters of a century ago, the group of younger men, among whom were prominent Woodhouse, Herschel, Peacock, and Babbage, threw themselves into the task of introducing the Continental Analysis into the course of study at the Universities of this country. The range of subjects to which the Newtonian geometrical methods were at that time applied in the schools, was for the most part limited to portions of the *Principia* itself, and to those portions in which the genius of its author had left least to be gleaned, except in matters of detail, by his followers. The problems then demanding solution, in Celestial Mechanics, were of unrivalled precision; but numerical calculation by unaided geometrical methods had been pushed about as far by Newton himself as it was possible for human ingenuity to carry it. What was chiefly wanted was the development of a new Calculus, which could compute the arithmetical results that flow from the theory of Universal Gravitation, and find out whether they corresponded, or were in discord, with the precise facts of Astronomy. It was, in fact, necessary that the direct geometrical exposition of Newton should give place to a method in which abstraction was made of the actual phenomena in order that the mind might be concentrated solely on the process of calculation. The Analysis of the Continental mathematicians was thus historically, to a great extent, a product of the necessities of this problem of Physical Astronomy. It became a tremendous engine for computation; and its very success for that purpose often somewhat obscured the deeper scientific aspects of its methods, and so left a rich harvest to be gleaned

an indirect product of the planetary dynamical astronomy.

from the writings of the analysts by subsequent investigators, whose more geometrical method of training incited them to penetrate further beneath the symbolism.

Contrast with strides in constructive physics and chemistry:

The progress attained by this analytical Calculus must have seemed almost miraculous to minds in this country that were familiar only with the purely geometrical methods, and had grown used to the limitations of their power in the direction in which the efforts of theoretical astronomers then lay. But the notion sometimes expressed that the eighteenth century was one of mathematical stagnation in Great Britain is one for which there is perhaps not much warrant, at any rate, on the Physical side. The foundations of the science of Electricity were, at that very time, being steadily and quietly laid by Cavendish, and those of the science of Heat by Black and others. And when at the very beginning of the present century theoretical Optics awoke to new life, which heralded the introduction of the modern Molecular Physics, the first long steps were taken by a Newtonian, whose erudition in the Continental Analysis was only less wonderful than the strength of his preference for the more direct geometrical methods of the Newtonian school. And when, again, the science of Electricity, and with it the whole method of modern Mathematical Physics, had to be reconstructed in the presence of a now wide and keenly interested scientific opinion, the place of honour came to the lot of an investigator whose mathematical equipment was confined to a keen perception of space-relations, who had the advantage of knowing nothing of the methods of algebraical Analysis, as at that time developed, and so was not biassed by the spell of their power, which, however effective in Astronomy, tended to lead into a wrong track in the newer subjects. The discovery of new principles in Natural Philosophy by the purely abstract process of algebraical reasoning requires indeed the highest qualities of intellect, and has perhaps been achieved to any considerable extent only by the great masters in Science. In the case of ordinary minds the study of Algebra usually only leads to a capacity to follow their reasoning and apply it to particular instances. In the geometrical method it is the thing itself that is before the mind instead of a numerical symbol of the thing; the training is in the direct survey of the relations and connections of different things, and in the simplest expression of these relations, as witness the subjects of Cartography and Graphical Mechanics. It has even happened that many of the most fertile ideas in modern algebraic Analysis have been directly transplanted from descriptive physical theories.

connected with national geometric mode of thought:

It would seem that the line of development of abstract mathematical thought in its higher branches has at length begun to turn aside from the purely analytical operations of differentiation and integration into which it was guided by the calculations of Physical Astronomy. The present aim is rather to establish a broader basis for the results of Analysis, and to reach wider points of view, by taking a survey of the distribution of the quantities it deals with, by the aid derived from representing them as mapped out in space, and thus exploring their relations through the geographical connections of the regions with which they become associated. It is perhaps safe to assert that the fabric of higher Pure Mathematics could hardly have its present vital qualities of growth apart from the help it receives from geometrical intuition. It has happened that when the more artificial analytical methods were in danger of sinking under the weight of their accumulated results, the powerful resources of space-perception have come to the rescue, and started a new line of progress.

its recent revival.

Each successive development of symbolical calculation seems to be limited in scope, and finally to reach a stage in which the mind cannot follow it further until it is revivified by being brought back to the fountain of direct intuition. The artificial ideas of the Calculus again give place to contemplation of the actual aspects of phenomena in time and space, but in the light of increased knowledge; and when a right orientation has thus been secured, a new harvest of numerical or quantitative results is once more within the power of Analysis. The algebraical method in this sense never supersedes the geometrical. The more vigorous and fully developed our geometrical ideas become, the more effectively shall we be able to coordinate the phenomena of experience, and thereby gain fruitful ground for the operations of the Calculus. But to be efficient for this purpose the geometrical ideal must be kept pure, and in constant relation to intuitive perception: a highly specialized geometrical theory is as feeble an instrument for tracing the wide general relations of things as abstract Algebra itself. The recognition of this informing power of Geometry as distinguished from the numerical or computing power of Analysis has in recent years received wide exemplification in the domain of the engineer. Owing to the somewhat vague and roughly approximate nature of the data usually at his disposal, the mechanical engineer very rarely wants exact numerical solutions of his problems; and the general views that can be derived from the simple inspection of geometrical constructions very often carry with them information which is sufficient for his purposes, in cases where a numerical solution would be far beyond the range of analytical calculation.

Inspiration of analysis by geometry.

The broad features of the distinction between the geometrical and the analytical methods of mathematical reasoning reveal themselves strikingly in the two main types of treatises on general Physics. In

The physical method in dynamics:

this country the model, as illustrated by such books, of different epochs, as Young's *Lectures on Natural Philosophy*, and Thomson and Tait's *Natural Philosophy*, has usually been a description and correlation of facts and laws over a wide field, enforced and illustrated by the application of such analytical methods as are capable of being presented in compact and handy form, and are therefore presumably within the reach of most people who are qualified to cultivate the subject. There is another type of treatise, intended only for the small body of competent specialists, which first surveys the limited field that it is to cover, then assigns a symbol to represent each variable in the problem, forms its equations and sails away into the ocean of algebraic Analysis, ultimately attaining results, perhaps few, but usually important, which for a long period may have to be taken on faith by the great majority of students. Even here, however, the algebraic method is not so self-centred as might appear. As already mentioned, its processes were originally invented mainly in order to deal with the problem of the inequalities in the movements of the bodies constituting the Solar System. A very precise descriptive knowledge of these inequalities was already in existence, as the result of centuries of astronomical observation: and the formal or geometric aspect of the phenomena, as thus crystallized by observation and reflexion, was the best possible guide to the character of the Analysis which would be most appropriate to explain and verify them. And afterwards it was not so difficult to pass on from the extremely simple and precise relations of the Sun and planets, to the generalized conditions of any dynamical system; and with the help of the geometrical principle of Virtual Work, and the semi-geometrical principle of Least Action, which had long formed part of the philosophy of descriptive Mechanics, to finally arrive at the great generalizations of Lagrange and Hamilton in Analytical Dynamics, whose application and verification throughout the range of physical phenomena is now going on.

in contrast with Lagrange's analytical method:

This view of the growth of Analytical Dynamics is in striking contrast to Lagrange's final presentation of its results in the *Mécanique Analytique*. A cardinal announcement in his Preface is the famous and occasionally much censured sentence, *On ne trouvera point de Figures dans cet ouvrage*. The framework of Dynamics has been cast by him into such a form, that when once the coordinates sufficing to express the positions and properties of the bodies composing the dynamical system are given, its subsequent history is deduced by a regular analytical process. "Les méthodes que j'y expose ne demandent ni constructions, ni raisonnements géométriques ou mécaniques, mais seulement des opérations algébriques, assujéties à une marche regulière et uniforme." This sentence represents the crowning success of an analytical method, to reduce everything to a calculus in which no further examination or independent consideration of the data is required. And here it is to be observed, on the other hand, that such an analytical generalization may lead into new regions that might not have been discovered at all by methods which kept closer to observation. The coordinates of a dynamical system may have their meaning widely generalized; and a comparison of the results of such a process with the facts of Nature may lead to yet more sweeping generalizations, and thus gather into the domain of pure Dynamics phenomena whose dynamical aspect would hardly have been unravelled by direct methods. It is in this kind of way that a dynamical basis is gradually being evolved for the phenomena of Heat and Electricity and Chemical Action, and will possibly some day be worked out for the secrets of the constitution of Matter itself.

an instrument for generalization:

A striking instance of the generalizing aspect of Analysis is in fact furnished by Lagrange's own treatment of the subject of Kinematics, or the pure geometrical theory of Motion. As has been already mentioned, he discards altogether the use of diagrams to help in the representation of the movements of solid bodies. So he is confined to conceiving motion as simply change in the values of his coordinates as the time passes; and he finds that some kinds of change are possible, representing motions of translation and rotation, but that others are impossible, if the body is rigid, and the distances between its parts are thus to remain unaltered. The various points of the moving body are known to his Analysis only by coordinates or measurements of position referred to some ideal framework with respect to which it moves; and the test which is to decide whether a given mode of change of these coordinates represents a possible motion of the rigid body, is that the square of the distance between each two points, represented to him only as a certain quadratic function of their relative coordinates, is to remain constant. If the idea had lain in Lagrange's path, it would have been an easy matter to examine the character of those changes of position that do not satisfy this condition, and thus investigate how far this notion of distance is involved in an idea of space, which regards it as merely a simple *continuum*, in which movement or change of position can go on. This question has actually received its answer at a much later date from von Helmholtz.

in his treatment of kinematics,

and of geometry:

But if in highly abstract fundamental discussions of this kind, the release from current habits of thought, which is conferred by purely analytical reasoning, is a powerful factor in Analysis and generalization, we may recall that on the other hand the first account which appeared, of the possibilities of wider laws of space-relation than the ones that belong to our experience, came from Lobatchewsky of

generalized.

Kasan, whose centenary has just been celebrated, and that they proceeded on the lines of strictly Euclidean development.

In these remarks we have allowed ourselves to stray somewhat from the subject we began with, which was to survey the characteristics of the geometrical method of investigation, and take note of the qualities that belong to it. Our conclusion will perhaps be that

The contrast.

there is no strict line of demarcation between Geometry and Analysis; that each method flourishes by means of the aid derived from the other, that when the labyrinth of Analysis becomes too complicated to be threaded by any finite clue, we must hail back to a geometrical standpoint for new fundamental conceptions; that on the other hand when the possibilities of wider geometrical domains than we are accustomed to are under investigation, we can remove our prepossessions as to space-relationship, however deep-seated, by casting the processes into a severely algebraic mould. But when all is counted up on both sides, there remains the fact that Analysis itself approaches perfection by coming closer to geometrical ideas, so that the geometrical method is the master-key in discussions about continuously varying magnitude.

Practical conclusion.

And this supplies a sufficient reason why the mode of formulation of the fundamental principles of Geometry, whether for philosophical or for educational purposes, has always been considered so essential a part of the general subject of scientific method. It may also allow us to entertain the view that in geometrical instruction there are two distinct ends to be aimed at; we must not neglect the severe training in the formulation of the logical basis of knowledge which is involved in a valid presentation of the principles of the Science; nor, on the other hand, must a too strict regard for logical form prevent the acquisition of that almost intuitive familiarity with the properties of figures in space, which is derived from practice in easy problems of Pure Geometry, and forms one of the most valuable trainings for work in most branches of exact knowledge.

44

ON THE THEORY OF MOVING ELECTRONS AND ELECTRIC CHARGES[1].

[*Philosophical Magazine*, August 1896.]

IN an interesting paper by Mr W. B. Morton, communicated by the Physical Society to the *Philosophical Magazine* for June, there is a criticism of a portion of my paper on "A Dynamical Theory of the Electric and Luminiferous Medium[2]," which if valid would affect its whole tenor. As, however, the formulae of that paper were to a considerable extent obtained by two independent trains of reasoning, it would have to be shown that both were wrong before an error could be fully substantiated. As a matter of fact, the criticism arises from reading into the analysis assumptions which are not there, but which had been used, with due limitations, in another place in the previous part of the paper. As the point is really fundamental, and as the analytical statement in the memoir is no doubt too brief to convey at once a grasp of the procedure employed, without somewhat detailed consideration on the reader's part, I beg leave to offer the following general explanation.

Faraday's electron:

The facts of chemical physics point to electrification being distributed in an atomic manner, so that an atom of electricity, say an electron, has the same claims to separate and permanent existence as an atom of matter. The fundamental question then is, how far the conception of separate isolated electrons, pervading the aether of free space, can provide an explanation of electrodynamic and optical phenomena. In the paper referred to I have gone further back, and have considered the question how far such a simple underlying scheme is able by itself to provide an explanation of physical phenomena in general; for it will obviously not be permissible to import into our dynamical notion of an atom of matter more than simple electric properties, unless these latter prove to be insufficient to include all actual knowledge of its relations. The conclusion arrived at in the memoir is that there is nothing in the ascertained laws of general physics which points to insufficiency in that scheme; while there are some experimental results which somewhat militate against the

applied to electro-dynamics and optics.

Gives a scheme wide enough for general physics:

[1] Communicated by the Physical Society.

[2] *Phil. Trans.* 1894 (A), pp. 719–822, and 1895 (A), pp. 695–743: as *supra*.

and even pointed to. existence of interatomic forces of any kind other than those included in it.

Aether not analogous to matter: The main feature of the theory referred to is that the aether is not matter, as ordinarily assumed, nor in any way like matter; it is the uniform substratum (analytical basis, if one is disposed to use that term, for it can never be the direct object of perception) in which the atoms of matter consist as permanent configurations of strain and but simpler: motion. As was to be expected, the relations of inertia and elasticity of this uniform medium are simpler than those of matter, which is merely a molecular aggregate involved in its constitution. In fact, the only way to arrive at a scheme of the relations between aether and matter which shall be a complete dynamical theory and not merely descriptive, is to abolish the apparent duality in the pheno- while matter is an atomic aggregation inherent in it. mena, either by taking as here the molecules of matter to constitute singularities (in the mathematical sense of the theory of functions*) in the uniform aether, or else by trying to make out the aether to be ordinary matter, and so giving up any attempt to explain why matter Necessity for atoms. is molecularly constituted. This molecular constitution of matter is essential to the former theory, just as it is to all other theories or illustrations, like the vortex theories, which hypothecate a uniform underlying medium; it is quite unintelligible—or rather quite unexplained—on the latter type of theory.

But however these things may be, the point criticized does not involve any considerations so refined, or—as possibly may be said—so ambiguous. The sections to which objection is taken claim to be a reconstruction of ordinary electrodynamic theory on the basis of Formal electro-dynamics does not require an ultimate model for the aether. permanent electrons associated with the atoms of matter. Whatever view one may entertain as to the presence of qualities other than electric in the atom, all are, I think, nowadays agreed that the electron is there. And whatever view one may have as to the validity and sufficiency of an aether with simple rotational elasticity, the formal equations to which that theory leads *for free space* are just those equations of Maxwell which Hertz's experimental work has fully verified. The problem of electrodynamics is then that of the free aether, whose properties are represented analytically by these acknowledged equations, disturbed by the action of the electrons of material atoms moving about in it. The original Amperean electrodynamics, proceeding by consideration of elements of current, has not proved valid or sufficient in matters involving electric radiation, or even

* It seems to have been sometimes supposed that this is meant to be taken literally: the meaning intended is that this course is sufficient for the problems in hand, which are independent of the unknown internal structures of the electrons or atoms.

ordinary electrodynamic force. A most successful modification of it was that proposed by Weber, in which elements of current were replaced, as the fundamental object of consideration, by moving electric particles which acted on each other *at a distance* according to a law of force involving their velocities. This theory was, however, shown long ago by Lord Kelvin and Professor von Helmholtz to be untenable, on account of its violating the principles of the modern theory of energy: now, of course, direct action at a distance is altogether out of court. The present question is whether a theory of electrons which act on each other, not directly according to a law of force, but mediately by propagation of the effect across the intervening aether, suffices to avoid the discrepancies of earlier theories and give a consistent account of electrical and optical phenomena: and it is maintained that the answer is altogether in the affirmative. This question is, presumably, sufficiently important and fundamental to justify the present detailed explanation.

Ampère: Weber: Kelvin: Helmholtz: culminate in a Maxwellian electron theory.

At the end of the first of the two papers referred to, building chiefly on the analytical results of previous theorists, the steady aethereal disturbance carried along by a moving electron had been investigated, and also the law of the force exerted on each other by two moving electrons through the intervention of the aether between them. This was on the hypothesis that each electron carried along with it a steady trail of aethereal disturbance, but that no sensible derangement of this steady motion ever occurred such as would lead to loss of the energy of the system by the starting of waves. If the velocities of the electrons remain always small compared with that of radiation, then, however their mutual influences alter their motions, this steady trail will instantaneously adjust itself to the new conditions without sensible excitation of radiation, and the theory will apply. But if any of the electrons are moving with velocities comparable with that of radiation, a change in velocity will involve derangement of this steady trail of aethereal strain and motion, giving rise to wave-motion which will carry off some of the energy by radiation. Accordingly in such a case it is altogether nugatory to speak of laws of action between electrons: the complete theory must then take account not only of the positions and velocities of each of the electrons at each instant, but also of the state of each volume element of the surrounding aether. And the theory of mutual actions of electrons as expressed in the memoir was in fact thus restricted to cases in which their velocities were small compared with that of radiation: unless that condition is satisfied there is no such theory at all.

Electrodynamics of motions slow compared with light: does not involve radiation, so need not involve an aether directly.

In the second paper (§ 15, *seqq.*) the general problem is attacked: it is now not a question of a set of electrons by themselves, each with a definite steady trail, but of the aethereal medium in general, *including* such electrons as exist in it. The analysis there given determines from foundations which all who adopt Maxwell's electrical scheme for free aether must allow, expressions for the force (P', Q', R') which acts on an element of volume of the free aether, and for the force e (P, Q, R), ordinarily called electric force, which acts on an electron e; and it uses these forces for further development of the theory. What computation virtually does is to assume that the trail of each electron is steady, and then to transfer to the electron itself the forcive due to (P', Q', R') acting on this aethereal trail. In the special case of no radiation, and of velocities small compared with that of radiation, this forcive can, as above explained, be transmitted through the aether to the electron itself, and be supposed there applied. But to so transmit it in general is to miss the point of the theory, and, as Mr Morton himself remarks, to reach the absurdity that the force on a moving charge depends not only on the state of the surrounding aether but on the state of the aether at a distance.

Aethereal force distinct from electric.

As regards the main subject of Mr Morton's paper, it may be of interest to state the following general theorem. Suppose a system of charged conductors is in steady translatory motion through the quiescent aether with velocity u, and let v represent the velocity of radiation in free aether: consider a correlative system of conductors obtained by uniform geometrical elongation of the actual system along the direction of motion in the ratio of $(1 - u^2/v^2)^{-\frac{1}{2}}$ to unity, and find the electrostatic distribution of the same charges on this system supposed at rest: then the actual distribution of the charges on the moving system will be exactly correlative, viz. equal charges will exist on all corresponding elements of the two systems. This proposition is, however, limited to the case in which none of the bodies of the moving system are dielectrics, but all are conductors*.

Simple effect of convection on a system of charged conductors, in free space.

* It is the exact Lorentz transformation for the case of steady motions.

45

ON THE THEORY OF OSMOTIC PRESSURE.

[*Proceedings of the Cambridge Philosophical Society*, Vol. IX. (1897) pp. 240–242.]

As osmotic theory is now attracting general attention in this country, it seems desirable that all the positions that are maintained in regard to it should be clearly set forth. The excuse for offering the following remarks is that for some time I have paid attention to the subject in its relation to general molecular theory, both in the thermal and the electrical aspects. I fail to recognize how the validity of the thermodynamic basis of the law of osmotic pressure can be shaken: and though the idea of ionic dissociation in solutions is an additional hypothesis which must be judged separately by the extent of its agreement with the facts, it appears to me that in some form—possibly not at all in the chemical imagery with which it is at present often expounded—it holds the field. It is difficult in fact to see how the hypothesis that the same chemical element can have different valencies in different series of compounds, which is now usually accepted, is fundamentally any whit less paradoxical than the hypothesis of ionic dissociation [in explanation of osmotic anomalies]; anything that throws light on the real nature of the one must also illuminate the other.

Its foundation independent of hypothesis.

In his recent note on this subject[1], Lord Kelvin appears to allow, within certain limits, the cogency of the argument which bases the law of osmotic pressure on Henry's empirical law of solubility for gases, an argument which has recently been carefully restated by Lord Rayleigh, having previously been employed, as he remarks, in forms more or less explicit, by van 't Hoff, Nernst and other investigators. The connection thus established, however, hardly amounts to a physical demonstration, because it only deduces one empirical relation from another. Yet it seems desirable to draw attention to the fact, which I have not seen anywhere remarked, that this method had been employed by von Helmholtz in 1883, two years before van 't Hoff announced his theory of the correlation between osmotic and gaseous pressures; and that the principles given by him in an investigation of the work equivalent of gaseous solution, made in

Involved in Henry's law of solubility of gases:

the connection already latent in Helmholtz's earlier theory.

[1] *Proc. R. S. Edin.* Jan. 1897; *Nature*, Jan. 21, 1897, p. 272.

connection with the theory of galvanic polarization[1], involve in fact an implicit prediction of the osmotic law. This circumstance, that the law of osmotic pressure, as regards dissolved gases, is tacitly involved in von Helmholtz's equations, does not of course confer on him a position in the actual development of the subject.

General theory of dilute systems:

But the theory of osmotic pressure can, I think, be placed on a purely abstract basis independently of the law of solubility of gases, which would then assume the form of a deduction from it. The broad principles on which this is to be done have been in fact laid down in a precise and very general manner, but without special application, by Willard Gibbs as early as 1875, in his fundamental development of the laws of mechanical availability of energy[2]. The following position is, I believe, sound*. Each molecule of the dissolved substance forms for itself a *nidus* in the solvent, that is, it sensibly influences the molecules around it up to a certain minute distance so as to form a loosely connected complex, in the sense not of chemical union but of physical influence. The laws of this mutual molecular influence are unknown, possibly unknowable; but provided the solution is so dilute that each such complex is, for very much the greater part of the time, out of range of the influence of the other complexes, as for instance are the separate molecules of a free gas, then the principles of thermodynamics necessitate the osmotic laws. It does not matter whether the nucleus of the complex is a single molecule, or a group of molecules, or the entity that is called an ion: the pressure phenomena are determined merely by the number of complexes per unit volume. To determine the osmotic forces, we must know the change in *available* energy that is involved in dilution of the solution by further transpiration of the pure solvent into it. In finding that change, the laws of mutual action between molecules of the dissolved substance are not required: for there is actually no action between them, and as soon as the solution becomes so concentrated that such mutual action between the complexes comes in, the theory is no longer exact. Nor are the laws of mutual action between the molecules of the dissolved substance and those of the solvent required, because the effect of transpiration of more of the solvent into the solution is not in any way to alter the individual complexes. The change in available energy

simple so long as the spheres of influence are independent:

no other condition enters:

for the change of available energy is due only to wider separation in space:

[1] H. von Helmholtz, "Zur Thermodynamik chemischer Vorgänge iii"; in collected papers Vol. III. pp. 105–114, especially his equation (4), and the theory of diffusion at the end. (The law had also been formulated explicitly in this manner by Willard Gibbs as early as 1876, *loc. cit. infra*, p. 227.)

The law goes back to Gibbs.

[2] *Trans. Connecticut Academy*, Nov. 1875, p. 138, "Effect of a Diaphragm (Equilibrium of Osmotic forces)."

* Contrast with Willard Gibbs' different abstract treatment of "dilute systems" in general.

of the system, on dilution, thus solely arises from the expansion of the complexes into a larger volume; and it can be traced into exact correlation with the change of available energy that occurs in the expansion of a gas. This argument meets the objection that a true theory should involve a knowledge of the molecular actions between the various molecules. It would seem that with just the same cogency it might be argued that a real investigation of the connection of the alteration of the freezing point of a liquid by pressure and its change of volume on freezing should involve a knowledge of the individual molecular actions in the liquid: and so it would, had we not the means of evading considerations of molecular constitution that is afforded by Lord Kelvin's great principle of dissipation, which is for this very reason at the basis of all physical theory.

thus a universal law must exist:

There is, however, one point to be remembered, namely, that the theoretical osmotic pressure is a limiting value which may not be reached by an actual arrangement, unless we can be certain that it works reversibly and so without heating effects.

but as an *optimum* law.

The remark has been made by Lord Kelvin, that the connection between Henry's law and the osmotic law must break down when the solution of the gas is accompanied by change in its state of molecular aggregation. It is also probable from the fundamental ideas as to dissociation and aggregation, that such change would usually be partial, and not uniform over all the dissolved molecules; so that it is not to be expected that Henry's law would in such circumstances hold good. The point in which the argument, as set forth in precise form by Lord Rayleigh[1], becomes then inapplicable, is that the gas expelled from solution by the osmotic process must be considered as emerging in the actual state of aggregation differing from that of its free condition, and its return to the latter state involves further change of available energy.

Complication of partial ionization: its cause.

If the considerations above stated, which will be most suitably developed in detail in another connection, are valid, it follows that Professor Poynting's[2] recent suggestion with a view to evading the necessity of the ionic dissociation hypothesis cannot avail, as it would not lead to the desired value for the osmotic pressure: that pressure depends on the number of molecular complexes involving the dissolved substance that exist on the dilute solution, but not on their individual degrees of complexity*.

[1] *Nature, loc. cit.* p. 254.

[2] *Phil. Mag.* Oct. 1896.

* The sphere of influence of an ion would, however, probably be much larger, and the simpler law would begin to fail at smaller concentration.

46

THE INFLUENCE OF A MAGNETIC FIELD ON RADIATION FREQUENCY.

[*Proceedings of the Royal Society*, Vol. LX. p. 514. Received and read February 11, 1897.]

IN the course of the development of a dynamical hypothesis[1] I have been led to express the interaction between matter and aether as wholly arising from the permanent electrons associated with the matter; and reference was made to von Helmholtz (1893) and Lorentz (1895) as having followed up similar views. A footnote in Dr Zeeman's paper* has drawn my attention to an earlier memoir[2] of Lorentz (1892), in which it was definitely laid down that the electric and optical influences of matter must be formulated by a modified Weberian theory, in which the moving electrons affect each other, not directly by action at a distance but mediately by transmission across the aether in accordance with the Faraday-Maxwell scheme of electric relations. The development of a physical scheme in which such action can be pictured as possible and real, not merely taken as an unavoidable assumption which must be accepted in spite of the paralogisms which it apparently involves, was a main topic in the papers above mentioned.

The Lorentz electronic theory.

The experiments of Dr Zeeman verify deductions drawn by Lorentz from this view. It might, however, be argued that inasmuch as a magnetic field alters the index of refraction of circularly polarized light, which depends on the free periods of the material molecules, it must therefore, quite independently of special theory, alter the free periods of the spectral lines of the substance. But the actual phenomena do not seem to be thus reciprocal. On the electric theory of light it is only the dispersion in material media that arises from direct influence of the free molecular periods, the main refraction arises from the static dielectric coefficient of the material, which is

Zeeman not the converse of Faraday effect:

but both are of gyrostatic type,

[1] *Phil. Trans.* 1894, A, pp. 719–822; 1895, A, pp. 695–743.

* It was the improbability of an observable effect if electrons were of mass comparabie to atoms that induced the writer to suggest to Lodge the importance of verifying the result, as soon as it became known through an abstract of a single sentence in *Nature*, Dec. 24, 1896.

[2] H. A. Lorentz, "La Théorie Electromagnétique de Maxwell, et ses Applications aux Corps Mouvants," *Archives Néerlandaises*, 1892. Cf. especially § 91.

not connected with the periods of molecules[1]. From the phenomena of magneto-optic reflexion it may be shown that, on the hypothesis that the Faraday effect is due to regular accumulated influences of the individual molecules, it must be involved in the relation between the electric force (PQR) and the electric polarization of the material ($f'g'h'$), of type*

$$f' = \frac{K-1}{4\pi} P - c_3 \frac{dQ}{dt} + c_2 \frac{dR}{dt},$$

where ($c_1 c_2 c_3$) is proportional to the impressed magnetic field. This relation, interpreted in the view that the electric character of a molecule is determined by the orbits of its electrons, simply means that the capacity of electric polarization of the molecule depends on its orientation with regard to the imposed magnetic field, that, in fact, the static value of K, depending on the molecular configurations just as much as do the free periods, is altered by the magnetic field. This relation agrees with the main feature of rotatory dispersion, namely, that it roughly follows the law of the inverse square of the wave-length. The specific influence of the molecular free periods, that is, of the ordinary dispersion of the material, on the Faraday effect, is presumably a secondary one; though it, too, follows the same law for different wave-lengths, in the case of substances for which Cauchy's dispersion formula holds good. It is this latter part of the Faraday effect that is reciprocal to Dr Zeeman's phenomenon†.

smoothed out:

and effect of dispersion is residual.

The question is fundamental how far we can proceed in physical theory on the basis that the material molecule is made up of revolving electrons and of nothing else. Certain negative optical experiments of Michelson almost require this view; at any rate, they have not been otherwise explained. It may be shown after the manner of *Phil. Trans.* 1894, A, p. 813 (and Dr Zeeman's calculation, in fact, forms a sufficient indication of the order of magnitude of the result), that in an ideal simple molecule consisting of one positive and one negative

Atoms are orbital electronic.

[1] *Loc. cit.*, *Phil. Trans.* 1894, A, p. 820; and 1895, A, p. 713.

* Cf. *supra*, p. 339.

† On the rudimentary Lorentz model, or illustration, of an electron describing an orbit under a central force varying as distance, expressions for both the Faraday effect and the Zeeman effect were deduced by FitzGerald (*Roy. Soc. Proc.* LXIII. 1898; *Scientific Papers*, p. 464), thus exhibiting their mutual connection. A similar analysis will perhaps be applicable to the general precessing model for Zeeman normal triplets, having one central nucleus, *Phil. Mag.* Dec. 1897: or *infra*. In neither case does the excitation of free vibrations come directly into the analysis.

These rotatory effects do not appear to have been formally subsumed as yet under the principle of electrodynamic relativity, an attempt in *Aether and Matter*, Ch. XIII. involving errors of algebra, as Lorentz pointed out. This perhaps points to absence of any simple mode of analytical expression.

Indication that an electron is much less massive than any atoms.

electron revolving round each other, the inertia of the molecule would have to be considerably less than the chemical masses of ordinary molecules, in order to lead to an influence on the period, of the order observed by Dr Zeeman. But then a line in the spectrum may be expected to arise rather from one of the numerous epicycles superposed on the main orbits of the various electrons in the molecule than from a main orbit itself.

47

A DYNAMICAL THEORY OF THE ELECTRIC AND LUMINIFEROUS MEDIUM.—PART III: RELATIONS WITH MATERIAL MEDIA.

(*Abstract: with a general discussion.*)

[*Proceedings of the Royal Society*, Vol. LXI. pp. 272–285.
Received April 21, read May 13, 1897.]

1. It was shown by Maxwell that the theory of electric current systems flowing in rigid conductors could be formulated dynamically if the current in each circuit, which is supposed constant all round it, is taken as a generalized velocity: a formal development in his manner of the properties of the current system, from the single dynamical foundation of Least Action, was given in the first of the present series of papers. It is implied in such a view that the positions of the linear conductors and the intensities of the currents that are flowing in them control completely, it may be in an entirely unknown manner, the motions that are going on in the surrounding aether; just as in the cyclic irrotational motion of a perfect fluid the positions of the rigid cores round which the circulations take place, and the amounts of these circulations, determine the motion. But it was pointed out that, whereas the vorticity of such a core is dynamically a momentum, on the other hand in order to satisfy the facts the currents in the circuits must be considered dynamically as velocities; so that there is not any real analogy between the two cases.

Maxwell's dynamics of linear currents:

its implications,

not of hydrodynamic type:

This simple dynamical formulation is no longer available when attention is not confined to complete rigid circuits, for example when the mechanical forces acting on a portion of a circuit are in question. It is insufficient also when the electromotive force (integrated electric force) between two points of a circuit carrying a current is to be discussed, that quantity being realizable and measurable by bridging the two points through the incomplete conducting circuit of an electrometer.

cannot apply to segments of circuits,

A knowledge of the electric force at a point, as well as of its integrated value round a circuit, is also essential when the current in the circuit is in part conducted and in part arises from changing electric displacement: it is thus essential in the theory of electric

***i.e.* to problems involving radiation.**

waves and radiation, though unimportant in ordinary electrodynamic applications; and for this reason that part of electric theory remained unsettled until Hertz showed how to produce and investigate free electric waves.

Previous to that time, the theory had been completed hypothetically in various ways, by extending the range of the dynamical formulae (with such generalizations as were admissible) from the complete rigid circuit and the complete circuital current to the geometrical elements of which they are composed. The way was prepared for this by F. Neumann, by his discovery of the electrodynamic potential formula, from which he deduced by a uniform analytical process the mechanical and electric forces in complete rigid circuits, as previously formulated by Ampère and Faraday. Then Helmholtz took up and extended Neumann's theory, introducing Maxwell's principle that the formula of Neumann represents the kinetic energy of the system of current elements, to which Lagrange's general dynamical equations, or more precisely the Action method, can be applied. His main aim was to include in this potential theory the conception of dielectric currents introduced by Faraday: but he was at the same time led to generalize Neumann's formula so as to include the possibility of waves of electric compression as well as transverse electric waves in dielectric media: the existence of the former type of waves Hertz's discoveries have disproved.

Neumann's potential of two current elements: generalized by Helmholtz in view of polarization currents, and involving compressional waves.

This theory of von Helmholtz thus originated by way of elucidating, from the older standpoint, the aether scheme advanced by Maxwell in the shape of a system of analytical relations, which were the mathematical formulation of Faraday's views, and involved also a unification of the electrical and optical functions of the aether. The equations of Maxwell include implicitly a definite theory of all electromotive phenomena in open circuits *at rest*, a fact which is not very obvious for the form into which they were thrown in Maxwell's *Treatise*, but which is plainer for the form of two conjugate circuital relations into which he had previously cast them as being the simplest and most direct formal expression of the theory, but which he left aside in favour of a conception of electrodynamic momentum or vector potential that was intended to connect the equations directly with dynamical principles*: this simple formal specification of aethereal relations has since been restated and utilized by Heaviside and Hertz. These Maxwellian fundamental equations, being purely electromotive, gave no direct explanation of ponderomotive forces: for that purpose the theory had of necessity to take on a dynamical form. A formula for the distribution of the electric energy throughout the

Maxwell's two circuital relations of the field.

* Cf. *Aether and Matter* (1900), Ch. VI.

aether was suggested by various considerations, and from it an attempt was made by the methods of general dynamics to establish the laws of the force exerted by the aether on the different parts of conductors conveying currents: and it was natural that the same procedure should be extended to an attempt to place the fundamental formal equations of the aether itself on a dynamical basis. But what was lacking for the satisfactory accomplishment of this purpose was a definite and consistent idea of how the electric charges and currents in the matter established a hold on the aether. In the absence of this, Maxwell had to rely, when actions on portions of circuits were under consideration, on the notion of a current element, and to derive the formulae from that conception with such degree of definiteness as was possible. It was this imperfect dynamical method which it was the aim of Helmholtz to discuss and elucidate in the analysis above mentioned.

Mechanical forces had to revert to current elements:

In so far as these theories were dynamical they all involved current elements: but the criticism of the second of the present memoirs is held to show that a current element is not a legitimate dynamical entity, in that the forces strictly derivable from that assumption, both mechanical and electric, are in disagreement with experimental knowledge: and the reason is indicated, namely that the method of current elements forms an incomplete specification of the phenomena, inasmuch as it gives no account of how a current is induced within an element of volume of the matter by separation of the two electricities under the action of the electric force. This led to the introduction of the mobile electron or atomic charge of electricity as the true physical element, and to a dynamical theory of molecular type which is held to be self-consistent and in full agreement with experimental knowledge, and which may be regarded as in a manner a final development of the Weberian notion of moving electric particles. The *true* electric current of moving electrons is thus made up of a current of conduction and a current of dielectric polarization: it does not flow in closed circuits, but if there is added on to it a quantity called the aethereal displacement current, which is not a flow of electricity at all but a flux of elastic displacement of the aether, there is obtained the *total* circuital current* of Faraday and Maxwell. This more precise three-fold specification of the total current, in place of the two-fold specification of Maxwell which ignores the physical distinction between the polarization current in the dielectric matter and the displacement current in the aether occupying the same space,

with results against experiment.

The moving electron introduced as the true physical element:

its beginnings with Gauss and Weber.

Currents now not closed unless on Maxwell's special hypothesis.

Polarization current separated out.

* This idea of total current has usually lapsed in the Continental treatises, a circumstance which probably makes British procedure on Maxwell's lines more difficult to follow.

introduces notable differences as regards ponderomotive forces, moving material media, electric radiation, and in other respects. In the main the results correspond with Faraday's geometrical mode of specification by means of tubes of induction; they agree more closely with the scheme derived by Maxwell from direct consideration of his mechanical model of the electric field than with the later analytical theory of his *Treatise*.

The simplification in the *Treatise* a backward step.

The general equations of the electric field, when it contains moving material bodies, dielectric or conducting, are formulated on these principles. They are applied to the problem of the uniform rotation of a charged conductor, or a dielectric body, in a magnetic field: to the influence of motion of a material medium on the velocity of radiation that is passing across it: and to the influence of steady translation of a material system through the aether on its configuration and the distribution of its electric charges.

Special problems treated.

2. The object of the present memoir is to further develop and apply this conception of the relations of aether and matter, which has thus been shown to form a working basis for optics and electrodynamics. According to this scheme the aether is the seat of the elastic transmission of electric and magnetic force: but the ponderomotive forces acting on its electrons or strain centres, which represent matter, are derived directly from the energy without any reference to such transmission; and this has been taken as an objection to the theory. We are therefore led to a critique of the sufficiency of the principle of step-by-step transmission or contact action as a basis for a complete connected representation of the play of physical agencies.

Theory of material media.

The ponderomotive forces.

With a view to thus attaining more precise ideas as to what constitutes ultimate physical explanation, the properties of a material elastic solid medium involving in its constitution centres from which intrinsic strain spreads out into the surrounding parts, are noticed. The theory of fluid vortices, irrespective of any claim to form a natural representation of material phenomena, has been of fundamental service as an illustration of what kind of interconnection it is conceivable to assume between matter and the universal aether: in vortex-atom theories the fundamental reality as regards pressure and inertia is transferred to the continuous fluid aether, and the properties of the atomic matter, which had been the original source of dynamical suggestion, reappear as secondary or derivative. Now just as a vortex atom is a state of intrinsic motion which can travel or flit through the fluid, independently of, and in addition to, motion of the fluid itself, so also can a strain centre in an elastic solid medium move about

Dynamics of vortex atoms typical.

Intrinsic strain centres.

independently of the medium[1]. Its motion forms an element beyond and in addition to the changing strain impressed on the parts of the solid medium by external influence. It would be easy to construct a solid body involving such strain centres, and this circumstance gives an impression of actuality which may be felt as wanting in an abstract theory: the material of the solid would illustrate aether, the strain form would represent an essential of matter. Such a representation is a step beyond and outside the ordinary idea of an elastic medium as the mere vehicle for transmission of forces: that notion, effective in its own sphere, would not give any account of where or how these forces originate, or what is the nature of the connection by which the matter gets a grip upon the impalpable aether. Nor was it to be expected that the mere idea of contact action, which has come to be interpreted as an idea of elastic transmission, could of itself give account of the actions of which it is only the mechanism of transmission. It is here maintained that the whole of the action between different portions of matter cannot possibly be represented as transmitted by the aether in that manner. It would even be a fair defence of this position to claim that the *onus* of demonstration lies on those who assert the opposite: but the material elastic medium, above mentioned, pervaded by intrinsic strain centres, furnishes a crucial illustration. The parts of that medium itself are in motion on account of the changing strain; while in addition the intrinsic strain centres put each other into motion across the medium (as vortices do across an intervening fluid) by their mutual actions. The theory of contact action so called, or elastic transmission, is not wide enough to include all this: the dynamical interactions between the strain centres are no more transmitted by elastic action than the interactions between fluid vortices without solid cores are transmitted by fluid pressure: though in each case that agency is a necessary concomitant to the dynamical effect. The really absolute thing in dynamical explanation, that on which this principle of elastic transmission or contact action has itself been built up by Lagrange and Green, and without which it could not have assumed a precise mathematical form, is the scheme of fundamental dynamical notions connecting inertia and force, in their modern generalized aspect which formulates them under the energy principle. The main problem of transcendental physics is to assign the nature of the ultimate medium or scheme of relations which combines physical phenomena into a unity, in whose relations

Solid with intrinsic strains.

Back from aether strain to its sources.

The dynamics of matter as subsisting in the aether:

is wider than direct contact transmission.

The Action scheme wide enough:

[1] The analogy is not quite complete; for the vortices move so that their cores [and a volume around them] are always made up of the same portions of the medium, while movement of a strain centre does not to any extent carry the medium with it.

these dynamical notions have their scope: and it is only the prejudice of education that would keep, in this wider field, too close to the ideal of mechanical transmission in a homogeneous elastic solid*.

In so far as we can collect the various dynamical principles into a simple formula, we shall attain the security that all the trains of results that come from that formula are consistent among themselves, and form parts of a single interlaced scheme. Such a formula was dimly foreshadowed by Maupertuis as the Principle of Least Action, and after various elucidations by Euler was finally established in its generality by Lagrange for ordinary material systems. The ultimate unification of physical theory, which transcends and includes ideas of contact action and other partial explanations, would thus lie in the formulation of the energy-function of the aether, including matter, in a manner suitable for the application of the Action analysis to the correlation of the observed phenomena. To assume that this will ever be accomplished absolutely and completely is nearly the same as to assume infinite capacity in the human understanding: but exact knowledge of inanimate nature comes by a process of analysing and classifying into types the main physical agencies as they present themselves in bulk to our senses, and it is a legitimate and feasible problem to seek out these aspects of the underlying unity which are the cause of the interactions and correlations of these agencies, so far as they have yet been unravelled. The mere recognition of the precise order that reigns in the larger workings of the unlimited diversity that constitutes matter is itself the strongest argument that the common basis of the varieties of matter involves something simple and universal in its relations, with which we are really in physics more intimately concerned than with the infinity of arrangements and collocations that the molecules of matter can individually and collectively assume.

[Margin: and necessarily self-consistent. — The problem of inanimate nature: — demands an underlying universal medium.]

It follows from the analysis of the second of these memoirs that the Action formula can be completely expressed and developed in a molecular theory sufficiently refined to take account of each electron separately; also that the main outline of the ordinary electrodynamic theory, for finite systems of bodies treated as continuous aggregates, can be developed from the Action formula transformed so as to be expressed in terms of matter in bulk, when the currents are specified as circuital and the different regions are homogeneous as regards electric and magnetic polarization. But in questions of details of

[Margin: Action theory for a system of electrons:]

* It would be closer to the Action scheme to assert that the vortex or strain centre involved a permanent associated collocation of motion or strain; two of them overlap and therefore influence each other immediately, but in a way which involves adjustment of their fields by aethereal propagation.

mechanical action on matter, especially when in motion, and in questions involving the distinction between the true current carried by the matter and the total circuital current, and also involving heterogeneity in the dielectric, the pure dynamics does not suffice, and recourse must be had to direct processes of averaging, such as are necessary in other domains of molecular theory, as for example the theory of gases. to be combined with statistical theory.

3. The utility of an elastic solid model of the kind above described is not to represent the aether, but to enlarge our ideas: for the optical phenomena show that the elasticity of the actual aether is of rotational, not distortional or elastic solid, type. It is, however, also explained in the memoir how, out of matter gyrostatically dominated, it is theoretically possible to construct a model which will represent the aether itself and its electrons for any assignable time, though not for ever: the properties of the model there described would gradually fade away, just as if matter were not eternal. A main difficulty in designing such a representation lies in the circumstance that it must possess the property of perfect fluidity for irrotational motions. Models necessarily imperfect: such as the gyrostatic model of the aether.

The distinction between the features of elastic transmission in the aether and in material elastic substances is brought out. The former medium is a pure *continuum* of which elasticity, inertia, and continuity of motion, are the sole ultimate and fundamental properties. Matter, on the other hand, is made up of discrete atoms or singular points in the aether; it has the inertia which belongs to these singularities individually; it has elasticity on account of their interactions through the aether; its continuity of motion is, in the case of fluids, of limited character, being maintained only by viscosity and other such causes. The elasticity of matter in bulk is to be based on the distribution of organized material energy per unit volume: the material energy depends on the relative positions of the atoms, therefore this organized or mechanical part of it depends on the change of their mutual configurations expressed with reference to the deformation of the element of volume, that is on the strain. It follows, as Green was the first to show with logical simplicity and precision, that the stress in an element of volume is self-conjugate. But none of these ideas have any application to the aethereal *continuum*: in it there is no question of change of mutual configuration of physical parts; the energy of strain is not thus restricted to be a function of deformation only, in fact after MacCullagh it assumes the geometrically more simple form of a function of the absolute position, more precisely of the rotational displacement, of the element; the resulting stress Aether the ideally simple medium. The elastic energy-function for matter: is less ultimate than for aether.

need not be self-conjugate,—nor is it in fact so even in a material medium that has gyrostatic quality. An electric field would consist of rotational strain in the aether, a magnetic field of irrotational flow, each in actual cases extremely slight: the motion involved in a permanent magnetic field combined with a permanent electric field would not become jammed in course of time, because it can relieve itself by a slight separation of free electrifications, which will again neutralize each other after this object has been attained.

Does a magneto-electric field gradually become jammed?

The Solar system type of atom:

vibrations and secular precessions.

4. An atom of matter has been represented by a collocation of electrons describing stable orbits round each other. The discussion of the internal vibrations of such an atom and the consequent radiation will follow the lines of Laplace's general analysis of the oscillations about steady motion of a system of connected bodies like the Solar system. When the gyrations or orbital motions are sufficiently rapid, there will be two types of vibrations produced by disturbance of the system; very rapid ones which radiate light, and very slow ones like the precession of a spinning top which do not involve appreciable internal deformation of the system. In gases it is only these latter that would be excited sensibly by the comparatively gentle encounters between the molecules: these are in relation to the thermal energy, but are only in indirect connection with radiation*. The difficult outstanding problem of the theory of gases, that namely of the connection between temperature and internal thermal energy, involving the relation of the two specific heats, would on this view take on a form different from the usual one. In various other respects, a recognition that the motions which constitute heat are not the vibrations which feed radiation seems to extend and improve the capabilities of molecular theory.

Do secular effects give out radiation?

Bearing on the difficulty of equi-partition.

Another field in which the influence of a gyratory character in the molecule might be expected to be prominent is that of optics, more particularly the influence of matter as it appears in refraction and reflexion. As an introduction to this subject, Lorentz's law of the relation between refraction and density is worked out; the argu-

The refraction invariant.

* Expressed from another angle, if the structure of an atom provides configurations of minimal energy, they cannot radiate, unless when some external cause of loosening of internal constraint permits a jump across to a lower minimum. There must also be a restoring process of some different kind if radiation is to recur. Experiment now associates these interactions with release and absorption of fast electrons in the structural system of the atom. In modern theories the orbital atom of Rutherford has been the source of a surprisingly wide range of tentative development at the hands mainly of Bohr and Sommerfeld. The orbital scheme now trends, however, towards the status of a symbolic model, with undeciphered, probably continuous, structure behind it.

ment is purely statical and independent of the constitution of the molecule, and closely follows a cognate investigation of Clausius as I afterwards discovered; it is however retained, as it appears to be exempt, within its proper scope, from the objections which are valid against other modes of demonstration of that law that have been proposed. When we pass on to discuss dispersion, the forced vibrations of the molecules come in: and these will be of different type according as the molecule is taken to be a system vibrating about a position of rest as has hitherto been tacitly done in optical theory, or a system vibrating about a state of steady motion as it is here required to be. The main result, for a medium devoid of non-selective opacity such as would arise from conduction, is that the Lorentz refraction equivalent $(\mu^2 - 1)/(\mu^2 + 2)\,\rho$ (not $(\mu^2 - 1)/\rho$ as in the usual dispersion theories) is an additive physical constant, equal for each simple medium to $\Sigma g_r/(p_r^2 - p^2)$, where p_r is a natural vibration frequency for the molecule and g_r is a related constant. The only simplification that comes in, when the gyratory quality is absent, is that then g_r is necessarily positive: in the present case the reasons for taking it to be positive are not so conclusive. If in any term g_r were negative the character of the anomalous dispersion near the corresponding absorption band would be the opposite to that indicated by Kundt's law, which has hitherto always been observed to hold good.

Dispersion formulae for static molecules: for kinetic molecules.

General formula:

its relation to Kundt's law.

The only way that is *à priori* unexceptionable for determining the complex index of refraction of a strongly absorbing medium such as a metal is Kundt's method of deviation by thin prisms: this gives only the real part of the index, but it is shown that by taking advantage of oblique as well as of normal incidence approximate values might be obtained for the other part as well. The fair agreement of Kundt's values with those derived from experiments on polarization by reflexion is however a confirmation that surface films are not seriously operative in the latter method, which has yielded both parts of the index. If there were no non-selective absorption, the curve representing the real part of the index would rise to infinity near each absorption band, then fall straight down to the axis, coincide with the axis for an interval, and finally again rise above it. But when there is also general absorption the curve will turn back before reaching infinity, and it will not descend as far as the axis, while the sharp corners will be eased off. These characteristics are precisely those of anomalous dispersion, for example of Pflüger's recent dispersion curves for solid fuchsin and other selectively absorbent solid substances; but there is absolutely nothing in these general features that would not fit one theory of dispersion as well as another,—they

Metallic refraction:

formula for a prism:

graph of dispersion near the band of selective absorption.

all arise from the mere notion of sympathetic vibration. The complete values of the index along the spectrum, not merely those of its real part, must be available before any preference can in this way be established.

5. A main feature of the interaction between aether and matter consists in the bodily mechanical forcives exerted on electrically and magnetically polarized material media. In the theory a transition has here to be made between the mere aggregate or sum of the varying energies of the individual molecules of a medium, and the co-ordinated and averaged part of this sum which is the energy pertaining to the element of volume of the medium in bulk: this further involves the definite enunciation of a general principle in molecular mechanics which has hitherto found an application, and that a restricted one, only in the theory of capillary attractions. In a polarized medium, a distinction has thus to be drawn between the averaged individual energies from which the forces polarizing the separate molecules are derived, and the organized mechanical energy of the medium as a whole, which is the aggregate of these energies after the local parts arising from the neighbouring molecules have been excluded. In this mechanical energy the bodily forces on the medium in bulk are involved: it must therefore be expressible as an analytical function of the configuration of the medium in bulk, for otherwise perpetual motions could supervene. In the study of the mechanical actions in material media, consideration of the properties of the molecules is available as a guide towards the mathematical form of the function which represents the distribution of the mechanical energy of the forces acting on the element of mass: but the province of molecular theory is ended in this general survey, and the actual values of the coefficients in the energy-function must be determined by observation and experiment*.

A continuous energy-function extracted: its properties: how ascertained.

In an electrically polarized material medium an expression for the distribution of this mechanical energy is obtained, and the bodily applied forces in the material are derived from it, the tractions exerted on an interface between two media being deduced from the forces that would act on a layer of gradual transition which in the limit is taken indefinitely thin. The result is that in a medium whose molecules are polarized to intensity i' by a field of electric force F, and isotropic so that i' and F are in the same direction, there is a bodily force $(d/dx, d/dy, d/dz)\int i'dF$, which could be balanced by a hydrostatic pressure $-\int i'dF$, and there is also a normal traction on each interface equal to the difference in the values of $-2\pi n'^2$ towards

Forces on matter in the electric field: expressed in stress form, fluid.

* Conducted on the medium, whose stability is assured by experience alone.

each side, where n' is the component of i' along the normal. Thus when a fluid medium is in equilibrium there must exist in it a hydrostatic pressure $\int i'dF$, and in addition on each interface a traction $2\pi n'^2 + \int i'dF$ along the normal towards each side, arising from other than electric causes and balancing the electric forcive: so that to maintain mechanical equilibrium, extraneous normal tractions $2\pi n'^2$ [$+ \int i'dF$] towards each side of each interface are alone required.

Contrasts with Helmholtz's analysis.

6. This result differs from that of von Helmholtz's investigation, also based on the method of energy: the origin of the discrepancy is traced to the circumstance that a single continuous energy-function cannot serve for the complex medium aether *plus* matter. This difference goes to the root of things, especially in optical theory, even in cases where the resulting expressions present no difference in form. Variation of the physical constants of the medium arising from the strain involved in the virtual displacement is also included by von Helmholtz in the deduction of the mechanical forcive, thus introducing effects which are here held to be more consistently explained as physical changes arising from the molecular action of the polarization.

Structural change.

Material linear stress:

Of the purely local part of the total energy of a molecular medium, there is a regular or organized portion depending on the deformation of the material in bulk, which is the energy of the mechanical stress that compensates the applied mechanical forces: the remaining, usually wholly irregular, part finds its compensation in other interactions between neighbouring molecules, which may reveal themselves in the aggregate in alterations of the local physical constants of the material as well as of its volume and other dimensions.

and constitutional change,

But in the circumstances of a medium electrically polarized this *residuum* itself involves a part which is regular in each element of volume, arising from the regularity in the orientation of the molecules which act on each other in that element. The mutual forcive thence originating may be expressed, though there is not much object in doing so, as regards the interior of an isotropic medium, as an internal molecular stress related to the lines of polarization. When the distance between the effective poles of a molecule is small compared with that between neighbouring molecules this stress is a tension [$\frac{2}{3}$.] $\frac{4}{5}\pi i'^2$ along the lines of polarization together with a pressure $\frac{2}{3} \cdot \frac{4}{5}\pi i'^2$ uniform in all directions at right angles to them: it is to be considered as balanced locally by cohesive reaction*.

in part expressed as a quadratic stress,

Under all circumstances, the forces between neighbouring molecules produce and are compensated by change of the relative con-

* In a uniform medium it is, by Maxwell's theorem, always internally balanced.

figuration of these molecules; they thus produce change of the *local* physical constants of the material, and also *local* intrinsic change of volume and other dimensions, all which are proportional to the square of the polarization; but they contribute nothing directly to the mechanical stress transmitted by the material in bulk. In a solid material, however, these intrinsic changes of configuration of the elements of volume may not fit together consistently with the continuity of the substance, and thus secondary strains may be produced which will complicate the problem. But in fluids, in which alone experiment is feasible, no such complication can occur.

which is not transmitted, but may produce local striction.

Illustration by a gaseous medium.

The position is aptly illustrated by the ideally simple case of a perfect gas polarized in an electric field. The mechanical forcive due to the polarization of the gas as a whole is there compensated by change of pressure, which is transmitted. There is another regular part of the forcive which arises from actions between neighbouring molecules, so that those in the line of polarization attract and those in lines at right angles to it repel each other, after the manner of little magnets; and this, which differs from the former by being proportional to the square of the polarization instead of its first power, is not experienced as a mechanical force, because it is wholly compensated on the spot where it originates by slight change in the ordinarily fortuitous distribution of velocities of the molecules of the gas, by which its constitution acquires an axial character with reference to the line of polarization so that the pressure is no longer quite the same in all directions.

Thermodynamics.

Energy available (at constant temperature).

7. The relation is explained which exists between this *organized* or *mechanical* energy and the *available* or *free* energy of thermodynamics. The principle of available energy, which itself is a direct consequence of the negation of perpetual motions, or rather of the negation of the unlimited availability of diffuse thermal energy, is the single essential foundation of that science. It is pointed out, that if we had no direct perception of temperature through our senses, this negation of perpetual motions would necessitate the introduction of that quantity into physics, somewhat in the same way as potential is introduced into electrical theory, and would yield a demonstration of its fundamental property. Instead then of making attempts, by the aid of special molecular hypotheses of more or less problematical character, to obtain a purely dynamical definition of temperature and an analytical demonstration of Carnot's principle, it is suggested that it is more philosophical to recognize that no physical scheme of matter and molecular action is conceivable that would involve perpetual motions (in the above sense) of matter in bulk, and to base the pure

Temperature fundamental.

theory of thermodynamics and thermochemistry directly on this postulate.

The case of homogeneous fluid media, which acquire energy of polarization of any kind when in a field of force, is considered in a general manner: it is shown that such media will be in mechanical equilibrium provided an extraneous traction along the normal is applied over each interface between them, of intensity equal to the difference of the densities of the mechanical energy of polarization on the two sides of the interface. Equilibrium of polarized fluids.

Osmotic pressure is related to the total available energy, not merely to the mechanical part of it: the limiting or maximum value which it cannot exceed is equal to the change in available energy produced per unit volume of transpiration across the partition. An ultimate deduction of van 't Hoff's law of analogy between osmotic and gaseous pressure is offered, on the foundation of the principle of available energy, which is independent of any assumption as to the character of that pressure whether purely kinetic or otherwise: if this be accepted, it will follow that no inference as to the physical state of the dissolved substance, except as regards its degree of effective dissociation, is deducible from the osmotic law. It appears to have escaped general notice that what was virtually a prediction of the law for the cases of dissolved gases is involved in the equations of von Helmholtz's discussion of the influence of dissolved gas on electromotive force, in which however the argument is based on Henry's law of solution; the law itself had indeed been formulated explicitly on similar theoretical grounds by Willard Gibbs still earlier. The influence of an electric field on osmotic pressure between dielectric fluids is estimated: this involves by cyclic processes the influence of an electric field on the vapour pressure and on the freezing point of a dielectric liquid. Some considerations connected with the nature of the process of ionization are brought forward. The laws of chemical equilibrium, as developed by Guldberg and Waage and by van 't Hoff, are placed in relation to the principle of available energy. That method is also applied in a discussion of the electromotive force of a voltaic cell, and especially of the dissipative part which is established by steady finite diffusion between solutions of different concentrations. Osmotic energy: gas laws hold for dilute systems. Historical. Vapour pressure and freezing point. Method of available energy.

8. A thermodynamic application which possesses interest, both from the light it throws on the nature of magnetism and from the circumstance that in it the heat supply is calculated indirectly from the magnetic energy that runs down, is the relation between magnetic susceptibility and temperature in substances not in the very susceptible or ferromagnetic condition. According to the Weberian Its application to magnetism:

theory, which fits in with the present view, diamagnetic energy which is not compensated mechanically goes to the induction of Amperean currents in the molecules; while paramagnetic energy not thus compensated goes to orientating the molecules, and thus into heat. It follows that the diamagnetic coefficient is independent of temperature: on the other hand it is shown that the paramagnetic coefficient should vary inversely as the absolute temperature. These Curie's law. laws were discovered experimentally by Curie, who finds from a very extensive investigation that they have the same order of accuracy at sufficiently high temperatures as the ordinary gaseous laws: at lower temperatures and in ferromagnetic substances the control of the polarized molecules arises in appreciable part from the magnetic interaction of their neighbours, thus vitiating the law as well as introducing effects of hysteresis. The well-known model of Ewing would thus represent an ideal perfect ferromagnetic in which the control arises wholly from the latter cause.

Flux of energy. In application of the previous results as to how far physical actions can be considered as transmitted across the aether by elastic stress, the conditions are formulated under which the correlative principle utilized by Poynting is valid, that the actual rate of change with time of the organized or mechanical energy within any region is expressible explicitly as a surface integral over its boundary.

Radiation pressure. The mechanical effects of light-waves are reconsidered in the light of this molecular theory. The conclusion is reached [which when now corrected agrees with Maxwell's result]. Partial analogies are furnished by the mechanical effects of Hertzian radiation on a medium built up of conducting linear circuits, and of sound waves on a medium formed of a system of resonators.

Electro-striction in condenser: As an application of the law of the mechanical forcive on dielectrics, the changes of dimensions of a condenser under electrification are considered. The problem is found to admit of exact solution if the condenser layer consists of a closed sheet, of any form, but of uniform solved: thickness. In that case the mechanical stress in the material of the sheet proves to be simply of the type of the Faraday-Maxwell stress. relation to experiment. The theory is compared with Quincke's experimental results: their main features are verified, including those which led Quincke to assign a wholly non-mechanical origin to the effect: but something less than half the change of volume remains over as an intrinsic electric deformation, not due to the transmitted mechanical forces.

Various special problems solved. Finally a series of practical illustrations of the mechanical theory are treated, some of which have already been employed for experimental measurement, and which are capable of still further application. The mechanical circumstances attending the refraction of uniform fields of electric force by fluid media are developed. The theory of various arrangements for measuring electric tractions and pressures in fluid dielectrics is worked out. The effect of an electric field on the velocity of ripples on the surface of a conducting or a dielectric fluid is determined: as also are the relations of electric polarization to vapour tension and fluid equilibrium. The internal mechanical forces in a complete magnetic circuit are examined, and also the traction between the interfaces when it is divided: and the mode of calculation of the stress in a sphere of iron in a uniform magnetic field is indicated, agreeing for this case with Kirchhoff. The mutual influence of stress and magnetization is analysed, with reference to the experimental investigations of Bidwell.

Throughout the memoir care is taken to dispense, as far as possible, with detailed algebraic processes, which are essential for special computations and verifications, but are best evaded in the discussion of general principles. Most of the discussion is also independent of the rotational aether scheme: the great advantage of an interlacing hypothesis of that kind, which remains even when it is only provisional, is that it gives an insight into the character of the formal relations that are possible or probable between the actual physical quantities involved in it*.

Method.

Aether model only an auxiliary.

* Referring to § 3, p. 631, the model of a shell electron of 1894, *supra* p. 522, may be carried further. In § 42, p. 449, it is explained that if the fluid aether, elastically rotational, is incompressible, an instantaneous fluid pressure ϑ' can subsist in it, which in statical circumstances is uniform but can change its value in crossing an interface. Such an interface is the surface of the shell electron on which the rotational aether strain abuts, abruptly changing to absence of strain inside the shell. Equilibrium of this spherical shell system is secured automatically by fall of pressure from outside to inside compensating the repulsive force $2\pi\sigma^2$ on the surface, where σ is the abrupt change of rotational strain expressive of electric surface density. When the shell electron is in uniform convection through the medium it changes into an oblate spheroid: and, as Poincaré noted, a difference of uniform hydrostatic pressures, outside and inside, will still equilibrate the electric repulsion on the parts of the spheroidal surface. This is the only type of compensating force that can be provided: and its sufficiency is a point scored for the model. But the main question, postponed then and still awaiting scrutiny, is how the model can acquire sufficient stability against random shocks. It is in close analogy to the problem of the limits of stability of an electrified dewdrop, solved by Rayleigh with important applications, a difference being that here there is flow outside as well as inside the surface.

An aether pressure essential to the electron model:

and sufficient when it is convected:

subject to stability being ensured.

APPENDIX (1927).

(i) *Historical Note on Hamiltonian Action* (p. 70).

The Dublin school.

THE original ideas of William Rowan Hamilton, nurtured, like all the Dublin school of his time, on the writers of the great age in France, have constituted an epoch in fundamental mathematical physics. They began with consideration of the properties of what he at first named on historical grounds the "Action function" for a System of Optical Rays, in a memoir communicated in June 1824 to the Royal Irish Academy when he was an undergraduate of nineteen years of age. As printed three years afterwards in the *Transactions*, there are prefixed very copious analytical tables of contents for three parts; but only the first part appeared in print—very dishevelled in form, doubtless from the distraction of his wide range of philosophical and poetic interests, not to mention early settlement of his academic career as Astronomer Royal for Ireland—*Trans. R.I.A.* Vol. xv. (1828) pp. 81–174. The other two parts seem never to have appeared at all, though the tables of contents give a definite digest of their material, including already an adumbration of the extension to dynamics. Incidentally, among other partially related topics, the differential geometry of Systems of Lines, straight or curved, was developed in extensive but undigested fashion, long before it emerged from Plücker as a systematic and fertile branch of abstract mathematics. There followed three Supplements, enforcing the fundamental ideas, but perhaps also still further confusing their application by excess of detail only partially relevant*.

Sketch of Hamilton's memoirs.

Of these the First Supplement, dated 1830 and sliding away into geometry of ray-systems straight and curved, is in the next volume of *Trans. R.I.A.* pp. 1–61 (cf. footnote, *supra*, p. 148).

Hamilton's own optical applications.

The Second Supplement (1831), relating to expansions of the Characteristic Function such as are required for changes of the origin in optical systems, soon branches away into complex and hardly practicable mathematical formulas, but ends up with a workmanlike verification of Sir John Herschel's results for aberration of lenses. This application, and interesting short notes, extracted incidentally

* It appears from contemporary notices that the original paper, of undergraduate days, entitled "On Caustics," had been referred back to him by the distinguished board of Referees of the Royal Irish Academy for further explanation and development, which perhaps accounts better for this untoward mode of exposition. For, as indicated below, he could express his views in concise and masterly manner, when compelled by circumstances.

from Hamilton, *Phil. Mag.* 1833, on aberration in refraction of a pencil across a prism, and *Phil. Mag.* 1841, on the aberrations of a lens formed of uniaxial crystal such as quartz, alone seem to survive to testify to his own practical mastery, when prevented from digression, of the new formulation for geometrical optics.

The Third Supplement, *Trans. R. I. A.* Vol. XVII. (Jan.–Oct. 1832) pp. 1–141, again with detailed table of contents, once more restating the foundation principle, but then largely drifting into general analytical discussion of Complexes and Congruences of Rays and their Caustic Surfaces, finally diverges into Fresnelian Optics, and ends with the once famous incidental prediction of Conical Refraction in crystals. Conical Refraction.

A main field of application of the matured theory, there restricted perforce into more practical optical form, yet still liable to digression into more abstract topics, appeared in a note hidden away for half a century in *British Association Report, Cambridge* 1833, pp. 360–372, the essential part of which was at length disinterred and reprinted by Lord Rayleigh in two pages in *Phil. Mag.* 1908; *Scientific Papers*, Vol. V. pp. 456–464. In it the modern five types of aberrations of an axial optical system are already exhibited as lying naturally within the modified Characteristic Function—recovered in recent times in a special form as the "Eikonal"—which Hamilton advances as the compendious systematic foundation for all practical investigations on optical instruments. Another form, utilized by Rayleigh himself as a still more compact foundation for this purpose, semi-modified as involving coordinates of points at the image-end and directions of rays at the object-end of the system, had already been expounded in detail, but as usual branching away into abstract digressions, by Hamilton as the function W of the First Supplement*. Systematization of optical aberrations.

In the year following, these desultory developments, which had extended now over ten years from the undergraduate beginnings, converging toward what was to be essentially a new outlook in physical mathematics, culminated in a magnificent synthesis, "On a General Method in Dynamics," in two memoirs, *Phil. Trans.* 1834, 1835. This departure has expanded the algebraic structure of Abstract Dynamics so as to fit that science to be the essential symbolic guide and touchstone for progress in all departments of physical science. Here again, under limitation as regards space, as he had doubtless also been in the memoir itself in *Phil. Trans.*, he has provided a coherent and significant abstract of the guiding principles in *Brit. Assoc. Report, Edinburgh* 1834, pp. 513–518. The General Method for Dynamics;

After this date (1835) Hamilton seems to have parted company

* It may be suitably recorded here that §§ 13, 16, p. 189, and § 6, p. 269 *supra*, are wrong. Cf. T. Smith, *Trans. Optical Society*, Jan. 19, 1928.

with fundamental dynamics and optics, devoting the remaining twenty-five years of his life largely to the domain of symbolic algebras, and particularly to the development, and the detailed application over wide ranges of subjects, of his system of quaternions, being the precursor in this extensive and imaginative field, on the future practical importance of which he laid great stress. An interesting very simple account of the efforts which finally hit upon quaternions, and are embalmed in their name, is given by him in *Phil. Mag.* of date 1844.

Symbolic algebras.

Method of Action: Thomson and Tait:

The fundamental character of his culminating doctrine of Varying Action*, as implicating the whole range of physical science, seems to have been first emphasized for a wider audience in Thomson and Tait's classic book on *Natural Philosophy* (1867). It appears from surviving letters that Tait had a large share of the merit of this recognition, by insistance on his more physical colleague's study of the work of his own Irish friend, of which the development, apart from applications in dynamical astronomy natural to the time, had hitherto, following the lead of Jacobi, been in the direction of abstract analysis of the implications of systems of partial differential equations. Tait failed however in interesting Lord Kelvin in Hamilton's quaternions as a working instrument; though as an invariant or intrinsic mode of expression of results it had more success with Maxwell and naturally with Hamilton's compatriot FitzGerald. From Thomson and Tait the illumination passed on to Helmholtz, culminating in his extensive memoir "Ueber die physikalische Bedeutung des Princips der kleinsten Wirkung," Crelle's *Journ. für Math.* 1886: cf. Planck's historical lecture, already referred to (p. 31). Reference is relevant here to the general paper of 1884 on Least Action, reprinted *supra*, pp. 31–58, but of preliminary and analogical scope as contrasted with Helmholtz's bolder formulations developed soon afterwards: also to J. J. Thomson's book *Applications of Dynamics to Physics and Chemistry* (1888).

Helmholtz.

Its practical application.

The practical applications in the extended Action domain made by Hamilton himself, both in optics and in astronomy, rested on the simple plan of successive improvements, starting from an initial approximate form of the Action function with the suitable number of adjustable parameters—a function which, as was thus implied, must be expected to involve within itself alone the complete and immediate consummation of any problem in hand. This was the method suitable to the actual dynamical astronomy, which was then confined to working out the modifications of the motions of the stable Solar system due to slight perturbing causes. The formal verification that any function whatever, satisfying the characteristic partial

Extensions by Jacobi:

* Implying comparison of the Actions of adjacent dynamical systems.

differential equation and containing the proper number of arbitrary constants, does in fact provide within itself a complete conspectus of the solution of the dynamical equations, was supplied by Jacobi, who also developed the theory far into the recondite domain of the formal implications inherent in knowledge that is conditioned by partial differential equations. The classic exposition is his *Lectures on Dynamics* as published in 1843, reprinted in *Werke, Supplement-Band*. (Cf. also Cayley's "Reports on Theoretical Dynamics," *Brit. Assoc. Reports*, 1857, 1862, or *Collected Papers*, Vols. III., IV., historically very instructive, especially as regards the early evolution of general dynamics from the problems of planetary astronomy in the hands of the French analysts.) Hamilton was content to take this for granted in his own simpler astronomical problems. The far-flung physical interpretations, involving reciprocal relations between any two distant stages of the progress of the dynamical system, separated by finite time, that were doubtless indicated in their early beginnings by the familiar yet profound physical fact of image correspondences in the optical domain, and later illustrated in arresting manner by problems of dynamical aim in Thomson and Tait, forming the other aspect of the Hamiltonian theory, even more fundamental for the study of Nature, were first expanded systematically into general physical science by Helmholtz. That concise consolidation of the criterion for the coherent dynamical nature of physical systems is obviously still open to far-reaching developments.

already implied in Hamilton.

Direct physical interpretations.

Helmholtz.

As a matter of scientific history, a germ of the earlier Hamiltonian developments for rays may be found, already in Newton's day, in the principle of Cotes, expanded by R. Smith in his *System of Opticks* and recalled to modern attention by Rayleigh. It asserts that the apparent distance of an object at A, seen across any optical instrument by an observer at any other place B, is equal to the apparent distance of the same object located at B as seen from A. For the special case when A and B are conjugate foci, the relations of object and image, as earlier generalized for all systems by Huygens, emerge from this principle. When rays are replaced by their analogues, the paths of projectiles in a field of force, the principle becomes a reciprocal relation expressive as *supra* of the accuracies of aim from any position in order to strike a target at any other position whatever. Under this aspect the subject was explored, involving the conjugate pattern relations on any two targets anyhow situated and for the general non-symmetrical case, as *supra*, in Thomson and Tait's *Natural Philosophy* (1867) as an early overt illustration of the wide scope of the terminal relations inherent in the method of Varying Action, afterwards developed and utilized by Helmholtz and Rayleigh.

Cotes' optical apparent distance: relation to image properties of Huygens:

dynamical analogues.

APPENDIX.

(ii) *On Relativity in Relation to Convection.*

To reduce the equation of propagation by waves in a convected system, as above obtained (§ 14, p. 565), to the normal form appropriate to a resting system, *i.e.* to a form that is relative to its own internal convected frame of reference, which is

$$\left(\frac{d^2}{dx^2}+\frac{d^2}{dy^2}+\frac{d^2}{dz^2}-\frac{K}{c^2}\frac{d^2}{dt^2}\right)\mathfrak{B}=0,$$

Exact transformation, now consolidated:

the unit of time on the right requires to be altered as well as that of x. For the fundamental case of free aether, for which K is unity, this shrinkage in time is equal to that in length along x; namely, its factor is β, equal to $(1-v^2/c^2)^{-\frac{1}{2}}$ as is obtained later in the same paragraph. Combining this shrinkage in space and time with the previous transformation, there results the Lorentz transformation as the type which maintains the equation of free propagation invariant for uniform convection.

extended to the free electric field:

This transformation was developed for the complete scheme of the electric equations of the field, not merely the final resulting equation of pure propagation in three dimensions of space, in *Aether and Matter* (1900), Chap. XI., but only up to the second order of v/c; being so restricted on the tacit ground that the finite size of the electrons (10^{-13} cm.) compared with the atom (10^{-8} cm.) to which they belong must in any case introduce uncertainty beyond the order of 10^{-10} which is the ratio of the squares of these magnitudes. On the other hand, for the general electrodynamic field outside the atoms, the correspondence there developed is readily recognized to be valid exactly* to all orders, just as in the present partial discussion.

but to interiors of atoms only approximately.

This complete scheme for the electrodynamic field *outside the atomic sources* was obtained in exact form independently by Lorentz (1904), developed on a basis of electric volume-density; and the correspondence leading to it is appropriately called by his name, as having been the initiator of this criterion of correspondence (as, in fact, the test of physical identity in varied circumstances) in his treatment of the aberration of light up to the first order of v/c in 1892.

The Lorentz transformation.

Correspondence as criterion of identity.

Any isotropic shrinkage of space and the same shrinkage of time might however be superposed, and such indeed has been contemplated in order to obtain unchanging corresponding volumes, *e.g.* of the electrons regarded as distributions of charge. But Lorentz there pointed out that such a shrinkage, even when restricted to the type involving only the square of v/c, is not reversible, so that when a system acquires a convection, and then the opposite convection is

The transformations restricted if the system is to be permanent:

* As also appears in an earlier exploration of the correspondence in Part III., *Phil. Trans.* (1897), § 16, with correcting note added in Vol. II. *infra*, p. 39.

imposed so as to cancel it, there would be no return to its original scheme of configuration. This postulate of intrinsic permanence of material systems thus limits the transformation definitely to the Lorentz form.

Soon afterwards Poincaré extended this remark (*Rend. di Palermo*) by showing that the Lorentz transformations form a self-contained group, in that any uniform convection in space superposed on any previous one is equivalent as regards the resulting change of configuration to a single convection of the same type. So far as this holds, and it is established for free aether on Maxwellian principles but could only be postulated hypothetically within the atomic structure, there could be no reason for giving a preference to one frame of space and time belonging to the Lorentz group rather than any other, as a standard expressing absolute rest. All convection, uniform translatory motion, would then be indeterminate, there being no standard frame in which to locate it. At any rate one is impelled to explore the implications of this postulate as far as they will carry: it in no degree discredits the theory of an aether, unless the intrinsic atoms of matter can be abolished also.

they thus form a group,

when combined with an atomic postulate:

translatory uniform convection then inessential.

Thus arose the theory of unrestricted electrodynamic relativity, as regards uniform convections but not other motions of the system. It was reached independently by more universal reasoning, electrodynamics being only an illustration, by Einstein, fortified in his views strangely by the principles of E. Mach, that in science we must build on what we observe and not analyze beyond the range of our direct sensual perceptions. He founded his general argument on a mysterious postulate, that translation applied to the system does not affect at all the velocity of radiation relative to it, which remains the same as when measured in the original resting frame of reference. This might well appear to be a pure paradox, until it acquires an interpretation as the statement of an algebraic correspondence, between frames as above, but masquerading in the language of kinematics.

The relativist general philosophy.

A new trend was soon imparted to the group theory by a remark of Minkowski. Briefly the differential operator of propagation

$$\left(\frac{d}{dx}\right)^2+\left(\frac{d}{dy}\right)^2+\left(\frac{d}{dz}\right)^2-\left(\frac{d}{cdt}\right)^2$$

is conjugate to the differential expression

$$\delta x^2+\delta y^2+\delta z^2-(c\delta t)^2,$$

so that the one is invariant for all fourfold (x, y, z, t) linear transformations for which the other is so. The transformations which retain the former as invariant were worked out directly (much as here, p. 565) long previous by W. Voigt, in a discussion relating to the aberration of light. Now the latter form may be regarded as the square of the

Minkowski's fourfold partial map of history:

element of length in a pseudo-spatial fourfold extension ($x, y, z, \iota ct$). And if we imagine the history of the universe as spread out from remote past to remote future in this pseudo-space, which is a uniform one but differing *in toto* from a proper space, viz. if its flat sections of constant t represent the succession of configurations of the material cosmos at different times, then the Lorentz transformations are the analogues of the changes to various restricted sets of parallel sections, the new effective time, restricted to pure imaginary, being measured transverse to them. The group property would thus, but it would seem only to a partial degree, take on an intuitive aspect: inasmuch as times so postulated are restricted to be all pure imaginary, in the correlative general imaginary fourfold. To adapt to the reality of nature, it is much more direct and intelligible to group in a real fourfold, if so preferred, with its own internal *real* frame of reference (x, y, z, ct) that is essential to it. The transformations of frame are now real ones; but they are no longer expressible as simple rotations, once the imaginary form, as however above restricted, in an algebraic-symbolic pseudo-space in which it simulated rotation, has been got rid of.

its spatial group too wide for Nature:

reversion to the real group.

A standard frame, after Descartes.

The practical aspect as regards all such theories is that it would be a crude and complex procedure to envisage every pair of configurations in the world as related directly to each other: it suffices that they are all relative individually to the same convenient choice of a standard frame of reference. A pair of frames, within the permitted group of transformations as they are envisaged by the linear equations connecting them, differ by a uniform relative translatory convection: but they also differ by FitzGerald-Lorentz shrinkages in space and time, combined with the great complication (a feature fortunately not relevant for our usual types of knowledge, though the aberration of light has to be put upon it) of changing epochs, in analogy with our terrestrial civil times differing in epoch with the longitude. We have it, then, that when a group of phenomena, local in space and time, of the cosmic history as referred to its own intrinsic frame, is transferred to another frame with respect to which it is uniformly convected, the natural procedure is to modify the frame so as to maintain the cosmic system expressed within its framework invariable, as proves to be possible without very complicated processes. The frames differ: while the cosmic history, framing itself in related ways in all of them, is absolute, not itself subject to change. The members of the group of such permitted frames are related to each other, or, more simply, to any convenient standard frame of the group, by the Lorentz transformation, expressible as regards any pair of them as uniform shrinkage in space and time, *of each relative to the other*, combined with changing local epoch of measurement of time.

An unvarying world referred to convected modified frames:

which are the Lorentz group.

The ultimate formulation thus contemplates a permanent unchanging world: it is referable to any of a permitted group of travelling frames, each travelling uniformly relative to the others: in each of them the cosmos falls identically into similar mode of expression, as presented to an observer attached to the frame, in terms of signals travelling with the finite speed of radiation. When the frame changes, the relative values of features of the cosmos measured in its units change inversely.

Astronomers' assurance of a permanent cosmos.

And it may be claimed to be fortunate for the simplicity of our natural knowledge that it can be so expressed. As an astronomer is carried round the Sun in the Earth's orbital motion, his personal and indispensable frame of reference in space and time is constantly changing: but he is now assured that, if he constructs in his own landscape an appropriate frame, and changes it by definite rule on account of his changing motion, the presentation of the universe, as inserted in the lattice of the frame, will not require any alteration other than the Bradley correction for aberration of which he has been aware since Newtonian times, and which is an affair mainly of relative convection of one frame through another.

Universal absolute space and time merely not convenient.

It is the concatenation of times of events, which has to be practically measured by aid of light-signalling, that creates the complexity. If astronomers could employ instantaneous signals, their use would provide an absolute space and an absolute time as the unique universal frame of reference. Such absolute space and absolute time are coherent as a logical scheme because we can imagine instantaneous signalling: but it is an impracticable one when we have to depend on light. The practicable synthesis then has to be found in this coherent group of equivalent frames, associated with the various possible speeds of convection. This scheme is complete and is verified for the range of electrical phenomena, and so for all of physical science that is of electrical nature. The problem how far it can be made consistent with Newtonian gravitation has opened out new domains, involving further interesting complexity in the relations of their intrinsic times for different observers.

For hitherto the convection is taken as uniform, and so must be one of pure translation: therefore each frame is universal, and uniform throughout all space. But as observers in different parts of the cosmos are actually being convected with differing speeds, it is natural to try the effect of frames restricted to be uniform only locally. Instead of an unlimited quadratic invariant in the fourfold,

$$\Delta s^2 = (x_2 - x_1)^2 + (y_2 - y_1)^2 + (z_2 - z_1)^2 - (ct_2 - ct_1)^2,$$

there is now only a local or differential invariant δs^2, after Einstein,

Restriction to universal convection removed: expressible in terms of any scheme of coordinates (x_1, x_2, x_3, x_4), one of them being in special relation to time. This invariant form may determine an auxiliary formal history, a generalization from the Gaussian surface, with its parts inter-connected in *quasi*-spatial manner, now however non-uniform, of which each differential element of extension stands in direct correspondence with an observer's local convected element of the actual cosmos; to this auxiliary therefore the regions of the actual world may be referred in definite correspondence, which need not however be complete. The correlation between them includes in particular the positions of atoms and electrons. The framing of this auxiliary partial model would have to be itself determined in some intrinsic invariant way by the forms within it that are correlatives with the paths of the actual atoms and electrons, these forms being for this purpose regarded as singularities in its own constitution. The simplest way in which this dependence of the varying space on the atoms and electrons is possible, turns out, after Einstein, to introduce a picture of gravitation into the auxiliary construct.

thus making room for gravitation.

Natural spaces and times: Thus each observer has the intrinsic space and time of his own landscape, which is absolute space and time, the same everywhere, because the spectroscope has revealed that nowhere in the universe is there matter of unfamiliar structure. But he cannot directly connect his own system, at any rate with feasible facility, with the systems as presented to other remote potential observers. That connection has to be made indirectly, by first transferring his local system into the auxiliary cosmos, itself an incomplete and artificial model as not being in any sense in space and time so not representative of direction or motion, then making the cross-connection in that auxiliary cosmos, and finally transferring from it to the actual landscape of the other observer. Compare the use of the globes—or else of a flat Riemannian surface with connections ruled by spherical trigonometry—as an intermediary, by a geographer wishing to establish connections with the antipodes which are outside his range of direct light-signals. The question arises, is this unique auxiliary cosmos the real world, thus static in four dimensions? Then the local world of each convected observer, with its atoms and electrons, its plants and animals, and all that it contains, would be merely a conventional and imperfect simplification of a small region of reality considered by itself, introducing ideas of position and motion as being capable of reference to absolute time the same for all potential observers. Like most such questions, it may perhaps be regarded as one largely of the conventions that are adopted: but the absence of space and time and motion in the auxiliary construct is against reality. Anyhow, a key position, resolving many questions, would be that in any

made coherent by aid of an auxiliary cosmos of reference,

representing incompletely the actual world-history.

case there is one definite world postulated: however different may be the appearances it presents, by means of their light-signals, to observers differently situated and differently convected, they must come into agreement by the proper choice of frames*. This position, that there exists an external world, and that the aspects presented to various observers on account of the delay of light must of necessity be related so as to be in keeping with its permanence, resolves itself into an affair only of different subjective frames of reference distorted relative to each other, and may seem at first sight remote from the usual postulations in relativity.

A definite real world but under changing aspects.

* Cf. *Nature*, supplement of April 9, 1927: where, however, the solution adopted for verifying the values of the optical effects of gravitation, as observed, must be abandoned. Cf. also *Nature*, Sept. 3; and *Proc. R. S. Edin.*, July 4, 1927.

APPENDIX.

(iii) *Do Accelerated and Retarded Electrons emit Radiation?*

Radiation from orbital electrons: excessive, except for partial compensation. Do therefore electrons radiate? Crucial for Maxwellian theory.

THE conception of an orbital electronic structure for atoms of matter, adumbrated in pp. 514–525, pointed directly to a mechanism, on first view plausible, for their free radiation. An orbital electron must radiate on account of its acceleration; but at a rate far too great for the atom to survive for any appreciable time, unless* the correlated motions of the electrons are so balanced that the vector sum $\Sigma e\dot{v}$ vanishes where $\dot{v}$ is acceleration, the wave-length being here supposed great compared with the atomic diameter. This condition is very nearly satisfied (cf. *loc. cit.*) in the usual nuclear model of an atom, provided it is symmetrical. But outstanding difficulties have conduced to a prevalent view that an electron in fact ought not to radiate, which amounts virtually to a denial that such a structure can be legitimately postulated at all. It will now be claimed that direct experiment has in fact recently given a decisive answer on this direct question.

Tested by recent experiments with arrested cathode torrent, with result suggesting a precise general principle,

The main and practically only effect at a sufficient distance, due to arrest of an electron, is the pulse of radiation that travels out from it. This shell of propagated disturbance involves within itself an electric force F, such as might produce an appreciable electromotive force $\int F ds$ when integrated round a galvanometer circuit situated in the region. This idea has been advanced recently†, and carried through by means of delicate experimentation, by S. R. Milner and J. S. Hawnt, by making use of the arrested cathode stream in a suitable electric bulb. They estimated that if the pulse, in part consisting of x-rays, were the whole effect of this source, their tube should give $4 \cdot 10^{-3}$ volts round a circuit which includes a cylindrical part surrounding the cathode. In the end, after spurious effects had been detected and eliminated, the observed electromotive force was less than one-half per cent. of that amount. This indicates a null result: which being precise suggests the question whether or not the Maxwellian theory would have predicted a general conclusion of that kind. If on that classic theory the motion of the electron would produce nothing but the pulse of radiation, the answer as stated above would be in the negative: and confirmation would be given to the view now widely held that an arrested electron cannot emit radiation, a view which really would involve complete breakdown of the theory of the Maxwellian field, and not of that alone.

On the other hand, the pulse of radiation is by the theory only a

* *Phil. Mag.* Dec. 1897, § 11, as *infra*.

† *Phil. Mag.* May 1927, pp. 1185–95. [The experiments described by Joseph Henry, *Trans. Amer. Phil. Soc.* (1840), "Scientific Writings," Vol. II. p. 165, §§ 32–93, also in part in *Phil. Mag.* June 1841, "On apparently two kinds of Electro-dynamic Induction," appear to be relevant in an interesting way.]

part of the disturbance of the local field due to the arrest of the electron, a part ultimately conspicuous because it travels free right away from it. If exact general electrodynamic theory proves to be able to show that the effect of the pulse is neutralized by a compensating effect of the unpropagated change of the intrinsic local field, the inference to be drawn from the experiments is thereby reversed. The order of possible effects is found to be substantial, but experiment discovers less than one per cent. of that order: while classic theory, which does involve radiation from an electron with changing speed, in fact requires, on the argument which follows, as the present writer at any rate holds, that there should be no such effect*. The issue is thus sharply set: and the experiments quoted may be claimed to be a direct verification that the classic theory involving Maxwellian radiation from accelerated or retarded electrons is sustained in actuality.

which proves to be involved in Maxwellian theory.

What, then, is to be made of the phenomena of radiation from atoms, and its relation to the phenomena of absorption and ejection of electrons? The view that atomic radiation arises from the accelerations of the orbital electrons was a lame one from the first, as leading to an amount of radiation in general impossibly great: though for an atom with its orbital electrons symmetrically balanced round its nucleus so that the latter remains at rest, the radiation might be extremely small, as regards waves of light which are very long compared with the diameter of the atom†. Yet even then it would not vanish, so the atom would during aeons fade away: unless there is some delicate feature, not necessarily other than a slight one, but as yet unrecognized, in the scheme of linkage of the electrons of the atomic system, which just annuls the tendency to such residual radiation. That would conform to the conclusion to which modern spectroscopic relations point, namely, that the atomic model with orbital electrons is an imperfect, but in various respects an effective and approximate, symbolical expression of its real structure.

Atoms must be usually balanced against radiating.

Standard electrodynamic theory, involving the radiation from free electrons, would stand firm: but atomic structure would be such that there is compensation prohibiting a permanent drain upon the atomic energy, unless in circumstances of exceptional internal excitement.

The proof that there could be no galvanometer effect in any closed resting circuit, adjacent to an electric bulb, due to starting and arrest of a torrent of electrons or of ions of any kind, as above asserted, is

* Expressed the other way round, if we contemplate with some theorists a process which estops radiation on Maxwellian lines from the arrested outside electrons at the cathode, the experiment ought to show more than a hundred times the observed result.

† Cf. *Phil. Mag.* Dec. 1897, as reprinted, Vol. II. *infra*.

immediate on Maxwell's principles. The electric field (P, Q, R) produced is due to a vector potential (F, G, H) and a scalar potential V, by his formulae of type

Violent electronic disturbance does not affect an adjacent galvanometer circuit.

$$P = -\frac{\partial F}{\partial t} - \frac{\partial V}{\partial x}.$$

It is not material for the argument* whether these potentials are the static potentials of Maxwell's total current, or the retarded potentials of the moving electrons alone.

The two components of the electric field.

As regards the first terms, $\int P dt$ vanishes when taken from an initial resting state of the system to a final resting state: thus on the slow-acting galvanometer there would be no effect. As regards the second terms, that part of the field is derived from a potential V, which is single-valued, so at each instant the tangential electric force integrated around the galvanometer circuit is null. For it is to be noted that the electric flow adjusts itself to uniformity all round the circuit, by electric waves travelling on it, with the speed of light: so there can be no question of electric force from an incident pulse acting on the part of the circuit containing the galvanometer while the rest of it may be screened and immune. It is to be concluded, then, that a slow indicator in the circuit, such as a galvanometer, would show nothing†: though in circumstances originating electric waves such a circuit could well be active as a secondary radiator, but without any cyclic current around it.

Source of energy of cathode torrent.

The immense though now familiar speed of the cathode torrent of electrons arises mainly from the unimpeded fall down the electric field. Whence is its energy derived? On analysis it will be found to be traceable to the generator or battery that drives the current in the bulb circuit, which has to work at slightly higher electric pressure.

* Cf. Maxwell, *Treatise*, Vol. II. § 598: or directly in terms of the flying electrons, from Least Action, in *Camb. Phil. Trans.* 1899, as *infra*, or in *Aether and Matter*, §§ 49–59.

Result involved in Faraday's law.

† More ultimately, this negative result rests on the Faraday law, on which Maxwell and his successors built, that the electromotive force (integrated electric force) round any circuit is in all circumstances $-dN/dt$, where N is the number of magnetic tubes it encloses: as N cannot increase or diminish indefinitely the force must be alternating; in the absence of any artificial commutating device, such as Faraday was himself the first to introduce in his model of a direct-current electric motor which reversed would be a dynamo.

Twofold effect of arrest of an electron.

The explanation in detail for an isolated electron coming on from a distant source appears to be as follows. A uniformly moving charge carries along a steady magnetic field. When it is suddenly arrested, it emits the appropriate sharp pulse of radiation. But in addition its steady magnetic field has then lost its nucleus, so has to collapse, spreading in waves partly outward, and partly inward, but ultimately reflected outward from the centre. This provides an additional pulse of radiation, which carries far less energy, because it is less concentrated, and energy involves square of the field; but it can effect compensation as regards electromotive force.

APPENDIX.

(iv) *Minimal Action and Electrodynamic Potentials.*

THE fundamental consolidating *rôle* of Minimal Action is well brought out in connection with the electrodynamic potentials of moving charges. The important formulation by Liénard, and later by Wiechert, though found to be necessary for a consistent analysis, seems hardly intelligible by itself: in this entangled domain mere following out of algebra, unchecked by interpretation, may lead anywhere. Opportunity is here taken to extend the Action formulation of the dynamics of electrons, as *infra, Camb. Phil. Trans.* (1899) or *Aether and Matter* (1900), Ch. VI, to cover the case when they are in rapid motion. The electric part of the Action is there reduced to the form

$$A = \tfrac{1}{2}\Sigma e \int (F dx + G dy + H dz - V dt),$$

where the contribution of an electron say e' to the potential is expressed as

$$F = \frac{e'\dot{x}'}{r}, \quad V = \mathrm{c}^2 \frac{e'}{r}.$$

This potential refers to a point in the aether, or in the standard frame of reference of the problem: it has no relation to a moving charge at that point, except through A in which the velocity of that charge appears as terms of type dx/dt. Specification of potentials:

If disturbance is propagated out from a moving electron into the medium in shells which remain concentrated, the shell which reaches the point at which the potential is to be reckoned is the one that emerged at a prior time, so that in the formulae for the potentials there is to be substituted for r a reduced value $|r|$ equal to $r - v_r'/\mathrm{c}$ when v_r' is the component along r of the velocity of this influencing electron relative to the frame. But Liénard found that coherence of the theory required further modification, from $|r|$ to $|r|(1 - v_r'/\mathrm{c})$, so that, if this is rightly quoted, the latter factor occurs twice. The ultimate explanation with which one has had to be content* is that a shell of disturbance issuing from e' in the interval dt is operative for a longer time dt', equal to $dt\,(1 - v_r'/\mathrm{c})^{-1}$ in passing over the locality in the frame. But the force of this explanation depends on the Action: the contribution to it from e' over a time dt is extended to moving sources:

$$e\int\left\{\frac{e'\dot{x}'}{|r|}dx + \frac{e'\dot{y}'}{|r|}dy + \frac{e'\dot{z}'}{|r|}dz - \mathrm{c}^2\frac{e'}{|r|}dt\right\},$$

* Cf. E. Cunningham, *The Principle of Relativity*, Ch. IX. § 6: H. A. Lorentz, *Theory of Electrons*, Appendix.

in which, in a frame travelling with the electron e', dt belongs to that electron which originates the Action. But, on changing into the standard frame of the problem, it is prolonged to $dt\,(1 - v_r'/c)^{-1}$, while dx is $\dot{x}\,dt$; and that may be held to be the origin of the Liénard factor attached to $|\,r\,|$. That factor thus would not belong to the potential, but to the Action: but things are *as if* the factor were transferred, the Action at a point in the standard frame being left as it was.

a factor transferred from the Action.

It is to be observed that propagation of effects in shells that remain concentrated is essential to all these arguments. They would fail when applied to the line sources of two-dimensional propagation, for then a trail is left behind the main shell of travelling disturbance. Nor, strangely, would they be valid for propagation in an anisotropic medium such as a crystal, as Lamé long ago discovered: though it was just that case which led Huygens to his elucidation of the law of crystalline refraction on the basis of disturbance spreading out in compact shells. See *infra*, Vol. II, from *Nature*, April 1927.

Restriction of idea of a source.

APPENDIX.

(v) *Maxwell's Stress and Radiation Pressure.*

The stress-tensor of the field has had to be introduced, or alternatively the Lorentz relation between frames of which it is an expression as in Math. Congress Lecture (1912) *infra*. The reason is that the usual formulation takes account only of electric currents and magnetism: yet the field refuses to be wholly merged. The relations of the current are formulated as made up by the influence of the field on electrons taken separately and additively: but there may be a part not simply additive. **Source of radiation pressure.**

This part ought to be involved in the Action formulation expressed primarily by an Action density over the field; that is invariant for transformation between permitted frames, and in fact from the Action specification alone this invariance determines their group, the Action being here also the ultimate condensed formulation of everything. Reference may be made for detailed analysis to the Stokes volume, *Camb. Phil. Trans.* (1899) as *infra*, Vol. II, or to *Aether and Matter* (1900), Ch. VI. The way in which a momentum outstanding in the field asserts itself directly out of simple dynamics is illustrated in the paper "The electromagnetic force on a moving charge in relation to the energy of the field," *Proc. Lond. Math. Soc.* (1913) as *infra*, Vol. II. From its point of view, which dispenses with the auxiliary formal stress-tensor of Maxwell, we may say that the play of the field Action in the background compensates itself by means of the standard forces affecting the separate electrons, to which it gives rise, all except a part that remains over by itself as a field integral and cannot be so transferred to *individual* electrons. This part, operating as a residual momentum in the aether, exhibits itself in simple manner only against electrons *in the aggregate* where they abruptly alter the field, by a material stress, including as a special case the pressure of radiation: against a single one the result depends on how it disperses the radiation. If, further, an ideal interface is imagined isolating a region of the field and excluding the Action of the part beyond, the result of its interaction is represented by a stress over that interface. But all this part of the field activity cannot readily be transferred into a form of expression in terms of interaction between the electrons as individuals. **Field momentum illustrated:** **more directly:** **a residue operative for volume specifications only:** **thus expressed in a stress-tensor.**

The reason why there is no sign of radiation pressure in the equations of the field, on which Kelvin insisted to their detriment, is not that

Field equations thus incomplete.

these are in any way imperfect dynamically, but that they are an incomplete expression of what is involved, as regards material manifestations, in the essential activity of the medium as formulated by its Action and controlled by its electrons.

Illustrated by a reflecting current sheet of electrons, or an equivalent dielectric medium.

Referring back to pp. 584–6, from the electronic point of view, the ideal perfect conductor of Maxwell, which turns back radiation by a surface current sheet in phase with the incident electric force, loses its simplicity. The current sheet is composed of electrons, and as there is no loss of energy they ought to encounter resistance only such as inertia would cause. But if that be so, the current is in the opposite phase to the electric and magnetic fields, and radiation pressure would be alternating instead of steady: for such a medium the dielectric constant K is very small, and the velocity of phase very great compared with that of light in free space: therefore by Fresnel's laws for transparent bodies the reflexion coefficient $\left(\frac{1-K}{1+K}\right)^2$ is nearly unity, and that of transmission $\frac{2K}{(1+K)^2}$ very small. The absence of any steady radiation pressure on the interface is perhaps connected with the necessity of some sort of ideal constraint to confine the electrons from escaping, which sustains an equivalent force*. Such considerations as these show up the very great convenience of the Maxwell stress, as a mode of specification of the mechanical forces of electric systems which evades all local features.

Radiation pressure in a material dielectric.

For a beam of light incident in a dielectric substance on a perfect reflector, say directly, we can calculate the pressure exerted by it in three ways. The magnetic field along the reflector is $2H$, the electric field null. (i) The reflexion produces a current sheet of intensity $2H/4\pi$ which is subject to a mean magnetic field $\frac{1}{2} \cdot 2H$, giving a pressure $(2H)^2/8\pi$ which is the density of energy in front of it, being twice the intensity of energy in the incident beam. (ii) There is involved a reversal with velocity c' of a distribution of aethereal momentum density DH, where D is the aethereal displacement $F/4\pi c^2$ and H is F/c': this gives a pressure $2 \cdot F^2/4\pi c^2$ or $K^{-1} \cdot (2H)^2/8\pi$, which *disagrees* with (i). There is something more involved: it must be a force on the electrons of the substance. Note that the momentum density in a beam is the energy density multiplied by c'/c^2. (iii) Again we may surround the place of reflexion by a surface: the momentum inside it is steady, therefore the force on its content is

Discrepancy.

* We may however assert that it is the secondary radiations from the surface electrons that form the reflected beam, which again introduces resistance to their motions. See p. 664 *infra*.

to be derived from the Maxwell stress over the boundary. The calculation depends on how the boundary crosses the stationary beam, say whether as a node or a loop: in the one case it is $\frac{K(2F)^2}{8\pi c^2}$, in the other $\frac{(2H)^2}{8\pi}$, both in agreement with (i).

We may probe the discrepancy of (ii) by the simplest case. Suppose a limited train of radiation is travelling within a region of the dielectric, entering normally across the boundary. The boundary stress is a pressure $\frac{1}{2}(1+K^{-1})E$ inwards: the interior gains from it *aethereal* momentum $K^{-1}E$ per unit time: the difference $\frac{1}{2}(1-K^{-1})E$ must be momentum gained by the electrons of the *material*. Thus momentum $\frac{1}{2}(1-K^{-1})E/c'$ would have to be communicated to the transmitting *material* per unit volume, remaining while the waves are passing through it and abstracted as they leave it; and this must be presumed to be the case.

Extra momentum required to initiate a train of waves.

APPENDIX.

(vi) *Relations between Electric Unitary Systems.*

The field equations: THE Maxwellian circuital equations of the electrodynamic field, with E for intensity of electric force and H of magnetic, are

$$\text{curl } H = \frac{1}{c^2}\frac{dE}{dt} + 4\pi\Gamma, \quad \text{curl } E = -\frac{dH}{dt},$$

implying $\quad \text{div } E = 4\pi c^2 \rho, \qquad \text{div } H = 0.$

Hence the vector Γ expresses $(\Sigma e\dot{x}, \Sigma e\dot{y}, \Sigma e\dot{z})$ the summation being per unit volume, as the electrons whatever they may be are small enough to be regarded even within the atoms as mere point-charges e; while the electric density ρ expresses Σe. Thus Γ is the flux of true electricity, which in special theory is analyzed into currents of conduction, of polarization, of convection of steady charges, and what not.

These equations may be transferred to a new unitary system indicated by accented symbols by a substitution

$$E' = \frac{k}{c}E, \quad H' = kH, \quad \Gamma' = 4\pi k c\Gamma, \quad e' = 4\pi k c e, \quad \rho' = 4\pi k c \rho,$$

transferred to other units: giving $\quad c \text{ curl } H' = \frac{dE'}{dt} + \Gamma', \quad c \text{ curl } E' = -\frac{dH'}{dt},$

implying $\quad \text{div } E' = \rho', \qquad \text{div } H' = 0.$

In these units electric density is the divergence of electric force. Magnetic force has no divergence.

So far k remains arbitrary: so another condition can be satisfied. further restricted. The Amperean electrodynamic force on an electron is evH: if this is not to be altered by mere change of electric units, eH must be equal to $e'H'$. This requires that $k^{-1} = (4\pi c)^{\frac{1}{2}}$.

The electric inertia of an electron, or of its model as an electric shell of radius a, is altered

Formulae in "rational" units:

$$\text{from } \frac{1}{2}\frac{e^2}{a} \text{ to } \frac{e'^2}{8\pi c a}:$$

the rate of radiation of any accelerated charge e

$$\text{from } \frac{2}{3}\frac{e^2}{c}\dot{v}^2 \text{ to } \frac{e'^2}{6\pi c^2}\dot{v}^2.$$

The numerical value of an electron becomes

$$e', \text{ equal to } (4\pi c)^{\frac{1}{2}} e, \text{ where } e \text{ is } 16 \,.\, 10^{-21}.$$

In this rational unitary system, as employed by Heaviside and Lorentz, the volt is of value $10^8/(4\pi c^3)^{\frac{1}{2}}$, the ampere $10^{-1}(4\pi c)^{\frac{1}{2}}$, the ohm $10^9/c$, and so on: where c is $3 . 10^{10}$.

There is a further simplification. The density of field energy is thus altered field energy and its flux:

$$\text{from } \frac{E^2}{8\pi c^2}+\frac{H^2}{8\pi} \text{ to } \tfrac{1}{2}E'^2+\tfrac{1}{2}H'^2,$$

the energy flow in the field

$$\text{from } \frac{1}{4\pi}[EH] \text{ to } c^2[E'H'].$$

The field Action is the specification by Action.

$$A'=\int dt\int(\tfrac{1}{2}H'^2-\tfrac{1}{2}E'^2)\,d\tau,$$

from which alone the new system can be developed, with divergence of electric force as charge and of magnetic force null.

The vector potential U (components usually denoted by F, G, H) and the electrostatic potential V (equal to $\Sigma c^2 e/r$) may be introduced without alteration of their relations to the field: thus Vector potentials

$$H'=\text{curl } U,\quad cE'=-\frac{dU}{dt}+\text{grad } V^*.$$

For a single charge moving with velocity v, then of moving sources.

$$U=\frac{e'/4\pi c\,.\,v}{|r|(1-v_r/c)},\quad V=\frac{e'/4\pi c}{|r|(1-v_r/c)},$$

where, as *supra*, p. 653, the second factor in the denominator really belongs to the Action.

But this determination of k is not really necessitated, as suggested above. The value of the electrostatic mechanical force on an electron is $(4\pi k^2)^{-1}e'E'$: also the value of the electrodynamic mechanical force on a moving electron is $(4\pi k^2 c)^{-1}e'vH'$: and k may be chosen in any way that is convenient. The choice in this system of rational units makes k^2 equal to $(4\pi c)^{-1}$: and then these mechanical forces are $ce'E'$ and $e'vH'$. It is a hybrid system, which would conform, were it not for the Heaviside factor involving 4π, to Helmholtz's proposal to measure electric force and charge in electrostatic units and the magnetic force in magnetostatic units. But as both the standard systems of units, the electrostatic and electromagnetic, are completely self-contained, to get a mixed system some link in their chain of relations must be broken. It is naturally the link connecting electric current and magnetism: in free aether a current circuit is now the Choices as regards mechanical forces: involving modification of some link in the unitary chain,

* The symbol grad here and elsewhere means the downward gradient, which is not now the universal convention.

equivalent of a magnetic shell of strength equal to that of the current multiplied by a dimensional factor $(4\pi c)^{-1}$. Complications involving K and μ arose for a material medium regarded as simply polarizable, which were discussed at the time between Helmholtz and Clausius: but as the total electric polarization is complex, partly ionic and partly aethereal, and the former part would give magnetic effect by convection, they are now obsolete except formally for stationary media.

not permissible in material dielectrics.

The dielectric and magnetic moduli as physical entities:

The electrostatic and electrodynamic may be merged into one system, after Rücker, by assigning dimensions to the product $K\mu$: thus it might be presumed that any third system might also be merged by assigning dimensions to K and μ separately. This could apply, only formally, to the case of stationary dielectrics: then the systematic development which would show how the differences are to be absorbed by making K and μ dimensional, would start from the distribution of Action

taking their place in the Action.

$$A = \int dt \int d\tau \left(\frac{K}{8\pi c^2} E^2 + \frac{\mu}{8\pi} H^2 \right)$$

transformed to new potential variables by relations regarded as constitutive

$$H = \text{curl } U, \quad \text{curl } H = \frac{1}{c^2} \dot{E} + 4\pi \Sigma ev$$

and carried through as in *Camb. Phil. Trans.* 1899, Vol. II, *infra*, or *Aether and Matter*, Ch. VI.

APPENDIX.

(vii) *Repulsion of Electrons, Ions and Molecules by Radiation.*

It has already been noted that, in order to derive a pressure of radiation (as *supra*, p. 586, footnote) from reflexion at a perfect conductor, the surface current must be in phase with the electric force, whereas if the electrons were free and isolated, as absence of any loss of the energy of their motion into heat suggests, the phase would be half a period in advance and there would be no steady pressure. It appeared that, under such circumstances, the pressure must be regarded as coming from deflection of the momentum of the radiation by the ions. This position may be further followed out.

The concept of a perfect conductor.

For an ion, under conditions which allow it to be treated as isolated, the process can be traced in detail. Its equation of motion, where subject to a train, or it may be merely a pulse, of radiation travelling along z, polarized so that its electric field F is along x, and its magnetic field H equal to F/c along y, also subject to reaction R of its own secondary radiation, is

The reaction of emitted radiation.

$$m\ddot{x} = eF - R.$$

The energy lost into its secondary radiation is, if k represent $2e^2/3\mathrm{c}$,

$$\delta E = \int k\ddot{x}^2 dt = |\, k\ddot{x}\dot{x} \,| - \int k\dddot{x}dx.$$

The first term, being the time gradient of $|\, \tfrac{1}{2}k\dot{x}^2 \,|$, is merely alternating; the second term expresses the work of a resisting force $k\dddot{x}$, which is thus the value of R. Or better, we can be exact, and add to this value a term R' to represent the alternating part which, owing to its phase relation as yet unexplored, might conceivably have some influence. Thus

$$R = k\dddot{x} + R'.$$

The magnetic field of the radiation, operating on the current element $e\dot{x}$, imposes on the ion a force Z transverse to the wave-front, given by

$$Z = e\dot{x}H = e\dot{x}F/\mathrm{c}$$
$$= \mathrm{c}^{-1}\dot{x}\,(m\ddot{x} + k\dddot{x} + R'),$$

so that

$$\int Zdt = \mathrm{c}^{-1}\left\{ |\, \tfrac{1}{2}m\dot{x}^2 \,| + |\, k\dot{x}\ddot{x} \,| + \int R'dt - \int k\ddot{x}^2 dt \right\}.$$

The first three terms in this expression are alternating. Therefore the steady repulsion on the ion is expressed by a loss of momentum along the direction of the ray equal per unit time to $k\mathrm{c}^{-1}\ddot{x}^2$: that is, its

The repulsion of an ion:

motion loses momentum per unit time equal to its secondary radiation per unit time multiplied by c^{-1}. This is in agreement with the Maxwell law of repulsion by radiation, which is expressed in terms of the fall of density of the incident radiation in passing across the substance.

The directed momentum that disappears from the incident radiation by scattering, and must be communicated to the ion which scatters, is thus $\delta E/c$, where δE is the energy of the radiation that is scattered. This is a sufficient foundation for the inference that free radiation carries momentum of some kind of amount E/c.

It is to be observed that m is not involved in the result. The repulsion by radiation is thus additive for all mixtures of ions, and arises solely from disorganization of its directed momentum. The function of the mass of an ion is to determine the amount which that ion scatters*.

due to deflected momentum. Mutual support of ions.

This deflection of momentum of radiation from a single isolated ion has been here determined indirectly, by aid of the loss of scalar energy: but for a reflecting stratum of mutually supporting free ions there is no such scattering of energy, while the momentum is simply turned back along with its ray. A mass of ionized gas, even very highly rarefied, should reflect back all the incident light, though diffusely: the supposed high reflecting power of some gaseous nebulae is recalled.

Application to atmospheres.

The results above for an isolated ion are directly applicable to a gaseous medium with its ions entirely free, especially to an ionized Solar or stellar atmosphere usually of extremely small density of

Scattering by free ions:

molecules. The question of transparency first invites attention. The rate of radiation from a free ion is $\frac{2}{3c}(e\dot{v})^2$, where here $m\dot{v} = eF_0 \cos pt$: as the flux of incident radiation is $\epsilon = c\frac{F_0^2}{8\pi c^2}$, the mean value is $\frac{8}{3}\pi\left(\frac{e}{m}\right)^2 e^2\epsilon$. If there are N' such free ions per cubic cm. in a medium, the loss of incident energy by their independent scattering is, per cm. of depth, N' times this amount.

by polarized atoms. Polarization equivalent of a molecule:

But a molecule or an atom also scatters radiation by its polarization: and a comparative estimate is necessary. In attempting to form it, we are brought up against a specific property of the atom, independent of temperature and in fact absolute, which may be named its polarization equivalent. As it belongs to the isolated molecule, it

its values for different substances.

will arise specially for gases: it is the standard value of $K - 1$ for the pure gas divided by the numerical density N' of the molecules

* The true absorptions and emissions of the Bohr theories (p. 665) are beyond the range of these illustrations of the imprisonment of radiation by diffuse matter.

which is $26 \cdot 10^{18}$. As K is μ^2 for gases where μ is refractive index, it appears as the so-called refraction equivalent. In that form Lorentz showed long ago that it can be recognized also in the solid and liquid states, with the suitable additional factor $3/(K+2)$. It is even found to persist to a considerable extent for atoms combined into a molecule, notwithstanding the mutual disturbance due to their proximity. For air μ is 1·0003; thus the molecular polarization equivalent, which may be denoted by k, and is here a mean between nitrogen and oxygen not far from the value for nitrogen, is $6 \cdot 10^{-4}/26 \cdot 10^{18}$, or $23 \cdot 10^{-24}$.

The polarity induced in a molecule in the electric field $F_0 \cos pt$ of the radiation is $\delta = \dfrac{k}{4\pi C^2} F_0 \cos pt$. As the electrons within it all oscillate coherently in step, the waves being long compared with its dimensions, this moment radiates in the manner of a vibrating ion for which ev is its time gradient. The scattering per cm. of depth of the medium, thus taken as due to the secondary atomic sources vibrating incoherently, is

$$\frac{2}{3C} N' \left(\frac{d^2\delta}{dt^2}\right)^2, \quad \text{which is} \quad \frac{2}{3CN'} \left(\frac{K-1}{4\pi C^2} p^2\right)^2 F_0{}^2 \cos^2 pt,$$

where $p = \dfrac{2\pi C}{\lambda}$, and the flux of incident radiation is $\epsilon = C \dfrac{F_0{}^2}{8\pi C^2}$. On substituting $\frac{1}{2}$ as the mean value of $\cos^2 pt$, and introducing ϵ and λ, this becomes

Rayleigh's formula for the blue sky.

$$\frac{8}{3} \frac{\pi^3}{N'} \left(\frac{K-1}{\lambda^2}\right)^2 \epsilon,$$

which is the Rayleigh formula, closely verified for the blue sky, and further developed for denser media by Raman and his colleagues. The extinction modulus of the light of the sky, which is the factor of ϵ, comes out for air at standard pressure and yellow light of wave-length $6 \cdot 10^{-5}$ about 10^{-7}, reducing the radiation to one-half in pure air in about 80 kilometres.

Electrons are the effective scatterers:

The scattering power of a single electron is as above $\dfrac{8\pi}{3} \left(\dfrac{e}{m}\right)^2 e^2$, which is $64 \cdot 10^{-26}$; thus it compares with about $3 \cdot 10^{-27}$ for a nitrogen molecule, for yellow light as derived from the Rayleigh formula, so is six hundred times greater. But for an atomically loaded electron the power is far smaller, the reducing factor being $7 \cdot 10^8$ for a nitrogen atomic ion: thus a nitrogen atom scatters 10^6 times more yellow light by its alternating polarization than by its ionic swing.

thus can hold together vast aggregates of free radiation,

The agency in holding together the aggregation of enormously intense free radiation within the extremely tenuous material frame of a giant star, and preventing it from flashing out into space as one would at first glance expect, in fact for producing opacity, must be

themselves held down by gravitation. predominantly the distribution of free electrons within it, themselves held down along with the free atoms by gravitation towards its centre.

Repulsion at Solar surface: At the Solar surface, a repulsion of each atom or electron is involved equal to c^{-1} times the energy it disperses per unit time. The justification is that by Maxwell's stress theorem the total repulsion of a layer is measured by the radiation it diverts per unit time from the beam passing across it divided by c, while by the verified Rayleigh principle this diversion is to be reckoned as due to the ions and atoms scattering independently. Taking the intensity of radiation at the Solar surface to be $7 . 10^{10}$ and the intensity of gravity $3 . 10^4$, and $K-1$ for hydrogen at standard density to be $3 . 10^{-4}$, it appears, taking the mean wave-length λ_0, which should be given by

$$\int \lambda^{-4} d\epsilon = \lambda_0^{-4} \epsilon,$$

of atoms: their repulsion compared with gravitation. to be $6 . 10^{-5}$, that a hydrogen molecule at the Sun would be repelled, owing to the dispersal by its vibrating polarization, by a force of order 10^{-26}, while its gravitation is 10^{-19}. For free electrons the repulsion by radiation would be $\frac{1}{45}$ of their gravitation.

Additional radiation in selective *quanta*. These figures, so far as they may be correct, illustrate the small part played, except for short waves, by general dispersal of radiation by the atoms treated on the analogy of the dispersal of energy of sound by resonators. But modern views contemplate beyond this theory that the atoms take in and emit packets of energy restricted to their own special periods, which are held to be far more potent as regards momentum (Eddington, Milne) and viscosity (Jeans), and even competent to sustain stellar atmospheres against gravitation*.

* The train of ideas presented above may admit of more definite general statement. The radiation of the Sun, and of stars of its type, comes from a gaseous photospheric layer which scatters light mainly in non-selective manner, so presumably as the same molecules in a terrestrial atmosphere would do. Proximate source of Solar radiation. Thus a depth measured in atoms corresponding to the order of 100 km. of air as estimated, p. 663, now with gradient of pressure due to Solar gravity, would be fairly opaque: so that most of the actual emergent Solar radiation would be emitted within such atomic depth, however its energy may have come there: which affords a rough idea of the contribution to it from each radiating atom. Above this layer is the chromosphere, which scatters almost wholly Chromosphere: selectively, in the bright arcs revealed in the flash spectrum. In it the atoms are thus, perhaps, very sparse, and their power of selective scattering, and therefore opacity, very intense. But the latter feature would hold also for the photosphere, composed of like atoms, so that little of this selective radiation would come up from it to be scattered, unless somehow a Stewart-Kirchhoff balance of exchanges becomes established. If the atoms of the chromosphere and the prominences are thus sustained, practically isolated, against Solar opaque if held up by radiation. gravity by repulsion exerted by the Sun's radiation, one atom of hydrogen would have to disperse to an amount equal to the whole incident Solar radiation over a front of $2 . 10^{-20}$ cm.^{2}, or to its own selective radiation say over 10^3 times

It is to be noted that in these estimates for a polarized atom, all the mobile electrons within it are taken as radiating coherently in step, thus increasing the scattering power proportionally to their number; this is on account of the large wave-length in comparison with the atomic dimensions. On the other hand, free ions are taken as scattering energy independently.

The problem of coherence for radiation.

The mechanism of this imprisonment of vast energies of free radiation in giant stars is that radiation travelling away in any direction is very soon broken back and scattered by the free electrons. The question emerges here as to the mode of statistical specification of natural radiation, after all regularity structurally inherent in wave-pulses has been thus shivered without limit: but that is a different subject.

To sustain steadily the lower strata of the rarefied gaseous system constituting a star, an increasing density of radiation inward would be required, after Eddington, rising to very high values. Equilibrium can be regarded in a rough way as attained by compression of the confined natural radiation by the gravitation of the layers above, until a balance is reached, after the manner of the discussion, by aid of ideal compression of radiation, of Wien's spectral law. There might be even gigantic radial pulsations such as have been invoked to account for the special type of variability of Cepheid stars. Reacting gas pressure of the atoms and electrons would play its part, usually more important unless where their density is very small: the density of the radiation thus compressed determining the temperature of the gas by the law of Stefan-Boltzmann. The astronomical evidence that, after Eddington, at enormous temperatures all atoms by loss of electrons shrink, in whatever manner, into extremely small effective dimensions, belongs to a different subject, still obscure in so far as the operation has to be reversible.

Compression of natural radiation. Stellar structure.

We may thus perhaps hold that the very dense free radiation is prevented from passing out of the rarefied stellar medium largely

Predominant influence of free electrons.

this front—and proportionately more for heavier atoms: this would make a stratum of atoms, comparable in number only to 10^{-2} cm. of air, strongly opaque to its own radiation. The flash spectrum would then not come from any considerable tangential depth at the Sun's rim. If however the chromospheric atoms attain to high speed under the repulsion of radiation, they would then disperse in their own relative period, so owing to Doppler effect would be fed by light slightly outside the absorption band: thus, following Milne, a more prolonged supply would be available for scattering, and greater heights would be attained, but possibly at the cost of extensive widening of the Fraunhofer lines. In any case a measure of the intensity of the flash spectrum should give some estimate of the actual intensity of selective radiation per atom, about which the theories of *quanta* have wide and definite views of their own, as also indeed ordinary electrodynamic theory in cases where the anomalous refraction close to the band has been measured.

by the obstruction of its free electrons: that its compression determines the temperature: and that the temperature determines the material gas pressure by some kind of mechanism the nature of which has been a problem ever since the beginning of spectrum analysis.

Theory of Solar spectroscopy.

Modern theory of the Solar atmospheric spectrum takes brilliant advantage, following Saha, of the hypothesis long familiar in other connections, that the free electrons are themselves to be regarded as one of the types of atomic constituents of the stellar atmosphere: subsisting in a state of balanced dissociation depending on temperature, such as has been reduced to law and practically confirmed for all time by Willard Gibbs' classical application* of his thermodynamic principles to the experimental data for nitric oxide and other freely dissociating gases.

The main obstacle to the application of statistical gas theory, in the manner of Maxwell-Boltzmann-Rayleigh-Gibbs, to an ionized gas, in fact to treating each ionic constituent as an independent component, has been that the forces between ions are not of molecular type effective only in close encounters. It has been an implication in gas theory that the encounters between atoms are practically all binary, whereas under inverse square law all the atoms at any rate in the neighbourhood are involved in every encounter. Cf. "On the Physical Aspect of the Atomic Theory: Appendix": in *Manchester Memoirs* 1908, as *infra*, Vol. II. The success, in application to relations of spectroscopy to density in stellar atmospheres (Saha), thus seems to bring to the fore the question whether and how far these ionic forces operative at a distance can be ignored in a reconstructed theory in comparison with the cohesive forces operative in intimate encounters.

Thermo-dynamics for ionized gas.

The classical resonance theory:

All that precedes is on the foundation of the resonance view of atomic radiation, now sometimes called classical. It starts from Stokes' (and Young's) explanation by analogy for the spectrum of bright and dark lines, such as might envisage a train of sound waves travelling through a medium strewn with a system of tuned violins or other resonant vibrators, representing the atoms. The theory attained to formal mathematical expression very briefly in the hands of Maxwell; it was developed independently soon after into detail by Sellmeier and Helmholtz; and finally it has tried to attach itself to the new foundation of the electric orbital atom, at first by astronomical methods (cf. *Phil. Trans.* 1897, as *infra*). The facts of anomalous dispersion strongly supported such a theory: perhaps its most striking

* "On the Vapour Densities of Peroxide of Nitrogen, Formic Acid, Acetic Acid, and Perchloride of Phosphorus," *Amer. Jour. Sci.* 1879, *Scientific Papers*, Vol. I. pp. 372–403.

verification is in the Rayleigh-Stokes theory of the blue sky as due to secondary radiation scattered by the molecules of the air, which has led even as far as an exact count of their number.

But a different account of the absorption and emission of radiation, at high temperatures, has been coming to the front on the basis of a new order of facts as interpreted under the lead of N. Bohr. Flying free electrons play an essential part, stimulating the atoms by imparting their own energy, but only in definite *quanta* which are stored in the atom and in time can break out again by some unimagined internal process in the different form of a limited train of pure radiation: and the completed process is postulated to be reversible, so that a statistical theory can be built up. The requisite *data* are mainly those of *quantum* numbers. The justification of this empirical scheme is found, apparently with much precision, in the excitation of individual spectral lines by regulated electronic impact in vacuum bulbs: now largely also in the beautiful application to elucidation of many perplexing phenomena of Solar and stellar spectroscopy advanced by Saha. Whether some formal scheme of this type is to be superposed on the classical resonance theory, or is to supersede it in some way not yet in evidence, is perhaps left open. In any case, it is of necessity on a different plane from the present straightforward discussions, which connect definitely the pressure of radiation with its scattering, on normal electrodynamic principles.

contrast with theory of *quanta*.

Delicate questions as to the propagation of natural radiation, at any rate on a wave theory, are relevant here, which will arise formally in another connection*. It can be held that it is on account of very slight Doppler differences in periods of scattered radiation, cumulating after travelling a sufficient distance into large differences in phases, that the molecules of a gas scatter light independently: such differences would arise by virtue of their irregular translatory motions. The same mixing of phases would arrive at greater distances even for radiation from a solid aggregation of molecules, on account of their thermal vibrations. But the Rayleigh analysis of dispersal of radiation by a particle of dust, of a smaller order than the wave-length, namely partly by reflexion and diffraction at its surface and partly by scattering from its interior, implies that all its molecules scatter energy by coherent oscillation, thus with energy in proportion to the square of their number instead of its first power as in a gas. Are we then to regard this radiation, deflected by the particle, as a local field of organized radiant energy through which the smaller amount that

The problem of natural radiant energy.

Dust nebula in contrast with gaseous nebula.

* "On the Constitution of Natural Radiation," *Phil. Mag.* Nov. 1905; as in Vol. II, *infra*.

is irrevocably lost makes its outward way? Even for a crystalline particle this question persists. However close the periods of component trains may be, their phases will ultimately become discordant. Within their own environment the particles of a fog scatter each of them coherently and so deflect much of the incident light. But does the energy of the broken light all get away as radiation into remote space? For instance, taking the case of a remote dark nebula illuminated by an adjacent star situated in front of it, is it certain that a dust nebula would shine any more intensely than a gas nebula of the same density of molecules? Mathematically this question belongs to the refined theory of the Fourier analysis*.

* Cf. "The Fourier Harmonic Analysis...," *Proc. Lond. Math. Soc.* 1916: as in Vol. II, *infra*. Reconciliation is perhaps to be sought in the principle, enforced *supra*, pp. 558, 563, that refraction of light is distinct from dispersion. The $K-1$ for a distribution of molecules is a static property, even rather closely so for light because the molecule can adapt itself without shock, so as a whole, to the slowly alternating field of its comparatively long waves: but it is not at all static for the far greater frequencies of X-ray radiation.

Coherence of radiation from an atom.

APPENDIX.

(viii) *The Residues of Atomic Mass regarded as Inertia of Aggregation.*

IT has been held by the writer on grounds of general dynamics, cf. p. 671, that if the Action of any orbital system, such as an atom is usually taken to be, is expressed as $\int L dt$, then its structure, statical and motional, makes a contribution to its inertia equal to $-L/c^2$. Here $-L$ is $W-T$ the excess of the positional over the kinetic energy, whereas the conserved energy E of the system is $W+T$. This inertia is a convective property, tested only in translatory motion of the system: the L that enters must be the mean value of the Action gradient internal to the atom, and the virial theorem (cf. Vol. II. p. 134) finds a useful opening. It is to be noted that in a complex system it is the true Action gradient that is involved: the modified Action (*supra*, p. 48), which as empirically determined may have to ignore latent motions, would demand closer discussion. If this contention is valid the mass has a variable part which may be sensible for a compound nucleus of an atom, but it is not expressive of atomic energy: its value is not E/c^2, but is the average of $(W-T)/c^2$ which is $(E-2T)/c^2$ or $(2W-E)/c^2$. Running down of potential energy diminishes mass: so does presence of internal kinetic energy. It is only when the system is static with no kinetic energy that the contribution of state of aggregation to mass is E/c^2. When the orbital motions are purely cyclic, the modified Action density reduces to $-E$; but a diminution of mass of amount E/c^2 thereby suggested would be unwarranted, the value being $(E-2T)/c^2$, where T is now constant as well as E on account of the unchanging cyclic configuration.

Action density contributes to convective inertia:

only for a static system it reduces to a contribution from energy.

But an atom, as already remarked, is an orbital system for which the value of the internal Action gradient L is probably variable. The inertia of aggregation must depend on its average: and fortunately there is a general result in dynamics that precisely fits the case. The virial theorem, applied by Clausius to aggregates of free atoms, asserts that in every self-contained system the mean value $(\overline{T})$ over time of the kinetic energy is equal to the mean of a virial function expressed as $-\frac{1}{2}\Sigma\,(Xx+Yy+Zz)$. The proof neglects only the deviations of $\Sigma m\,(x^2+y^2+z^2)$ from the mean. Applied to an orbital system of unchanging point masses, where mutual forces are restricted to attractions or repulsions according to inverse square, the virial is

Simplified value for an orbital atom.

equal to half the potential energy that has run down in the assembling of the orbital system from its widely separated constituents. Thus for any such system $2\overline{T} = -\overline{W}$, a relation derived long ago by Jacobi for dynamical astronomy. It will apply to our atom model, if we neglect as we may the electrodynamic features discussed *supra*, p. 524, and also the dependence of inertia on speed*. Thus the residual mass is $-3\overline{T}/c^2$, so always negative.

Interpretation in terms of nuclear structure. The facts revealed by Aston's fundamental discovery are held by him provisionally to fit with the Rutherford hypothesis that his atomic nuclei are built up out of primordial atoms on protons, to be identified with atoms of hydrogen; thus reverting in a way towards the original naïve guess of last century (*supra*, p. 523), except that now this proton takes the place of the missing positive electron. The nucleus of an atom on the Rutherford-Bohr theories would be constituted of an integer number of positive protons held together by the predominating attractions of a number of negative electrons, this "binding" being a main function served by electrons. These protons and electrons, somehow concentrated into a space astonishingly small†, are yet regarded as possibly in part describing free mutual orbits and in part limited by structural constraints. In either case there is a sensible fraction of the mass to be ascribed to their extreme propinquity in the atomic nucleus: and the present considerations indicate that its value should be negative.

The Lavoisier law. Last century it was a bulwark of the early electron theory that the mass of a molecule comes out practically the sum of the masses of its constituent atoms, thus avoiding a chief difficulty encountered by the simple Kelvin vortex-atom scheme: now the question for exploration is turned round, into the extent to which such equality may not be exact.

The Aston residues. When the mass m' of an atom is expressed as an integer number m combined with a very small residue δm, Aston's recent determinations (Bakerian Lecture, *Roy. Soc. Proc.* 1927, p. 510) make the residue positive for the lighter elements and negative for the heavy ones, instead of being always negative. As he remarks, the unit of measurement for m is alterable at choice: on this basis he indicates, without apparently putting weight on the relation, that if the unit were increased so as to make oxygen 16 *minus* ·02 instead of exactly 16, the residue becomes for all atoms a constant ·025. From another

* It is easy however to include the latter.

† In this connection Prof. Lorentz remarks (*Pasadena Lectures*, 1922–7) that the theories of relativity abandon the view (*supra*, p. 523) that matter is constituted of singularities in an aether, which formed after the ideas of Kelvin so satisfying an account of the necessity for an atomic theory: as no other physical conception of atomic matter takes its place the imagination there becomes entirely unrestricted.

point of view, the proton mass-unit might be increased as to satisfy the condition of the residue being negative; especially as, the atomic number being there different from mass-number, deviation from that principle may be permissible for the heavy atoms. Possibly however the residue would then be too great for the facts of atomic disintegration. The tabulated masses, though they are relative to an origin on their mass-spectrum diagrams, become absolute when checked against standard chemical weighings: thus a constant term, such as ·025 above, is not merely due to change of origin: but this of course does not prevent their numerical fractions from changing with change of the value assigned for the proton mass-unit.

Internal Action alone is concerned.

The virial of a travelling system is an affair of mutual internal forces: thus the $\overline{T}$ in these expressions for inertia of aggregation is the kinetic energy of the internal motions of the system. If translatory motion were included, the condition of steadiness which limits the virial theorem would not be satisfied. This elucidates how it is that translatory motion and its energy have nothing to do with the intrinsic inertia m_0 of an atomic system: though the momentum relative to the frame is modified (in the reckoning of p. 672 *infra*) from $m_0 v$ to $m_0 (1 - v^2/c^2)^{-\frac{1}{2}}.v$. However complex the orbital atomic system may be, it possesses this intrinsic mass, in part aggregational, invariant for all time because inherent in an invariant dynamical Action.

An invariant atomic mass exists.

Proof of dependence of convective inertia on Action:

The demonstration of the relation asserted between mass and Action may here be conveniently indicated. The Action being the fundamental dynamical entity, it is naturally invariant for the optical group of permitted frames of reference of the convected system. In its own frame, which travels with the system, its value is

$$A = \int (T_0 - W_0)\, dt_0,$$

where $T_0 - W_0$ belongs to the system referred to itself and dt_0 is an element of its own time. This time is absolute, therefore as A is invariant, so should be $T_0 - W_0$, say L_0, for change of spatial frame: it is in fact a constant in so far as the atomic system is steady*. We have to transfer from the system's own frame $(x', \ldots t')$ into the one $(x, \ldots, t)$ with regard to which its convective velocity v is expressed. The reversed Lorentz transformation is

$$dx = \epsilon^{\frac{1}{2}} (\delta x' + v\delta t'),\quad \delta t = \epsilon^{\frac{1}{2}}\left(\delta t' + \frac{v}{c^2}\,\delta x'\right),\quad \epsilon^{-1} = 1 - \frac{v^2}{c^2}.$$

Here $v\delta x'/c^2$ is negligible compared with $\delta t'$ because $\delta x'$ is slight

* This argument as to frame requires expansion: see *infra*, p. 674.

internal displacement of the steady system in its own frame. Hence we may substitute $\epsilon^{-\frac{1}{2}}dt$ for dt'. Thus in the frame $(x, \ldots t)$ of the convection the Action is expressed as

$$A = L_0 \int \left(1 - \frac{1}{c^2}\frac{dx^2}{dt^2}\right)^{\frac{1}{2}} dt,$$

$$\text{giving } \delta A = \int -\frac{L_0}{c^2}\left(1 - \frac{1}{c^2}\frac{dx^2}{dt^2}\right)^{-\frac{1}{2}} \frac{dx}{dt}\frac{d\delta x}{dt} dt$$

$$= \left| -\frac{L_0}{c^2}\left(1 - \frac{v^2}{c^2}\right)^{-\frac{1}{2}} v\delta x \right| + \int \frac{d}{dt}\left\{\frac{L_0}{c^2}\left(1 - \frac{v^2}{c^2}\right)^{-\frac{1}{2}} v\right\} \delta x\, dt.$$

The latter term implies working kinetic reaction as derived from a translatory momentum

$$M = -\frac{L_0}{c^2}\left(1 - \frac{v^2}{c^2}\right)^{-\frac{1}{2}} v,$$

which it is legitimate to describe as arising from a varying inertia

$$m = -\frac{L_0}{c^2}\left(1 - \frac{v^2}{c^2}\right)^{-\frac{1}{2}}.$$

result. This represents for ordinary speeds a definite contribution to mass, $-L_0/c^2$ or $(W_0 - T_0)/c^2$ in which W_0 and T_0 fluctuate round mean values, as above asserted.

Illustration from the self-contained spherical electron model: For the spherical shell model of a static electron, self-contained because sustained from disruption by differential pressure of the fluid aether, established between inside and outside, as *supra*, pp. 459, 500, this aggregation term in the mass is, for small values of v/c,

$$\frac{W}{c^2} \text{ or } \frac{1}{2}\frac{e^2}{a}.$$

The convective energy of the system is then made up from energy

analysed. of the magnetic field $\frac{1}{3}\frac{e^2}{a}v^2$ as calculated *supra*, p. 522, less work run down in it by this *uniform* pressure* concentrating its volume by the convective shrinkage, readily seen to be $\frac{1}{12}\frac{e^2}{a}v^2$: the difference is in all $\frac{1}{4}\frac{e^2}{a}v^2$, corresponding to an inertia $\frac{1}{2}\frac{e^2}{a}$, which thus verifies.

The convective invariance of Action: The issue here involved, that the mass of an atom or other self-contained system is an affair of its Action which is intrinsic, not of its energy which alters with the frame of reference, is fundamental. Ideas may be cleared by following out the case of a system of perfect conductors which are to represent the electrons, carrying charges and

* This difference of uniform hydrostatic pressures, subsisting of necessity between inside and outside for the electron model in a rotational fluid aether, and exact for all values of v/c, was later recognized as a mathematical result by Poincaré.

in mutual orbital motion. The activities are now all located in the intervening fields, which alone need be considered. The very special case where the conductors are rigidly connected was developed at length in an early stage, *Aether and Matter* (1900), Ch. IX. The atomic system with its internal orbital motions is convected with velocity v in the frame $(x, ..., t)$ to which its motion is referred. Its intrinsic Action and energy are to be calculated in its own frame $(x', ... t')$. A correlation arises as expressed above: it is *as if* $(x', ... t')$ were shrunk to $(x, ... t)$ in the ratio $\epsilon^{-\frac{1}{2}}$ or $(1 - v^2/c^2)^{\frac{1}{2}}$ along with change of epoch of time, the combined effect being to maintain $c^2t^2 - x^2$ unaltered and therefore also $\iint dx\,dt$. The correlation involves also (cf. Vol. II, *infra*, p. 39, or *Aether and Matter*, § 112), using accented letters to express the field in its own frame, confirmed on a special atomic model:

$$(f', g', h') = \epsilon^{\frac{1}{2}}\left(\epsilon^{-\frac{1}{2}}f,\ g - \frac{v}{4\pi c^2}c,\ h + \frac{v}{4\pi c^2}b\right),$$

$$(a', b', c') = \epsilon^{\frac{1}{2}}\left(\epsilon^{-\frac{1}{2}}a,\ b + 4\pi v h,\ c - 4\pi v g\right).$$

Thus we have for the Action the form $\int dt'.L'$, where, $d\tau'$ being the spatial element $dx'\,dy'\,dz'$,

$$L' = \int d\tau' \left\{\frac{1}{8\pi}(a'^2 + b'^2 + c'^2) - \frac{1}{8\pi c^2}(f'^2 + g'^2 + h'^2)\right\}$$

which transforms to

$$\int d\tau' \left\{\frac{1}{8\pi}(a^2 + b^2 + c^2) - \frac{1}{8\pi c^2}(f^2 + g^2 + h^2)\right\}.$$

The integrand being thus invariant, the Lagrangian function L' equal to $T' - W'$ transforms into the spatial frame to which the motion is referred by a factor $d\tau/d\tau'$ which is $\epsilon^{\frac{1}{2}}$ or $(1 - v^2/c^2)^{-\frac{1}{2}}$: but the energy of the system equal to $T' + W'$ changes in a complicated way which emphasizes that energy is not intrinsic but relative. The Action is thus $\int(T' - W')\,dt'$, equal to $\int(T - W)\,dt$, and so its invariance is verified.

This isolates dt', put equal to $\epsilon^{-\frac{1}{2}}dt$ as in p. 672, thus neglecting the slight internal changes of epoch. But in the optical group of frames $\iint dx\,dt$ is invariant exactly, and the Action is invariant for this group of correlations. Therefore the equations of the field, which are derivable from it alone, are also unconditionally invariant: which, combined with the invariance of the electronic charge which is also involved, is the expression of optical relativity. compared with electrodynamic relativity:

But what we are now specially concerned with is the inertia of the moving system regarded as a whole. In so far as it is steady, its Lagrangian function L' in its own frame has a constant mean value it involves that mass is not atomic energy.

which is not mixed up with its changing convection v: thus the Action is expressed by

$$A = \int L'dt' = L' \int \left(1 - \frac{v^2}{c^2}\right)^{\frac{1}{2}} dt,$$

from which the dynamics of its convective motion, expressible in terms of the idea of varying inertia, follows as above: the inertia of aggregation of an atom, namely the excess of its $W - T$ over the sum of those of its isolated constituents, being $-\frac{3}{2}$ of the potential energy that runs down in their coming together, or -3 times the resulting kinetic energy. These are statements of mean: the defect of inertia due to aggregation is -3 times the total energy that has run down: but that energy also is a mean, and depends on the convection by a factor $(1 - v^2/c^2)^{-\frac{1}{2}}$.

Varying mass implies a hybrid frame,

This energy of the system has been familiar after Lorentz for the case of a single electron. But is there not involved reference to a hybrid frame, one outside the group in which adjustments are all made by optical means? For, as has just been seen, it is the Action $\int Ldt$, not its integrand L, that is unchanged in transfer from one optical frame to another. Therefore, employing the Michelson experiment to fix ideas, the law of change of mass involves spatial measurements in the observer's frame travelling with his apparatus, without any shrinkage, combined with time measured in the astronomical frame of the Solar system, thus shrunk in the ratio dt/dt', now with sensible change of epoch with locality: or, more simply, it might be spatial measurements in the Solar frame, involving shrinkage along the direction of the terrestrial motion, combined with local time determined spectroscopically from local vibrating atoms. If this is right, cosmical arguments based on varying mass imply the latter alternative, which is Newtonian time.

with Newtonian time.

APPENDIX.

(ix) *On Gyromagnetics: and on Electrons and Atoms having Helicoidal Inertia.*

CONSIDER an electron revolving rapidly in a steady orbit. Its average magnetic effect is as if it were distributed round the orbit with density at each point inversely proportional to the velocity there. Cf. pp. 504, 516 *supra*. That gives a uniform electric current round the circuit, which integrates, in the periodic time of the orbit, to the value of the electron. Its magnetic moment is current multiplied by area, the latter in the vector sense that the component in any direction is its projection transverse to that direction. The magnetic moment is thus expressible as rate of description of area: but that changes with choice of origin, and to obtain a definite value it must be averaged over the periodic time of the orbit. The exception is when the origin is a centre of the only force acting in the electron: we may thus assert that then the magnetic moment is the angular momentum of the electron multiplied by $\frac{1}{2}e/m$, and this will always be the effective mean value for any orbital system of electrons. **The Amperean atom.**

In the Bohr symbolic scheme the orbital momentum of an electron is a quantum constant $h/2\pi$: thus the scheme provides a scalar quantum of magnetic moment, equal to $e/m.h/4\pi$, which has been named the Bohr magneton. **The Bohr magneton.**

Magnetism seems to promise clues towards atomic dynamics. Emergence of a resultant magnetic moment by partial alignment of the molecules thus involves emergence of angular momentum of the electrons of proportional amount, which would exert a material reaction. This reaction has been detected and measured carefully and repeatedly, but with results nearly exactly half the amount thus suggested. **Gyro-magnetic discrepancy.**

Now consider a perfectly conducting linear circuit carrying a current in a magnetic field, or, to gain precision, consider a number of such currents interacting across the aether. The Maxwellian dynamics will hold for such a system, as specified for changes slow compared with radiation by positional functions of self and mutual induction: for ultimately, the system will be subject to Minimal Action for the aethereal field, which in the present case reduces to that form. In each perfectly conducting circuit the correlated electric momentum will remain constant: that is, the current will vary inversely as the number of tubes of magnetic force the circuit embraces, which depends on the geometrical configuration, and also on the currents linearly. But if the currents change by continuous induction in this way, which is the accepted explanation of diamagnetism, the magneton could not remain a constant. Such implication of induction necessarily involves (p. 515) that the currents are made up of discrete travelling electrons, notwithstanding the hyperconductance here assumed, and now realized in actuality at very low temperatures by Kamerlingh Onnes and his associates. On the other hand, on a continuous vortex ring (imperfect) analogy each current would remain constant; it would itself be of the nature of a conserved momentum: cf. *supra*, p. 57. But there again no constant magneton would arise.

Paradoxes as regards magneton.

The processes are reversible on account of the hyperconductance. One may enlarge the possibilities by imagining the circuits to be expansible, like vortex rings. A system of such currents, circular on a common straight axis, will influence each other with forces equal and opposite to those of a system of vortex rings: so their motion could be traced graphically after the manner developed by Helmholtz and Kirchhoff for an axial vortex system: for instance, two such free currents in opposite directions would pass through each other alternately as they progress. But there do not seem to be any constant magnetons so long as a dynamical framework is retained.

Vortical parallels.

Consider for precise illustration two permanent circular currents, parallel to a plane in which they are mutual images. If they circulate in like directions they will attract one another, and diminish in value as they approach; and they will remain circular and equal, even if contracting in radius. But the concomitant material rotational momentum, of gyromagnetic type, of the electrons of each of them will change. Energy associated with attraction or translatory momentum in each of them is thus transferable into energy associated with angular momentum round its line of action, and conversely: though only slightly for large circuits.

Interchange of momenta.

This is analogous to the case of propelling screws of ships or aeroplanes. It implies helicoidal type of inertia: an illustration, the case of two ideal mutually influencing screws working in perfect fluid on a common axis, has already been examined, *supra*, p. 81.

Analogy of screw propeller.

Now consider the original spherical model (1894) of an electron and its field, *supra*, p. 520, constituted as subsisting in a rotationally elastic fluid aether. Its translational inertia was there determined as the factor in kinetic energy of a motion *restricted* to uniform translation: it came out $2e^2/3a$, corrected afterwards into $e^2/2a$ by virtue of the relativity shrinkage during convection. But the helical radial twist in an aethereal medium, which there expresses the actual screwing character of the electrodynamic field round the electron, points to helicoidal quality in its inertia, in Lord Kelvin's phrase, rather strongly. In the Parson type of model of a magnetic electron, an Amperean vortical current, the helical transformation of translational into rotational momentum is explicitly involved as above.

Electron as an isotropic helicoidal structure.

This suggests that an atomic model consisting of a number of electrons circulating in the same orbit with axial symmetry, constituting practically an Amperean current, would have inertia in some degree helicoidal, of type symmetrical around the axis of the orbit. The formulae determining the motion of helicoidal solids such as screws in perfect fluid, as initiated by Kelvin*, are here applicable in illustration: see *supra*, p. 76. In the case of axial symmetry, around the axis of z, with origin symmetrically chosen on the axis of symmetry, the kinetic energy is (p. 77, § 3) of the form

Atom as axially helicoidal:

form of its energy.

$$T = \tfrac{1}{2}A\,(u^2+v^2) + \tfrac{1}{2}Cw^2 + \tfrac{1}{2}A'\,(p^2+q^2) + \tfrac{1}{2}C'r^2 + L\,(up+vq) + Nwr,$$

in which for a system with small nuclei A', C' may perhaps be presumed to be small compared with the other inertial constants. It has been found that such a solid could progress through the perfect fluid in spiral motion round an axis which itself progresses spirally round a central axis. The result is

Its permanent modes of free translation,

$$\frac{u}{p} = \frac{v}{q} = \epsilon\frac{w}{r} = k, \text{ constant,}$$

where† $\epsilon = \dfrac{A\,(A-C)\,k^2 + A\,(2L-N)\,k + L\,(L-N) + CA'}{NAk^2 + AC'k + LC'A - NA'}$, constant;

also w is constant, and the radial component u is oscillatory so that u^2+v^2 is constant, and similarly for r and p^2+q^2; reducing to the result worked out in Lamb's *Hydrodynamics*, § 129, in agreement with Kirchhoff, when there is no helicoidal quality so that L, N are null.

* The general equations of motion of a solid in fluid, as published by Lord Kelvin in *Phil. Mag.* 1871 (also Baltimore Lectures, Appendix G), under the title "Hydrokinetic Solutions and Observations," still exist in his handwriting in one of the long series of "green books," under the date Jan. 6, 1858, as there stated. They are derived in Eulerian manner from the components of the impulse, obtained as gradients of the energy relative to the components of the motion, thus ten years before the suggestion of explicit treatment on those lines in Thomson and Tait, *Nat. Phil.*, 1867, systematized independently by Kirchhoff in 1869.

Historical.

† This formula is taken from a letter from Prof. Orr of date 26 Feb. 1910.

are helical even when the inertia is not helicoidal, unless it is translationally isotropic.

These formulae express a permanent motion of the system relative to itself, in which its central point traverses a spiral path. This spirality persists even when helicoidal quality is absent; unless A and C are equal and the translatory inertia thus isotropic, when the solution reduces to the familiar Euler-Poinsot precession.

Magnetic moment and rotational momentum.

If we imagine this system to represent an electronic atom, as suggested above, it would have a mean magnetic moment $\frac{1}{2}e/m.C'r$ and a mean rotational momentum $C'r + Nw$; the ratio, of latter to former, is a function of k, of a value twice too great for the facts when the translatory velocity w is null, but smaller if Nw is negative as it will be if the electrons are negative. Though in a solid the molecule cannot have free translation w, yet its orientation by magnetic influence, possibly even when it is not helicoidal, may produce an effort towards it, which would be held up by reaction of the aether affecting the momentum*.

Isotropic helicoidal electron:

Further discussion may be restricted to the tractable case of Lord Kelvin's isotropic helicoid, such as would represent an electron if it were helicoidal. Then

$$T = \tfrac{1}{2}A\,(u^2 + v^2 + w^2) + \tfrac{1}{2}A'\,(p^2 + q^2 + r^2) + L\,(up + vq + wr).$$

The natural paths of such electrons would be spirals, say of radius ϖ: the momentum is referred to axes in the body, that of z parallel to the central axis of the spiral, an instantaneous axis of x radial from that axis; and the translational and rotational components of the impulse are M, G along the axis. The inertial coefficients are now the same for every set of axes in the system, so that the equations of conservation of momentum (following Lamb, *Hydrodynamics*, § 130) are by the Kelvin principle

$$Au + Lp = 0 \qquad Av + Lq = 0 \qquad Aw + Lr = M$$
$$A'p + Lu = 0 \qquad A'q + Lv = \varpi M \qquad A'r + Lw = G$$

The three related pairs of component speeds thus change, each independently. The first pair u, p vanish: the other two pairs determine the motion, now completely steady. The pitch of the spiral path is determined by the Kelvin formula

its spiral free paths:

$$\frac{v/\varpi}{w} = -\frac{LM}{A'M - LG};$$

* For a particle describing an orbit round a centre, another invariant addition to the (scalar) energy is possible besides this helicoidal inertial term, one proportional to the triple determinantal product of translatory and rotatory momenta of the particle and its distance. It appears (cf. C. G. Darwin, *Roy. Soc. Proc.* Dec. 1927) that in wave mechanics this term by itself, without helicoidal inertia, is competent to put things in order as a scheme for an atom or electron in a magnetic field.

so that the linear period of the spiral path is determined solely by the pitch M/G of the impulse, up to the limit when it is straight.

its magnetic moment:

An electron of this helicoidal type travelling free would have a magnetic moment $\frac{1}{2}ewv$ along its translatory axis, and a rotational momentum G around it, where r/w is determined in terms of M/G by the last pair of equations.

its rotational momentum.

Nothing definite has here emerged in the present connection—whatever relation if any there may be to the *corpus* of spectral *quanta*; the possibilities of the model are wide, in contrast with the precise ratio of gyromagnetic discrepancy which suggested it. But suspicion appears to have been entertained before now that free electron paths in the Wilson condensation chamber do show spiral features.

Printed in the United States
By Bookmasters